CAMBRIDGE LIBRARY COLLECTION

Books of enduring scholarly value

Technology

The focus of this series is engineering, broadly construed. It covers technological innovation from a range of periods and cultures, but centres on the technological achievements of the industrial era in the West, particularly in the nineteenth century, as understood by their contemporaries. Infrastructure is one major focus, covering the building of railways and canals, bridges and tunnels, land drainage, the laying of submarine cables, and the construction of docks and lighthouses. Other key topics include developments in industrial and manufacturing fields such as mining technology, the production of iron and steel, the use of steam power, and chemical processes such as photography and textile dyes.

Electromagnetic Theory

Oliver Heaviside (1850–1925) was a scientific maverick and a gifted self-taught electrical engineer, physicist and mathematician. He patented the co-axial cable, pioneered the use of complex numbers for circuit analysis, and reworked Maxwell's field equations into the more concise format we use today. In 1891 the Royal Society made him a Fellow for his mathematical descriptions of electromagnetic phenomena. Along with Arthur Kennelly, he also predicted the existence of the ionosphere. Often dismissed by his contemporaries, his work achieved wider recognition when he received the inaugural Faraday Medal in 1922. Published in 1912, this is the last of three volumes summarising Heaviside's enormous contribution to electromagnetic theory. It includes a review of his work on waves from moving sources, and an appendix on vector analysis that compares its merits to quaternions.

Cambridge University Press has long been a pioneer in the reissuing of out-of-print titles from its own backlist, producing digital reprints of books that are still sought after by scholars and students but could not be reprinted economically using traditional technology. The Cambridge Library Collection extends this activity to a wider range of books which are still of importance to researchers and professionals, either for the source material they contain, or as landmarks in the history of their academic discipline.

Drawing from the world-renowned collections in the Cambridge University Library, and guided by the advice of experts in each subject area, Cambridge University Press is using state-of-the-art scanning machines in its own Printing House to capture the content of each book selected for inclusion. The files are processed to give a consistently clear, crisp image, and the books finished to the high quality standard for which the Press is recognised around the world. The latest print-on-demand technology ensures that the books will remain available indefinitely, and that orders for single or multiple copies can quickly be supplied.

The Cambridge Library Collection will bring back to life books of enduring scholarly value (including out-of-copyright works originally issued by other publishers) across a wide range of disciplines in the humanities and social sciences and in science and technology.

Electromagnetic Theory

VOLUME 3

OLIVER HEAVISIDE

CAMBRIDGE
UNIVERSITY PRESS

CAMBRIDGE UNIVERSITY PRESS

Cambridge, New York, Melbourne, Madrid, Cape Town,
Singapore, São Paolo, Delhi, Tokyo, Mexico City

Published in the United States of America by Cambridge University Press, New York

www.cambridge.org
Information on this title: www.cambridge.org/9781108032179

This edition first published 1912
This digitally printed version 2011

ISBN 978-1-108-03217-9 Paperback

ELECTROMAGNETIC THEORY.

BY

OLIVER HEAVISIDE.

VOLUME III.

LONDON:
"THE ELECTRICIAN" PRINTING AND PUBLISHING COMPANY,
LIMITED,
SALISBURY COURT, FLEET STREET, E.C,
And all Booksellers at Home and Abroad.

Printed and Published by
"THE ELECTRICIAN" PRINTING AND PUBLISHING CO., LTD.,
1, 2 and 3, Salisbury Court, Fleet Street,
London, E.C.

PREFACE TO VOL. III.

LONG ago I had the intention, if circumstances were favourable, of finishing the third volume of this work about 1904, and the fourth about 1910. But circumstances have not been favourable. That is all that need be said about it here, save to add that I have excluded parts of the third volume, and included parts of the fourth.

It would be as wrong to love your enemies as to hate your friends. Nevertheless, "the way of life is wonderful; it is by renunciation." Especially when prodding is no longer necessary. If my life is spared, I hope to be able to present a bust of the eminent electrician who invented everything worth mentioning to the Institution over which he once ruled, to be placed under that of Faraday.

OLIVER HEAVISIDE.

August 23, 1912.

IN MEMORY OF

GEORGE FRANCIS FITZGERALD, F.R.S.

" We needs must love the highest when we" know him.

CONTENTS OF VOLUME III.

[The dates within brackets are the dates of first publication, or of printing. All up to p. 158 was published in " The Electrician " ; the rest of the articles in the Journals indicated ; but if only printed at the dates given, they are now first published, 1912.]

CHAPTER IX.

WAVES FROM MOVING SOURCES.

(Pages 1 to 130.)

SECTION PAGE

APPENDIX J.

APPENDIX K.

CHAPTER X.

WAVES IN THE ETHER.

(Pages 144 to 519.)

CHAPTER IX.

WAVES FROM MOVING SOURCES.

Adagio. Andante. Allegro moderato.

§ 450. The following story is true. There was a little boy, and his father said, "Do try to be like other people. Don't frown." And he tried and tried, but could not. So his father beat him with a strap; and then he was eaten up by lions.

Reader, if young, take warning by his sad life and death. For though it may be an honour to be different from other people, if Carlyle's dictum about the 30 millions be still true, yet other people do not like it. So, if you are different, you had better hide it, and pretend to be solemn and wooden-headed. Until you make your fortune. For most wooden-headed people worship money; and, really, I do not see what else they can do. In particular, if you are going to write a book, remember the wooden-headed. So be rigorous; that will cover a multitude of sins. And do not frown.

There is a time for all things: for shouting, for gentle speaking, for silence; for the washing of pots and the writing of books. Let now the pots go black, and set to work. It is hard to make a beginning, but it must be done.

Electric and magnetic force. May they live for ever, and never be forgot, if only to remind us that the science of electromagnetics, in spite of the abstract nature of the theory, involving quantities whose nature is entirely unknown at present, is really and truly founded upon the observation of real Newtonian forces, electric and magnetic respectively. I cannot appreciate much the objection that they are *not* forces; because they *are* the forces per unit electric and magnetic pole. All the same, however, I think Dr. Fleming's recent proposal that electric force and magnetic force shall be called the voltivity and the gaussivity a very good one; not as substitutes for with abolition of the old terms, but as alternatives; and beg to recommend their use if found useful, even though I see no reason for giving up my own use of electric and magnetic force until they become too antiquated.

Having thus got to the electric and magnetic forces, it is only a short step farther to near the end of the book—namely, to the simple cases in which they occur simultaneously. It does not follow that the matter which comes towards the end of a treatise—for instance, Maxwell's great work—is harder than that in the first chapter of his Vol. I. On the contrary, some parts of it are easier out of all comparison. In the course of the next generation many treatises on electromagnetics will probably be written; and there is no reason whatever (and much good reason against it) why the old-fashioned way of beginning with electrostatics (unrelated to the general theory) should be followed. After all, should not the easier parts of a subject come first, to help the reader and widen his mind? I think it would be perfectly practical to begin the serious development of the theory with electromagnetic waves of the easy kind. First of all, of course, there should be a good experimental knowledge all round, not necessarily very deep. Then, considering the structure of a purely theoretical work to co-ordinate the previous, a general survey is good to begin with, with consideration in more detail of the properties of circuits and the circuital laws. Then, coming to developments, start with plane electromagnetic waves in a dielectric non-conductor. The algebra thereof, even when pursued into the details of reflections, &c., is perhaps more simple than in any other part of the science, save Ohm's law and similar things; and the physical interest is immense. You can then pass to waves along wires. First the distortionless theory in detail, and then make use of it to establish the general nature of the effects produced by practical departures from such perfection, leaving the difficult mathematics of the exact results for later treatment. Now, all this and much more is ever so much easier than the potential functions and spherical harmonics and conjugate transformations with which electrostatics is loaded, and there is more exercise for the brains in the electromagnetic than in the electrostatic problems. The subsequent course may be left open. There are all sorts of ways.

Simple Proof of Fundamental Property of a Plane Wave.

§ 451. At present, in dealing with some elementary properties, the object is to smooth the road to the later matter. First of all, how prove the fundamental property of a plane wave, that it travels at constant speed undistorted, if there be no conductivity, or, more generally, no molecular interference causing dispersion and other disturbances? We have merely

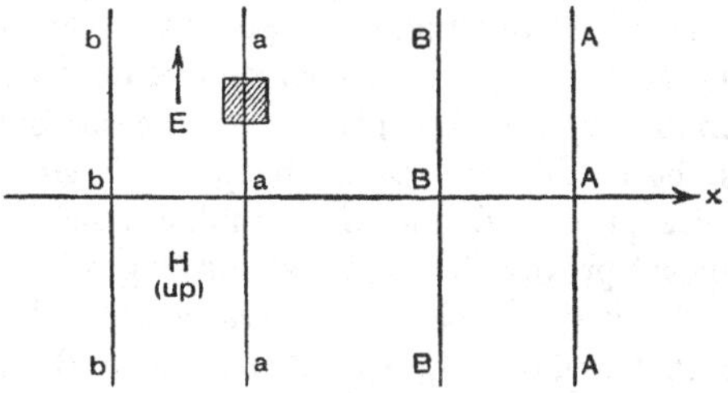

FIG. 1.

to show that the two circuital laws are satisfied, and that can be done almost by inspection. Thus, let the region between two parallel planes *aaa* and *bbb* be an electric field and a magnetic field at the same time, the electric force **E** being uniform and in (and parallel to) the plane of the paper, whilst H is also uniform, perpendicular to **E** and directed up through the paper. Also let their intensities be connected by $E = \mu v H$, or $H = c v E$; c being the permittivity and μ the inductivity, whilst v is defined by $\mu c v^2 = 1$. This being the state at a given moment, such as would be maintained stationary by steadily acting impressed forces $\mathbf{e} = \mathbf{E}$ and $\mathbf{h} = \mathbf{H}$ in the slab between the planes, what will happen later, in the absence of all impressed force?

Apply the two circuital laws. They are obviously satisfied for all circuits which are wholly between the planes a and b, or else wholly beyond them to right or left. There are left only circuits which are partly inside and partly outside the slab. Consider a unit square circuit in the plane of the paper, as shown in the figure. The circuitation of E is simply E, and by the second circuital law this must be the rate of decrease of induction $B = \mu H$ through the circuit downward, or its rate of increase upward. Then turn the square circuit at right angles to the paper. The circuitation of H is then simply H, and by the first circuital law this must be the rate of increase of displacement $D = Ec$ through the circuit. Now let the plane *aaa* move to the right at speed *v*. The two rates of increase are made to be $v\mu H$ and vcE respectively. That is, $E = \mu v H$ and $H = cvE$ express the circuital laws. They are harmonised by the definition of *v*. We prove that the circuital laws are satisfied in the above way. That there is no other way of putting induction and displacement in the two circuits may be seen by considering circuits two of whose sides are infinitely near the plane *a* on opposite sides. The fluxes must be added on just at the plane itself, extending the region occupied by E and H. Similar reasoning applied to the plane *bbb* proves that it must also move to the right at speed *v*. Thus the whole slab moves bodily to the right at speed *v*, so that *a* moves to A and *b* moves to B in the time given by $vt = a\text{A}$ or $b\text{B}$.

The disturbance transferred in this way constitutes a pure wave. It carries all its properties with it unchanged. The density of the electric energy, or $U = \frac{1}{2}cE^2$, equals the density of the magnetic energy, or $T = \frac{1}{2}\mu H^2$. The flux of energy is $\mathbf{W} = \mathbf{v}(U + T)$, the simplest case of the general formula $\mathbf{W} = \mathrm{V}(\mathbf{E} - \mathbf{e})(\mathbf{H} - \mathbf{h})$. *See* Vol. I., § 70.

The General Plane Wave.

§ 452. What is proved for a discontinuity is proved for any sort of variation. For the slab may be of any depth and any strength, and there may be any number of slabs side by side behaving in the same way, all moving along independently and unchanged. So $E = \mu v H$ expresses the general solitary wave, where, at a given moment, E may be an arbitrary function of *x*, real and single-valued of course, but without any necessary continuity in itself or in any of its differential coefficients. Denoting it by $f(x)$ when $t=0$, it becomes $f(x-vt)$ at the time t.

If we change the sign, and make $E = -\mu v H$, this will represent a negative wave, going from right to left. There may be a positive and a negative wave coexistent, separate in position, or superimposed. This constitutes the complete solution for plane waves with straight lines of E and H. If E_0 and H_0 are given arbitrarily (with no connection) at the moment $t=0$, the two waves at that moment are

$$\text{(positive)}\quad E_1 = +\mu v H_1 = \tfrac{1}{2}(E_0 + \mu v H_0),$$

$$\text{(negative)}\quad E_2 = -\mu v H_2 = \tfrac{1}{2}(E_0 - \mu v H_0),$$

as may be immediately verified. Move E_1 to the right, and E_2 to the left, at speed v, to produce the later states.

Since every slab is independent of the rest, there need be no connection between the directions of E in one slab and the next. The direction may vary anyhow along the wave. This makes a mathematical complication of no present importance, the behaviour of individual slabs being always the same.

The overlapping of positive and negative waves should be studied to illustrate the conversion of electric to magnetic energy, or conversely. For two equal waves moving oppositely, which fit when they coincide, there is a complete temporary disappearance and conversion of one or the other energy, according as E is doubled, leaving no H, or H is doubled, leaving no E. If E_0 exists alone initially, it makes two equal oppositely-going waves, of half strength as regards E. Similarly as regards initial H_0.

The reversal of sign of both E and H in either a positive or a negative wave does not affect the direction of motion. But if only one be reversed, it is turned from a positive to a negative wave, or conversely. Slabs of uniform strength should be studied, not simply periodic trains of waves, for simplicity of ideas. Only when there is dispersion, and the wave speed varies, is it necessary to consider a train of waves of given frequency or given wave-length; because then a slab spreads out behind as it travels, producing a diffused wave of difficult mathematical representation, by reason of the partial reflection of its different parts as it progresses.

Generation of Waves by a Plane Source of Induction.

§ 453. Consider next how to generate plane waves by impressed electric or magnetic force. Say by **e** first. It must obviously be of the same type as **E**, *i.e.*, in uniform slabs. But it is not **e** itself, but its curl, say **f**, that is the real source of the waves. A plane surface of **f** is the simplest case. Let **e** be uniform on the right side and zero on the the left side of the plane AA, beginning to act at the moment $t=0$, and continuing steady later. What will happen? Here

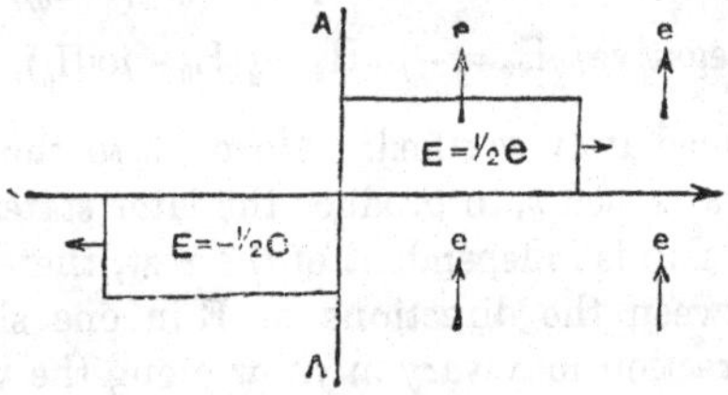

FIG. 2.

f the curl of **e** is uniform on the sheet AA; its density is $f=e$, and its direction is upward through the paper. It generates induction at the rate f on the plane AA per unit area. Or we may say that the strength of the source of B is f. Once generated, the induction divides fairly to right and left, and since the speed is v, the amount $\frac{1}{2}f$ is spread over the distance v in a second. Therefore

$$B=\frac{f}{2v}, \qquad \text{or} \qquad H=\frac{f}{2\mu v}= \pm cv\mathrm{E} \tag{1}$$

expresses the full connection between f and the waves generated. The plus sign is for the positive wave, and the minus sign for the negative wave. We introduce E in this way because the disturbance, once started, makes free waves. The solution, of course, only holds good up to the two wave fronts, which are at distance vt from the source. H is up through the paper in both waves. Here we see the inner meaning of the impedance $2\mu v$ of the doubly infinite unit tube of flux of energy. It depends essentially upon the speed with which the medium can carry away from the source the induction supplied there. The value of E is $\frac{1}{2}e$ on the right and $-\frac{1}{2}e$ on the left side of the plane source. Since **E** is the force of the flux **D**, the force of the field, or **E** − **e**, is of value $-\frac{1}{2}e$ all the way between the two wave fronts.

When **f** varies anyhow in time, it is just the same as regards the generation of **B**. If **f** is impulsive, it makes two impulsive waves. Equations (1) may be used when **f** is variable, if we understand that the elementary slab of **E** and **H** referred to, say at distance x, belongs to the **f** at the source at the moment earlier by the amount x/v.

It is **e** that does the work, though, but only where there is electric current. That is, only at the wave front on the positive side, when the source is steady, and intermediately only when f varies; because $-dH/dx$ measures the current density. The flux of energy **W** is V(**E** − **e**)**H**. It is entirely from the right to the left wave front when **f** is steady. It may be tested that

$$\mathbf{e}\dot{\mathbf{D}} = \dot{U} + \dot{T} + \operatorname{div}\mathbf{W} \tag{2}$$

is the equation of activity in any case.

In the case of a uniform slab of **e** of finite depth, there are two plane sources of **f**, acting oppositely, or one as a source in the above manner and the other simultaneously as an equal sink of induction. There are therefore two positive and two negative waves, separate up to a certain time, and then superimposed. Four stages are shown in the figure.

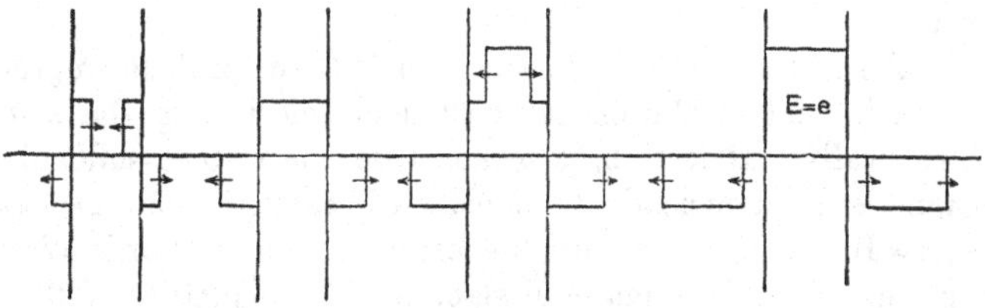

Fig. 3.

First, before the initial waves have begun to overlap inwardly; secondly, when overlapping has just commenced; thirdly, before the overlapping is completed; fourthly, a little while after completion of overlapping, showing the emergence of two pure waves. It is the **E** that is referred to. Every wave has its corresponding **H** according to $E = \pm\mu v H$.

The final result is $E = e$ in the slab, and two pure waves in which $E = -\frac{1}{2}e$. The energy of these free waves together equals that of the steady flux of **D** without **B** which is established in the slab. The work is done by **e** where **e** and **D** coexist, that is, at the wave fronts in the region of **e**. The total induction is zero, because there are two opposite **f**'s. The total displacement is also zero, for another reason.

Generation of Waves by a Plane Source of Displacement.

§ 454. In order to generate displacement finitely in total amount, we require another kind of source. Let the impressed force be magnetic, say **h**. Let its negative curl be **g**. Then **g** generates **D** exactly in the same way as **f** generates **B**, as before described in detail. Thus, considering a single plane source, g is the total displacement generated per unit time per unit area of the plane, and $\frac{1}{2}g$ the amount going each way, spreading over the distance v in a second. The result is that a steady source makes

$$D = \frac{g}{2v}, \quad \text{or} \quad E = \frac{g}{2cv} = \pm \mu v H \qquad (3)$$

between the plane of **g** and the two wave fronts. The same Fig. 2 will do for this case, only for E must be understood H in the two waves. Otherwise stated, $H = \frac{1}{2}h$ on the right side and $-\frac{1}{2}h$ on the left; whilst $H - h = -\frac{1}{2}h$ on both sides. The work is done by **h** only where there is magnetic current.

If the impressed force in the slab is **e**, the problem represented is that of the effect of suddenly electrising the slab intrinsically. If it is **h**, it means that the slab is suddenly magnetised intrinsically to density of magnetisation $I = \mu H = \mu h = B$. This represents the complete and full induction possible. If it is a material slab, and the inductivity differs from that outside, the waves will not be quite so simple. But it is only when the slab is of finite depth that we can get the full induction; when of infinite depth one way, the final result is only half as much; no steady state is reached; the flux of energy continues indefinitely. And if the slab is infinitely extended both ways, a uniform **h** cannot produce any induction in it. There is no **g** source.

As regards the situations of e and h to suit given f and g, they may be altered in any way we like, provided f and g do not change. Thus, considering the magnetisation of a slab, h may be shifted outside it, provided we reverse its direction. It is now the external region that is the seat of the intrinsic magnetisation. Nevertheless, the induction will be set up in the slab, whilst only the free waves will traverse the magnetised region on their way out.

If the sheet of f, when there is but one sheet, is of finite breadth, instead of extending over an infinite plane, all the above results regarding the generation of waves are true at a given point up to the moment when disturbance reaches that point from the nearer of the two edges of the strip of f. That is, there is immediate failure at the edges, involving two waves of a fresh kind, which "interfere" with the two plane waves. This case will be given later. Reducing the width of the strip brings the source down to a single straight line of f. This generates a single cylindrical wave. [*See* § 466.]

Returning to the infinitely extended plane sheet source of disturbance, let it be the seat of both f and g simultaneously. Then we can adjust them so as to generate a single wave, going either to the right or to the left of the sheet. This may be shown by using the previous results for f and g separately, and superposing them. Take $e = \pm \mu v h$, or $f = \pm \mu v g$; and a single wave will result. It will be of full strength, instead of half strength.

Comparison of Electromagnetic with Aerial Waves.

§ 455. It is perhaps worth while pointing out the corresponding properties in sound waves in air. Let s be the condensation and u the speed in a plane wave, that is, the speed of the air perpendicular to the plane, or in the direction of motion of the wave, whose own speed is v. Then the connections of the two variables s and u are

$$f - v^2\frac{ds}{dx} = \dot{u}, \qquad g - \frac{du}{dx} = \dot{s}, \tag{4}$$

where the new quantities f and g represent sources. If ρ is the density, and $\mathrm{F} = \rho f$, then F represents impressed moving force per unit volume. What it generates is momentum.

And if $G = \rho g$, then what G generates is matter; in another form, it is a source, real or effective, of condensation. Comparing with the corresponding equations for electromagnetic plane waves, viz.,

$$f - \frac{d\text{E}}{dx} = \mu\dot{\text{H}}, \qquad g - \frac{d\text{H}}{dx} = c\dot{\text{E}}, \tag{5}$$

where f is the measure of the curl of e, and $-g$ that of h, we may conclude that $u = \pm vs$ is the relation corresponding to the electromagnetic $\text{E} = \pm \mu v \text{H}$. That is, in a plane sound wave the air speed and condensation are in constant ratio, and the wave always moves the same way as the air in its condensed parts. (Rayleigh, "Sound," Vol. II., § 245.)

Also, if there be a plane source F generating momentum, F acting normal to the plane,

$$\rho u = \frac{\text{F}}{2v}, \quad \text{or,} \quad u = \frac{\text{F}}{2\rho v} = \pm vs \tag{6}$$

shows the u in the two waves leaving the source, first by splitting of the momentum, and then by its transfer away from the source.

Similarly, if G is the amount of fluid supplied at the plane per unit area per second,

$$\rho s = \frac{\text{G}}{2v}, \quad \text{or,} \quad s = \frac{\text{G}}{2\rho v} = \pm \frac{u}{v} \tag{7}$$

shows the condensation produced in the two waves which result.

If F and G are coexistent, and $\text{F} = \text{G}v$, the result is

$$u = vs = \frac{\text{F}}{\rho v},$$

on the right side only. Take $\text{F} = -\text{G}v$ to bring it to the left side.

Of course we cannot have a continuous, steadily-acting source of condensation in a fluid; it must fluctuate. There is no such restriction necessary in the electromagnetic case.

Waves ending perpendicularly upon a Conductor. Conduction and Convection.

§ 456. If a perfectly conducting sheet of metal be placed so that the axis of a plane wave lies in the sheet, it will usually interfere with the continuance of the wave in its simple state. The exception is when the sheet (infinitely thin) is turned round so as to be perpendicular to D in the wave. Then the wave will run on unchanged. It will run on to the sheet at P (Fig. 4), traverse the strip, and run off at Q, and go on as it was before. At the same time, it is to be remarked that the wave is really split into two waves, which are quite independent of one another save in the accident of being timed together. The upper side of the strip is positively electrified, the lower side negatively, and the electrifications cancel one another as the wave runs off.

Either of the waves may be abolished, so far as the possibility of existence of the other is concerned (Fig. 5). It will run along the strip by itself. But when it comes to the edge of the strip, what then? The wave goes right on. But

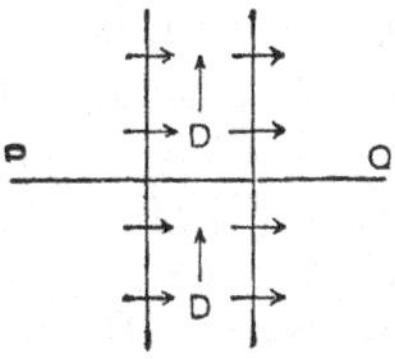

FIG. 4.

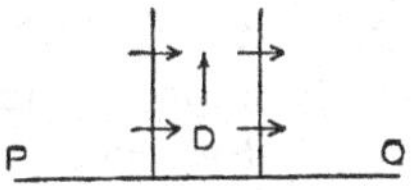

FIG. 5.

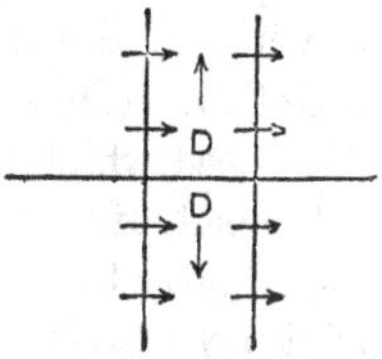

FIG. 6.

since* the electrification cannot leave the sheet, a new wave is generated, beginning the moment the wave reaches the edge. It is cylindrical, having its axis at the edge. It, together with the plane wave, form the complete new state of D. But this is too complicated for present discussion.

To simplify matters, we may prevent the generation of the cylindrical wave by convection. Directly the front of the wave reaches the edge, or at any rate not later than that moment, set the conducting sheet moving with the wave. Then the plane wave will go on unchanged. Fig. 5 will do for this case also, PQ being now imagined to move instead of being stationary. The current on the strip is no longer a conduction current, but a convection current. Also, it no

* "Since" had better be "if," in view of modern experiments relating to the discharge of conductors. Such secondary effects are not in question here.

longer matters whether the sheet is conducting or non-conducting. It must be something that can be charged. In addition, there is the assumption that its motion does not disturb the ether. For merely descriptive purposes we may speak of the motion of electrification. The surface density being D, the convection current is Dv per unit area. It is continuous with the electric currents on the front and back of the slab wave. The current is the curl of the magnetic force, as usual.

We may have another wave below the one in question, and there need be no connection between their magnitudes. An extreme case is reached by letting D be of the same strength, but oppositely directed in the two waves, as in Fig. 6. As there is similar electrification on both sides of the strip, the electrification density is 2D if we ignore the strip. This example is of some significance, because the state represented is what would be tended to if the electrification, given initially at rest, were set in motion in its own plane along PQ at a speed which ultimately became v.

Instead of one plane of electrification, as in Fig. 6, there may clearly be any number. In the limit we come to any arbitrary distribution of electrification. For instance, D may be any function of y, the distance measured parallel to D. Its divergence is then the density of electrification, which must be moved with the wave of course, or it will break up.

Returning to Fig. 5, if the reader thinks, he may observe a curious thing. The current is the curl of H, both when the wave runs along the stationary conductor and when the conductor is carried along with the wave. In the latter case Dv is the surface current density, and it is a convection current. But when the conductor is stationary the electrification is there all the same, and it moves at the same speed. Why is it not a convection current in this case also? Because the conductor is not moved. This is not a mere matter of convention. To illustrate: Suppose the conductivity is not quite perfect, so that there is a thin film of conduction current near the surface. H is now continuous there, and there is no surface current, in spite of the transfer of electrification. The variation of H which measures the current is distributed throughout a thin film under the electrification belonging to the wave. This remains true however great the conductivity is made. In the limit, with perfect conductivity the current becomes a surface current. But there is not the slightest reason for supposing that it suddenly, in the limit, becomes a convection current. To obtain the convection current, abolish the conduction current by moving the material supporting the electrification.

In the true physics of the matter, the current in the conductor itself may be partly of the convective nature, of course, by the electrons being set moving by the wave penetrating the conductor from its surface. But that is another matter. A theory of matter is now in course of evolution, but it is not made yet. In any case, however, the supposed interior convection currents represent a secondary action due to the electromagnetic wave running along the surface of the conductor.

Oblique Reflection at a Conducting Surface with H tangential. Transformation to a Convection Problem.

§ 457. Returning to a perfectly conducting plane surface PQ at rest, let the H in a plane wave be tangential, and perpendicular to the plane of incidence, but let the D be no longer perpendicular to the conductor. Fig. 7 shows an unstable

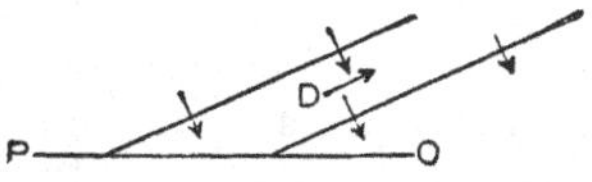

FIG. 7.

state of affairs, because E has a tangential component, which would be instantly destroyed by the generation of a new wave. Passing over the complicated details of the process, consider the ultimate result that is tended to, shown in Fig. 8.

Here we have a pure wave on the right side moving obliquely towards the conductor, and an exactly similar wave on the left side moving away from it. They are represented by plane slabs. They join together perfectly where they overlap, at PAQ and just above. If we consider a single tube of flux of energy in the incident wave, with its continuation in the reflected wave, the two together make an incident and reflected ray (or thin beam). For instance, X is incident, and Y reflected, at the point A. The energy travels along XA in the right wave, and then along AY in the left wave. In the overlapping region H is doubled. It is up through the paper in both waves. Also, D is vertical. If θ is the angle of incidence, PAQ travels along the surface from left to right at speed u connected with v by $v = u \sin \theta$.

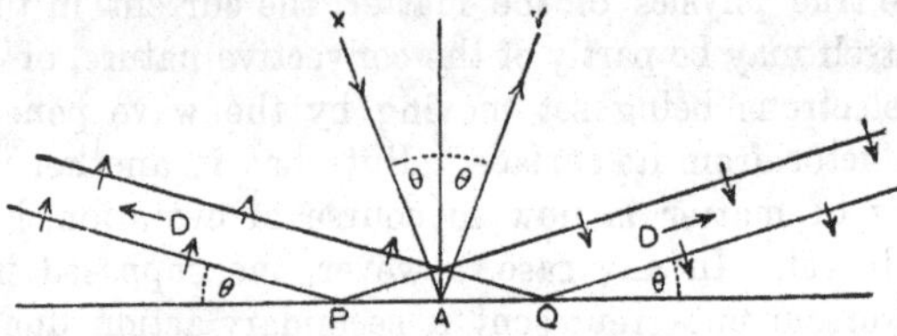

FIG. 8.

Thus, $u = v$ at grazing incidence. That is, when XAY is horizontal. The two waves then make one doubled simple wave, 2D being perpendicular. But at perpendicular ray incidence $D = 0$. This is the case of direct reflection of a plane slab striking flush against a perfect conductor. The electric force is destroyed, and H is doubled during coincidence of the incident and reflected waves. The reflected wave has the same H but reversed E. Five stages are shown in Fig. 9.

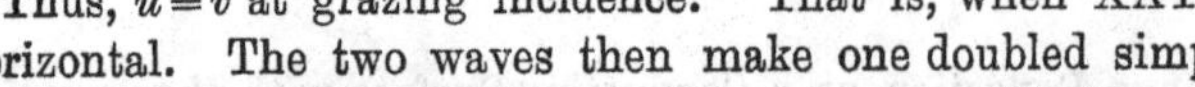

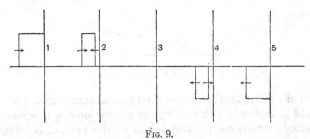

FIG. 9.

Stage 1 shows E (by the ordinate) at the moment the front of the slab strikes the conductor; 2, when it has gone in one-fourth; 3, when half gone in; 4, when three-fourths gone in; and 5, all gone in, and replaced by a pure negative wave. "Gone in" does not here imply entrance.

Now consider the case of oblique incidence, represented by Fig. 8. The density of electrification is $2D \sin \theta$, and its speed of transference is $u = v/\sin\theta$; so $2Dv$, that is, 2H, is the equivalent electric current per unit surface of conductor. Now 2H is the curl of H under the circumstances, but, as before explained, it is a conduction current. We may, however, easily turn the problem into one relating to convection. When, or before the electrification reaches the end of the sheet, move the sheet itself in its own plane along with the electrification. Then the two waves will go on behaving exactly as before. We may dismiss the idea of a conductor, and consider PAQ to represent a plane strip of electrification, density $= 2D\sin\theta$, moving in its own plane perpendicularly to its edges at the speed u, accompanied by the two electromagnetic waves. An upward pressure on the strip is required to maintain the state, but no work is done by it. Energy is being constantly transferred from one wave to the other. In fact, we may say that the one wave is being continuously turned into the other.

Generation of a Pair of Inclined Plane Waves by Motion of an Electrified Strip; $u>v$. The applied force required.

§ 458. This state of things is obviously not that which could arise from putting the strip, given charged and at rest, into motion at the speed u, because there is a wave in advance of the strip, which is moving faster than the speed v of propagation

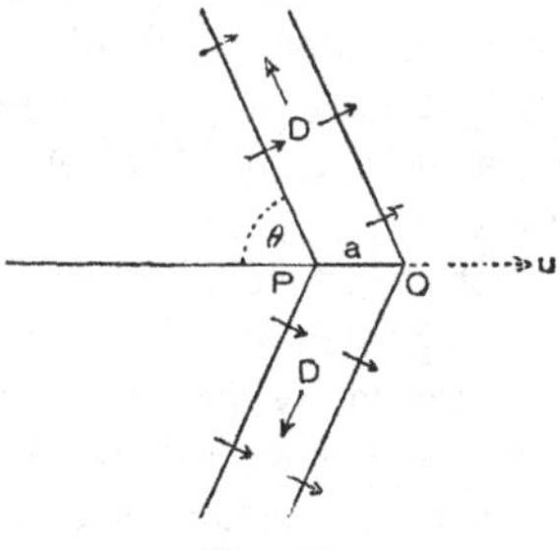

FIG. 10.

of disturbances. But we may readily construct the proper solution of the new problem stated. In Fig. 8, keep the reflected wave in its place, but turn the incident wave over to the other side and start afresh, as in Fig 10. The electrified strip PQ moves along the x-axis at any speed $u>v$. The final state that is tended to consists of two plane waves as shown, inclined backwards. D is away from the strip in both waves, which are exactly similar, and the angle between either D and the axis is the θ given by $v=u\sin\theta$. This solution is only true when $v<u$, or in the limit $v=u$. When $v>u$, the displacement spreads out over all space, obviously.

This is a very important case theoretically, in spite of its simplicity, so some more details may be given. There is no conversion of one wave into another, as in Fig. 8, but a real generation of the waves by the motion of the electrification. For the waves move normal to their planes, as usual, and additions are being continuously made to them at PQ. If the breadth of the strip is a, the depth of the two waves is $b=a\sin\theta$. The energy density is 2U; and the speed of elongation of the waves is $u\cos\theta$. Therefore the rate of increase of energy in the upper wave (per unit distance perpendicular to the paper) is

$$2\mathrm{U}a\sin\theta\times u\cos\theta=\mathrm{U}ua\sin 2\theta,$$

and its double is therefore the rate of increase of the energy of both waves. Therefore, by the law of activity,

$$\mathrm{F}=2\mathrm{U}a\sin 2\theta$$

measures the applied force required to maintain the electrification in steady motion. This force F acts along x. That is, it is a push from behind or a pull in front. If it be removed, the electrification will at once begin to slow down, and this will continue until its speed is v. The state represented in Fig. 6 will be tended to.

As Newton is not yet superseded, we must have equal action and reaction. The companion to the impressed F is the force exerted on the electrification by the stress in the two waves. The stress in a pure wave is a pressure 2U acting along the ray. So $2\mathrm{U}a\cos\theta$ is the slanting back pressure upon PQ from above. Similarly in the lower wave. Their up and down components cancel one another. Their tangential components are additive, and together make $2\mathrm{U}a\sin 2\theta$, as before got by a consideration of the energy added to the waves.

It is not necessary for the waves to be externally unbounded. The displacement may be terminated upon stationary perfectly conducting plane surfaces. So we get the state of things shown in Fig. 11, where the two planes are AB and AC. Everything else is as before. Here the energy (per unit depth normal to the paper) is always finite, increasing at a uniform rate. To make more complete, we may show how to initiate the waves. Starting with zero displacement everywhere, let an impulsive voltage act in the direction of u all along the line A, generating the double charge $+\rho$ and $-\rho$ there, and let the line of $+\rho$ be immediately carried away at speed u. The line of $-\rho$ will then split into two halves, one of which will run along AC and the other along AB at the speed v. They will be joined to the $+\rho$ by two impulsive plane waves, as in Fig. 11, save that the waves there are slabs. To have slab waves, the initial voltage must continue for a finite interval of time,

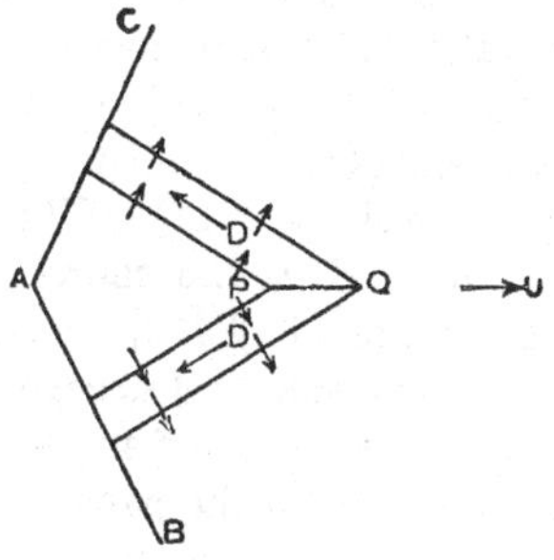

FIG. 11.

Now go back to Fig. 10. If the width of the strip PQ be reduced to nothing, whilst the charge upon it is maintained the same, we come to the case of an electrified line moving transversely at any speed greater than v, accompanied by impulsive plane waves. No separate figure is needed. But the force F, although it is finite when PQ is finite, becomes infinitely great by the concentration of electrification assumed. To show this, put F in terms of the total charge on PQ. If Q be this charge (per unit length along the strip), we have

$$\mathrm{Q} = 2\mathrm{D}a \sin\theta, \quad \text{and} \quad \mathrm{U} = \tfrac{1}{2}\mathrm{ED} = \tfrac{1}{2}c\mathrm{E}^2.$$

These make F become

$$\mathrm{F} = \frac{\mathrm{Q}^2}{2ca}\cot\theta = \frac{\mathrm{Q}^2}{2ca}\left(\frac{u^2}{v^2} - 1\right)^{\frac{1}{2}},$$

which increases infinitely with a^{-1} and also with u. It is, however, zero when $u=v$. This means that the two waves are in one plane. It is necessary, therefore, to have finite surface density of electrification, in order to be able to move the charge through the ether at speeds greater than light by the application of finite force.

Generation of a Pair of Inclined Plane Waves by a Moving Source of Induction, when $u>v$.

§ 459. Several modifications of these results will come in due course. One may be given here, being closely related to Fig. 10. In the arrangement there represented, reverse **D** in the lower wave. In order that the wave shall move the same way, **H** must also be reversed. The two displacements, in the upper and lower waves, are now continuous. PQ is therefore unelectrified. What then must it represent, in order that the problem may be really electrical, with growth of the waves by external work done on PQ? To answer this, observe that **E** is discontinuous tangentially at PQ. This means that the strip is a strip of **f** or curl **e**, the source which generates induction as before considered, § 453, whilst generating displacement positively on one side and negatively on the other. Remembering that **E** is towards the f surface below and from it above, whilst **H** is up through the paper in both, we have

$$\mathrm{H}=\frac{f}{2\mu v\cos\theta}, \qquad \mathrm{E}=\frac{f}{2\cos\theta},$$

and the total displacement in each wave is

$$\mathrm{Q}=\tfrac{1}{2}cfa\tan\theta.$$

In the case of a filament, put $af=f_0$. That is, f_0 is the line source. Then we have two impulsive waves, in which

$$\mathrm{Q}=\tfrac{1}{2}cf_0\tan\theta.$$

But if f_0 is finite, there is the same infiniteness of applied force involved as previously.

Two ways of Dividing a Pair of Crossing Pure Plane Waves to make Convection Problems; $u>v$.

§ 460. An extension of the case represented by Fig. 10 may be arrived at in a different way. In Fig. 12 are shown two pure impulsive waves crossing one another. The magnetic force H is up through the paper in one wave and down in the other; both are equally intense. There is nothing abnormal. Remembering, however, that the waves move perpendicular to their planes at speed v, it follows that their line of intersection moves from left to right at speed u given given by $v=u\sin\theta$, where θ is the angle between D and the line of motion. Next, we may regard the line of intersection as consisting of two coincident equally and oppositely electrified lines, charged to linear density ± 2D, if D is the strength of the impulsive waves. This is merely a truism, because they annul one another. What is not a truism, however, but a striking fact, is that if we separate the two charges and the waves from one another, producing the two states shown in Figs. 13, 14, these states are separately possible.

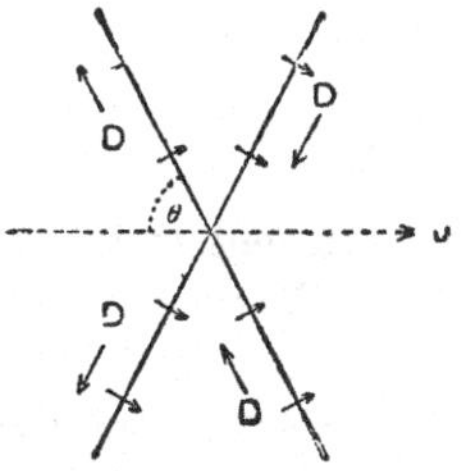

FIG. 12.

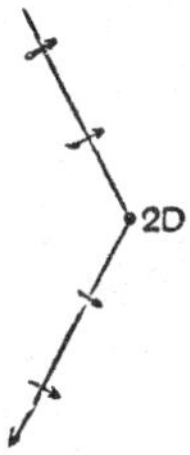

FIG. 13.

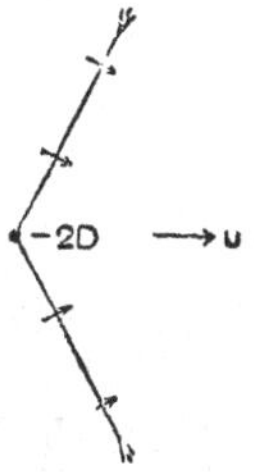

FIG. 14.

Here it may be recognised that Fig. 13 shows the case before considered, Fig. 10, involving the generation of plane waves by an electrified line (or plane) moving at speed u, by force applied to the moving electrification, acting from left to right. In Fig. 14, on the other hand, the waves advance towards one another, and are continuously destroyed at the moving electrified line. The force due to the stress tends to push the electrification along faster; so, to maintain the state steadily, an applied force pulling back the electrification is required. Without this applied force the tendency will be to make the two waves into one, since the force due to the stress will decrease. But this will not take place in any simple manner.

In Fig. 13 the charge may be negative, and in Fig. 14 it may be positive. Both E and H are then to be reversed, u remaining the same. The important distinction between Figs. 13 and 14 is merely that the former arrangement can arise from rest naturally, as before explained, whereas the latter needs some more artificial initial conditions.

Here is another way of splitting the crossed pure waves of Fig. 12:—In Fig. 15 there is no electrified line in either case. To interpret, consider the upper pair of waves, and expand to slab waves, as in Fig. 16. We see that there is no H in the

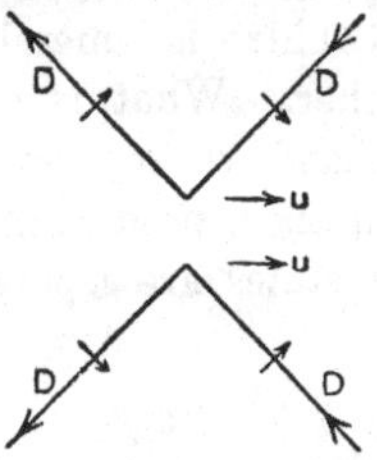

FIG. 15.

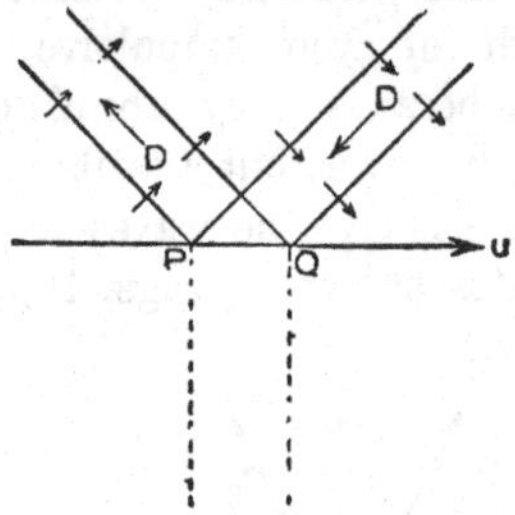

FIG. 16.

overlapping region, and that the displacement is tangential therein. The tangential electric force is $2E \cos \theta$ from right to left. This is the measure of f, the curl of e on the surface PQ. That is, PQ must be a plane f source (say due to intrinsic electrisation), moving at speed u in its own plane. Similar remarks apply to the lower pair of waves in Fig. 15. But since one wave is in advance of PQ, neither of the cases in Fig. 15 can arise from rest in static condition.

Slanting Plane Waves Generated by a Moving Magnet, and by an Electrified Strip moving normally to its Plane.

§ 461. Closely connected with Fig. 16 is the case arising by turning D to B therein. No separate figure is needed. There is now no E in the overlapping region, and the induction is tangential, since it goes down one wave and up the other. The magnetic force at PQ is $2H \cos \theta$. There are two interpretations. First, PQ may be a portion of a stationary perfect electric conductor. Then the right wave is incident and the left wave reflected. Secondly, PQ may be a plane g source (negative curl of h, or intrinsic magnetisation), moving from left to right at speed u. The dotted lines may represent the two sides of a slab magnetised parallel to **u**. Here, however, one wave is in advance of the moving slab, so some special initial state is required.

To substitute a state which could arise from rest naturally, have both waves behind the **g** source, as in Fig. 17. First of all, consider the top part only. The two waves have B normally continuous at PQ, which is right and proper. D is fully continuous there, but the tangential discontinuity in H at PQ is measured by $2H\cos\theta$. This is the measure of the surface density of the moving **g** source. There are many ways of developing this problem. The one shown in Fig. 17 concerns a magnetised slab moving in the direction of

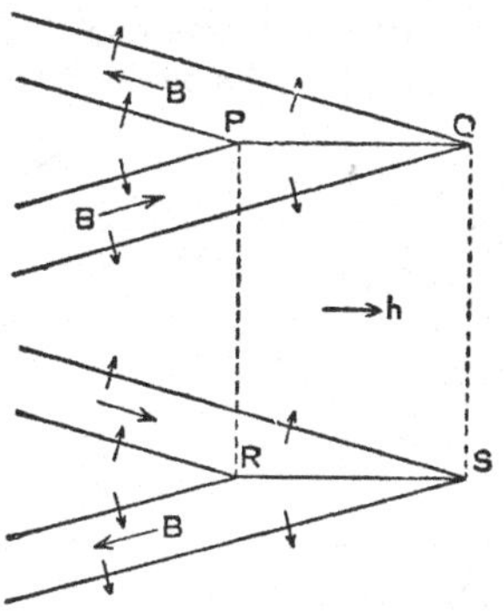

FIG. 17.

its magnetisation at speed u. The dotted lines show the forward and backward faces of the slab, magnetised from left to right, $h = 2H\cos\theta$. PQ and RS are its upper and lower sides, the seat of the **g** sources. They are equal but oppositely directed—viz., down through the paper on PQ and up on RS.

Such a magnetised slab, moved in the way indicated, will ultimately tend to have its induction distributed in the way shown—viz., in two pairs of pure electromagnetic waves, which are being continuously generated by the g sources through the agency of force (Newtonian) applied to the slab. The amount of this force has been given before equivalently, since it is determined by the speed and the density of energy (or the stress) and the depth of the wave. Double as much is wanted here, because there are two pairs of waves. The induction is shifted altogether out of a part of the magnet. That is nothing. More important are the reservations. The inductivity μ of the magnet must be the same as that of the ether. If different, the waves are not so simply calculated, though the general ideas are the same. And no allowance at all is made for any disturbance from rest of the ether itself due to the motion of the magnet. We do not know that such disturbance exists. And the magnet must be non-conducting.

An interesting modification is got by decreasing the depth PR or QS till the magnetised slab becomes a plate moving in its own plane, and the waves in Fig. 17 overlap and cancel so much as to produce an apparently different case. The result is shown in Fig. 18. The plate, magnetised from left

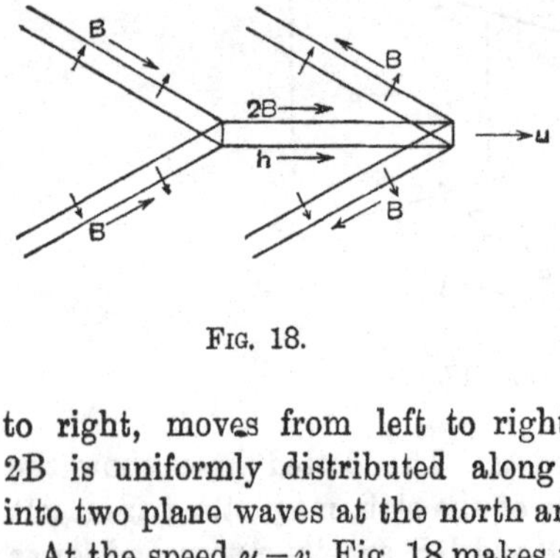

FIG. 18.

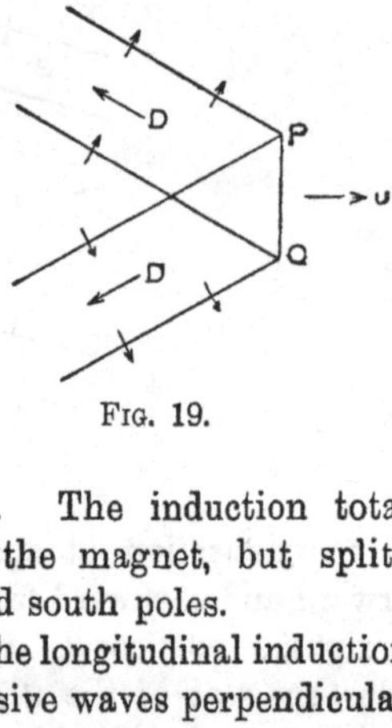

FIG. 19.

to right, moves from left to right. The induction total 2B is uniformly distributed along the magnet, but splits into two plane waves at the north and south poles.

At the speed $u=v$, Fig. 18 makes the longitudinal induction 2B be accompanied by terminal impulsive waves perpendicular to it. On the other hand, Fig. 17 apparently gives nothing when $u=v$. But take u a little greater than v, and it will be found that the overlapping of the waves in Fig. 17 leads to exactly the same result as Fig. 18, both inside and outside the magnet.

We must not take one pair of waves by itself and ignore the rest of the arrangement, because the induction has no divergence in reality. But we may equivalently realise this process by turning B to D. We then obtain Fig. 19, wherein

PQ is an electrified strip moving normally to its plane. This, of course, can be considered a development of the case of the transverse motion of a electrified line. There are, additionally, some interesting points to notice. Thus, there is no H in the overlapping region, where the displacement is normal to PQ. That is, there is no H all along PQ. Yet there is convection current there, of course. Also the current density is the curl of the magnetic force. There is, superficially, a contradiction.

The true current is $\dot{\text{D}}+\text{u}\,\text{div}\,\text{D}$, the sum of the displace ment current and the convection current. This is zero along the strip PQ, because one term is the negative of the other.

If a is the breadth of the strip, the applied force is

$$\text{F}=\tfrac{1}{2}a\frac{(2\text{D}\cos\theta)^2}{c}=\frac{\text{Q}^2}{2ca}$$

per unit depth perpendicular to the paper. Notice that this is finite at any speed, even up to $u=\infty$, when the two waves merge into one, containing no H. The tubes of displacement then simply grow longer at their ends upon PQ.

Generation of a Single Plane Wave by Motion of an Electrified Electret. Also of Two Separated Plane Waves.

§ 462. In the previous, the moving source generates two plane waves, or else is accompanied by these without generation. But we may easily arrange matters so that only one terminated plane wave is generated. This brings us to a new class of problems—viz., concerning double sources with one wave, instead of two waves with one source.

Consider Fig. 20 for example. There is supposed to be a single plane pure wave terminated at PQ, where it is continuously growing. What sources are required? The normal displacement is D sin θ. Therefore this is the surface density of electrification on PQ moving at speed u. But, in addition, the tangential electric force is E cos θ. This indicates the surface density of an f source. So PQ may be the end of a non-conducting slab (sides indicated by dotted lines) electrised to density ce, where $e=\text{E}\cos\theta$; the direction of e being parallel to PQ. By itself this moving slab would produce a pair of waves. The electrification on PQ, however, by itself would also produce two waves; the same as f does above, but oppositely below. So the resultant effect is one wave only.

In Fig. 21 the electrification is the same, but the electrisation is reversed. Also, as the wave is being destroyed, and is

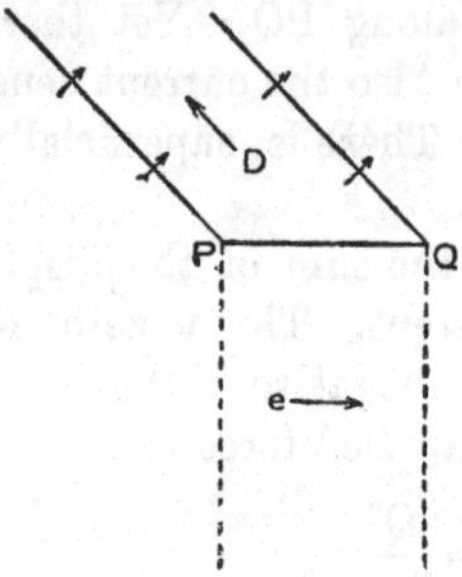

FIG. 20.

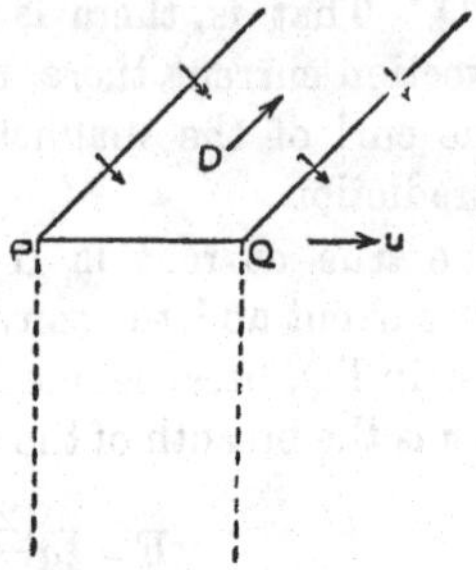

FIG. 21.

in advance of PQ, a special initial state is required, whereas the case of Fig. 20 could arise naturally from rest, if no infinite resistance is met with in the transition from speed 0 to u.

If the waves are underneath PQ, there are similar interpretations.

If we change **D** to **B**, then e is changed to h, or the slab must be intrinsically magnetised. But then the electrification is changed to magnetification. So this imagined single wave is physically unreal.

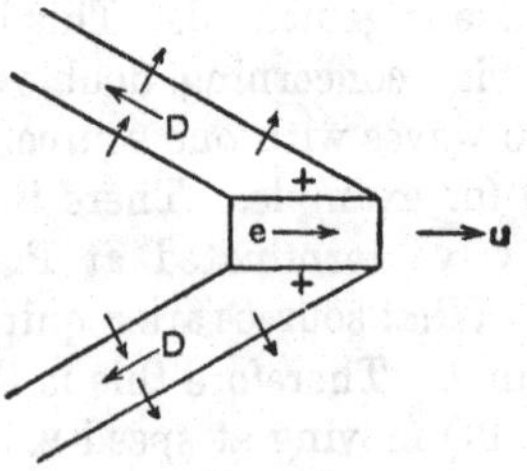

FIG. 22.

When the slab is of finite depth let it be positively electrified on the top and also positively beneath, as in Fig. 22, besides being electrised in the direction of motion. There is no **E** or **H** inside the slab. By reducing the depth till the slab becomes a mere thin sheet the result is the same as Fig. 10, concerning the motion of a single sheet of electrification in its own plane. This is because the effect of e ultimately vanishes, and the two electrifications unite to make one. In fact, Fig. 22 shows how to separate the two waves in Fig. 10.

Although nearly all the above has been descriptive, yet the demonstrations of the various propositions have not been omitted. They all rest upon the simple properties of plane waves which were proved at the beginning, together with the general connections of **E** and **D**, **H** and **B**, **e** and **h**, &c., direct and crosswise. These are to be understood all along, in the way explained in Chap. II., Vol. I., especially the early part. When understood, the satisfaction of the proper conditions at the junctions becomes visible by inspection, and no repeated formal demonstrations are needed.

In electrostatics the lines of force are perpendicular to the equipotential surfaces. But if the charges be set in steady motion, the distribution of displacement is so changed as to reduce the angle between the lines of force and the new equipotential surfaces, the electric force being derived from the potential in an eolotropic manner. As the speed increases the angle between the lines of force and the surfaces decreases, until at the speed of light the lines lie in the surfaces. And after that? In all the above diverging plane waves the lines of force lie in the equipotential surfaces. But this requires some mathematics for its explanation.

Theory of the Steady Rectilinear Motion of a Point-Charge or "Electron" through the Ether when $u<v$ and when $u>v$.

§ 463. In Vol. I., § 164, it was shown that a point charge q in steady rectilinear motion along the axis of z, at a speed u, less than v, was associated with the potential

$$V=\frac{q}{4\pi c\{z^2+\kappa^2(x^2+y^2)\}^{\frac{1}{2}}}, \qquad (1)$$

where $\kappa^2=1-u^2/v^2$, a fraction ranging from 1 to 0, as u increases from 0 to v. Here it is to be understood that V is the potential at the point x, y, z when the origin is at the charge, so that V accompanies q in its motion. It is further to be understood that the electric force is derived from the potential in the eolotropic manner specified by

$$\mathbf{E} = -(\mathbf{i}\nabla_1 + \mathbf{j}\nabla_2 + \kappa^2\mathbf{k}\nabla_3)\mathrm{V}. \tag{2}$$

The displacement is obtained by multiplying by c, as usual. The result, as regards the displacement, is therefore the same as if the charge were at rest, but the permittivity in the direction of z both ways were reduced from c to $\kappa^2 c$. This produces a lateral concentration of the displacement, with the ultimate result that when u is increased to v, the displacement is entirely in the equatorial plane, forming an impulsive plane wave with circular magnetic force. In any case, the magnetic force is the vector product of the velocity **u** and the displacement **D**.

The question now to be considered is what occurs when u is greater than v. Are the formulæ still valid? We can see immediately that some reservations are necessary, even though no change of formula may be required. For κ^2 is now negative; and V, and also **E** and **H** are made imaginary when

$$z^2 < s^2(x^2 + y^2), \tag{3}$$

where $$s^2 = u^2/v^2 - 1 = \operatorname{cosec}^2\theta - 1 = \cot^2\theta. \tag{4}$$

This means that V is real inside the two cones to right and left of the moving charge whose angles are 2θ, connected with

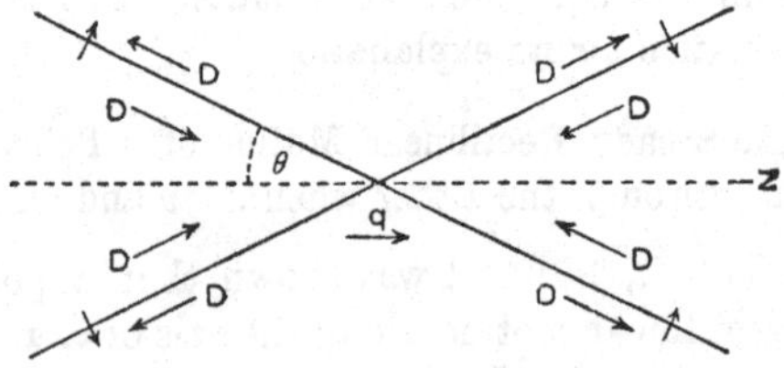

FIG. 23.

s^2 by equation (4), but unreal in the intermediate region outside the cones. We must certainly reject the unreal V, because this is a matter of physics ultimately, not of possible mathematics. In Vol. II. were numerous formulæ relating to waves which had no existence when they assumed unreal meanings. Certainly it is the same thing here.

But next, seeing that disturbances are propagated only at speed v, whilst the charge q moves at the greater speed u, the locus of the spherical waves sent out by the charge as it moves along forms the *left* conical surface only. So we must reject the right cone altogether, if we are considering a charge brought from rest up to speed u.

So far is rejection without change. But closer consideration will make it probable, if not certain, that a change in the formula is wanted as well. For, assuming that equation (1) is correct when κ^2 is negative, provided we keep to real values, it still belongs to both cones. Now it was standardised so as to make the total displacement leaving the charge be q. This was with $u<v$, when the displacement emanated in all directions. As we employ now the same formula, the same property should hold good, keeping to the real values, however. But V is symmetrical. At corresponding points in the two cones V is the same. So the displacement leaving q for the right cone can be only $\frac{1}{2}q$, and similarly for the left cone. The practical meaning is that if we reject the right cone, and still have the charge at the apex of the left cone represented by q, we must double the right side of equation (1). That is to say,

$$\mathrm{V}=\frac{q}{2\pi c\{z^2-s^2(x^2+y^2)\}^{\frac{1}{2}}}\tag{5}$$

should be the proper solution. Similarly, the right members of the formulæ for E and H, true when $u<v$, must be doubled when $u>v$. I gave this result in 1892, El. Pa., v. 2, p. 516, footnote, having obtained it by a straightforward operational process, to be given later. It is possible, however, that some readers may find the above reasoning sufficiently convincing. It is plausible enough, but by no means demonstrative, on account of the infinite values concerned, which evade evaluation. There are two ways of corroborating equation (5), viz., to use it to construct a known solution, and to prove it directly from the operational solution. The first way is much the easier.

At any point P inside the cone, we have

$$\mathrm{V}=\frac{q}{2\pi c}\frac{\tan\theta}{\{z^2\tan^2\theta-(x^2+y^2)\}^{\frac{1}{2}}},\tag{6}$$

or

$$\mathrm{V}=\frac{q}{2\pi c}\frac{\tan\theta}{(\mathrm{PQ}\times\mathrm{PR})^{\frac{1}{2}}}.\tag{7}$$

The last form shows the meaning, the second denominator expressing the geometrical mean of the two parts into which P divides the diameter QR, Fig. 24. So V is a minimum on the axis, and increases to infinity on the cone. Outside the cone V is zero. Deriving the electric force by equation (2), it will be found that E is radial, and is directed *towards* the charge. This is inside the cone. Its size is

$$E = \frac{q}{2\pi c} \frac{r \tan\theta}{l^{\frac{3}{2}}}, \tag{8}$$

at distance r from the apex, where l is the above-mentioned geometrical mean. The conical surface is the seat of a sheet of displacement away from the apex. This follows because V suddenly drops to zero outside the cone. The amount of

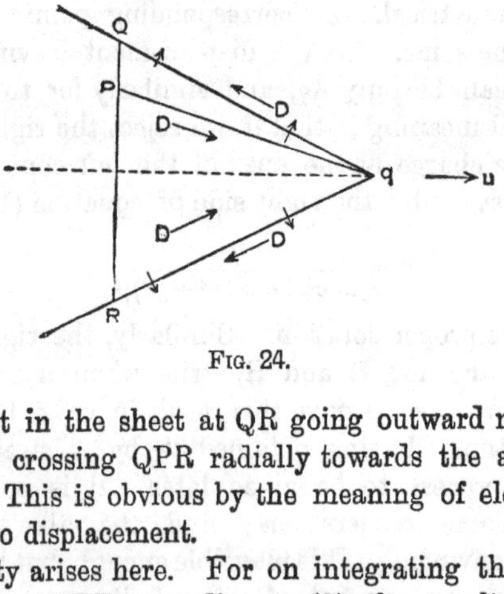

FIG. 24.

displacement in the sheet at QR going outward must exceed the amount crossing QPR radially towards the apex by the amount q. This is obvious by the meaning of electrification in relation to displacement.

A difficulty arises here. For on integrating the flux of D inwards across QPR, according to (8), the result is infinity, the infinite part arising at the boundary. So the outward flux on the outside must be also infinity, plus q. The meaning of this double infinity will appear later.

Construction of the Slanting Plane Waves generated by an Electrified Line by means of the Potential of an Electron.

§ 464. Now use equation (6) to calculate V by integration in the case of an electrified line moving transversely. We know the result already, so can disprove equation (6) if it does not work properly. Let σ be the linear density of the electrified line. Then, by (6),

$$V=\frac{\sigma}{2\pi c}\int\frac{\tan\theta . dx}{(z^2\tan^2\theta-x^2-y^2)^{\frac{1}{2}}}, \qquad (9)$$

where x is the distance of dx from the plane of the paper in Fig. 24, it being supposed that the electrified line passes through q, and y is measured upward on the paper. To find the limits, observe that every element of the electrified line has its cone to correspond, of angle 2θ, and that those portions only of the line contribute to V whose cones cover the point in question where V is wanted. The extreme cones therefore have the point on their surfaces. Therefore

$$z^2\tan^2\theta-x^2-y^2=0,$$

or

$$x=\pm(z^2\tan^2\theta-y^2)^{\frac{1}{2}} \qquad (10)$$

at the limits. The net result is that (9) includes all real values. The integral is an easy one, and makes

$$V=\frac{\sigma}{2\pi c}\tan\theta\left[\sin^{-1}\frac{x}{x_0}\right]_{-x_0}^{+x_0}, \qquad (11)$$

where x_0 is given by (10). Or, finally,

$$V=\frac{\sigma}{2c}\tan\theta. \qquad (12)$$

So $V=0$ everywhere outside the two planes which envelop the cones, and V has the constant value just written inside the planes. This means that there is no **E** or **H** anywhere save at the planes themselves. The total displacement they carry can be estimated by the sudden drop in V in passing through the planes to be $\frac{1}{2}\sigma$ on both; and the magnetic force to correspond satisfies $E=\mu v H$, as in all pure plane waves.

Now these results agree with the arrangement in Fig. 10, which was proved. So the proper standardisation of the potential formula for a point charge is assured. In the case of a moving strip of surface density σ and breadth a, the potential is

$$V_0=\frac{\sigma a}{2c}\tan\theta, \qquad (13)$$

constant, on the left side of the slab waves; $V=0$ on the right side; whilst in the waves themselves V rises uniformly from 0 outside to V_0 inside—that is, referring to Fig. 25,

$$V = V_0 \times \frac{PQ}{QR} = \frac{\sigma}{2c}(z \tan\theta - y) = \frac{\sigma}{2c} \times PS, \qquad (14)$$

where PS is the vertical distance of P, where V is reckoned, from the outer plane. Thus the equipotential surfaces are parallel to the wave fronts, and **E** lies in them. This property remains true until *u* is reduced to *v*. After that they split, separate, and widen out both ways into closed surfaces, and the disturbance spreads over all space.

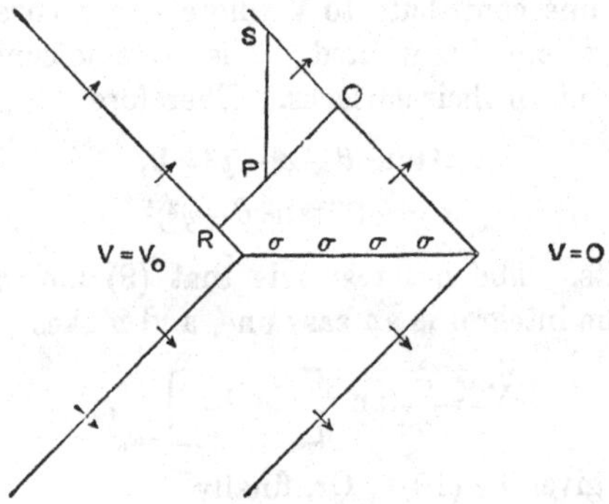

FIG. 25.

In the case of a point charge, the equipotential surfaces are hyperboloids of revolution. The conical surface is the extreme limit, to which the last hyperboloid goes infinitely near. These surfaces become infinitely compressed into one plane when $u=v$. After that, with $u<v$, they split and become flattened ellipsoids, and ultimately spheres, when *u* is reduced to nothing. The ellipsoidal theory was given in Vol. I., §§ 164 to 167. I should add that Mr. Searle has added interesting developments to the ellipsoidal part of the subject (*Phil. Trans.*, 1896, and *Phil. Mag.*, 1896).

The steady Rectilinear Motion in its own line of a Terminated Electrified Line when $u>v$, and interpretation of the Impure Conical Wave following an Electron.

§ 465. What is not clear in the above is the state of things at and just within the conical surface itself. To elucidate this, consider a more general case—viz., that of the motion of an electrified line of finite length, or at least infinite only one way, in its own line. The investigation of V will follow later. The formula is

$$V = \frac{\sigma}{2\pi c} \log \left\{ \frac{z \tan \theta}{h} + \left(\frac{z^2 \tan^2 \theta}{h^2} - 1 \right)^{\frac{1}{2}} \right\}, \tag{15}$$

where

$$h^2 = x^2 + y^2. \tag{16}$$

This is when the electrified line, linear density σ, is infinitely long one way, its termination being at the origin. The angle θ has the same meaning as before, in relation to the conical boundary of the region occupied by E and H. Now when the line is of finite length a, the V is the difference of two formulæ of the type (15) having their origins at the two ends of the line. That is,

$$V = \frac{\sigma}{2\pi c} \log \frac{z \tan \theta + (z^2 \tan^2 \theta - h^2)^{\frac{1}{2}}}{z_0 \tan \theta + (z_0^2 \tan^2 \theta - h^2)^{\frac{1}{2}}} \tag{17}$$

where $z_0 = z - a$. The quantity z_0 is always positive. The formula (15) is true in the region between the two cones, and (17) inside the inner one.

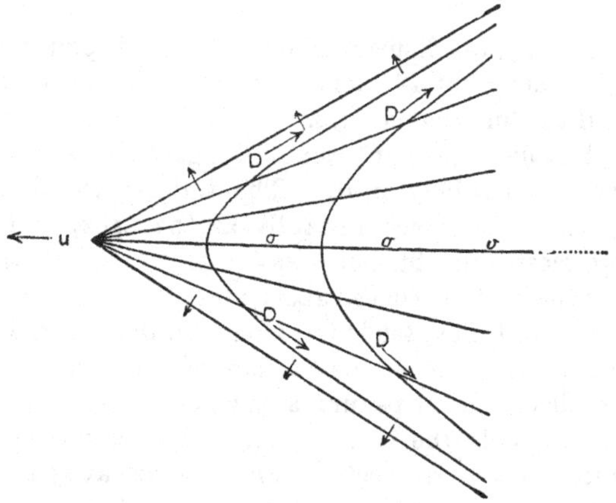

FIG. 26.

Considering (15) first, it may be seen that the equipotential surfaces are all cones with a common apex at the forward end of the line, and that V increases from 0 on the outer cone to infinity at the axis. Deriving the electric force by (2), the result is

$$\mathbf{E} = \frac{\sigma}{2\pi c h \tan\theta} \frac{z \tan^2\theta . \mathbf{h}_1 + h\mathbf{k}}{(z^2 \tan^2\theta - h^2)^{\frac{1}{2}}}, \tag{18}$$

where k and h_1 are unit vectors along and transverse to the axis. The ratio of the transverse to the axial component is

$$\frac{E_h}{E_z} = \frac{z \tan^2\theta}{h} \tag{19}$$

Now $z/h =$ constant is an equipotential surface. So, by (19), all the lines of force cut a given equipotential surface at the same angle. Observe that for a given distance from the axis, the ratio of the transverse to the axial component increases in proportion to the distance from the apex. On the axis itself the axial component is

$$E_z = \frac{\sigma}{2\pi c z \tan^2\theta}, \tag{20}$$

varying inversely as the distance from the apex. Also, when h/z is small,

$$E_h = \frac{\sigma}{2\pi c h}. \tag{21}$$

That is to say, the displacement at the axis is perpendicular thereto, spreading out all round from the electrified line precisely as if the line were at rest. But this does not continue, for all the displacement eventually goes to the conical boundary, or infinitely near. The turning round of the displacement takes place instantly at the apex, and more and more slowly the further away from it. As all the displacement ultimately concentrates about the boundary, the amount collected there tends to increase uniformly with z. At the same time, so does the surface of the cone. So the amount of displacement per unit area tends to become uniform, and since $\mathbf{H} = \mathbf{VuD}$, the state of affairs at the surface tends towards that of a pure electromagnetic wave, away from the apex, as the surface becomes more planar. A purely conical wave of the kind cannot exist alone. The lines of force are hyperbolas.

Fig. 26 will give some notion of the conical surfaces of potential and the hyperbolic lines of force, though, being done out of my head, I cannot vouch for the accuracy of shape of the curves.

Now let the electrified line be of finite length a. We must then superpose on the field of force in Fig. 26, the exact negative copy of the same, but shifted through the distance a to the right. The result of the superposition is that in the overlapping region D tends to become radial, towards the moving charge, weak about the axis, and very intense at the conical surface; whilst outside, between the inner and outer cones, is an impure conical electromagnetic wave, in which D is very intense. Pushing things to the limit, reducing a to

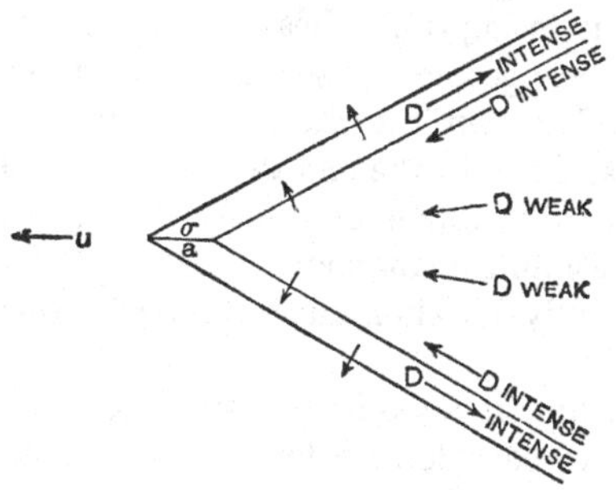

FIG. 27.

zero, and keeping the charge upon it finite, we come to the theory of the moving point charge, already given. It may seem to be much ado about nothing, but is there any easier way?

The Wave from a Straight Line Source of Induction in a Non-conducting Dielectric.

§ 466. In consequence of the circuitality of the currents, electric and magnetic respectively, all states of induction and displacement which are due to impressed electric and magnetic force depend solely upon the curl of e or of h. This was shown in Vol. I., § 87. Now, intrinsically electrised and magnetised bodies are brought into the theory by what are equivalent to e and h. The disturbances produced by moving such bodies therefore involve the consideration of the waves emitted from the seat of the curl of e or h. It is also true that the static states themselves may be regarded as arising from the superimposition of waves continuously emitted by the sources mentioned. But this is useless, though curious. On the other hand, when e and h vary in intensity, either in time or in place, or in both, the consideration of the waves in question is imperative. In Vol. I., § 89, was described generally the nature of the waves from a circular source; and above have been investigated several cases of moving magnets, &c., under such circumstances as give rise to simple plane waves. The results are sometimes rather incredible, but that soon wears off. The human mind is so made as to be capable of believing anything, no matter how silly, provided it is imprinted heavily enough. That is well known to all astute people—to those priests in particular who convert men and women into sheep by saying the same thing over and over again. In the end they have no judgment left. The incredibility of results is no evidence by itself of necessary error. The reasoning man may not believe till he sees it: that does not imply disbelief, but a reserved attitude, waiting for more light.

The electromagnetic wave from a straight line source when stationary I have considered before, *El. Pa.*, Vol. II., p. 456. It is wanted here, as a foundation for the theory of the same when moving; so for completeness I give what is necessary, in a somewhat different way, and with an addition. Let there be a straight line source of f or curl e. It generates induction at the rate f per second. That is how the wave originates. For the induction spreads all round at speed v, and along with it is electric force, according to the circuital laws. But it is not so simple a matter in the present case as with a plane source of f, already considered. There is a mixing up in the cylindrical wave.

By symmetry H is uniformly distributed round about the axis which is the straight line of f, so that the general characteristic $\nabla^2\mathrm{H} = q^2\mathrm{H}$, where $q = p/v = d/d(vt)$, reduces to the zeroth Bessel form

$$\left(\frac{d^2}{dr^2} + \frac{1}{r}\frac{d}{dr}\right)\mathrm{H} = q^2\mathrm{H}, \tag{1}$$

where H is the intensity of magnetic force at distance r from the axis.

The appropriate solution is

$$H = K_0(qr).A, \tag{2}$$

where A is a time function to be found to suit f given as a function of the time. To find A, note that the second circuital law

$$-\text{curl}\,(E - e) = \mu p H \tag{3}$$

takes the special form

$$\frac{1}{r}\frac{d}{dr}(rE) = f - \mu p H. \tag{4}$$

This is proved at once by considering two adjacent circles of E. In this equation f is the density of curl e. But let all the f be concentrated at the axis to make a line source, of strength f_0, say, and then integrate (4) over a circle of infinitesimal area centred upon f_0. The integral of $\mu p H$ vanishes, whilst that of f is f_0. This is the same as saying that the circuitation of E round f_0 tends to the value f_0 as r is reduced to 0. That is,

$$2\pi r E = f_0, \quad \text{when} \quad r = 0. \tag{5}$$

But the first circuital law

$$\text{curl}\,H = cpE,$$

reduces to

$$-\frac{dH}{dr} = cpE; \tag{6}$$

so, by (2) and (6), used in (5), we get

$$f_0 = -\frac{2\pi r}{cp}\frac{d}{dr}K_0(qr)A = -\frac{2\pi r}{cp}\frac{d}{dr}\left(-\frac{2}{\pi}\log\frac{qr}{2}\right) = \frac{4}{cp}A. \tag{7}$$

See Vol. II., p. 226, equation (4), or p. 250, equation (48), for the full formula for $K_0(qr)$, of which the only significant part at the axis is the log term here employed. So $A = \frac{1}{4}cpf_0$, and therefore

$$H = \tfrac{1}{4}cpK_0(qr)f_0 = \frac{1}{4\mu v}K_0(qr)qf_0 \tag{8}$$

is the operational solution for H, true for any varying f_0.

Let f_0 be steady, beginning at the moment $t = 0$. The algebrisation was given in Vol. II., § 436, p. 456, equation (119) equivalently. But interchange r and vt in that investigation to suit the present case. The result is

$$\mathrm{H}=\frac{f_0}{2\pi\mu v(v^2t^2-r^2)^{\frac{1}{2}}}, \tag{9}$$

and the corresponding E is

$$\mathrm{E}=-\frac{1}{cp}\frac{d\mathrm{H}}{dr}=\frac{vtf_0}{2\pi r(v^2t^2-r^2)^{\frac{1}{2}}}. \tag{10}$$

These are valid between the axis and the wave front at $r=vt$, beyond which E and H are zero. At the axis H varies as t^{-1}. If the value of H at the axis had been known the above investigation would have been unnecessary, because H at r is derivable from H_0 at the axis by

$$\mathrm{H}=\mathrm{I}_0(qr)\,\mathrm{H}_0=\mathrm{I}_0(qr)\frac{cf_0}{2\pi t}, \tag{11}$$

which produces the formula (9). Compare with equation (118*e*), §436, Vol. II., in which Δ and q, r and vt should be interchanged to suit the present case.

In the above the divergent K_0 function was employed to effect the algebrisation. But we may obtain the same result by the convergent Bessel function. Thus,

$$\mathrm{I}_n(qr)q1=\frac{(\frac{1}{2}qr)^nq}{\underline{|0}\,\underline{|n}}+\frac{(\frac{1}{2}qr)^{n+2}q}{\underline{|1}\,\underline{|n+1}}+\dots \tag{12}$$

$$=\frac{1}{vt}\left\{\frac{(r/2vt)^n}{\underline{|0}\,\underline{|n}\,\underline{|-(n+1)}}+\frac{(r/2vt)^{n+2}}{\underline{|1}\,\underline{|n+1}\,\underline{|-(n+3)}}+\dots\right\}, \tag{13}$$

the transition from (12) to (13) being made by turning p^n1 to $t^{-n}/\underline{|-n}$, according to the theory given in Vol. II., with numerous examples. Now

$$\mathrm{K}_0(qr)=-\frac{2}{\pi}\frac{d}{dn}\mathrm{I}_n(qr), \quad (n=0). \tag{14}$$

This applied to (13) makes

$$\mathrm{K}_0(qr)q1=\frac{1}{vt}\frac{2}{\pi}\left\{g'(-1)+\frac{(r/2vt)^2g'(-3)}{\underline{|1}\,\underline{|1}}+\frac{(r/2vt)^4g'(-5)}{\underline{|2}\,\underline{|2}}+\dots\right\} \tag{15}$$

$$=\frac{2}{\pi vt}\left\{1+\tfrac{1}{2}\left(\frac{r}{vt}\right)^2+\frac{1.3}{2.4}\left(\frac{r}{vt}\right)^4+\dots\right\} \tag{16}$$

$$=\frac{2}{\pi}\frac{1}{(v^2t^2-r^2)^{\frac{1}{2}}}. \tag{17}$$

This corroborates equation (9). For the values of the g' functions used see Vol. II., p. 451, equations (78).

The Wave from a Straight Line Source of Induction in a Conductor.

§ 467. Being in the way of it, it is worth while giving the corresponding results when the medium is a conductor instead of a non-conductor, and its permittivity is ignorable. The operator cp in the first circuital law becomes k, the conductivity. This makes $q^2 = \mu kp$. Therefore

$$\mathrm{H} = \tfrac{1}{4}k\mathrm{K}_0(qr)f_0 \tag{18}$$

is the operational solution, instead of (8). To algebrise, proceed thus.

$$\mathrm{I}_n(qr)1 = \frac{(\frac{1}{4}\mu kpr^2)^{\frac{1}{2}n}}{\underline{|0}\,\underline{|n}} + \frac{(\frac{1}{4}\mu kpr^2)^{\frac{1}{2}n+1}}{\underline{|1}\,\underline{|n+1}} + \dots$$

$$= \frac{(\mu kr^2/4t)^{\frac{1}{2}n}}{\underline{|0}\,\underline{|n}\,\underline{|-\frac{1}{2}n}} + \frac{(\mu kr^2/4t)^{\frac{1}{2}n+1}}{\underline{|1}\,\underline{|n+1}\,\underline{|-(\frac{1}{2}n+1)}} + \dots \tag{19}$$

To this apply (14), giving

$$\mathrm{K}_0(qr)1 = -\frac{2}{\pi}\Big\{\log \mathrm{Z}^{\frac{1}{2}} + g'(0) - \tfrac{1}{2}g'(0) - \tfrac{1}{2}\mathrm{Z}g'(-1) - \tfrac{1}{2}\mathrm{Z}^2\frac{g'(-2)}{\underline{|2}\,\underline{|2}} - \tfrac{1}{2}\mathrm{Z}^3\frac{g'(-3)}{\underline{|3}\,\underline{|3}} - \dots\Big\}$$

$$= -\frac{1}{\pi}\Big\{\log \mathrm{Z} + \gamma - \mathrm{Z} + \frac{\mathrm{Z}^2}{2\underline{|2}} - \frac{\mathrm{Z}^3}{3\underline{|3}} + \dots\Big\}, \tag{20}$$

where $\mathrm{Z} = \mu kr^2/4t$, and $\gamma = g'(0) = 0{\cdot}5772$. Therefore

$$\mathrm{H} = -\frac{kf_0}{4\pi}\Big\{\log \mathrm{Z} + \gamma - \mathrm{Z} + \frac{\mathrm{Z}^2}{2\underline{|2}} - \frac{\mathrm{Z}^3}{3\underline{|3}} + \dots\Big\} \tag{21}$$

expresses H at r at time t due to the source f_0 at the axis beginning at $t = 0$. The series is convergent, and is therefore suitable when Z is not large. To suit larger values of Z we may turn (21) to a divergent series. Use equation (93), p. 452, Vol. II. It produces

$$\mathrm{H} = \frac{kf_0}{4\pi \mathrm{Z}}\epsilon^{-\mathrm{Z}}\Big\{1 - \frac{1}{\mathrm{Z}} + \frac{\underline{|2}}{\mathrm{Z}^2} - \frac{\underline{|3}}{\mathrm{Z}^3} + \dots\Big\}, \tag{22}$$

which is proper when Z is too large for the practical use of (21). It is curious to see that the series following $\epsilon^{-\mathrm{Z}}$ is that exponential series itself with every term turned upside down. It is also curious to see that we cannot get away from the divergent series, either operationally or numerically. They seem to have come to stay, like other things before.

If f_0 is impulsive, we get the corresponding H by time-differentiation. Let the impulse be B_0, then we put pB_0 for f_0 in (18). This makes

$$H = \tfrac{1}{4}kK_0(qr)pB_0 = \frac{kB_0}{4\pi t}\epsilon^{-z}, \tag{23}$$

by differentiating (21). This, observe, or test to corroborate, is the same as

$$H = I_0(qr)\frac{kB_0}{4\pi t}, \tag{24}$$

where the operand is the value of H at the axis. But if you try to turn (23) to a divergent series by differentiating (22), you will come to the convergent solution (23) again.

Observe that (23) is the same as

$$2\pi r(B = \mu H) = -B_0\frac{d}{dr}\epsilon^{-z}. \tag{25}$$

It follows that the total B over all space is B_0, constant. This is because we impulsively generate B_0, and it then simply diffuses outwards, but cannot be destroyed by the electric conductivity of the medium. It requires magnetic conductivity to destroy it. See Vol. I., §§ 194, 195.

Similarly, in case of the continued source f_0, the total B at time t amounts to f_0t. We must integrate over all space to find it, because in pure diffusion the speed v is infinite. On the other hand, in the nonconducting dielectric, the induction to the amount f_0t is to be found always between $r=0$ and vt. An easy integration of (9) will show that the induction between $r=0$ and mvt, where m is any fraction, is proportional to f_0t. Here the radius of the circle considered is always a constant fraction of the radius of the wave front. But within any circle of constant radius, the total induction falls to zero, because B tends to zero, everywhere save at the wave front, where it is always infinite. This infinity, however, is of no consequence, being empty.

The electric field tended to is specified by

$$E = \frac{f_0}{2\pi r}. \tag{26}$$

It is the electric field due to an electric shell of strength f_0 having a single straight edge at the axis, and therefore extending to an infinite distance. But if it is stopped by another straight edge parallel to the first at any distance, a second source is introduced, which is the negative of the first. There are then two waves produced, according to the above formulæ. The total induction (algebraically reckoned) is now always zero; although before the waves overlap the steady generation at a rate of $+f_0$ by one source and $-f_0$ by the other, and their existence in separate regions is evident.

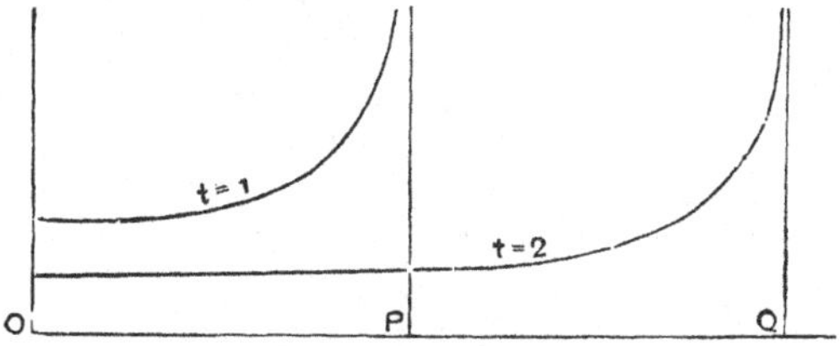

FIG. 28.

Fig. 28 will give a general idea of the spread of the H wave from a line source. Distance from the axis at O is reckoned along the base line, and the two curves show the intensity of H in the wave at two moments of time, say $t = 1$ and $t = 2$, when the wave front has reached P and Q respectively. The scale is quite arbitrary, because one curve is a copy of the other. At the initial moment, there is just the solitary infinite value of H at the commencing wave front, at O. The infiniteness is merely because f_0 is assumed to start action with infinite suddenness.

The Waves due to a Growing Plane Sheet of Sources of Induction, and to Travelling Filaments, at any Speed.

§ 468. Now consider the state of things due to a line source in motion, or to a plane source steadily extending itself by the continuous addition of line sources at its straight edge. To fix ideas and formulæ, refer to Fig. 29. Let z_1 be the actual position of the edge of the plane source at time t, whilst z and r are the co-ordinates of the point P where H is to be determined. The directions of f and H are perpendicular to the paper. The origin of z may be conveniently situated at a distance to the left, to avoid negative values. Let the plane sheet be of density f per unit length along z, and start at z_3 at the moment t_3, and grow along z at speed u. The position z_2 indicates the limit of action at P. That is, P at time t receives H from the sheet between z_3 and z_2, but not from the part between z_2 and z_1, where its edge is situated. As for z_0, that is any intermediate position to be employed in the integral.

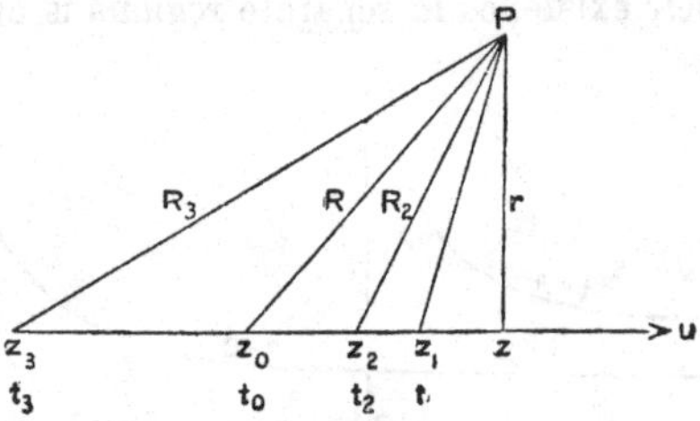

FIG. 29.

With the above conventions, the value of H at P is given by

$$\mathrm{H}=\frac{fu}{2\pi\mu v}\int_{t_3}^{t_2}\frac{dt_0}{\{v^2(t-t_0)^2-r^2-[z-z_1+u(t-t_0)]^2\}^{\frac{1}{2}}},\qquad(27)$$

this being derived from (9) above by putting in the special distance concerned. The diagram is drawn for the case $v>u$, so that the time of going from z_2 to P at speed v equals the time of going from z_2 to z_1 at speed u.

The above is the same as

$$\mathrm{H}=\frac{fu}{2\pi\mu v}\int_{\tau_2}^{\tau_3}\frac{d\tau}{(v^2-u^2)^{\frac{1}{2}}\sqrt{\tau^2-a^2}},\qquad(28)$$

where

$$\tau=t-t_0-\frac{u(z-z_1)}{v^2-u^2},\qquad a=\frac{\sqrt{v^2(z-z_1)^2+r^2(v^2-u^2)}}{v^2-u^2}.\qquad(29)$$

Integrating (28), we get

$$\mathrm{H}=\frac{fu}{2\pi\mu v(v^2-u^2)^{\frac{1}{2}}}\Big[\log(\tau+\sqrt{\tau^2-a^2})\Big]_{\tau_2}^{\tau_3},\qquad(30)$$

provided $v>u$. But (28) may also be written

$$\mathrm{H}=\frac{fu}{2\pi\mu v}\int_{\tau_2}^{\tau_3}\frac{d\tau}{(u^2-v^2)^{\frac{1}{2}}(a^2-\tau^2)^{\frac{1}{2}}};\qquad(31)$$

so, when $u>v$, we have the result

$$\mathrm{H}=\frac{fu}{2\pi\mu v(u^2-v^2)^{\frac{1}{2}}}\left[\sin^{-1}\frac{\tau}{a}\right]_{\tau_2}^{\tau_3}. \tag{32}$$

At present consider the case of $v>u$. Then the limits τ_2 and τ_3 are got by putting t_2 and t_3 for t_0 in the expression for τ in (29) above. At the lower limit in (28) the radical vanishes. So $\tau_2=a$. The upper limit depends upon the moment t_3 of starting action. So

$$\mathrm{H}=\frac{fu}{2\pi\mu v(v^2-u^2)^{\frac{1}{2}}}\log\frac{\tau_3+(\tau_3^{\,2}-a^2)^{\frac{1}{2}}}{a} \tag{33}$$

is our solution. Note that H begins at P when $\mathrm{R}_3=v(t-t_3)$. This makes $\tau_3=a$. So $\mathrm{H}=0$ initially.

If we differentiate (33) with respect to t_3 we get a formula of the type (9), expressing the effect of the introduction of a stationary filament at z_3. But if we differentiate (33) to z_1, we obtain the formula for the H due to a filament of f introduced at z_3 at time t_3, and then moved at speed u to the right, being at z_1 at time t. In carrying out the differentiation, which is complicated, both τ_3 and a must be differentiated, because they both contain z_1. The result is

$$\frac{d\mathrm{H}}{dz_1}=\frac{f}{2\pi\mu v}\left[\frac{v^2(v^2-u^2)^{-1}}{\{v^2(t-t_3)^2-\mathrm{R}_3^{\,2}\}^{\frac{1}{2}}}+\frac{uv^2(z-z_1)(v^2-u^2)^{-2}}{\{v^2(t-t_3)^2-\mathrm{R}_3^{\,2}\}^{\frac{1}{2}}\{\tau_3+(\tau_3^{\,2}-a^2)^{\frac{1}{2}}\}}\right.$$
$$\left.+\frac{u(z-z_1)(v^2-u^2)^{-\frac{1}{2}}}{(z-z_1)^2+r^2(1-u^2/v^2)}\right]. \tag{34}$$

In this formula R_3 is defined in the last figure. It is connected with τ_3 by

$$(v^2-u^2)(\tau_3^{\,2}-a^2)=v^2(t-t_3)^2-\mathrm{R}_3^{\,2}. \tag{35}$$

If f is understood to be the strength of the filament, then $d\mathrm{H}/dz_1$ means the magnetic force to correspond. When $u=0$ we get the cylindrical wave from the stationary filament at z_3. But when $t_3=-\infty$, we get the steady wave of H travelling with the filament when it has been in motion for a sufficiently long time. Calling this H′, it is expressed by

$$\mathrm{H}'=\frac{f_0}{2\pi\mu v}\frac{u}{(v^2-u^2)^{\frac{1}{2}}}\frac{z-z_1}{(z-z_1)^2+r^2(1-u^2/v^2)}. \tag{36}$$

The eolotropic effect is now clearly visible.

When u/v is very small,

$$\mathrm{H}' = \frac{f_0}{2\pi\mu v}\frac{u}{v}\frac{z-z_1}{(z-z_1)^2+r^2} = \frac{f_0 u}{2\pi\mu v^2}\frac{\cos\theta}{\mathrm{R}}, \qquad (37)$$

where R and θ are the polar co-ordinates of P where H′ is reckoned with respect to the filament's position and line of motion. This formula may be readily checked by the static method.

There is always vanishing of H′ at the plane $\theta=\frac{1}{2}\pi$. At the same time, increasing u causes a concentration of H′ towards that plane. So, in the limit, when $u=v$, the result is two plane electromagnetic waves of reverse natures, both travelling the same way.

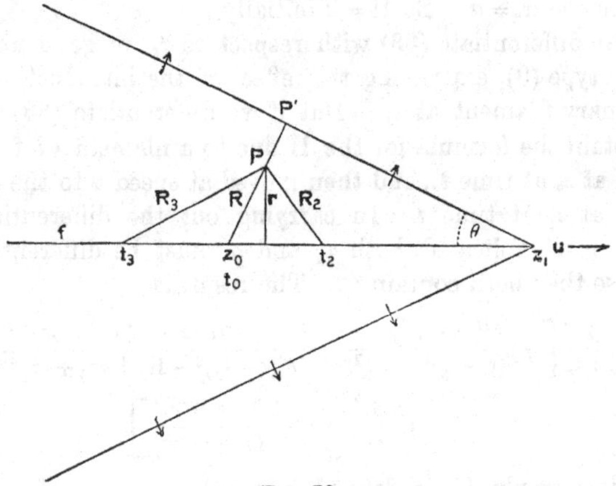

FIG. 30.

Now pass to the case $u>v$, equation (32) above, in which τ and a are known, whilst the limits have to be considered carefully. We require a separate figure. As the edge of the sheet advances the envelope of the cylindrical waves emitted by it make the wave front consist of two planes inclined to one another at the angle 2θ, the edge of the sheet being their meeting line. The angle θ is given by $\sin\theta=v/u$. Now if P is outside these planes, there is no H at all. When it is just on the boundary, say at P′, the cylinder from z_0 just reaches it, and $t_2=t_3$. When it is inside the planes, as at P in Fig. 30, thers are two limiting points, marked t_2 and t_3.

whose distances from z_1 are u/v times the distances of the same points from P. In the figure $u/v=2$. So the range of integration is from z_3 to z_2, or τ_2 to τ_3 determined by giving τ the corresponding values. That is to say, the H at P is entirely due to the filaments from z_3 to z_2. The waves coming from filaments to the left of z_3 have not reached P; nor have those coming from filaments between z_2 and z_1, which last is the position of the foremost filament at the moment in question. Here, of course, it is assumed that the sheet began to form before the time t_3. The range is, of course, less at the beginning. The important matter is the final result when the limits are full. Then we have

$$v^2(t-t_0)^2 - R^2 = v^2(t-t_0)^2 - r^2 - \{u(t-t_0) - (z_1 - z)\}^2, \quad (38)$$

and also

$$= (u^2 - v^2)(a^2 - \tau^2). \quad (39)$$

So the full limits are $\pm\ a$ for τ in equation (32). This makes

$$H = \frac{fu}{2\mu v(u^2 - v^2)^{\frac{1}{2}}} \quad (40)$$

constant everywhere between the two planes.

Reversion to Divergent Plane Waves. Is the Ether Fixed?

§ 469. The last result brings us back to the simple theory of pairs of inclined terminated plane waves. It was, in the first place, from the theory of generation of a single complete plane wave that we concluded that a straight line of curl of e of strength f generated induction at the rate f. This property has now been applied to the wave generated by a plane sheet of f continuously growing at one of its straight edges, thus introducing new filaments of f, both when $u < v$ and when $u > v$. The first case includes, for a steadily moving state, the theory of Vol. I., § 166. The second case is much simpler in results. The H can be determined from the first moment of starting the action of f, including the initial partially cylindrical cap. Behind this cap H soon becomes constant, equation (40). This signifies slanting divergent plane waves. For, by taking differences, we can derive the H due to a plane strip of f of constant width started and set in motion at any moment in its own plane at speed u. Except as regards the cap, or rather, with the addition of the cap, we get a pair of diverging plane waves, as described at the end of § 456 above, according to the formulæ there given. Likewise the result for a moving filament of f. These results were easily obtained, but the initial effects required closer consideration. Besides that, we have had the advantage of seeing in detail that the results of moving magnets or electrets are really due to the waves coming from the sources f and g.

But when we pass from the abstract problems of moving sources to those connected with material bodies, it must be so done that the conditions assumed are not changed by the materialisation if the results are to be the same. In particular, the ether must not be disturbed by the motion of the matter. This is the assumption I have always made in my investigations of the waves arising from moving electrification or other sources. (See, for example, *El. Pa.*, Vol. II., p. 504; and E.M.T., Vol. I., p. 269 to 282, and the above investigations.) The reasons are tolerably obvious. First, simplicity. Next, the changes due to any dragging along of the ether with the matter cannot be determined without some assumptions as to the extent of the drag, and its nature. This is one of the unsettled and very speculative matters in electrical science applied to moving bodies. I would rather have no drag at all, or else complete drag, if I had the management of it. But it may be that a partial drag is an essential condition of the connection between ether and matter, even though that connection be entirely through electrification, that is, by ions or electrons. Even a full drag is not inconceivable. For consider a single charge or electron moving through the ether without any assumed drag externally. It might really be the limiting case of a moving ether; that is, the electrification might be inexorably associated with a definite part of the ether. Or again, with perfect slipping possible, resultant effects of moving collections of matter and electrification might give an apparent motion to the ether in the circuital equations.

Drag of Matter upon Ether. Modified Circuital Equations, and the Wave-speed resulting.

§ 470. The question of the drag of a moving body on the ether has occupied the attention of physicists for a very long time, especially from the optical point of view, and it is open to question whether the last word has been said about it by the latest investigators. The following comparisons of different results will serve to illustrate some aspects of the subject.

Divide the displacement D and the induction B each into two parts, thus,

$$D = cE = (c_0 + c_1)E = D_0 + D_1, \tag{1}$$

$$B = \mu H = (\mu_0 + \mu_1)H = B_0 + B_1, \tag{2}$$

Here B_0 and D_0 belong to the ether, and B_1 and D_1 are the extra parts due to the matter, that is, the two polarisations, magnetic and electric.

Next, write the two circuital laws for a moving medium in the form

$$\text{curl}\,(\mathbf{H} - \mathbf{VD}_0\mathbf{q}_0 - \mathbf{VD}_1\mathbf{q}_1) = \mathbf{C} + \dot{\mathbf{D}} + \mathbf{u}\rho, \tag{3}$$

$$-\text{curl}\,(\mathbf{E} - \mathbf{Vw}_0\mathbf{B}_0 - \mathbf{Vw}_1\mathbf{B}_1) = \dot{\mathbf{B}}. \tag{4}$$

Compare with § 71, Vol. I. We leave out impressed forces, except motional, also the hypothetical magnetic conduction current and magnetification. As regards the motional electric and magnetic forces, instead of one velocity, there are now four, one for each part of the two fluxes concerned. This is required for our comparisons. As for u, the velocity of the electrification ρ or divD, it need not be assumed to be the same as that of the matter.

To ease matters, let there be a simple plane wave traversing a non-conducting dielectric without volume electrification, and let E || x, H || y, VEH || z, and let all the q's and w's be || z. Then $\mathbf{Vq}_0\mathbf{D}_0$ || H; also $\mathbf{Vq}_1\mathbf{D}_1$; and $\mathbf{VB}_0\mathbf{w}_0$ || E; also $\mathbf{VB}_1\mathbf{w}_1$. These simplifications bring the circuital equations to

$$-\Delta(H + q_0 D_0 + q_1 D_1) = pD, \tag{5}$$

$$-\Delta(E + w_0 B_0 + w_1 B_1) = pB, \tag{6}$$

where Δ is d/dz and p is d/dt. All scalars now.

But the type of a plane wave is $f(z - Ut)$, where U is the wave speed. So $p = -U\Delta$, and therefore

$$H + q_0 D_0 + q_1 D_1 = UD, \tag{7}$$

$$E + w_0 B_0 + w_1 B_1 = UB. \tag{8}$$

These are the same as

$$H=(U-q)D, \tag{9}$$

$$E=(U-w)B, \tag{10}$$

where q and w are the speeds given by

$$q=\frac{c_0q_0+c_1q_1}{c}, \qquad w=\frac{\mu_0w_0+\mu_1w_1}{\mu}. \tag{11}$$

So there are only two effective speeds of motion. In this notation the circuital equations become

$$\text{curl}\,(\mathbf{H}-\mathrm{V}\mathbf{D}\mathbf{q})=\mathbf{C}+\dot{\mathbf{D}}+\mathbf{u}\rho, \tag{12}$$

$$-\text{curl}\,(\mathbf{E}-\mathrm{V}\mathbf{w}\mathbf{B})=\dot{\mathbf{B}}. \tag{13}$$

When q and w are zero, $U=v$, given by $\mu cv^2=1$. And in general, (9) and (10) lead to

$$v^2=(U-q)(U-w); \tag{14}$$

of which the solution is

$$U=\tfrac{1}{2}(q+w)+\sqrt{v^2+\tfrac{1}{4}(q-w)^2}; \tag{15}$$

and further, when q and w are small compared with v, we get

$$U-v=\tfrac{1}{2}(q+w). \tag{16}$$

This represents the increased wave speed (referred to stagnant ether at a distance) due to the motion of the matter. It is the mean of the effective speeds associated with **D** and **B**. Now compare different cases.

Comparison of Wave-speeds in special Cases.

§471. (1). When there is symmetry, so that $c_1/c=\mu_1/\mu$, and also $q_0=w_0$, $q_1=w_1$, then

$$q=w=\frac{q_0c_0+q_1c_1}{c}, \tag{17}$$

and

$$U-v=q, \tag{18}$$

exactly; that is, without limitation to small speeds. This is a fanciful case, though interesting.

(2) Let $q_0=q_1=w_0=w_1$; then also $=q$, and

$$U-v=q, \tag{19}$$

again exactly. Here q is the speed of the matter and ether moving together, equivalent to a single medium theory. There is full superposition of velocities; that is, the wave speed *through the body* is the same whether it is stationary or moving. This is the result we ought to have in any rational theory of one medium, in a primary theory, that is to say. (See Preface to Vol. 1, E.M.T., for remarks on this subject.)

(3). Consider an unmagnetisable body, $\mu_1 = 0$. Then

$$\mathrm{U} - v = \frac{c_0 q_0 + c_1 q_1}{2c} + \tfrac{1}{2} w_0. \tag{20}$$

If we can agree that w_0 should be the same as q_0, then

$$\mathrm{U} - v = \tfrac{1}{2} q_0 \left(1 + \frac{c_0}{c}\right) + \tfrac{1}{2} q_1 \frac{c_1}{c}. \tag{21}$$

(4). If the ether is entirely left behind as the matter moves through it, we may take $q_0 = 0$ in (21); then

$$\mathrm{U} - v = \tfrac{1}{2} q_1 \frac{c_1}{c} \tag{22}$$

Here q_1 is the speed of the matter. The result is therefore just one-half of Fresnel's result, according to which, in the present notation, the extra wave speed is $q_1 c_1/c$, which means that the wave traverses the moving body more slowly than when it is at rest by the amount $q_1 c_0/c$.

(5). Let q_0 and q_1 be both zero in the first circuital law, and $\mathrm{w}_0 = \mathrm{w}_1$ in the second. Then

$$\mathrm{U} - v = \tfrac{1}{2} w, \tag{23}$$

where w is the velocity of the matter. This result was obtained by J. J. Thomson by using Maxwell's theory in its unamended form. In my amendment of the same, given in 1885, I concluded that the motional argument which led Maxwell to the motional electric force $\mathbf{e} = \mathrm{V}\mathbf{w}\mathbf{B}$ involved a similar one leading to a motional magnetic force $\mathbf{h} = \mathrm{V}\mathbf{D}\mathbf{w}$, using the full flux in the case of $\mathbf{h}$ quite as much as in the case of $\mathbf{e}$; for an obvious gastronomical reason relating to sauce. Whether right or wrong, it rationalises Maxwell's theory, and I maintain, as before, that it is the proper expression of his theory, because of his form of the motional electric force. It leads to a full superposition of velocities, as in example (2) above. But it does not follow that either $\mathbf{e}$ or $\mathbf{h}$ is correct in form universally. That is a distinct question.

(6). Larmor, "Ether and Matter," p. 60, takes $q_0 = 0$ in the first law, and $w_0 = w_1 = q_1$ in the second. This produces

$$U - v = \tfrac{1}{2}w\left(1 + \frac{c_1}{c}\right), \tag{24}$$

where w is the speed of the matter. This is the sum of J. J. Thomson's and one half of Fresnel's results. The ether is assumed to be stagnant, or not dragged at all. Nevertheless, the full $e = VwB$ is used, but the full $h = VDw$ is replaced by VD_1w. This pictures the larmorial bearings correctly, I think.

(7). But Larmor (*l.c.*) does not obtain the result (24), but Fresnel's wc_1/c instead. This is, in fact, the reason why he adopts VD_1w, though it may be questioned whether the optical result is sufficient to determine the theory. His equations are equivalent to

$$-\Delta(H + wD_1) = cpE, \tag{25}$$

$$-\Delta(E + wB) = \mu pH, \tag{26}$$

in my notation.* I derive (24) in the above way, getting

$$v^2 = \left(U - w\frac{c_1}{c}\right)(U - w) \tag{27}$$

for the quadratic, a special case of (14) above.

It may perhaps fairly be asked, if the assumed fact that the ether is stationary causes the influence of the electric polarisation to introduce $h_1 = VD_1w$, why should not the influence of the magnetic polarisation introduce $e_1 = VwB_1$? I do not dogmatise, but wait for more light.

(8). Let us see what connection is needed between q_0 and q_1 to obtain Fresnel's result. That depends upon w_0 and w_1 also. If we take, as seems the fairest at first sight, $w_0 = q_0$, and $w_1 = q_1$, then equation (21) supplies the answer, for an unmagnetisable medium. The right member has to be q_1c_1/c; so

$$\frac{q_0}{q_1} = \frac{c_1}{c + c_0} = \frac{c_1}{2c_0 + c_1}, \tag{27A}$$

which may be grossly imagined to represent the drag of the matter on the ether. It is here conditioned by the electric polarisation, and cannot exist without it. It is small when

* *See* § 476 later for correction of Larmor's investigation.

c_1/c_0 is small, but may be nearly unity with large c_1/c_0. The corresponding e is

$$\mathbf{e} = \mathrm{V}(\mathbf{w}_0\mathbf{B}_0 + \mathbf{w}_1\mathbf{B}_1) = \mathrm{V}\mathbf{w}\mathbf{B}$$

where
$$w = \frac{\mu_0 q_0 + \mu_1 q_1}{\mu} = \left(\frac{\mu_1}{\mu} + \frac{\mu_0}{\mu}\frac{c_1}{2c_0 + c_1}\right)q_1 \tag{28}$$

So e is zero in an unpolarisable medium, for then $\mu_1 = 0$, $c_1 = 0$. But this e is the electric force impressed in a wire carried perpendicularly across the lines of magnetic force in a stationary ether. The vanishing of e seems wrong at first; but there is no necessary absurdity. For by assumption the body is unpolarisable, and is unaffected by either magnetic or electric force. Being perfectly neutral, why should there be an e impressed upon it when it just slips through the ether without disturbing it? Then why, on the present hypothesis, is there e in a conductor? Because c_1 is not zero. Conduction in the physical part of Maxwell's theory implies polarisation, because it is a breaking down of the displacement, with external renewal of the same, maintaining an average steady state of **D**. The value of c/c_0 can only be measured in very poor conductors. Even then it may be very large: for example, the value 60 to 80 in water has been found. It is likely to be much larger in good conductors. This would make $\mathbf{e} = \mathrm{V}\mathbf{w}\mathbf{B}$ closely, with $w = q_1$ nearly, with nearly full drag of the ether, or what is equivalent thereto.

If, on the other hand, we assume $w_0 = q_1$, and inquire into the relation between q_0 and q_1 to obtain Fresnel's result in an unmagnetisable medium, we have

$$\frac{q_0 c_0 + q_1 c_1}{2c} + \frac{q_1}{2} = \frac{q_1 c_1}{c}, \tag{29}$$

which makes q_0 be negative; nothing more need be said about that.*

It is to be noted that should c_1/c be very large, the idea and the measure of $\mathbf{D}_1$ as an elastic phenomenon become indistinct. It is then also of little importance to preserve it in the equations, because conduction, unless very feeble, works a revolution in the manner of propagation, turning elastic waves to diffusive waves. The precise nature of the failure to

* Prof. G. F. FitzGerald asks me if I am wise to so hastily dismiss this case. I do not think I am.

support elastic displacement without a continuous supply of energy, in relation to the ions, electrons, or corpuscles, is a matter of speculation at present. I am rather prejudiced against the view that in a solid conductor the conduction consists of a continuous flow of electrons or of J. J. Thomson's corpuscles either one or both ways, and think it more likely to be a local phenomenon in the main.

(9). A very simple way of obtaining Fresnel's result is to assume $e = VqB$, $h = VDq$, that is, the standard forms of e and h, but with the velocity of q so chosen as to give the required result. We require

$$q = \frac{c_1}{c} q_1, \qquad (30)$$

if q_1 is the real velocity of the matter. This applies to both circuital equations. But an electrical argument founded upon the absence of drag which would justify this procedure in the first circuital law, would surely fail in the second. However, there is this remarkability, that Fresnel's result becomes exact, instead of for small speeds only of the matter. I do not know that this is to be considered a recommendation. But it may be possible to invent some hypothesis to justify the procedure in the second circuital equation as well as in the first.

There is a difficulty in applying optical evidence to the circuital laws which has not been mentioned, viz., dispersion and change of wave-speed with the frequency of vibration. In applications involving assumed definite equations of motion or of connection between the etherial E and the molecular atomic electrical arrangements, the symbols c and μ, constants for the ether, become differential operators (functions of p), and the permittivity and inductivity can only be effectively constants at a given frequency of simply periodic vibration. This must necessarily affect in some degree all arguments of the above kind.

Effect of Modified Circuital Equations on Electrical Distributions.

§ 472. The general nature of the modification of electrical distributions due to the introduction of the four velocities q_1, q_2, w_1 and w_2 in the circuital equations (3), (4) above may be readily seen in the manner of § 164, Vol. 1, by considering the effect of moving a single charge in a homogeneous medium. Using them in the form (12), (13), impress upon the whole system the velocity $-\mathbf{u}$. This makes

$$\text{curl}\,\{\mathbf{H} - \mathrm{V}\mathbf{D}(\mathbf{q} - \mathbf{u})\} = \dot{\mathbf{D}}, \tag{31}$$

$$-\text{curl}\,\{\mathbf{E} - \mathrm{V}(\mathbf{w} - \mathbf{u})\mathbf{B}\} = \dot{\mathbf{B}}. \tag{32}$$

As we have not altered the relative velocities by thus bringing the electrification to rest, these are the proper equations when we travel with the electrification. In a steady state, that is to say, moving forward steadily at speed u, the right members vanish. Equation (31) then gives

$$\mathbf{H} = \mathrm{V}\mathbf{D}(\mathbf{q} - \mathbf{u}), \tag{33}$$

and (32) makes

$$\text{curl}\,\{\mathbf{E} - \mathrm{V}(\mathbf{w} - \mathbf{u})\mathbf{B}\} = 0. \tag{34}$$

Insert (33) in (34), and suppose $\mathbf{u}$ and $\mathbf{w}$ are parallel to $\mathbf{k}$. Then the vector

$$\begin{aligned} &\mathbf{E} + \mu c(u - q)(u - w)\mathrm{V}\mathbf{k}\mathrm{V}\mathbf{k}\mathbf{E} \\ &= \mathbf{E} + \mu c(u - q)(u - w)(\mathbf{k}\mathrm{E}_3 - \mathbf{E}) \\ &= (1 - m^2)\mathbf{E} + m^2\mathrm{E}_3\mathbf{k} \quad \text{if} \quad m^2v^2 = (u - q)(u - w). \end{aligned} \tag{35}$$

By (34) it is polar. So the vector

$$\mathbf{E} + \frac{m^2}{1 - m^2}\mathbf{k}\mathrm{E}_3 = \mathbf{i}\mathrm{E}_1 + \mathbf{j}\mathrm{E}_2 + \frac{\mathbf{k}\mathrm{E}_3}{1 - m^2} \tag{36}$$

is polar. Therefore

$$\mathbf{D} = -c\{\mathbf{i}\nabla_1 + \mathbf{j}\nabla_2 + \mathbf{k}(1 - m^2)\nabla_3\}\mathrm{P}, \tag{37}$$

where P is the potential. The result is, when m^2 is positive, to effectively reduce the permittivity in the line of motion both ways, to an amount governed by the size of m^2, so as to cause a concentration of displacement equatorially. This concentration is complete when $m^2 = 1$. Then $u = \mathrm{U}$. That is, the previously investigated U is the speed with which the electrification must move to make a plane wave. There is no distortion when either $u = q$, or $u = w$, if u and w are unequal. But m^2 is not necessarily positive. If u lies between q and w, m^2 is negative, and the distortion is a concentration in the line of motion. An eminent authority once remarked that there is a lot of humbug in mathematical papers. He knew, having done it himself several times. To obviate a remark of that sort, I remark that the cases of vanishing distortion are unlike as regards H. For, if $u=q$, then $\mathrm{H}=0$; but if $u=w$, then $\mathrm{H}=\mathrm{VD(q-w)}$, showing magnetic force proportional to the difference of the speeds associated with D and B. In general, H is positive or negative according as q exceeds or is less than w. Unless D is reversed. When can that happen? When $m^2>1$. Then, as in the case of a charge moving through Maxwell's ether at a speed greater than that of light, there is a generation of an impure conical wave within which D is reversed.

Lorentz's Equations of a Moving Dielectric.

§ 473. Another way of getting Fresnel's wave-speed was described to me by Prof. G. F. FitzGerald. Write the second, or Faraday's circuital law in the form

$$-\mathrm{curl}\,\mathrm{F}=\mu p\mathrm{H}, \tag{38}$$

and the first, or Maxwell's, in the form

$$\mathrm{curl}\,(\mathrm{H}-\mathrm{V}c_1\mathrm{Fw})=p\mathrm{D}=p\mathrm{D}_0+p\mathrm{D}_1, \tag{39}$$

where D_0 belongs to the ether, and D_1 is the electrisation or electric polarisation added on by the matter. Then let

$$\mathrm{D}_0=c_0\mathrm{F}, \qquad \mathrm{D}_1=c_1(\mathrm{F}+\mathrm{e})=c_1\mathrm{E}, \tag{40}$$

where $\mathrm{e}=\mathrm{VwB}$. These equations give Fresnel's wave speed when w is small, that is, $\mathrm{U}=v+(c_1/c)w$.

It occurred to me that the term $\mathrm{V}c_1\mathrm{Fw}$ should be $\mathrm{V}c_1\mathrm{Ew}$, because in the additional part thus introduced, or $\mathrm{V}c_1\mathrm{ew}$, all the factors are related to the matter. This change makes no difference in the approximate result, and was not objected to. But Prof. FitzGerald believed the above way was equivalent to Lorentz's.

To find Lorentz. After profound research I succeeded in discovering Prof. H. A. Lorentz's "Versuch einer Theorie der electrischen und optischen Erscheinungen in bewegten Körpern," Leiden, 1895. This important application of Maxwell's theory to optical phenomena ought to have been done into English at once to save repetitional labour. Though sad, it is a fact that few Britons have any linguistic talent. This is not due to laziness, but mainly to a real mental incapacity, combined with the feeling that one language is quite enough. Foreigners, on the other hand, seem to be gifted linguists quite naturally, so much so that they have invented a large number of lingos, and are commonly skilled in several at once. Very well; I would say let them give us poor islanders the benefit of their skill, by doing all their best work into English. And why not make English the international scientific language? It would be all the same to the foreigners, and a great boon to Great-Britain-and-Irelanders, and the other English-speaking people.

On examination, I find that Lorentz's equations do lead to the Fresnel wave-speed, and in the above way. As his work is not so well known in England as it deserves to be, I will give a short sketch of Lorentz's way of treating the circuital equations, with reference to the matter in hand, but without any particular confinement in details. I observe with some personal gratification that Lorentz employs Vector Analysis (not Quaternions), and that he works entirely upon the circuital equations and their accessories, without the vector-potential or the Principle of Least Action, which last, in spite of its name, has the remarkable property of increasing the amount of the work to be done.

Let H and F be the magnetic and electric forces in ether away from matter and electrification, then

$$\text{curl H} = p\text{D}_0 = c_0 p\text{F}, \tag{41}$$

$$-\text{curl F} = p\text{B}_0 = \mu_0 p\text{H}, \tag{42}$$

are the fundamental circuital equations. I think it proper to have two ethereal constants. I do not consider the velocity of propagation to be a fundamental constant. It is a compound idea. Propagation is caused by some properties of the ether, I believe, and all known cases of propagation arise from the interaction of two properties. Lorentz has $\mu_0 = 1$, or ignores μ altogether, and $c_0 = \text{V}^{-2}$. Of course I also rationalise his units.

If there is electrification, measured by its volume density $\rho = \text{div D}_0$, the above equations suffer no change unless the electrification moves. Let the velocity of ρ be $\text{u} + \text{w}$, where the part w is constant, the same for all the electrification, and u is variable. As is well-known, $\rho(\text{u} + \text{w})$ is a part of the

Maxwellian true current, the missing link in the chain of the displacement current $p\mathbf{D}_0$. So

$$\text{curl}\,\mathbf{H} = p\mathbf{D}_0 + \rho(\mathbf{u}+\mathbf{w}) \tag{43}$$

takes the place of (41). The reference space in these equations is the ether itself, supposed fixed. But let the reference space have the same common motion $\mathbf{w}$ as the matter carrying the electrification. Let p' be the time differentiator for the moving matter, then

$$p' = p + \mathbf{w}\nabla \tag{44}$$

shows its connection with p. The circuital equations become

$$\text{curl}\,\mathbf{H} = (p' - \mathbf{w}\nabla)\mathbf{D}_0 + \rho(\mathbf{u}+\mathbf{w}), \tag{45}$$

$$-\text{curl}\,\mathbf{F} = (p' - \mathbf{w}\nabla)\mathbf{B}_0. \tag{46}$$

Now, in general,

$$\text{curl}\,\mathrm{V}\mathbf{wB} = \mathrm{V}\nabla\mathrm{V}\mathbf{wB} = \mathbf{w}.\nabla\mathbf{B} + \mathbf{B}\nabla.\mathbf{w} - \mathbf{B}.\nabla\mathbf{w} - \mathbf{w}\nabla.\mathbf{B}, \tag{46A}$$

(Vol. I., p. 200, equation 181).

Applying to (46), we have

$$-\mathbf{w}\nabla.\mathbf{B}_0 = \mathrm{V}\nabla\mathrm{V}\mathbf{wB}_0,$$

and applying to (45),

$$\mathbf{w}.\nabla\mathbf{D}_0 - \mathbf{w}\nabla.\mathbf{D}_0 = \mathrm{V}\nabla\mathrm{V}\mathbf{wD}_0,$$

so (45), (46) become

$$\text{curl}\,(\mathbf{H} - \mathrm{V}\mathbf{wD}_0) = p'\mathbf{D}_0 + \rho\mathbf{u}, \tag{47}$$

$$-\text{curl}\,(\mathbf{F} - \mathrm{V}\mathbf{wB}_0) = p'\mathbf{B}_0, \tag{48}$$

when we travel with the matter. The velocity $\mathbf{w}$ may be imagined to be the translational velocity of the earth through the ether at the place in question, it being assumed that the ether is stagnant. Lorentz makes a further transformation for the consideration of aberrational questions, but that is not wanted here. What is wanted is the introduction of the electric moment or the electrisation in a non-conducting material. There is no volume electrification, owing to molecules being equally endowed with positive and negative ions. These ions are displaced under the action of electric force, producing a sort of polarisation; and the first approximation is to assume an equilibrium theory between the moving force on an ion and the displacement. This takes no count of inertia, and the resultant effect is to make the electrisation simply proportional to the moving force. That is

$$\mathbf{D}_1 = c_1\mathbf{E} = c_1(\mathbf{F}+\mathbf{e}), \tag{49}$$

where $e = VwB_0$, is the result. Perhaps the averaging is the obscurest part of the process. The moving force on ρ at a point is ρE, not ρF, the additional part ρe being Maxwell's electromagnetic force on the current element ρw.

The circuital equations (47), (48) therefore become

$$\text{curl}\,(H - VwD_0) = p'(D_0 + D_1), \qquad (50)$$
$$= p'(cF + c_1 e),$$
$$= p'(cE - c_0 e),$$
$$-\text{curl}\,(F + VwB_0) = p'B_0. \qquad (51)$$

These are the equations employed by Lorentz for a moving dielectric. But I think it is somewhat plainer to let the reference space be the fixed ether. To transform, put $p' = p + w\nabla$, and use (46A), or rather the special simpler forms following it. We get

$$\text{curl}\,(H - VD_1 w) = pD = p(D_0 + D_1), \qquad (52)$$
$$= p(cF + c_1 e),$$
$$-\text{curl}\,F = pB_0 = \mu_0 pH. \qquad (53)$$

Or, in terms of E and H,

$$\text{curl}\,(H - VD_1 w) = p(cE - c_0 e), \qquad (54)$$
$$-\text{curl}\,(E - e) = p\mu_0 H. \qquad (55)$$

Perhaps the most significant form is got by taking out the $VD_1 w$ in (52). Then

$$\text{curl}\,H = pD_0 + p'D_1, \qquad (56)$$
$$-\text{curl}\,F = pB_0, \qquad (57)$$

express the relations. As the ether is supposed to be stationary the time differentiator is simply p for D_0. But the matter moves at speed w, so the time differentiator is p' instead. The distinction is most important, and the activities of p and p' must not be confounded. In the other equation (57), B_0 belongs to the ether, the matter being regarded as unmagnetisable, so the time differentiator is p simply. In this way of regarding (56), (57) we can eliminate the ionic hypothesis, if we can trust that D_0 and D_1 can be separated in the manner assumed.

The Wave-speed according to Lorentz.

§ 474. Lorentz deduces the wave speed from equations equivalent to (50), (51) only after a transformation of variables, including a changed time variable, making the zero of time different for different parts of space. But there is no need for this complication. The mathematics of plane waves is quite simple. Thus, using (52), (53), let $\mathbf{w} \parallel \mathbf{z}$, $\mathbf{F} \parallel \mathbf{x}$, $\mathbf{H} \parallel \mathbf{y}$ They reduce to

$$-\Delta\{\mathrm{H} + c_1 w(\mathrm{F} - w\mu_0\mathrm{H})\} = p(c\mathrm{F} - c_1 w\mu_0\mathrm{H}), \tag{58}$$

$$-\Delta\mathrm{F} = \mu_0 p\mathrm{H}, \tag{59}$$

where Δ is d/dz, and the quantities are now all scalars.

Also $p = -\mathrm{U}\Delta$, U being the wave speed; so

$$\mathrm{H} + c_1 w(\mathrm{F} - w\mu_0\mathrm{H}) = \mathrm{U}(c\mathrm{F} - c_1 w\mu_0\mathrm{H}), \tag{60}$$

$$\mathrm{F} = \mu_0\mathrm{UH}, \tag{61}$$

giving the quadratic

$$\mathrm{U}^2 - 2\mathrm{U}\frac{c_1}{c}w = v^2 - \frac{c_1}{c}w^2, \tag{62}$$

where $\mu_0 c v^2 = 1$, and, therefore, approximately

$$\mathrm{U} - v = \frac{c_1}{c}\,w, \tag{63}$$

the required result.

That the moving force on $\rho\mathbf{w}$ due to its motion in the field $\mathbf{B}_0$ is $\rho\mathbf{e}$ may be most simply seen thus. The magnetic force due to $\rho\mathbf{w}$ is the ordinary magnetic force of a current element. The moving force on a magnetic pole in its field is therefore known. Conversely, the force on $\rho\mathbf{w}$ due to the pole is known. It is the $\rho\mathbf{e}$ in question. That the pole gives a special kind of magnetic field is immaterial, because it is only the value of $\mathbf{B}_0$ at ρ that is concerned. [This simple argument is valid only when w/v is very small.]

But I think the above theory is possibly too simple to be true. In the first place, can the optical result be considered to be so well established as to make it be necessary to get the Fresnel wave-speed? And, if so, is it proper that the result should come out direct from elementary equations which take no count of dispersion? There is a very wide difference between light vibrations, billions per second, and those concerned in long waves. The above is really a propagational theory, without any reflex actions. On the other hand, U should strictly be the speed of a simply periodic wave-train, itself the resultant of the forward and backward actions. Now it is true that when U does not vary much within a given range of frequency, any wave which is compounded of simply periodic waves within that range is propagated with little distortion, so that a simple pulse theory is approximately true. But when U varies very widely with the frequency, and the arbitrary wave when analysed involves widely discrepant values of U, for example, an impulsive wave itself, an impulsive theory cannot be even approximately true. For this reason I conceive that a more proper theory should be of a secondary character, involving the differential connection between the ethereal disturbance and the internal motions.

Possible Equations for a Moving Magnetised Substance.

475. And yet it is suggested that the relations (56), (57) may be proper for long waves. But then the effect of magnetisation is not exhibited. If the above reasoning about the meaning of the time differentiators is correct, independently of the ionic hypothesis, it will be equally correct to extend them to

$$\text{curl}\,\mathbf{H} = p\mathbf{D}_0 + p'\mathbf{D}_1, \qquad (64)$$

$$-\text{curl}\,\mathbf{F} = p\mathbf{B}_0 + p'\mathbf{B}_1, \qquad (65)$$

where the new $\mathbf{B}_1$ is the magnetisation, existent when μ differs from μ_0, and of course assumed to be of elastic character. The question then arises as to the relation of $\mathbf{D}_1$ to $\mathbf{F}$, and $\mathbf{B}_1$ to $\mathbf{H}$. If we write

$$\mathbf{D}_1 = c_1(\mathbf{F} + \mathbf{e}), \qquad (66)$$

$$\mathbf{B}_1 = \mu_1(\mathbf{H} + \mathbf{h}), \qquad (67)$$

it is still matter for argument whether e should be the former $\text{V}\mathbf{w}\mathbf{B}_0$, or should become $\text{V}\mathbf{w}\mathbf{B}$, using the complete induction. And as regards h, whether it should be $\text{V}\mathbf{D}_0\mathbf{w}$, or $\text{V}\mathbf{D}\mathbf{w}$, or existent at all. In the theory of a movable ether of variable c and μ, the variable parts representing matter, everything can be done symmetrically. But with matter moving through ether, and special hypotheses in connection therewith, there would appear to be no longer a necessity for symmetry.

Prof. FitzGerald says there are electrons, but there are no magnetons, therefore there will be a want of symmetry, other than that arising from putting zero for quantities which might exist. But I think some salt is required here. For, first, it is known not that there are no magnetons, but rather that if there be any, they are in pairs like the ions, and do not dissociate as far as can be concluded from $\nabla B = 0$. Next, granting that there are no real magnetons, the theory of the magnetic shell remains. It shows that a closed electric current, which may consist of moving electrons, may be replaced by a double sheet of magnetons so far as external magnetic force is concerned, Thirdly, there are considerations concerning space and time variation of vectors which are quite independent of the physical nature of the vectors. So, altogether, I think it will be wise not to be overhasty in rejecting terms in the circuital equations because magnetons do not exist. Let us try to keep to symmetry to aid understanding any want of symmetry that may be forced upon us by special hypotheses. Say, therefore,

$$\mathrm{e} = n\mathrm{VwB}_0, \qquad \mathrm{h} = m\mathrm{VD}_0\mathrm{w}, \tag{68}$$

where m and n are numerics. This will give considerable latitude. What is now the wave-speed?

First, considering a plane wave travelling the same way as w, (64), (65) make

$$-\triangle(\mathrm{H} + \mathrm{D}_1 w) = p\mathrm{D}, \tag{69}$$

$$-\triangle(\mathrm{F} + \mathrm{B}_1 w) = p\mathrm{B}, \tag{70}$$

and therefore

$$\mathrm{H} + \mathrm{D}_1 w = \mathrm{UD}, \tag{71}$$

$$\mathrm{F} + \mathrm{B}_1 w = \mathrm{UB}. \tag{72}$$

Now introduce (66), (67), with the meanings given by (68), we get

$$\mathrm{H} + c_1 w(\mathrm{F} - wn\mu_0\mathrm{H}) = \mathrm{U}(c\mathrm{F} - c_1 wn\mu_0\mathrm{H}), \tag{73}$$

$$\mathrm{F} + \mu_1 w(\mathrm{H} - wmc_0\mathrm{F}) = \mathrm{U}(\mu\mathrm{H} - \mu_1 wmc_0\mathrm{F}), \tag{74}$$

which give U for any value of w. Neglecting squares, the quadratic is

$$\mathrm{U}^2 - \mathrm{U}w\left[\frac{c_1}{c}\left(1 + n\frac{\mu_0}{\mu}\right) + \frac{\mu_1}{\mu}\left(1 + m\frac{c_0}{c}\right)\right] = v^2; \tag{75}$$

where $c = c_0 + c_1$, $\mu = \mu_0 + \mu_1$, and $\mu c v^2 = 1$, That is

$$\frac{\mathrm{U} - v}{w} = \frac{c_1}{2c}\left(1 + n\frac{\mu_0}{\mu}\right) + \frac{\mu_1}{2\mu}\left(1 + m\frac{c_0}{c}\right). \tag{76}$$

To revert to the former case, $\mu_1 = 0$, and $n = 1$. But if both m and n are unity, then

$$\frac{\mathrm{U} - v}{w} = \frac{c_1}{2c}\left(1 + \frac{\mu_0}{\mu}\right) + \frac{\mu_1}{2\mu}\left(1 + \frac{c_0}{c}\right); \tag{77}$$

and if we employ e = VwB, h = VDw as another extreme case, then $n = \mu/\mu_0$ and $m = c/c_0$, producing

$$U - v = \left(\frac{c_1}{c} + \frac{\mu_1}{\mu}\right)w. \qquad (78)$$

Here magnetisation acts similarly to electrisation. Notice that this way of introducing μ_1 allows the coefficient of w to be greater than unity. Then the speed of the wave is increased from v to more than $v + w$ by moving the matter at speed w. This is not impossible, however improbable it may appear.

If we ask how still to obtain Fresnel's wave-speed when μ_1 is finite, examination of (76) shows that $m = -1$ will do it, along with the former $n = +1$. But if $n = \mu/\mu_0$, so as to use the full e, then we require $m = -c/c_0$. But it seems very unlikely that μ_1 should have no influence on the wave-speed in the part depending upon w.

There are many other ways of modifying the circuital equations so as to always lead to Fresnel's wave-speed when $\mu_1 = 0$, but it is of no use discussing them without some particular reason.

Larmor's Equations for a Moving Body.

§ 476. In § 471 above I referred to Dr. Larmor's investigation of the wave-speed. He has kindly asked me to delete the § 37 of his stimulating work "Ether and Matter," being convinced of its erroneousness, and has referred me to his § 73 later on. There is no investigation of the wave-speed there, but on making the necessary reductions I find (subject to an indeterminateness) that the general equations there given do lead to Fresnel's speed, and in the same way as Lorentz's. But there is some indefiniteness about the meaning of the vector (α, β, γ), which he identifies with Maxwell's magnetic force, so I give the equations here in vectorial form and rationalised, that the matter may be made more distinct.

Let $(P, Q, R) = \mathbf{E}$; $(a, b, c) = \mathbf{B}$; $(p, q, r) = \mathrm{w}$. Then Larmor's second circuital law is

$$-\operatorname{curl}(\mathbf{E} - \mathrm{e}) = p\mathbf{B}, \qquad \mathrm{e} = \mathrm{V}\mathrm{w}\mathbf{B}. \qquad (79)$$

This requires no remark.

Let $(A, B, C) = \mathbf{I}$; $(A_1, B_1, C_1) = \mathbf{I}_1$; $(A_0, B_0, C_0) = \mathbf{I}_0$;
$(u, v, w) = \mathbf{J}$; $(f, g, h) = \mathbf{D}_0$; $(f', g', h') = \mathbf{D}'$;
$(f'', g'', h'') = \mathbf{D}''$; $(f'_0, g'_0, h'_0) = \mathrm{d}_0$; $(\alpha, \beta, \gamma) = \mathbf{H}$.

Then Larmor's first circuital law is

$$\text{curl}\,\mathbf{H} = \mathbf{J}, \tag{80}$$

where J is what he terms Maxwell's total current

$$\mathbf{J} = \sigma\mathbf{E} + \mathbf{w}\rho + pc_0(\mathbf{E} - \mathbf{e}) + p(c_1\mathbf{E} + \mathbf{d}_0) \tag{81}$$

i.e., the sum of the conduction current, convection current, rate of time increase of the ethereal electric displacement, and ditto of the material electric displacement. For it is given that

$$\mathbf{J} = \sigma\mathbf{E} + \mathbf{w}\rho + p\mathbf{D}'', \qquad \mathbf{D}'' = \mathbf{D}_0 + \mathbf{D}' \tag{82}$$

$$\mathbf{D}_0 = c_0(\mathbf{E} - \mathbf{e}) = c_0\mathbf{F}, \qquad \mathbf{D}' = \mathbf{d}_0 + c_1\mathbf{E}. \tag{83}$$

Here w is the speed of the matter and of ρ the electrification, and $\mathbf{d}_0$ is intrinsic electrisation, analogous to $\mathbf{I}_0$, which is intrinsic magnetisation, whilst σ is the electric conductivity.

Now as regards H, it is given that

$$\mathbf{H} = \mathbf{B} - \mathbf{I}_1, \qquad \mathbf{I}_1 = \mathbf{I} + \mathrm{V}\mathbf{D}'\mathbf{w}, \qquad \mathbf{I} = \mathbf{I}_0 + \kappa\mathbf{H}, \tag{84}$$

which lead to

$$\mathbf{H} = \mathbf{B} - \mathbf{I}_0 - \kappa\mathbf{H} - \mathrm{V}\mathbf{D}'\mathbf{w}. \tag{85}$$

There are no μ's; that is, the units are "electromagnetic," only rationalised. Remembering this, we have in Maxwell's theory

$$\mathbf{H} = \mathbf{B} - \mathbf{I}_0 - \kappa\mathbf{H}, \tag{86}$$

because $1 + \kappa = \mu$ ("electromagnetic," numerics). This makes $\mathrm{V}\mathbf{D}'\mathbf{w} = 0$, which is obviously not meant. On the other hand, if we ignore (86), and use only (85), we express H in terms of B and VD′w; viz.,

$$(1 + \kappa)\mathbf{H} = \mathbf{B} - \mathrm{V}\mathbf{D}'\mathbf{w}. \tag{87}$$

Perhaps this is not meant either. It occurs to me to suggest that the H in equation (80) is not meant for Maxwell's H; call it H′ instead, and then in (85) we shall have H′ on the left side and H, subject to (86), on the right side. This will give

$$\text{curl}\,(\mathbf{H} - \mathrm{V}\mathbf{D}'\mathbf{w}) = \mathbf{J}, \tag{88}$$

which I think it probable is what Dr. Larmor really means. Because, transferring VD′w to the other side, by using (46A), this equation is the same as

$$\text{curl}\,\mathbf{H} = \sigma\mathbf{E} + p\mathbf{D}_0 + p'\mathbf{D}', \tag{89}$$

which is equivalent to Lorentz's equation, with the addition of conduction current and intrinsic displacement; because D′ reduces to D_1 when $d_0 = 0$.

It took some trouble to effect this reconciliation, and I could wish that Dr. Larmor had himself brought his work into line with that of other people; but as he has not done so, he may perhaps be grateful to me for trying to do it. If, however, I have not correctly interpreted his (α, β, γ) or H or H′, in the above way, I do not see how to harmonise matters. I should call curl H the total current, as in (89); not the J in (80).

Theory of Moving Electrified Cones; $u>v$. The Moving Force upon them and upon an Electrified Line.

§477. The case above considered, §465, of an electrified line moving in its own line casts light upon that of a moving point-charge, because it explains how the infinities arise that rendered the interpretation so obscure. In the theory of the moving line there is nothing in the distribution of the displacement that is seemingly impossible to understand. Referring to Fig. 26, we have a continuous transition from plane waves to a limiting conical wave, through intermediate hyperboloidal waves.

But how about the applied force needed to maintain the motion of the electrified line? We might expect it to be infinite because of the linear concentration of the electrification. Or it might be infinite only at the end of the line, and finite elsewhere. But we could not expect it to be zero. And yet the distribution of the displacement, being always perpendicular to the electrified line, does not indicate any moving force in the line of motion, whilst transversely there is a balance. The moving force due to the induction also seems ineffective. What is the explanation of this conflict between expectation and appearances?

It is the appearances that are deceptive. The moving force due to the stress (electric and magnetic) is indeed perpendicular to the electrified line in any plane containing the line; but since it is infinite, any finite force along the line compounded with it is lost sight of. We get no information immediately as to whether there is or is not a moving force along the line. To find out, we must generalise the problem, reduce the infiniteness to finiteness, see what the force along the line then is, and finally revert to the original problem and see what the limiting value of the force turns out to be. This means that we must examine the moving force upon an electrified cone instead of a line.

Referring to Fig. 26, here reproduced for convenience, there are two intermediate conical equipotential surfaces exhibited. Let the inner one be the electrified surface. Abolish the electric and magnetic disturbance inside this cone, but have everything else the same as before.

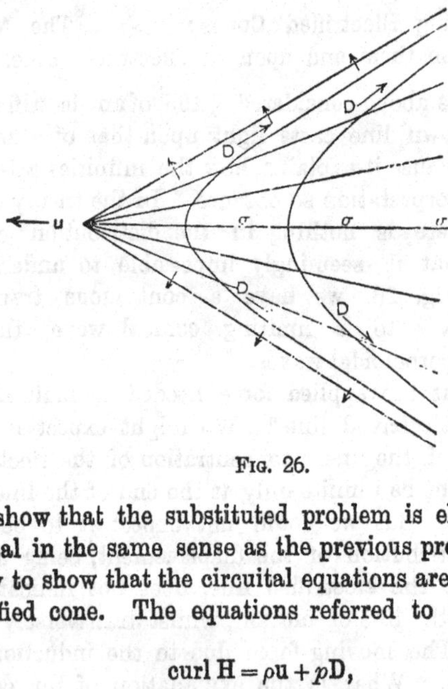

FIG. 26.

To show that the substituted problem is electromagnetically real in the same sense as the previous problem, we have merely to show that the circuital equations are satisfied at the electrified cone. The equations referred to the fixed ether being

$$\text{curl}\,\mathbf{H} = \rho\mathbf{u} + p\mathbf{D}, \tag{1}$$

$$-\text{curl}\,\mathbf{E} = p\mathbf{B}, \tag{2}$$

become, when referred to reference space moving with the cone,

$$\text{curl}\,(\mathbf{H} - \mathrm{V}\mathbf{u}\mathbf{D}) = p'\mathbf{D} = 0, \tag{3}$$

$$-\text{curl}\,(\mathbf{E} - \mathrm{V}\mathbf{B}\mathbf{u}) = p'\mathbf{B} = 0. \tag{4}$$

So $\mathbf{H} = \mathrm{V}\mathbf{u}\mathbf{D}$, because $\text{div}\,\mathbf{H} = 0$; and at the conical surface

$$\mathrm{V}\mathbf{N}(\mathbf{E} + \mathrm{V}\mathbf{u}\mathbf{B}) = 0 \tag{5}$$

expresses the second circuital law. That is, curl is $V\nabla$, and this is turned to VN at a surface, N being unit normal from the surface (see § 131, Vol. I.).

This equation means that the moving force on the electrification is perpendicular to the surface. The other moving force, pVDB per unit volume (§ 85, Vol. I., equation (30)) on the medium is inoperative, because by assumption the medium is fixed. So, by using the expression for H, we get

$$-\mathrm{VNE} = \mathrm{VNVuB} = \mathrm{u.NB} - \mathrm{Nu.B}.$$

Or, since NB = 0,

$$-\mathrm{VNE} = -\mathrm{Nu.B}. \tag{6}$$

This expresses the same as (2), when that equation is applied to a unit square circuit in the plane of the paper fixed in the ether and traversed by the electrified cone. For + Nu is the speed of the cone normal to itself, and − Nu . B is the rate of increase of induction through the circuit.

Similarly equation (1) is the same as

$$\mathrm{VNH} = \mathrm{u.ND} - \mathrm{Nu.D}, \tag{7}$$

by turning ρ to ND, the surface density, and p to − Nu. This is an identity because H = VuD. Therefore both circuital laws are satisfied at the moving electrified cone. Its situation may be anywhere between the axial line and the limiting cone, which is the seat of a free wave. I give these details of transformation in order to encourage timid readers to study the Vector Analysis in Vol. I. of this work. I repeat that there are no quaternions in it, so there is nothing to be afraid of.

The formula for the electric force is (18), § 465, or

$$\mathbf{E} = \frac{\sigma}{2\pi ch \tan\theta} \frac{z \tan^2\theta \,.\, \mathbf{h}_1 + h\mathbf{k}}{(z^2\tan^2\theta - h^2)^{\frac{3}{2}}}, \tag{8}$$

and from this is derived

$$\mathrm{VuB} = -\frac{1}{\sin^2\theta} \frac{\sigma}{2\pi ch \tan\theta} \frac{z \tan^2\theta \,.\, \mathbf{h}_1}{(z^2\tan^2\theta - h^2)^{\frac{3}{2}}}. \tag{9}$$

Therefore, by addition,

$$\mathbf{E} + \mathrm{VuB} = \frac{\sigma}{2\pi ch \tan\theta} \frac{h\mathbf{k} - z\mathbf{h}_1}{(z^2\tan^2\theta - h^2)^{\frac{3}{2}}}. \tag{10}$$

Here σ is not the surface density, but the linear density when the electrification is on the axis of the cone. The size of **E** is

$$\mathbf{E} = \frac{\sigma}{2\pi ch \tan\theta} \frac{(z^2 \tan^4\theta + h^2)^{\frac{1}{2}}}{(z^2 \tan^2\theta - h^2)^{\frac{1}{2}}}, \tag{11}$$

and the angle between **E** and **N** is given by

$$\cos \widehat{\mathbf{EN}} = \frac{z^2 \tan^2\theta - h^2}{(z^2 \tan^4\theta + h^2)^{\frac{1}{2}}(z^2 + h^2)^{\frac{1}{2}}}; \tag{12}$$

so, by (11) and (12),

$$\mathrm{ND} = \frac{\sigma}{2\pi h \tan\theta} \frac{(z^2 \tan^2\theta - h^2)^{\frac{1}{2}}}{(z^2 + h^2)^{\frac{1}{2}}}. \tag{13}$$

This being the surface density, the moving force on it is, by (10) and (13),

$$\mathrm{ND}(\mathbf{E} + \mathrm{V}\mathbf{uB}) = c\left(\frac{\sigma}{2\pi ch \tan\theta}\right)^2 \frac{h\mathbf{k} - \mathbf{h}_1 z}{(z^2 + h^2)^{\frac{1}{2}}}, \tag{14}$$

of which the size is

$$\frac{1}{c}\left(\frac{\sigma}{2\pi h \tan\theta}\right)^2. \tag{15}$$

This is the normal pressure per unit area of cone. Multiply it by h/r to get the z-component, and then by $2\pi h$ to get the total for a circular hoop round the cone. The result is

$$\mathrm{F} = \frac{\sigma^2}{2\pi cr \tan^2\theta}, \tag{16}$$

where F is the size of the resultant moving force of the electromagnetic field on the circular hoop. The direction of this force is from left to right, or against the motion. So F measures the applied force required to maintain the motion, acting in the direction of **u**. It varies inversely as the distance r of the hoop from the apex. It is always finite, save at the apex, which is a point where some modification of the shape of the V surfaces is required to produce finiteness, by rounding off the sharpness.

Finally, we come to the initial question of the moving force on the electrified line. F remains finite (save at the apex again) when h is reduced to zero, which makes r become z. Then

$$\mathrm{F} = \frac{\sigma^2}{2\pi cz \tan^2\theta} \tag{17}$$

is the force required to maintain the state, per unit length of the electrified line. If the linear density be reduced from σ to zero gradually, instead of suddenly at the apex, it will do away with the difficulty there, I think.

The flux of energy VEH is perpendicular to the hyperboloidal surfaces of D. This requires special interpretation at the electrified line, like the above. Substitute the electrified cone, and it is all right. The work done by the applied force is accounted for.

Suppose we have a second electrified cone outside the first. Referring to the figure again, abolish the E and H disturbances outside the second cone as well as inside the first. The second cone is then negatively electrified. Its theory is just like that above given for the first cone. There is no free extreme conical wave now, of course. The moving force due to the field on the second or outer cone is with the motion, and the applied force must be against it. So now energy goes from the inner cone to the outer through VEH. External work is done on the inner cone, and is returned to externality at the outer one.

Theory of Electrified Line of Finite Length Moving Transversely; $u>v$.

§ 478. Now consider the theory of an electrified line moving transversely to itself, not when of infinite length, as done in § 464, and previously, but of finite length. When of infinite length, the effect is to produce a pair of divergent plane waves, and there are only two forms of solutions, V = 0 in one region, and V = constant in the rest of space. But when the electrified

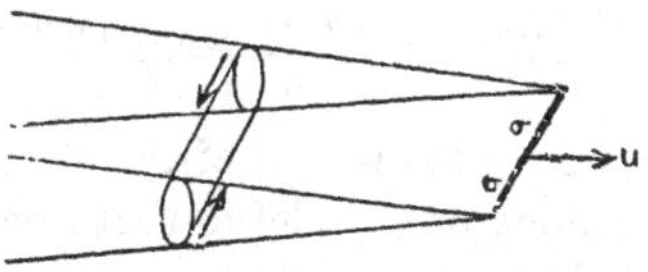

Fig. 31.

line is of finite length, there are five formulæ for the potential. To see this, from the two ends of the line as apexes construct cones of angle 2θ, given by $u \sin \theta = v$, whose axes are in the direction of u reversed. Join these cones together by a pair of planes touching them at opposite ends of diameters. These planes diverge from the electrified line and are inclined at angle 2θ. The Fig. 31 will give an idea of what is meant.

The thick line $\sigma\sigma$ is the electrified line moving from left to right. The two planes are plain enough. The ovals are supposed to represent the circular sections of the two cones at a certain distance, seen sideways.

The formula for V is (9), § 464; that is, when integrated, (11), with proper limits, or

$$V = \frac{\sigma}{2\pi c}\tan\theta\left[\sin^{-1}\frac{x}{\sqrt{z^2\tan^2\theta - y^2}}\right]. \tag{1}$$

The only trouble is to find the limits properly, to be done by consideration of the geometry of the wedge and two cones in Fig. 31. Let for convenience

$$V = \frac{\sigma\tan\theta}{2\pi c}W, \tag{2}$$

and consider the values of W, the potential measured in special units.

(1). Outside the planes and cones, $W = 0$. (3)

(2). Between the planes, but outside the cones, $W = \pi$. (4)

(3). Inside the left cone, but not also inside the right cone,

$$W = \tfrac{1}{2}\pi + \sin^{-1}\frac{a-x}{h}. \tag{5}$$

(4). Inside the right cone, but not also inside the left cone,

$$W = \tfrac{1}{2}\pi + \sin^{-1}\frac{a+x}{h}. \tag{6}$$

(5). Inside both cones at once

$$W = \sin^{-1}\frac{a-x}{h} + \sin^{-1}\frac{a+x}{h}. \tag{7}$$

Understand that $2a$ is the length of the electrified line, and that x is measured along or parallel to that line, with origin at its middle point. Also

$$h = \sqrt{z^2\tan^2\theta - y^2}, \tag{8}$$

where z is measured against u, or to the left, and y upwards from the median plane in which the electrified line is moving.

The single formula (7) will cover the whole on certain understandings common in the mathematics of waves—viz., the three geometrical variables x, y, z are to receive all values which allow W to be real; and then when, for instance, x or y

passes out of these bounds, and makes one of the inverse sine functions impossible, the value of W is to be considered to remain what it was at the moment of leaving bounds. In this way the first four cases are derivable from the fifth. The displacement from the electrification divides equally between the two planes, just as if the line were infinitely long, but there is an auxiliary system of displacement in the two cones, to be mentioned presently. The induction in the planes, too, is the same as if they did not terminate—namely, parallel to the electrified line, from the top of the right to the top of the left cone in the upper plane, and from the bottom of the left to the bottom of the right cone in the lower plane. The question now is, how the induction gets across from one plane to the other, and back again to complete its circuit. Fig. 32 is a section of the wedge in the plane of the induction, as seen

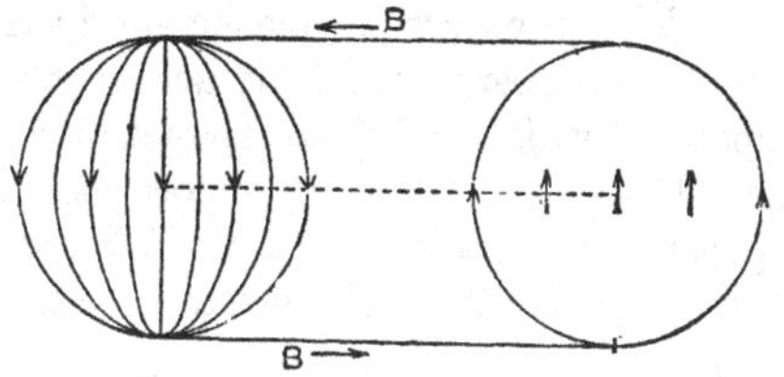

FIG. 32.

from the electrified line, whose position is vertically above the dotted line at distance z. The radius of the circles is $z \tan \theta$. If now we measure x in left circle from its centre to the right, and y upwards on the paper, the formula for W becomes

$$W = \tfrac{1}{2}\pi + \sin^{-1} \frac{x}{(z^2 \tan^2 \theta - y^2)^{\frac{1}{2}}}, \qquad (9)$$

inside the left circle. W falls from π on the right semicircle to 0 on the left semicircle, being $\frac{1}{2}\pi$ on the vertical diameter. The equipotential lines are semi-ellipses with a common major axis.

Now $H = VuD$. The z-component of D is here inoperative, so H is perpendicular to the lines of D in the plane of the paper (ignoring the z-component). But these are given by the slope of the potential in the plane, so they are perpendicular to the lines of V. It is only the z-component of D that follows a different law. Therefore the lines of H are the equipotential lines themselves. The flux of induction is therefore from the effective point source at the top of the circle to the effective sink at the bottom along the semi-elliptic paths. This applies to all planes parallel to the paper, from the electrified line itself to any distance beyond. As regards the intensity of H, it is smallest along the central vertical diameter and increases to infinity on the two extreme semicircles. This infinity is empty, of course, the total induction passing from one plane to the other (per unit distance along z) being finite. It is easy to see that if the electrified line be turned into a rod of finite size, carrying the same amount of electrification, the sheets of induction will become of finite depth, and then the density of the induction will be finite everywhere.

As regards the electric displacement in the cones, that is more difficult to follow, because it is in three dimensions instead of two. The transverse displacement in the plane of the paper, Fig. 32, goes perpendicularly across the lines of V; but the z-component, which is differently reckoned, and varies as $+\nabla V$ instead of $-\nabla V$, is very important. It is down through the paper in the outer half of the left one, and up in the inner. The full formula for the electric force is

$$\mathbf{E} = -\frac{(z^2 \tan^2\theta - x^2 - y^2)^{\frac{1}{2}}}{z^2 \tan^2\theta - y^2}\mathbf{i} - \frac{x\mathbf{r}}{(z^2 \tan^2\theta - y^2)\,(z^2 \tan^2\theta - x^2 - y^2)^{\frac{1}{2}}} \qquad (10)$$

where $\mathbf{r}$ is the vector from the origin at the end of the electrified line.

The cones overlap one another at some finite distance depending upon the length of the electrified line and the ratio u/v settling the angle θ. In the overlap the E and H are the resultants of those for the two cones. Go far enough away from the electrified line and the circles in a plane section as in Fig. 32 tend to coincide, when viewed through a diminishing glass; or, viewed from the line, by natural perspective. Then the left crescent in Fig. 33 is all that is left of the left circle, and the induction is down in it from top to bottom in nearly circular paths. The right crescent shows what is left of the right circle, and the induction is up in it. The resultant H in the overlap, which is nearly a circle, is nearly circular, oppositely directed to that in the two crescents. It is intense at the boundary, and falls off towards the centre. The short horizontal line in the middle is the projection of the electrified line. We approximate in this way to the conical wave with reversed internal E and H due to a moving point-charge.

The voltages employed by Trowbridge are so greatly in excess of those which give estimated European speeds to electrons comparable with the speed of light, that it seems very probable that in his experiments electrons do have speeds given to them exceeding that of light. They cannot maintain them. That is one thing. The other principal peculiarity is the reversal of their action upon other electrification. Thus an electron moving much faster than light does will draw after it other slowly-moving electrons of the same sign,

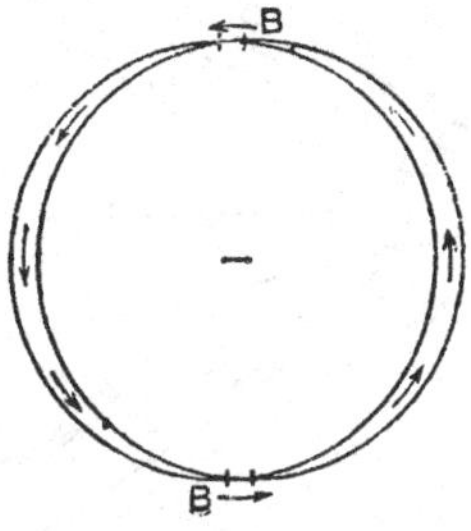

FIG. 33.

instead of repelling them, unless they get into the conical sheet, or rather, into the more diffused real disturbance corresponding to the conical sheet of the abstract theory.

It is to be noted that, by the theory of § 465, to which Fig. 27 refers, the inward displacement does not go to the point-charge first, and then come out again along the conical sheet, when the point-charge is enlarged to finite dimensions. It turns round into the conical sheet all the way along it; that is, forcing an electron along faster than light goes produces a sort of vortex of displacement in its wake.

Motion of Electrified Hyperboloids; $u>v$.

§ 479. The motion of electrified hyperboloids comes under the theory of a moving electron merely because the equipotential surfaces inside the conical wave are hyperboloids of revolution. As the essential mathematics has been given, only some special points need be noticed here. Fig. 26 will serve to illustrate, only changing its meaning. The electron is now to be at the apex. The outer cone is the wave front. The two internal cones are sheets of electric force. The two hyperboloids are equipotential. The equations are, by § 463,

$$V=\frac{Q\tan\theta}{2\pi cl}, \qquad E=-\frac{rQ\tan\theta}{2\pi cl^3}, \qquad H=VuD, \tag{1}$$

where l is given by

$$l^2=z^2\tan^2\theta-y^2. \tag{2}$$

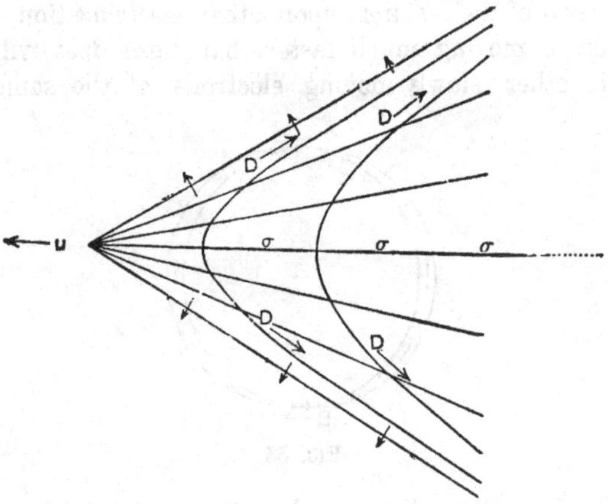

FIG. 26.

The apex is the origin, z being measured to the right, and y upwards on the paper, whilst $\mathbf{r}$ is the vector from the origin.

Consider the inner hyperboloid. Abolish E and H outside it. It then becomes an electrified hyperboloid moving parallel to its axis. Its field of force is internal, and is given by the above formulæ. E is directed everywhere towards a certain external point, the apex of the asymptotic cone. The electrification is negative, surface density ND, if N is the inward normal. By differentiating (2), remembering that l is constant for the hyperboloid, we find

$$N=\frac{z\tan^2\theta-y}{(y^2+z^2\tan^4\theta)^{\frac{1}{2}}}, \tag{3}$$

therefore, by the formula for E, we get

$$\mathrm{ND} = -\frac{\mathrm{Q} \tan \theta}{2\pi l\,(y^2 + z^2 \tan^4\theta)^{\frac{1}{2}}}. \tag{4}$$

The total electrification on the hyperboloid is infinite, because if ds is an element of the curve in the plane of the paper,

$$2\pi y \frac{ds}{dz} = 2\pi y \frac{(y^2 + z^2 \tan^4\theta)^{\frac{1}{2}}}{y}, \tag{5}$$

which makes

$$\iint \mathrm{ND}d\mathrm{S} = \frac{\mathrm{Q} \tan \theta}{l} \int dz, \tag{6}$$

so that the total electrification on the hyperboloid is proportional to the distance from its vertex.

The moving force per unit of electrifica tion is E + VuB, or

$$\mathrm{E} + \frac{\mathrm{VuVuE}}{v^2} = \left(1 - \frac{u^2}{v^2}\right)\mathrm{E}_2 + \mathrm{E}_3,$$

$$= -\mathrm{E}_2 \cot^2\theta + \mathrm{E}_3 = -\frac{\mathrm{Q} \cot \theta}{2\pi c l^3}(\mathrm{z} \tan^2\theta - \mathrm{y}); \tag{7}$$

of which the tensor is

$$\frac{\mathrm{Q} \cot \theta}{2\pi c l^3}(y^2 + z^2 \tan^4\theta). \tag{8}$$

So, by (4) and (8),

$$\mathrm{ND}(\mathrm{E} + \mathrm{VuB}) = \frac{1}{c}\left(\frac{\mathrm{Q}}{2\pi l^2}\right)^2 \mathrm{N}. \tag{9}$$

This is constant all over the hyperboloid. It is somewhat remarkable that the normal pull of the electromagnetic field upon the electrification should be everywhere of the same intensity. As in previous cases, applied force in the direction of motion is required to maintain it. The question arises whether the force may be of finite total in this case. It is unlikely to be so, on account of the association of the theory with that of a point charge, but as there are no places of infinite E and H now, it is worth looking to see. We have

$$\cos \widehat{\mathrm{zN}} = \frac{z \tan^2\theta}{(y^2 + z^2 \tan^4\theta)^{\frac{1}{2}}}, \qquad \cos \widehat{\mathrm{z}ds} = -\frac{y}{(y^2 + z^2 \tan^4\theta)^{\frac{1}{2}}}, \tag{10}$$

and therefore

$$2\pi y \cos \widehat{\mathrm{zN}} \frac{ds}{dz} = 2\pi y \frac{\cos \widehat{\mathrm{zN}}}{\cos \widehat{\mathrm{z}ds}} = 2\pi z \tan^2\theta; \tag{11}$$

consequently the total moving force in the direction of motion is

$$\frac{1}{c}\left(\frac{Q}{2\pi l^2}\right)^2 \int 2\pi z \tan^2\theta \,.\, dz, \tag{12}$$

where the lower limit is finite, and the upper infinity. The total for the complete hyperboloid is therefore infinite.

Now look at Fig. 26 again. If we extend the electric field until it reaches the outer hyperboloid, we shift the electrification to it, and the same theory applies, with a different value given to l. Next, abolish the **E** and **H** inside the inner hyperboloid. We then have two charged hyperboloids moving along together, the inner one being positively, and the outer one negatively electrified. The electromagnetic field exerts the normal traction given by (9) on both hyperboloids. The outer surface is pulled back, the inner one forward. The necessary applied forces required to maintain the motion are with the motion on the outer surface and against it on the inner.

If the direction of motion is reversed, without changing **E**, the effect is merely to reverse **H**. The moving forces remain the same.

Growing Plane Source of Induction. Transition from $u>v$ to $u<v$.

§ 480. In considering (§ 468) the waves sent out from a growing plane source of induction, two distinct cases arose, according as the speed of growth was less or greater than that of light. The question of the transition from the one case to the other presents itself; whether there is any difficulty in the intermediate case of equal speeds, and what is the state of affairs in the cylindrical cap which completes the diverging plane waves generated when the speed of growth exceeds that of light.

One figure will serve for both cases in connecting the two solutions. The circle is of radius vt. Its centre is the place where the **f** source began, say at the moment $t=0$. When $u<v$, the circle (section of cylinder) is the wave front, and the region occupied by the source extends from the centre to some distance less than vt, say to a, when $u=\frac{1}{2}v$. Then any point P within the circle receives induction not from the existent source between O and a, but between O and some

point not so far as a, except at a itself. If P is on the circle itself it is only just beginning to receive induction, namely, from O. The intensity of H is greater in the right than in the left semicircle, because of the source being to the right

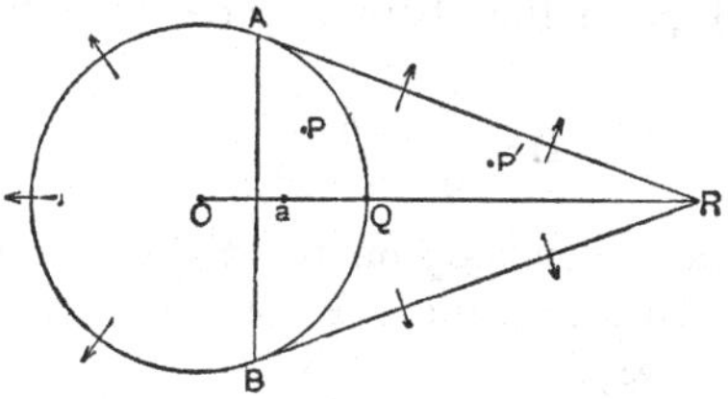

FIG. 34.

of O. The formula for H is (29), § 468. Put $t_3 = 0$, and give τ_2 the value there determined. The result may be written in the form

$$H = \frac{f}{2\pi\mu v\,(v^2 - u^2)^{\frac{1}{2}}} \log\left(\frac{1+\lambda}{1-\lambda}\right)^{\frac{1}{2}}, \tag{1}$$

where λ is given by

$$\lambda = \frac{(v^2 - u^2)^{\frac{1}{2}}(v^2t^2 - z^2 - y^2)^{\frac{1}{2}}}{v^2t - uz}. \tag{2}$$

Here z is measured from O to the right, and y upward on the paper. Note that the denominator in λ cannot vanish, provided we keep inside the circle of disturbance, so that λ is real positive and < 1.

In the limiting case $u = v$, the solution reduces to

$$H = \frac{f}{2\pi\mu v}\,\frac{(v^2t^2 - z^2 - y^2)^{\frac{1}{2}}}{vt - z}. \tag{3}$$

The f source now extends from O to Q. The magnetic force is still zero on the circle, save at the point Q, where it is $f/2\pi\mu v$.

Increasing u above v, the wedge begins at Q, with sides tangential to the cylinder. In the case shown, $u = 3v$, so that R is the extreme limit of the source at time t. We cannot expect the formula (1) to be valid outside the circle when taken literally. But it may be suitably interpreted. It is equivalent to

$$H = \frac{fu}{2\pi\mu v}\,\frac{1}{(u^2 - v^2)^{\frac{1}{2}}} \tan^{-1}\frac{(u^2 - v^2)^{\frac{1}{2}}(v^2t^2 - y^2 - z^2)^{\frac{1}{2}}}{v^2t - uz}, \tag{4}$$

and when $u>v$, the vanishing of the denominator separates the circle into two regions, to right and left of the chord AB. On the left side, the angle $\tan^{-1}$... increases from 0 on the circular boundary to $\frac{1}{2}\pi$ on the chord. On the right side it increases from $\frac{1}{2}\pi$ on the chord to π on the remainder of the circle. That is

$$\mathrm{H}=\frac{fu}{2\mu v(u^2-v^2)^{\frac{1}{2}}}, \tag{5}$$

on the cylindrical boundary on the right side, one-half this value on the plane AB, and zero on the remainder of the cylindrical boundary.

Now the last formula shows the constant value of H inside the wedge as found by a different kind of integration, § 468, suitable for the case $u>v$. See Fig. 30, § 468, indicating the limits of integration when the point P is shifted outside the circle, say to P′. So, by using the formula appropriate to the $u<v$ case, the extension to $u>v$ requires us to keep H at the circular boundary value, and continue it up to the plane wave fronts.

We must next show that the complete integral in the $u>v$ case passes properly into the above form when we leave the region of constant H and enter the circle. This can be done. Thus, by (31), § 468,

$$\mathrm{H}=\frac{fu}{2\pi\mu v}\frac{1}{(u^2-v^2)^{\frac{1}{2}}}\int\frac{-d\tau}{(\beta^2-\tau^2)^{\frac{1}{2}}}$$

$$=\frac{fu}{2\pi\mu v}\frac{1}{(u^2-v^2)^{\frac{1}{2}}}\left[-\sin^{-1}\frac{\tau}{\beta}\right]; \tag{6}$$

where $$\beta=\left\{\left(\frac{v^2t-uz}{u^2-v^2}\right)^2+\frac{v^2t^2-y^2-z^2}{u^2-v^2}\right\}^{\frac{1}{2}}, \tag{7}$$

and $$\tau=\frac{v^2t-uz}{v^2-u^2}-t_0. \tag{8}$$

At the limits, when complete,

$$t_0=\frac{v^2t-uz}{v^2-u^2}\pm\beta, \tag{9}$$

using the + sign with the larger value of t_0, which indicates two values of the time variable. So the τ limits are $\mp\beta$, which used in (6), produce the preceding result (5). But this is only true provided the lower t_0 limit does not pass the centre of the circle, because the source does not exist on the left of O. But when P′ is shifted to the circle, the lower t_0 limit is sent down to zero. After that, when P′ is shifted inside the circle, the lower limit for t_0 must remain zero. Therefore, applying this to (6), by (8) we get

$$\mathrm{H} = \frac{fu}{2\pi\mu v(u^2 - v^2)^{\frac{1}{2}}}\Big[-\sin^{-1}\frac{\tau}{\beta}\Big]_{\frac{v^2t-uz}{v^2-u^2}}^{-\beta} \tag{10}$$

That is,

$$\mathrm{H} = \frac{fu}{2\pi\mu v(u^2 - v^2)^{\frac{1}{2}}}\left\{\frac{\pi}{2} - \sin^{-1}\frac{v^2t - uz}{(u^2 - v^2)\beta}\right\}. \tag{11}$$

This gives the full value (5) on the right portion of the circular boundary, half this value on the line AB, and zero on the left circular boundary. Just as before, in fact. So formula (11) should mean the same as (4). This is the case, for they are mutually transformable. But pay proper attention to the sign of $v^2t - uz$. Otherwise these inverse circular functions are treacherous. The radical must be kept to its + value always.

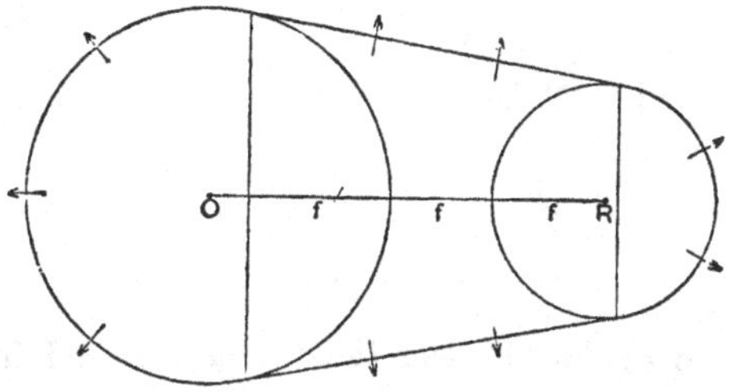

FIG. 35.

We have therefore a complete harmony. We can pass only suggestively from the $u<v$ to the $u>v$ case. But the latter is more comprehensive, and we can include in it both cases. We have then two forms of solution, one outside, the other inside the circle; and if v is made less than u, the outside solution disappears and leaves the inside one alone.

Now go on further. When the source has reached R, stop its further extension, and see what happens. The former state of things will continue, but with an exactly similar negative state superposed, with origin at R instead of at O. After a little while therefore, we shall have things as in Fig. 35.

The plane source extends from O to R, beginning at O at time $t=0$. Inside the left circle the value of H is given by (11). Between the two circles and between the tangent lines it has the constant value (5), this being the region of the plane waves. In the right circle, the formula for H is like (11), with suitable change of origin and time reckoning. H falls from the full value to half value in passing to the vertical chord, and then to zero on the right circular boundary. As time goes on the circles overlap, and the region of constant **H** and **E** disappears.

The Waves from a Plane Strip Source of Induction suddenly started.

§ 481. If we increase u to infinity, we obtain the result due to the immediate starting (all over at the same moment) of the plane strip of f extending from O to R. Fig. 32 suits this case excellently.

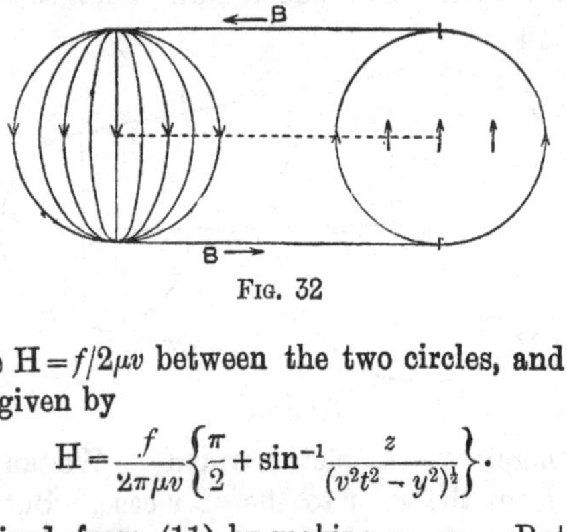

FIG. 32

We have $H=f/2\mu v$ between the two circles, and in the left circle it is given by

$$H=\frac{f}{2\pi\mu v}\left\{\frac{\pi}{2}+\sin^{-1}\frac{z}{(v^2t^2-y^2)^{\frac{1}{2}}}\right\}. \tag{12}$$

This is derived from (11) by making $u=\infty$. But it may, of course, be obtained from the reduced form of the integral leading to (11). H falls to half the constant value on the vertical diameter, and then to zero on the left semicircle. Similarly in the right circle.

It may be asked what the semi-elliptic lines now represent. In a previous problem, § 478, they were the lines of the induction entering at the top and leaving at the bottom, their continuations being plane sheets of induction. But in the present case the plane sheets are sheets of electric current. Does the current divide over the circles in the same way in the present problem? Yes, just the same. The y and z components of the current, say C_2 and C_3, are

$$C_2 = \frac{dH}{dz}, \qquad C_3 = -\frac{dH}{dy}; \tag{13}$$

that is, the rate of increase of H along any line in the plane measures the transverse current density from right to left across the line. To fix ideas, x is downward through the paper, y upward on the paper, and z to the right. Origin at centre of left circle. So, if f is down through the paper, so is H all over, whilst E is from left to right above f, and from right to left below it, between the circles. The electric current, therefore, enters at the bottom and leaves at the top of the left circle. Using (13), applied to (12), we have

$$C_2 = \frac{f}{2\pi\mu v}\frac{1}{(v^2t^2 - r^2)^{\frac{1}{2}}}, \qquad C_3 = \frac{f}{2\pi\mu v}\frac{-yz}{(v^2t^2 - y^2)(v^2t^2 - r^2)^{\frac{1}{2}}}, \tag{14}$$

where $r^2 = y^2 + z^2$.

Now the displacement cE is the time integral of the current. Outside the circles this means simply first $E_3 = 0$, until $vt = y$, and then $E_3 = \pm\frac{1}{2}f$, from $vt = y$ until $vt = r$. At this moment the circle vt reaches the point, and later still the time integrals of (14) must be added on. Thus,

$$E_2 = \frac{1}{c}\int C_2 dt = \frac{f}{2\pi}\Big[\log\{vt + (v^2t^2 - r^2)^{\frac{1}{2}}\}\Big]_r^{vt}$$

$$= \frac{f}{2\pi}\log\frac{vt + (v^2t^2 - r^2)^{\frac{1}{2}}}{r}. \tag{15}$$

This is the complete formula for the upward component of the electric force in the circle.

As regards the other, we have by differentiation,

$$\frac{d}{d(vt)}\sin^{-1}\frac{vtz}{r(v^2t^2 - y^2)^{\frac{1}{2}}} = -\frac{z\sqrt{y^2}}{(v^2t^2 - y^2)(v^2t^2 - r^2)^{\frac{1}{2}}}. \tag{16}$$

Particularly attend to the $\sqrt{y^2}$. If it is written y, everything will go wrong in the subsequent work, and the displace ment will not be continuous in the circle. Having found this out by troublesome experience, it follows that in integrating (14), which has the numerator $-yz$, we, of course, use the left member of (16) when y is +, but must change its sign when y is −.

The final results are

$$E_3 = \pm \frac{f}{2\pi}\left(\frac{\pi}{2} + \sin^{-1}\frac{vtz}{r(v^2t^2 - y^2)^{\frac{1}{2}}}\right) \tag{17}$$

inside the left circle, using the + or − sign according as y is + or −, that is above or below the plane of the source of induction. The initial displacement due to the plane waves, already referred to, is included in (17). The two equations (15) and (17) give a complete account of how the two oppositely directed uniform horizontal displacements above and below the plane of f are joined together through the terminal circles. The nature of a line of E differs according to whether it strikes the f plane in the cylinder or goes round its extremity in the outer part of the cylinder. If it goes round it does so in a continuous curve which is naturally vertical across the continuation of the f plane. This applies to the outer lines of force. But the inner ones strike the f plane sideways above and below it, in the right part of the left circle and in the left part of the right circle, so as to produce closed lines of displacement possessing a pair of convex cusps.

Impressed Current along a Straight Axis. The Operational Solution in General.

§ 482. When electrification moves along a straight line, the magnetic force generated is in circles centred upon that line, in planes perpendicular to it. One datum, the intensity of magnetic force at distance r from the axis and distance z along it is sufficient to specify H. The characteristic $v^2\nabla^2 E = p^2 E$ also suffers a reduction in the number of variables by the symmetry, E being representable by an axial component and a radial component. Say z is along the axis, x and y transverse; then $r = (x^2+y^2)^{\frac{1}{2}}$, and the characteristic of the z component is

$$\frac{1}{r}\frac{d}{dr}r\frac{dE}{dr} + \frac{d^2E}{dz^2} = \frac{p^2}{v^2}E. \tag{1}$$

This allows the immediate use of Bessel functions. Let

$$q^2 = \frac{p^2}{v^2} - \frac{d^2}{dz^2}, \tag{2}$$

then the characteristic is

$$\frac{1}{r}\frac{d}{dr}r\frac{dE}{dr}=q^2E, \tag{3}$$

of which the solution is

$$E=I_0(qr)A+K_0(qr)B, \tag{4}$$

where A and B are any functions of t and z.

It being supposed here that E means the axial component of **E**, let F be the transverse component; then the second circuital law is fully given by

$$\frac{dF}{dz}-\frac{dE}{dr}=-\mu pH, \tag{5}$$

if H is the intensity of **H**. But the first circuital law has two representatives, namely

$$\frac{1}{r}\frac{d}{dr}rH=cpE, \qquad -\frac{dH}{dz}=cpF. \tag{6}$$

These equations (5), (6) are best obtained by immediate application of the circuital principles to elementary circuits It follows that

$$H=\frac{cp}{q^2}\frac{dE}{dr}, \qquad F=-\frac{1}{cp}\frac{dH}{dz}. \tag{7}$$

The second of these is an obvious consequence of the second of (6). The first one is proved by inserting E according to (4), and then using (3).

These general equations (4) and (7) apply to other problems besides the present one, to straight round wires and tubes for example. (El. Pa., Vol. II., p. 175). But we do not require such complication as is there involved. Let there be no external interferences or boundaries or changes of medium. Then we have only one set of solutions, and we do not want the I_0 function in (4), but only the K_0. Or, if there was first an external boundary, its removal to an infinite distance would cause reflections to disappear and the I_0 function to drop out. That is, all we want now is given by

$$E=K_0(qr)A, \qquad F=-\frac{1}{cp}\frac{dH}{dz}, \tag{8}$$

$$H=\frac{cp}{q^2}\frac{d}{dr}K_0(qr)A, \tag{9}$$

with one function A, which must be found to suit given conditions at the axis.

Now the motion of a point-charge or electron along the axis obviously comes under this theory. Also, I have previously shown how to algebrise the special forms assumed in this case, without direct reference to a point-charge, however. (El. Pa., Vol. II., p. 456; E. M. T., Vol. II., p. 456). So, by now pointing out how to construct the operational solution, and previously how to algebrise it, I might consider the matter essentially completed, and leave the working out of details to my readers. Perhaps, however, they will be more satisfied if I continue to elaborate the subject. Besides, there are various other matters to be noticed in connection. So I continue.

To find A. If C is the electric current impressed at the axis, the circuitation of H must be C when r is reduced to 0. That is,

$$2\pi r\frac{cp}{q^2}\frac{d}{dr}\mathrm{K}_0(qr)\mathrm{A} = \mathrm{C}, \quad \text{when} \quad r = 0. \tag{10}$$

Or,

$$2\pi r\frac{cp}{q^2}\left(-\frac{2}{\pi r}\right)\mathrm{A} = \mathrm{C}, \text{ therefore } \mathrm{A} = -\tfrac{1}{4}\frac{q^2}{cp}\mathrm{C}. \tag{11}$$

Using this in (8), (9), we get

$$\left.\begin{aligned} \mathrm{E} &= -\tfrac{1}{4}\frac{q^2}{cp}\mathrm{K}_0(qr)\mathrm{C}, \qquad \mathrm{F} = -\frac{1}{cp}\frac{d\mathrm{H}}{dz}, \\ \mathrm{H} &= -\tfrac{1}{4}\frac{d}{dr}\mathrm{K}_0(qr)\mathrm{C}, \end{aligned}\right\} \tag{12}$$

giving E, F, and H in terms of C, which may be any function of z and t. What we have to find is $\mathrm{K}_0(qr)\mathrm{C}$. The algebrisation may be effected in a good many cases, immediately in simply periodic cases, but more difficultly in more interesting cases. If C is a convection current, and it is written $=\sigma u$, then σ may mean the linear density of electrification moving along the line at speed u. But there may be two convection currents. In either case, it is the second of equations (8) that finds the linear density of electrification, which is $2\pi rc\mathrm{F}$.

Regarding the use of the $\mathrm{K}_0(qr)$ operator alone, a caution is needed which is of some significance. $\mathrm{I}_0(qr)$ is finite at the axis, infinite at ∞, when qr is numerical; whereas $\mathrm{K}_0(qr)$ is infinite at the axis, and vanishes at infinity. This seems a sufficient reason for excluding I_0 when the source of disturbance is at the axis, and there is no external interference. For do not the waves expand, and so decrease in intensity as they go out to infinity? Certainly they do, but that is not the real reason for using K_0. The true reason is because $K_0(qr)$ is the operator appropriate to an outward going wave. The disturbance need not always vanish at infinity. There may be a wave front of infinite intensity, and which remains always of infinite intensity. (See Vol. II., pp. 240, 257, for the parts played by H_0 and K_0 in inward and outward waves.) There is a similar property in plane waves, when there is no attenuation at all. Thus, $\epsilon^{-qx}f(t) = f(t - x/v)$; so ϵ^{-qx} is the operator for a positive wave, q meaning p/v. Now ϵ^{-qx} vanishes at infinity, but the disturbance does not, if it can be imagined ever to get there, which is a separate question.

The $K_0(qr)$ operator is the cylindrical analogue of ϵ^{-qx} for the plane. The analogue of ϵ^{qx} for a negative plane wave is $H_0(qr)$. This is appropriate when a wave travels towards the axis from an external source.

Nevertheless, the $I_0(qr)$ function can be properly used instead of $K_0(qr)$ to obtain a wave coming from a source at the axis, provided we know what the disturbance there is. This is not the same thing as the strength of source. The I_0 function is compounded of both H_0 and K_0. The disturbance at the axis is due to inward and outward waves. Knowing it, then $I_0(qr)E_0 = E_r$ finds E_r from E_0. In the purely mathematical aspect, these relations of H_0 and K_0 to I_0 are rather obscure, but the wave theory throws light upon them, and makes the two divergent functions useful working agents.

Algebrisation of the Operational Solution in the case of Steady Motion of an Electron or of an Electrified Line; $u > v$.

§ 483. Returning to equations (12), there are two differentiators in the operator q. But in the important case of steady motion, there is an effective reduction to one differentiator, which makes the algebrisation quite easy for an electron. Thus, let $C = \epsilon^{-ut\Delta}C_0$, and C_0 be a function of z only. Then

C is the same as C_0 in shape, but travels at speed u along the z-axis, making $C=f(z-ut)$. In another form, we may say $C=\epsilon^{-pz/u}F(t)=F(t-z/u)$. This way is sometimes more convenient. $F(t)$ is the value of C at the fixed origin $z=0$, and may be the given datum. In either case, if $d/dz=\Delta$, $d/dt=p$, we have

$$u\Delta=-p, \qquad q^2=\lambda^2\Delta^2=\lambda^2\frac{p^2}{u^2}, \tag{13}$$

if

$$\lambda^2=\frac{u^2}{v^2}-1, \tag{14}$$

and both Δ and q are expressed in terms of p. Equations (12) become

$$E=\tfrac{1}{4}\epsilon^{-pz/u}\frac{\lambda^2\Delta}{cu}K_0\left(\frac{\lambda pr}{u}\right)C_0, \tag{15}$$

$$H=-\tfrac{1}{4}\epsilon^{-pz/u}\frac{d}{dr}K_0\left(\frac{\lambda pr}{u}\right)C_0, \tag{16}$$

$$F=H/cu, \tag{17}$$

where C_0 is the value of C at $z=0$, a function of the time.

Now let

$$U=\frac{1}{4cu}K_0\left(\frac{\lambda pr}{u}\right)C_0. \tag{18}$$

then the preceding equations assert that

$$E=\epsilon^{-pz/u}\lambda^2\frac{dU}{dz}, \quad F=-\epsilon^{-pz/u}\frac{dU}{dr}, \qquad H=ucF. \tag{19}$$

These equations may be compared with § 463. They show that U is the travelling potential, and that the electric force is derived from it in the eolotropic manner there described, and that H = VuD. The operator $\epsilon^{-pz/u}$ merely does the translation.

We have therefore to determine U according to (18). First of all, for a travelling electron—a charge Q moving at speed u. Suppose that it passes the origin at the moment $t=0$, then $C_0=Qp1$ operationally expressed. This makes

$$U=\frac{Q}{4cu}K_0(qr)p1. \tag{20}$$

which has been already algebrised. But as it is shortly done, it may be here repeated, for completeness.

Let q be any differentiator, say with respect to the variable l, then

$$\mathrm{K}_0(qr)q1 = \epsilon^{-qr}\left(\frac{2}{\pi qr}\right)^{\frac{1}{2}}\left(1 - \frac{1^2}{8qr} + \frac{1^2 3^2}{\underline{|2}(8qr)^2} - \dots\right)q1$$

$$= \epsilon^{-qr}\left(\frac{2}{\pi r}\right)^{\frac{1}{2}}\left(\frac{l^{-\frac{1}{2}}}{\underline{|-\tfrac{1}{2}}} - \frac{1^2 l^{\frac{1}{2}}}{8r\underline{|\tfrac{1}{2}}} + \frac{1^2 3^2 l^{\frac{3}{2}}}{\underline{|2}(8r)^2\underline{|\tfrac{3}{2}}} - \dots\right)$$

$$= \epsilon^{-qr}\left(\frac{2}{rl}\right)^{\frac{1}{2}}\frac{1}{\pi}\left\{1 - \frac{1}{2}\left(\frac{l}{2r}\right) + \frac{1.3}{2.4}\left(\frac{l}{2r}\right)^2 - \dots\right\}$$

$$= \epsilon^{-qr}\frac{1}{\pi}\left(\frac{2}{rl}\right)^{\frac{1}{2}}\left(1 + \frac{l}{2r}\right)^{-\frac{1}{2}} = \frac{1}{\pi}\left(\frac{2}{r(l-r)}\right)^{\frac{1}{2}}\left(1 + \frac{l-r}{2r}\right)^{-\frac{1}{2}}.$$

That is, finally,

$$\mathrm{K}_0(qr)q1 = \frac{2}{\pi}\frac{1}{(l^2 - r^2)^{\frac{1}{2}}}. \tag{21}$$

Applying this to (20), we have $q = \lambda p/u$, so $l = ut/\lambda$. This makes

$$\mathrm{U} = \frac{\mathrm{Q}}{2\pi c}\frac{1}{(u^2t^2 - \lambda^2 r^2)^{\frac{1}{2}}}. \tag{22}$$

This is the potential at the plane of the origin. It is zero before the moment $t=0$, and later is given by (22), but only provided ut is not less than λr. We assume u to be positive; then the operand 1 in (20) begins at the moment $t=0$; and then again the operator ϵ^{-qr} turns ut to $ut - \lambda r$, so that $ut - \lambda r$ begins with the value zero.

Finally, if V is the potential at the point z, r, we have

$$\mathrm{V} = \epsilon^{-pz/u}\mathrm{U} = \frac{\mathrm{Q}}{2\pi c}\frac{1}{\{(ut-z)^2 - \lambda^2 r^2\}^{\frac{1}{2}}}, \tag{23}$$

provided $(ut - z)$ is not less than λr. This confines us to the cone behind the travelling electron, outside which we have $\mathrm{V} = 0$.

The above argument appears to be perfectly distinct in all respects. But it is necessary to carefully note the underlying assumption, and its consequence. Thus, at the beginning, the assumption $\mathrm{C} = \mathrm{F}(t - z/u)$ implies that the electron has been in motion for an infinitely long time before it reached the origin. That is why we obtain a travelling steady state of **E** and **H**. No information is given as to how it was arrived at a long time previously. To show that, a different operand for C is required, showing how the electron passes from rest to the state of uniform motion.

Closely connected with this problem is that of a moving infinitely long line of electrification. Let σ be the linear density and u the speed (positive), and let its free end reach the origin at the moment $t=0$. Then C_0 jumps from 0 to the value σu at the moment $t=0$, and remains at that value for ever after. Therefore

$$U=\frac{\sigma}{4c}K_0\left(\frac{\lambda rp}{u}\right)1_t \qquad (24)$$

is the potential at the plane of the origin. It only differs from (20) in the absence of p/u. We have, instead of (21),

$$K_0(qr)1=\frac{2}{\pi}\log\left\{\frac{l}{r}+\sqrt{\frac{l^2}{r^2}-1}\right\}. \qquad (25)$$

This may be proved in the same way as (21) was proved. But it is unnecessary to go through the work, because having the operand 1_t now instead of $p1_t$ only requires us to integrate the former U solution from 0 to t. So

$$U=\frac{\sigma}{2\pi c}\log\left\{\frac{ut}{\lambda r}+\sqrt{\left(\frac{ut}{\lambda r}\right)^2-1}\right\}, \qquad (26)$$

and then V is got by changing ut to $ut-z$, as before. This is the solution used in § 465.

Application of Simply Periodic Analysis. The Transition from $u<$ to $u>v$.

§ 484. Now for a change. Apply harmonic analysis to the operational solution (20), and see what we come to, and whether the definite integrals are recognisable. We have

$$p1=\frac{1}{\pi}\int_0^\infty \cos nt\, dn. \qquad (27)$$

Use this operand in (20), and work by differentations. Assume that $p=ni$ is true in the K_0 function, making

$$U=\frac{Q}{4\pi cu}\int_0^\infty K_0\left(\frac{\lambda nir}{u}\right)\cos nt\, dn, \qquad (28)$$

where i is the differentiator p/n really, but has to be worked according to $i^2=-1$. Now we at once note that if λ^2 is negative, that is, $u<v$, the argument of the K_0 function becomes real. Say $\kappa=\lambda i$, and $\kappa=(1-u^2/v^2)^{\frac{1}{2}}$, then

$$U=\frac{Q}{4\pi cu}\int_0^\infty K_0\left(\frac{\kappa nr}{u}\right)\cos nt\,dn, \tag{29}$$

which is a real integral without further change.

Now we know, § 463, that the solution is

$$U=\frac{Q}{4\pi c}\frac{1}{(u^2t^2+\kappa^2r^2)^{\frac{1}{2}}}, \tag{30}$$

because changing ut to $ut-z$ produces the proper ellipsoidal solution about which there is no difficulty. We conclude that

$$\int_0^\infty K_0\left(\frac{\kappa nr}{u}\right)\cos nt\,dn=\frac{u}{(u^2t^2+\kappa^2r^2)^{\frac{1}{2}}}. \tag{31}$$

On consulting Gray and Mathews' "Bessel Functions," p. 227, I find this integral given, ascribed to Basset. So far good. Then how does the doubling of the formula take place when $u>v$? The 4π in (30) has then to become 2π.

Go back to (28) with λ real. We have

$$K_0(xi)=G_0(x)-iJ_0(x), \tag{32}$$

when x is real. (Vol. II., p. 253.) So (28) is

$$U=\frac{Q}{4\pi cu}\int_0^\infty\left\{G_0\left(\frac{\lambda nr}{u}\right)\cos nt+J_0\left(\frac{\lambda nr}{u}\right)\sin nt\right\}dn, \tag{33}$$

which is a real integral, λ being real. Now our U in this case is zero when t is negative, and since the second part changes sign, whilst the first does not, when t is made positive, it follows that

$$\int_0^\infty J_0\left(\frac{\lambda nr}{u}\right)\sin nt\,dn=\frac{u}{(u^2t^2-\lambda^2r^2)^{\frac{1}{2}}}, \tag{34}$$

$$\int_0^\infty G_0\left(\frac{\lambda nr}{u}\right)\cos nt\,dn=\frac{u}{(u^2t^2-\lambda^2r^2)^{\frac{1}{2}}}. \tag{35}$$

The first of these (34) I find in Gray and Mathew ascribed to Weber. The second (35) I fail to find, so for the present I will ascribe it to myself.

Thus, as u passes from less than v to greater than v, the value of the integral doubles itself. This would not perhaps be a correct way of putting it if the integral were purely algebraical, for we must consider that the integral concerned is really (28), in which ni stands for d/dt. Strictly it takes two forms, one of which gives double the result of the other.

Observe, too, that the symbol i has had two distinct meanings. In (32) it is algebraical. In (28) it is a differentiator. Both follow $i^2 = -1$. Although they have not come into conflict, it may very easily happen that such will happen in investigations of this class, involving definite integrals and harmonic analysis; and to avoid error, it is desirable to be guided by the conditions of the physical problem concerned. That will serve to counteract the ambiguity of the purely mathematical machinery.

If we employ the other method referred to above, making $C = f(z - ut) = \epsilon^{-ut\Delta}C_0$ where C_0 is a function of z at the initial moment, and C what it becomes at time t later, we eliminate p and use Δ instead. The reasoning all through is quite similar to the above, and the results are the same. But this way of working I did not find so easy to follow in the particular problem of a moving electron on account of purely technical difficulties of the kind just mentioned, and the fact that the solution admits of a double interpretation. The conical wave may be either expanding or contracting, although only the expanding wave can be imagined to arise naturally.

Train of Simply Periodic Forced Waves along an Axis. The Work done and Waste of Energy.

§ 485. The simply periodic train of waves utilised in the last paragraph possesses some properties which deserve notice, especially as regards the activity of the forces and the waste of energy. There is no waste at all when the speed of the wave train is less than v, but there is waste when it is greater. Say that the impressed current at the axis is

$$C = C_0 \cos m(ut - z), \tag{36}$$

existent permanently. Then

$$V = \frac{C_0}{4cu}\left\{G_0(\lambda rm)\cos + J_0(\lambda rm)\sin\right\}m(ut - z) \tag{37}$$

is the potential when $u > v$, and λ is real; whilst when $u < v$ it is

$$V = \frac{C_0}{4cu}K_0(\kappa rm)\cos m(ut - z), \tag{38}$$

and κ is real.

Start with $u=0$, and go right through to show the effect of the motion. With stationary electrification, density $\sigma\cos mz$, so that $C_0=\sigma u$,

$$V=\frac{\sigma\cos mz}{4c}K_0(mr). \tag{39}$$

It is only the wavy distribution of the electrification that prevents V from being infinite everywhere—namely, when $m=0$, a well-known electrostatic result, though the electric force is finite save at the axis. At a distance from it, or more generally, when mr is large,

$$V=\frac{\sigma\cos mz}{4c}\epsilon^{-mr}\left(\frac{2}{\pi mr}\right)^{\frac{1}{2}}\left\{1-\frac{1}{8mr}+\dots\right\}; \tag{40}$$

and close to the axis, or when mr is small,

$$V=\frac{\sigma\cos mz}{2\pi c}\left\{-I_0(mr)\left(\log\frac{mr}{2}+0\cdot5772\right)+\left(\frac{mr}{2}\right)^2+\dots\right\}, \tag{41}$$

which is nearly the same as

$$V=-\frac{\sigma\cos mz}{2\pi c}\log mr. \tag{42}$$

This makes

$$E=-\frac{dV}{dz}=-\frac{\sigma\sin mz}{2\pi c}m\log mr, \qquad F=-\frac{dV}{dr}=\frac{\sigma\cos mz}{2\pi cr}. \tag{43}$$

When $m=0$ the axial component vanishes, leaving the simple state $F=\sigma/2\pi rc$.

Now the effect of increasing u from zero turns $K_0(mr)$ to $K_0(\kappa mr)$, where $\kappa=(1-u^2/v^2)^{\frac{1}{2}}$. The result is, as may be seen by (40), to decrease the rapid rate at which V falls off as the axis is receded from. But there is still no change of sign of V between the axis and $r=\infty$, for any particular value of z, because $K_0(\kappa mr)$ is always positive. This action continues until $\kappa=0$, or $u=v$. The potential solution is then useless, but it is the case of plane progressive waves travelling at the natural speed v. E is zero again, and

$$F=\frac{\sigma\cos m(ut-z)}{2\pi rc}, \tag{44}$$

without any function of r as a factor, save the usual r^{-1}.

Increasing u above v brings in a different state of things, as in (37), where $\lambda=(u^2/v^2-1)^{\frac{1}{2}}$. It makes V oscillatory along **r**, for any special value of z; and the larger λ is made, the shorter the wave length. By sufficient increase of u/v we may pack the regions of positive and negative V as closely as we please. These results will be understood on remembering the conical nature of the wave fronts corresponding to the different elements of electrification, and that its arrangement upon the axis is alternating.

As regards the electric and magnetic forces, we have

$$\mathrm{F} = -\frac{d\mathrm{V}}{dr} = -\frac{\sigma}{4c}\Big\{\mathrm{G}_0'(\lambda mr)\cos + \mathrm{J}_0'(\lambda mr)\sin\Big\} m(ut-z), \quad (45)$$

$$\mathrm{E} = \lambda^2\frac{d\mathrm{V}}{dz} = \frac{\sigma\lambda^2 m}{4c}\Big\{\mathrm{G}_0(\lambda mr)\sin - \mathrm{J}_0(\lambda mr)\cos\Big\} m(ut-z), \quad (46)$$

and $\mathrm{H} = cu\mathrm{F}$, as before. The big accent means d/dr.

Now H is perpendicular both to E and to F. So

$$\mathrm{FH} = cu\mathrm{F}^2 = cu\left(\frac{\sigma}{4c}\right)^2\Big\{\mathrm{G}_0'^2\cos^2 + \mathrm{J}_0'^2\sin^2 + 2\mathrm{G}_0'\mathrm{J}_0'\cos\sin\Big\} \quad (47)$$

Averaging, we obtain

$$\overline{\mathrm{FH}} = \tfrac{1}{2}cu\left(\frac{\sigma}{4c}\right)^2(\mathrm{G}_0'^2 + \mathrm{J}_0'^2). \quad (48)$$

This is the mean flux of energy per unit area along z or with the wave. It does not involve any waste.

As regards the flux outward along r it is

$$-\mathrm{EH} = -cu\mathrm{EF} = -cu\lambda^2 m\left(\frac{\sigma}{4c}\right)^2\Big\{\mathrm{J}_0\mathrm{G}_0'\cos^2 - \mathrm{G}_0\mathrm{J}_0'\sin^2 + (\mathrm{J}_0\mathrm{J}_0' - \mathrm{G}_0\mathrm{G}_0')\sin\cos\Big\} \quad (49)$$

Averaging, we obtain

$$-cu\overline{\mathrm{EF}} = -\tfrac{1}{2}cu\lambda^2 m\left(\frac{\sigma}{4c}\right)^2(\mathrm{J}_0\mathrm{G}_0' - \mathrm{G}_0\mathrm{J}_0'),$$

where $$\mathrm{J}_0\mathrm{G}_0' - \mathrm{J}_0'\mathrm{G}_0 = -2/\pi r. \quad (50)$$

(Vol. II., p. 257.) So the averaged outward flux per unit area is

$$\tfrac{1}{2}cu\lambda^2 m\left(\frac{\sigma}{4c}\right)^2\frac{2}{\pi r}. \quad (51)$$

This must be multiplied by $2\pi r$ to obtain the waste per unit length of axis. It is

$$2cu\lambda^2 m(\sigma/4c)^2, \quad (52)$$

independent of the distance. This is real waste of energy. We conclude that the averaged applied force (Newtonian) needed at the axis to keep the electrification moving in the way specified is (52) divided by u, or

$$\lambda^2\sigma^2 m/8c. \tag{53}$$

This increases infinitely with u or λ. It is zero when $u=v$, or $\lambda=0$. It remains zero when $u<v$.

To confirm this result, consider the mechanical reaction of the electromagnetic field upon the electrification. It is

$$\mathrm{E}\sigma \cos m(z-ut) \tag{54}$$

at any point on the axis. Or, by (46),

$$\frac{\sigma^2\lambda^2 m}{4c}\{\mathrm{G}_0 \sin\cos - \mathrm{J}_0 \cos^2\}. \tag{55}$$

The average of the first part is zero. Of the second it is the negative of (53). So (53) represents the average regarded as impressed, and (52) the average activity.

There are several cautions to be expressed regarding the above. First the investigation has no reference to ordinary waves along wires. They do not behave in the above way, even if all resistance were done away with. The above waves are forced waves, whether u be less or greater than v, although only in the latter case is there permanent activity on the average. If we want to represent waves of this type along a wire, we require a continuous distribution of impressed electric force along the wire, or something equivalent. That is, the wire is to be a source of energy, instead of a sink, as is usually the case with waves along them, for the loss of energy by radiation of the heat is a separate matter, which does not come in question. Prof. G. F. FitzGerald was the first to calculate loss by radiation from a wire, but I do not know the precise circumstances he contemplated.* Whether there is any loss depends materially upon the circumstances.

There is waste when the impressed current is everywhere in the same phase, say $\mathrm{C}=\mathrm{C}_0 \cos nt$, with C_0 constant. This is a reduced case of the above—viz.,

$$u=\infty, \quad m=0, \quad mu=n, \quad \sigma u=\mathrm{C}_0. \tag{56}$$

* George Francis FitzGerald is dead. The premature loss of a man of such striking original genius and such wide sympathies will be considered by those who knew him and his work to be a national misfortune. Of course the "nation" knows nothing about it, or why it should be so.

These reduce the expression (52) for the waste to

$$\frac{mu\sigma^2}{8c}\left(\frac{u^2}{v^2}-1\right)=\frac{nC_0^2}{8cv^2}=\frac{\mu nC_0^2}{8}. \tag{57}$$

This is per second. But $2\pi/n$ is the period, so the waste per period is $\frac{1}{4}\pi\mu C^2{}_0$, at any frequency, provided the amplitude of the axial current is kept the same. This independence is also true in the former case, down to $u=v$. But we should not apply it to the extreme of zero frequency, or infinitely long period, because then we cannot have reversal of current to make a period.

Also notice that in the calculation of the waste by averaging at the axis, the neglected part, which goes out on averaging, is infinitely large compared with the retained part; which may seem absurd. But the infiniteness is of no consequence at all, when its reason is considered. It is to simplify results that the impressed current is condensed in a line instead of being distributed in a rod. The condensation causes the infinite values at the axis. But the effective results away from the axis are not materially affected by finitising all results by spreading the axial current throughout a small rod. So there is nothing to be alarmed about. It is nothing like so curious as the result in § 465 and § 477, where we found a finite moving force (and electric force) acting quite perpendicularly to the lines of electric force.

Construction of the Simply Periodic Wave Train from the Two Electronic Steady Solutions.

§ 486. Having employed in § 484 the simply periodic wave formulæ to build up the two kinds of solution for an electron in steady rectilinear motion, the converse problem presents

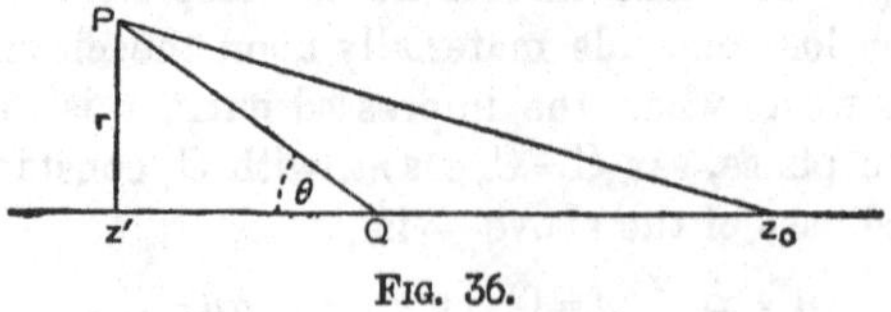

FIG. 36.

itself for completeness—namely, to build up the simply periodic wave formulæ out of the two electron solutions. If this cannot be done, there must be something wrong.

Required U at the point P due to the distribution of electrification of linear density $\sigma \cos mz'$ along the axis, this z' being the former $z-ut$. We suppose that P also travels along at speed u. The U required is the sum at P of the steady potentials of all the elements of electrification which produce potential at P. There are two cases. First, if $u<v$, every element of electrification operates at P, because the electric force of an element extends over all space. So, by equating the integral of the electron solution to the simply periodic solution we obtain

$$\frac{\sigma}{4\pi c}\int_{-\infty}^{\infty}\frac{\cos mz_0\, dz_0}{\{(z_0-z')^2+\kappa^2 r^2\}^{\frac{1}{2}}}=\frac{\sigma}{4c}\mathrm{K}_0(m\kappa r)\cos mz', \tag{58}$$

where κ is real. Put $z_0-z'=y$, and there results

$$\frac{2}{\pi}\int_0^{\infty}\frac{\cos my}{(y^2+\kappa^2 r^2)^{\frac{1}{2}}}dy=\mathrm{K}_0(m\kappa r), \tag{59}$$

which is a known definite integral.

But in the other case, with $u>v$, the total U at P is derived only from the electrification to the right of Q, if the angle PQz' be the θ before used making $u\sin\theta=v$, and the electrification is moving from left to right; becanse the electrification on the left side of Q does not act at P. So the lower limit must be $h=z'+r\lambda$, where $r\lambda$ is the distance z'Q in the figure, instead of $-\infty$. Therefore, by equating the new integral to the new simply periodic solution, we have, with λ real,

$$\frac{\sigma}{2\pi c}\int_h^{\infty}\frac{\cos mz_0\, dz_0}{\{(z_0-z')^2-\lambda^2 r^2\}^{\frac{1}{2}}}=\frac{\sigma}{4c}\{\mathrm{G}_0(\lambda mr)\cos-\mathrm{J}_0(\lambda mr)\sin\}\, mz'. \tag{60}$$

Putting $z_0-z'=y$, reduces the left side to

$$\mathrm{U}=\frac{\sigma}{2\pi c}\int_{r\lambda}^{\infty}\frac{\cos m(y+z')}{(y^2-r^2\lambda^2)^{\frac{1}{2}}}dy, \tag{61}$$

which, by comparison with the right side of (60), requires

$$\frac{2}{\pi}\int_{r\lambda}^{\infty}\frac{\cos my\, dy}{(y^2-r^2\lambda^2)^{\frac{1}{2}}}=\mathrm{G}_0(\lambda mr),\qquad \frac{2}{\pi}\int_{r\lambda}^{\infty}\frac{\sin my\, dy}{(y^2-\lambda^2 r^2)^{\frac{1}{2}}}=\mathrm{J}_0(\lambda mr), \tag{62}$$

which are also known Bessel integrals if I correctly understand the formula given by Gray and Mathews, p.230, ascribed to Weber, viz.,

$$\int_1^{\infty}\frac{\epsilon^{iz\lambda}}{(\lambda^2-1)^{\frac{1}{2}}}d\lambda=-\{\mathrm{Y}_0(z)+(\gamma-\log 2)\mathrm{J}_0(z)\}+\frac{i\pi}{2}\mathrm{J}_0(z).$$

Finally, it may be remarked that the application of Fourier's theorem to the first of (62) leads to the formula (35) above. (See also Vol. II., p. 106.) We have, therefore, a good harmonisation of results all round by all methods. This is particularly desirable, because no intelligible and valid mechanical analogue has yet been invented which will enable us to see clearly what the behaviour of electrons in motion ought to be according to the circuital laws, without resort to troublesome mathematics. From another point of view, having established the two steady and the two periodic solutions independently of the various integrals, we may consider the solutions to establish the values of the integrals.

To my mind, proofs of this character are more convincing than those involving the theory of functions.

Connection between Moving Electrification and Moving Electrisation. Transition from Cylindrical to Conical Wave.

§ 487. Some light is cast upon the effects due to moving electrification by the connected theory of moving electrisation. It is not necessary that the impressed current in the above investigations should be moving electrification, or $\mathrm{C} = \sigma u$, say. We may exhibit the axial source in terms of impressed electric force. I have before pointed out that the effects due to convection current $\rho\mathbf{u}$, ρ being the volume density, are the same as those due to impressed $\mathbf{e}$, such that $\rho\mathbf{u} = pc\mathbf{e}$ (*El. Pa.*, Vol. II., p. 509). Here $c\mathbf{e}$ may be regarded as intrinsic electrisation, the electric analogue of intrinsic magnetisation, and its time rate of increase is equivalent to impressed electric current. Applying this to a rod reduced ultimately to a mere line, as ρ is turned to σ, the linear density, $\mathbf{e}$ must be multiplied by the cross-section of the rod to preserve equivalence. Thus, put

$$\mathrm{C} = cpe_0, \tag{63}$$

instead of convection current, in the equations (12). They become

$$\mathrm{E} = -\tfrac{1}{4}\, q^2 \mathrm{K}_0(qr)e_0, \qquad \mathrm{F} = -\frac{1}{cp}\frac{d\mathrm{H}}{dz}, \tag{64}$$

$$\mathrm{H} = -\tfrac{1}{4}\frac{d}{dr}\mathrm{K}_0(qr)cpe_0. \tag{65}$$

There are the equations expressing the electric and magnetic forces generated by a straight filament of e_0 along the z-axis, where e_0 may be regarded as the product of section and impressed voltivity; or, equivalently, ce_0 may be regarded as the total strength of intrinsic electrisation. These equations may be obtained independently, as before done. (See *El. Pa.*, Vol. II., p, 457, where various results due to a filament of e_0 are given. The G function there employed is the negative of the present one. A consideration of the measure of an axial source and the transition from K_0 to $G_0 - iJ_0$ will show that the present positive reckoning is proper.)

We may therefore translate results due to varying impressed force to those due to moving electrification. To show explicitly by an example, let PQ be an electrised rod. If at rest, it is known by statical considerations that the electric

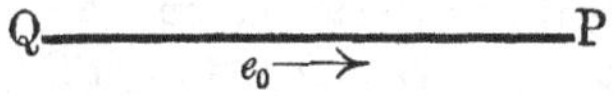

displacement outside the rod is the same as that due to a positive charge at P and a negative charge at Q, both of size ce_0. That is, the convergence of the intrinsic electrisation is the measure of the fictive electrification. There is no real electrification, because the displacement is made circuital by the flux of displacement ce_0 along the rod. There may be electrons, but that is a different matter, involving a finer kind of space division to find them. If we put a negative charge at P and a positive one at Q, both of size ce_0, the external displacement disappears. There is left only the displacement ce_0 along the rod from the positive charge to the negative. The rod is then a charged condenser which will not discharge itself even though its poles are conductively connected. To discharge the charges, e_0 must be relaxed. (See *El. Pa.*, Vol. I. pp. 466 to 515 for details about real and fictive electrification, and intrinsic displacement. For the benefit of the uninitiated, I should explain that *El. Pa.* means my "Electrical Papers." They can be picked up cheap, because the remainder was sold off in quires for a few pence per volume, on account of the deficiency in storage room. So look in the fourpenny boxes. Though somewhat vexed at first by this disposal of my laboured lucubrations, it has, later, given me and others occasion for much laughter.)*

* I do not delete this remark under the sad circumstances of the last footnote, for FitzGerald was the first man to see the humour of the proceedings of those unhumorous publishers.

Now if the rod PQ, electrised but uncharged, be moved at speed u from left to right, the external effects, electric and magnetic, are the same as those due to a pair of equivalent charges, positive at P, negative at Q, both moving the same way at speed u. The effects due to a single moving charge are therefore the same as those due to an infinitely long electrised rod, ending at P, and pushing forward. The speed u may be anything from 0 to ∞. It is not necessary for the rod to move. The region occupied by the electrisation has to grow at the P end. We may imagine little demons putting on fresh electrisation at any rate we like, and whether P effectively travels at a speed greater or less than that of propagation in the medium makes no difference to the demons individually.

If $u = \infty$, this is the same as putting on e_0 all over at the same moment. The result is a cylindrical wave. (*El. Pa.*, Vol. II., p. 460.) D is of two sorts. At the wave front is a shell in which D is oppositely directed to e_0; and inside the shell D is directed the same way as e_0. The same reversal occurs with H. At the wave front it is so directed as to make $E = \mu v H$ with outward expansion, as usual. Inside the shell it is reversed.

Now, suppose u is not infinite, but still very great compared with v. The cylinder becomes a cone of very small angle, nearly the same cylinder in fact, if we consider only a part of it at a time. It has the same characteristics as before. The cylindrical outer sheet becomes a conical sheet, in which D is directed away from the apex, from right to left. Inside it, D goes the other way, in the direction of e_0 at the axis. H is similarly reversed.

These will be seen to be the properties found for a moving electron, when its speed is greater than v. For the equivalent electrification is situated at the end of the e_0 region where it is growing.

In *El. Pa.*, Vol. II., p. 461, is worked out in some detail the case of finite section and $u = \infty$, and I meant to adapt that investigation here to illustrate the electron theory. But in looking through the troublesome calculations I find some numerical errors, so the matter must be postponed. But a general idea of the cause of the reversal in the theory of e may be shortly given. An electrised rod of finite size corresponds to a disc of electrification at its end. Now the waves are initiated at the surface of the rod, the seat of the curl of e, the impressed voltivity. Induction is generated there at the rate e per second, in a cylindrical sheet. It spreads both ways, outward and inward. Along with it the outer displacement is against e, and the inner with e, just as in the case of plane waves, § 453 above. The inner wave of D condenses to infinite intensity at the axis, but does not change sign in expanding again, being longitudinal or axial. But the corresponding H does reverse effectively. Not in reality, because an element of a circle of H in condensing and crossing the axis preserves its direction. But when the circle expands that element is on the opposite side to its old place. So there is effective reversal of H, as well as of D, compared with those at the outer wave front, save at the beginning. This property continues true when the rod is condensed to a line, and also when u is reduced, so that the wave front turns from a cylinder to a cone.

Spherical Impulsive Wave due to sudden Displacement of an Electron.

§ 488. The equivalence in external results of convection current ρu and current of intrinsic electrisation $c p$e, both being impressed currents, enables us to immediately interpret worked out problems in impressed e in terms of the equivalent convection current. An important fundamental problem concerns the wave generated by the sudden establishment of e at a point, or rather, in a spherical portion of the ether. This will be found fully worked out and interpreted in *El. Pa.*, v. 2, pp. 409 to 412. Only a small part is required here for present purposes. Changing from cpe to ρu is the same as changing from ce to $p^{-1}\rho$u If, then, e is suddenly established in a sphere of radius a, the magnetic wave sent out is the same as that due to the impulsive time integral of ρu throughout the sphere—that is, to the displacement of ρ through the distance $p^{-1}u = dx$. So $ce = \rho dx$ is the equivalence.

The intensity H of the circular magnetic force at the point r, θ, at time t, is

$$\mathrm{H} = \frac{a \sin\theta}{2\mu v r}\left\{\epsilon^{-q(r-a)}\left(1 - \frac{1}{qa}\right) + \epsilon^{-q(r+a)}\left(1 + \frac{1}{qa}\right)\right\}\left(1 + \frac{1}{qr}\right)e, \quad (1)$$

the origin being at the centre of the sphere, and θ being measured from the direction of e. (*El. Pa.*, v. 2, p. 409, equation (139)). Here e may vary anyhow, it being the operational solution that is written. Making e simply periodic produces the spherical wave trains used by Hertz as a first approximation to the theory of his vibrator, together with an impulsive wave at the front. By taking e constant, beginning when $t=0$, by an easy algebrisation we obtain an inward and an outward wave from the spherical boundary of e, leading ultimately to a shell wave of H of depth $2a$, given by (*loc. cit.*, p. 410, equation (142))

$$\mathrm{H}=\frac{e\sin\theta}{4\mu v}\left(1-\frac{v^2t^2-a^2}{r^2}\right). \tag{2}$$

This formula expresses the magnetic force from the first moment between the limits $r=a\pm vt$, until $vt=a$; but after that, between the limits $r=vt\pm a$. It is a double shell of H, positive outside, negative inside, vanishing at the distance $r=(v^2t^2-a^2)^{\frac{1}{2}}$.

The time-integral of H during the passage of the wave past the point r, θ is

$$p^{-1}\mathrm{H}=\frac{ce_0}{4\pi r^2}\sin\theta=\frac{\mathrm{Q}dx}{4\pi r^2}\sin\theta, \tag{3}$$

where we write ce_0 for the product of ce and the volume of the sphere. It is the moment of the electrisation. Its equivalent is $\mathrm{Q}dx$, if Q is the total charge, or density × volume.

But the formula is somewhat deceptive, because it is only a difference. If H_1 and H_2 refer to the outer (positive) and inner (negative) shell, their time integrals are

$$p^{-1}\mathrm{H}_1=\frac{e\sin\theta}{6\mu v^2}r\left\{-1+\frac{a^3}{r^3}+\left(1+\frac{a^2}{r^2}\right)^{\frac{3}{2}}\right\}$$

$$=\frac{cer\sin\theta}{6}\left(\frac{a^3}{r^3}+\frac{3}{2}\frac{a^2}{r^2}+\dots\right), \tag{4}$$

$$p^{-1}\mathrm{H}_2=\frac{e\sin\theta}{6\mu v^2}r\left\{+1+\frac{a^3}{r^3}-\left(1+\frac{a^2}{r^2}\right)^{\frac{3}{2}}\right\}$$

$$=\frac{cer\sin\theta}{6}\left(\frac{a^3}{r^3}-\frac{3}{2}\frac{a^2}{r^2}-\dots\right). \tag{5}$$

The sum depends on r^{-2} as just seen. But when a/r is small, the two impulses are sensibly equal and opposite, and depend upon r^{-1}. Their values are nearly

$$= \pm ce_0 \frac{3 \sin\theta}{16\pi ar} = \pm Qdx \frac{3 \sin\theta}{16\pi ar}, \qquad (6)$$

in terms of the electric moment. With $a=0$ we get ∞. That is nothing unusual. We cannot have a finite charge at a *point*. If it existed it would be immovable by a finite force.

The electric fields of ce_0 and Qdx only differ in this respect. There is no E to begin with in the case of the electrisation. There is in the case of the electrification—say the stationary polar force of Q. So add this on to the electric force produced by establishing ce_0 to get the electric force which exists when Q is shifted. Now, what ce_0 does is to set up circuital displacement similar to the induction of a spherical uniform magnet. This state is fully formed right up to the rear of the wave of magnetic force at any moment. So when Q is shifted its polar displacement persists at any point until the front of the shell wave reaches it, and when the wave has passed it is turned into the slightly changed polar displacement of Q in its new position. It is the electric force in the shell wave itself that lays down the change in the internal electric force (see also v. 1, p. 314). At a great distance $E = \mu v H$ in the wave.

The magnetic energy in the shell being always positive need not be divided into two parts. It amounts to (*El. Pa.*, v. 2, p. 412)

$$T = \frac{ce_0^2}{8\pi a^3} = \frac{(Qdx)^2}{8\pi ca^3}, \qquad (7)$$

and is constant on the whole journey, as soon as the wave is fully formed. Compare with the usual static energy of Q. It is

$$\frac{Q^2}{8\pi ca}, \text{ external, and } \frac{Q^2}{40\pi ca}, \text{ internal.} \qquad (8)$$

So $$T = \left(\frac{dx}{a}\right)^2 \times \frac{5}{6} \text{ of stationary energy.}$$

This energy is wasted. But the total waste is 2T, because there is equal electric energy in the shell ultimately. 2T is, therefore, the work done in shifting Q.

I expect, in passing, that Maxwell's characteristic expression, "electric displacement," may need to be changed. It did not formerly present any trouble that I know of. I never heard of anyone mistaking it for the spacial displacement of electrification, until one day I heard that I had been supposed to have done so myself. But in reading Dr. Larmor's book it occurs to me that trouble may very easily arise in the way mentioned, particularly because the spacial displacement of charges is so much in evidence at present. Maxwell's electric displacement D was (and I suppose is still) primarily a phenomenon in the ether, not necessarily connected with a spacial displacement in the direction of D. In matter D is increased. The excess is also not necessarily conditioned by the spacial displacement of positive and negative charges, but it is a reasonable hypothesis that it may be so, and the facts of electrolysis show that it is a probable hypothesis. The resultant makes up the excess displacement, to be coupled with the other and counted with it to make the total D called by Maxwell the electric displacement. Here the word displacement has no reference to the supposed spacial displacement of charges. So a new word, and a good word, is wanted for D. This would also be the case if an ether theory were elaborated in which D in the ether itself involved spacial displacement of + and − charges. There is room in the ether for much speculation.

Spherical Impulse due to sudden change of Velocity of an Electron. Röntgen Rays.

§489. There is another sort of impulsive magnetic wave, not to be confounded with the above, which is supposed by J. J. Thomson to represent the elementary portion of Röntgen rays. Presumably also, previously by Sir G. Stokes, though he did not express himself in magnetic language (Wilde Lecture, 1898). This is the magnetic pulse arising when a moving electron suddenly changes its velocity. By "electron" I only mean an electric charge, without hypothesis that it is really a "singular point" in the ether. I do not think Larmor has established that point. I wish he may, for the matter of ether and matter is in a very obscure state. Personal opinions at second-hand count for little, but I may as well say that I incline at present to J. J. Thomson's idea that his corpuscles are electrified *matter*. This does not exclude altogether the idea that the matter may be conditioned by ether and electrification. To show this pulse go back to (1). Put $a=0$ and reduce. The operational solution becomes

$$\mathrm{H} = -\sin\theta \frac{d}{dr}\frac{\epsilon^{-qr}}{4\pi r}cpe_0 = -\sin\theta\frac{d}{dr}\frac{\epsilon^{-qr}}{4\pi r}\mathrm{Q}u. \tag{9}$$

Or, in full,

$$\mathrm{H} = \sin\theta\left(q+\frac{1}{r}\right)\frac{\epsilon^{-qr}}{4\pi r}\mathrm{Q}u. \tag{10}$$

So, if u at the moment $t=0$ suddenly changes by the amount u_0, the impulsive H, say, H_0, the time integral of H at $t=r/v$ is,

$$\mathrm{H}_0 = \sin\theta\frac{\epsilon^{-qr}}{4\pi rv}\mathrm{Q}u_0. \tag{11}$$

This shows a spherical pulse out from the place where the change of velocity occurred. The operator ϵ^{-qr} only fixes the position of the pulse. Strike it out to express the magnitude. That there are such pulses produced by the collision of electrons with obstacles can hardly be doubted. That they fully express the Röntgen rays is not so clear. For there is the body struck to be considered, and the special vibrations emitted by it when struck.

To see the state of things in the pulse itself, we must take a finite. Use the full operational solution (1). It is the same as

$$\mathrm{H} = \frac{a\sin\theta}{2qr}\left(1+\frac{1}{qr}\right)\left\{\epsilon^{-q(r-a)}\left(1-\frac{1}{qa}\right)+\epsilon^{-q(r+a)}\left(1+\frac{1}{qa}\right)\right\}\rho u. \tag{12}$$

Here let u at the moment $t=0$ jump from 0 to u_0. An easy immediate algebrisation makes

$$\mathrm{H} = \frac{\rho u_0 a\sin\theta}{2r}\left\{vt_1 - \frac{v^2t_1^2}{2a}+\frac{v^2t_1^2}{2r}-\frac{v^3t_1^3}{6ar}\right\}$$
$$+\frac{\rho u_0 a\sin\theta}{2r}\left\{vt_2+\frac{v^2t_2^2}{2a}+\frac{v^2t_2^2}{2r}+\frac{v^3t_2^3}{6ar}\right\}. \tag{13}$$

The algebrisation is done by $q^{-n}1 = (vt)^n/\underline{|n}$, and then Taylor's theorem to translate, making

$$vt_1 = vt - r + a, \qquad vt_2 = vt - r - a, \tag{14}$$

confined to positive values only of t_1 and t_2.

We have assumed that the charge has not sensibly shifted its position. Now the first line in (13) expresses the wave out from the surface of the sphere. The second line is the wave going in, and then out again, following the first, and cancelling it partly by overlapping. If, therefore, we take the sum, it expresses not the impulse, but the internal dregs. The sum makes

$$\mathrm{H} = \frac{\mathrm{Q}u_0 \sin\theta}{4\pi r^2}, \tag{15}$$

which is recognised to be correct under the circumstances considered, that the position of the origin with respect to the point r,θ does not sensibly change, for (15) is the magnetic force due to the steady convection current $\mathrm{Q}u_0$. But the impulse is expressed by the first line only of (13), for it is the part of the primary wave, between the limits $vt = r \pm a$ which is free of the second wave. To verify, take the time integral between these limits. We get

$$p^{-1}\mathrm{H} = \frac{\rho u_0 a \sin\theta}{2rv}\frac{2a^2}{3}\left(1 + \frac{a}{r}\right) = \frac{\mathrm{Q}u_0 \sin\theta}{4\pi rv}, \tag{16}$$

which agrees with the result (11). The first line of (13) may be used to calculate the energy of the pulse.

Wave Train due to Damped Vibrations.

§ 490. If it is desired to draw the lines of displacement, we may make use of

$$2\pi r \sin\theta.\mathrm{H} = p\mathrm{D}_0, \tag{17}$$

where D_0 is the total displacement through the circle r, θ, as is obvious by the circuitation of H. Then, in terms of D_0, we have (17) to derive H, and

$$\mathrm{D}_r = \frac{1}{2\pi r^2 \sin\theta}\frac{d\mathrm{D}_0}{d\theta}, \quad \mathrm{D}_\theta = -\frac{1}{2\pi r \sin\theta}\frac{d\mathrm{D}_0}{dr} \tag{18}$$

to find the two components of D along and perpendicular to the radius vector r. And $\mathrm{D}_0 =$ constant expresses a line of D

If e_0 is a damped vibration, say $= f_0 \epsilon^{-mt} \sin nt$, beginning when $t = 0$, then

$$\mathrm{D}_0 = -\frac{cf_0}{2} r \sin^2\theta \frac{d}{dr} \frac{\epsilon^{-m(t-r/v)} \sin n(t - r/v)}{r}, \tag{19}$$

by equation (17) and (9) above. There is an impulsive wave

at the front. Ignoring that, we obtain for the state of things in the rear of the wave front,

$$H = cf_0 \sin\theta\, \epsilon^{-mt_1} \frac{n}{4\pi vr}\left\{\left(\frac{v}{r} - 2m\right)\cos + \left(\frac{m^2}{n} - n - \frac{mv}{nr}\right)\sin\right\} nt_1, \tag{20}$$

$$D_\theta = \frac{H}{v} + \frac{cf_0 \sin\theta}{4\pi r^3}\epsilon^{-mt_1}\sin nt_1, \tag{21}$$

$$D_r = \frac{cf_0}{r}\epsilon^{-mt_1} 2\cos\theta \frac{n}{4\pi vr}\left\{\cos + \left(\frac{v}{nr} - \frac{m}{n}\right)\sin\right\} nt_1, \tag{22}$$

where $t_1 = t - r/v$, and D_r, D_θ are the components of D along r and $\perp$ r in the direction of increasing θ.

To find the speed of a particular phase, let

$$H \propto \sin n(t - r/v + \beta),$$

then
$$\tan n\beta = \frac{v/r - 2m}{m^2/n - n - mv/nr}. \tag{23}$$

Let $t - r/v + \beta$ be constant, and $r = U$. Then

$$1 - \frac{U}{v} + \frac{d\beta}{dr}U = 0; \tag{24}$$

leading to

$$\frac{v}{U} = 1 - \frac{(v/nr)^2}{1 + (v/nr - m/n)^2}. \tag{25}$$

Similarly, let $D_r \propto \sin n(t - r/v + \beta_1)$, then

$$\tan n\beta_1 = \frac{1}{v/nr - m/n}, \tag{26}$$

and
$$\frac{v}{U_1} = 1 - v\frac{d\beta_1}{dr}, \quad \text{leading to} \quad U = U_1. \tag{27}$$

So the speeds of phases of H and E_r are the same—viz., U, and so is the speed of the part μvH of E_θ. The other part of E_θ has speed v. U becomes negative near the origin. But the idea of U being a wave speed becomes very artificial then; there is such huge distortion. A paper by Pearson and Lee, *Phil. Trans.*, 1899, on this subject contains a series of diagrams showing how the tubes of displacement behave as the vibrations decay. The behaviour is, of course, not so simple as in the maintained vibrations assumed by Hertz in his theory of a vibrator.

Investigation of the Electromagnetic Field due to an Impressed Electric Current growing in a Straight Line. The Solutions in Sphere and Cone.

§ 491. Now consider comprehensively the state of things arising when a charge is brought from rest suddenly to rectilinear motion at any steady speed. Previous investigations discuss the steadily travelling electromagnetic field or else the impulses. But there is a variable field as well. Its investigation is by no means so difficult as might be anticipated. We can ease matters in two ways. First, by discussing the easier theory of a growing straight line of impressed current. The results for a solitary charge will follow by differentiation. Secondly, by working it out for the case $u>v$. This will save a separate

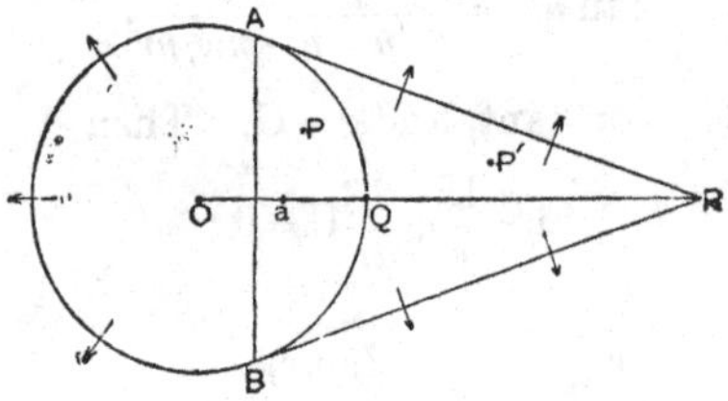

FIG. 34.

investigation for the case $u<v$, which might be expected to be easier, but is not. The results when $u>v$ may be used to find those when $u<v$. But we cannot pass the other way. The process is similar to that employed in § 468 and § 480 to find the magnetic force due to a growing plane source. The formulæ are all different, but the same diagram will serve, with a fresh interpretation.

Let it be given that there is an impressed current C along the line $OaQR$, whose length is ut, increasing uniformly with the time at the end R, and beginning at the origin O when $t=0$. Find the magnetic force at any time t.

Let $u>v$. Then the wave front is the cone ARB with a spherical cap on the left side, the radius of this sphere being vt. To find H at any point P′ inside the cone, but outside the sphere, we must sum up the magnetic forces at P′ due to all the elements of C which are within speaking distance of P′ at the moment. They are comprised between those two points on the line of C, whose distances from R are u/v times their distances from P′. That is, if z and h are the co-ordinates of P′, and z_1 the distance of the apex from the origin, the equation

$$\frac{h^2+(z-z_0)^2}{v^2}=\left(\frac{z_1-z_0}{u}\right)^2 \tag{28}$$

finds the values of z_0 required. This quadratic gives

$$z_0=z\sec^2\theta-z_1\tan^2\theta\pm\sec\theta\sqrt{(z_1-z)^2\tan^2\theta-h^2} \tag{29}$$

These are the limiting values of z required, provided they are both positive. But shifting P′ towards the sphere sends the lower z_0 down to 0. This is proved by equating z_0 to 0 in (29) when there results

$$z_1{}^2\sin^2\theta=h^2+z^2, \tag{30}$$

which belongs to the sphere.

Formula (9) above shows the H due to any element. Writing it in the form

$$\mu\mathrm{H}=-\frac{d\mathrm{A}}{dh}, \tag{31}$$

we have

$$\mathrm{A}=\int\frac{\mu\mathrm{C}d(z_0-z)}{4\pi\{h^2+(z-z_0)^2\}^{\frac{1}{2}}}=\frac{\mu\mathrm{C}}{4\pi}\Big[\log\{z_0-z+[h^2+(z_0-z)^2]^{\frac{1}{2}}\}\Big]. \tag{32}$$

By (29), the limits are

$$z_0-z=(z-z_1)\tan^2\theta\pm\sec\theta\sqrt{(z_1-z)^2\tan^2\theta-h^2}, \tag{33}$$
$$=m\pm n,\ \text{say},$$

outside the sphere. This makes

$$\mathrm{A}=\frac{\mu\mathrm{C}}{4\pi}\log\frac{m+n+\{h^2+(m+n)^2\}^{\frac{1}{2}}}{m-n+\{h^2+(m-n)^2\}^{\frac{1}{2}}}. \tag{34}$$

But inside the sphere the limits for z_0-z are $-z$ and $m+n$, making

$$\mathrm{A}_0=\frac{\mu\mathrm{C}}{4\pi}\log\frac{m+n+\{h^2+(m+n)^2\}^{\frac{1}{2}}}{-z+(h^2+z^2)^{\frac{1}{2}}}. \tag{35}$$

Here m and n are defined in equation (33), so the two A solutions look very complicated. But they may be much simplified by some troublesome algebraic reductions which I omit for brevity. We come finally to

$$A = \frac{\mu C}{2\pi} \log \frac{l + \sqrt{l^2 - h^2}}{h}, \tag{36}$$

outside the sphere, where l is the semidiameter of the cone, that is

$$l = (z_1 - z) \tan \theta. \tag{37}$$

And inside the sphere, we come to

$$A_0 = \frac{\mu C}{4\pi} \log \left[\left(\frac{u - v}{u + v}\right)^{\frac{1}{2}} \frac{l + \sqrt{l^2 - h^2}}{-z + \sqrt{z^2 + h^2}} \right]. \tag{38}$$

Though much work remains to derive the electrical results, the solution of the problem is now virtually complete. For, if we reduce u, the cone dwindles and disappears when $u = v$; and after that we have merely the spherical solution (38). As, however, it is expressed in terms of l, l should be got rid of in discussing the case of $u < v$. Put it in terms of u and v, by (37) and the relation $\sin \theta = v/u$. Then (38) takes the form

$$A_0 = \frac{\mu C}{4\pi} \log \frac{(z_1 - z)v + \sqrt{(z_1 - z)^2 v^2 + h^2(v^2 - u^2)}}{(u + v)\{-z + \sqrt{h^2 + z^2}\}}. \tag{39}$$

The magnetic force inside the cone is given by (31) applied to (36). It gives

$$H = \frac{C}{2\pi h} \frac{l}{(l^2 - h^2)^{\frac{1}{2}}}, \tag{40}$$

in agreement with § 465. This remarkable result follows, that the state of E and H in the cone which was previously investigated is established *instantly*; that is, it is not the final result of a variable state, but begins at the moment $t = 0$. We may also see this in (36), which is independent of the time in the sense required. On the other hand, there is a variable state of things inside the sphere which has to be elucidated.

The above A and the travelling potential called V in the previous investigation are in constant ratio, for we obtained by the operational process (§ 483, equation (26),)

$$V = \frac{\sigma}{2\pi c} \log \frac{l + \sqrt{l^2 - h^2}}{h}, \tag{41}$$

when $C = \sigma u$. That is, the relation of the two potentials is

$$A = \mu c u V = \frac{u}{v^2} V. \qquad (42)$$

The electric field demands separate consideration, to follow

The Ellipsoidal and Conical Equipotential Surfaces.

§ 492. Digging out the details of the problem under consideration deserves to be done with care, for this reason. The results of the two important cases of the steady motion of an electron at a speed either less or greater than that of light, hitherto treated separately, though abundantly verified, do not exhibit much compatibility. But in the present investigation the spherical and the conical states are both in action together, and we shall see how they are harmonised. It is done by means of the spherical pulse due to the sudden change of velocity at the initial moment. Or, if the change is not quite sudden, or the charge is of finite size, then by a shell wave of finite depth. In fact, the wave on the surface of the cone is continued upon the surface of the sphere and, by leakage therefrom, connection is made between the displacement inside the sphere and that inside the cone, which are of different densities.

In the first place, as regards the potential A in the cone and A_0 in the sphere, equations (36) and (38). The outermost equipotential surface is the wave front itself, partly the surface

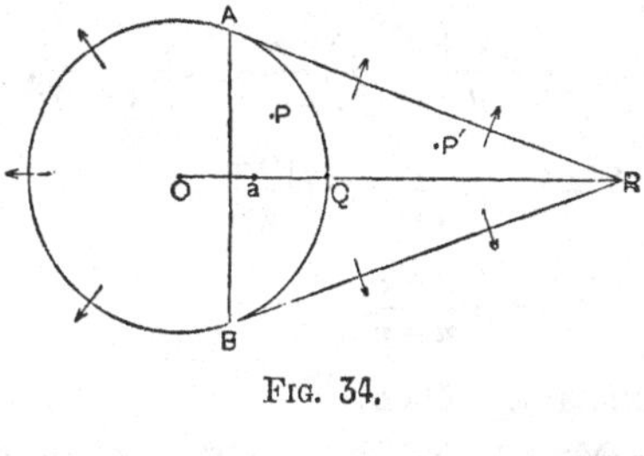

FIG. 34.

of the cone, partly of the sphere ; $A = 0$ on the one, and $A_0 = 0$ on the other. Also $A = A_0$ upon the portion AQB of the sphere, Fig. 34. Prove by (38). We have identically

$$l^2 - h^2 = \frac{(v^2 t - uz)^2}{u^2 - v^2} + v^2 t^2 - h^2 - z^2, \qquad (43)$$

and when $v^2t^2 = h^2 + z^2$, we get

$$A_0 = \frac{\mu C}{4\pi} \log \left\{ \left(\frac{u-v}{u+v}\right)^{\frac{1}{2}} \frac{(ut-z)v \pm (uz-v^2t)}{(u^2-v^2)^{\frac{1}{2}}(vt-z)} \right\} \tag{44}$$

on the circle. Here $uz - v^2t$ is + on the right side and − on the left side of the plane AB, and the value of the radical $(l^2 - h^2)^{\frac{1}{2}}$ in (38) must be always taken +. Allowing for this, $A_0 = 0$ on the left side, and

$$A_0 = \frac{\mu C}{4\pi} \log \left(\frac{vt+z}{vt-z} \frac{u-v}{u+v} \right) \tag{45}$$

on the right side of AB. This also agrees with A derived from (36) in the cone.

The innermost equipotential surface is a surface enclosing infinitely closely the axial current C, on the line OQR. There $A = \infty$, $A_0 = \infty$. Now we know already that the equipotential surfaces in the cone are co-axial cones, § 477, and since the outermost cone is made complete by a spherical continuation, it is an easy guess that the internal cones are completed within the sphere by means of ellipsoids of revolution. That is, the equipotential lines in the plane of the figure are ellipses with their major axes on the line OQR. In fact, it is quite easy to draw them roughly by the mere look of things. They must all pass round the point O on the left side. For the help of anyone who may wish to go further in this respect, I give the equation of the ellipses, without the complicated work. Let

$$X = \left(a^2 \frac{u+v}{u-v} - 1\right) \frac{v}{u-v} - a \frac{2v^2 - u^2}{(u-v)^2}, \tag{46}$$

$$Y = 1 + 2a \frac{v}{u-v} - a^2 \frac{u+v}{u-v}, \tag{47}$$

$$Z = 1 + a^2 \frac{u+v}{u-v}, \tag{48}$$

where a is a constant. Then,

$$\left(z + \frac{uvtY}{2(u-v)X}\right)^2 + \frac{Z^2h^2}{4aX} = \frac{au^2v^2t^2}{(u-v)^2X} + \left(\frac{uvtY}{2(u-v)X}\right)^2 \tag{49}$$

is the equation required, in a form showing the position of the centre upon the axis of C, and the size of the semi-axes. The co-ordinates are z and h, origin at O; and ut is the distance OR.

The position of the centre, from the values of Y and X, may be represented by

$$z = \frac{\frac{1}{2}vt}{\frac{v}{u} + \frac{b}{1 - b^2 + (v/u)(1-b)^2}}, \tag{50}$$

if $b = a^{-1}$. From this we can see that when $u < v$ the position of the centre varies between $z = 0$ and $\frac{1}{2}ut$. When $u = v$,

$$z = \frac{1-b}{2-b}vt, \tag{51}$$

going from $\frac{1}{2}vt$ to 0 as b goes from 0 to 1, and all the ellipses touch at the point Q, that being the extremity of their major axes. The line of impressed current is then OQ without the extension QR. The ratio of the minor to the major axis is $\sqrt{b(2-b)}$, and the semi-axis major is $vt/(2-b)$.

That is enough about the nature of distribution of A, for its only utility here is to lead to the physically significant magnetic force by equation (31). It happened to be easier to obtain H through A than directly. We may also use it to derive the electric force, at least partly, and fully by the use of an auxiliary potential function, but for electrical clearness there is a better way than that, which will be used presently.

The Magnetic Force and Electric Current in the Cone and Sphere. The Spherical Current Sheet.

§ 493. Denoting by H_0 and H the intensities of magnetic force inside the sphere and the cone (outside the sphere), we have

$$H = \frac{C}{2\pi h} \frac{l}{(l^2 - h^2)^{\frac{1}{2}}}, \tag{52}$$

$$H_0 = \frac{C}{4\pi h}\left\{\frac{l}{(l^2 - h^2)^{\frac{1}{2}}} + \frac{z}{(z^2 + h^2)^{\frac{1}{2}}}\right\}. \tag{53}$$

These are valid whether u be $>$ or $<$ v. But in the latter case, l, which is the semidiameter of the cone when $u > v$, should be put in terms of u and v, by $l = (ut - z)\tan\theta$, and $\sin\theta = v/u$. This makes

$$H_0 = \frac{C}{4\pi h}\left\{\frac{ut - z}{\{(ut-z)^2 + h^2(1 - u^2/v^2)\}^{\frac{1}{2}}} + \frac{z}{(z^2 + h^2)^{\frac{1}{2}}}\right\}, \tag{54}$$

which is equivalent to (53), and is valid at any speed, provided we keep inside the sphere. But this H_0 is not continuous with H. We have

$$H - H_0 = \frac{C}{4\pi h}\left\{\frac{l}{(l^2 - h^2)^{\frac{1}{2}}} - \frac{z}{R}\right\}, \tag{55}$$

at the junction $R = (z^2 + h^2)^{\frac{1}{2}} = vt$ of the sphere and inside of cone. This indicates a current sheet upon the part AQB of the sphere. It has no energetic significance, but makes continuity with the current elsewhere. It may be made to be of finite volume density by having a not quite sudden start.

The value of $2\pi h H$ on the axis is C between Q and R, and the value of $2\pi h H_0$ is C between Q and O, but is zero on the other side of O. These results are necessitated by the initial datum about impressed current. H is infinite at the conical wave front, where $l = h$, as well as at the axis. There is a minimum value in the cone when $h^2 = \frac{1}{2}l^2$.

There is no impulsive wave. On the spherical boundary H increases from 0 at the extreme left, to

$$H = \frac{C}{4\pi h}\frac{u}{v} \tag{56}$$

at the top (and bottom, of course, negatively). At the junction of the spherical and conical portions of the wave front it is infinite.

When ut is very large, we should approximate, round about O, to the magnetic force due to an infinitely long (one way) straight current. The formula (53) or (54) leads to

$$H_0 = \frac{C}{4\pi h}\left(1 + \frac{z}{h}\right), \tag{57}$$

which is a well-known elementary result, provided there be a proper radial continuation at the origin of the axial current C. The conditions of the present problem necessitate this as ultimate result.

In general, the current density is got by the first circuital law, making

$$C_h = -\frac{dH_0}{dz}, \qquad C_z = \frac{1}{h}\frac{d}{dh}hH_0, \tag{58}$$

and these, applied to (53), give

$$C_h = \dot{D}_h = -\frac{C}{4\pi}\left(\frac{h}{R^3} + \frac{h \tan\theta}{(l^2 - h^2)^{\frac{3}{2}}}\right), \tag{59}$$

$$C_z = \dot{D}_z = -\frac{C}{4\pi}\left(\frac{z}{R^3} - \frac{l}{(l^2 - h^2)^{\frac{3}{2}}}\right), \tag{60}$$

the transverse and axial components of the current density—the rate of increase of the displacement. Here the first parts represent the components of the inward radial current. The rest ultimately vanishes. It is directed radially towards the other end, R, of the impressed current, though not equably, like the first part, but with increasing density outwards from the axis.

Similarly, it will be found that in the cone outside the sphere

$$C_h = -\frac{C}{2\pi}\frac{h\tan\theta}{(l^2-h^2)^{\frac{3}{2}}}, \qquad C_z = +\frac{C}{2\pi}\frac{l}{(l^2-h^2)^{\frac{3}{2}}}, \tag{61}$$

and this represents a radial current towards R exactly like that last mentioned, but with 2π in the denominator instead of 4π. The current towards the apex therefore doubles itself in crossing over from the sphere into the cone, without changing its direction.

To account for this singular behaviour, we have the current sheet before mentioned, equation (55). It may be reduced to

$$H - H_0 = \frac{C}{4\pi R}\cdot\frac{\frac{u}{v}\left(1-\frac{z^2}{R^2}\right)^{\frac{1}{2}}}{\frac{u}{v}\frac{z}{R}-1}, \tag{62}$$

where $R = vt$, the radius of sphere. The direction of both H and H_0 is positive, or up through the paper above the axis, and down below, and H is the greater. So, by the rule of curl applied to a surface, if β is the polar angle, reckoned from the forward pole of the sphere, and $C\beta$ the surface density of the current along the lines of longitude, reckoned positive from the forward pole, the value of $C\beta$ is

$$C\beta = \frac{C}{4\pi R}\quad\frac{\frac{u}{v}\sin\beta}{1-\frac{u}{v}\cos\beta}, \tag{63}$$

between $\beta = 0$ and the value of β making the denominator vanish. That is at the circle of contact with the cone.

Beyond this, on the left part of the sphere, the surface density is given by the value of H_0 alone, because there is no external magnetic force. It works out to the same formula. That is, (63) is the proper formula all over the sphere, and this is true whether u is less or greater than v. If $u/v<1$, $C\beta$ is positive everywhere, from the forward to the backward pole. But if $u/v>1$, $C\beta$ is positive behind the cone only, and negative forward, or from the circle of contact to the axis. Continuity is made at the circle of contact by the current at the boundary of the cone.

The reader who is practised in making translations of application of electrical formulæ (by interchanges of electric and magnetic quantities, or by differentiations) will perhaps be able to see the application of (63) to the problem of a moving electron. It is this. Let a charge ρ, initially at O, be suddenly started and moved along the axis at speed u, then the spherical impulse is fully represented by

$$\overline{\mathrm{H}}_0 = v\overline{\mathrm{D}}_0 = \frac{\rho}{4\pi\mathrm{R}} \frac{u \sin\beta}{1 - \frac{u}{v}\cos\beta}, \tag{64}$$

all over the sphere, at any speed. This is important in the theory of Röntgen rays, and is the generalisation of the result (11) in § 489 for small speeds. $\overline{\mathrm{H}}_0$ and $\overline{\mathrm{D}}_0$ mean the values of H and D multiplied by the depth of the pulse, positively reckoned in direction in the same way as for $C\beta$ above. But as the translation of meaning is not very obvious, I will give a demonstration when the present problem of an axial impressed current is completed.

The Manner of Continuity of the Electric Current.

§ 494. So, returning to (63), consider its connection with the internal currents. Multiply by $2\pi\mathrm{R}\sin\beta$, making

$$\tfrac{1}{2}\mathrm{C}\frac{u}{v} \frac{\sin^2\beta}{1 - \frac{u}{v}\cos\beta}, \tag{65}$$

This is the total current in the sheet crossing the circle R, β, from right pole to left when positive. Its rate of decrease with $\mathrm{R}\beta$ is therefore the leakage from the sheet between two circles at unit distance apart. The result is

$$\frac{\mathrm{C}}{2\mathrm{R}} \sin\beta \left\{ 1 - \frac{1 - \frac{u^2}{v^2}}{(1 - \frac{u}{v}\cos\beta)^2} \right\} \tag{66}$$

Now by (59), (60), the outward radial component of the internal urrent, say C_R, is

$$C_h\frac{h}{R}+C_z\frac{z}{R}=\frac{C}{4\pi R}\left\{\frac{zl-h^2\tan\theta}{(l^2-h^2)^{\frac{3}{2}}}-\frac{1}{R}\right\}, \qquad (67)$$

in which $R=vt$. Putting in the values of l and $\tan\theta$, this reduces to

$$C_R=\frac{C}{4\pi R^2}\left\{\frac{\left(1-\frac{u^2}{v^2}\right)\left(1-\frac{u}{v}\cos\beta\right)}{\left\{\left(1-\frac{u}{v}\cos\beta\right)^2\right\}^{\frac{3}{2}}}-1\right\}; \qquad (67\text{A})$$

and, since there is a radical concerned, we must be careful about the sign. When $u<v$, there is no trouble. For (67A) is then the negative of (66) divided by $2\pi R\sin\beta$. That is, the leakage per unit area from the current sheet is equal to the sum of the two internal currents reckoned from the sheet, thus making continuity. The same applies precisely when $u>v$, provided we keep on the left side of the circle of contact with the cone.

But it is different on the right side. Then $1<(u/v)\cos\beta$, because $\cos\beta=\sin\theta=v/u$ at the circle of contact. The radical is always to be positive in these physical problems (a rigorous demonstration of the necessity might, however, be lengthy and obscure). So the denominator in (67A) is positive whilst $1-(u/v)\cos\beta$ in the numerator is negative. So (67A) becomes

$$C_R=\frac{C}{4\pi R^2}\left\{\frac{\frac{u^2}{v^2}-1}{\left(1-\frac{u}{v}\cos\beta\right)^2}-1\right\}. \qquad (68)$$

This is inside and towards the surface. On the outside, there is the current (61). Its radial component outward is double the first part of (67A). That is

$$C=\frac{C}{2\pi R^2}\,\frac{\frac{u^2}{v^2}-1}{\left(1-\frac{u}{v}\cos\beta\right)^2} \qquad (69)$$

The excess of (69) over (68) therefore represents the leakage per unit area from the sheet. This excess is exactly the quantity in (66) divided by the circumference $2\pi R \sin\beta$. So there is continuity all over, except that we have not considered the circle of contact, where the current divides; but the conical sheet current joins on there and makes things right. The current, being the curl of the magnetic force, is necessarily circuital when the discontinuities and junctions are properly attended to. The value of C_β is infinite only at the circle of contact, and the leakage from it also. The current in the conical sheet is infinite, too; but just under it there is a reverse current of infinite value going the other way, which makes matters right, for, of course, the infinite current at the conical wave front could not be continuous with finite currents in the spherical sheet.

The Electric Force, and Time Integral of Magnetic Force.

§ 495. So far nothing has been done as regards the electric field. An impressed current C of any kind has been assumed, and the magnetic force and external current due to it have been established. There was no magnetic force initially. To fix the electric force similarly let there be none initially also. Two principal suppositions may be made about the nature of C. It may be the impressed current of increasing intrinsic electrisation. If so, there will be no electrification at any time. Or, it may be supposed that there is a continuous electric separation going on at the point O, generating equal amounts of positive and negative electrification, and that the positive moves along the axis at speed u, so that $C = \sigma u$, where σ is the linear density of electrification, whilst the negative remains stationary, so that there is a negative charge $-Ct = -\sigma ut$ at the origin at time t. In either case the external displacement will be the same. In the case of intrinsic electrisation there will be fictive electrification of amount $-\sigma ut$ at the origin, indicating the convergence of displacement there, but this displacement will pass along the axis and leak out uniformly all the way along, producing the same effect as if there was electrification σ along the axis.

For distinctness, take the case of $C = \sigma u$, or real convection of electrification, with stationary compensating charge. Now, we had two formulæ for H; but it is easy to see that by time integration they will lead to three for D, the displacement.

For let P be a point where D is required. From the origin O draw a cone passing through the circle of contact. Call it the secondary cone. Now, if P lies to the left of this secondary cone, it is never inside the primary cone at any moment. It will be traversed first by the spherical shell, and then by a portion of the sphere. This makes one formula for D. But if P is on the right side of the secondary cone, it will be first traversed by the conical wave front and then be within the primary cone for a time. This makes a second formula. But later on it will be traversed by the spherical wave front and then be inside the sphere. This makes the third formula.

But to avoid the complications of the wave front and the shell, we had better calculate results a little differently. Since the electric current is the curl of the magnetic force, the displacement outside the axis is the curl of the time-integral of the magnetic force. Let

$$Z = p^{-1}H, \quad \text{then} \quad \text{curl } Z = D,$$

where D is the time-integral of curl H, therefore the electric displacement required. This would be also true on the axis in the case of electrisation; but with convection current there, since it exists as current in the first circuital law, its time-integral also counts. The theorem of curl has no exceptions. This Z is also the vector potential of the magnetic current (*El. Pa.*, v. 1, p. 467), but we need not employ that curious property.

If the point P is inside the cone at time t,

$$Z = \int H dt = \int H \frac{dl}{u \tan\theta} = \frac{C \cot\theta}{2\pi h u}\Big[(l^2 - h^2)^{\frac{1}{2}}\Big]_h^l.$$

That is,

$$Z = \frac{C \cot\theta}{2\pi h u}(l^2 - h^2)^{\frac{1}{2}}, \tag{70}$$

by the H formula (52), remembering that $l = (ut - z)\tan\theta$. This state of thing lasts until the sphere reaches P. The time limits are then t_1 and t_2, given by

$$(ut_1 - z)\tan\theta = h, \qquad (vt_2)^2 = h^2 + z^2. \tag{71}$$

This value of t_2 used in l in (70) gives Z at time t_2. Later on, we must add on to this Z the integral of H_0 according to

(53) in the sphere from t_2 to t. This makes

$$Z_1 = \frac{C}{4\pi h}\frac{z(t-t_2)}{(z^2+h^2)^{\frac{1}{2}}} + \frac{C\cot\theta}{4\pi hu}\left\{\sqrt{l^2-h^2}+\sqrt{l_2^{\,2}-h^2}\right\}, \qquad (72)$$

where l_2 is l with t_2 in it. This formula continues valid.

Thirdly, when P is on the left of the secondary cone the time limits for the integral of H_0 are from t_2 to t, without any extra. The result is

$$Z_2 = \frac{C}{4\pi h}\frac{z(t-t_2)}{(z^2+h^2)^{\frac{1}{2}}} + \frac{C\cot\theta}{4\pi hu}\left\{\sqrt{l^2-h^2}-\sqrt{l_2^{\,2}-h^2}\right\}. \qquad (73)$$

Comparing Z_1 with Z_2, and calling them Z_0 for the sphere, we may write it

$$Z_0 = \frac{C}{4\pi h}\frac{z(t-t_2)}{(z^2+h^2)^{\frac{1}{2}}} + \frac{C\cot\theta}{4\pi hu}\left\{\sqrt{l^2-h^2}\pm\sqrt{l_2^{\,2}-h^2}\right\}. \qquad (74)$$

The complete Z is now given by (70) in the cone and Z_0 in the sphere, Z_0 being really two formulæ. Use the + sign on the right and the − sign on the left side of the secondary cone. It may be inferred that $\sqrt{l_2^{\,2}-h^2}$ vanishes on the secondary conical surface, and the fact may be verified. It is then suggested that (74) may be converted to a form not involving any $\pm$. This is the case, for the quantity under the radical sign is the perfect square of a function which changes sign in passing through the secondary cone. We get

$$Z_0 = \frac{C}{4\pi h}\left\{\frac{zt}{R} - \frac{R}{u} + \frac{\sqrt{l^2-h^2}}{u\tan\theta}\right\} \qquad (75)$$

anywhere inside the sphere, at distance R from its centre.

We may now derive **D** from Z in the same easy way as the current from the magnetic force, since Z is like **H**, in circles round the axis, though unlike **H**, it is continuous at the spherical boundary between sphere and primary cone.

The Distribution of Displacement.

§ 496. The displacement components D_h perpendicular to, and D_z along the axis are found by

$$D_h = -\frac{dZ}{dz}, \qquad D_z = \frac{1}{h}\frac{d}{dh}Zh. \qquad (76)$$

Applying these to (70) and (75), we obtain

$$D_h = \frac{Cl}{2\pi uh\sqrt{l^2-h^2}}, \qquad D_z = -\frac{C\cot\theta}{2\pi u\sqrt{l^2-h^2}}, \qquad (77)$$

which apply inside the cone, and

$$\left.\begin{aligned} D_h &= -\frac{C}{4\pi h}\left\{\frac{h^2 t}{R^3} - \frac{z}{Ru} - \frac{l}{u\sqrt{l^2 - h^2}}\right\}, \\ D_z &= -\frac{C}{4\pi h}\left\{\frac{hzt}{R^3} + \frac{h}{Ru} + \frac{h \cot\theta}{u\sqrt{l^2 - h^2}}\right\}, \end{aligned}\right\} \quad (78)$$

inside the sphere. These are exhibited in a form admitting of a ready interpretation.

According to (77), and as previously seen, the displacement in the cone starts from the axis, to the amount C/u per unit length, spreads uniformly all round, perpendicularly to the axis at first, and then turns round in hyperbolic curves, ultimately tending to parallelism with the conical wave front. This displacement, of total amount C/u multiplied by the distance of the apex from the sphere, all reaches the sphere, and enters it.

Now the terms at the end of (78), containing l, are of the same nature as those in (77). Therefore this part of D in the sphere represents the continuation of the process just described, originating along the axis all the way from the right to the left side of the sphere, but weakened, as if the electrification density were $\frac{1}{2}C/u$.

Next, the middle terms in (78) represent displacement in semi-circles centred at O, the centre of the sphere. Consider a spherical shell of radius $R=(h^2+z^2)^{\frac{1}{2}}$ and of unit depth, centred at O. Then $\frac{1}{2}C/u$ of displacement starts from the pole of this shell on the axis on the right side, spreads out uniformly in the shell, follows the lines of longtitude all round, and terminates at the pole on the axis on the left side. (Both poles here are really unit lines). This distribution of displacement, combined with the former, makes the linear density of electrification be fully C/u on the right side of O, and zero on the left side.

Finally, the terms containing t in (78) represent a radial distribution of displacement such as is due to the charge of amount $-Ct$ situated at the origin, and $+Ct$ upon the surface of the sphere of radius vt. The charge at the origin is the necessary complement to the charge along the axis (real or fictive as the case may be). The apparent charge

$+\mathrm{C}t$ on the sphere does not exist, for the displacement has no divergence there.

To examine the state of things on the spherical surface. The radial component of D, say $\mathrm{D_R}$, anywhere in the sphere is

$$\mathrm{D_R} = \frac{h\mathrm{D}_h + z\mathrm{D}_z}{\mathrm{R}} = -\frac{\mathrm{C}}{4\pi}\left\{\frac{t}{\mathrm{R}^2} - \frac{1}{\mathrm{R}u}\frac{l - z\cot\theta}{\sqrt{l^2 - h^2}}\right\}. \tag{79}$$

Now, when $\mathrm{R} = vt$, we have $l = l_2$, and

$$l^2 - h^2 = (\mathrm{R}u/v - z)^2 \tan^2\theta - h^2 = (\mathrm{R}\tan\theta - z\sec\theta)^2$$

and
$$l - z\cot\theta = \frac{\mathrm{R}\tan\theta - z\sec\theta}{\sin\theta}.$$

So, taking the + value of the radical in both cases, we get $\mathrm{D_R} = 0$ on the left side of the circle of contact with the cone, and

$$\mathrm{D_R} = -\frac{\mathrm{C}t}{2\pi\mathrm{R}^2} \tag{80}$$

on the right side. This formula (80) also represents $\mathrm{D_R}$ in the cone, according to (77), at $\mathrm{R} = vt$, and completes the verification.

It should be noticed, because it is rather remarkable, that the displacement in the cone reaches the sphere everywhere at the proper angle to make the normal component the same. That is, the displacement enters the sphere equably all over that portion which is to the right of the circle of contact. Multiply (80) by the area $2\pi\mathrm{R}^2(1 - v/u)$, and we get $\mathrm{C}t(1 - v/u)$, which is the charge upon the portion of the axis outside the sphere. The rest of the axial charge is joined to the charge at the origin by tubes of displacement entirely inside the sphere, of course.

Although the formulæ are necessary for precision, yet the general nature of things may be easily seen without them, guided by the radial displacement just round O, and the hyperbolic in the cone, and the fact that the wave front is the outermost sheet of displacement. The hyperbolic tubes of displacement, as soon as they enter the sphere, begin to be deflected towards the axis. They all converge to the centre ultimately, though the outer ones only do so by making a wide sweep round the other side of O.

Using the two potentials A_0 and P_0 in the sphere, defining P_0 by

$$P_0 = -\frac{Ct}{4\pi Rc} + \frac{v^2}{u}A_0 \tag{81}$$

we may write the electric force thus

$$E = -\dot{A}_0 - \nabla P_0, \tag{82}$$

as may be verified. Knowing already that curl $A_0 = \mu H_0$, the form of expression has become quite antiquarian, or at least classical. P_0 is the potential of the electrification within range of the point considered. Now the direction of A_0 is that of the impressed current, which happens to be all the same way. Hence the appearance of A_0 in P_0, equation (81). In general, P_0 and A_0 have no simple connection.

Solutions for an Electron Jerked Away from a Stationary Compensating Charge. The Spherical Pulse.

§ 497. The passage from the above theory of a moving line of electrification to that of a moving point-charge is tolerably obvious. First, it is easy to construct the formulæ for a short piece of electrified line moving in its own line. For imagine at the moment $t=0$ electric separation to begin at O, the negative electrification remaining there, whilst the positive is moved along the axis so as to constitute the linear current. The resulting E and H are completely known by the above. Next, at the moment $t=\tau$, stop the electric separation, without other change. Then we have a line charge of amount $C\tau$ moving in its own line at speed u, running away from a fixed charge $-C\tau$ at the origin. The results are also completely known, because the stoppage of the first supply is equivalent to the superposition upon the first supply (without stopping it) of a new supply, the negative of the first, only differing in starting at the moment $t=\tau$. So the results for E and H are, after the moment $t=\tau$, the same as before in a certain forward region, and behind it the results are the difference of two similar states of E and H which overlap. If τ is very small, we have two nearly coincident cones and spheres. In the region common to both is a differential state of affairs, though in the space between the cones is the original state, and at the rear the hinder part of the sphere shows the negative of the original state. The previous formulæ make the state of E, H and electric current intelligible everywhere.

Now shorten the interval τ to 0, and at the same time magnify σ in proportion. In the limit we come to this problem. At the moment $t=0$, a finite charge ρ is projected from the origin at speed u, leaving behind the compensating charge $-\rho$. The solutions for E and H are obtained by differentiating the previous solutions for a line charge with respect to the time, with alteration of the constant factor.

Thus, in the sphere, if we put $\rho u = \mathrm{C}$, so that C is the moment of the impressed electric current, we shall have

$$\mathrm{A}_0 = \frac{\mu \mathrm{C} \tan\theta}{4\pi(l^2 - h^2)^{\frac{1}{2}}}; \tag{83}$$

or, expanding l,

$$\mathrm{A}_0 = \frac{\mu \mathrm{C}}{4\pi\{(ut - z)^2 + h^2(1 - u^2/v^2)\}^{\frac{1}{2}}}, \tag{84}$$

and the corresponding magnetic force is

$$\mathrm{H}_0 = \frac{-\mathrm{C}h \tan\theta}{4\pi(l^2 - h^2)^{\frac{3}{2}}}, \tag{85}$$

or, expanded,

$$\mathrm{H}_0 = \frac{\mathrm{C}}{4\pi} \frac{(1 - u^2/v^2)h}{\{(ut - z)^2 + h^2(1 - u^2/v^2)\}^{\frac{3}{2}}}. \tag{86}$$

Again, inside the cone, but outside the sphere,

$$\mathrm{A} = \frac{\mu \mathrm{C} \tan\theta}{2\pi(l^2 - h^2)^{\frac{1}{2}}}, \qquad \mathrm{H} = \frac{-\mathrm{C}h \tan\theta}{2\pi(l^2 - h^2)^{\frac{3}{2}}}. \tag{87}$$

These are entirely residual formulæ due to the near coincidence of similar opposite states. Thus (86) holds good only up to the inside of the spherical surface, $\mathrm{R} = vt$, and (87) holds good from the axis up to the inside of the conical surface, and the outside of the sphere. The conical and spherical sheets require separate consideration.

The magnetic force is now negative in both the cone and the sphere (when the cone exists, that is to say) or in the contrary direction to the natural magnetic force of the impressed current under usual circumstances. Its intensity just outside the spherical shell is double that just inside. But this is no guide to the magnetic force of the shell itself. This shell is bounded by sheets of oppositely directed electric current. But the shell is really an electromagnetic pulse, consisting of tangential E and H, to be presently noticed. The conical surface is also the seat of a sheet of displacement and magnetic force, and their directions are unreversed, that is, D is from the apex of the cone, and H is as usual. But just underneath it D and H are reversed—viz., at the boundary of the internal conical region.

The complicated distribution of electric current in the previous line-problem is now replaced by a similar distribution of displacement, also obtained by time-differentiation. The results are

$$\left.\begin{aligned} D_h &= -\frac{\rho}{4\pi}\left\{\frac{h}{R^3} + \frac{h\tan\theta}{(l^2-h^2)^{\frac{3}{2}}}\right\}, \\ D_z &= -\frac{\rho}{4\pi}\left\{\frac{z}{R^3} - \frac{l}{(l^2-h^2)^{\frac{3}{2}}}\right\}, \end{aligned}\right\} \qquad (88)$$

in the sphere. Similar to (59), (60), in fact.

Here the first part represents the radial displacement due to the charge $-\rho$ at the origin, but extending only as far as $R = vt$. The rest represents displacement directed towards the apex of the cone. Calling this part d, its components are

$$d_h = -\frac{d}{dh}\frac{\rho}{4\pi}\frac{\tan\theta}{(l^2-h^2)^{\frac{1}{2}}}, \qquad d_z = \cot^2\theta\,\frac{d}{dz}\frac{\rho}{4\pi}\frac{\tan\theta}{(l^2-h^2)^{\frac{1}{2}}}. \qquad (89)$$

This exhibits the displacement d in the sphere derived in eolotropic manner from the scalar potential used in previous investigations of steadily travelling states. This derivation (89) is valid whether $u <$ or $> v$. But should the cone be non-existent ($u < v$), then the moving charge ρ is inside the sphere, and the displacement is in its natural direction, away from ρ. The complete displacement is then the sum of two radial distributions—one isotropic, centred at the fixed charge, the other eolotropic, centred at the moving charge. They do not cancel at the spherical sheet unassisted, but by leakage from the sheet.

The cone formulæ for the displacement are

$$D_h = -\frac{d}{dh}\frac{\rho}{2\pi}\frac{\tan\theta}{(l^2-h^2)^{\frac{1}{2}}}, \qquad D_z = \cot^2\theta\frac{d}{dz}\frac{\rho}{2\pi}\frac{\tan\theta}{(l^2-h^2)^{\frac{1}{2}}}, \qquad (90)$$

similar to (89), but doubled. The apex is the point of convergence, as before, so there is no change in direction of the displacement.

The complete course of the displacement is thus: First the conical sheet of displacement from the apex. When this reaches the circle of contact it divides right and left in the spherical shell. Now the conical sheet carries an infinite amount of displacement. So the amount entering at the circle of contact is also infinite. But the leakage from the shell is also infinite close up to the circle of contact, and this forms the reversed flux of displacement before mentioned just under the real conical wave front. The rest of the displacement entering at the circle of contact makes a spherical shell of finite intensity, with finite leakage. Continuity is made between the displacement in the sphere and in the cone by the leakage from the spherical shell. It is defined by

$$\overline{\mathrm{H}}_0 = v\overline{\mathrm{D}}_0 = \frac{\rho}{4\pi \mathrm{R}} \frac{u \sin\beta}{1-(u/v)\cos\beta} \tag{91}$$

if $\overline{\mathrm{H}}_0$ and $\overline{\mathrm{D}}_0$ mean the product of H and D in the shell and its depth, and β is the polar angle of the sphere, radius $\mathrm{R}=vt$. The proof is that this formula for $\overline{\mathrm{D}}_0$ harmonises the displacement in the sphere and cone according to (88) and (90). Full details were given of the continuity of the current in the line-problem. The details are exactly the same here as regards the displacement, so they need not be repeated in the new application.

Solutions for a Jerked Electron Without Compensating Charge.

§ 498. To treat the case of $u<v$, we have merely to omit the cone formulæ. The rest remain valid, including the spherical pulse. I pointed out in the line-problem that the state of things in the cone was established instantly. The same is true in this point-problem in a more complicated manner. But even when $u<v$, there is a similar instantaneous establishment of the steady state understood in a particular way. Thus, let there be no compensating charge at the origin. The problem is then this: The charge ρ, initially at rest at the origin, is suddenly projected along the axis at speed u, leaving nothing behind. No change is made in the magnetic force, or in the spherical pulse. To find the change in the displacement, we have merely to superimpose upon the previous D

the polar displacement of ρ in its initial position. The new displacement is therefore the d of equations (89) inside the sphere. Outside the sphere is the remains of the initial polar displacement in any case, and the cone displacement in addition when $u>v$.

Now let the charge be inside the sphere. As it changes its position, and the sphere expands, there is a variable state of things. But if we travel with the charge and keep near it, there is no change at all. That is, the steady state is established instantly at the charge itself, when its motion is altered. It is the spherical pulse alone that does all the work of establishing E and H, or else the spherical and conical pulses together. The initial act of changing the motion generates the pulse. The external work is, if $u<v$, done momentarily too. It is easily seen that the distribution of displacement round the charge is then symmetrical and involves no inertial resisting force. But if $u>v$, continuous working is needed to maintain the velocity, because the cone is being constantly regenerated at the apex.

But the finite point-charge must not be taken literally. Even in electrostatics it is a make-believe, because its energy is infinite. In its motion there is a higher degree of make-believe, and infinities of less easy interpretation. We should spread out the charge to finitize results. In special cases of plane and line distributions the energetic relations becomes quite plain and in accordance with Newtonian dynamics.

The apparent inertia of a charge due to electromagnetic causes arises from the lag involving bending of the displacement just round it when its velocity changes. There is no bending there when u becomes constant. It is then further away, being propagated outward. But this apparent inertia is not a constant. It is a function of u. The continuous working required when $u>v$ is something like frictional resistance. Change of velocity should be understood in Newton's sense. Thus an electron revolving at constant speed in a circular orbit wastes energy continuously on account of the curvature.

On the question of inertial resistance to change of motion I take the opportunity of making a correction to the Note, App. G., vol. 2., p. 533. That Note agrees generally with the investigations of this volume, but the paragraph on p. 534, beginning "Where does the energy come from?" requires some emendation, to be made in the sense of the present and following section. Mr. G. F. C. Searle called in question my statement that the force on a charged body set moving, whilst tending to vanish when $u<v$, does not when $u=v$. He said it did tend to vanish. I agreed. But he said further that the force vanished because it varied as t^{-1}. On consideration I differed, and said only the initial impulse was wanted after all.

Comparison of two Cases of Motion of Electrification at the Speed of Light.

§ 499. The reason for this conclusion was the existence of a class of easily-understandable cases in which no resisting force whatever is experienced after the initial impulse required to start the waves, although the speed of motion of the electrification is the same as that of propagation of the waves. One simple example will be sufficient.

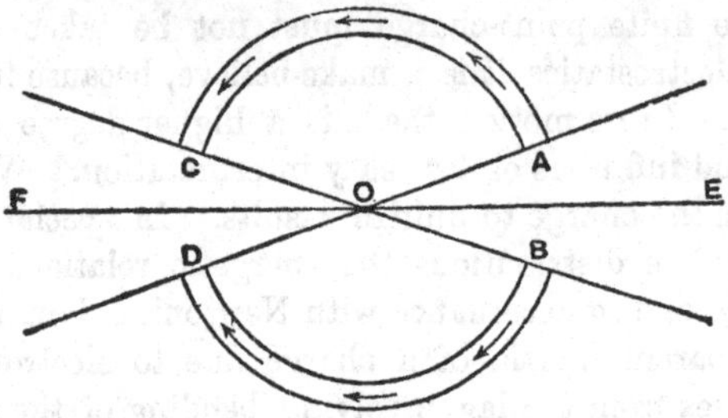

FIG. 37.

Let BOA and DOC be two cones having a common apex at O. Let them be perfect conductors first. Then a voltaic impulse acting at O will generate a spherical shell wave bounded by the conical surfaces. AB is a band of positive electrification and CD a band of negative electrification going round the cones. The displacement is in circular arcs, from A to C and B to D, symmetrically all round. In the wave itself $E=\mu vH$, and H is in circles round the axis of the cones. The wave expands uniformly, so as to make the bands of electrification run along the cones at speed v.

Now, there is no mechanical force on the conductors under the circumstances, because the stress is a pressure in the wave perfectly radial from O, and therefore tangential to the conical surface. Or, we may say that the moving force **E**ρ on the electrification element ρ is exactly neutralised by the moving force ρ**VvB** on the current element ρ**v**, the same ρ moving with velocity **v**. At A, for example, the first force is upward, the second downward; both are equal, and perpendicular to the conical surface.

If we take away the conical conductors altogether no change whatever will be made in the condition of the wave, provided the two bands of electrification move through the ether in the same way as before. The energy is finite and constant; therefore nothing more than an initial force impulse is required to set things going, and no force is needed to keep them going.

The angle of the two cones may be reduced to nothing. In the limit we come to the case of positive electrification moving along OE, connected through a spherical shell with equal negative electrification moving along OF. That the energy becomes infinite in the limit is nothing. The point in question is the balance of forces.

Many other cases of this sort concerning the projection of electrification radially from a centre at the speed of light may be constructed, having the same property of not requiring energetic assistance after the initiation of the waves.

But now consider the matter from a slightly different point of view, through a simple solution of another kind.

Let O (Fig. 38) be a continuous source of separation of + and − electrification, the + to move to the right along OA, and the − to the left along OB, both moving at speed v. Let σ be the linear density, then there is a linear convection current $C = \sigma v$ extending all the way from B to A. The com plete solution is, at distance r from the axis,

$$H = cvE = \frac{C}{2\pi r}, \tag{92}$$

with the additional information that **H** is in circles round the axis in the usual way, and that **E** is in semicircles as shown, from the + electrification to the −, and that the sphere within which (92) is valid is of radius vt, if the current begins at O at the moment $t=0$.

The supply of electrification need not be at a constant rate. The quantities σ and C may vary anyhow with the time. Still, (92) will be true for the particular spherical shell associated with a particular pair of $+\sigma$ and $-\sigma$. So this example looks exactly like the last, with the angle of the cones reduced to nothing. If there were no difference the force on the electrification would be zero.

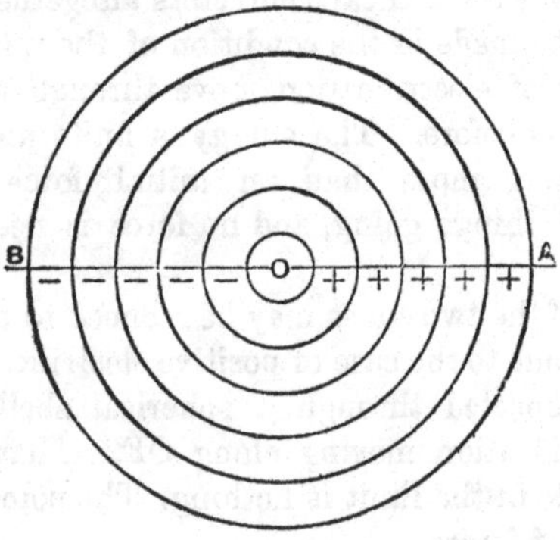

FIG. 38.

But there is a difference. In Fig. 38 the electromagnetic force on σv is perpendicular to the line of σ all round, and does not produce any resistance. The electric force E is also perpendicular to σ all round, but it does produce a resistance to motion. It has a finite component along the axis in the limit.

Thus, considering any point near the axis between O and A, the z-component of E, at distance z from O, is

$$E_z = -\frac{\sigma}{2\pi rc} \times \frac{r}{(z^2+r^2)^{\frac{1}{2}}} = -\frac{\sigma}{2\pi c(z^2+r^2)^{\frac{1}{2}}}. \qquad (93)$$

In the limit, with $r=0$, this is $-\sigma/2\pi cz$. The mechanical force on any linear element σ is, therefore,

$$F = -\frac{\sigma^2}{2\pi cz} = -\frac{\sigma^2}{2\pi cvt}. \qquad (94)$$

Applied force is therefore required, not merely at the moment of projection from O, but later as well. It varies as t^{-1}.

The reason of the contradiction of results is that the infinitely fine line of electrification is treated as a cone in one case, no matter how slowly its diameter may increase, but as a cylinder in the other case. The axial components of the electric and electromagnetic moving forces are both finite and balance one another in one case. In the other case only the electric part exists, unbalanced. It is the greatness of the electric and magnetic forces at small distances that makes the usually insensible difference between a fine cylinder and a fine cone of insensible angle become of significance.

Peculiarities at the Speed of Light.

§ 500. The above theory, (93), (94), seems to me to contain the essential part of Mr. Searle's argument that the moving force on a charged sphere jerked into motion at speed v varies as t^{-1} ultimately, his calculation being professedly a first approximation. It is here put, perhaps, in its simplest form, in an exact solution.

Now, as regards the reason why there should be a residual force when u is exactly equal to v, whereas there is none when u is less than v by any amount, however small. If a single electron, of radius a, is suddenly jerked into motion at any speed making u/v very small, the spherical pulse generated, which is of depth $2a$, is clear of the electron almost at once, in a time interval just over $2a/v$. Similarly, any distribution of electrification similarly jerked, has the whole of the impulsive part clear of the electrification in a time interval a little over $2a/v$, if $2a$ is the greatest diameter of the electrified region. After that, the steady state is fully formed there, and there is no resisting force left on the electrification as a whole. The mutual forces that remain balance. But when u is increased it takes longer for the impulsive wave to get clear of the electrification, and so bring about the state of no resisting force. In the limit, when $u=v$, the electrification travels with the impulsive waves, which, therefore, never get clear. There may now be a residual resisting force, or there may not, according to circumstances. In the spherical waves of the class to which Fig. 37 refers, the electrification is projected radially from the origin, and therefore spreads out as it travels. In Mr. Searle's case there is no spreading.

We have next to ask what bearing the residual force varying as t^{-1} has upon the question whether it is possible to move electrification through the ether at a speed equalling or exceeding that of light? Assuming that a charged body, say a sphere, is really moving at speed v, it travels with what is practically a plane wave. If the wave were fully formed, extending to an infinite distance, its energy would be infinite, if it travelled alone (though finite if it travelled with an equal compensating charge). But this is really no reason why there should be an infinite resisting force, or why it should not be possible to increase u above v. For the assumed final state with infinite energy would never be reached. And the t^{-1} theory shows that the increase of energy in the wave tends to go on infinitely slowly. A further increase of speed will then bring in the resisting force of the conical theory, which is a different thing altogether, and requires special calculation in particular cases.

There is, however, one reservation to be made about the $u=v$ case. An impulse is an instantaneous affair, and is, therefore, ideal. It is convenient for calculation and for general simplification of reasoning. If the work done by an impulse is finite, there is no harm in replacing practical impulses by ideal instantaneous ones. This is the case when $u<v$. But if $u=v$, the energy in the impulsive wave, if really impulsive, is infinite. Then we should not assume the existence of an ideal impulse; the change of speed from 0 to v must be done gradually, making a diffused impulsive wave of finite energy.

The importance of this point will be plainly exhibited by a consideration of how the steady final state comes about. Suppose $u<v$; then first we generate a spherical pulse at an electron, after which the change in the field is made by the pulse itself. So the complete addition to the energy of the final field comes from the pulse as it travels. This will work right when the energy of the pulse is finite, but not when it is infinite. There is always some energy wasted in a pulse, carried away to infinity, after supplying the field. How much is wasted depends upon the rapidity of the change of velocity. In the limit, with the ideal instantaneous change, the wasted energy equals the work done in the field, or twice as much work has to be done initially as is accounted for in

the form of stored energy. So in the limiting case, when $u=v$, instantaneous change of velocity of a charged sphere from 0 to v will cause infinite resistance merely because it is instantaneous. A calculation of the energy wasted when finite will follow presently.

In case of a distribution of separated electrons impulsively moved, there will be a highly complicated state of affairs in the very short interval before the steady state is established just round the distribution. For the first effect is a spherical pulse from every one of the electrons. Every pulse has to cross every other electron, and in doing so there is external energetic action (by the force $\mathbf{E}\rho$). Not only the initial work done, but all the later additions (or reductions) have to be taken into account in order to fully account for the energy of the field which results.

The Energy Wasted in the Spherical Pulse from a Jerked Electron, and the Energy left behind.

§ 501. We are in possession of the formula for the total flux of **B** and of tangential **D** in the spherical pulse from an electron when jerked. But that is not enough. In order to calculate the energy in the pulse we require to know the distribution of **B** and **D** in the depth of the pulse. That depends upon the constitution of the electron in detail. Not to make a complicated matter of it, I shall calculate the magnetic energy and the energy wasted on the simple supposition that **H** is ultimately constant in the depth of the pulse. The results are interesting enough and bring us to Searle's and Morton's formula for the potential of a moving charged sphere.

Thus, by the pulse formula (91) above, we have

$$2a\mathrm{H}=\frac{\mathrm{Q}}{4\pi\mathrm{R}}\frac{u\sin\beta}{1-(u/v)\cos\beta}, \tag{95}$$

if $2a$ is the depth of the pulse and Q the charge. This constancy of depth implies sphericality of the electron. The magnetic energy is, therefore,

$$\Sigma\tfrac{1}{2}\mu\mathrm{H}^2=\tfrac{1}{2}\mu\left(\frac{\mathrm{Q}u}{4\pi\mathrm{R}\,.\,2a}\right)^2\int_{-1}^{1}\frac{\sin^2\beta}{\{1-(u/v)\cos\beta\}^2}2\pi\mathrm{R}^2.\,2a.\,d\,(\cos\beta) \tag{96}$$

$$=\frac{\mu\mathrm{Q}^2u^2}{32\pi a}\int_{-1}^{1}\frac{1-\nu^2}{\{1-(u/v)\nu\}^2}d\nu. \tag{97}$$

Put $1-(u/v)\nu=y$, then we get

$$\frac{\mu Q^2u^2}{32\pi a}\int_{1-u/v}^{1+u/v}\left(\frac{1-v^2/u^2}{y^2}-\frac{v^2}{u^2}+\frac{2v^2}{u^2y}\right)\frac{v}{u}dy\,; \tag{98}$$

integrable at sight, and reducing to

$$\frac{Q^2}{8\pi ca}\left\{\frac{v}{2u}\log\frac{1+u/v}{1-u/v}-1\right\} \tag{99}$$

$$=\frac{Q^2}{8\pi ca}\left\{\frac{1}{3}+\frac{1}{5}\frac{u^2}{v^2}+\frac{1}{7}\frac{u^4}{v^4}+\dots\right\}\frac{u^2}{v^2}. \tag{100}$$

To show the meaning, let

$$P=\frac{Q}{4\pi ca}\left(1+\frac{1}{3}\frac{u^2}{v^2}+\frac{1}{5}\frac{u^4}{v^4}+\dots\right), \tag{101}$$

and let P_0 be the value of P when $u=0$. Then the magnetic energy in the pulse is $\frac{1}{2}(P-P_0)Q$. Here $\frac{1}{2}P_0Q$ is the electric energy before the impulse acted, P_0 being the usual electrostatic potential of a sphere of radius a charged superficially, with Q in equilibrium. And P is the potential of the same after the impulse has worked. For it agrees with the potential of the final state as calculated by Searle* and by Morton,† and this final state is assumed on the spot as soon as the pulse is clear of the electrification. The total waste, in the case of an ideal impulse is, therefore,

$$W=Q(P-P_0)=U+T-U_0, \tag{102}$$

if U_0 is the initial, U the final electric energy, and T the magnetic energy of the field that ultimately results, assuming equality of the waste to the increase of stored energy. We can now find U and T separately in terms of P.

For if $F=-\nabla P$, and $f=\kappa^2F=(1-u^2/v^2)F$, we have

$$f=E+VuB, \tag{103}$$

inside the spherical pulse, by § 164, vol. 1., p. 271. Multiply by D. Then

$$fD=ED+DVuB=ED-BVuD=ED-HB, \tag{104}$$

because $H=VuD$. Now, here $\frac{1}{2}ED$ is the density of the electric energy, and $\frac{1}{2}HB$ the density of the magnetic energy, so their space integrals are

$$U-T=\Sigma\,\tfrac{1}{2}fD=\kappa^2\Sigma\,\tfrac{1}{2}P\rho, \tag{105}$$

* G. F. C. Searle, *Phil. Trans.*, 1896, and *Phil. Mag.*, Oct. 1897.

† W. B. Morton, *Phil. Mag.*, June, 1896.

if $\rho = \text{div D}$. The "space" extends up to the inner surface of the pulse, and in general ρ has to include the apparent electrification there. But when the pulse has gone out far enough, only the electronic charge counts, so

$$U - T = \tfrac{1}{2}PQ\kappa^2, \tag{106}$$

a relation proved by Searle for the ultimate steady field. Thus, U being known in terms of T and P, the formula (102) of the pulse gives us T in terms of P. We get

$$T = \tfrac{1}{4}Q\left\{\left(1 + \frac{u^2}{v^2}\right)P - P_0\right\}, \tag{107}$$

for the final magnetic energy. Or, in series form for plainness,

$$T = \frac{\mu Q^2 u^2}{16\pi a}\left\{\left(1 + \frac{1}{3}\right) + \left(\frac{1}{3} + \frac{1}{5}\right)\frac{u^2}{v^2} + \left(\frac{1}{5} + \frac{1}{7}\right)\frac{u^4}{v^4} + \dots\right\}. \tag{108}$$

This agrees with Searle's formula obtained by space integration. I have much pleasure in arriving at his results, because I could not understand his λ. It seemed academical, not electrical. He said it was not lemma, but lambda. That might be, but was unconvincing.

The Potential of a Charged Spheroid moving along its Axis.

§ 502. We may readily obtain the formula for P without using the academical λ integral, for which I have, nevertheless, the profound respect of ignorance. For a given distribution of

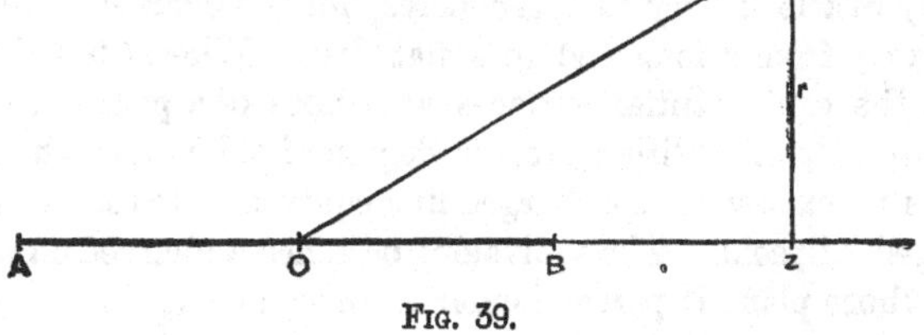

FIG. 39.

electrification, P is the integral of the potential of a point charge. In case of a sphere this will be an integral over its surface. But it is easier to integrate along a line in an equivalent manner. Let the charge Q be spread uniformly along a line of length $2l$, which moves in its own line at speed u.

Let AOB be the charged line, and let the potential at P be required. Co-ordinates z and r. Origin at O. By (1), § 463, it is

$$P=\frac{Q}{2l}\int_{-l}^{l}\frac{dy}{4\pi c\{(z-y)^2+\kappa^2r^2\}^{\frac{1}{2}}},\qquad(109)$$

where dy is any element of length. This makes

$$P=\frac{Q}{8\pi cl}\log\frac{z+l+\sqrt{(z+l)^2+\kappa^2r^2}}{z-l+\sqrt{(z-l)^2+\kappa^2r^2}}.\qquad(110)$$

The equipotential surfaces are ellipsoids of revolution round the line AOB. To find the relation of the axes, put first $r=0$, $z=a$, and then again $z=0$, $r=b$, and equate the two P's. Then

$$\frac{a+l}{a-l}=\frac{l+\sqrt{l^2+\kappa^2b^2}}{l-\sqrt{l^2+\kappa^2b^2}},\qquad(111)$$

which reduces to $$l^2=a^2-\kappa^2b^2,\qquad(112)$$

giving a in terms of b, when a and b are the semi-axes of any equipotential surface. One of them is a sphere. Put $a=b$, then $l/a=u/v$ gives its radius a. Shift the charge Q from the line to this sphere, then P in (110) becomes the potential of the charged sphere at any point outside it, when we put $l=au/v$. Further, put $z=a$ to obtain the potential of the sphere itself, as in equation (101) above.

When the charge Q is not on a sphere, but on the ellipsoid a, b, the magnetic energy is not given by (108), but by the more general formula obtained by writing l/a instead of u/v inside the { }, l being given by (112). This general formula got by Searle is very comprehensive, for the charged ellipsoid may vary from a long rod to a flat disc. When l is reduced to 0, the equipotential surfaces are those of a point charge, and are oblate. With a greater degree of oblateness, the line $2l$ has no existence as a charged line equivalent to the external charged ellipsoid. The equivalent becomes a charged circular disc whose plane is perpendicular to the previous line.

Since $$P=\frac{Q}{8\pi cl}\log\frac{a+l}{a-l}\qquad(113)$$

is the surface potential of the ellipsoid a, b, when in motion, its potential P_0 when at rest is the same with $\kappa=1$. These values should be used in the waste formula (102) when a charged ellipsoid is concerned.

APPENDIX J.

NOTE ON THE SIZE AND INERTIA OF ELECTRONS.

When an electron is set in motion its energy is increased by the amount

$$\tfrac{1}{2}m_0u^2 + \mathrm{W},$$

the first term being the ordinary kinetic energy of the mass m_0 of the electron, if it has mass, and the second being the addition made to the energy of the field, that is, $\mathrm{U} - \mathrm{U}_0 + \mathrm{T}$. The expression for W is (E.M.T., vol. 3, § 501),

$$\mathrm{W} = \mathrm{Q}(\mathrm{P} - \mathrm{P}_0) = \frac{\mathrm{Q}^2}{4\pi ca}\left\{\frac{v}{2u}\log\frac{1+u/v}{1-u/v} - 1\right\}$$

$$= \frac{\mathrm{Q}^2}{4\pi ca}\frac{u^2}{v^2}\left\{\frac{1}{3} + \frac{1}{5}\frac{u^2}{v^2} + \frac{1}{7}\frac{u^2}{v^2} + \dots\right\}.$$

The series is useful for small values of u/v, but large ones are now to be the subject of calculation, so use the log formula.

Since $cv^2 = \mu^{-1} = 1$ in electromagnetic units, and $\mathrm{Q}^2 = 4\pi e^2$, if e is the charge in common irrational units, we may write

$$\mathrm{W} = \frac{e^2u^2}{a}\mathrm{X}, \quad \text{if} \quad \mathrm{X} = \left(\frac{v}{2u}\log\frac{1+u/v}{1-u/v} - 1\right)\frac{v^2}{u^2};$$

and therefore, if M is the effective mass of the electron,

$$\tfrac{1}{2}\mathrm{M}u^2 = \tfrac{1}{2}\left(m_0 + \frac{2e^2}{a}\mathrm{X}\right)u^2;$$

from which we derive

$$\frac{\mathrm{M}}{e} = \frac{m_0}{e} + \frac{2e}{a}\mathrm{X},$$

where X is a known function of u/v.

In *The Electrician*, March 14, 1902, are reproduced data given by W. Kaufmann (*Gött. Nach.*, 2, 1901). Five values of u/v and the corresponding five values of M/e are given. I am not competent to express any opinion upon the accuracy of the data. But, taking them to really express what is professed, we can employ them to find the ratio m/e and the size of an electron, in so far as it is expressible by the quantity a, the radius of the sphere upon which the charge e is distributed in the above theory, in terms of e itself. The data are:

$$u/v = 0{\cdot}94, \quad 0{\cdot}91, \quad 0{\cdot}86, \quad 0{\cdot}83, \quad 0{\cdot}79.$$
$$e/\mathrm{M} = 630^5, \quad 770^5, \quad 9750^4, \quad 1170^5 \quad 1310^6.$$

From u/v calculate the values of 2X. They are:

$$2\mathrm{X} = 2{\cdot}16, \quad 1{\cdot}64, \quad 1{\cdot}36, \quad 1{\cdot}24, \quad 1{\cdot}16.$$

Use these in the formula for M. The results are:

$$\begin{aligned} \mathrm{M}/e = 159 \times 10^{-9} &= 2{\cdot}16 e/a + m_0/e, \\ 130\ldots \quad &= 1{\cdot}64\ldots + \ldots \\ 103\ldots \quad &= 1{\cdot}36\ldots + \ldots \\ 86\ldots \quad &= 1{\cdot}24\ldots + \ldots \\ 76\ldots \quad &= 1{\cdot}16\ldots + \ldots \end{aligned}$$

Taking these in pairs, so as to eliminate m_0/e, the values of e/a are rather wild. We cannot tell whether m_0 exists at all from these data. But from the rough proportionality of the figures on the two sides it may be concluded that m_0 is a small fraction of M. If we ignore m_0 altogether, we get

$$\frac{e}{a} = \frac{73,\ 79,\ 76,\ 69,\ 66}{10^9}; \quad \text{mean } \frac{e}{a} = \frac{73}{10^9}.$$

A remarkable thing about the figures for M/e is their smallness, compared with what might be expected from the generally-accepted size of m/e and the largeness of u/v. If we put $u = 0$ in the equation of M, and write m instead of M, we have

$$\frac{m}{e} = \frac{m_0}{e} + \frac{2e}{3a},$$

because $\mathrm{X} = \frac{1}{3}$ when $u = 0$. This m is the effective mass at small speeds, m_0 is common mass, and $2e^2/3a = m_1$ is the addition due to electromagnetic inertia.

A. Schuster was the first to try to measure m/e. Later experimenters, J. J. Thomson and many others, have done better. J. J. Thomson made it 10^{-7}. Schuster (*Phil. Mag.*, February, 1901) adopts Wiechert's $1{\cdot}3 \times 10^{-7}$. There does not seem to have been any experimental separation of m into m_0 and m_1. If $m_0 = 0$, then m_1 or $2e^2/3a$ is the effective mass.

Now, using the above high-speed estimate of e/a we find $m/e = 49 \times 10^{-9}$. This is only half of J. J. Thomson's value.

In fact, his value is that of M/e at the high speed $u=0{\cdot}86v$, which doubles the effective mass.

Which, then, are the more probably correct, the high-speed or the low-speed experiments? I should think the latter, but do not profess to know anything about it.

As regards the values of e, m and a separately, e and m are given by electrolytic measurements, combined with the estimate of the mass of a hydrogen atom founded upon the kinetic theory of gases and the experimental estimate of m/e. This mass is, therefore, subject to uncertainty probably much greater than that in the ratios m/e and e/a.

Thus, take $e/m=1{\cdot}3\times10^7$, and $e/m'=10^4$ for hydrogen. Schuster calculates $m'=2\times10^{-24}$ and $e=2\times10^{-20}$, and these lead by the low-speed formula to $a=1{\cdot}8\times10^{-13}$ centim., "which is 30,000 times smaller than the molecular distance in solids."

Now we have

$$\frac{m}{a}=\frac{2}{3}\left(\frac{e}{a}\right)^2=\frac{3}{2}\left(\frac{m}{e}\right)^2, \quad \text{and} \quad ma=\frac{2}{3}e^2;$$

so, if we use the high-speed value of e/a, giving the low value of m/e, we get

$$\frac{m}{a}=\frac{3}{2}\left(\frac{49}{10^9}\right)^2=\frac{\frac{3}{8}}{10^{14}}.$$

Use also the datum $e=2\times10^{-20}$, which does not depend upon the electronic experiments, and we get

$$m=10^{-27}\text{ gram}; \quad a=2{\cdot}6\times10^{-13}\text{ centim.}$$

In short, the high-speed data about halve the value of m and double that of a as compared with the low-speed data. It is a matter of fundamental philosophical and scientific importance to know whether an electron has mass independently of the electromagnetic inertia. No doubt the experimental difficulties are great, but there are some remarkably able experimenters at work. That is one thing. Another thing of importance is a perfectly reliable estimate of e/m, and its harmonisation with e/M at any speed. There is a difference between m and M which is not merely numerical. The common m is a constant, whilst M varies with the speed, though it is only in nearing the speed of light that it increases very fast. This variability may necessitate some modification in the interpretation of experimental observations.

[Postscript, July 7, 1902. The above $m/e = 49 \times 10^{-9}$, or say $e/m = 2 \times 10^7$, is not much different from the latest results relating to cathode rays at low speeds, and to the Zeeman effect. Thus (*The Electrician*, June 13, 1902, p. 303, and W. Seitz, *Ann. der Physik*, No. 6, 1902), e/m has been put at $1{\cdot}3 \times 10^7$ by Wiechert, 2×10^7 by Wien, and $1{\cdot}77 \times 10^7$ by Kaufmann, and W. Seitz found that Kaufmann and Simon's method gave the best and most consistent results, making $e/m = 1{\cdot}87 \times 10^7$ or $m/e = 53 \times 10^{-9}$.

Also, (*The Electrician*, June 27, 1902, and E. Riecke, *Phys. Zeitschr.* June 15, 1902), E. Riecke points out that in Zeeman's experiment, the isolated mercury lines, which form true triplets, give $e/m = 1{\cdot}86 \times 10^7$, according to Lorentz's theory.

So the formula for M/e used above gives results which harmonize roughly with the accepted low speed results. But the formula is not invariable, because there is no invariable connection between M and u.]

APPENDIX K.

VECTOR ANALYSIS.*

More than a third part of a century ago, in the library of an ancient town, a youth might have been seen tasting the sweets of knowledge to see how he liked them. He was of somewhat unprepossessing appearance, carrying on his brow the heavy scowl that the "mostly-fools" consider to mark a scoundrel. In his father's house were not many books, so it was like a journey into strange lands to go book-tasting. Some books were poison; theology and metaphysics in particular; they were shut up with a bang. But scientific works were better; there was some sense in seeking the laws of God by observation and experiment, and by reasoning founded thereon. Some very big books bearing stupendous names, such as Newton, Laplace, and so on, attracted his attention. On examination, he concluded that he could understand them if he tried, though the limited capacity of his head made their study undesirable. But what was Quaternions? An extraordinary name! Three books; two very big volumes called Elements, and a smaller fat one called Lectures. What could quaternions be? He took those books home and tried to find out. He succeeded after some trouble, but found some of the properties of vectors professedly proved were wholly incomprehensible. How could the square of a vector be negative? And Hamilton was so positive about it. After the deepest research, the youth gave it up, and returned the books. He then died, and was never seen again. He had begun the study of Quaternions too soon.

Perhaps there was no cause for wonder. Great mathematicians have even done worse, by wholly misconceiving the essential nature of the quaternion itself. Such, at least, is what I gather from one of the latest papers written by Prof. Tait, professedly devoted to an exposure of Prof. Klein's errors

* [This was written as a review of E. B. Wilson's work upon Vector Analysis, founded upon the lectures of Prof. J. Willard Gibbs.]

on the subject. The quaternion, regarded as the ratio of two vectors in Hamiltonian fashion, is a quite distinct idea. It is the logical deduction of vectorial properties that is so difficult.

My own introduction to quaternionics took place in quite a different manner. Maxwell exhibited his main results in quaternionic form in his treatise. I went to Prof Tait's treatise to get information, and to learn how to work them. I had the same difficulties as the deceased youth, but by skipping them, was able to see that quaternionics could be employed consistently in vectorial work. But on proceeding to apply quaternionics to the development of electrical theory, I found it very inconvenient. Quaternionics was in its vectorial aspects antiphysical and unnatural, and did not harmonise with common scalar mathematics. So I dropped out the quaternion altogether, and kept to pure scalar and vectors, using a very simple vectorial algebra in my papers from 1883 onward. The paper at the beginning of vol. 2 of my *Electrical Papers* may be taken as a developed specimen; the earlier work is principally concerned with the vector differentiator ▽ and its applications, and physical interpretations of the various operations. Up to 1888 I imagined that I was the only one doing vectorial work on positive physical principles; but then I received a copy of Prof. Gibbs's Vector Analysis (unpublished, 1881-4). This was a sort of condensed synopsis of a treatise. Though different in appearance, it was essentially the same vectorial algebra and analysis to which I had been led. That is, it was pure vectorial algebra, and the method of treating ▽ by the operations potential, curl, divergence and slope was practically the same. Not liking Prof. Gibbs's notation so well as my own, I did not find it desirable to make any change, but have gone on in the old way. My chapter on Vectorial Algebra and Analysis (vol. 1, E.M.T.) was written to meet a want, but specially for my own readers. Regarded as a treatise, it was a stop-gap.

Prof. Tait's attitude towards quaternionic innovations was originally that of the extremest conservatism. Anyone daring to tamper with Hamilton's grand system was only worthy of a contemptuous snub. Even Prof. Gibbs should be regarded as a retarder of quaternionic progress, he said. Yet it is a fact, that if he had considered Gibbs's work fairly, he would have found that Gibbs had actually extended the field of operation of Hamilton's ϕ, miscalled a quaternionic operator. As for myself, I was too small to be seen, at first. However, things changed as time went on, and after a period during which the diffusion of pure vectorial analysis made much progress, in spite of the disparagement of the Edinburgh school of scorners (one of whom said some of my work was "a disgrace to the Royal Society," to my great delight), it was most gratifying to find that Prof. Tait softened in his harsh judgments, and came to recognise the existence of rich fields of pure vector analysis, and to tolerate the workers therein. Besides those impertinent tamperers, Tait had to stick up for quaternionics against Cayley, for quite different reasons. There was danger of a triangular duel, or perhaps quadrangular, at one time, but I would not engage in it for one. I appeased Tait considerably (during a little correspondence we had) by disclaiming any idea of discovering a new system. I professedly derived my system from Hamilton and Tait by elimination and simplification, but all the same claimed to have diffused a working knowledge of vectors, and to have devised a thoroughly practical system. This system, like that of Gibbs, does not profess to be a grand system like Hamilton's Quaternions. But it has, nevertheless, a wide field of application. First, to geometry in three dimensions, and then to most of the physical mathematics of real space. In this respect it is like the old Cartesian mathematics done in terms of the really essential quantities concerned, the vectors and functions thereof, instead of coordinates and components. The work is simplified, and one keeps in touch with physical ideas throughout.

Dr. Wilson's book is, I believe, the first formal treatise on the subject. It is founded upon Gibbs's pamphlet before referred to, and his lectures, with plenty of explanatory matter and examples in illustration and for practice. In some small matters Dr. Wilson follows me, but on the whole his work stands for Gibbs. Personally, I should have liked to have seen Gibbs's pamphlet reprinted as an appendix. That he did not publish it was unfortunate, but it should not, therefore, be overlooked. Its publication would prevent the possibility of its being thought that Gibbs explains or uses my algebra. Is such a calamity possible? Yes, for I find it stated upon p. 573 of Whitehead's "Universal Algebra" that Gibbs's algebra is explained and used by me; and I am sure Mr. Whitehead would not have said so if he had not thought so or been told so.

I think Dr. Wilson has done his work exceedingly well, and has turned out a book that will be useful to students of physical mathematics. There is much to praise and little to blame. But as space is very limited here, I shall leave out the praising and confine myself to a few critical remarks. They are likely to be more useful in the long run.

Dr. Wilson's remark (p. 13) on the impossibility of adding a scalar to a vector is, I think, liable to be misunderstood by early students. It is really quite legitimate to add together all sorts of different things. Everybody does it. My washerwoman is always doing it. She adds and subtracts all sorts of things, and performs various operations upon them (including linear operations), and at the end of the week this poor ignorant woman does an equation in multiplex algebra by equating the sum of a number of different things received in the basket at the beginning of the week to a number of things she puts in the basket at the end of the week. Sometimes she makes mistakes in her operations. So do mathematicians. I think Dr. Wilson should explain the real point thus:—

If $$A_1 + B_1 + C_1 + \ldots = A_2 + B_2 + C_2 + \ldots,$$

and the A's are of one sort, the B's of another, and so on, then we have also $A_1 = A_2$, $B_1 = B_2$, and so on, separately, each to each, as well as all together. Just like the washerwoman. There is a moral. Mathematicians who want to extend multiplex algebra should go to the washerwoman, and observe her ways. The ant is an idiot in comparison, as Mark Twain has noticed.

Dr. Wilson, in making the extension from the space vector to physical vectors, does not explain any criterion for the identification of vectors or enter into any detailed justification of the validity of treating the physical like the space vectors, in addition, multiplication, and so on. Nor should I consider it necessary. The recognition of physical magnitudes as vectors is usually pretty obvious. Yet it seems that some people do really want some argument about it. There was a letter in *Nature* lately on this matter. I think the following will be a sufficient descriptive chain, for the identification of vectors, to make quite sure. Start with the space vector. The circuital property (polygon of vectors) is obvious. It is just a fact, which nobody can deny; any so-called proof must be hanky-panky. Next, differences of vectors are vectors, of course. Therefore, differentiants of vectors are vectors. Because $d\mathrm{a}/dt$ is the limit of $\Delta\mathrm{a}/\Delta t$, and $\Delta\mathrm{a}$ is a vector. In fact, $d\mathrm{a}/dt$ is a very small space vector very largely magnified by the multiplier $1/dt$. So velocity is a vector, and so are acceleration and momentum and force; and (multiplying by scalars) we see that electric force and magnetic force, and electric current are vectors; and rotational velocity, and so on, all over the field. Spacial displacement is a vector, and so are they all—all vectors. They are infinitesimal space vectors, with scalar factors making them finite and endowing them with the proper physical dimensions. They do not go from place to place like the space vector, but vectorially they are all similar, and each may have its own diagram. The parallelogram of velocities and of forces is true because it is true for the space vector, and for no other reason that I can see. Of course, you can make up proofs of the parallelogram of forces; you can do anything by logic if you are artful enough. Dr. O. W. Holmes made some excellent remarks about logic in his "Autocrat."

Dr. Wilson's book is, on the whole, commendably free from that straining after rigour by the invention of difficulties which the Autocrat would have laughed at. But I observe a remarkable exception, and it is so typical of what is done in other books that I shall hold it up to view as a shocking example. It is about linear vector operators. You have first to get to know what a linear operator is. Then comes this property, $n\phi(\mathrm{r}) = \phi(n\mathrm{r})$; that is, multiplying the linear function of r by the scalar n is the same as multiplying r itself. Dr. Wilson says it is tolerably obvious, and then proceeds to *prove* it, presumably for the benefit of unhumorous people. He first proves it when n is a positive integer; then extends n to a rational positive number; then passes to negative numbers, and finally extends n to an incommensurable number by the method of limits. Now I declare that the whole of this is really what FitzGerald used to call "hugger-mugger." But I may be asked, "Is not the proof right, then?" I don't know. I dare say it is. I would not dispute it. "Then how would you prove it?" Just by inspection, of course. Write down any formula for $\phi(\mathrm{r})$, say

$$\phi\mathrm{r} = \mathrm{ra.l} + \mathrm{rb.m} + \mathrm{rc.n},$$

and multiply it by n. Then

$$n\phi\mathrm{r} = n\mathrm{ra.l} + n\mathrm{rb.m} + n\mathrm{rc.n},$$

that is, $n\phi\mathrm{r} = \phi n\mathrm{r}$, instantly, and without any argument. It is like the sum in arithmetic that Alice did for Humpty Dumpty. And when he sees that it seems right, he may ask *why* it goes. Just because n times the vector r is the vector nr. And the rationality or irrationalty of n has nothing to do with it. Besides, n might be a complex number, and Dr. Wilson has not proved that case.

It seems to me that the demonstration I have poked fun at is typical of a lot of work made up by the brain-torturers who write books for young people and college students who are going to be Senior Wranglers, perhaps. Let mathematics be humanised if possible. The best of all proofs is to set out the fact descriptively, so that it can be seen to be a fact.

The vector algebra, considered as algebra, does not take very much space in Dr. Wilson's book. In fact, very little is wanted. It is more a matter of applications than abstract theory. But there is a good deal of space given to the theory of the space-differentiator ∇, called del here (it will do), and divergence, curl, slope and potential. The method of operating is something like that of my early work (1883), but I have no space for detailed notice. Further on, there is a very full account of the properties of linear vector operators. This is Prof. Gibbs's speciality. His treatment of ϕ is somewhat different from that of the quaternionists. He makes ϕ act backward as well as forward, by algebraical processes; of this useful extension of operation I have sometimes been glad to avail myself. Unless I am mistaken, Hamilton was the inventor of the matrix, for that is what ϕ is. There is nothing quaternionic about it. But Hamilton was so wedded to quaternions that he probably failed to see the full power of ϕ, as he might have done by purely vectorial treatment. There may be any number of unitary symbols instead of the three i, j, k, making generalised vectors which are obviously treatable in the same way in direct products and linear functions as the space vectors. Keeping to three only, what I write

$$\phi\mathrm{r} = \mathrm{a}.\mathrm{lr} + \mathrm{b}.\mathrm{mr} + \mathrm{c}.\mathrm{nr},$$

or

$$\phi\mathrm{r} = \mathrm{a}(\mathrm{lr}) + \mathrm{b}(\mathrm{mr}) + \mathrm{c}(\mathrm{nr}),$$

is what Hamilton wrote as

$$\phi\rho = -\alpha S\lambda\rho - \beta S\mu\rho - \gamma S\nu\rho.$$

Here ϕ itself, when isolated, is

$$\phi = \mathrm{a}.\mathrm{l} + \mathrm{b}.\mathrm{m} + \mathrm{c}.\mathrm{n};$$

and the conjugate function of r is

$$\mathrm{r}\phi = \mathrm{ra}.\mathrm{l} + \mathrm{rb}.\mathrm{m} + \mathrm{rc}.\mathrm{n},$$

which is equivalent to Hamilton's $\phi'\rho$. Now Gibbs writes

$$\phi = \mathrm{al} + \mathrm{bm} + \mathrm{cn},$$

$$\phi.\mathrm{r} = \mathrm{al}.\mathrm{r} + \mathrm{bm}.\mathrm{r} + \mathrm{cn}.\mathrm{r}.$$

He calls ϕ a dyadic, and al is called a dyad. This brings me to the somewhat important questions of notation, and what should be called a product. Gibbs's scalar product is a.l (= my al); his vector product is a × l (= my Val); and al itself without any mark between (except a space, which, however, does not seem to be essential, because these spaces between letters occur all through the book, giving the formulæ a rather loose appearance), he calls the indeterminate product, and says it is the most general product. I have great respect for Professor Gibbs, and have therefore carefully read what Dr. Wilson says in justification of regarding the dyad as a product. But I have failed to see that it is a product at all. The arguments seem very strained, and I think this part of Gibbs's dyadical work will be difficult to students. In what I write a.l, the dot is a separator; a and l do not unite in any way. With a vector operand r we get a.lr, or lr times a. That is plain enough, but I do not see any good reason for considering the operator a.l to be the general product, in whatever notation it may be written.

As regards notation, the following comparisons will illustrate.

(Mine) $\mathrm{VaVbc} = \mathrm{b}.\mathrm{ca} - \mathrm{c}.\mathrm{ab} = (\mathrm{b}.\mathrm{c} - \mathrm{c}.\mathrm{b})\mathrm{a}.$

(Gibbs) $\mathrm{a} \times (\mathrm{b} \times \mathrm{c}) = \mathrm{bc}.\mathrm{a} - \mathrm{ca}.\mathrm{b} = (\mathrm{bc} - \mathrm{cb}).\mathrm{a}.$

Looking into my last article for the most developed product, I find it is $V(V \nabla VDq)B$. This will be Gibbs's $[\nabla \times (D \times q)] \times B$.

In short, I amalgamate the members of products; Gibbs separates them.

Now I once devoted a good deal of thought to the question of notation. What I had particularly in view was poor human nature, and the needs of elementary workers. Generally speaking, it may be said that at first glance, all sorts of ways of denoting the scalar and vector products, as $\overline{mn}$ and $\overline{\overline{mn}}$, or (mn) and [mn], &c., are equally good. But in developments they work out differently, and are by no means on equal terms. I arrived at these conclusions:—(1) The members of products should not be separated, but should be amalgamated. (2) The marks in use in common algebra should continue to have the same meanings. (3) Any marks needed to specialise the nature of a product should not be on both sides of the product. (4) The ideas of common algebra should be closely followed. (5) Only one special mark is wanted in elementary vector algebra and analysis.

Dr. Wilson agrees with (1), on his p. 269, but the odd thing is that the application is made to the dyad, the members of which are not multiplied together. As regards (2), this applies to dots and crosses, and bracket marks, used for grouping and distinctness. They should not have special vectorial meanings. Regarding (3), we get into mixtures difficult to read and manipulate algebraically by having initial and terminal marks. Number (4) is important because common algebra is fundamental and well understood, and very few men will go beyond the elements of vector algebra. I do not, for one, not finding it necessary. Considering (5), there are two special products, the scalar and the vector. But the scalar product is the proper generalisation of the product in common algebra, so no special mark is wanted. It is properly Fv, for example, the activity of a force. The vector product is a sort of extra special; it wants a mark, therefore. After consideration, I concluded that Hamilton's prefix V, making Vab, was, perhaps, as good as any way, and a great deal better than many other ways. There is a very remote contingency. It might come into contact or conflict with a scalar called V. I have not known it do so, but even that can be avoided by using

Dr. Föppl's specially shaped letter, long and slender. Now it seems to me that my notation complies pretty well with the desiderata mentioned, and that Gibbs's does not. I am naturally very sorry that Dr. Wilson (who was given *carte blanche*) did not adopt my notation, and sincerely hope he may be able to see his way to do so in his second edition.

Dr. Wilson gives a full account of the different classes of dyadics, and their effects as strainers. Whilst about it, he might well have given a few details regarding the more general matrix of Cayley. It is not difficult. Say

$$\mathbf{A} = A_1\mathbf{i}_1 + A_2\mathbf{i}_2 + \ldots,$$

subject to $\mathbf{i}_1^2 = 1$, $\mathbf{i}_2^2 = 1$, $\mathbf{i}_1\mathbf{i}_2 = 0$, &c. Then the direct product of **A** and **B** is

$$\mathbf{AB} = A_1B_1 + A_2B_2 + \ldots,$$

and a general form of linear operator is

$$\phi = \mathbf{a}_1 . \mathbf{1}_1 + \mathbf{a}_2 . \mathbf{1}_2 + \ldots.$$

I find this notation work much more easily than what is given in the "Universal Algebra" I have already referred to. If Heinrich Hertz had used it in his Mechanics, I think it would have simplified his arguments considerably. But I hope that Prof. Boltzmann will not see this, for fear he may be led to tell another story.*

* [I spoke of Hertz's experimental establishment of Maxwell's electromagnetic waves as a "great hit." Prof. Boltzmann turned this into "lucky hit," and then fell upon me!]

CHAPTER X.

WAVES IN THE ETHER.

Matter, Electricity, Ether and the Pressure of Radiation.

§ 503. On the very borders and extreme limits of Maxwell's theory of the ether in its electromagnetic functions we come to definite indications that a wider theory is wanted. Not a narrower one, involving retrogression, but a wider one. Gravitation is left out in the cold, for one thing, although I have shown (v. 1, p. 455) that the propagation of the gravitational influence may be done at finite speed, which may be that of light, by relations which imitate the circuital equations of electromagnetics, and that the old idea that the speed of gravitation must be an enormous multiple of the speed of light in order to avoid aberration is only moonshine. Then there is the unknown nature of the connecting link between ether and matter. This is Maxwell's electrification. To call it an electron does not help much. It is true that Maxwell's idea of fundamental atoms of electricity has received very important experimental support, which may lead to better knowledge not merely of electrification, but of the nature of matter itself; but at present there is little definitely certain about the actual nature of any one of the three unknowns—matter, electrification, ether.

Then there is the moving force on the ether under electromagnetic influence. Maxwell's theory, to my mind, *proves* the existence of this force, by sound dynamics, and I must consider any theory that ignores it to contain unsound dynamics. (This, of course, assumes that the Maxwellian equations are true. If the force be denied, the circuital equations must be altered, or the distribution of energy.) But as this force exists, what does it do, what motions result? It does not follow necessarily that any motion results. The medium cannot be compressed if it be incompressible, or distorted if rigid. So an incompressible rigid ether would simply oppose the force due to the electromagnetic cause by its own natural reaction. The moving force will do nothing in the way of motion or work.

But I think a rigid incompressible ether is an exceedingly difficult idea. How move matter and electrification through it without elastic or frictional resistance if it is, by hypothesis, rigid and incompressible? I think a deformable ether is far more probable. By the consideration of electromagnetic effects in a deformable medium I was led to my dynamic theory of the forces and stresses in the ether, which should not be confounded with Maxwell's static theory, as people will find out if they will take the trouble to read it (*El. Pa.*, v. 2, p. 524). There is nothing to show the necessity or even probability of incompressibility. If necessary, I should rather prefer to go to the other extreme as a hypothesis, and consider that the ether does not oppose compression elastically. This would do away with the difficulty about the motion of matter through the ether. Besides, there is some theoretical evidence, which counts for what it is worth. The only mechanical imitation of the circuital laws known is that furnished by the ideal rotational ether, when done for infinitely small motions. Now I have shown (v. 2, p. 503) that in this analogy the existence of finite compressibility is inadmissible, if electrification is to keep its identity and not be dissipated in wave fashion, and that it is by the assumption of no elastic resistance to compression that we come to Maxwell's purely transverse waves, with persistence of electrification. Even an analogy counts for something. But what about radiation from the sun and stars? This constitutes progressive plane waves, practically. If Maxwell's "pressure of radiation" exists, its space variation along the ray is the moving force, so it must move a compressible ether and compress it, and it would appear from the circuital laws taken in the form given by me for a deformable medium that there must be a reaction upon the radiation itself, distorting the light waves and altering the speed of propagation. Now it is a conclusion from observation that the speed of light in ether is constant, to a very high degree of constancy. Any variation of speed with the amplitude that is permissible must be very minute. By consideration of stellar distances, therefore, it is possible to obtain rough ideas of the velocity permissibly produced in a compressible ether so as not to cause more than a stated small fraction of a wave length of displacement of the antinodes in a wave train by the changed and variable value of the speed of propagation. In large scale electromagnetics, on the other hand, there might be large motions of the ether allowable, since the speed of propagation is so great as to allow of a sensibly instantaneous re-adjustment. But in making these calculations concerning disturbed radiation, which were rather troublesome and involved quite speculative data concerning the density and compressibility of the ether, I was induced to ask myself, Why have the speed of radiation variable at all? Why not make it an absolute constant? It will save all this bother caused by a variable speed. The following investigation was the result.

The Moving Force Acting on a Deformable Ether.

§ 504. As regards the force on the ether, I pointed out long ago that it was the sum of the electromagnetic force and the magnetoelectric force, which are equal in a solitary wave, and that the force has an interesting form—viz.,

$$\mathrm{F} = \frac{d\mathrm{M}}{dt}, \quad \text{where} \quad \mathrm{M} = \mathrm{VDB} = \frac{\mathrm{VEH}}{v^2}. \tag{1}$$

This is the first approximation, for a practically non-moving medium. By remembering Newton's equation of motion, M is momentum density, either the momentum produced, or a part thereof. J. J. Thomson tried to make this go. He assumed M to be the momentum itself. But I cannot follow his investigation ("Recent Advances," p. 22), and believe the result is erroneous.

My own interpretation (*El. Pa.*, v. 2, p. 558) is this. The Newtonian equation of motion of any material (or inertial) medium is

$$\mathrm{F} + \mathrm{X} = m\frac{\delta \mathrm{q}}{\delta t}, \tag{2}$$

where m is the density and q the velocity, F impressed force, and X the force due to the strain of the medium itself, or to its resistance to strain. Here F is known by electromagnetics, but X is quite unknown. It depends upon the nature of the ether considered as a substance.

To go further, we require to use the expression for F corrected for the motion of the ether. The stress vector on the plane whose normal is N is

$$\mathrm{PN} = \mathrm{E}\,.\,\mathrm{DN} + \mathrm{H}.\,\mathrm{BN} - \mathrm{N}(\mathrm{U}+\mathrm{T}), \qquad (3)$$

(v. 1, p. 84). The force per unit volume may be got by changing N to ∇, according to the theorem of divergence extended (v. 1, p. 190), giving

$$\mathrm{F} = \mathrm{E}\,.\,\mathrm{D}\nabla + \mathrm{H}\,.\,\mathrm{B}\nabla - \nabla(\mathrm{U}+\mathrm{T}). \qquad (4)$$

Here ∇ has to act on all quantities. So

$$\mathrm{F} = (\mathrm{E}\rho + \mathrm{H}\sigma) + (\mathrm{D}\nabla\,.\,\mathrm{E} + \mathrm{B}\nabla.\mathrm{H}) - \nabla(\mathrm{U}+\mathrm{T}), \qquad (5)$$

if ρ and σ are the divergences of D and B. Assuming isotropy, so that there is no torque, this is transformable to (v. 1, p. 107),

$$\mathrm{F} = (\mathrm{E}\rho + \mathrm{H}\sigma) + \mathrm{V}\,\mathrm{curl}\,\mathrm{H}\,.\,\mathrm{B} + \mathrm{V}\,\mathrm{curl}\,\mathrm{E}\,.\,\mathrm{D} - (\tfrac{1}{2}\mathrm{E}^2\nabla c + \tfrac{1}{2}\mathrm{H}^2\nabla\mu). \qquad (6)$$

In a medium of constant μ and c, and with $\rho = 0$, $\sigma = 0$, as in pure ether, only the middle terms exist, the two vector products.

There is another interesting form, the corrected form of (1) above. Use the circuital equations (v. 1, p. 71) for a non-conducting single medium,

$$\mathrm{curl}\,\mathrm{H} = \dot{\mathrm{D}} + \mathrm{u}\rho + \mathrm{curl}\,\mathrm{VDq}, \qquad (7)$$

$$-\,\mathrm{curl}\,\mathrm{E} = \dot{\mathrm{B}} + \mathrm{w}\sigma + \mathrm{curl}\,\mathrm{VBq}, \qquad (8)$$

and use them in the vector products in (6). Then

$$\mathrm{V}(\mathrm{V}\nabla\mathrm{H})\mathrm{B} + \mathrm{V}(\mathrm{V}\nabla\mathrm{E})\mathrm{D} = \mathrm{V}(\dot{\mathrm{D}} + \mathrm{V}\nabla\mathrm{VDq})\mathrm{B} - \mathrm{V}(\dot{\mathrm{B}} + \mathrm{V}\nabla\mathrm{VBq})\mathrm{D} + \rho\mathrm{VuB} - \sigma\mathrm{VwD}. \qquad (9)$$

Here the deflecting forces on ρ and σ moving through the medium are

$$\mathrm{g} = \rho\mathrm{V}(\mathrm{u}-\mathrm{q})\mathrm{B} + \sigma\mathrm{V}(\mathrm{q}-\mathrm{w})\mathrm{D}. \qquad (9\text{A})$$

Ignoring them, the rest on the right side is f say, where

$$\mathrm{f} = \frac{\delta\mathrm{M}}{\delta t} + \mathrm{VD}(\mathrm{B}\,.\,\nabla\mathrm{q} - \mathrm{B}\nabla.\,\mathrm{q}) - \mathrm{VB}(\mathrm{D}\,.\,\nabla\mathrm{q} - \mathrm{D}\nabla.\,\mathrm{q}), \qquad (10)$$

which, by the formula (55) in the chapter on vector algebra (v. 1, p. 163) is the same as

$$\mathrm{f} = \left(\frac{\delta}{\delta t} + \nabla\mathrm{q}\right)\mathrm{M} + \nabla_q(\mathrm{Mq}). \qquad (11)$$

Here ∇_q means that ∇ differentiates q only. This is important. Prof. G. Mie has been investigating this subject lately,* but he has employed a wrong formula, derived from Helmholtz, equivalent to

$$\mathrm{f} = \frac{\delta \mathrm{M}}{\delta t} - \nabla_{\mathrm{M}}(\mathrm{Mq}). \qquad (12)$$

The absence of ∇q merely means the assumption of incompressibility. There is also a changed sign. But the important matter is that M is to be differentiated, not q. This is fatal to the vitality of developments. So I have given above the details.

The complete F is therefore (6), with f + g put for the sum of the vector products, f being as in (11), and g being the ρ, σ terms in (9A). First consider (11), no electrification, &c., and constant c and μ. Then

$$\mathrm{F} = \left(\frac{\delta}{\delta t} + \nabla \mathrm{q}\right)\mathrm{M} + \nabla_q(\mathrm{Mq}), \qquad (13)$$

$$\mathrm{F} = \left(\frac{\delta}{\delta t} + \nabla \mathrm{q}\right) m\mathrm{q} - \mathrm{X}, \qquad (14)$$

where (14) is a changed form of (2). The suggestion is again that M = mq. But if so, X is determined. That is,

$$\mathrm{M} = m\mathrm{q}, \qquad \mathrm{X} = -m\nabla(\tfrac{1}{2}\mathrm{q}^2), \qquad (15)$$

make (13), (14) identical. Or, if X = $-\nabla p$, then

$$\frac{dp}{dq} = mq, \qquad (16)$$

is the relation between the pressure p (not the Maxwellian pressure) and the momentum, if q is the size of q, as usual.

Now what sort of a medium can it be in which the moving force due to the strain is the slope of a pressure connected with the momentum by the relatio n (16)? A fluid, of course, but known fluids do not behave like that. However, it has to be remembered that it is not a mere question of a fluid, but a fluid under electromagnetic influence, or something more than a fluid. The suggestion that M = mq is so inevitable by (13), (14) (though not inexorably inevitable), that I have tested various relations between p and m to try to make a decent fit. The speed of propagation through space always varies with the amplitude of the electrical disturbances. This

* *Phys. Zeitschrift*, 2, No. 21, p. 319.

causes accumulation at the nodes, given time enough—thousands or millions of years, for example—according to the size of m and q. I may return to this part of the subject. But at present it is much easier to change the data somewhat, and make the speed through space constant.

How to have Constant Speed through Space of Plane Radiation Traversing a Moving Compressible Ether.

§ 505. Let u be the speed of radiation in plane waves, v the speed of propagation through the medium, and q its own speed in the same direction, and let

$$u = v + q, \tag{17}$$

then u is to be later on an absolute constant, whilst v varies to suit the variation in q. This cannot be done with constant μ and c, for we have

$$\mu c v^2 = 1, \tag{18}$$

but since the medium is compressible, we can vary μ and c.

Consider a solitary plane wave. The common and necessary equation of continuity of mass

$$-\operatorname{div}(m\mathrm{q}) = \dot{m}, \tag{19}$$

becomes

$$-(mq)' = \dot{m}, \tag{20}$$

if the accent, here and below, means d/dx, the wave travelling along x. That is, the matter entering any fixed unit volume through its boundary is not lost, but stored therein.

There is a similar magnetic property. The induction, moving transversely, is not lost, but persists in amount. That is,

$$-\{(v+q)\mathrm{B}\}' = \dot{\mathrm{B}}, \tag{21}$$

if v is the speed of the tubes of B without reflection through the ether, which itself moves at speed q. Here $(v+q)\mathrm{B}$ is the rate of flux of B transversely. Similarly as regards D. We have

$$-\{(v+q)\mathrm{D}\}' = \dot{\mathrm{D}}. \tag{22}$$

In a plane common electromagnetic wave B and D are perpendicular to one another and to the direction of motion of the wave, and $\mathrm{E} = \mu v \mathrm{H}$ connects their size. But in order that the two equations (21), (22) shall be consistent in general, they must become identical by a relation between μ, c, v.

Let

$$\tfrac{1}{2}c\mathrm{E}^2 = \tfrac{1}{2}\mu\mathrm{H}^2, \quad \text{and} \quad \mu c v^2 = 1. \tag{23}$$

These make $E = \mu v H$, and then (21), (22) become

$$-\{(v+q)\mu H\}' = \dot{\overline{\mu H}}, \tag{24}$$

$$-\{(v+q)c\mu v H\}' = \dot{\overline{c\mu v H}}. \tag{25}$$

These become identical by cv = constant, or μv = constant. I showed some 14 years ago that this is the property required to prevent reflection in transit, when μ and c and v are variable. Every plane slice of B and D keeps to itself, without mixing. So

$$-\{(v+q)B\}' = \dot{B} \tag{26}$$

is the single equation necessary, understanding that $E = \mu v H$, and μv = constant.

Now go further, and let $v + q = u$ be itself constant, not previously assumed.

Then,
$$-uB' = \dot{B}, \tag{27}$$

or B and D travel at constant speed u, unreflected and without distortion of the wave shape. It is easy to *say* so, but can it be done? First, we have

$$\mu = \frac{\mu v}{u - q}, \qquad c = \frac{cv}{u-q}, \tag{28}$$

showing how μ and c depend upon q.

Next, let the momentum mq have the same property (27), viz.,

$$-u(mq)' = \dot{\overline{mq}}. \tag{29}$$

Comparing with (20), we have

$$u\dot{m} = m\dot{q} + q\dot{m}, \quad \text{or} \quad (u-q)\dot{m} = m\dot{q} = v\dot{m}. \tag{30}$$

Therefore, mv is also constant, and

$$m = \frac{mv}{u-q}, \tag{31}$$

like (28). That is, c and μ are both proportional to the density, and all three vary inversely as v or $u - q$.

Now, if we introduce the condition $M = mq$, we must find whether it is possible to harmonise with the moving force F, without absurdity, and find what X is. We have

$$F + X = m(\dot{q} + qq'), \tag{32}$$

and this is the same as

$$F + X = -mvq', \tag{33}$$

because $uq' = -\dot{q}$, and $v + q = u$.

As regards F we must use a form allowing for variability of μ and c. Go back to (5). Applying to a plane wave, $D\nabla.E$ and $B\nabla.H$ are zero, and U and T are equal, so

$$F = -(\mu H^2)' = -(mvq)' = -mvq', \tag{34}$$

because $M = DB = B^2/\mu v = mq$, and mv is constant.

Comparing with (33) we see that $X = 0$. This is unique. It cannot be zero in general, but is made zero by the relations of μ, c, m and q. All conditions are fully satisfied. It is like magic. The result is, that radiation is transmitted by a compressible ether at absolutely constant speed through space if its inductivity and permittivity vary as its density, and if the ether behaves as a substance subject to the moving force arising from the electromagnetic cause, whilst the electrical relations are the general ones introduced by me for a deformable medium.

There is no restriction to small velocities, for q may have any value less than u. The ether always moves the same way as the wave, and that is how we get the constant speed. We have

$$q = \frac{uB^2}{B^2 + \mu mv^2}, \tag{35}$$

where μmv^2 is constant. So q varies as B^2 at first, then more slowly. It requires $B = \infty$ to make $q = u$, so there is plenty of room. But, of course, other considerations make it desirable to keep q/u very small.

One relation between the magnetic energy T and the kinetic energy is

$$T^2 = (\tfrac{1}{2}mq^2)(\tfrac{1}{2}mv^2); \tag{36}$$

that is, when q/u is small, T is the geometrical mean of the small energy of m at the actual speed and the large energy of the same at speed v. The proportionality of c, μ and m suggests that they are different aspects of the one property of inertia.

The vanishing of X means no elastic resistance to compression. There is, of course, resistance to compression in another, not the usual sense—viz., inertial resistance. It is strictly translational inertial resistance, but compression is not possible without translation.

In the general case, $X=0$ requires

$$0=\nabla(\mathrm{qM}-\tfrac{1}{2}c\mathrm{E}^2-\tfrac{1}{2}\mu\mathrm{H}^2), \tag{37}$$

on the understanding that ∇ acts upon q, c, μ only.

Or,
$$0=\mathrm{M}d\mathrm{q}-\tfrac{1}{2}\mathrm{E}^2dc-\tfrac{1}{2}\mathrm{H}^2d\mu. \tag{38}$$

But $dc/c=d\mu/\mu=-dv/v$; so

$$-\frac{dv}{v}=\frac{\mathrm{M}d\mathrm{q}}{\mathrm{U}+\mathrm{T}}. \tag{39}$$

Here $\mathrm{M}=\mathrm{VDB}$. To have $v+q$ exactly constant, we want $\mathrm{M}\parallel d\mathrm{q}$ and $\mathrm{M}v=\mathrm{U}+\mathrm{T}$. This means the plane waves before considered, or ordinary radiation, except close to the sources.

Connection Between the Compressed Electromagnetic Wave and Rankine's Wave of Compression.

§ 506. I have shown (v. 2, p. 498) that when the circuital equations are appropriately extended, three sorts of waves arise—namely, (1) the Maxwellian, or electromagnetic wave, (2) the compressional electric wave, and (3) the compressional magnetic wave. But I gave reasons why the compressional waves should not be allowed to appear as possible extensions of Maxwell's system. Now we have just had a wave of compression under consideration. But it must not be confounded with either of the rejected compressional waves. It is a pure electromagnetic wave. It is also compressed. It may, therefore, be fairly described as a compressed electromagnetic wave.

Subject to some reservation, the compressed electromagnetic wave might also be considered to be an electromagnetic wave accompanied by a wave of mechanical compression, such as Rankine's classical solitary wave of compression in a fluid, which may, however, be solid. This works by means of a tension varying as the rarity of the medium, the tension being the negative of the excess of the greatest over the actual pressure. Our wave resembles the Rankinian wave in respect to the compression and its connection with the actual motion. But the pressure which in Rankine's theory is imagined to have presumably a purely mechanical origin, must be abolished. The place of the excess pressure in Rankine's theory must be taken by the Maxwellian pressure of radiation.

If the pressure of radiation bears the same relation to the density of the ether as the mechanical pressure in the Rankinian waves bears to the density, and if the variations of the pressure of radiation constitute moving force in the same way as the variations of pressure in the Rankinian wave, then it it is clear that the behaviour of the two mediums as regards motion and compression will be identically similar. This is the mechanical part of it. That the electromagnetic circuital equations are also obeyed, by making the ether be compressed in its permittivity and inductivity as well as in its materiality, is a separate matter.

The compressed electromagnetic wave is essentially accompanied by the wave of material compression. Neither can exist alone. For the Rankinian wave will not work without the pressure, and the electromagnetic wave will not work in the manner required without the proper compression of the Rankinian wave. The result is undistorted propagation at constant speed through space, even though the medium may be itself moving at any speed not exceeding that of radiation.

The wave-speed u is the sum of the speed v of propagation through the ether, and of q, the speed of motion of the ether, and this sum is constant. What happens is this. If the induction B increases, so do q and m, the density. Along with increased density come increased μ and c (the electromagnetic constants). This lowers the value of v, since $\mu c v^2 = 1$, and to the same extent as q was increased. The displacement D and induction B may have any values, but there is a greatest pressure and a greatest electric and magnetic energy density. The total energy density is $2(\mathrm{T}+\mathrm{T}_0)$, and the energy flux is $2(\mathrm{T}+\mathrm{T}_0)\mathrm{u}$, if T is the magnetic energy and T_0 the energy of motion.

In the propagation of an electromagnetic wave through a compressible medium it is necessary in general to consider the moving force to be $\mathbf{F}+\mathbf{X}$, where $\mathbf{F}$ is Maxwellian and X is due to the strain of the medium as ordinarily considered. Now we found $\mathbf{X}=0$. But then F performs exactly the same functions as the abolished X of the Rankinian wave. There is, therefore, an alternative way of viewing the matter. The Maxwellian pressure may be regarded as being *the* pressure in the medium. This may be somewhat a matter of convention.

To understand fully would need an explanation of how the Maxwellian pressure arises. That is going too far at present. Yet it is necessary to allow for the work done by p and F in translating and compressing the ether just as if they were the abolished p and X. That is, not only are the electric and magnetic energies equal, but the energies of translational motion and of compression, as in Rankine's wave.

Theory of the Rankinian Wave of Compression.

§ 507. The best way to understand the connections is to take the Rankinian wave alone first. The electromagnetics can be tacked on afterwards. Let q be the speed of the fluid (along the x axis); then the equations of persistence of mass and momentum are

$$-(mq)' = \dot{m}, \qquad -p' = m(\dot{q} + qq'). \tag{41}$$

If there is a solitary wave at constant speed u, then $d/dt = -u\,d/dx$, so the first of (41) becomes

$$(mq)' = um', \quad \text{therefore} \quad m(u-q) = \text{const.} = mv, \tag{42}$$

if $u = v + q$, so that v is the speed of propagation through the fluid.

Similarly, eliminating q from the second of (41), we get

$$p' = mvq', \quad \text{or} \quad \frac{dp}{dq} = mv, \quad \therefore \quad p - p_0 = mvq, \tag{43}$$

if p_0 is the least pressure, when $q = 0$. So the working pressure varies as the speed. Also, since mv is constant,

$$m'v = -mv' = mq', \quad \text{because} \quad u = v + q. \tag{44}$$

Put this value of q' in the first of (43). Then

$$p' = m'v^2, \quad \text{or} \quad v^2 = \frac{dp}{dm}. \tag{45}$$

This makes

$$p_1 - p = mv^2 = \frac{m^2v^2}{m}, \tag{46}$$

if p_1 is the greatest pressure, when $m = \infty$, or $q = u$, and $v = 0$. The deficit from this pressure varies as v, and as m^{-1}.

If m_0 is the least density, when $q = 0$, then $m_0u = mv$. Then also $p = p_0$, so

$$p_1 - p_0 = m_0u^2 = \frac{m^2v^2}{m_0}, \tag{47}$$

and

$$p - p_0 = m^2v^2\left(\frac{1}{m} - \frac{1}{m_0}\right) = mv^2 - m_0u^2. \tag{48}$$

The activity of the working stress is $(p-p_0)q$. The convective flux of energy is $q(\mathrm{U}_0+\mathrm{T}_0)$. Their sum is the flux of energy. Its convergence is the rate of increase of energy density. That is,

$$-\{(p-p_0+\mathrm{U}_0+\mathrm{T}_0)q\}'=\dot{\mathrm{U}}_0+\dot{\mathrm{T}}_0. \qquad (49)$$

This applies in general, without specifying the law of pressure or that u is constant. But making $d/dt=-ud/dx$, (49) becomes

$$-\{(p-p_0)q\}'=-\{v(\mathrm{U}_0+\mathrm{T}_0)\}',$$

if $v=u-q$. Therefore

$$(p-p_0)q=v(\mathrm{U}_0+\mathrm{T}_0) \qquad (50)$$

if $\mathrm{U}_0=0$ when $q=0$. Putting in the value of $p-p_0$, it follows that $\mathrm{U}_0=\mathrm{T}_0$. This value of U_0 may also be directly calculated. Thus, if $m^{-1}=\mathrm{V}$, we have

$$\mathrm{U}_0=-\frac{1}{\mathrm{V}}\int_{\mathrm{V}_0}^{\mathrm{V}}(p-p_0)d\mathrm{V}=m\int_{m_0}^{m}\frac{p-p_0}{m^2}dm. \qquad (51)$$

Put in the value of $p-p_0$ in terms of m, and integrate. The result is

$$\mathrm{U}_0=\tfrac{1}{2}m\,.\,m^2v^2\left(\frac{1}{m_0^{\,2}}-\frac{1}{m^2}\right)=\tfrac{1}{2}mq^2=\mathrm{T}_0, \qquad (52)$$

because

$$q=mv\left(\frac{1}{m_0}-\frac{1}{m}\right). \qquad (53)$$

Another expression is

$$\mathrm{U}_0=\tfrac{1}{2}(p-p_0)\frac{q}{v}, \qquad (54)$$

as may be seen by (50). So $2\mathrm{T}_0u$ is the flux of energy, including the equal stored energies moving at speed q, and the activity of the working pressure.

Crossing of Two Waves. Riemann's Solution.

§ 508. The wave may travel either way. So there may be two waves approaching one another, both travelling at speed u without distortion. On emergence from one another after crossing, they will also follow the above theory. But when coincident, there is mutual influence, and consequently change of type. A classical investigation of Riemann allows us to see what happens in crossing. Go back to equations (41), and put $\lambda=\log m$, and $v^2=dp/dm$. Then they become

$$-v^2\lambda'=\dot{q}+qq', \qquad -q'=\dot{\lambda}+q\lambda'. \qquad (55)$$

Multiply the second equation by v and add it or subtract it from the first. Thus,

$$-v(v\lambda'+q')=\frac{\delta q}{\delta t}+v\frac{\delta\lambda}{\delta t}, \qquad -v(v\lambda'-q')=\frac{\delta q}{\delta t}-v\frac{\delta\lambda}{\delta t}. \tag{56}$$

These are the same as

$$\left\{\frac{d}{dt}+(q+v)\frac{d}{dx}\right\}(q+\int v\,d\lambda)=0, \tag{57}$$

$$\left\{\frac{d}{dt}+(q-v)\frac{d}{dx}\right\}(q-\int v\,d\lambda)=0; \tag{58}$$

which show that

$\frac{1}{2}(q+\int v\,d\lambda)$ travels positively at speed $v+q$; (59)

$\frac{1}{2}(q-\int v\,d\lambda)$ travels negatively at speed $v-q$. (60)

The sum and difference are q and $\int v\,d\lambda$, which may be the given data, since the latter is a function of m.

Now in Rankine's fluid mv is constant, where $v=(dp/dm)^{\frac{1}{2}}$, so

$$\int v\,d\lambda = mv\int\frac{dm}{m^2}=mv\left(\frac{1}{m_0}-\frac{1}{m}\right)=u-v. \tag{61}$$

So, by Riemann's results just got,

$\frac{1}{2}(q+u-v)$ is the positive wave; speed $v+q$; (62)

$\frac{1}{2}(q-u+v)$ is the negative wave; speed $v-q$. (63)

Of course, it is not assumed here that $u=v+q$, but merely that $mv=$constant. The quantities q and v may be the initial data, v being a known function of m. Then (62), (63) show the division into the instantaneous positive and negative waves. Now $q=u-v$ makes the negative wave disappear; and then (62) shows that both q and v travel at speed u. This is Rankine's wave again.

Modification of the Rankinian to make a Compressed Maxwellian Wave.

§ 509. Having got back to the solitary wave, we can now introduce the electromagnetic part. We must first abolish $p-p_0$ and substitute the Maxwellian pressure of radiation, say p_r. Or else, suppose that $p-p_0$ is p_r. Or equivalently, put $p_0=0$, and then let p itself stand for the Maxwellian pressure. Say

$$p=\mu H^2=2T, \tag{64}$$

where H is magnetic force transverse to the direction of motion of the wave. Let there be also E likewise transverse, and perpendicular to H. If the intensities are connected by $\mathrm{E}=\mu v\mathrm{H}$, we have a plane electromagnetic wave. It is the same as $\mathrm{B}=\mu v\mathrm{D}$, if $\mathrm{B}=\mu\mathrm{H}$, $\mathrm{D}=c\mathrm{E}$. We also have $\mathrm{U}=\mathrm{T}$. Considering the state of things in a thin plane slab, the connections between the electric and magnetic quantities and the quantities q, m, v, u in the Rankinian wave are definite. But in order that the electromagnetic wave should keep up with the wave of compression, not only must v be the same in both, that is,

$$\mu c v^2=1, \qquad \text{as well as} \qquad v^2=\frac{dp}{dm}, \tag{65}$$

but we require μv and cv to be constant, as well as mv. That is, μ, c and m all vary together, or the compression of the "matter" of the ether (its inertia of the usual kind) involves simultaneous equal compression of the two properties associated with the magnetic induction and the electric displacement. The electromagnetic wave now travels with the Rankinian, at the constant speed u, without change of type.

Remembering that $p=\mu\mathrm{H}^2$, we can derive

$$q=\frac{\mu\mathrm{H}^2}{mv}=\frac{c\mathrm{E}^2}{mv}=\frac{u\mathrm{B}^2}{\mathrm{B}^2+\mu m v^2}=\frac{u\mathrm{D}^2}{\mathrm{D}^2+cmv^2}, \tag{66}$$

showing that D and B may go from 0 to $\pm\infty$, but q can only go from 0 to u. In the limit, $v=0$, or there is no propagation through the medium. The pressure is then the greatest possible, for

$$p_1-p=mv^2 \qquad \text{makes} \qquad p_1=m_0u^2=mvu. \tag{67}$$

This p_1 shows also the limiting value of the sum of the equal electric and magnetic energies. But the limiting values of E and H are zero. For

$$\mathrm{H}^2=\frac{p}{\mu}=\frac{mvq}{\mu}, \qquad \mathrm{E}^2=\frac{p}{c}=\frac{mvq}{c}, \tag{68}$$

and the ratios m/μ and m/c are constant, whilst $q=u$ makes $v=0$.

An important relation is

$$\mathrm{T}_0 v=\mathrm{T}q. \tag{69}$$

The complete flux of energy is

$$\mathrm{W}=2(\mathrm{T}+\mathrm{T}_0)u=2\mathrm{T}\frac{u^2}{v}=\frac{pu^2}{v}. \tag{70}$$

This tends to ∞ as q tends to u, making v tend to 0. It is the energy of motion that causes this (as in the Rankinian wave), because m has no limiting value. Considering this, it is striking that the electric and magnetic energies should have no similar tendency, but tend to a finite limit.

The above union of the compressed electromagnetic wave and the compressed etherial wave might be suspected to be merely formal, due to superposition or coincidence. But it is more than that. The dynamics of the electromagnetic wave necessitate the mechanical force $-(2\text{T})'$, and this is precisely what is required in the etherial wave of compression.

The electromagnetic equation of activity has also to be harmonised with the above. It is (*El. Pa.*, v. 2., p. 546, equation (121)),

$$\text{conv}\{\text{VEH}+(\text{U}+\text{T}+p)\text{q}\}=\dot{\text{U}}+\dot{\text{T}}+\text{Fq}+\frac{\delta}{\delta t}(\text{U}_c+\text{T}_\mu), \quad (71)$$

where p is the pressure operator, and **F** the corresponding force. The last terms are required because $\delta c/\delta t$ and $\delta\mu/\delta t$ do not vanish here. In a plane wave this becomes

$$-\{\text{EH}+(\text{U}+\text{T}+p)q\}'=\dot{\text{U}}+\dot{\text{T}}-p'q+\frac{\text{U}}{c}\frac{\delta c}{\delta t}+\frac{\text{T}}{\mu}\frac{\delta\mu}{\delta t}, \quad (72)$$

and now p is a pure pressure, and $\text{F}=-p'$.

Putting in the values $\text{EH}=pv$; $\text{U}+\text{T}=p$; $q+v=u$, constant; $\mu v=$ constant, there is a perfect satisfaction But for the full interpretation of the equation, especially as regards the terms depending upon the variation of c and μ, I must say a few words about the stress or pressure operator, and its significance in electromagnetics. I have already written much about it. What is now wanted concerns the concrete application at present in question.

Concerning Riemann's solutions (57), (58) above, it may be shown that if the operands $q\pm\int vd\lambda$ are replaced by the electromagnetic $\text{E}\pm\mu v\text{H}$, the solutions thus obtained are the corresponding ones in electromagnetics, but *only* when μ and c vary as the density of the ether, not in general.

The Waste of Energy from a Moving Electron.

(*Nature*, Oct. 30, 1902, p. 6; Nov. 6, 1902, p. 32.)

§ 510. The subject of the dynamics of a moving charge being of considerable interest now, I have thought the following may be useful. I have shown [§193, 497, &c.], that a charge Q on a sphere of radius a, when suddenly jerked into motion at speed u, generates a spherical electromagnetic shell of depth $2a$, in which the magnetic force H tends to the value given by

$$2a\mathrm{H}=\frac{\mathrm{Q}}{4\pi\mathrm{R}}\frac{u\sin\theta}{1-\frac{u}{v}\cos\theta},\tag{1}$$

when R, the distance from the initial centre of Q, is great. Along with this H, we have perpendicular electric force in the shell, according to $\mathrm{E}=\mu v\mathrm{H}$, or vectorially, $-\mathbf{E}=\mathrm{V}\mathbf{v}\mathbf{B}$, if $\mathbf{v}$ is the vector velocity of the shell. The angle θ is that between $\mathbf{u}$ and $\mathbf{R}$.

The energy wasted by this shell equals the energy left behind, that is, $\mathrm{U}-\mathrm{U}_0+\mathrm{T}$, if U_0 is the initial, U the final electric energy in the field, and T the final magnetic field energy. On its first formation **H** and **E** in the shell are different; they then include in accumulated form all the **H** and **E** which are left behind by the shell as it expands.

The applied force impulse follows from my formula for the force on the ether, viz.,

$$\mathbf{F}=\frac{d}{dt}\mathrm{V}\mathbf{D}\mathbf{B}$$

per unit volume. Denoting the time-space integral by **M**, then

$$\mathbf{M}=\mathbf{M}_1+\mathbf{M}_2,$$

where $\mathbf{M}_2$ belongs to the shell ultimately, and is lost, whilst $\mathbf{M}_1$ is left behind in the field. We have

$$\mathrm{T}=\tfrac{1}{2}\mathrm{M}_1u,\qquad\text{and}\qquad\mathrm{U}-\mathrm{U}_0=\tfrac{1}{2}\mathrm{M}_2u;$$

so that altogether,

$$\tfrac{1}{2}\mathrm{M}u=\mathrm{U}-\mathrm{U}_0+\mathrm{T}.\tag{2}$$

Both $\mathbf{M}_1$ and $\mathbf{M}_2$ are parallel to $\mathbf{u}$.

If, now, a second impulse acts, changing the velocity from $\mathbf{u}_1$ to $\mathbf{u}_2$ say, another spherical shell is generated. Disregarding the part left behind, (1) above shows that the magnetic force in it is given by

$$2a\mathrm{H}=\frac{\mathrm{Q}}{4\pi\mathrm{R}}\left(\frac{u_2\sin\theta}{1-\frac{u_2}{v}\cos\theta}-\frac{u_1\sin\theta}{1-\frac{u_1}{v}\cos\theta}\right),\tag{3}$$

when the direction does not change. More generally, substitute the vector change in the quantity on the right side of (1) properly vectorised. Then the change in θ will be allowed for as well.

The energy lost in this second shell may be calculated by (3). It amounts to

$$\left\{\frac{u_2 P_2 - u_1 P_1}{u_2 - u_1}\left(1 - \frac{u_1 u_2}{v^2}\right) - P_0\right\}Q, \tag{4}$$

where P is the potential function

$$P = \frac{Q}{4\pi a c}\left(1 + \frac{1}{3}\frac{u^2}{v^2} + \frac{1}{5}\frac{u^4}{v^4} + \dots\right), \tag{5}$$

investigated by Searle and Morton. Take $u = 0$, u_1, and u_2, to obtain P_0, P_1, P_2.

It may be shown that the substitution of two impulsive changes in the same direction for a single one reduces the waste; that is, the one impulse u_2 wastes more energy than the two successive impulses u_1 and $u_2 - u_1$. In fact, the saving is great, and ten equal partial impulses in succession waste not much more than one-tenth part of that wasted by a single impulse of size equal to their sum. There is a residuum, however, and that is what appears as continuous waste when u varies continuously.

When Δu is small, by (3),

$$2aH = \frac{Q}{4\pi R}\frac{\sin\theta\,\Delta u}{\left(1 - \frac{u}{v}\cos\theta\right)^2}, \tag{6}$$

and now the waste of energy in the shell wave corresponding to Δu is, by (4),

$$\frac{\mu Q^2}{12\pi a}\frac{(\Delta u)^2}{\left(1 - \frac{u^2}{v^2}\right)^2}. \tag{7}$$

The magnetic force in the above shells is uniform in the depth of the shell, when the impulse acts strictly at the front of a shell. But if Δu be distributed uniformly over the time $2a/v$, the shell will be doubled in depth, and H will rise at uniform rate from 0 to the same full value in the middle of the shell, and then fall similarly to zero in the second half. Now if a second Δu acts in the same way, beginning as soon as the first Δu has made H reach full strength, H will continue of that full strength. And so on with a third Δu. Finally, if $2a = v\Delta t_1$, and $\Delta u/\Delta t_1$ is steady, and allowing for the variable depth of the shell according to (11) below, we come to

$$\Delta t_1 \frac{\mu Q^2}{6\pi v}\left(\frac{\Delta u}{\Delta t_1}\right)^2 \frac{1}{\left(1-\frac{u^2}{v^2}\right)^3}, \tag{8}$$

to represent the waste in time Δt_1. Or, if W is the rate of waste,

$$W = \frac{\mu Q^2}{6\pi v}\left(\frac{du}{dt_1}\right)^2 \frac{1}{\left(1-\frac{u^2}{v^2}\right)^3}. \tag{9}$$

This holds when the acceleration and the velocity are parallel. By the manner of construction, it is necessary that du/dt_1 should not vary sensibly in the time taken by light to traverse the diameter $2a$.

By a fuller analysis, allowing for change of direction of motion, I find that the waste of energy per second from a charge Q with velocity **u** and acceleration **A** is

$$W = \frac{\mu Q^2 A^2}{6\pi v} \frac{1-\frac{u^2}{v^2}\sin^2\theta_1}{\left(1-\frac{u^2}{v^2}\right)^3}, \tag{10}$$

where θ_1 is the angle between the velocity and the acceleration (absolute). The dimension a does not appear. W is the same for any size, subject to the restriction mentioned. The smaller a the better, of course. It is exactly true with $a=0$, only then the motion would be impossible.

This calculation of the waste may be confirmed by following up my investigation of the electric and magnetic field by the method I gave in 1889 ("Electrical Papers," vol. 2, p. 504).

The waste is greatest when the velocity and acceleration are parallel, and least when perpendicular. There is another reservation, viz., u must be less than v. If not, special treatment is required, after the manner I have already published.

The meaning of waste is this. When Q moves through the distance udt_1, it casts off a spherical shell of depth

$$vdt_1\left(1-\frac{u}{v}\cos\theta\right), \tag{11}$$

and the energy of this shell when it has gone out to an infinite distance is Wdt_1.

When at a finite distance, **E** and **H** in this elementary shell are given by

$$\mathbf{E} = \mathbf{E}_1 + \mathbf{E}_2, \qquad \mathbf{H} = \mathbf{H}_1 + \mathbf{H}_2, \qquad \mathbf{H}_1 = \mathrm{V}\mathbf{u}\mathbf{D}_1, \qquad \mathbf{H}_2 = \mathrm{V}\mathbf{v}\mathbf{D}_2, \tag{12}$$

$$\mathbf{E}_1 = \frac{Q}{4\pi R^3 c}\,\frac{\left(\mathbf{R} - \frac{\mathbf{u}}{v}R\right)\left(1 - \frac{u^2}{v^2}\right)}{\left(1 - \frac{u}{v}\cos\theta\right)^3}, \tag{13}$$

$$\mathbf{E}_2 = -\frac{\mu Q}{4\pi R}\left\{\frac{\mathbf{A}}{\left(1 - \frac{u}{v}\cos\theta\right)^2} - \frac{A\left(\mathbf{R} - \frac{\mathbf{u}}{v}R\right)\cos\phi_1}{R\left(1 - \frac{u}{v}\cos\theta\right)^3}\right\}. \tag{14}$$

Here the part $\mathbf{E}_1$, $\mathbf{H}_1$ belongs to the steady travelling state of steady **u**, whilst the other part $\mathbf{E}_2$, $\mathbf{H}_2$ is electromagnetic, and represents the waste. The angle between the acceleration **A** and **R** is ϕ_1. The waste part has $\mathbf{E}_2$, $\mathbf{H}_2$ tangential, that is, perpendicular to **R**. $\mathbf{H}_1$ is also tangential to the sphere, but $\mathbf{E}_1$ is radially directed from the point which Q would reach at the moment in question (belonging to the sphere R) if it were not accelerated at all. This means the steady travelling state (see "Electrical Papers," vol. 2, p. 511, equation (29)).

There is another way of treating the question, viz., by the vector and scalar potentials. The vector potential of the impressed current Q**u** is not Q**u**/4πR, but *(loc. cit.)*

$$\mathbf{A} = \frac{Q\mathbf{u}}{4\pi R\left(1 - \frac{u}{v}\cos\theta\right)}. \tag{15}$$

This is referred to origin at the virtual position of the charge, not the actual. The actual is best for the steady state, the virtual to show the waves emitted. The factor

$$\{1 - (u/v)\cos\theta\}^{-1}$$

expresses the Doppler effect. Divide by *u*c to obtain the scalar potential Φ. Then

$$\mathbf{H} = \operatorname{curl}\mathbf{A}, \qquad \mathbf{E} = -\mu\dot{\mathbf{A}} - \nabla\Phi,$$

in Maxwell's manner. The trouble here is the differentiations, which require great care, since **u**, **R** and θ all vary in a rather complicated way as Q moves. The relations (12) exhibit the field clearly.

For an infinitely small sphere of Q, the energies in the shell at distance R corresponding to the displacement udt_1 of Q are

$$T = T_1 + T_2 + 2T_{12}, \qquad U = U_1 + U_2 + 2U_{12},$$

where $_1$ relates to the $\mathbf{E}_1$, $\mathbf{H}_1$ part, and $_2$ to the other part, whilst $_{12}$ refers to the mutual energy. They are connected thus:

$$U_2 = T_2, \quad U_{12} = T_{12}, \quad U_1 = T_1 + \frac{Q^2 v dt_1}{8\pi R^2 c}, \tag{16}$$

$$T_1 = \frac{Q^2 v dt_1}{12\pi R^2 c} \frac{u^2/v^2}{\kappa^2}, \qquad T_{12} = \frac{\mu Q^2 A}{12\pi R} \frac{udt_1 \cos\theta_1}{\kappa^4}, \tag{17}$$

$$T_2 = \frac{\mu Q^2 A^2 dt_1}{12\pi v} \frac{1 - \frac{u^2}{v^2}\sin^2\theta_1}{\kappa^6}, \tag{18}$$

where $\kappa^2 = 1 - u^2/v^2$.

The corresponding "momenta," or force-impulses, say $\mathbf{M}_1 = \Sigma \mathrm{V}\mathbf{D}_1\mathbf{B}_1$, $\mathbf{M}_2 = \Sigma \mathrm{V}\mathbf{D}_2\mathbf{B}_2$, $\mathbf{M}_{12} = \Sigma \mathrm{V}\mathbf{D}_1\mathbf{B}_2$, $\mathbf{M}_{21} = \Sigma \mathrm{V}\mathbf{D}_2\mathbf{B}_1$, are given by

$$\mathbf{M}_1 = \frac{2T_1}{\mathbf{u}}, \quad \mathbf{M}_2 = \frac{2T_2\mathbf{u}}{v^2}, \quad \mathbf{M}_{21} = \frac{2T_{12}\mathbf{u}}{v^2}. \tag{19}$$

These are all parallel to $\mathbf{u}$. But $\mathbf{M}_{12}$ is not, though it is in the plane of $\mathbf{u}$ and $\mathbf{A}$. Its components parallel to $\mathbf{u}$ and to $\mathbf{A}$ are

$$\frac{2T_{12}}{u}, \quad \text{and} \quad \frac{2T_{12}}{u} \frac{1 - \frac{u^2}{v^2}\sin^2\theta_1}{\cos\theta_1}. \tag{20}$$

With the previous restriction, these are independent of the size of the sphere of Q. But to obtain exact formulæ without this restriction, either a very difficult integration must be effected over the surface of the sphere of Q, every element of which will usually have (effectively) a different velocity and acceleration, on account of the Doppler effect, or we may derive the resulting formulæ by a differentiating operator. Thus, for example, exhibiting it for Φ only, let Φ_0 be the formula when $a = 0$, then the real Φ is, by a previous investigation,

$$\Phi = \frac{\operatorname{shin} qa}{qa} \Phi_0, \tag{21}$$

outside the sphere, and

$$\Phi = \frac{\operatorname{shin} qR}{qR} \Phi_{0a}, \tag{22}$$

inside the sphere, where q is the differentiator $d/d(vt)$, and Φ_{0a} is the common value of both Φ's at $R=a$. But this t is not the same as the previous t_1; it is the corresponding value; the place where the differentiations are performed is at the end of R. The differentiations are troublesome. Thirdly, we may calculate the time-integral of Φ_0, and then apply Taylor's theorem. Nearly all the trouble in the electronic theory is connected with the necessity of making a finite to have finite energy (though this does not apply to the waste) and finite moving forces, with the consequent resulting two superposed waves, one outward from the surface of Q, the other inward, and then outward again. The results for impulses work out easily enough, but not for continuous accelerations.

Details of the above will be published in vol. 3 of "Electromagnetic Theory" (and perhaps elsewhere), which is, as the advertisement says, "in preparation."

Returning to the waste formula, an electron revolving in a circular orbit of radius r has $\theta_1 = \frac{1}{2}\pi$, and $u^2/r = A$. So we want an applied force along **u** varying as u^3 to maintain the motion, since the waste varies as u^4. This revolving electron has sometimes been supposed to be a circular current. But it is really a vibrator. The free path followed under decay of energy without fresh supply would perhaps be difficult to follow completely. It is rather hard for the "explanation" of magnetism.

The kinetic energy of molecules is the natural source of the radiation, but the connection between them and the electrification is very obscure, and how the electrons get knocked off is harder still, and what they are is hardest of all. Larmor thinks they run through the ether like knots on a string. If they do, as they may, *how* do they do it? Connections are wanted.

Sound Waves and Electromagnetics. The Pan-potential.

(*Nature*, Jan. 1, 1903, p. 202.)

§ 511. The photographs taken some years ago by Prof. Boys of flying bullets showed the existence of a mass of air pushed along in front of the bullet. Is there anything analogous to this in the electromagnetics of an electron? Suppose, for example, that an electron is jerked away from an atom so strongly that its speed exceeds that of light. Then it will slow down by reason of the resisting force to which I have shown it is subjected. So long as its speed is greater than that of light, it is accompanied in its motion by a conical wave. The question is whether there is any disturbance ahead of the electron, close to it, as in the case of a bullet moving through the air. It is a question of fact, not of theory. When Maxwell's theory shows that there is no disturbance in front of the electron, that is only because it is virtually assumed to be so at the beginning, by the assumption that the ether continues fixed when the electron traverses it.

Apart from this detail, the analogy between the conical sound wave and the conical electromagnetic wave is interesting in connection with C. A. Bjerknes's theory of pulsations in a liquid, as developed by V. Bjerknes's in his "Vorlesungen über hydrodynamische Fernkräfte nach C. A. Bjerknes' Theorie." The liquid is incompressible, and is set into a pulsating state by pulsating sources, and the result shows remarkable analogies with electric and magnetic phenomena when they are static.

Now, if the liquid is compressible, the results must be approximately the same provided the pulsations are not too quick. But if very rapid, and the compressibility be sufficient to lower the speed of propagation sufficiently, new phenomena will become visible with pulsating sources, like sound waves, and the question is how far they are analogous to electromagnetic phenomena?

Here, for example, is an interesting case. Let f be the density of the source, such that (if $q = d/d(vt)$)

$$(\nabla^2 - q^2)\mathrm{V} = -f \qquad (1)$$

is the characteristic of the velocity potential V, so defined that $-\nabla\mathrm{V}$ is the velocity. Then f signifies the amount of fluid (unit density) generated per unit volume per second and diverging outward. Then, for a point source of strength Q, the V it produces is

$$\mathrm{V} = \frac{\epsilon^{-qr}}{4\pi r}\mathrm{Q} = \frac{\mathrm{Q}_{t-r/v}}{4\pi r}, \qquad (2)$$

at distance r. This is equivalent to Rayleigh's account of Helmholtz's spherical waves from a centre ("Theory of Sound," vol. 2), except in the interpretation of f or Q, which I do not altogether understand in that work.

Q is a fluctuating function of the time in the above in the acoustic application, though, of course, fluctuation is not necessary in the ideal theory. Now if the source Q moves through the air with velocity **u**, the potential becomes

$$V = \frac{Q}{4\pi r\{1 - (u/v)\cos\theta\}}, \qquad (3)$$

if θ is the angle at Q between **r** and **u** at the proper moment.

This equation therefore expresses the theory of a very small pulsating source moved through the air, and is so far very like that of an electric charge Q (which does not pulsate) moved through the ether. The analogy does not continue in details, when, for example, we compare velocity with electric displacement. The electromagnetic theory is more involved.

When u exceeds v, equation (3) is no longer the complete solution. If u is less than v, there is just one and only one position of Q at a given moment where it is, so to speak, in communication with P, the point where V is reckoned. But when $u > v$, there may be just one point, or two, or there may be any number. Thus, if the source Q starts at moment $t = 0$ from a certain point, and then moves steadily in a straight line, the wave front is conical, with a spherical cap, or spherical, with a conical spike, Q being at the apex. If P is inside the sphere, there is only one position for Q. But if P is inside the cone, there are two. The value of V at P is the same for both, given by (3), reckoned positive always. So the real V at P is double as much.

If the speed varies, the values of u will usually be not the same in the two positions, so the two partial V's must be separately reckoned. But the speed and path may vary in such a way that there are more than two positions of Q which are the centres of waves which all arrive at P at the same moment.

When there are any number of electrons moving about in given paths, the following will give a broad idea of the nature of the problem. To find V at a fixed point P at the moment t. Let at that moment a spherical surface expand from P at speed v, not forward in time, but backward. In expanding from radius 0 to ∞, it will cross the electrons one after another. Take note of the times of passage, t_1, t_2, &c. (less than t), of the charges, and their velocities. Then

$$V = \Sigma \frac{Q}{4\pi R\{1 - (u/v)\cos\theta\}}, \tag{4}$$

where $R_n = v(t - t_n)$, and θ_n is the angle at Q_n between R_n and $\mathbf{u}_n$. Similarly as regards the vector-potential.

When u is allowed to exceed v, the effect is to increase the number of crossings of electrons. An electron crossed twice counts as two electrons.

The value of $\dot{t}_n$ is given by

$$\dot{t}_n = \frac{1}{1 - \frac{u_n}{v}\cos\theta_n}.$$

The vector $\mathbf{u}_n$ is the real velocity of Q_n at the moment t_n. Its apparent velocity, as viewed from P at the moment t, is $\mathbf{u}_n\dot{t}_n$ or $-\dot{\mathbf{R}}_n$. It has no necessary resemblance to the real velocity, and may be positive or negative. The dot here signifies differentiation to t at P.

Talking of potentials, I am tempted to add a few words about their King, the Pan-potential. In equation (1) above, let q be not $d/d(vt)$, but any sort of complex time-differentiator, for example, if $p = d/dt$,

$$q^2 = (k + cp)(g + \mu p),$$

which is the special form for electromagnetic waves in a conductor. Then (2) is still the solution for a point source, and in general,

$$V = \Sigma \frac{\epsilon^{-qr}f}{4\pi r} = \text{pan} f \tag{5}$$

is the pan-potential due to the distributed source f. It is not the complete solution, because ϵ^{qr} has not been counted, but that is not wanted when there is no barrier to reflect.

For instance, if **C** is *impressed* electric current, in a conductor, the characteristic of **H**, magnetic force, is

$$(\nabla^2 - q^2)\mathbf{H} = -\text{curl}\,\mathbf{C}. \tag{6}$$

It follows by the above that

$$\mathbf{H} = \text{pan curl}\,\mathbf{C}; \tag{7}$$

that is, the magnetic force is the pan-potential of the curl of the impressed current. The operations pan and curl are interchangeable, so

$$\mathbf{H} = \text{curl pan } \mathbf{C}, \tag{8}$$

that is,

$$\mathbf{H} = \text{curl } \mathbf{A}, \qquad \text{if} \qquad \mathbf{A} = \text{pan } \mathbf{C}. \tag{9}$$

(Similarly, ∇ pan = pan ∇, and div pan = pan div.)

I worked out this problem for a fixed point source of impressed current some time ago ("Electrical Papers," vol. 2, p. 432) without reference to the pan-potential. The operational solution there given, equation (258), represents either (7) or (8). The algebrisation was also done. There is no advantage in using the **A** function in this particular case; it is, in fact, more difficult to find **A** first and then derive **H** than to obtain **H** without **A**. Similarly, as regards **E**, the electric force. The second circuital law derives it from the **H** equation, so that it is not required to introduce Φ to supplement **A**.

If the point-source is in motion, the pan-potential requires dopplerisation as well as the ordinary potential. But this does not require explicit representation for continuously distributed sources. For example, the electromagnetic circuital equations

$$\text{curl}\,(\mathbf{H} - \mathbf{h}) = \mathbf{u} \text{ div } c\mathbf{E} + (k + cp)\mathbf{E}, \tag{9}$$

$$\text{curl}\,(\mathbf{e} - \mathbf{E}) = \mathbf{w} \text{ div } \mu\mathbf{H} + (g + \mu p)\mathbf{H}, \tag{10}$$

where **u**, **w**, **e**, **h**, are functions of position and time, have the solutions

$$\mathbf{E} - \mathbf{e} = \text{pan } \mathbf{X}, \qquad \mathbf{H} - \mathbf{h} = \text{pan } \mathbf{Y}. \tag{11}$$

To prove this, and determine the nature of **X** and **Y**, it suffices to put the characteristics of $\mathbf{E} - \mathbf{e}$ and $\mathbf{H} - \mathbf{h}$ in the form (1), q^2 having the more general later meaning. Now (9) and (10) lead to

$$(q^2 - \nabla^2)(\mathbf{E} - \mathbf{e}) = -\nabla\rho - \text{curl } \mathbf{G} - (g + \mu p)\mathbf{C}, \tag{12}$$

$$(q^2 - \nabla^2)(\mathbf{H} - \mathbf{h}) = -\nabla\sigma + \text{curl } \mathbf{C} - (k + cp)\mathbf{G}; \tag{13}$$

where

$$\rho = \text{div}(\mathbf{E} - \mathbf{e}), \qquad \sigma = \text{div}(\mathbf{H} - \mathbf{h}), \tag{14}$$

$$\mathbf{C} = \mathbf{u} \text{ div } c\mathbf{E} + (k + cp)\mathbf{e}, \tag{15}$$

$$\mathbf{G} = \mathbf{w} \text{ div } \mu\mathbf{H} + (g + \mu p)\mathbf{h}. \tag{16}$$

So **X** and **Y** are the right members of (12) and (13) as defined. **C** is the impressed electric current, **G** the impressed magnetic current. It will be seen that no separate determination of scalar potentials is required, because they are already included in X and Y.

The Radiation from an Electron describing a Circular Orbit.

(*Nature*, Jan. 28, 1904, p. 293.)

§ **512.** The complete formula for the radiation may be useful to some of those who are now indulging in atomic speculations. It is derived from the general formula I gave a year ago in *Nature* (Oct. 30, Nov. 6, 1902, pp. 6, 32), expressing the electromagnetic field everywhere due to an electron moving anyhow. Put in the special value of R required, which is a matter of elementary geometry, and the result is the complete finite formula. But only the part depending on R^{-1} is required for the radiation; and, in fact, we only want the r^{-1} term (if $r =$ distance from the centre of the orbit), if the ratio of the radius of the orbit to the distance is insensible; and that, of course, is quite easy, on account of the extreme smallness of electronic orbits. The magnetic force is given by

$$H_\phi = \frac{Qun}{4\pi rv}\alpha^3 \cos\theta \cos\phi_1, \qquad (1)$$

$$H_\theta = \frac{Qun}{4\pi rv}\alpha^3(\sin\phi_1 - \beta), \qquad (2)$$

subject to

$$\alpha = \frac{1}{1-\beta\sin\phi_1}, \qquad \beta = \frac{u}{v}\sin\theta, \qquad (3)$$

$$\phi_0 = \phi_1 + \beta\cos\phi_1 = \phi - nt + nr/v. \qquad (4)$$

There is no limitation upon the size of u/v, save that it must be less than 1. But there is a limitation regarding the acceleration. If the change in the acceleration is sensible in the time taken by light to traverse the diameter of the electron, it will sensibly alter the results. The size of the electron itself will then have to be considered. But this is very extreme.

To explain the symbols: The (surface) charge is Q moving at speed u and angular speed n in a circle in the plane perpendicular to the axis from which θ is measured. It revolves positively round this axis, and its position when $t=0$ is $\phi=0$. Also, r, θ, ϕ are the usual spherical coordinates of the point of observation, and H_ϕ, H_θ are the ϕ and θ components of the magnetic force at that point at the moment t. The coefficient α^3 shows the Doppler effect on H. The difference between ϕ_0 and ϕ_1 must be noted.

It will be readily seen what an important part the Doppler effect plays if, as has been sometimes assumed, subatomic motions of electrons involve values of u which are not insensible fractions of v. For instance, in the plane of the orbit, $H_\phi = 0$, and

$$\dot{H}_\theta = \frac{Qun}{4\pi rv} \frac{\sin\phi_1 - u/v}{[1-(u/v)\sin\phi_1]^3}. \qquad (5)$$

The effect is to compress H in one half and expand it in the other half of a period, with corresponding strengthening and weakening of intensity, and also with a shifting of the nodes towards the compressed part. When u/v is made large, there is a great concentration at $\phi_1 = \phi_0 = \frac{1}{2}\pi$, $2\frac{1}{2}\pi$, $4\frac{1}{2}\pi$, &c., with only a weak disturbance of opposite sign between them. That is, there is a tendency to turn the original simply periodic vibration into periodic pulses, which become very marked as u increases towards v. The radiation of energy is very rapid. It involves (*l.c.*) the factor $(1-u^2/v^2)^{-2}$. This becomes so great as seemingly to shut out the possibility of more than momentary persistence of revolution. But there might be a solitary partial revolution, or nearly complete, in cometary fashion, which would generate a single pulse, if there cannot be a sequence of several at speeds nearly equal to that of light.

Three suggestions have been made about the X-rays. Röntgen suggested a longitudinal ether disturbance. This has not found favour, because it requires a new theory of electricity. Schuster suggested very rapid vibrations. This is tenable, because in the inside of an atom rudimentary calculations show that vibrations much more frequent than light are easily possible with revolving electrons. Stokes suggested collisional pulses. This is tenable, too, for the collisions must produce electromagnetic pulses. I think X-rays are mixed Stokes pulses and Schuster vibrations, the latter arising from the atoms of the body struck. Now, a pulse is not the same as a continued vibration, though it may be analysed into the sum of various sorts of continued vibrations, just as the distorted simply periodic vibration in (5) above may be. There ought, then, to be a physical difference between the effects of collisional pulses and continued very rapid vibrations.

Apart from the emission of electrons and matter, there might be six sorts of radiation at least; say, light vibration, below light, above light, collisional pulses, cometary pulses, and possibly periodic pulses. The last may have to be excluded for the reason mentioned. The cometary pulses would resemble the collisional pulses, though less dense. The above-light vibrations need not require u/v to be more than a small fraction, though even then their maintenance is a difficulty. They require renewal again and again, perhaps in a collisional manner.

There is a good deal to be found out yet in the relations of electricity to matter. There is also sometimes a good deal of misconception as to the relations of theory to fact. A purely dynamical theory of electricity, like Maxwell's, can give no information about the connection between electricity and matter. For example, Zeeman's experiment, as interpreted by Lorentz, brought out the striking fact that it was the negative electricity that revolved, not seemingly the positive, and the fact harmonises with J. J. Thomson's negative corpuscles. Theory could never predict such a fact, because it is not in the theory. It could not be there, because it has no dependence upon the dynamics of electricity in the theory. The same may be said of various other new facts much discussed of late. Now, though the theory cannot predict such facts, it is useful, of course, as a guide in framing hypotheses to account for the new facts, for it is no use flying in the face of solid theory. Whether the solid theory itself (not meaning that the ether is solid) will need to be altered remains to be seen. There is no sign of it yet, though I cannot believe the etherial theory is complete.

To analyse the dopplerised vibrations expressed by (1), (2) into simply periodic vibrations seemed to involve very complicated work at first, save just for two or three terms. But there is a trick in it, which, when found, allows the complete expansions to be developed in a few lines. First show that (this is the trick)

$$\left.\begin{aligned} a^3 \cos \phi_1 &= -\frac{d^2}{d\phi_0^2} \cos \phi_1, \\ a^3 (\sin \phi_1 - \beta) &= -\frac{d^2}{d\phi_0^2} \sin \phi_1, \end{aligned}\right\} \qquad (6)$$

Next, by the theorem known as Lagrange's, sin ϕ_1 can be at once put in the form of a series involving the derivatives of various powers of cos ϕ_0. Do not find the derivatives from them, but put $\cos^n \phi_0$ in terms of the sum of first powers of cosines by the well-known circular formula. The full differentiations, not forgetting those in (6), may then be done at sight in one operation. The result is

$$a^3(\sin\phi_1-\beta)=\sin\phi_0-\beta\,.\,2\cos 2\phi_0-\tfrac{3}{8}\beta^2(9\sin 3\phi_0+\sin\phi_0)$$
$$+\tfrac{4}{3}\beta^3(4\sin 4\phi_0+\cos 2\phi_0)+\frac{\beta^4}{\lfloor 4}\frac{1}{2^4}(5^5\sin 5\phi_0+5\,.\,3^4\sin 3\phi_0+10\sin\phi_0)$$
$$-\frac{\beta^5}{\lfloor 5}\frac{1}{2^5}(6^6\cos 6\phi_0+6\,.\,4^4\cos 4\phi_0+15\,.\,2^4\cos 2\phi_0)-\dots, \qquad (7)$$

and so on to any extent. Then, to find the other one, differentiate the series in (7) with respect to ϕ_0, and divide the nth term by n. Thus,

$$a^3\cos\phi_1=\cos\phi_0+2\beta\sin 2\phi_0-\frac{\beta^2}{8}(27\cos 3\phi_0+\cos\phi_0)-\dots, \qquad (8)$$

and so on. This analysis of the vibrations is useful in some special developments, but, of course, the original distorted simple vibration is the most significant. In fact, the result of the analysis exhibits the common failing of most series developments that the resultant meaning is not evident.

Another way. Use Bessel's series for the sine and cosine of ϕ_1, and then carry out (6). It is remarkable that the relation between the eccentric and mean anomaly in a planetary orbit should be imitated, for the dynamics is quite different.

When I was a young child I conceived the idea of infinite series of universes, the solar system being an atom in a larger universe on the one hand, and the mundane atom a universe to a smaller atom, and so on. I do not go so far as that now, but only observe that there is a tendency to make the electrons indivisible, and all exactly alike. But they must have size and shape, and be therefore divisible; unless, indeed, they are infinitely rigid. Or they may vary in shape without dividing. There are infinite possibilities in the unknown.

Kaufmann's measurements go to show that the mass of an electron, if there is any, is only a small fraction of its effective electromagnetic mass, although that is not a definite quantity subject to the Newtonian second law. But it is too soon to say that the electron has no mass at all, that is, to be quite sure that negative electricity is absolutely separable from matter, though it seems likely. It would be well to have, if possible, similar measurements made on positive electricity. If permanently attached to matter, it should not exhibit the increased inertia with increased speed in a sensible manner.

The Radiation from an Electron moving in an Elliptic, or any other Orbit.

(*Nature*, Feb. 11, 1904, p. 342.)

§ 513. I have been looking for a tolerably simple way of expressing the radiation at a distance from an electron, to avoid the work involved in reducing the general formulæ (*Nature*, Oct. 30, Nov. 6, 1902) in special cases. The result is

$$|\mathbf{E}| = \frac{\mu Q}{4\pi r}|\ddot{\mathrm{s}}| \sin \gamma, \tag{1}$$

subject to

$$\mathrm{R} = v(t - t_1). \tag{2}$$

Here understand that Q is the charge moving in the path defined by the vector s from the origin at the moment t_1; and E the electric force at the corresponding moment t at the point P at the end of the vector r from the origin, at distance R from Q; and γ is the angle between r and $\ddot{\mathrm{s}}$. That is, the electric force is the tangential part of the vector $-\ddot{\mathrm{s}}\mu Q/4\pi r$, or the part perpendicular to r. The magnetic force is perpendicular to **E**, given by $\mathrm{E} = \mu v \mathrm{H}$. It is assumed that s/R is very small, but no assumption has been made about u/v, so the waves are fully dopplerised. The dot indicates time-differentiation at P.

Example. Elliptic orbit. Let

$$\mathrm{s} = \frac{1}{n}(\mathrm{i}u_2 \cos nt_1 + \mathrm{j}u_1 \sin nt_1). \tag{3}$$

Then Q describes an ellipse in the plane x, y, axes u_2/n and u_1/n, where $n/2\pi$ is the frequency. It is the spring or pendulum kind of elliptic motion. Describe a spherical surface with centre at the centre of the ellipse, and project s upon the surface, and insert the result in (1). Then we get

$$\mathrm{E}_\theta = \frac{\mu Q}{4\pi r n} \cos \theta \frac{d^2}{dt^2}(u_2 \cos \phi \cos nt_1 + u_1 \sin \phi \sin nt_1), \tag{4}$$

$$\mathrm{E}_\phi = \frac{\mu Q}{4\pi r n} \frac{d^2}{dt^2}(u_2 \sin \phi \cos nt_1 - u_1 \cos \phi \sin nt_1), \tag{5}$$

expressing the θ and ϕ components of **E** at the point r, θ, ϕ, if θ is measured from the z axis, and ϕ from the plane z, x.

Yet one thing more. The connection between t and t_1 is

$$nt_1 = n\left(t - \frac{r}{v}\right) + \frac{\sin\theta}{v}(u_2 \cos\phi \cos nt_1 + u_1 \sin\phi \sin nt_1), \qquad (6)$$

which gives

$$\dot{t}_1 = \left\{1 - \frac{\sin\theta}{v}(u_1 \sin\phi \cos nt_1 - u_2 \cos\phi \sin nt_1)\right\}^{-1}, \qquad (7)$$

which is required when (4), (5) are differentiated. This process introduces the factor $\dot{t}_1^3$, and so, at high speeds, converts the radiation into periodic pulses, as in the case of a circular orbit (*Nature*, Jan. 28, 1904, p. 293). Put $u_1 = u_2 = u$ in the present formulæ to reduce to the circular. The analysis to simply periodic vibrations may be done in a similar way. If the motion in the elliptic orbit is of the planetary kind, the equation (3) is replaced by a much less manageable one. Electrons can conceivably vibrate in both these ways, according as the centre of force is condensed positive electricity, or is the centre of diffused positive electricity.

This is not the place for detailed proofs, but I can indicate one way of representing the matter which has some interest apart from the speciality of orbital motion. Given that Q is moving anyhow, it may be shown that my general formula for **E** may be converted to

$$\mathbf{E} = \frac{\mu Q}{4\pi}\ddot{\mathbf{R}}_1 + \frac{\mu Q v}{4\pi R^2}(\dot{\mathbf{R}} - 3\dot{R}\mathbf{R}_1 + v\mathbf{R}_1). \qquad (8)$$

This gives **E** at P, at distance R from Q, and $\mathbf{R}_1$ is the unit vector **R**/R. The centre varies as we shift P, because Q is moving. It is always to be understood that Q and P are at every moment of time uniquely connected when $u < v$. Any value given to t fixes a corresponding value t_1 for Q, and its position as well. This formula (8) is a very curious way of representing **E**, and physically very unnatural. But the form of the first part is such that it leads easily to the radiational formula above given. Reject the second part of **E** in (8), because it varies as R^{-2}. Then carry out d^2/dt^2, and reject the R^{-2} part again. There is left

$$\mathbf{E} = \frac{\mu Q}{4\pi R}(\ddot{\mathbf{R}} - \ddot{R}\,\mathbf{R}_1). \qquad (9)$$

Lastly, put $R = r - s$; then $\ddot{R} = -\ddot{s}$; and if s/r is very small, $\ddot{R} = -\ddot{s}r_1$. So we come to formula (1) above, as required. I hope this will be satisfactory. If not, there are lots of other much more complicated ways of doing the work.

The Principle of Least Action. Lagrange's Equations.

(*Nature*, Jan. 29, 1903, p. 297.)

§ **514.** Whether good mathematicians, when they die, go to Cambridge, I do not know. But it is well known that a large number of men go there when they are young for the purpose of being converted into senior wranglers and Smith's prizemen. Now at Cambridge, or somewhere else, there is a golden or brazen idol called the Principle of Least Action. Its exact locality is kept secret, but numerous copies have been made and distributed amongst the mathematical tutors and lecturers at Cambridge, who make the young men fall down and worship the idol.

I have nothing to say against the Principle. But I think a good deal may be said against the practice of the Principle. Truly, I have never practised it myself (except with pots and pans), but I have had many opportunities of seeing how the practice is done. It is usually employed by dynamicians to investigate the properties of mediums transmitting waves, the elastic solid for example, or generalisations or modifications of the same. It is used to find equations of motion from energetic data. I observe that this is done, not by investigating the actual motion, but by investigating departures from it. Now it is very unnatural to vary the time-integral of the excess of the total kinetic over the total potential energy to obtain the equations of the real motion. Then, again, it requires an integration over all space, and a transformation of the integral before what is wanted is reached. This, too, is very unnatural (though defensible if it were labour saving), for the equation of motion at a given place in an elastic medium depends only upon its structure there, and is quite independent of the rest of the medium, which may be varied anyhow. Lastly, I observe that the process is complicated and obscure, so much so as to easily lead to error.

Why, then, is the P. of L. A. employed? Is not Newton's dynamics good enough? Or do not the Least Actionists know that Newton's dynamics, viz., his admirable Force = Counterforce and the connected Activity Principle, can be directly applied to construct the equations of motion in such cases as above referred to, without any of the *hocus-pocus* of departing from the real motion, or the time integration, or integration over all space, and with avoidance of much of the complicated work? It would seem not, for the claim is made for the P. of L. A. that it is a commanding general process, whereas the principle of energy is insufficient to determine the motion. This is wrong. But the P. of L. A. may perhaps be particularly suitable in special cases. It is against its misuse that I write.

Practical ways of working will naturally depend upon the data given. We may, for example, build up an equation of motion by hard thinking about the structure. This way is followed by Kelvin, and is good, if the data are sufficient and not too complicated. Or we may, in an elastic medium, assume a general form for the stress and investigate its special properties. Of course, the force is derivable from the stress. But the data of the Least Actionists are expressions for the kinetic and potential energy, and the P. of L. A. is applied to them.

But the Principle of Activity, as understood by Newton, furnishes the answer on the spot. To illustrate this simply, let it be only small motions of a medium like Green's or the same generalised that are in question Then the equation of activity is

$$\mathrm{div}\,\mathrm{qP} = \dot{\mathrm{U}} + \dot{\mathrm{T}}; \qquad (1)$$

that is, the rate of increase of the stored energy is the convergence of the flux of energy, which is $-\mathrm{qP}$, if q is the velocity and P the stress operator, such that

$$\mathrm{Pi} = \mathrm{P}_1 = \mathrm{iP}_{11} + \mathrm{jP}_{12} + \mathrm{kP}_{13} \qquad (2)$$

is the stress on the **i** plane. Here qP is the conjugate of Pq.

By carrying out the divergence operation, (1) splits into two, thus,

$$\mathbf{F}\mathrm{q} = \dot{\mathrm{T}}, \qquad \mathbf{G}\mathrm{q} = \dot{\mathrm{U}}. \qquad (3)$$

Here **F** is a real vector, being the force, whilst G is a vector force operator. Both have the same structure, viz., $\mathrm{P}\nabla$; but in **F** the differentiators in ∇ act on P, whereas in **G** they are free, and act on q, if they act at all.

Now when U is given, $\dot{\text{U}}$ becomes known. It contains $\mathbf{q}$ as an operand. Knock it out; then $\mathbf{G}$ is known; and therefore $\mathbf{F}$; and therefore the equation of motion is known, viz.,

$$\mathbf{F}=m\frac{d\mathbf{q}}{dt}$$

where m is the density, or the same generalised eolotropically, or in various other ways which will be readily understood by electricians who are acquainted with resistance operators.

Of course, P becomes known also. So the form of U specifies the stress, the translational force and the force operator of the potential energy. To turn $\mathbf{G}$ to $\mathbf{F}$ is the same as turning $\text{A}(d/dx)$ to $d\text{A}/dx$.

If, for example, the displacement is $\mathbf{D}$, the potential energy is a quadratic function of the nine differentiants, $d\text{D}_1/dx$, &c., of the components. Calling these r_{11}, r_{12}, &c.,

$$\text{U}=\tfrac{1}{2}r_{11}\frac{d\text{U}}{dr_{11}}+\tfrac{1}{2}r_{12}\frac{d\text{U}}{dr_{12}}+\dots, \tag{4}$$

by the homogeneous property. Therefore, since $\dot{r}_{12}=dq_1/dy=\mathbf{i}d\mathbf{q}/dy$,

$$\dot{\text{U}}=\left(\frac{d\text{U}}{dr_{11}}\mathbf{i}\frac{d}{dx}+\frac{d\text{U}}{dr_{12}}\mathbf{i}\frac{d}{dy}+\dots\right)\mathbf{q}=\mathbf{G}\mathbf{q}; \tag{5}$$

therefore, writing P_{21} for $d\text{U}/dr_{12}$,

$$\mathbf{F}=\mathbf{i}\left(\frac{d\text{P}_{11}}{dx}+\frac{d\text{P}_{21}}{dy}+\frac{d\text{P}_{31}}{dz}\right)+\dots \tag{6}$$

$$=\frac{d\mathbf{P}_1}{dx}+\frac{d\mathbf{P}_2}{dy}+\frac{d\mathbf{P}_3}{dz}. \tag{7}$$

It is clear that the differentiants in (4) (which involve the large number 45 of coefficients of elasticity in the general case of eolotropy) are the nine components of the conjugate of the stress operator. Of course, vector analysis, dealing with the natural vectors concerned, is the most suitable working agent, but the same work may be done without it by taking the terms involving q_1, q_2, q_3 separately.

Another expression for $\dot{\text{U}}$ is $\tfrac{1}{2}\mathbf{G}\mathbf{D}$, which shows how to find $\mathbf{F}$ from U directly.

Another claim made for the P. of L.A. is that it leads to Lagrange's equations of motion. That is not remarkable, seeing that both are founded upon Newtonian ideas. I suppose Lagrange's equations can be made to lead to the P. of

L.A. But the practical way of proving Lagrange's form is to derive it immediately from Newton's Principle of Activity. Thus, when there are n independent coordinates x, with velocities v, the kinetic energy T is a homogeneous quadratic function of the v's, with coefficients which are functions of the x's. This makes

$$2\mathrm{T} = v_1\frac{d\mathrm{T}}{dv_1} + v_2\frac{d\mathrm{T}}{dv_2} + \dots ; \qquad (8)$$

therefore

$$2\dot{\mathrm{T}} = \frac{d}{dt}\frac{d\mathrm{T}}{dv_1}v_1 + \frac{d\mathrm{T}}{dv_1}\dot{v}_1 + \dots \qquad (9)$$

But also, by the structure of T,

$$\dot{\mathrm{T}} = \frac{d\mathrm{T}}{dx_1}v_1 + \frac{d\mathrm{T}}{dv_1}\dot{v}_1 + \dots \qquad (10)$$

So by subtraction of (10) from (9),

$$\dot{\mathrm{T}} = \left(\frac{d}{dt}\frac{d\mathrm{T}}{dv_1} - \frac{d\mathrm{T}}{dx_1}\right)v_1 + \dots ; \qquad (11)$$

and therefore, by Newton, the force on x_1 is the coefficient of v_1, and similarly for the rest.

Some people who had worshipped the idol did not altogether see that the above contained the really essential part of the establishment of Lagrange's form, and that the use of the activity principle is proper, instead of *vice versâ*. To all such, the advice can be given, Go back to Newton. There is nothing in the P. of L.A., or the P. of L. Curvature either, to compare with Newton for comprehensive intelligibility and straight correspondence with facts as seen in Nature. It must, however be said that Newton's Third Law is sometimes astonishingly misconceived and misapplied, perhaps because it is badly taught.

The Principle of Activity and Lagrange's Equations. Rotation of a Rigid Body.

(*Nature*, Feb. 19, 1903, p. 368.)

§ 515. Prof. Orr's opening remarks* perhaps indicate that the want of appreciation of Newton's dynamics is even greater than I supposed. My authority for Newton is that stiff but thorough going work, Thomson and Tait. On comparison, I

* [See a letter by Prof. Orr in *Nature*, Feb., 1903. Also another a little later.]

find that Prof. Orr's "some people" seem to overlook the vitally important Third Law, without which there could be no dynamics resembling the reality, and also the remarkable associated scholium "Si æstimetur . . . ," which is of such universal and convenient application, both by practicians and by some theorists. In my short outline of the beginning of the theory of Lagrange's equations, my argument "by Newton," referred to the activity principle.

The example of failure given by Prof. Orr is remarkable in more than one way. If the three coordinates specified the configuration, then the equations of motion would come out in the way indicated. It is clear, therefore, from the failure, that in the concrete example of a rotating rigid body, the coordinates employed, which are the time-integrals of the angular velocities about three moving axes, are not proper Lagrangian coordinates within the meaning of the Act. If we use coordinates which do fix the configuration (Thomson and Tait, § 319), there is no failure.

But it is quite easy to avoid the usual complicated trigonometrical work, and obtain the proper equation of motion by allowing for the motion of the axes. Thus, if **a** is the angular velocity, the angular momentum is

$$\mathbf{i}\frac{dT}{da_1} + \ldots = Aa_1\mathbf{i} + Ba_2\mathbf{j} + Ca_3\mathbf{k},$$

and the torque is its time-differentiant; that is

$$\mathbf{F} = A\dot{a}_1\mathbf{i} + B\dot{a}_2\mathbf{j} + C\dot{a}_3\mathbf{k} + Aa_1\frac{d\mathbf{i}}{dt} + Ba_2\frac{d\mathbf{j}}{dt} + Ca_3\frac{d\mathbf{k}}{dt}.$$

Here **i**, **j**, **k** are unit vectors specifying the directions of the principal axes. They only vary by the rotation, so $d\mathbf{i}/dt = \mathrm{V}\mathbf{a}\mathbf{i}$, &c., and this makes

$$\begin{aligned}\mathbf{F} &= Aa_1(\mathbf{j}a_3 - \mathbf{k}a_2) + Ba_2(\mathbf{k}a_1 - \mathbf{i}a_3) + Ca_3(\mathbf{i}a_2 - \mathbf{j}a_3) + A\dot{a}_1\mathbf{i} + \ldots \\ &= \mathbf{i}\{A\dot{a}_1 - a_2a_3(B-C)\} + \mathbf{j}\{\ldots\} + \mathbf{k}\{\ldots\}.\end{aligned}$$

This exhibits Euler's three well-known equations of motion round the three principal moving axes.

In general, $T = \frac{1}{2}\mathbf{a}\mathrm{M}\mathbf{a}$, when M is a vectorial matrix (or linear vector operator) fixed in the body. Then the momentum is Ma, and the torque is

$$\mathbf{F} = \mathrm{M}\dot{\mathbf{a}} + \dot{\mathrm{M}}\mathbf{a} = \mathrm{M}\dot{\mathbf{a}} + (\mathrm{V}\mathbf{a}\mathrm{M})\mathbf{a}.$$

This allows M to be specified with respect to any axes fixed in the rotating body. Of course, the principal axes are the best. I may refer to my "Electrical Papers," vol. 2, p. 547, footnote, for details of a similar calculation relating to the torque (and activity thereof) produced in an eolotropic dielectric under electric stress.

The following concisely exhibits the necessity of allowing for variation of M, and how it is done in the case of n independent variables:—Let

$$\mathrm{T} = \tfrac{1}{2}\mathbf{v}\mathrm{M}\mathbf{v} = \tfrac{1}{2}\mathbf{p}\mathbf{v}.$$

Then $\mathbf{v}$ is a "vector," or complex of n velocities, and $\mathbf{p} = \mathrm{M}\mathbf{v}$ is the corresponding momentum, whilst M is a symmetrical matrix. By differentiation to t,

$$\dot{\mathrm{T}} = \mathbf{v}(\mathrm{M}\dot{\mathbf{v}} + \tfrac{1}{2}\dot{\mathrm{M}}\mathbf{v}) = \mathbf{F}\mathbf{v}, \qquad \text{(Hamilton)}$$

or

$$\dot{\mathrm{T}} = \mathbf{v}(\dot{\mathbf{p}} - \tfrac{1}{2}\dot{\mathrm{M}}\mathbf{v}) = \mathbf{F}\mathbf{v}, \qquad \text{(Lagrange)}$$

Here $\mathbf{F}$ is the force on the system, in the same sense as $\mathbf{v}$ is the velocity of the system. For $\dot{\mathrm{M}}$ substitute $v(d\mathrm{M}/dx)$, to come to the usual forms by breaking up into n components. But the above are more general, because M may vary independently of x. Activity should be the leading idea.

The Undistorted Cylindrical Wave.

(*Nature*, May 21, 1903, p. 54.)

§ **516**. The receipt of a paper by Prof. H. Lamb, "On Wave Propagation in Two Dimensions" (*Proc.* London Math. Soc., vol. 35., p. 141), stimulates me to publish now a condensation of a portion of a work which will not be further alluded to. I once believed that there could not be an undistorted cylindrical wave from a straight axis as source. But some years ago the late Prof. FitzGerald and I were discussing in what way a plane electromagnetic wave running along the upper side of a plane conducting plate and coming to a straight edge, managed to turn round to the other side. Taking the wave as a very thin plane slab, one part of the theory is elementary. The slab wave itself goes right on unchanged. Now Prof. FitzGerald speculatively joined it on to the lower side of the plate by means of a semi-cylindrical slab wave. I maintained that this could not possibly work, because the cylindrical wave generated at the edge was a complete one, causing backward waves on both sides of the plate. Moreover, it was not a simple wave, for the disturbance filled the whole cylindrical space, instead of being condensed in a slab. It was in the course of examining this question that I arrived at something else, which I thought was quite a curiosity, namely, the undistorted cylindrical wave.

Maxwell's plane electromagnetic wave consists of perpendicularly crossed straight electric and magnetic forces, in the ratio given by $E=\mu v H$. Thinking of a thin slab only, it travels through the ether perpendicularly to itself at speed v, without any change in transit. I have shown that this may be generalised thus. Put any distribution of electrification in the slab, and arrange the displacement **D** in the proper two-dimensional way, as if the medium were non-permittive outside the slab. Then put in **H** orthogonally according to the above-mentioned rule, and the result is the generalised plane wave, provided the electrification moves with the wave. Otherwise, it will break up. Another way is to have the electrification upon fixed perfectly conducting cylinders arranged with their axes parallel to the direction of propagation.

Now the first kind of plane wave has no spherical analogue, obviously. But I have shown that the other kinds may be generalised spherically. Put equal amounts of positive and negative electrifications on a spherical surface, arranged anyhow. Distribute the displacement in the proper way for a spherical sheet, as if constrained not to leave it. Then put in **H** orthogonally, as above. The result constitutes an undistorted spherical electromagnetic wave, provided the electrification moves radially with the wave, and attenuates in density as its distance from the centre increases, in the proper way to suit **E** and **H**. This attenuation does not count as distortion. Similarly, the other sort of generalised plane wave may be imitated spherically by having conical boundaries.

But when we examine the cylinder, there is apparently no possibility of having undistorted waves. For with a simple axial source it is known that if it be impulsive, the result is not a cylindrical impulse, but that the whole space up to the wave front is filled with the disturbance. It is easy to see the reason, for any point within the wave front is receiving at any moment disturbances from two points of the source on the axis, and there is no cancellation. And if the source be on a cylindrical surface itself, producing an inward and an outward wave, the whole space between the two wave fronts is filled with the disturbance.

How, then, is it possible to have an undistorted wave from a straight line source? By not arguing about it, but by showing that it can be done. The reason will then come out by itself. As the solution can be easily tested, it is only necessary to give the results here. Take plane coordinates r and θ. Let the magnetic force be perpendicular to the plane, of intensity H. Let Z be its time-integral, then

$$Z = \frac{\cos \frac{1}{2}\theta}{vr^{\frac{1}{2}}} f(vt - r), \qquad H = \frac{\cos \frac{1}{2}\theta}{r^{\frac{1}{2}}} f'(vt - r), \tag{1}$$

expresses the magnetic field, f being an arbitrary function. Now the displacement **D** is the curl of **Z**. So if E_1 is the radial component of E, and E_2 the tangential component, in the direction of increasing θ, we have the electric field given by

$$E_1 = \frac{-\mu v \sin \frac{1}{2}\theta}{2r^{\frac{3}{2}}} f, \qquad E_2 = \frac{\mu v \cos \frac{1}{2}\theta}{r^{\frac{1}{2}}} f' + \frac{\mu v \cos \frac{1}{2}\theta}{2r^{\frac{3}{2}}} f. \tag{2}$$

The attenuation factor $r^{-\frac{1}{2}}$ in (1) does not count as distortion.

The wave may go either way, and various cases can be elaborated. If the wave is outward, the axis $r = 0$ is the source. The plane $\theta = 0$ is a perfect electric conductor. The electrification is of the same sign on its two sides. Other details may be got from the formulæ.

I give an example to show the not very obvious electrical meaning. Let the infinite plane conductor with the straight edge be one pole of a condenser, and a straight wire placed parallel to the edge, and close to it, be the other pole. Join them by a battery, charging the plate and the wire. Bring the wire right up to the edge, and reduce its magnitude to a mere line. (This is done in order to attain the ideal simplicity of the formulæ.) Take away the battery. Then the electric field is given by

$$cvE_1 = -\frac{\sin \frac{1}{2}\theta}{2r^{\frac{3}{2}}} f_0, \qquad cvE_2 = \frac{\cos \frac{1}{2}\theta}{2r^{\frac{3}{2}}} f_0, \tag{3}$$

where f_0 is a constant, and c is the permittivity.

Finally, discharge the condenser by contact between edge and wire. Then the result at time t later is that outside the cylinder of radius $r=vt$ the above field (3) persists, whilst inside the cylinder there is no E or H. An electromagnetic wave separates these regions. It started from the axis at the moment of contact, and as it expands it swallows up the whole energy of the field, and carries it to infinity. Similarly as regards the charging of the plate, only the "battery" should, to have the same formulæ, be an impressed force acting at the axis, between the edge and the wire. At time t after contact, the electric field is established fully within the cylinder $r=vt$. On its boundary is the impulsive wave which is laying down the remainder. It also, if the contact be instantaneous, wastes an equal amount of energy at infinity.

Similarly, by varying the impressed voltage anyhow with the time, the emission of an arbitrary wave of H results. With a real plate and real wire, the main features would no doubt be the same. The use of the line wire introduces infinite voltage.

What somewhat disguises the electromagnetics is the existence of the steady electric force, or parts thereof, along with the electromagnetic E and H, particularly when f is arbitrary. There is a similar complication in the spherical wave when the total electrification in any thin shell is not zero. There is then an auxiliary internal or external electric force to make continuity.

We cannot have an undistorted wave from a simple line source. But in the example the apparent line source will be found to be a doublet. For the curl of e (impressed force) is the source of the wave. It is double, positive on one side, negative on the other.

Solutions of the type

$$\mathrm{H}=\Sigma\frac{\mathrm{A}r^n \cos(n\theta+a)}{(v^2t^2-r^2)^{n+\frac{1}{2}}}, \tag{4}$$

or the same with r and vt interchanged in the denominator, are not distortionless, save for the solitary term in which $n=-\frac{1}{2}$. The above distortionless cylindrical wave (1) is unique. Prove by the characteristic.

Extension of Kelvin's Thermoelectric Theory.

(*Nature*, May 28, 1903, p. 78.)

§ **517.** Lord Kelvin's thermoelectric theory has always seemed to me to be one of his best works. Since its enunciation the scope of the electric current has been extended, as in Maxwell's theory. It is now the curl of the magnetic force of the field always and everywhere. A corresponding extension of the thermoelectric theory is needed. I do not know whether it has been done, but it may be shortly stated, and contains some striking results. As regards the necessity, the following case will show it plainly. Make up a circuit of two parallel wires of different materials, both thermoelectrically neutral, say one of lead, the other of one of Tait's alloys. The places of thermoelectric force in the circuit are then the terminals. Now send short waves along the circuit, in the way so often done of late years. There need be no current at all in the circuit at one end to pair with that at the other. So there is complete failure of the theory of metallic circuits.

But the needed extension is easily made by following Lord Kelvin's method, and using the enlarged meaning of electric current. Let **e** be the intrinsic voltage per unit length due to reversible thermal action, and let **C** be the current density. Then **eC** is the heat per unit volume absorbed per second, and the second thermodynamic law requires that $\Sigma\, \mathbf{eC}/\theta = 0$, if θ is temperature, the summation to be complete as regards **e** Here **C** may be any circuital current, so $\mathbf{e}/\theta$ is polar; that is, $\mathbf{e} = -\theta \nabla p$, where p is a scalar, the thermoelectric power. In a homogeneous conductor, p is a function of the temperature only, to suit Magnus's results. But it is also a function of the material. In what way is not known, but it shows itself at the junction of different metals. Then p changes, say from p_1 to p_2, so the intrinsic voltage at the junction is $P_{12} = \theta(p_1 - p_2)$. This is the Peltier force from the first to the second metal. So far is all that is necessary for steady currents. But when the current varies, part of it leaves the metals. Now at a metal-air junction, the thermoelectric power falls from p to 0, so there is an additional thermoelectric force $P\mathbf{N}$ or $p\theta\mathbf{N}$ acting outwards, **N** being the unit normal. It is here assumed that the thermoelectric power of air is zero. It does not seem likely that its value is important compared with p in a metal. This $P\mathbf{N}$ multiplied by the current leaving the conductor measures the reversible thermal effect of the boundary. The system is now complete, provided there is no external **e**. But should there be, then it must be counted too, if, for instance, a current is induced in an external conductor. In any, case $e = -\theta \nabla p$ will be valid, with the usual proper interpretation of discontinuities, and the Maxwellian meaning of the current.

It will be sufficient to suppose that $p=0$ outside a circuit of two metals. Then there is the Thomson force in the metals, the Peltier force at the metal junctions, and the metal-air force PN of variable intensity all over the circuit. In the extreme case with which I commenced, there may be only one Peltier force in operation, or even none at all, but just the metal-air force alone. If so, there is reversible evolution of heat at some parts, and absorption at other parts of the boundary.

As regards the application of the second thermodynamic law, it seems to be justified by experimental results with steady currents. I see no reason why it should not be applied to variable currents, even when varying very rapidly. For p is a property of the material and its temperature at any place, and has nothing to do at the moment with what is going on at other places. Yet a reservation is necessary. For the second law results from averages. So there must be some limit to the rapidity with which the current at any spot may vary, if the second law is to be fully valid there.

The Volta contact-force must not be forgotten in connection with the metal-air thermoelectric force. Mr. J. Brown has lately made the Volta force disappear by heating it away in oil. If this is fully confirmed, it perhaps proves that chemical action between the metal and an electrolytic film of moisture is the real source of the energy of the transient Volta current, as Mr. Brown maintains. How will this affect the thermal force? If we allow properly for the change in p in passing through the film from the metal to the air, it seems likely that the thermoelectric effects will be simply superposed upon the Volta effects, because the sources of energy are different. Yet they might have to be combined in some unknown way.

Returning to the steady current in a circuit of two metals, Lord Kelvin showed that the complete intrinsic voltage amounted to $\Sigma p d\theta$. This does not express the real distribution of intrinsic force in the circuit, and seems to have no meaning. But it has a curious interpretation which is of importance in the extended theory. The necessity of the metal-air force is shown in another way. I have shown that the source of **H** in varying states is the curl of e everywhere. Here this is $f = V\nabla p \nabla \theta$. It is zero in a homogeneous conductor, and also at the metal junctions, but has the boundary value $VN\theta\nabla p$, which would represent the source of **H** if there were no metal-air force. But add on the curl of the metal-air force PN or $p\theta N$. It is $-VN\nabla p$; and the sum of the two is $-VNp\nabla\theta$. Now this is also the curl of the fictitious intrinsic force referred to, that is, $p\nabla\theta$ in the metals only. So we come to this striking result, that Lord Kelvin's $\Sigma\, p d\theta$ in the metal circuit alone is a fictitious distribution which not only gives the same steady current as the real distribution of intrinsic force, but also gives the true **E** and **H** everywhere in variable states as well, provided the real intrinsic forces include the metal-air forces along with the Peltier and Thomson forces.

The Pressure of Radiation.

(*Nature*, March 9, 1905, p. 439.)

§ 518. The success of Lebedeff and Nichols and Hull in recognising and measuring the pressure of radiation has aroused much interest in radiation pressure generally, real or apparent. It has some interesting and sometimes somewhat difficult theoretical aspects. In the first place, if the ether is really absolutely at rest (this rigidity is an exceedingly difficult idea), the moving force on it has no activity, and its time-integral VDB can only be called momentum out of compliment. The force becomes active in a moving ether, with interesting consequences not now under examination. The present question is rather how to interpret the pressure of radiation on the assumption of a fixed ether, in the measure of its effects on matter which is either fixed or moving through the ether.

The following is striking in what it proves. Let plane radiation fall flush upon a perfect reflector moving in the same direction at speed u, a case considered by Larmor. Let the energy density be $p = p_1 + p_2$, the incident being p_1, the reflected p_2. Assume, which seems reasonable at first, that p_3, the pressure in the reflector, is zero; then the moving force $p_1 + p_2 - p_3$ reduces to $p_1 + p_2$. Therefore

$$p_1(v-u) - p_2(v+u) = (p_1 + p_2)u, \tag{1}$$

because the left side is the rate of loss of energy from the waves, and the right side the activity of the force on the reflector. So

$$\frac{p_2}{p_1} = \frac{1 - 2u/v}{1 + 2u/v} = s^2, \text{ say;} \tag{2}$$

and $s = H_2/H_1$ is the ratio of magnetic forces in the electromagnetic case. Now (2) asserts that the reflected wave gets smaller as the mirror goes faster, and vanishes when $u = \frac{1}{2}v$. Or if the mirror be pushed against the radiation, the reflected wave gets stronger, and the resisting force stronger until $u = -\frac{1}{2}v$, when it is infinite. The mirror could not be pushed against the radiation faster than $\frac{1}{2}v$.

An immediate objection is that when u has risen to $\frac{1}{2}v$, if the mirror be maintained at that speed it acts like a perfect absorber to the incident energy. Moreover, since there is the pressure p_1 left, why should it not accelerate the mirror? But if it does, p_2 becomes negative, and s becomes imaginary. Considered mechanically only, say by $F = m\dot{u}$, the motion of m is quite determinate when $u > \frac{1}{2}v$, up to v, in fact. But electromagnetically it means that the energy in the reflected wave is negative. Now although there is nothing to object to quantitatively in a continuous transition from a Maxwellian stress consisting of a tension along an axis combined with an equal lateral pressure, to its negative, a pressure along the axis with equal lateral tension, still the negativity of the reflected energy causes difficulty. The stress for both the electric and magnetic energy becomes of the gravitational type. That is, like electrifications attract, and unlike repel, or matter is imaginary electrification in this comparison. The moving forces and energies are real. But let a real charge and an unreal one coexist, the energy density becomes imaginary. That is out of all reason in a real universe.

We should, I think, regard (2) as a demonstration that (1) is untrue, in that $(p_1 + p_2)u$ is not the activity of the force on the mirror, although $p_1 + p_2$ may be actually the pressure of the radiation. In fact, in the electromagnetic case, the variation of p constitutes a force on the ether itself. We must find the force on the mirror in another way.

Let radiation fall flush upon the plane surface of a dielectric, which call glass, moving the same way at constant speed u, and let the circuital equations in the glass be

$$-\frac{d\mathrm{H}}{dx}=c\dot{\mathrm{E}}+\frac{\delta\mathrm{I}}{\delta t}, \qquad -\frac{d\mathrm{E}}{dx}=\dot{\mathrm{B}}=\mu\dot{\mathrm{H}}; \tag{3}$$

that is, the same as for the ether, with the addition of the electric current of polarisation $\delta\mathrm{I}/\delta t$. The reference space is the fixed ether, and $\delta/\delta t$ is the moving time-differentiator. Now if the relation between I and E is such as to permit of an undistorted plane wave, we shall have

$$\underset{\text{(incident.)}}{\mathrm{E}_1=\mu v\mathrm{H}_1}, \qquad \underset{\text{(reflected.)}}{\mathrm{E}_2=-\mu v\mathrm{H}_2}, \qquad \underset{\text{(transmitted.)}}{\mathrm{E}_3=\mu w\mathrm{H}_3}, \tag{4}$$

if v is the speed in the ether, and w the wave speed, referred to the ether, in the glass. This w is a function of u. Also, the boundary conditions

$$\mathrm{E}_1+\mathrm{E}_2=\mathrm{E}_3, \qquad \mathrm{H}_1+\mathrm{H}_2=\mathrm{H}_3, \tag{5}$$

combined with (4), give

$$\frac{\mathrm{H}_2}{\mathrm{H}_1}=\frac{v-w}{v+w}, \qquad \frac{\mathrm{H}_3}{\mathrm{H}_1}=\frac{2v}{v+w}. \tag{6}$$

An incident pulse of unit depth is stretched to depth $(1-u/v)^{-1}$ in the act of reflection; the reflected pulse is of depth $(v+u)(v-u)^{-1}$, and the transmitted pulse of depth $(w-u)(v-u)^{-1}$.

The rate of loss of energy from the waves in the process of reflection is

$$p_1(v-u)-p_2(v+u)-p_3(w-u), \tag{7}$$

where the p's are the energy densities. But, by the above,

$$p_1v=p_2v+p_3w; \tag{8}$$

therefore the rate of loss of energy is

$$(p_3-p_1-p_2)u, \tag{9}$$

and the moving force on the mirror is

$$\mathrm{F}=p_3-p_1-p_2. \tag{10}$$

This is, in its expression, exactly the *negative* of the previous pressure difference. It is in the direction of the rise of energy density. Its amount is

$$\mathrm{F}=2\mu\mathrm{H}_1\mathrm{H}_2=2p_1\frac{v-w}{v+w}=\tfrac{1}{2}\mu\mathrm{H}_3^2-\tfrac{1}{2}c\mathrm{E}_3^2=\mathrm{U}_0. \tag{11}$$

The first form in terms of H_1, H_2 is useful. The second is in terms of the wave speeds. The third is in terms of the etherial energy inside the glass.

All these come out of the ratios H_2/H_1, etc. Now the electric energy equals the magnetic energy in the transmitted wave. Consequently U_0 means the energy of the polarisation I. And the activity is $U_0 u$, the convective flux of energy.

These properties are true for various relations between I and E. The first approximation is $I = c_1 E$. The second, introduced by Lorentz, is $I = c_1 (E - uB)$, that is, the polarisation is proportional to the moving force on a moving ion. Other forms allowing of undistorted wave propagation may be proposed. In Lorentz's case

$$U_0 = \tfrac{1}{2} c_1 E_3^2 (1 - u/w)^2. \tag{12}$$

To pass to perfect reflection, reduce w to u, its least value. U_0 does not vanish, but has the value given by (10), (11) still, with $w = u$. The moving force on the glass is now

$$F = 2p_1 \frac{v - u}{v + u}, \tag{13}$$

and finally, if $u = 0$, $F = 2p_1$.

Here we come right back to the pressure of radiation. It does measure the force on the glass when at rest, when it reflects perfectly, and it looks as if (13) were merely the form $p_1 + p_2$ a little modified by the motion. But appearances are very deceitful here, for (10) above is the proper formula.

[But observe that in (8) above it is assumed that the electric and magnetic energy densities in the transmitted wave are equal. So the further results are limited in application by this condition. It is obeyed in the case to which (12) refers. For further results see the next §.]

As regards the distribution of F. With an actual transmitted wave consisting of a pulse of uniform intensity all through, F is entirely at the wave front. So, with total reflection, it is just under the surface of the glass. Again, if E_3 varies continuously in the transmitted wave, F is distributed continuously, to the amount $B(\delta I/\delta t)$ per unit volume. What F means in (11) now is the total of this volume force, *i.e.*, the integral from the surface up to the wave front, expressed in terms of the momentary surface state.

After a pulse has left the surface there is an equal opposite force at its back, so there is no further loss of energy or moving force on the glass. The obscurities and apparent contradictions arise from the assumption that the ether is quite motionless. If we treat the matter more comprehensively, and seek the forces in a moving ether, with moving polarisable matter in it as well, if this is a complication one way, it is a simplification in another, viz., in the ideas concerned. There is harmony produced with the stress theory. To illustrate, $(\delta/\delta t)$VDB is the moving force per unit volume when the ether and polarisable matter have a common motion, **D** and **B** being the complete displacement and induction. (The variation of **u** is ignored here). But if we stop the ether, a part of this force becomes inactive. If the matter is unmagnetisable, the only active part is that containing the polarisation current, for that is carried along.

Besides this electromagnetic force, there is a force due to a pressure of amount U_0. But it does not alter the reckoning of the moving force on the glass, because the pressure acts equally and oppositely at the front and back of a pulse.

Some other illustrations of the curious action between electromagnetic radiation and matter can be given. [The polarisation energy is calculated by (12) above.] For example, two oppositely moving plane pulses inside moving glass. Say $E_1 = \mu w_1 H_1$ one way with the glass, and $E_2 = -\mu w_2 H_2$ against the glass. If $H_1 = -H_2$, work is done upon the glass when they cross, ceasing the moment they coincide, so that the energy of the momentary electric field is less than the wave-energy. On separating, the loss is restored. If, on the other hand, $E_1 = -E_2$, work is done by the glass on the waves when uniting, so that the momentary magnetic energy, together with the polarisation energy, is greater than the wave energy. In this second case, too, it is noteworthy that the solitary waves are of unequal energy, whereas they are equal in the first case. But details must be omitted, as this communication is perhaps already too long.

Electromagnetics in a Moving Dielectric.

(*Nature*, April 27, 1905, p. 605.)

§ **519.** Some time ago, when considering the assumption that the ether inside a body is quite stationary when a body is moved, and that in the application to Maxwell's etherial equation this involves the use of a fixed time differentiation for the ether, and a moving one for the matter, I argued that the same applied not only to the electric polarisation, as done by Lorentz and by Larmor, but also to the magnetic polarisation. I told the late Prof. FitzGerald that to make the extension seemed to be a sort of categorical imperative. For it involves no assumption as to how the magnetic polarisation is produced. At the time I made the application to plane waves only. Since then I have extended it to the general case. The principal interest at present lies in the mechanical activity, fundamentally involved in the question of the pressure of radiation, and electromagnetic moving forces in general. The results confirm the desirability of applying similar reasoning to the magnetic and to the electric polarisation, in so far as they are relatively simple, and cast light upon the subject.

Thus, let $\mathbf{M}=\mathrm{V}\mathbf{DB}$ be the complete *quasi*-momentum per unit volume, and $\mathbf{M}_0=\mathrm{V}\mathbf{D}_0\mathbf{B}_0$ the etherial part. Then if the velocity of the matter is $\mathbf{u}$, and of the ether $\mathbf{q}$, the motional activity (in the absence of free electrification, or variation of the electrical constants in space) comes to

$$\{\mathbf{u}(d/dt)+\nabla(\mathbf{u}.\mathbf{u})\}(\mathbf{M}-\mathbf{M}_0)+\{\mathbf{q}(d/dt)+\nabla(\mathbf{q}.\mathbf{q})\}\mathbf{M}_0; \qquad (1)$$

or, in a more developed form,

$$\mathbf{u}\{d/dt+\mathbf{u}\nabla+\nabla\mathbf{u}+\nabla_u.\mathbf{u}\}(\mathbf{M}-\mathbf{M}_0)$$
$$+\,\mathbf{q}\{d/dt+\mathbf{q}\nabla+\nabla\mathbf{q}+\nabla_q.\mathbf{q}\}\mathbf{M}_0. \qquad (2)$$

Here the factor of $\mathbf{u}$ is the moving force on the matter, and that of $\mathbf{q}$ the force on the ether. It will be seen that in the material part we simply deduct that part of the complete $\mathbf{M}$ that does not move with the matter. This makes a great simplification of ideas. To avoid misconception, the ∇ in (1) acts on all that follows, whereas in (2) the first ∇ acts on the M's, but the second and third on the velocities only, as may be seen on comparison with (1).

It is necessary, however, to point out distinctly the data involved in the above, as the simplification comes about in a special way. Divide the displacement $\mathbf{D}$ into

$\mathbf{D}_0=c_0\mathbf{E}$ in the ether, and $\mathbf{D}_1=c_1\mathbf{E}_1$ in the matter,

where $\mathbf{E}_1=\mathbf{E}+\mathbf{e}$, and $\mathbf{e}=\mathrm{V}(\mathbf{u}-\mathbf{q})\mathbf{B}_0$.

Similarly, divide the induction $\mathbf{B}$ into

$\mathbf{B}_0=\mu_0\mathbf{H}$, and $\mathbf{B}_1=\mu_1\mathbf{H}_1$,

where $\mathbf{H}_1=\mathbf{H}+\mathbf{h}$, and $\mathbf{h}=\mathrm{V}\mathbf{D}_0(\mathbf{u}-\mathbf{q})$.

The electric energy is

$$\mathrm{U} = \mathrm{U}_0 + \mathrm{U}_1 = \tfrac{1}{2}\mathbf{ED}_0 + \tfrac{1}{2}\mathbf{E}_1\mathbf{D}_1,$$

and the magnetic energy is

$$\mathrm{T} = \mathrm{T}_0 + \mathrm{T}_1 = \tfrac{1}{2}\mathbf{HB}_0 + \tfrac{1}{2}\mathbf{H}_1\mathbf{B}_1.$$

Also, let there be four eolotropic pressures, of Maxwellian type, say P_0, P_1 electric, and Q_0, Q_1 magnetic. For example,

$$\mathrm{P}_1 = \mathrm{U}_1 - \mathbf{E}_1.\mathbf{D}_1\,;$$

meaning a tension U_1 parallel to $\mathbf{E}_1$ combined with equal lateral pressure. The rest are similar. Finally, the two circuital equations are

$$\mathrm{V}\nabla(\mathbf{H} - \mathbf{h}_0 - \mathbf{h}_1) = \dot{\mathbf{D}} \qquad -\mathrm{V}\nabla(\mathbf{E} - \mathbf{e}_0 - \mathbf{e}_1) = \dot{\mathbf{B}}, \qquad (3)$$

where the motional electric and magnetic forces are defined by

$$\mathbf{h}_0 = \mathrm{V}\mathbf{D}_0\mathbf{q}, \qquad \mathbf{h}_1 = \mathrm{V}\mathbf{D}_1\mathbf{u}, \qquad \mathbf{e}_0 = \mathrm{V}\mathbf{q}\mathbf{B}_0, \qquad \mathbf{e}_1 = \mathrm{V}\mathbf{u}\mathbf{B}_1.$$

This completes the data, and from them may be derived the equation of activity

$$-\nabla\{\mathrm{V}\mathbf{EH} + \mathbf{q}(\mathrm{U}_0 + \mathrm{T}_0 + \mathrm{P}_0 + \mathrm{Q}_0) + \mathbf{u}(\mathrm{U}_1 + \mathrm{T}_1 + \mathrm{P}_1 + \mathrm{Q}_1)\}$$
$$= \dot{\mathrm{U}} + \dot{\mathrm{T}} + (\mathrm{U}_0/c_0)\dot{c}_0 + (\mathrm{U}_1/c_1)\dot{c}_1 + (\mathrm{T}_0/\mu_0)\dot{\mu}_0 + (\mathrm{T}_1/\mu_1)\dot{\mu}_1 + \mathbf{F}_0\mathbf{q} + \mathbf{F}_1\mathbf{u}, \qquad (4)$$

where $\mathbf{F}_0$ and $\mathbf{F}_1$ are the forces displayed in (2).

The meaning is that the left side of (4) is the convergence of the flux of energy made up of the Poynting flux, the convective flux, and the activity of the pressures, whilst the right side shows the result in increasing the stored energy, *and* in work done upon the ether and matter, either, both, or neither, according to the size of the two velocities.

The terms involving $\dot{c}$, etc., in (4) represent residual activity which may be of different sorts. The commonest is when the constants vary in space, especially at a boundary. For example, $\dot{c}_1 = -\mathbf{u}\nabla.\,c_1$, if c_1 does not vary in the moving matter. This means a moving force $-(\mathrm{U}_1/c_1)\nabla c_1$. But if there is compression, c_1 probably always varies intrinsically as well.

It will be found that the omission of the auxiliary $\mathbf{h}$ has the result of complicating instead of simplifying the force formulæ. Similarly the omission of $\mathbf{e}$ complicates them. Now the use of $\mathbf{e}$ is founded upon the idea that the electric polarisation is produced by a separation of ions under the action of $\mathbf{E}$, for $\mathbf{E}_1$ is the moving force on a moving unit electric charge

Analogously, $\mathbf{H}_1$ is the moving force on a moving unit magnetic charge or magneton. If there are really no such things, the interpretation must be made equivalent in other terms. But the categorical imperative is not easily to be overcome.

The application to plane waves I described in a recent letter (*Nature*, March 9, 1905) will be found to harmonise with the above in the special case. But a correction is needed. In the estimation of the moving force on glass receiving radiation, the assumption was made that the electric and magnetic energies in the transmitted wave were equal. So the result is strictly limited by that condition. But the conditions $E=wB$ and $U=T$ are not coextensive in general, though satisfied together in Lorentz's case. When U not $=T$, we have, instead of (8), [§518, above],

$$p_1v-p_2v-p_3w=w(T_3-U_3),$$

and the rate of loss of electromagnetic energy is

$$2\mu H_1H_2u+(w-u)(T_3-U_3).$$

Now this is zero when $e=0$, or the polarisation is proportional to the electric force. The question is raised how to discriminate, according to the data stated above, between cases of loss of energy and no loss. To answer this question, let $\mathbf{e}$ and $\mathbf{h}$ in the above be unstated in form; else the same. Then, instead of (4), the activity equation will be

$$-\nabla\mathbf{W}=\dot{U}+\dot{T}+\{\tfrac{1}{2}E^2(\delta c_0/\delta t)+\ldots\}+(\mathbf{f}_0\mathbf{q}+\mathbf{f}_1\mathbf{u})-(\mathbf{e}\mathbf{J}_1+\mathbf{h}\mathbf{G}_1), \quad (5)$$

where $\mathbf{W}$ is as in (4), whilst $\mathbf{f}_0$, $\mathbf{f}_1$ are the forces derived from the stresses specified (not the same as $\mathbf{F}_0$ and $\mathbf{F}_1$), and $\mathbf{J}_1$, $\mathbf{G}_1$ are the electric and magnetic polarisation currents, thus,

$$\mathbf{J}_1=\dot{\mathbf{D}}_1+V\nabla\mathbf{h}_1, \quad \text{etc.} \quad (6)$$

It follows that it is upon $\mathbf{e}$ and $\mathbf{h}$ that the loss of energy depends in plane waves, when $\mathbf{u}$ and $\mathbf{q}$ are constant. For the stresses reduce to longitudinal pressures, so that by line integration along a tube of energy flux we get

$$\Sigma(eJ_1+hG_1)=\Sigma(\dot{U}+\dot{T}). \quad (7)$$

Thus, when a pulse enters moving glass from stationary ether, the rate of loss of energy is $\Sigma(-eJ_1)$. If e is zero, so is the loss, as in the special case above. There is also agreement with the calculated loss in the other case.

That the moving force on the glass should be controlled by e is remarkable, for it is merely the small difference between the electric force on a fixed and a moving unit charge. The theory is not final, of course. If the electromagnetics of the ether and matter could be made very simple, it would be a fine thing; but it does not seem probable.

The Charging of a Cable through a Condenser and Resistance.

(*The Electrician*, Dec. 23, 1904, p. 394.)

§ 520. In vol. 2 of "Electromagnetic Theory" I have described in considerable detail how waves are sent along a cable, and how to calculate them. Also how to find the voltage and current impressed upon a cable by an impressed voltage in the terminal apparatus at its beginning, from which the waves mentioned are derived. There is a good deal of numerical work to be done to determine the curves of voltage and current in specified cases. It is not difficult, though sometimes laborious. In a recent article (*The Electrician*, vol. 53, pp. 905, 954, 994 and 1,019), Mr. W. Gaye has done some of this work in a way that ought to be interesting to some of the many men all over the world who ought to be interested in this subject. The object of the present article is to promote the study of special cases numerically and graphically, especially as regards the inclusion of the self-induction of the cable, which is by no means a negligible quantity even now, whilst it ought to be a paramount agent. If it is love that makes the world go round, it is self-induction that makes electromagnetic waves go round the world.

But it is well to begin with simple cases, and then go on to the more difficult. So, first, considering no self-induction, the current sent into a cable at its beginning through a resistance r and condenser s by the impressed force e is

$$\mathrm{C}=\frac{e}{r+\dfrac{1}{sp}+\left(\dfrac{\mathrm{R}}{\mathrm{S}p}\right)^{\frac{1}{2}}},\qquad \mathrm{V}=\left(\frac{\mathrm{R}}{\mathrm{S}p}\right)^{\frac{1}{2}}\mathrm{C}, \tag{1}$$

where the denominator under e is the sum of the resistance operators of the terminal resistance, condenser, and cable, the last being necessarily very long, since no allowance is made for the effect of terminal reflection at the far end upon the state of things at the beginning. R and S are the resistance and condensance of the cable per unit length, and p is the time-differentiator d/dt. I am here using condensance on trial for the electrostatic capacity of a condenser, because people do not seem to care much about permittance, which I think a good name. Along with condensance goes condensivity, of course. It is by trial and error that terminology becomes established. International Conventions and Committees are not much good.

If $s=\infty$, the condenser is equivalently short-circuited, and if $r=0$ the resistance is similarly cut out. In either case only one curve requires to be calculated, as exhibited in a table by Mr. Gaye, because by altering the time-scale all other cases of a resistance alone or a condenser alone may be included. But it is different when the resistance and condenser are both on at once. Thus, put

$$z=\frac{Rs}{rS}, \qquad q=\left(\frac{z}{rsp}\right)^{\frac{1}{2}}; \tag{2}$$

then

$$\frac{rC}{e}=\frac{1}{1+q+z^{-1}q^2}. \tag{3}$$

We see that by varying the constant z we change the type of the curve. Mr. Gaye has drawn the curves for several values of z; if he had given the sets of numbers as well, it would have been better still. Not to repeat too much methods described in "Electromagnetic Theory," vol. 2, it may be said that in these cases of simply formed operators, the way of algebrisation is merely to divide the numerator by the denominator, and so obtain a power series in terms of p, and then turn p^{-n} to $t^n/\underline{|n}$.

Now there is a certain range for z in which we may make use of the table referred to to calculate C in the present more advanced case. This can be shown very simply.

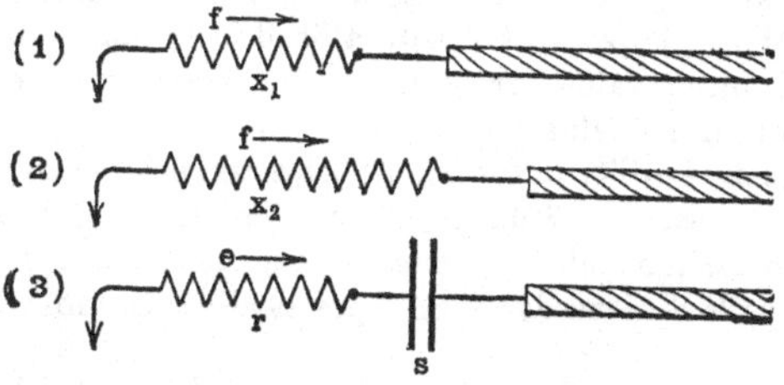

In (1) we have resistance x_1, no condenser, and impressed force f. In (2) the resistance is changed to x_2, else the same. In (3) the impressed force is e, the resistance r, and the condenser s is put in. Let C_1, C_2, and C be the currents in the three cases. Then

$$C_1-C_2=\frac{f}{x_1+\left(\frac{R}{Sp}\right)^{\frac{1}{2}}}-\frac{f}{x_2+\left(\frac{R}{Sp}\right)^{\frac{1}{2}}}$$

$$=\frac{(x_2-x_1)f}{x_1x_2+\left(\frac{R}{Sp}\right)^{\frac{1}{2}}(x_1+x_2)+\frac{R}{Sp}}; \tag{4}$$

and in the third case

$$C=\frac{e}{r+\frac{1}{sp}+\left(\frac{R}{Sp}\right)^{\frac{1}{2}}}. \tag{5}$$

By inspection it can be seen that the constants can be adjusted so as to make $C = C_1 - C_2$. We require

$$\frac{e}{r} = \frac{x_2 - x_1}{x_1 x_2} f, \qquad es = \frac{f(x_2 - x_1)S}{R}, \qquad e = \frac{(x_2 - x_1)f}{x_1 + x_2}; \tag{6}$$

or

$$e = f\frac{x_2 - x_1}{x_2 + x_1}, \qquad r = \frac{x_1 x_2}{x_1 + x_2}, \qquad s = \frac{S}{R}(x_1 + x_2). \tag{7}$$

Any two values of x being given, the corresponding r and s become known. So the required curve of current through the condenser is the difference of the curves of current in two cases of terminal resistance only. We also have

$$z = \frac{(r_1 + x_2)^2}{x_1 x_2}. \tag{8}$$

It follows that the least value of z is 4, or this reduction to simplicity ranges from $z = 4$ to $z = \infty$. Or thus,

$$\frac{e}{f} = \left(1 - \frac{4}{z}\right)^{\frac{1}{2}}, \qquad \frac{x_1}{x_2} = \frac{z}{2r}\left\{1 \pm \left(1 - \frac{4}{z}\right)^{\frac{1}{2}}\right\}. \tag{9}$$

The range from $z = 0$ to 4 requires imaginary x's and f.
Example,

$$x_1 = 100, \quad x_2 = 200, \qquad \frac{e}{f} = \tfrac{1}{3}, \qquad r = 66\tfrac{2}{3}, \qquad z = 4\tfrac{1}{2}.$$

Since z means Rs/rS or $(R/S)/(r/s)$, the ratio r/s must be less than one-fourth of the cable ratio. So, with a fixed s, we may range r from 0 up to a limiting value; or, with a fixed r, we may range s from ∞ down to a limiting value.

The principal utility of the condenser is to curb the signals. The curbing is increased by reducing the condensance. But as this also reduces the charge and the strength of the signal, compensation by increased battery is needed. So a practical compromise has to be made.

The above being a practical simplification, what follows next relates to the general theory to exhibit the effect of z above and below the critical value and connect with other matters. Go back to equation (3). The roots of the denominator equated to zero are

$$\frac{q_1}{q_2} = \tfrac{1}{2}(-z \pm \sqrt{z^2 - 4z}),$$

$$\therefore \quad q_1 q_2 = z, \quad q_1 - q_2 = \sqrt{z^2 - 4z}, \quad q_1^2 = \frac{z}{2}(z - 2 - \sqrt{z^2 - 4z}). \tag{10}$$

We may therefore write

$$\frac{rC}{e} = \frac{z}{q_1 - q_2}\left(\frac{1}{q_2 - q} - \frac{1}{q_1 - q}\right)$$

$$= \frac{z}{q_1 - q_2}\left\{\frac{1}{q_2}\left(1 + \frac{q}{q_2} + \frac{q^2}{q_2^2} + \dots\right) - \frac{1}{q_1}\left(1 + \frac{q}{q_1} + \frac{q^2}{q_1^2} + \right)\right\} \tag{11}$$

The result is algebrisable at sight, as the difference of two convergent series. Now, when z is greater than 4, it represents the already described differential problem. Or, which is the same,

$$\frac{rC}{e} = \frac{z}{q_1 - q_2}\left[\frac{1}{q_2}\epsilon^{zt/rsq_2^2} + \frac{q}{q_2^2} + \frac{q^3}{q_2^4} + \frac{q^5}{q_2^6} + \dots\right]$$
$$- \frac{z}{q_1 - q_2}\left[\frac{1}{q_1}\epsilon^{zt/rsq_1^2} + \frac{q}{q_1^2} + \frac{q^3}{q_1^4} + \frac{q^5}{q_1^6} + \dots\right]. \qquad (12)$$

By pairing terms, and using the values of $q_1^{2n} - q_2^{2n}$, this form may be converted to the convergent series used by Mr. Gaye, obtained by division. But as the table has been calculated to suit this case, with any two positive values of x_1 and x_2, which mean real negative values of q_1 and q_2, it is not necessary to transform further. If, however, $z < 4$, it is possible that this form may also be useful, noting that it is only the real parts of the two series that are effective. The value of q^n is

$$q^n = \left(\frac{z}{rsp}\right)^{\frac{n}{2}} = \frac{\left(\frac{zt}{rs}\right)^{\frac{n}{2}}}{\underline{|\tfrac{1}{2}n}}. \qquad (13)$$

When we transform (12) to divergent form, it makes

$$\frac{rC}{e} = \frac{z}{q_1 - q_2}\left[\left(\frac{1}{q} + \frac{q_1^2}{q^3} + \frac{q_1^4}{q^5} + \dots\right) - \left(\frac{1}{q} + \frac{q_2^2}{q^3} + \frac{q_2^4}{q^5} + \right)\right] \qquad (14)$$

This is done by using the generalised exponential formula, remembering that q_1 and q_2, when real, are negative. It goes equally well when they are imaginary, with real parts negative.

It is remarkable that the series in (12) involving the odd powers of q, contain an oscillating part when $z < 4$. For we may write (12) thus :—

$$\frac{rC}{e} = \epsilon^{\frac{(z-2)t}{2rs}}\left[\cos + \frac{z}{(4z - z^2)^{\frac{1}{2}}}\sin\right]\frac{t}{2rs}(4z - z^2)^{\frac{1}{2}}$$
$$+ \frac{z}{q_1 - q_2}\left[\frac{1}{z^2}(q_1^2 - q_2^2)q + \frac{1}{z^4}(q_1^4 - q_1^4)q^3 + \dots\right]. \qquad (15)$$

But since there is no self-induction, there can be no oscillations in C. So the second line really involves an oscillating part which is the negative of the first line, the remainder being non-oscillatory. It is by conversion to the divergent form (14) that we get rid of the redundant mutually cancelling oscillations. Furthermore, by pairing terms to get rid of the powers of q_1 and q_2, (14) may be converted to the divergent series used by Mr. Gaye.

The singular case $z=4$ makes $q_1 = q_2$, and $x_1 = x_2$. There is another case of simplification, viz. $z=2$. Then $q^1_2 = -1 \pm i$; $q_1^2 = -2i = -q_2^2$; $q_1^4 = q_2^4 = -4$; $q_1^6 = 8i = -q_2^6$; $q_1^8 = q_2^8 = 16$; $q_1^{10} =$

$-32i=-q_2^{10}$ and so on. Also $q_1-q_2=2i$. These reduce (15) to

$$\frac{rC}{e}=\sqrt{2}\cos\left(\frac{t}{rs}-\frac{\pi}{4}\right)-\sqrt{2}\left\{\left(\frac{1}{rsp}\right)^{\frac{1}{2}}-\left(\frac{1}{rsp}\right)^{2\frac{1}{2}}+\left(\frac{1}{rsp}\right)^{4\frac{1}{2}}-\dots\right\}, \tag{16}$$

where

$$\left(\frac{1}{rsp}\right)^{\frac{n}{2}}=\frac{\left(\frac{t}{rs}\right)^{\frac{n}{2}}}{\lfloor\tfrac{1}{2}n}.$$

Now this special formula is an interesting example of the generalised cosine series, namely

$$\cos\left(x-\frac{n\pi}{2}\right)=\dots-\frac{x^{2n-2}}{\lfloor 2n-2}+\frac{x^{2n}}{\lfloor 2n}-\frac{x^{2n+2}}{\lfloor 2n+2}+\dots, \tag{17}$$

the series to be made complete both ways, x being real positive and n any number from $-\infty$ to $+\infty$. We are concerned at present with the value $n=\frac{1}{2}$. It makes

$$\cos\left(x-\frac{\pi}{4}\right)=\dots-\frac{x^{-1\frac{1}{2}}}{\lfloor -1\frac{1}{2}}+\frac{x^{\frac{1}{2}}}{\lfloor \frac{1}{2}}-\frac{x^{2\frac{1}{2}}}{\lfloor 2\frac{1}{2}}+\dots. \tag{18}$$

Give to x the value t/rs, then

$$\cos\left(\frac{t}{rs}-\frac{\pi}{4}\right)=\dots-\left(\frac{1}{rsp}\right)^{-1\frac{1}{2}}+\left(\frac{1}{rsp}\right)^{\frac{1}{2}}-\left(\frac{1}{rsp}\right)^{2\frac{1}{2}}+\dots. \tag{19}$$

This transforms (16) to

$$\frac{rC}{e}=-\sqrt{2}\{(rsp)^{1\frac{1}{2}}-(rsp)^{3\frac{1}{2}}+(rsp)^{5\frac{1}{2}}-\dots\}$$

$$=\frac{1}{(2\pi)^{\frac{1}{2}}}\left(\frac{rs}{t}\right)^{1\frac{1}{2}}\left\{1-3.5\left(\frac{rs}{2t}\right)^2\left(1-7.9\left(\frac{rs}{2t}\right)^2\left(1-\dots\right.\right.\right. \tag{20}$$

Finally, this is what (14) reduces to when $z=2$. Half the terms cancel out.

The cosine transformation occurs in a variety or physical problems. It is not fancy work, as some ignorant people have supposed, to use divergent series, but is highly practical in general, and is sometimes the only way to work. Here is an example—deep water plane waves. Given the initial state of vertical displacement at the surface to be D_0 condensed at the origin, and everywhere at rest, what is the vertical displacement D at time t on the surface at distance x from the origin? The usual theory when worked out gives

$$D=\frac{D_0}{\pi a^2t^2}\left\{\frac{y}{\lfloor 2}-\frac{\lfloor 3}{\lfloor 6}y^2+\frac{\lfloor 5}{\lfloor 10}y^3-\frac{\lfloor 7}{\lfloor 14}y^4+\dots\right\} \tag{21}$$

where $y=a^4t^4/x^2$. The constant a means $\sqrt{g}$. Although mathematically perfect, the result is only directly calculable within a small range. Proceeding from $x=\infty$ towards the origin, D rises from zero up to a hump at a certain distance, depending upon the value of y. Going nearer, the hump becomes a hollow. After that, the labour of calculation is too great. But by the look of the formula there is an infinite series of alternate humps and hollows between the outermost hump and the origin. Now the above generalised cosine formula (which may be obtained from

$$\epsilon^x = \Sigma \frac{x^r}{\underline{|r}}, \tag{22}$$

by $x=ai$, where a is real positive), when applied to (21) above, turns it to

$$D = \frac{D_0}{2\sqrt{\pi}} \frac{at}{x^{1\frac{1}{2}}} \cos\left(\frac{a^2t^2}{4x} - \frac{\pi}{4}\right) - \frac{2D_0}{\pi a^2 t^2}\left(1 - \frac{\underline{|5}}{\underline{|2}} \frac{1}{y} + \frac{\underline{|9}}{\underline{|4}} \frac{1}{y^2} - \ldots\right). \tag{23}$$

The second part is divergent, so the formula is only closely calculable when y is large enough to make the second term in the divergent series small compared with the first. Now this is just when the previous formula is unusable. So the changed form allows us to calculate the oscillatory region.

Although this is the fundamental solution in the usual incompressible theory of deep water surface waves, it must be taken with much salt, being physically absurd. Nor do we arrive at intelligibility by allowing for the compressibility of the water. In the incompressible case the wave speed varies as the square root of the wave length. It is infinity for infinitely long waves. This is why the initial disturbance is immediately communicated to any distance. It is also zero for infinitely short waves. This is why there is such a packing of infinitely rapid vibrations close to the origin, where there was initially a discontinuity in the vertical displacement. Now, allowing for compressibility, the relation between wave speed and vibration frequency is exhibited in the periodic solution

$$D = A \sin n\left[t - \frac{x}{v}\left(1 + \frac{n^2v^2}{a^4}\right)^{\frac{1}{2}}\right], \tag{24}$$

where v is the speed of sound in water. So the wave speed w is

$$w = \frac{v}{\left(1 + \frac{n^2v^2}{a^4}\right)^{\frac{1}{2}}}. \tag{25}$$

As n goes from 0 to ∞, w goes from v to 0. There is a remarkable improvement in so far as the long waves are concerned. The greatest wave speed is v, so there is a real proper wave front at distance vt, at time t after a disturbance begins at the origin. But still the shortest waves have speed 0, so we do not get rid of the infinitely closely packed vibrations by the improvement. We must go deeper, and see what happens underneath the surface. Define a point by the coordinates r, ϕ, where $r = \sqrt{z^2 + x^2}$ is its distance from the origin, and ϕ the angle between $\mathbf{r}$ and the vertical downward. Then if D_z is the vertical displacement

$$D_z=\frac{1}{\pi}\left\{\frac{\cos\phi}{r}-\frac{a^2t^2}{\underline{|2}}\frac{\cos 2\phi}{r^2}+\underline{|2}\,\frac{a^4t^4}{\underline{|4}}\frac{\cos 3\phi}{r^3}-\underline{|3}\,\frac{a^6t^6}{\underline{|6}}\frac{\cos 4\phi}{r^4}+\dots\right\} \tag{26}$$

is the convergent solution corresponding to (21). There is a similar formula for D_x the horizontal displacement. Both are contained in

$$D_z-iD_x=\frac{1}{\pi R}\left\{1-\frac{a^2t^2}{\underline{|2}R}+\frac{\underline{|2}}{\underline{|4}}\frac{a^4t^4}{R^2}-\frac{\underline{|3}}{\underline{|6}}\frac{a^6t^6}{R^3}+\dots\right\}, \tag{27}$$

where $R=r\epsilon^{\phi i}$. This shows the helical nature of motion. There is nothing unintelligible about it, except at the surface, where $z=0$, or $\phi=\frac{1}{2}\pi$. It is a limiting surface of absurdity. It is therefore the boundary condition assumed in this theory that is imperfect. The condition

$$\dot{D}_z=g\frac{dD_z}{dz} \tag{28}$$

is indeed approximately true at the plane $z=0$, when this is underneath the surface, but is not true at the surface itself. But if we give z a small finite value the initial state of vertical displacement will be the hump represented by

$$D_z=\frac{zD_0}{\pi(z^2+x^2)}, \tag{29}$$

and now the initiation, growth, and decay of the wave trains to right and left can be followed.

It is also interesting to see that in the vertical plane $x=0$ the generalised ϵ^{-z} formula turns up. That is

$$\epsilon^{-z}\cos r\pi=\Sigma\pm\frac{z^r}{\underline{|r}} \tag{30}$$

when z is positive. For put $\phi=0$ in (26). Then

$$D_z=\frac{1}{\pi z}\left\{1-\frac{a^2t^2}{\underline{|2}z}+\frac{\underline{|2}}{\underline{|4}}\frac{a^4t^4}{z^2}-\frac{\underline{|3}}{\underline{|6}}\frac{a^6t^6}{z^3}+\dots\right\}. \tag{31}$$

This is the convergent result. As t increases from 0 upward, D falls to zero, reverses itself, and then rises again. Is there any oscillation? It appears not, but calculation becomes difficult. Now, the generalised ϵ^{-z} formula turns (31) to

$$D_z=-\frac{2D_0}{\pi a^2t^2}\left(1+\frac{\underline{|3}}{\underline{|1}}\frac{z}{a^2t^2}+\frac{\underline{|5}}{\underline{|2}}\frac{z^2}{a^4t^4}+\dots\right), \tag{32}$$

and this shows the subsidence to equilibrium from the negative maximum without oscillation. The vertical plane $x=0$ must be uniquely non-oscillatory. By actual numerical calculation it may be shown that the convergent and divergent formulæ join properly in the middle region where both are calculable. The effect of surface tension does not appear in the deep water wave theory; only gravity.

Returning to the submarine cable, I have, in "Electromagnetic Theory," vol. 2, pp. 346, 348, &c., given a general method of expanding the required formulæ when the self-induction is taken into account (and also leakage if desired), which saves a great deal of experimental work, and puts the results in the form suitable for showing the progress of the waves along the cable. Thus, with a condenser and resistance at the beginning, to illustrate, if there is no leakage the current entering the cable is given by

$$\frac{C}{e}=\frac{1}{r+\frac{1}{sp}+\left(\frac{R+Lp}{Sp}\right)^{\frac{1}{2}}}. \tag{33}$$

Let $R/2L=\sigma$, $LSv^2=1$, then

$$\frac{C}{e}=\frac{1}{r+\frac{1}{sp}+Lv\left(\frac{p+2\sigma}{p}\right)^{\frac{1}{2}}}. \tag{34}$$

Introduce the prefactor $\epsilon^{-\sigma t}$. To do this, change p to $p-\sigma$, and put on the postfactor $\epsilon^{\sigma t}$ or $p/(p-\sigma)$. So we get

$$\frac{C}{e}=\epsilon^{-\sigma t}\frac{\frac{p}{p-\sigma}}{r+\frac{1}{s(p-\sigma)}+Lv\left(\frac{p+\sigma}{p-\sigma}\right)^{\frac{1}{2}}}. \tag{35}$$

Expand this in $I_n(\sigma t)$ functions. To do this, put

$$p=\frac{\sigma}{2}(a+a^{-1}), \tag{36}$$

and introduce the postfactor $(1-a^2)/(1+a^2)$. See "Electromagnetic Theory," vol. 2, p. 348. We come to

$$\frac{C}{e}=\frac{\epsilon^{-\sigma t}}{r+Lv}\,\frac{1-a^2}{1-2a\frac{r-1/s\sigma}{r+Lv}-\frac{Lv-r}{Lv+r}a^2}. \tag{37}$$

Or

$$\frac{C}{e}=\frac{\epsilon^{-\sigma t}}{r+Lv}\,\frac{1-a^2}{1-ma-na^2}. \tag{38}$$

This is, in a condensed form, the algebraical solution of the problem. To exhibit fully, expand in powers of a, and then turn a^n to $I_n(\sigma t)$. The expansion is easily done. Thus,

$$\begin{aligned}\frac{C}{e}=\frac{\epsilon^{-\sigma t}}{r+Lv}(1-a^2)\,[1&+a(m+na)+a^2(m^2+2mna+n^2a^2)\\&+a^3(m^3+3m^2na+3mn^2a^2+a^3)+a^4(m^4+4m^3na+6m^2n^2a^2\\&+4mn^3a^3+n^4a^4)+\ldots]\\=\frac{\epsilon^{-\sigma t}}{r+Lv}\{1&+am+a^2(m^2+n)+a^3(m^3+2mn)+a^4(m^4+3m^2n+n^2)+\ldots\\&\qquad\;\,-a^2\qquad\qquad\;\;-a^3m\qquad\qquad\;\;-a^4(m^2+n)\qquad\quad-\ldots\}\end{aligned}\tag{39}$$

Or, finally

$$\begin{aligned}\frac{C}{e}=\frac{\epsilon^{-\sigma t}}{r+Lv}\{1&+am+a^2(m^2+n+1)+a^3(m^3+2mn-m)\\&+a^4(m^4+3m^2n+n^2-m^2-n)+\ldots\}.\end{aligned}\tag{40}$$

To carry this on to the extent desired for calculation, all that is wanted is the law of the coefficients. It is easiest to do it arithmetically thus,

1						
1						
1	1					
1	2					
1	3	1				
1	4	3				
1	5	6	1			
1	6	10	4			
1	7	15	10	1		
1	8	21	20	5		
1	9	28	35	15	1	
1	10	36	56	35	6	1

The first two columns explain themselves. All the rest follow this rule :—The sum of the first n terms in any column $= n^{th}$ term in the next column. Taking the figures in the rows, they are the coefficients, as in the first line of (39) above.

Another way of expanding (38) is similar to (11) above, in terms of a_1 and a_2, the roots of

$$1 - ma - na^2 = 0,$$

thus making a difference problem, meaning the same as (40).

For small values of σt the convergent formula is

$$a^n = I_n(\sigma t) = \frac{(\frac{1}{2}\sigma t)^n}{\underline{|0}\,\underline{|n}} + \frac{(\frac{1}{2}\sigma t)^{n+2}}{\underline{|1}\,\underline{|n+1}} + \frac{(\frac{1}{2}\sigma t)^{n+4}}{\underline{|2}\,\underline{|n+2}} + \dots \qquad (41)$$

But for larger values,

$$\epsilon^{-\sigma t} I_n(\sigma t) = \frac{1}{(2\pi\sigma t)^{\frac{1}{2}}}\left\{1 + \frac{1^2 - 4n^2}{1.8\sigma t}\left(1 + \frac{3^2 - 4n^2}{2.8\sigma t}\left(1 + \frac{5^2 - 4n^2}{3.8\sigma t}\left(1 + \dots\right\}\right.\right.\right. \qquad (42)$$

So the problem is brought to complete numerical realisation. There are tables of $I_n(\sigma t)$ in Gray and Mathew up to $\sigma t = 5$ or so. But the divergent formula may be used for much smaller values in some cases.

To find V, the potential, go thus :—

$$V = \left(\frac{R + Lp}{Sp}\right)^{\frac{1}{2}} C = Lv\epsilon^{-\sigma t}\frac{1 + a}{1 - a} C\epsilon^{\sigma t} \qquad (43)$$

$$= \frac{Lv}{r + Lv}\epsilon^{-\sigma t}\frac{(1 + a)^2}{1 - ma - na^2}. \qquad (44)$$

The net result is that the multiplier $1 - a^2$ in the C expansion becomes $(1 + a)^2$. See how equation (39) is constructed from the preceding one. Do similarly with the changed multiplier to find V.

Details of these processes will be found in the work referred to. Also how to include the influence of leakage by having $\epsilon^{-\rho t}$ instead of $\epsilon^{-\sigma t}$ as factor. Also more primitive methods, obtaining power series directly. But the systematic development is best in advanced work. How to turn the terminal solutions into progressive waves need not be considered here.

The values of m and n are

$$m=2\frac{r-1/s\sigma}{r+\mathrm{L}v}, \quad n=\frac{\mathrm{L}v-r}{\mathrm{L}v+r}. \tag{45}$$

They may be positive or negative. One or other of them may be zero. When both are zero we have a remarkable reduction to simplicity. We require

$$r=\mathrm{L}v=(s\sigma)^{-1}.$$

These give the values of r and s in terms of the cable constants. The terminal resistance r equals the initial impedance of the cable, and the terminal condensance s is $2\mathrm{L}/\mathrm{R}r$. That is, the time constant of the condenser and resistance on short-circuit equals twice the time constant of the cable itself on short-circuit when supporting a current of uniform strength, meaning C uniform all along the cable, varying with the time as $\epsilon^{-\mathrm{R}t/\mathrm{L}}$.

The terminal potential and current are

$$\mathrm{V}=\tfrac{1}{2}e\epsilon^{-\sigma t}(\mathrm{I}_0+2\mathrm{I}_1+\mathrm{I}_2)(\sigma t), \tag{46}$$

$$\mathrm{C}=\frac{e}{2\mathrm{L}v}\epsilon^{-\sigma t}(\mathrm{I}_0-\mathrm{I}_2)(\sigma t)=\frac{e}{\mathrm{L}v}\epsilon^{-\sigma t}\frac{\mathrm{I}_1(\sigma t)}{\sigma t}. \tag{47}$$

Now, if the cable were direct to earth, $r=0$, $s=\infty$, we should have

$$\mathrm{C}=\frac{e}{\mathrm{L}v}\epsilon^{-\sigma t}\mathrm{I}_0(\sigma t); \tag{48}$$

and if, further, $\mathrm{L}=0$, this reduces to

$$\mathrm{C}=e\left(\frac{\mathrm{S}}{\pi \mathrm{R}t}\right)^{\frac{1}{2}}. \tag{49}$$

Now these formulæ are all very easily calculated and graphed. So the influence of the L of the cable, and also the influence of the terminal resistance and condensance in a special case are easily obtainable. For example, comparing (48) with (47), the effect of r and s is to turn $\mathrm{I}_0(\sigma t)$ to $\mathrm{I}_1(\sigma t)/\sigma t$, or

$$1+(\tfrac{1}{2}\sigma t)^2+\frac{(\frac{1}{2}\sigma t)^4}{2.2}+\frac{(\frac{1}{2}\sigma t)^6}{\lfloor 3\lfloor 3}+\dots \text{ to } \tfrac{1}{2}\left(1+\frac{(\frac{1}{2}\sigma t)^2}{1.2}+\frac{(\frac{1}{2}\sigma t)^4}{\lfloor 2\lfloor 3}+\dots\right). \tag{50}$$

The factor $\frac{1}{2}$ comes in because $r=\mathrm{L}v$. The current is halved at the first moment. But of course the curves are not otherwise similar. Introducing the factor $\epsilon^{-\sigma t}$, we may write

$$\mathrm{C}=\frac{e}{2r}\left\{1-\sigma t\left(1-\frac{5}{8}\sigma t\left(1-\frac{7}{15}\sigma t\left(1-\frac{9}{24}\sigma t\left(1-\frac{11}{35}\sigma t\left(1-\dots\right\}\right.\right.\right.\right.\right., \tag{51}$$

or

$$\mathrm{C}=\frac{e}{r\sigma t(2\pi\sigma t)^{\frac{1}{2}}}\left\{1-\frac{3}{2}\frac{1}{4\sigma t}\left(1+\frac{5}{4}\frac{1}{4\sigma t}\left(1+\frac{21}{6}\frac{1}{4\sigma t}\left(1+\frac{45}{8}\frac{1}{4\sigma t}\left(1+\dots\right\}\right.\right.\right.\right.. \tag{52}$$

Use the first when σt is small and the second when work by the first becomes too long. Or, the tables may be used in the first case. *E.g.*, $\sigma t=5$ makes the first sum very long; so use the second, and a short sum makes $rC/e=\cdot 032$. The value $\sigma t=2$ seems about the neutral place; by both formulæ, $rC/e=\cdot 108$. My general practice is to count half the smallest term in the divergent series.

But let us see if the critical values of r and s are practical. Say $R=2\times 10^4$, or 2 ohms per kilom., and $S=2\times 10^{-21}$, or $\frac{1}{5}$ microfarad per kilom. Then

$$r=\sqrt{\frac{L}{20}}\times 10^{11} \quad =100\sqrt{\frac{L}{20}}\text{ ohms}.$$

Giving L values from 2 to 100 for example, it is seen that the values are practical. Also

$$s=\frac{2L}{Rr}=\frac{2L}{2\times 10^4\times\sqrt{L/20}\times 10^{11}}=\frac{\sqrt{20L}}{10^{15}}=\sqrt{20L}\text{ microfarads.}$$

This again gives practical values. It is certainly striking that the values of r and s should be just in the proper region. There is a Providence in these things.

It would be possibly of interest to cable electricians if someone would work out the curves for various values of L; say, 0, $2\frac{1}{2}$, 5, 10, 20, etc., to show the effect of L on the sent currents, not merely with the critical values of r and s, but with various values. The general effect is not difficult to see.

If leakage is allowed for, Sp becomes $K+Sp$ in (33). Then let

$$\rho=R/2L+K/2S, \qquad \sigma=R/2L-K/2S. \tag{53}$$

This makes

$$\frac{C}{e}=\epsilon^{-\rho t}\frac{\dfrac{p}{p-\rho}}{r+\dfrac{1}{s(p-\rho)}+Lv\left(\dfrac{p+\sigma}{p-\sigma}\right)^{\frac{1}{2}}}, \tag{54}$$

which by the a transformation leads to

$$\frac{C}{e}\epsilon^{\rho t}=\frac{\frac{1}{2}s\sigma(1-a)(1-a^2)}{A_0+A_1a+A_2a^2+A_3a^3}, \tag{55}$$

where

$$\begin{aligned}A_0&=\tfrac{1}{2}s\sigma(r+Lv), & A_1&=1-\rho s(r+Lv)+\tfrac{1}{2}s\sigma(Lv-r),\\ A_2&=\tfrac{1}{2}s\sigma(r+Lv)-1-s\sigma(Lv-r), & A_3&=\tfrac{1}{2}s\sigma(Lv-r).\end{aligned} \tag{56}$$

Here (55) is the complete solution, on expansion in rising powers of a. But it is very complicated in general. An exception, of course, is the distortionless case, which should not be done through a at all; but direct from the primitive operational solution, making

$$C=\frac{e}{r+Lv}\epsilon^{-\frac{t}{s(r+Lv)}}. \tag{57}$$

Nevertheless a great simplification of (55) occurs in the critical case. For $A_3=0$ when $r=Lv$; and then $A_2=0$ if $rs\sigma=1$ as well.

But we cannot make $A_1=0$ with possible values of cable constants. In this critical case,

$$\frac{C}{e}\epsilon^{\rho t}=\frac{1}{2r}(1-a^2)\frac{1-a}{1-ca},\qquad \text{if}\quad c=\frac{2\rho}{\sigma}-1, \tag{58}$$

$$=\frac{1}{2r}\{1+(c-1)a+(c^2-c-1)a^2+\dots\}. \tag{59}$$

Although making an infinite series of $I_n(\sigma t)$ functions, the coefficients are very simply formed.

Other Critical Cases. Mathematical Excursion.

§ **521.** There is a second critical case in which, without leakage, the formulæ for the potential and current reduce to simple form when a condenser and resistance are put in terminally. Say

$$r=Lv,\qquad 1=2rs\sigma;\qquad \text{then}\qquad m=-1,\qquad n=0.$$

Equations (38) and (44) reduce to

$$\frac{C}{e}=\frac{\epsilon^{-\sigma t}}{2r}(1-a)=\frac{\epsilon^{-\sigma t}}{2r}(I_0-I_1)(\sigma t), \tag{60}$$

$$\frac{V}{e}=\tfrac{1}{2}\epsilon^{-\sigma t}(1+a)=\tfrac{1}{2}\epsilon^{-\sigma t}(I_0+I_1)(\sigma t). \tag{61}$$

The value of the terminal condensance is one half the value in the first critical case. Being in the practical region, the present case makes a substantial addition to the simply calculable cases possible without employing an infinite series of I_n functions.

If the resistance r is next the cable, and e is on the other side of the condenser, without extra resistance, the potential V_1 at the junction of the resistance and condenser is

$$V_1=V+rC=e\,\epsilon^{-\sigma t}I_0(\sigma t), \tag{62}$$

by the above. Compare with (48). The curve of V_1 in this second critical case is the same as the curve of C in the standard case ot terminal short-circuit. If, however, e is directly associated with r, there is the usual modification of potential distribution, whereas C is the same through the resistance and condenser whichever way they are put, and wherever e is put.

If $r=Lv$, and $s=\infty$, we have a third critical case,

$$\frac{C}{e}=\frac{\epsilon^{-\sigma t}}{2r}(I_0+I_1). \tag{63}$$

As the condenser is cut out, this curve is of a quite different nature. In fact the curve of C is the same as that of V in the second critical case. The curve of V, however, requires an infinite series.

A fourth critical case is worth noticing, though it does not reduce to finite terms. It is that of equal roots of the a equation. Let

$$m^2+4n=0,\qquad \text{then}\qquad a_1=a_2=2m^{-1}.$$

We may eliminate n, and reduce (38) to

$$\frac{\mathrm{C}}{e}=\frac{\epsilon^{-\sigma t}}{r+\mathrm{L}v}\frac{1-a^2}{(1-\frac{1}{2}ma)^2}=\ldots\ldots(1-a^2)\{1+2(\tfrac{1}{2}ma)+3(\tfrac{1}{2}ma)^2+\ldots\} \quad (64)$$

There do not seem to be any more finite reductions than those described.

Although the method of treating the a fraction similar to (11) ultimately only leads to the result (40), there are some views by the way worth looking at. We have

$$\frac{1-a^2}{1-ma-na^2}=\frac{1-a^2}{a_2-a_1}\left(\frac{a_2}{1-\frac{a}{a_1}}-\frac{a_1}{1-\frac{a}{a_2}}\right), \quad (65)$$

where
$$a_1 \text{ or } a_2=-\frac{m}{2n}\pm\sqrt{\frac{m^2}{4n^2}+\frac{1}{n}}, \quad (66)$$

and the question is what electrical problems the two simpler ones represent. We have to find the meaning of

$$\frac{1}{1-ca}=\mathrm{I}_0+c\mathrm{I}_1+c^2\mathrm{I}_2+c^3\mathrm{I}_3+\ldots, \quad (67)$$

where c is a constant, and of the same multiplied by $1-a^2$. Take (67) first.

Let
$$\mathrm{Z}=\left(\frac{\mathrm{R}+\mathrm{L}p}{\mathrm{S}p}\right)^{\frac{1}{2}}, \quad \text{then} \quad \mathrm{Z}^2=\mathrm{L}^2v^2\left(1+\frac{2\sigma}{p}\right). \quad (68)$$

That is, the square of the resistance operator of the cable is the product of the resistance $\mathrm{L}v$, and of $r+(sp)^{-1}$, where r and s have the special values $\mathrm{L}v$ and $(2\mathrm{L}v\sigma)^{-1}$. This is what takes place in the second critical case above described. But in the more general case, we can put a in terms of p by means of

$$p=\tfrac{1}{2}\sigma(a+a^{-1}), \quad \text{or} \quad a=\frac{\left(\frac{p+\sigma}{p-\sigma}\right)^{\frac{1}{2}}-1}{\left(\frac{p+\sigma}{p-\sigma}\right)^{\frac{1}{2}}+1}. \quad (69)$$

It follows that
$$\epsilon^{-\sigma t}\frac{1}{1-ca}=\frac{\frac{\mathrm{L}v+\mathrm{Z}}{1+c}}{\mathrm{Z}+\frac{\mathrm{Z}^2}{\mathrm{L}v}\frac{1-c}{1+c}}. \quad (70)$$

To verify this, it may be done the other way, Given the right side, solve it in terms of I_n functions with the exponential factor in the way (38) was obtained. The result is the left side. As regards the meaning, the denominator signifies a cable, resistance and condensance in sequence, with special values of r and s, which are positive when $c<1$. The numerator is a constant × the sum of the resistance operator of the cable and of a resistance $\mathrm{L}v$. So the left side of (70) is the sum of two solutions, viz., the potential at the cable terminal due to a steady e of a certain value, and of the current into the cable multiplied by a certain resistance, not the same as r. If $c>1$, this problem is electrically impossible, owing to the negativity of r and s.

If we turn c to its reciprocal, we obtain

$$\epsilon^{-\sigma t}\frac{1}{1-c^{-1}a}=\frac{\dfrac{Lv+Z}{1+c^{-1}}}{Z-\dfrac{Z^2}{Lv}\dfrac{1-c}{1+c}}, \tag{71}$$

which is electrically possible when $c>1$.

By uniting (70) and (71) we get

$$\epsilon^{-\sigma t}\left(\frac{1}{1-ca}+\frac{1}{1-c^{-1}a}-1\right)=\epsilon^{-\sigma t}(I_0+(c+c^{-1})I_1+(c^2+c^{-2})I_2+\dots), \tag{72}$$

on the left side; and

$$\frac{\dfrac{Lv+Z}{1+c}}{Z+\dfrac{Z^2}{Lv}\dfrac{1-c}{1+c}}+\frac{c\dfrac{Lv+Z}{1+c}}{Z-\dfrac{Z^2}{Lv}\dfrac{1-c}{1+c}}-\frac{Lv}{Z}, \tag{73}$$

on the right side. Here note that the 1 in the a function always means a^0; so that $\epsilon^{-\sigma t}a^0=Lv/Z$. This expression (73) reduces to

$$\frac{1-x^2}{1-\dfrac{Z^2}{L^2v^2}x^2}=\frac{1-x^2}{1-x^2\left(1+\dfrac{2\sigma}{p}\right)}, \quad \text{if } x=\frac{1-c}{1+c}. \tag{74}$$

This can be algebrised at sight, and makes

$$\epsilon^{\frac{\sigma t(1-c)^2}{2c}}=\epsilon^{-\sigma t}\times\epsilon^{\frac{\sigma t}{2}\left(c+\frac{1}{c}\right)}. \tag{75}$$

Comparing with (72), we corroborate a well-known besselian formula.

We can also exhibit it as the difference of two cable problems without condensers, but with resistances only. For the left side of (74) is the same as

$$\frac{Lv}{2x}(1-x^2)\left\{\frac{1}{Z+Lv/x}-\frac{1}{Z-Lv/x}\right\}, \tag{78}$$

which expresses the differences of the current from a special e into a cable through the resistances Lv/x and $-Lv/x$; one of which, unfortunately, is negative. But that is of no consequence in the momentary application, which is to elucidate the treatment of operators in the solution of differential equations, rational or fractional. This matter has a great future, not merely in electromagnetics, but in all mechanical applications, as academical mathematicians will find out in time, if they live long enough.

In a similar way, it may be shown that

$$\epsilon^{-\sigma t}\frac{1-a^2}{1-ca}=\frac{2\mathrm{L}v}{\mathrm{Z}+\mathrm{L}v+\frac{(1-c)\mathrm{R}v}{2p}}, \tag{79}$$

so the problem represented is that of a cable with terminal resistance and condensance, the first of which has the critical value, but the second neither of the critical values, and moreover $c<1$ for electrical reality. Applying this to (65) above, we get

$$\frac{\mathrm{C}}{e}=\frac{\frac{2\mathrm{L}v}{r+\mathrm{L}v}\frac{a_2}{a_2-a_1}}{\mathrm{L}v+\mathrm{Z}+\frac{a_1-1}{2a_1}\frac{\mathrm{R}v}{p}}-\frac{\frac{2\mathrm{L}v}{r+\mathrm{L}v}\frac{a_1}{a_2-a_1}}{\mathrm{L}v+\mathrm{Z}+\frac{a_2-1}{2a_2}\frac{\mathrm{R}v}{p}}. \tag{80}$$

This exhibits C in the case of any values of r and s in terms of two C's, both with the critical $r=\mathrm{L}v$, but with different s's, which need not be real, however. The impressed forces are also different. I hoped this method might be as illuminating as in the corresponding case when $\mathrm{L}=0$, but it is not.

As another example, examine the meaning of

$$\frac{\mathrm{C}}{e}=\frac{1}{\mathrm{Z}+c\mathrm{Z}^2}, \quad \text{where} \quad \mathrm{Z}=\left(\frac{\mathrm{R}+\mathrm{L}p}{\mathrm{K}+\mathrm{S}p}\right)^{\frac{1}{2}}. \tag{81}$$

Here clearly c is a conductance. Put in terms of ρ and σ, equation (53). Then

$$\frac{\mathrm{C}}{e}=\frac{1}{\mathrm{L}v\left(\frac{p+\rho+\sigma}{p+\rho-\sigma}\right)^{\frac{1}{2}}+c\mathrm{L}^2v^2\left(1+\frac{2\sigma}{p+\rho-\sigma}\right)}. \tag{82}$$

We therefore have

$$r=c\mathrm{L}^2v^2, \qquad k+sp=\frac{\rho-\sigma}{2\sigma c\mathrm{L}^2v^2}\left(1+\frac{p}{\rho-\sigma}\right). \tag{83}$$

These show the value of the terminal resistance r, and also that the condenser is shunted, or has the conductance k; and shows the values of k and s.

To develop in terms of a, change p to $p-\rho$, at the same time putting on the prefactor $\epsilon^{-\rho t}$ and the postfactor $\epsilon^{\rho t}$ or $p/(p-\rho)$; then finally put $\frac{1}{2}\sigma(a+a^{-1})$ for p, and put on the postfactor $(1-a^2)/(1+a^2)$, which is the besselian development of unity. The result is

$$\frac{\mathrm{C}}{e}=\epsilon^{-\rho t}\frac{(1-a^2)(1-a)^2}{(1+a^2-2\rho\sigma^{-1}a)\,\mathrm{L}v\,\{(1-a^2)+c\mathrm{L}v(1+a)^2\}}. \tag{84}$$

Finally, this may be expanded by division. If $c\mathrm{L}v=1$, it simplifies, viz.:—

$$\frac{\mathrm{C}}{e}=\epsilon^{-\rho t}\frac{(1-a)^3}{2\mathrm{L}v(1+a^2-2a\rho\sigma^{-1})} \tag{85}$$

To destroy the leakage $\rho=\sigma$. We then reduce to the case of equation (61) above

The Curbing Effect of an Inductance Shunt.

§ 522. That an inductance will curb signals when used as a shunt to the receiving instrument, instead of a condenser in series, is obvious enough. But the inductance and the condenser curves of current in the receiver cannot be quite similar; and it is not easy to say, given an unlimited power of varying the size of the condenser and of the induction coil, which method would be preferable on the whole. I have been informed by Mr. W. Gaye, that in certain cases the inductance shunt has been found to be 12 to 14 per cent. better than the condenser, owing to the better definition of successive dots on the same side. But the working speed is the resultant of so many factors that this estimate should not be extended too far beyond present experience.

We may get some theoretical guidance by comparing the curves of current in the two cases of condenser and inductance under the same circumstances as regards impressed voltage. The condenser with resistance in series has been done. Now do the coil theory.

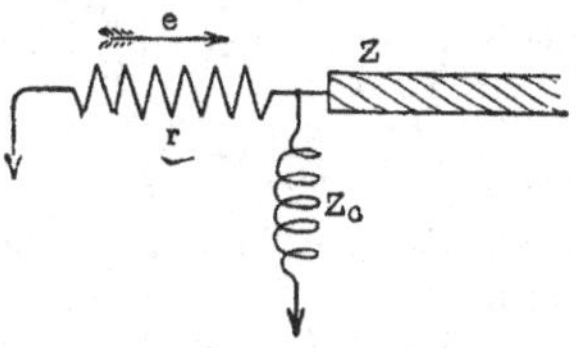

Let the impressed force e act in the resistance r, joined to the beginning of an infinitely long cable, with an inductance coil as shunt. The inductance should be very large, and the resistance of the coil small. The initial current is e/r, just as when a condenser is used in series with r, instead of the shunt coil. The final current is $e/(r+R_0)$ in the battery and the coil, but zero in the cable. This is again like the action of a condenser, as regards the cable current, though not otherwise. But the course of the current in the cable is not the same in the two cases.

The current in r at any moment is

$$C_r = \frac{e}{r + ZZ_0/(Z+Z_0)}, \qquad Z = (R/Sp)^{\frac{1}{2}}, \quad Z_0 = R_0 + L_0 p; \tag{1}$$

and the current entering the cable is

$$C = C_r \frac{Z_0}{Z} = \frac{e}{r + Z + rZ/Z_0}, \tag{2}$$

by the usual formulæ for conductors, but using the resistance operators instead of the resistances.

We must now algebrize this formula. Take $Z_0 = L_0 p$. The resistance R_0 is not material. Then

$$C = \frac{e}{r\left\{1 + \frac{a}{p^{\frac{1}{2}}}\left(1 + \frac{b}{p}\right)\right\}}; \qquad a = \frac{1}{r}\left(\frac{R}{S}\right)^{\frac{1}{2}}, \qquad b = \frac{r}{L_0}. \tag{3}$$

It can be seen at sight that long division will give a convergent solution, which is therefore *the* solution, and two others of mixed character. Thus, the convergent solution is

$$C = \frac{e}{r}\left\{1 - \frac{a}{p}\left(1 + \frac{b}{p}\right) + \frac{a^2}{p}\left(1 + \frac{b}{p}\right)^2 - \frac{a^3}{p^{1\frac{1}{2}}}\left(1 + \frac{b}{p}\right)^3 + \dots\right\}, \tag{4}$$

which is to be algebrized by turning p^{-n} to $t^n/\lfloor n$. Since this solution is in sets of terms alternately + and −, this may be the best arrangement for calculation, as it stands. The range of utility with moderate work must be found out. Since $b = r/L_0$, the larger L_0 is the easier the work.

The solution may also be regarded as the difference of two power series, one in integral powers only, the other fractional only. Thus,

$$C = \frac{e}{r}\left\{1 + \frac{a^2}{\mu}\left(1 + \frac{b}{p}\right)^2 + \frac{a^4}{p^2}\left(1 + \frac{b}{p}\right)^4 + \dots\right\}$$

$$- \frac{e}{r}\left\{\frac{a}{p^{\frac{1}{2}}}\left(1 + \frac{b}{p}\right) + \frac{a^3}{p^{1\frac{1}{2}}}\left(1 + \frac{b}{p}\right)^3 + \dots\right\}, \tag{5}$$

and it is easy, in (4) and (5), to collect the coefficients so as to have any power of t isolated.

The corresponding condenser formula is

$$C = \frac{e}{r + \frac{1}{Sp} + Z} = \frac{e}{r\left\{1 + \frac{a}{p^{\frac{1}{2}}}\left(1 + \frac{c}{p^{\frac{1}{2}}}\right)\right\}} \tag{6}$$

where a is as before, whilst c is a fresh constant. That is, c/p in the condenser case becomes $b/p^{1\frac{1}{2}}$ in the coil case. This suggests more effective curbing, but regular calculation is required to show it.

Since the denominator in (3) is a cubic in $p^{\frac{1}{2}}$, the divergent solution requires special treatment to harmonize with the convergent. This is reserved.

If we shift e to the other side of the junction, so as to be in the cable itself, C then represents the current in r. That is to say, C is current in the receiver at the end of a very long cable due to a quite special sort of signal, and curbed by the inductance shunt. This may be compared with the corresponding received current with a condenser inserted instead. So that the difference in the curbing action as calculated at the sending end serves to illustrate the same at the receiving end. The actual formula for a practical received signal which has traversed the cable is of course very complicated.

If the inductance shunt is better than the condenser in receiving, it may be expected to be better for sending as well. The proper way is to try it thoroughly. Perhaps earth currents may be detrimental.

The Transverse Momentum of an Electron.

[*Nature*, Aug. 31, 1905, p. 429.]

§ 523. When Newton's Third Law is applied to an electron it makes

$$\mathbf{F} = \dot{\mathbf{M}} + \dot{\mathbf{N}}, \tag{1}$$

where **M** is the "momentum" in the field, or that part of the time integral of the force on the ether which is in the field, or Σ **VDB**, and **N** is the momentum already wasted, whilst **F** is the applied force on the electron.

Similarly, Newton's Fourth Law (or the Scholium to the Third) makes

$$\mathbf{F}\mathbf{u} = \dot{\mathrm{U}} + \dot{\mathrm{T}} + \mathrm{W}, \tag{2}$$

if **u** is the velocity of the electron, U the electric and T the magnetic field energy, and W the rate of waste of energy.

Now, both W and $\dot{\mathrm{N}}$ are known in terms of the velocity and acceleration of the charge at any moment by formulæ I gave in *Nature*, Nov. 6, 13, 1902. But when applied to (1), (2), these equations do not let us determine **M** generally in terms of the velocity and acceleration, on account of the variability of the state of the field, and of the waste of energy and momentum. **M** is indefinite. But in long-continued forced circular motion of a charge, $\dot{\mathrm{U}} + \dot{\mathrm{T}} = 0$. So [p. 163, eq. (18)]

$$\mathbf{F}\mathbf{u} = \mathrm{W} = \frac{\mu \mathrm{Q}^2 \mathrm{A}^2}{6\pi v \kappa^4}, \tag{3}$$

(*loc. cit.*), where Q is the charge, and A the acceleration (or u^2/R, if R is the radius of the orbit). Also, $\kappa^2 = 1 - u^2/v^2$. The direct or **u** component of **F** is therefore known. We also have (*loc. cit.*) [p. 163, eq. (19),]

$$\dot{\mathrm{N}} = \frac{\mathbf{u}}{v^2}\mathrm{W}. \tag{4}$$

Using these in (1), along with (3), we come to

$$\kappa^2 \mathbf{F}_1 = \dot{\mathbf{M}}_2, \qquad \mathbf{F}_2 = \dot{\mathbf{M}}_1, \tag{5}$$

where $\mathbf{F}_1$ is the **u** component, and $\mathbf{F}_2$ the transverse component, towards the centre.

Thus only the part $\kappa^2\mathbf{F}_1$ of the direct force is associated with the transverse or centripetal force $\mathbf{F}_2$ in keeping up the revolving state; the rest of $\mathbf{F}_1$, that is, $(u^2/v^2)\mathbf{F}_1$, being the wasted part as regards momentum, although the whole of $\mathbf{F}_1$ is concerned in the waste of energy.

Now, $\dot{\mathbf{M}} = \mathrm{V}\mathbf{n}\mathbf{M}$, if $\mathbf{n}$ is the angular velocity. That is,

$$\dot{\mathbf{M}} = \dot{\mathrm{M}}_1\mathbf{u}_1 + \mathrm{M}_1\dot{\mathbf{u}}_1 + \dot{\mathrm{M}}_2\mathbf{A}_1 + \mathrm{M}_2\dot{\mathbf{A}}_1, \tag{6}$$

if $\mathbf{u}_1$ and $\mathbf{A}_1$ are unit vectors, making

$$\dot{\mathbf{u}}_1 = (u/\mathrm{R})\mathbf{A}_1, \qquad \dot{\mathbf{A}}_1 = -(u/\mathrm{R})\mathbf{u}_1. \tag{7}$$

Also, $\dot{\mathrm{M}}_1 = 0$, $\dot{\mathrm{M}}_2 = 0$, because the motion is steady. So we convert (5) to

$$\kappa^2\mathbf{F}_1 = \dot{\mathbf{M}}_2 = -\mathrm{M}_2(u/\mathrm{R})\mathbf{u}_1, \quad \mathbf{F}_2 = \dot{\mathbf{M}}_1 = \mathrm{M}_1(u/\mathrm{R})\mathbf{A}_1. \tag{8}$$

Finally, although we get no formula for M_1, we do obtain a complete formula for M_2, viz.,

$$\mathrm{M}_2 = -\frac{\mu\mathrm{Q}^2\mathrm{A}}{6\pi v\kappa^2}. \tag{9}$$

This is the transverse momentum of Q in steady circular motion, without any limitations upon the size of the velocity and acceleration, save the usual ones, $u<v$, and A not excessively great in regard to the diameter of the electron.

It would seem that an integration over the whole field, in which **E** and **H** are known (*loc. cit.*) [p. 162], is required to find M_1, the direct momentum. If, however, the acceleration is infinitesimal, the known formula for M_1 in steady rectilinear motion may be employed, viz., $\frac{1}{2}\mathrm{M}_1 u = \mathrm{T}$.

Finally, I have pleasure in saying that Mr. G. F. C. Searle, F.R.S., led me to see that my waste formulæ led to the formula (9) for the transverse momentum, by submitting to me a calculation of M_2 in the special case of infinitesimal acceleration and velocity. He made no use of the waste formula, not being aware of it, but since in the circumstances the waste is infinitesimal, it did not matter. In fact, $\frac{1}{2}\mathrm{M}_1 u = \mathrm{T}$ leads to the reduced special value of the transverse momentum when u and A are infinitesimal. The argument was somewhat obscure by the want of comprehensiveness, but the result agrees with (9).

Extension to Helixal Motion.

§ 524. To pass from a circular to a right helixal path, we have merely to add to the circular velocity $\mathbf{u}$ a constant rectilinear velocity $\mathbf{w}$ perpendicular to the circle. Since the acceleration is unaltered and the waste of energy goes on uniformly, it may be inferred that the formula for the transverse momentum is the same, with allowance made for the changed velocity. But there are some changes otherwise. Thus, let $\mathbf{q}=\mathbf{u}+\mathbf{w}$ be the actual velocity, then $\kappa^2=1-q^2/v^2$ now, and

$$W=\frac{\mu Q^2A^2}{6\pi v\kappa^4}, \qquad \dot{\mathbf{N}}=\frac{\mathbf{q}}{v^2}W, \tag{10}$$

represent the rate of waste of energy and momentum. Here $A=u^2/R$, as before, but now $\dot{\mathbf{N}}$ has an axial component.

Let $\mathbf{M}=\mathbf{M}_1+\mathbf{M}_2+\mathbf{M}_3$, where $\mathbf{M}_1$ is the circular component, $\mathbf{M}_2$ the radial component inward, and $\mathbf{M}_3$ the axial component. We have $\mathbf{M}_3=$constant, because its direction is constant, and the symmetry makes its size constant. So the equation

$$\mathbf{F}=\dot{\mathbf{M}}+\dot{\mathbf{N}}$$

splits up into

$$\mathbf{F}_1=\dot{\mathbf{M}}_2+(\mathbf{u}/v^2)W=-M_2(u/R)\mathbf{u}_1+(\mathbf{u}/v^2)W, \tag{11}$$

$$\mathbf{F}_2=\dot{\mathbf{M}}_1=M_1(u/R)\mathbf{A}_1, \qquad \mathbf{F}_3=(\mathbf{w}/v^2)W. \tag{12}$$

Also, the activity equation becomes, by (11) and (12),

$$W=\mathbf{Fq}=F_1u+F_3w=-M_2A+(q^2/v^2)W, \tag{13}$$

from which

$$\kappa^2W=-M_2A, \tag{14}$$

so that M_2, F_1 and F_3 are all known in terms of W; thus,

$$M_2=-\frac{\kappa^2}{A}W, \qquad F_3=\frac{w}{v^2}W, \qquad F_1=\frac{W}{u}\left(1-\frac{w^2}{v^2}\right), \tag{15}$$

where W is given in (10). But F_2, M_1, and M_3 require further work.

The transverse momentum is increased by the axial motion added to the circular (supposed unchanged) by reason of the changed value of the velocity and therefore of κ^2. It may be largely multiplied in this way. F_1 is also increased. The new force F_3 also mounts up with w. It would be zero, if the motion were strictly rectilinear at constant speed.

But if it is the actual speed q in the helixal path that is kept constant, then M_2 and F_1 fall off as w increases, whilst F_3 first increases and then falls off.

Deep Water Waves.

§ 525. This subject turned up in § 520 very strangely in the theory of a condenser and resistance placed in series with a cable. But some time before that, in Lord Kelvin's "Baltimore Lectures," my attention was directed to the water waves by the remark that a certain definite integral was "irreducible." This meant, I supposed, that it had not been algebrised. Whether so or not, I found that it was readily algebrisable operationally to a convergent series, which also showed the necessity of a second form, divergent, for purposes of calculation. I make no pretence of writing about deep water waves of various types from the practical point of view. What follows is to illustrate operational methods.

Let
$$D=\frac{D_0}{\pi}\int^{\infty}\cos qx\cos qvt\,dq, \tag{1}$$

What does D mean? Put $t=0$. Then $D=0$ everywhere save at $x=0$, where $\int D\,dx=D_0$. That is, D at time t results from the amount of D represented by D_0 condensed at the origin initially, and made to spread right and left in a manner controlled by the partial characteristic of D, modified also by the similar spreading of the initial states of D, $\ddot{D}$, &c., so far as they are concerned.

Say $v=aq^n$, or $v\propto\lambda^{-n}$, if λ is the wave length in a train. Then D is equivalently expressed by

$$D=\frac{D_0}{\pi}\int_0^{\infty}\frac{\cos qx\,dq}{1+a^2q^{2n+2}/p^2}=\frac{D_0}{\pi}\int_0^{\infty}\frac{\cos atq^{n+1}\,dq}{1+q^2/\Delta^2}, \tag{2}$$

where $p=d/dt$ and $\Delta=d/dx$. Or by

$$D=\frac{D_0}{\pi}\int_0^{\infty}\frac{dq}{(1+q^2/\Delta^2)\,(1+a^2q^{2n+2}/p^2)} \tag{3}$$

If any of these forms be integrated, the result is to express D in terms of D_0 operationally, with one differentiator concerned in the case of either of (2) and with two differentiators in the case of (3). This may perhaps be done for any value of n, though certainly it will be complicated. For special values of n, it can easily be done. If $n=0$, v is constant. This is an easy case. If $n=1$, $v=aq$. This occurs in the theory of lateral vibrations of an elastic rod. If $n=-1$, $v\propto\lambda$, so it is a singular case of stationary vibrations.

If $n=-\frac{1}{2}$, we come to the deep water surface wave theory. Then, selecting the form (3), we have

$$D=\frac{D_0}{\pi}\int_0^{\infty}\frac{dq}{(1+q^2/\Delta^2)\,(1+a^2q/p^2)}. \tag{4}$$

The indefinite integral is known, thus

$$\frac{D}{D_0}=\frac{1}{\pi}\frac{\Delta^2p^2/a^2}{\Delta^2+p^4/a^4}\left[\log\frac{q+p^2/a^2}{\sqrt{q^2+p^4/a^4}}+\frac{p^2}{a^2\Delta}\tan^{-1}\frac{q}{\Delta}\right]_0^{\infty} \tag{5}$$

So, finally,

$$\frac{D}{D_0}=\frac{1}{\pi}\frac{\Delta^2p^2/a^2}{\Delta^2+p^4/a^4}\left\{\frac{p^2}{a^2\Delta}\frac{\pi}{2}-\log\frac{p^2}{a^2\Delta}\right\}. \tag{6}$$

This being an operational solution, can be algebrised in the usual way.

The rational part of (6) makes

$$\tfrac{1}{2}\frac{p^2}{a^2}\sin\frac{p^2x}{a^2},\quad\text{by }\Delta\text{ first,} \tag{7}$$

or, $$\tfrac{1}{2}\Delta\left(1-a^4\Delta^2\frac{t^4}{\lfloor 4}+a^8\Delta^4\frac{t^8}{\lfloor 8}-\cdots\right),\ \text{by } p \text{ first.} \tag{8}$$

Next do the logarithmic part of (6). With Δ first we get

$$-\frac{1}{\pi}\frac{d}{dn_0}\frac{(p^2/a^2)\,(p^2/a^2\Delta)^n}{1+p^4/a^4\Delta^2}, \tag{9}$$

where n_0 means that we put $n=0$ after differentiation to n. Or,

$$-\frac{1}{\pi}\frac{d}{dn_0}\left\{\frac{x^n}{\lfloor n}\frac{(at)^{-(2n+2)}}{\lfloor -(2n+2)}+\frac{x^{n+2}}{\lfloor n+2}\frac{(at)^{-(2n+6)}}{\lfloor -(2n+6)}+\cdots\right\}$$

$$=\frac{2}{\pi}\Big[(at)^{-2}g'(-2)-\frac{x^2}{\lfloor 2}(at)^{-6}g'(-6)+\cdots\Big]$$

$$=-\frac{2}{\pi a^2t^2}\left\{1-\frac{\lfloor 5}{\lfloor 2}\frac{x^2}{a^4t^4}+\frac{\lfloor 9}{\lfloor 4}\left(\frac{x^2}{a^4t^4}\right)^2-\cdots\right\}. \tag{10}$$

But if we do p first in the logarithmic part of (6), we get

$$-\frac{1}{\pi}\frac{d}{dn_0}\frac{(p^2/a^2)^{n-1}\Delta^{-(n-2)}}{1+a^4\Delta^2/p^4} \tag{11}$$

$$=-\frac{1}{\pi}\frac{d}{dn_0}\left\{\frac{x^{n-2}}{\lfloor n-2}\frac{(at)^{-(2n-2)}}{\lfloor -(2n-2)}-\frac{x^{n-4}}{\lfloor n-4}\frac{(at)^{-(2n-6)}}{\lfloor -(2n-6)}+\cdots\right\}$$

$$=-\frac{1}{\pi}\left\{x^{-2}g'(-2)\frac{a^2t^2}{\lfloor 2}-x^{-4}g'(-4)\frac{(at)^6}{\lfloor 6}+\cdots\right\}$$

$$=\frac{1}{\pi}\left(\frac{at}{x}\right)^2\left\{\frac{1}{\lfloor 2}-\frac{\lfloor 3}{\lfloor 6}\frac{a^4t^4}{x^2}+\frac{\lfloor 5}{\lfloor 10}\frac{a^8t^8}{x^4}-\cdots\right\}. \tag{12}$$

The sum of (12) and (8) is one form of solution, the sum of (10) and (7) is another. But (12) is convergent and (10) divergent. Now it is a result of experience that when an algebrisation comes out in the form of a convergent series like (12) coupled with an operational series in integral positive powers of the differentiator, like (8), then the latter part means zero, just as any one would suppose, in fact. Therefore,

$$\frac{D}{D_0}=\frac{1}{\pi}\left(\frac{at}{x}\right)^2\left\{\frac{1}{\lfloor 2}-\frac{\lfloor 3}{\lfloor 6}\frac{a^4t^4}{x^2}+\frac{\lfloor 5}{\lfloor 10}\frac{a^8t^8}{x^4}-\cdots\right\} \tag{13}$$

is the convergent solution of our problem. But it is equally a result of experience that when we come to a *divergent* series like

(10) coupled with an operational series like (7), which would seem to mean zero, then it may do so, or it may not. That is, the divergent equivalent of (13) is

$$\frac{D}{D_0}=\frac{D_1}{D_0}-\frac{2}{\pi a^2t^2}\left\{1-\frac{\lfloor 5}{\lfloor 2}\frac{x^2}{a^4t^4}+\frac{\lfloor 9}{\lfloor 4}\frac{x^4}{a^8t^8}-\cdots\right\}, \tag{14}$$

where D_1 is an auxiliary function to be found from (7), or by other means.

Deferring this work, numerical examination of (13) shows it to be oscillatory, but only the extreme outer part, as said before, can be conveniently calculated. Give a constant value to t, and then vary x from ∞ downwards. D mounts up from zero at ∞ to the top of a hump at a finite distance and then falls to zero. Going nearer the origin, there is a negative hump, or a hollow, now of finite length, of course, and, moreover, its maximum is greater than that of the outermost hump. This is quite sufficient, along with experience of this kind of formula, to let one see that between the outermost hump and the origin is an infinite series of humps and hollows with amplitudes increasing and lengths decreasing towards the origin. But we also see the necessity of finding D_1 in (14), to enable the waves to be numerically estimated. Without that, (13) shows that if $y=a^4t^4/x^2$, for any constant value of y, x varies as t^2 and D varies as t^{-2}. Also $\dot{x}$ varies as t, so the velocity of a particular phase, as a node, varies as t. The distance between two nodes varies as t^2.

The following is one way of finding D_1. Put $q=a^2t^2/4x$ in (13), and $\Delta=d/dq$; then

$$\frac{D}{D_0}=\frac{4q^{1\frac{1}{2}}}{\pi^{\frac{1}{2}}a^2t^2}\left(\frac{q^{\frac{1}{2}}}{\lfloor\frac{1}{2}}-\frac{q^{2\frac{1}{2}}}{\lfloor 2\frac{1}{2}}+\frac{q^{4\frac{1}{2}}}{\lfloor 4\frac{1}{2}}-\cdots\right) \tag{15}$$

$$=\frac{4q^{1\frac{1}{2}}}{\pi^{\frac{1}{2}}(at)^2}\times\Delta^{-\frac{1}{2}}\cos q. \tag{16}$$

In this form we see that the transformation from (13) to (14) depends upon the generalised cosine function. Thus, let

$$\mathrm{Cos}\, q=\cdots+\frac{q^{-4}}{\lfloor -4}-\frac{q^{-2}}{\lfloor -2}+1-\frac{q^2}{\lfloor 2}+\frac{q^4}{\lfloor 4}-\cdots, \tag{17}$$

then Cos q is numerically the same as $\cos q$. But it does not behave the same when differentiated, for

$$\Delta^{-r}\,\mathrm{Cos}\, q=\cdots+\frac{q^{r-4}}{\lfloor r-4}-\frac{q^{r-2}}{\lfloor r-2}+\frac{q^r}{\lfloor r}-\frac{q^{r+2}}{\lfloor r+2}+\frac{q^{r+4}}{\lfloor r+4}-\cdots \tag{18}$$

Now $\Delta\cos q=\cos(q+\frac{1}{2}\pi)$, so the effect of Δ^{-r} is to add $-\frac{1}{2}r\pi$ to the argument. This makes

$$\Delta^{-r}\,\mathrm{Cos}\, q=\cos(q-\tfrac{1}{2}r\pi)=\cdots-\frac{q^{r-2}}{\lfloor r-2}+\frac{q^r}{\lfloor r}-\frac{q^{r+2}}{\lfloor r+2}+\cdots, \tag{19}$$

$$\Delta^{-r}\,\mathrm{Sin}\, q=\sin(q-\tfrac{1}{2}r\pi)=\cdots-\frac{q^{r-1}}{\lfloor r-1}+\frac{q^{r+1}}{\lfloor r+1}-\frac{q^{r+3}}{\lfloor r+3}+\cdots \tag{20}$$

The above is something like the way of constructing an integral for $x^n/\underline{|n}$, in vol. 2, p. 435. Or thus (if $\Delta = d/dx$), by the above,

$$\Delta^{-n} \cos mx = m^{-n} \cos (mx - \tfrac{1}{2}n\pi);$$

so
$$\Delta^{-n} \frac{1}{\pi}\int_0^\infty \cos mx\, dm = \frac{1}{\pi}\int_0^\infty \frac{\cos (mx - \frac{1}{2}n\pi)}{m^n} dm.$$

This is $\Delta^{-n}\Delta 1$, so

$$\frac{x^n}{\underline{|n}} = \Delta^{-n} 1 = \frac{1}{\pi}\int_0^\infty \frac{\sin (mx - \frac{1}{2}n\pi)}{m^{n+1}} dm. \tag{21}$$

Finally, we should show the connection with the generalised exponential function. Say

$$\epsilon^x = \cdots + \frac{x^{r-1}}{\underline{|r-1}} + \frac{x^r}{\underline{|r}} + \frac{x^{r+1}}{\underline{|r+1}} + \cdots. \tag{22}$$

Put $x = yi$. Then we get

$$\cos y + i \sin y = \Bigg[\left(\cdots + \frac{y^r}{\underline{|r}} - \frac{y^{r+2}}{\underline{|r+2}} + \cdots\right)$$
$$- i\left(\cdots + \frac{y^{r-1}}{\underline{|r-1}} - \frac{y^{r+1}}{\underline{|r+1}} + \cdots\right)\Bigg] i^r. \tag{23}$$

Next, use $i^r = (\cos + i \sin)\frac{1}{2}r\pi$, and assume that y is real positive. Then we may separate the real and imaginary parts of (23), and obtain

$$\cos y = \left(\cdots + \frac{y^r}{\underline{|r}} - \frac{y^{r+2}}{\underline{|r+2}} + \cdots\right)\cos\tfrac{1}{2}r\pi + \left(\cdots + \frac{y^{r-1}}{\underline{|r-1}} - \frac{y^{r+1}}{\underline{|r+1}} + \cdots\right)$$
$$\sin\tfrac{1}{2}r\pi, \tag{24}$$

and a similar equation. Both these results involve (19) and (20).

It is to be noted that (22) is numerically true when x is real positive, and that (19), (20) are numerically true when q is real positive. As a test, consider the case $r = \frac{1}{2}$, making

$$\cos (q - \tfrac{1}{4}\pi) = \cdots + \frac{q^{\frac{1}{2}}}{\underline{|\tfrac{1}{2}}} - \frac{q^{2\frac{1}{2}}}{2\frac{1}{2}} + \frac{q^{4\frac{1}{2}}}{\underline{|4\tfrac{1}{2}}} - \cdots. \tag{25}$$

I have worked out the long sum for the value $(2q)^2 = 100$, and obtain the value $-{\cdot}478$, whilst the tables make the left side come to $-{\cdot}477$. Also for the value $(2q)^2 = 50$. The series makes $-{\cdot}917$, whilst the tables make $-{\cdot}92$. These tests are good, and the errors are well within the admissible errors conditioned by the finite size of the last convergent term in the divergent series.

Now use (25) in (15). We see that

$$\frac{D}{D_0} = \frac{4q^{1\frac{1}{2}}}{\pi^{\frac{1}{2}}a^2t^2}\left\{\cos (q - \tfrac{1}{4}\pi) + \frac{q^{-1\frac{1}{2}}}{\underline{|-1\tfrac{1}{2}}} - \frac{q^{-3\frac{1}{2}}}{\underline{|-3\tfrac{1}{2}}} + \cdots\right\}; \tag{26}$$

or in terms of x and t,

$$\frac{D}{D_0} = \frac{at}{2\sqrt{\pi}x^{1\frac{1}{2}}}\cos\left(\frac{a^2t^2}{4x} - \frac{\pi}{4}\right) - \frac{2}{\pi a^2t^2}\left(1 - \underline{\left|\tfrac{5}{2}\right.}\,\frac{x^2}{a^4t^4} + \cdots\right). \tag{27}$$

This is the complete divergent form of solution. The auxiliary function D_1 is found to be

$$\frac{D_1}{D_0}=\frac{at}{2\sqrt{\pi}x^{1\frac{1}{2}}}\cos\left(\frac{a^2t^2}{4x}-\frac{\pi}{4}\right). \tag{28}$$

Here is another way of finding the auxiliary function. Go back to (16). Say

$$E=\Delta^{-\frac{1}{2}}\cos q=\frac{\Delta^{-\frac{1}{2}}}{1+\Delta^{-2}}. \tag{29}$$

If we algebrise this in a convergent series, we obtain (15), of course. But do it the other way. Thus, say E_2 now,

$$E_2=\frac{\Delta^{1\frac{1}{2}}}{1+\Delta^2}=\Delta^{1\frac{1}{2}}-\Delta^{3\frac{1}{2}}+\cdots=\frac{q^{-1\frac{1}{2}}}{\underline{|-1\frac{1}{2}}}-\frac{q^{-3\frac{1}{2}}}{\underline{|-3\frac{1}{2}}}+\cdots. \tag{30}$$

Comparing with (26), we see that the result is only the divergent part. The other part, say $E_1=\cos(q-\frac{1}{4}\pi)$, does not come. Now $E=E_1+E_2$, if E is the convergent series, E_2 the divergent, and $E-E_2$ represents the complete generalised series. A practical way is wanted of easily determining its value, in a great variety of problems, the present being only a special case. I do not know of a general method, but find the Expansion Theorem (vol. 2, p. 127) is often useful, either to give the true result, or else a result which is suggestive. Thus, to give examples, say

$$u=\frac{\Delta^{-r}}{1-\Delta^{-n}}. \tag{31}$$

When n and r are integers, the Expansion Theorem may be used, with caution or some modification, however, when $r>n$. It gives the value in a series of functions of x, of the power series which is obtainable at sight by division. This is the proper use of the Expansion Theorem. But when r is not integral, it gives the value of the generalised series sometimes. Let $\{...\}$ indicate the use of the theorem. Then, $n=1$, and let r be fractional, and less than n.

$$\left\{\frac{\Delta^{-r}}{1-\Delta^{-1}}\right\}=\left\{\frac{\Delta^{1-r}}{\Delta-1}\right\}=\frac{1^{1-r}\epsilon^x}{1}=\epsilon^x. \tag{32}$$

This is the value of the complete series $\Sigma x^r/\underline{|r}$. Again

$$\left\{\frac{\Delta^{-r}}{1+\Delta^{-1}}\right\}=\left\{\frac{\Delta^{1-r}}{\Delta+1}\right\}=\epsilon^{-x}(\cos r\pi-i\sin r\pi), \tag{33}$$

of which the real part is the value of the complete series $\Sigma\pm x^r/\underline{|r}$.

Passing to $n=2$, we have

$$\left\{\frac{\Delta^{2-r}}{\Delta^2+1}\right\}=\frac{i^{2-r}\epsilon^{ix}}{-2}+\frac{(-i)^{2-r}\epsilon^{-ix}}{-2}=\cos(x-\tfrac{1}{2}r\pi). \tag{34}$$

Here again the result is the value of the generalised series, as above determined. So far is very encouraging. But, passing to $n=3$,

$$\left\{\frac{\Delta^{3-r}}{\Delta^3-1}\right\}=\tfrac{1}{3}\epsilon^x+\tfrac{1}{3}\epsilon^{-\frac{1}{2}x}\{\cos(x\sqrt{\tfrac{3}{4}}-2\pi r/3)+\cos(x\sqrt{\tfrac{3}{4}}+4\pi r/3)\}$$
$$+\ldots\quad\{i\sin(\quad\ldots\quad)-i\sin(\quad\ldots\quad).\} \tag{35}$$

Now the real part of the right side, in the special case $r=\frac{1}{2}$, does not seem to represent the proper value of the generalised series, and so the suggested extension of the Expansion Theorem to fractional differentiations requires to be hedged round with restrictions. Leaving this matter, therefore, as being in an experimental stage, return to the main question.

Observe how the algebrisation of the log part of (6) was done, through (9) to (10), and then again through (11) to (12). This is the reason. Not knowing in advance exactly what the log of a differentiator means, we have to find out. Things of this sort are not considered in any respectable mathematical work, either because they are beneath notice, or for other reasons. But all mathematics, in the exploring stage, is observational and experimental, as I took pains to insist upon at the beginning of volume 2. Now here we have to find the meaning of log $(p^2/a^2\Delta)$, with unit operand, and of functions of the same. If x is a quantity, we know that $\log x=(d/dn_0)x^n$. The same is true formally for log $(p^2/a^2\Delta)$; so we turn it to $(d/dn_0)(p^2/a^2\Delta)^n$. Here $(p^2/a^2\Delta)^n$ can be algebrised in the usual way, so we algebrise first, and then proceed to the logarithm by differentiation. And the justification of faith is by work, for the process works. If it failed, then we should have to find some other way.

I have already given an example of this process in the present volume, p. 37, where it was applied to work out the new result due to a straight line source of magnetic force in an electrical conductor. The theory is the same as that of the diffusion of heat from a straight line source in a heat conductor.

Here is another way, essentially the same, but a little simpler.

$$\log\frac{p^2}{a^2\Delta}=\frac{d}{dn_0}\left(\frac{p^2}{a^2\Delta}\right)^n=\frac{d}{dn_0}\frac{x^n t^{-2n}}{a^{2n}\underline{|n}\,\underline{|-2n}}=\log\frac{x}{a^2t^2}+\gamma-2\gamma, \qquad (36)$$

where $\gamma=g'(0)$. So, by (6),

$$\frac{\text{D}}{\text{D}_0}=\frac{1}{\pi}\frac{p^2/a^2}{1+p^4/a^4\Delta^2}\left(-\log\frac{x}{a^2t^2}+\gamma\right). \qquad (37)$$

There are two ways of developing this, first with p as a differentiator, and Δ^{-1} as integrator, giving

$$\frac{\text{D}_2}{\text{D}_0}=\frac{1}{1+p^4/a^4\Delta^2}\left(-\frac{2}{\pi a^2t^2}\right)=(10)\text{ above, divergent,} \qquad (38)$$

and then by multiplying the numerator and denominator in (37) by $a^4\Delta^2/p^4$, so that Δ becomes a differentiator and p^{-1} an integrator. This makes

$$\frac{\text{D}}{\text{D}_0}=\frac{1}{1+a^4\Delta^2/p^4}\left(\frac{a^2t^2}{2\pi x^2}\right)=(13)\text{ above, convergent.} \qquad (39)$$

In connection with the treatment of the logarithm of a differentiator is the treatment of $\Delta^r \log x$, where r is any real number. It is rather similar. Put $(d/dn_0)x^n$ for $\log x$. Then

$$\Delta^r \log x = \Delta^r \frac{d}{dn_0} x^n = \frac{d}{dn_0} \frac{x^{n-r}}{\lfloor n-r} \lfloor n$$

$$= \frac{x^{-r}}{\lfloor -r}(\log x - g'(0)) + x^{-r} g'(-r). \tag{40}$$

When r is a positive integer, the result is $x^{-r}g'(-r)$. When a negative integer, the result is that of $-r$ complete integrations. The function $g(m)$ is the reciprocal of $\lfloor m$, and $g'(m)$ is its derivative. See vol. 2, p. 360, 438.

So much for logarithms. Now go back to the original definite integral. I selected the form (3) at first rather than the simpler ones in (2), because I happened to recognise the indefinite integral. Owing to two differentiators being concerned, as well as the log, it turned out to be instructive, various points of interest being present. But now examine (2), say the first form, with $n = -\frac{1}{2}$, and see how it goes.

$$\frac{D}{D_0} = \frac{1}{\pi}\int_0^\infty \frac{\cos qx\, dq}{1 + a^2 q/p^2}. \tag{41}$$

Not seeing the indefinite integral, convert it to operational form thus

$$\frac{D}{D_0} = \frac{1}{\pi}\int_0^\infty \frac{(1 - a^2 q/p^2)\cos qx}{1 - a^4 q^2/p^4} dq \quad = \frac{1}{\pi}\int_0^\infty \frac{\cos qx - a^2\Delta/p^2 \sin qx}{1 + a^4\Delta^2/p^4} dq. \tag{42}$$

The changes made are obvious, and enable us to transform to

$$\frac{D}{D_0} = \frac{1}{\pi}\frac{1}{1 + a^4\Delta^2/p^4}\int_0^\infty \cos qx\, dq - \frac{1}{\pi}\frac{a^2\Delta/p^2}{1 + a^4\Delta^2/p^4}\int_0^\infty \sin qx\, dq. \tag{43}$$

Here $\int \cos qx\, dq = \Delta 1$, and $\int \sin qx\, dq = x^{-1}$, so

$$\frac{D}{D_0} = \frac{\Delta}{\pi(1 + a^4\Delta^2/p^4)} + \frac{a^2/p^2}{1 + a^4\Delta^2/p^4}\frac{1}{\pi x^2}. \tag{44}$$

Comparing with (39), we can see that this gives the convergent result (13).

As regards the use of $x^{-1} = \int \sin qx\, dq$, or $1 = \int x \sin qx\, dq$, or

$$1 = \int_0^\infty -\frac{d}{dq}(\cos qx) \,.\, dq = \Big[-\cos qx \Big]_0^\infty, \tag{45}$$

it would be strictly true if $\cos qx = 0$, when $qx = \infty$. Really it is indefinite, with mean value zero. I think, therefore, that the truth of $x^{-1} = \int \sin qx\, dq$ requires reservations and caution in general application. Nevertheless, it is a very useful working formula in physical applications, like $\Delta 1 = \int \cos qx\, dq$.

Now there is the second form of (2) to be done. Similarly treated, it makes

$$\int_0^\infty \frac{\cos atq^{\frac{1}{2}}}{1+q^2/\Delta^2}dq = \frac{1}{\pi}\int_0^\infty \frac{\cos atq^{\frac{1}{2}}}{1+p^4/a^4\Delta^2}dq = \frac{1}{\pi(1+p^4/a^4\Delta^2)}\int_0^\infty \cos atq^{\frac{1}{2}}dq, \quad (46)$$

because $p^2 = -a^2q$ on the circular function. Here let $q = Q^2$, then $dq = 2QdQ$, and we come to

$$\frac{2p/a}{\pi(1+p^4/a^4\Delta^2)}\int_0^\infty \sin atQ\,dQ = \frac{p/a}{\pi(1+p^4/a^4\Delta^2)}\frac{2}{at}. \quad (47)$$

Comparing with (38), we see that this way produces the divergent series D_2/D, first got in (10). There is an evident imperfection in the use of $\int \sin atQ\,dQ = (at)^{-1}$ here. Perhaps the reservation to be made is that it can be used safely when the result is convergent, as in (44).

Next consider the meanings of the initial states of D, D_1 and D_2. They all represent solutions of the same characteristic, viz. :—

$$(\Delta^2 + p^4/a^4)D = 0, \quad (48)$$

but arise from different initial states. These may be obtained from the formulæ for D and D_1, and the results are that when $t=0$,

$$D_1 = 0, \qquad \dot{D}_1 = \frac{aD_0}{2(2\pi)^{\frac{1}{2}}x^{1\frac{1}{2}}}, \qquad \ddot{D}_1 = 0, \qquad \dddot{D}_1 = \frac{3a^3D_0}{4(2\pi)^{\frac{1}{2}}x^{2\frac{1}{2}}}, \quad (49)$$

$$\underset{(\text{at } x=0)}{\int D_2 dx = D_0}, \qquad \dot{D}_2 = \frac{-aD_0}{2(2\pi)^{\frac{1}{2}}x^{1\frac{1}{2}}}, \qquad \ddot{D}_2 = \frac{a^2D_0}{\pi x^2}, \qquad \dddot{D}_2 = \frac{-3a^3D_0}{4(2\pi)^{\frac{1}{2}}x^{2\frac{1}{2}}}, \quad (50)$$

$$\underset{(\text{at } x=0)}{\int D dx = D_0}, \qquad \dot{D} = 0, \qquad \ddot{D} = \frac{a^2D_0}{\pi x^2}, \qquad \dddot{D} = 0. \quad (51)$$

Thus D_1 arises from distributions of D_1 and $\ddot{D}_1$, but not of D_1 or $\ddot{D}_1$, whilst the actual D arises from the condensed D_0 and a distribution of $\ddot{D}$. The other one, D_2, being $D - D_1$, arises from condensed D_2 and distributions of the three derivatives.

Now the characteristic (48) has four independent initial states; for its general solution, got by the four roots of $p^4 = = -a^4\Delta^2$, is

$$D = \theta D_0 + (p^{-1}\theta)\dot{D}_0 + (p^{-2}\theta)\ddot{D}_0 + (p^{-3}\theta)\dddot{D}_0, \quad (52)$$

where D_0, $\dot{D}_0$, etc., are the functions of x representing the initial states of D and its first three derivatives, and θ is the operator

$$\theta = \cos bt \cosh bt = 1 - \frac{4b^4t^4}{\lfloor 4} + \frac{16b^8t^8}{\lfloor 8} - \cdots \quad (53)$$

$$= 1 - \frac{a^4t^4\Delta^2}{\lfloor 4} + \frac{a^8t^8\Delta^4}{\lfloor 8} - \cdots, \quad (54)$$

where $b = a\sqrt{\tfrac{1}{2}\Delta}$. In (52), $p^{-1}\theta$ means the time integral of θ, and so on. So in full, (52) is

$$D = \cos bt \cosh bt\, D_0 + (\cos bt \operatorname{shin} bt + \sin bt \cosh bt)\dot{D}_0/2b$$
$$+ \sin bt \operatorname{shin} bt\, \ddot{D}_0/2b^2 + (\sin bt \cosh bt - \cos bt \operatorname{shin} bt)\dddot{D}_0/4b^3. \quad (55)$$

This equation expresses D due to any given D_0, etc., given as functions of x. But if we try a condensed source by itself, say $D=\theta\Delta 1$, we find that $\theta\Delta$ is an operator which, when expanded in powers of the differentiators, is like the rational part of (6), so that D is either zero, or is indeterminate this way. The same applies to $(p^{-1}\theta)\Delta$, $(p^{-2}\theta)\Delta$, and $(p^{-3}\theta)\Delta$. That is, we determine nothing in this way from a condensed source of any one of D, or $\dot{D}$, $\ddot{D}$, $\dddot{D}$, taken by itself.

But work out $\epsilon^{bt}\sin bt\,\Delta 1$, which involves more than one initial state. In the usual way, it makes

$$\epsilon^{bt}\sin bt\,\Delta 1=\frac{2}{a^2t^2}\frac{(a^2t^2/2x)^{1\frac{1}{2}}}{\underline{|-1\frac{1}{2}}}\times$$
$$\left\{1-\frac{a^2t^2}{4x}-\frac{(a^2t^2/4x)^2}{\underline{|2}}+\frac{(a^2t^2/4x)^3}{\underline{|3}}+--+\dots\right\}, \qquad (56)$$

and the same result comes from $\epsilon^{-bt}\sin bt\,\Delta 1$. So, by combination,

$$\sin bt\,\text{shin}\,bt\,\Delta 1=0 \qquad \sin bt\cosh bt\,\Delta 1=\text{result in (56)}. \qquad (57)$$

If we change $\sin bt$ to $\cos bt$, we get

$$\epsilon^{bt}\cos bt\,\Delta 1=+\frac{2}{a^2t^2}\frac{(a^2t^2/2x)^{1\frac{1}{2}}}{\underline{|-1\frac{1}{2}}}\times$$
$$\left\{1+\frac{a^2t^2}{4x}-\frac{(a^2t^2/4x)^2}{\underline{|2}}-\frac{(a^2t^2/4x)^3}{\underline{|3}}++--\dots\right\}, \qquad (58)$$

and the negative of the same comes from $\epsilon^{-bt}\cos bt\,\Delta 1$. So, by combination,

$$\cos bt\cosh bt\,\Delta 1=0, \qquad \cos bt\,\text{shin}\,bt\,\Delta 1=\text{result in (58)}. \qquad (59)$$

It also follows that

$$(\cos bt\,\text{shin}\,bt+\sin bt\cosh bt)\Delta 1=\frac{4}{a^2t^2}\frac{(a^2t^2/2x)^{1\frac{1}{2}}}{\underline{|-1\frac{1}{2}}}\sin\frac{a^2t^2}{4x}, \qquad (60)$$

$$(\cos bt\,\text{shin}\,bt-\sin bt\cosh bt)\Delta 1= \quad \dots\dots\dots \quad \cos\dots \qquad (61)$$

Here (60) is due to an initial distribution $\dot{D}_0$ varying as $b\Delta 1$ or $\Delta^{1\frac{1}{2}}$ or $x^{-1\frac{1}{2}}$, and (61) is due to a distribution of $\ddot{D}_0$. But in both cases there is an auxiliary initial distribution.

To illustrate, examine what $u=\cos atq^{\frac{1}{2}}\cos qx$ means operationally. It is the same as

$$u=\left(1+\frac{a^4t^4}{\underline{|4}}q^2+\frac{a^8t^8}{\underline{|8}}q^4+\dots\right)\cos qx-\left(\frac{a^2t^2}{\underline{|2}}+\frac{a^6t^6}{\underline{|6}}q^2+\dots\right)q\sin qx; \qquad (62)$$

or, by putting $-\Delta^2$ for q^2,

$$u=\cos bt\cosh bt\cos qx-\sin bt\,\text{shin}\,bt\sin qx; \qquad (63)$$

and comparing this with (51), (52), it will be seen that the initial states of u and $\ddot{u}$ are correctly given. As for $\dot{u}$ and $\dddot{u}$ they are initially zero.

We are led to ask what the solution $u=\theta \cos qx$ means? It is

$$u=\left(1+\frac{a^4t^4}{\lfloor 4}q^2+\frac{a^8t^8}{\lfloor 8}q^4+\dots\right)\cos qx=\tfrac{1}{2}\cos qx \cos atq^{\frac{1}{2}} + \tfrac{1}{2}\cos qx \cosh atq^{\frac{1}{2}}. \tag{64}$$

The first part is a standing oscillation. The second part increases with the time. Now, initially u is finite, whilst $\dot{u}$, $\ddot{u}$ and $\dddot{u}$ are all zero. That is, a given single initial state out of the possible four produces a solution, a part of which increases with the time without limit. This shows the utility of the paired initial states. They produce solutions which remain finite. The partial characteristic is too large for the physical circumstances of waves in water limited by a free surface, flat when undisturbed. Lord Kelvin said that the existence of initial $\ddot{D}_0$ along with D_0, when $\dot{D}_0=0$ was required on account of the applied forcive all over the surface which is statically needed to balance D_0. We may say, in other words, that the water being assumed to be incompressible, there is an instantaneous transmission of pressure to the surface at any distance due to a surface displacement. When the surface is free, motion begins to be instantly produced everywhere, and therefore $\ddot{D}_0$ is finite at the first moment, and behaves like an initial state. I think it may be concluded that if D_0, $\dot{D}_0$, $\ddot{D}_0$ and $\dddot{D}_0$ are all given arbitrarily, then $\ddot{D}_0$ and $\dddot{D}_0$ will immediately adjust themselves to suit D_0 and $\dot{D}_0$, whenever we take care to exclude the solutions which increase with the time. But the need of paired solutions is not fully explained by the incompressibility, as will be seen later.

The other way of solving the characteristic, in terms of two time functions, gives

$$D=\cos\left(\frac{xp^2}{a^2}\right)E_0+\frac{\sin(xp^2/a^2)}{p^2/a^2}E_1, \tag{65}$$

where E_0 and E_1 are two time functions expressing the values of D and dD/dx at the origin. This gives a very easy way of constructing solutions, though they may not be practical, being more curious than useful if they indicate that D increases towards $+\infty$ or $-\infty$ or both. But they are not all like that. For instance, if $E_1=0$, and $E_0=e_0/\pi a^2t^2$, we get

$$D=\frac{e_0}{\pi a^2t^2}\left(1-\frac{\lfloor 5}{\lfloor 2}\frac{x^2}{a^4t^4}+\frac{\lfloor 9}{\lfloor 4}\frac{x^4}{a^8t^8}-\dots\right). \tag{66}$$

Comparing with (10), it is the same, if $e_0=-D_0$, and it may be turned to a convergent form.

In two dimensions the characteristic corresponding to $\Delta^2=-p^4/a^4$ is

$$\frac{1}{r}\frac{d}{dr}r\frac{dD}{dr}=-\frac{p^4}{a^4}D, \tag{67}$$

where $r^2=x^2+y^2$; and the solution corresponding to (65) is

$$D=J_0(p^2r/a^2)F_0+G_0(p^2r/a^2)F_1 \tag{68}$$

where J_0 and G_0 are the oscillating zeroth Besselian functions, and F_0, F_1 are two time functions, F_0 being the value of D at the origin, and F_1 proportional to $r(dD/dr)$ at that place. For example, if $F_1=0$, and $F_0=f_0/at$, then

$$D=\frac{f_0}{at}\left\{1-\frac{|\underline{4}(r/2a^2t^2)^2}{|\underline{1}\,|\underline{1}}+\frac{|\underline{8}(r/2a^2t^2)^4}{|\underline{2}\,|\underline{2}}-\dots\right\}, \tag{69}$$

and if $F_0=f_0/a^2t^2$, then

$$D=\frac{f_0}{a^2t^2}\left\{1-\frac{|\underline{5}(r^2/4a^4t^4)}{|\underline{1}\,|\underline{1}}+\frac{|\underline{9}(r^2/4a^4t^4)^2}{|\underline{2}\,|\underline{2}}-\dots\right\}. \tag{70}$$

Here (69) corresponds to (66) for plane waves, and like it, may be turned to a convergent form.

If $F_0=f_0t^n/|\underline{n}$, the solution is

$$\frac{D}{f_0}=\frac{t^n}{|\underline{n}}-\left(\frac{r}{2a^2}\right)^2\frac{t^{n-4}}{|\underline{1}\,|\underline{1}\,|\underline{n-4}}+\left(\frac{r}{2a^2}\right)^4\frac{t^{n-8}}{|\underline{2}\,|\underline{2}\,|\underline{n-8}}-\dots, \tag{71}$$

and the corresponding plane wave solution is, by (65),

$$\frac{D}{e_0}=\frac{t^n}{|\underline{n}}-\frac{x^2}{a^4}\frac{t^{n-4}}{|\underline{2}\,|\underline{n-4}}+\frac{x^4}{a^8}\frac{t^{n-8}}{|\underline{4}\,|\underline{n-8}}-\dots \tag{72}$$

In simple waves, either standing vibrations or progressive waves, or combinational, the symbol p^2 in (68) is replaced by a negative constant, *e.g.*,

$$D=J_0\left(\frac{n^2r}{a^2}\right)\sin nt \tag{73}$$

is a case of standing vibrations.

Returning to the original problem with its condensed oscillations proceeding from the origin, arising out of

$$\frac{D}{D_0}=\frac{-a^2\Delta/p^2}{1+a^4\Delta^2/p^4}\frac{1}{x\pi}, \tag{74}$$

it asks itself whether the condensed waves will disappear when we substitute for the condensed D_0 a continuous uniform initial $D=E_0$ on the left side of the origin, with no D on the right side. By line integration, the solution is

$$\frac{D}{E_0}=\frac{a^2/p^2}{1+a^4\Delta^2/p^4}\frac{1}{x\pi}=\frac{a^2t^2}{x\pi}\left(\frac{1}{2}-\frac{|\underline{2}}{|\underline{6}}\frac{a^4t^4}{x^2}+\frac{|\underline{4}}{|\underline{10}}\frac{a^8t^8}{x^4}-\dots\right). \tag{75}$$

This is the D at time t on the originally undisturbed side of the origin. Calculation shows that we do not get rid of condensed waves. This is due to the initial discontinuity in D at the origin. Examination of various cases led me to the conclusion that condensed waves make their appearance, not only when there is a discontinuity in the initial D, but also when dD/dx is discontinuous:

or, in fact, if any succeeding derivative is discontinuous. So the initial D, if condensed waves are to be avoided, should be represented by a perfect function continuous in itself and all its derivatives from $x=-\infty$ to $+\infty$. But functions of this sort are not usually practical functions for showing how disturbances are propagated, particularly when they go into still water from local disturbance.

I thought of imitating closely the curve $\cdots\overline{\qquad}|\underline{\qquad}\cdots$ by means of the "arrival curve" of current C at the end of a long cable. It is known, by the diffusion theory, that not only C, but all its time derivatives, up to the nth, where n is any finite integer, however large, are zero at the first moment (and before), and finite later. And we may make the rise of C as steep as we please by reducing the time-constant, and so approximate to a sudden jump from $C=0$ to constancy without any discontinuity in C or its derivatives. But this process turned out to be very complicated in the work, and was given up.

Now Lord Kelvin proposed this case ("Baltimore Lectures," p. 530)

$$\frac{D}{D_0}=\frac{1}{\pi}\int_0^\infty \epsilon^{-q^2t_1/4}\cos qx \cos atq^{\frac{1}{2}}dq. \tag{76}$$

First see what the initial state is. Put $t=0$, Δ for d/dx, and p_1 for d/dt_1, then we may at once transform to

$$\frac{D}{D_0}=\frac{1}{\pi}\frac{1}{1-4p_1/\Delta^2}\int_0^\infty\frac{dq}{1+q^2/4p_1}=\cdots\left[\tan^{-1}\left(\frac{q^2}{4p_1}\right)^{\frac{1}{2}}\right]_0^\infty$$

$$=\frac{p_1^{\frac{1}{2}}}{1-4p_1/\Delta^2}=p_1^{\frac{1}{2}}\left(1+\frac{4p_1}{\Delta^2}+\dots\right)=\frac{\epsilon^{-x^2/t_1}}{(\pi t_1)^{\frac{1}{2}}}. \tag{77}$$

This makes an interesting example of the solution of definite integrals. This function makes

$$\int_{-\infty}^{\infty} D\,dx=D_0. \tag{78}$$

So the initial state is the amount D_0 distributed continuously and symmetrically, with a hump at the origin. If we reduce t_1 to zero, it is all condensed there. So if t_1 is very small, D is practically zero save at and close to the origin, but being continuous, we may perhaps get rid of the infinitely condensed waves.

We can now solve (76). First in rising powers of t_1. Put Δ^2 for $-q^2$ in the exponential function. Then

$$\frac{D}{D_0}=\epsilon^{t_1\Delta^2/4}\frac{1}{\pi}\int_0^\infty \cos qx\cos atq^{\frac{1}{2}}dq$$

$$=\epsilon^{t_1\Delta^2/4}\frac{a^2t^2}{\pi x^2}\left(\frac{1}{2}-\frac{|\underline{3}}{|\underline{6}}\frac{a^4t^4}{x^2}+\frac{|\underline{5}}{|\underline{10}}\frac{a^8t^8}{x^4}-\dots\right), \tag{79}$$

by (13). The work is obvious. The commencement makes, if $y=a^4t^4/x^2$,

$$\begin{aligned}\frac{\mathrm{D}}{\mathrm{D}_0}=\frac{a^2t^2}{\pi x^2}\Bigg[&\left(\frac{1}{2}-\frac{\underline{|3}}{\underline{|6}}y+\frac{\underline{|5}}{\underline{|10}}y^2-\frac{\underline{|7}}{\underline{|14}}y^3+\dots\right)\\ &+\frac{t_1}{4x^2}\left(\frac{\underline{|3}}{\underline{|2}}-\frac{\underline{|5}}{\underline{|6}}y+\frac{\underline{|7}}{\underline{|10}}y^2-\frac{\underline{|9}}{\underline{|14}}y^3+\dots\right)\\ &+\frac{1}{\underline{|2}}\left(\frac{t_1}{4x^2}\right)^2\left(\frac{\underline{|5}}{\underline{|2}}-\frac{\underline{|7}}{\underline{|6}}y+\frac{\underline{|9}}{\underline{|10}}y^2-\frac{\underline{|11}}{\underline{|14}}y^3+\dots\right)\\ &+\frac{1}{\underline{|3}}\left(\frac{t_1}{4x^2}\right)^3\left(\frac{\underline{|7}}{\underline{|2}}-\frac{\underline{|9}}{\underline{|6}}y+\frac{\underline{|11}}{\underline{|10}}y^2-\frac{\underline{|13}}{\underline{|14}}y^3+\dots\right)+\text{etc.}\Bigg].\end{aligned} \tag{80}$$

This is sufficient to show all the rest. But the solution is obviously unpractical in the above form. For the first line above contains the condensed waves, and so presumably do all the succeeding lines. Some rearrangement, which is not obvious, is required.

Next, do it in descending powers of t_1. Equation (76) may be turned to

$$\frac{\mathrm{D}}{\mathrm{D}_0}=\frac{1}{\pi}\int_0^\infty\frac{\epsilon^{-q^2t_1/4}\,dq}{(1+q^2/\Delta^2)(1+qa^2/p^2)}=\frac{1}{\pi}\int_0^\infty\frac{(1-qa^2/p^2)\epsilon^{-q^2t_1/4}\,dq}{(1+q^2/\Delta^2)(1-q^2a^4/p^4)}, \tag{81}$$

and now we may substitute $-4p_1$ for q^2 in the denominator, making

$$\begin{aligned}\frac{\mathrm{D}}{\mathrm{D}_0}&=\frac{1}{\pi(1-4p_1/\Delta^2)(1+4p_1a^4/p^4)}\int_0^\infty\left(1-\frac{qa^2}{p^2}\right)\epsilon^{-q^2t_1/4}\,dq\\ &=\frac{\dfrac{1}{(\pi t_1)^{\frac12}}-\dfrac{a^2}{p^2}\dfrac{2}{\pi t_1}}{(1-4p_1/\Delta^2)(1+4p_1a^4/p^4)}.\end{aligned} \tag{82}$$

The auxiliary variable t_1 proves to be analytically useful. The operand is the numerator, a function of t and t_1. The denominator contains three differentiators, Δ, p, and p_1. Yet the algebrisation goes easily enough.

First of all, examine the state of things at the origin itself, where $x=0$. Put $\Delta=\infty$, then

$$\begin{aligned}\frac{\mathrm{D}_{x=0}}{\mathrm{D}_0}&=\left(1-\frac{4p_1a^4}{p^4}+\dots\right)\left(\frac{1}{(\pi t_1)^{\frac12}}-\frac{a^2t^2}{\underline{|2}}\frac{2}{\pi t_1}\right)\\ &=\frac{1}{(\pi t_1)^{\frac12}}\left\{1+\tfrac12\frac{a^4t^4}{\underline{|4}}\frac{4}{t_1}+\frac{1\,.\,3}{2\,.\,2}\frac{a^8t^8}{\underline{|8}}\left(\frac{4}{t_1}\right)^2+\dots\right\}\\ &\quad-\frac{2}{\pi t_1}\left\{\frac{a^2t^2}{\underline{|2}}+\frac{a^6t^6}{\underline{|6}}\frac{4}{t_1}+\frac{a^{10}t^{10}}{\underline{|10}}\left(\frac{4}{t_1}\right)^2\underline{|2}+\frac{a^{14}t^{14}}{\underline{|14}}\left(\frac{4}{t_1}\right)^3\underline{|3}+\dots\right\}.\end{aligned} \tag{83}$$

This is the difference of two convergent series. Taking $t_1=4$, and $a^2t^2=1$, 2, 5, 10, 20, D is positive at first and falls towards zero. It may reverse itself, but there is no sign of rapid oscillations; that is, at the origin.

To find D at any distance x, (82) makes

$$\frac{\mathrm{D}_x}{\mathrm{D}_0}=\left\{1+\frac{4p_1}{\Delta^2}+\left(\frac{4p_1}{\Delta^2}\right)^2+\dots\right\}\frac{\mathrm{D}_{x=0}}{\mathrm{D}_0}=\cos\frac{xp^2}{a^2}\,\frac{\mathrm{D}_{x=0}}{\mathrm{D}_0}. \tag{84}$$

The equivalence of the two forms shown arises from the equivalence of $4p_1$ to $-p^4/a^4$ when operating upon (83). That is, one differentiation to the auxiliary variable t_1 takes the place of four to t. Carrying out (84), let $x^n/\lfloor n$ be denoted by Δ^{-n}, and $a^n t^n/\lfloor n$ by P^{-n}, then the result is

$$\begin{aligned}\frac{\mathrm{D}}{\mathrm{D}_0}=\frac{1}{(\pi t_1)^{\frac{1}{2}}}\Big[1&+\tfrac{1}{2}\frac{4}{t_1}(\mathrm{P}^{-4}-\Delta^{-2})+\frac{1.3}{2^2}\left(\frac{4}{t_1}\right)^2(\mathrm{P}^{-8}-p^{-4}\Delta^{-2}+\Delta^{-4})\\&+\frac{1.3.5}{2^3}\left(\frac{4}{t_1}\right)^3(\mathrm{P}^{-12}-\mathrm{P}^{-8}\Delta^{-2}+\mathrm{P}^{-4}\Delta^{-4}-\Delta^{-6})+\dots\Big]\\-\frac{2}{\pi t_1}\Big[\mathrm{P}^{-2}&+\frac{4}{t_1}(\mathrm{P}^{-6}-\mathrm{P}^{-2}\Delta^{-2})+\lfloor 2\left(\frac{4}{t_1}\right)^2(\mathrm{P}^{-10}-\mathrm{P}^{-6}\Delta^{-2}+\mathrm{P}^{-2}\Delta^{-4})\\&+\lfloor 3\left(\frac{4}{t_1}\right)^3(\mathrm{P}^{-14}-\mathrm{P}^{-10}\Delta^{-2}+\mathrm{P}^{-6}\Delta^{-4}-\mathrm{P}^{-2}\Delta^{-6})+\dots\Big].\end{aligned} \tag{85}$$

Here we have a formula suitable for calculation, with all signs of condensed oscillations removed. Give to $4/t_1$ a convenient value for calculation, not too large, then this formula worked out at and near the origin, will show the transformation of the solitary hump into two wave trains, one to the right, the other to the left, and their transference outward.

But there is another sort of hump, of simpler nature, which will serve this purpose by less work. Instead of (76) above, let

$$\frac{\mathrm{D}}{\mathrm{D}_0}=\frac{1}{\pi}\int_0^\infty \epsilon^{-qt_1}\cos qx\cos atq^{\frac{1}{2}}\,dq=\frac{1}{\pi}\int_0^\infty\frac{\epsilon^{-qt_1}dq}{(1+q^2/\Delta^2)(1+qa^2/p^2)}. \tag{86}$$

Here p_1 is equivalent to $-q$, so we transform to

$$\frac{\mathrm{D}}{\mathrm{D}_0}=\frac{\dfrac{1}{\pi}\displaystyle\int_0^\infty\epsilon^{-qt_1}dq}{(1+p_1^2/\Delta^2)(1-p_1a^2/p^2)}=\frac{(\pi t_1)^{-1}}{(1+p_1^2/\Delta^2)(1-p_1a^2/p^2)}. \tag{87}$$

Comparison with (82) shows a notable simplification. At the origin,

$$\frac{\mathrm{D}_{x=0}}{\mathrm{D}_0}=\frac{(\pi t_1)^{-1}}{1-p_1a^2/p^2}=\frac{1}{\pi t_1}\left(1-\frac{a^2t^2}{\lfloor 2}\,\frac{1}{t_1}+\frac{a^4t^4}{\lfloor 4}\,\frac{2}{t_1^2}-\frac{a^6t^6}{\lfloor 6}\,\frac{\lfloor 3}{t_1^3}+\dots\right); \tag{88}$$

and initially, anywhere,

$$\frac{\mathrm{D}_{t=0}}{\mathrm{D}_0}=\frac{(\pi t_1)^{-1}}{1+p_1^2/\Delta^2}=\frac{1}{\pi t_1}\left(1-\frac{x^2}{\lfloor 2}\,\frac{\lfloor 2}{t_1^2}+\dots\right)=\frac{1}{\pi t_1(1+x^2/t_1^2)}. \tag{89}$$

This shows a nice practical hump for initial state. Like the other, it is symmetrical, and the total D comes to D_0.

Examine (88), to show the behaviour at the origin. By preliminary rough work, it is to be seen that D falls to zero, and becomes negative; then rises again towards zero. The work soon becomes too long; so turn (88) to a divergent series. Thus, put $z=a^2t^2/4t_1$, then (88) is the same as

$$\frac{\mathrm{D}}{\mathrm{D}_0}=\frac{(z\pi)^{\frac{1}{2}}}{\pi t_1}\left(\frac{z^{-\frac{1}{2}}}{\lfloor -\frac{1}{2}}-\frac{z^{\frac{1}{2}}}{\lfloor \frac{1}{2}}+\frac{z^{1\frac{1}{2}}}{\lfloor 1\frac{1}{2}}-\dots\right). \tag{90}$$

This series belongs to the type of the generalised ϵ^{-z} series, with z positive, so (90) is equivalent to

$$\frac{\mathrm{D}}{\mathrm{D}_0}=\frac{(z\pi)^{\frac{1}{2}}}{\pi t_1}\left(\frac{z^{-1\frac{1}{2}}}{\lfloor -\frac{1}{2}}-\frac{z^{-2\frac{1}{2}}}{\lfloor -2\frac{1}{2}}+\frac{z^{-3\frac{1}{2}}}{\lfloor -3\frac{1}{2}}-\dots\right)$$

$$=-\frac{1}{\pi}\left(\frac{2}{a^2t^2}+\frac{\lfloor 4}{\lfloor 2}\frac{t_1}{a^4t^4}+\frac{\lfloor 6}{\lfloor 3}\frac{t_1^2}{a^6t^6}+\dots\right). \tag{91}$$

This shows return towards equilibrium without oscillation. As a specimen of values, let $t_1=1$ and $\mathrm{D}_0/\pi=1$; then, by (88), the initial value of D is 1, which falls to $\frac{1}{2}$ in a little over $at=1$, passes 0 a little before $at=2$, and reaches $-\cdot 41$ at $a^2t^2=10$. Calculation by this convergent series now begins to be lengthy, but Providence comes to assist in the shape of the divergent series (91). It is just beginning to be useful, for $a^2t^2=10$ makes $-\cdot 38$. But when $a^2t^2=20$, a very long sum of 21 terms of the convergent series makes $-\cdot 1573$, whilst a short sum of five terms (counting half the last convergent term) makes $-\ 1602$. The difference, $\cdot 0029$, is only a small fraction of the l.c.t. This is an excellent test. After that, D rises to zero asymptotically. The beginning of the wave trains to right and left is shown during the first fall of the hump in its middle part. This creates two smaller humps, outward moving.

Now derive the value of D at x. By (87), (89),

$$\frac{\mathrm{D}}{\mathrm{D}_0}=\frac{1}{\pi}\,\frac{1}{1-a^2p_1/p^2}\,\frac{t_1}{t_1^2+x^2}$$

$$=\frac{1}{\pi}\left\{\frac{t_1}{t_1^2+x^2}+\frac{a^2t^2}{\lfloor 2}\,\frac{x^2-t_1^2}{(\ .\ )^2}+\lfloor 2\frac{a^4t^4}{\lfloor 4}\,\frac{t_1^3-3x^2t_1}{(\ .\ )^3}+\lfloor 3\frac{a^6t^6}{\lfloor 6}\,\frac{6t_1^2x^2-t_1^4-x^4}{(\ .\ )^4}\right.$$

$$+\lfloor 4\frac{a^8t^8}{\lfloor 8}\,\frac{t_1^5+5x^4t_1-10x^2t_1^3}{(\ .\ )^5}+\lfloor 5\frac{a^{10}t^{10}}{\lfloor 10}\Big(-t_1^6+15x^2t_1^4$$

$$\left.-15x^4t_1^2+x^6\Big)+\cdots\right\}. \tag{92}$$

If we put $t_1=0$, this reduces to the old familiar first solution (13) in which the waves are infinitely condensed, whereas (92) calculated for small values of x and t_1 will show them finitely spread. $(\ .\ )$ means $(t_1^2+x^2)$, for brevity.

Now if we use the divergent series (91) to obtain D at x by the same process, it is clear that the solution arising cannot be complete, because the auxiliary function previously called D_1, equation (28), makes no appearance. What we get is, calling this partial solution D_2,

$$\frac{D_2}{D_0}=-\frac{2}{\pi a^2t^2}\Bigg[1-\frac{\underline{|5}}{\underline{|2}}\frac{x^2}{a^4t^4}+\frac{\underline{|9}}{\underline{|4}}\frac{x^4}{a^8t^8}-\dots+\frac{t_1}{a^2t^2}\Big(\underline{|3}-\frac{\underline{|4}}{\underline{|2}}\frac{x^2}{a^4t^4}$$

$$+\frac{\underline{|11}}{\underline{|4}}\frac{x^4}{a^8t^8}-\dots\Big)+\frac{t_1^{\,2}}{\underline{|2}\,a^4t^4}\Big(\underline{|5}-\frac{\underline{|9}}{\underline{|2}}\frac{x^2}{a^4t^4}+\frac{\underline{|13}}{\underline{|4}}\frac{x^4}{a^8t^8}-\dots\Big)+\dots\Bigg]. \quad (93)$$

Returning to (92). Let $r=(x^2+t_1^{\,2})^{\frac{1}{2}}$, and $t_1/r=\cos\phi$, $x/r=\sin\phi$. Then by reference to a trigonometrical work for the expansion of $\cos n\phi$, it will be seen that (92) may be expressed as

$$\frac{D}{D_0}=\frac{1}{\pi r}\left\{\cos\phi-\frac{a^2t^2}{\underline{|2}}\frac{\cos 2\phi}{r}+\underline{|2}\frac{a^4t^4}{\underline{|4}}\frac{\cos 3\phi}{r^2}-\underline{|3}\frac{a^6t^6}{\underline{|6}}\frac{\cos 4\phi}{r^3}+\dots\right\}. \quad (94)$$

Now for any given values of t_1 and x, the cosine tables may be used to ease the full calculations. The explanation of this form is as follows: In the diagram let the straight line AA represent the undisturbed surface of the water, and the lines BB and CC represent

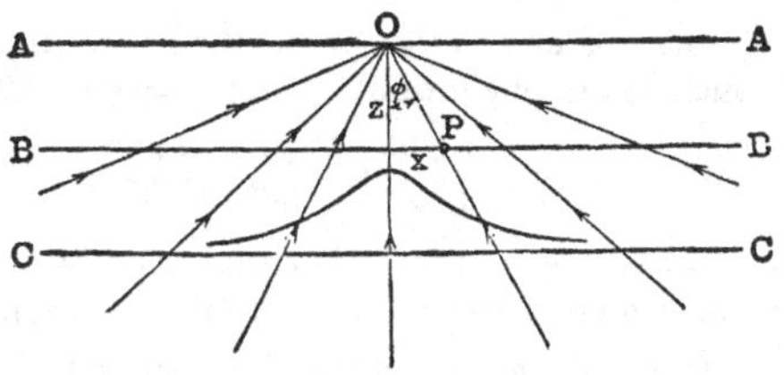

undisturbed lower levels. Then let the water everywhere be displaced along the radial lines (representing planes) leading to the origin O, the amount of this displacement at any point P being $D_0/\pi r$, at distance r from O. The total displacement through the semi-circle through P will be D_0. This is true at any distance. Moreover, the upward displacement at P is $(D_0/\pi r)\cos\phi$ or $D_0 z/(z^2+x^2)$, and by integration along the level through P, it will be seen that the total upward displacement through that level (and all other levels) is D_0.

The shape of the hump representing the upward displacement at any level is exactly the one last considered, when t_1 represents z, the depth of the undisturbed level in question below the top level. One of them is roughly shown for the level CC, but exaggerated of course. Going to higher levels the hump condenses, and at the top level becomes infinitely condensed at the point O. We then have the original hump with which we started.

If, then, we take the level CC to be the undisturbed surface of the water, and its upward displacement to be the hump corresponding to the distance below AA, then we have not only solved the problem of the subsequent history of the hump upon the CC level, but by increasing the value of z (the former t_1) for all other levels beneath it; *i.e.*, for the whole body of the water. The process may be continued upward, by having water above CC, and only when we reach the level AA, from which z is measured, do we come to condensed waves. It is to be carefully noted that when the hump is not condensed, and yet belongs to the level of the water surface, then z is not the distance below the level, but below an ideal level surface above the real one.

Knowing the vertical displacement upward, say D_z, or

$$D_z=\frac{D_0}{\pi r}\left\{\cos\phi-\frac{a^2t^2}{|\underline{2}}\frac{\cos 2\phi}{2}+|\underline{2}\frac{a^4t^4}{|\underline{4}}\frac{\cos 3\phi}{r^2}-\dots\right\} \tag{95}$$

throughout the water, it is easy to deduce the horizontal displacement from the condition of incompressibility, $dD_x/dx+dD_z/dz=0$. It makes

$$D_x=\frac{D_0}{\pi r}\left\{\sin\phi-\frac{a^2t^2}{|\underline{2}}\frac{\sin 2\phi}{r}+|\underline{2}\frac{a^4t^4}{|\underline{4}}\frac{\sin 3\phi}{r^2}-\dots\right\}. \tag{96}$$

This is D_x towards the origin, that is, to the left on the right side. Both may be included in one formula. Let $R=r\epsilon^{i\phi}$. Then

$$D_z-iD_x=\frac{D_0}{\pi R}\left\{1-\frac{1}{|\underline{2}}\frac{a^2t^2}{R}+\frac{|\underline{2}}{|\underline{4}}\frac{a^4t^4}{R^2}-\frac{|\underline{3}}{|\underline{6}}\frac{a^6t^6}{R^3}+\dots\right\}. \tag{97}$$

From D_x and D_z the displacement potential may be worried out. But perhaps better more directly thus. Go back to (87). We have

$$D_z=\{1-p_1a^2/p^2\}^{-1}\times\text{initial value of } D_z. \tag{98}$$

This must apply equally to the potential of which D_z is the derivative. Let the potential be Φ. Its initial value is $(D_0/\pi)\log r$, as it is obvious on differentiation that it produces the assumed initial state of radial displacement. Therefore

$$\Phi=\left\{1+\frac{a^2t^2}{|\underline{2}}p_1+\frac{a^4t^4}{|\underline{4}}p_1^2+\dots\right\}\frac{D_0}{\pi}\log r, \tag{99}$$

where $p_1=d/dz$. Carry this out. The result is, by the immediate use of $(-p_1)^n\log r=-|\underline{n-1}\,r^{-n}\cos n\phi$,

$$\Phi=\frac{D_0}{\pi}\left\{\log r+\frac{a^2t^2}{|\underline{2}}\frac{\cos\phi}{r}-\frac{a^4t^4}{|\underline{4}}\frac{\cos 2\phi}{r^2}+|\underline{2}\frac{a^6t^6}{|\underline{6}}\frac{\cos 3\phi}{r^3}-\dots\right\}; \tag{100}$$

and from this the above D_x and D_z may be derived. Also the radial and tangential components D_r and D_ϕ. Thus,

$$D_r=\frac{D_0}{\pi r}\left\{1-\frac{a^2t^2}{|\underline{2}}\frac{\cos\phi}{r}+\frac{|\underline{2}}{|\underline{4}}\frac{a^4t^4}{r^2}\cos 2\phi-\frac{|\underline{3}}{|\underline{6}}\frac{a^6t^6}{r^3}\cos 3\phi+\dots\right\}, \tag{101}$$

$$D_\phi=\frac{D_0}{\pi r}\left\{-\frac{a^2t^2}{|\underline{2}}\frac{\sin\phi}{r}+\frac{|\underline{2}}{|\underline{4}}\frac{a^4t^4}{r^2}\sin 2\phi-\frac{|\underline{3}}{|\underline{6}}\frac{a^6t^6}{r^3}\sin 3\phi+\dots\right\}, \tag{102}$$

where D_r and D_ϕ are reckoned in the directions of decrease of r and of ϕ. On this understanding,

$$D_r - iD_\phi = \frac{D_0}{\pi r}\left\{1 - \frac{a^2t^2}{\underline{|2}\,R} + \frac{\underline{|2}}{\underline{|4}}\frac{a^4t^4}{R^2} - \frac{\underline{|3}}{\underline{|6}}\frac{a^6t^6}{R^3} - \dots\right\}, \tag{103}$$

which only differs from (97) in the outer factor. Here, if we put $c = a^2t^2/4R$, the last equation becomes

$$D_r - iD_\phi = \frac{(c\pi)^{\frac{1}{2}}}{\pi r}\left\{\frac{c^{-\frac{1}{2}}}{\underline{|-\tfrac{1}{2}}} - \frac{c^{\frac{1}{2}}}{\underline{|\tfrac{1}{2}}} + \frac{c^{1\frac{1}{2}}}{\underline{|1\tfrac{1}{2}}} - \dots\right\} = \frac{(c\pi)^{\frac{1}{2}}}{\pi r}\Delta_c^{\frac{1}{2}}\,\epsilon^{-c}. \tag{104}$$

This c is in general complex. The series is a part of the generalised ϵ^{-c} series. If c is real, that is, in the vertical plane $\phi=0$ through the origin, we may convert to divergent form for numerical calculation of D_z in that plane, as done above. Also, in the horizontal plane, $\phi=\frac{1}{2}\pi$, we have converted D_z to divergent form by the generalised cosine series. The same may be done for the horizontal displacement. Put $\phi=\frac{1}{2}\pi$ in (96). Then

$$D_x = \frac{D_0}{\pi x}\left\{1 - \frac{\underline{|2}}{\underline{|4}}\frac{a^4t^4}{x^2} + \frac{\underline{|4}}{\underline{|8}}\frac{a^8t^8}{x^4} - \dots\right\}, \tag{105}$$

which is convergent, with the same defect practically as the similar D_z formula. Next, put $a^2t^2/4x = y$, then

$$\frac{D_x}{D_0} = \frac{(y\pi)^{\frac{1}{2}}}{\pi x}\left\{\frac{y^{-\frac{1}{2}}}{\underline{|-\tfrac{1}{2}}} - \frac{y^{1\frac{1}{2}}}{\underline{|1\tfrac{1}{2}}} + \frac{y^{3\frac{1}{2}}}{\underline{|3\tfrac{1}{2}}} - \dots\right\} = \frac{(y\pi)^{\frac{1}{2}}}{\pi x}\Delta^{\frac{1}{2}}\cos y$$

$$= \frac{(y\pi)^{\frac{1}{2}}}{\pi x}\left\{\cos\left(y + \tfrac{1}{4}\pi\right) + \frac{y^{-2\frac{1}{2}}}{\underline{|-2\tfrac{1}{2}}} - \frac{y^{-4\frac{1}{2}}}{\underline{|-4\tfrac{1}{2}}} + \dots\right\}, \tag{106}$$

This is done by using (20). Or, in terms of x and t,

$$\frac{D_x}{D_0} = \frac{at}{2\sqrt{\pi}\,x^{1\frac{1}{2}}}\left\{\cos\left(\frac{a^2t^2}{4x} + \tfrac{1}{4}\pi\right) + \frac{y^{-2\frac{1}{2}}}{\underline{|-2\tfrac{1}{2}}} - \frac{y^{-4\frac{1}{2}}}{\underline{|-4\tfrac{1}{2}}} + \dots\right\}, \tag{107}$$

which is the companion to (27).

Now, although the calculation difficulties of the convergent formulæ (95), (96) for any point not too near the limiting level through the origin are not comparable with those connected with the limiting level itself, it would be desirable to convert (95), (96) to divergent form as well. But it is not so easy to do it as at the limiting level, on account of difficulties in the numerical interpretation of complex divergent formulæ. As a preliminary, we may transform (104) so as to introduce a time factor of decay. Thus,

$$\Delta^{\frac{1}{2}}\epsilon^{-c} = \epsilon^{-c}(\Delta - 1)^{\frac{1}{2}} = \epsilon^{-c}\Delta^{\frac{1}{2}}(1 - \Delta^{-1})^{\frac{1}{2}}$$

$$= \epsilon^{-c}\Delta^{\frac{1}{2}}\left(1 - \tfrac{1}{2}\Delta^{-1} - \frac{1.1}{2.4}\Delta^{-2} - \frac{1.1.3}{2.4.6}\Delta^{-3} - \dots\right)$$

$$= \frac{\epsilon^{-c}}{(c\pi)^{\frac{1}{2}}}\left(1 - c - \frac{c^2}{3\underline{|2}} - \frac{c^3}{5\underline{|3}} - \frac{c^4}{7\underline{|4}} - \dots\right). \tag{108}$$

Here c is complex, $=y\epsilon^{-\theta i}$, if y is the real $a^2t^2/4r$. So, separating the two components, we convert (104) to

$$D_r - iD_\phi = \frac{\epsilon^{-y\cos\phi}}{\pi r}\left\{\begin{array}{l}X\cos(y\sin\phi) - Y\sin(y\sin\phi)\\ \quad + iX\sin(y\sin\phi) + iY\cos(y\sin\phi)\end{array}\right\}, \quad (109)$$

where

$$X = 1 - y\cos\phi - y^2/\underline{|2}\cos 2\phi - y^3/\underline{|3}\cos 3\phi - \dots$$

$$Y = y\sin\phi + y^2/\underline{|2}\sin 2\phi + y^3/\underline{|3}\sin 3\phi + \dots. \quad (110)$$

The factor of decay in (109) is good, by itself, and if X and Y were of such a nature as to be readily calculable for large values of y, the real meaning would be ascertainable. We should rather have X and Y expressed in descending powers of y. That is, the c series in (108) should be turned to an equivalent divergent series, numerically true for the real and unreal parts separately.

Try another way. Since the water is assumed to be incompressible, its state of displacement all through is entirely determined by its state at the surface level, and this connection is instantaneous. Thus, if $d/dx = \Delta$, and $d^2V/dx^2 + d^2V/dz^2 = 0$, then

$$V = \cos z\Delta . V_0 + \frac{\sin z\Delta}{\Delta}\frac{dV_0}{dz} \quad (111)$$

finds V at x, z in terms of the values of V and dV/dz at $z=0$. But the boundary condition in this theory is

$$\frac{dV}{dz} = \frac{\ddot{V}}{a^2} \qquad \text{at } z=0; \quad (112)$$

so $$V = \cos z\Delta . V_0 + \frac{\sin z\Delta}{\Delta}\frac{\ddot{V}_0}{a^2} = \text{real } \epsilon^{z\Delta i}\left(1 - \frac{ip^2}{\Delta a^2}\right)V_0. \quad (113)$$

If we know V_0 as a function of x and t, then also $\ddot{V}_0$. The data are complete, for the operator $\epsilon^{z\Delta i}$ turns x in V_0 to $x+zi$. Calling this S,

$$V = \text{real}\left(1 - \frac{ip^2}{\Delta a^2}\right)V_0, \quad (114)$$

if $\Delta = d/dS$, and V_0 contains S instead of x.

If we apply this process to D_z given convergently for the top level, (114) will give without difficulty the corresponding D_z at any depth. That is, (13) will be turned to (95). D_x may be similarly treated. But now go through the same process with the divergent formula for the surface elevation. Use (27) in (114). First find $p^2D_z/\Delta a^2$. We have

$$D_z = \frac{-\Delta a^2/p^2}{1+\Delta^2a^4/p^4}\frac{1}{x\pi} \quad \therefore \quad \frac{p^2D_z}{\Delta a^2} = \frac{-1}{1+\Delta^2a^4/p^4}\frac{1}{x\pi} = -D_x, \quad (115)$$

by inspection of (105), and so now we can use (107) combined with (27). The oscillating terms alone make, if $y=a^2t^2/4x$,

$$\frac{at}{2\pi^{\frac{1}{2}}x^{1\frac{1}{2}}}\left(\cos(y-\tfrac{1}{4}\pi)+i\cos(y+\tfrac{1}{4}\pi)\right)=\cdots(i^{\frac{1}{2}}\cos y+i^{-\frac{1}{2}}\sin y)$$
$$=\frac{at}{2\pi^{\frac{1}{2}}x^{1\frac{1}{2}}}i^{\frac{1}{2}}\epsilon^{-yi}. \tag{116}$$

Now change x to $x+zi=r\epsilon^{\theta i}$, and we get, with y turned to $a^2t^2/4r$,

$$\frac{at}{2\pi^{\frac{1}{2}}r^{1\frac{1}{2}}}\epsilon^{-i\left\{y(\cos\theta-i\sin\theta)-\frac{1}{4}\pi+1\frac{1}{2}\theta\right\}}; \tag{117}$$

of which the real part is

$$\frac{at}{2\pi^{\frac{1}{2}}r^{1\frac{1}{2}}}\epsilon^{-y\sin\theta}\cos(y\cos\theta-\tfrac{1}{4}\pi+1\tfrac{1}{2}\theta). \tag{118}$$

Doing the same for the divergent part of D_z, we get

$$-\frac{2}{\pi a^2t^2}\left(1-\frac{\lfloor 5}{\lfloor 2}\frac{r^2\cos 2\theta}{a^4t^4}+\frac{\lfloor 9}{\lfloor 4}\frac{r^4\cos 4\theta}{a^8t^8}-\ldots\right)$$
$$-\frac{1}{\pi}\left(\frac{\lfloor 4}{\lfloor 2}\frac{r\sin\theta}{a^4t^4}-\frac{\lfloor 8}{\lfloor 4}\frac{r^3\sin 3\theta}{a^8t^8}+\ldots\right)$$
$$=-\frac{2}{\pi a^2t^2}\left\{1+\frac{\lfloor 3}{\lfloor 1}\frac{r\sin\theta}{a^2t^2}-\frac{\lfloor 5}{\lfloor 2}\frac{r^2\cos 2\theta}{a^4t^4}-\frac{\lfloor 7}{\lfloor 3}\frac{r^3\sin 3\theta}{a^6t^6}+\ldots\right\}. \tag{119}$$

Now add together (118) and (119), and restore the old notation, through $\theta=\frac{1}{2}\pi-\phi$, and we obtain

$$\frac{D_z}{D_0}=-\frac{at}{2\pi^{\frac{1}{2}}r^{1\frac{1}{2}}}\epsilon^{-y\cos\phi}\sin(y\sin\phi-1\tfrac{1}{2}\phi)$$
$$-\frac{2}{\pi a^2t^2}\left(1+\frac{\lfloor 3}{\lfloor 1}\frac{r\cos\phi}{a^2t^2}+\frac{\lfloor 5}{\lfloor 2}\frac{r^2\cos 2\phi}{a^4t^4}+\frac{\lfloor 7}{\lfloor 3}\frac{r^3\cos 3\phi}{a^6t^6}+\ldots\right), \tag{120}$$

where $y=a^2t^2/4r$. This is to represent the vertical displacement at any depth. To test harmony with former results, put $\phi=0$. Then we obtain the formula (91) for the displacement in the vertical plane through the origin, which is numerically equivalent to (88) when z/a^2t^2 is sufficiently large. Again, put $\phi=\frac{1}{2}\pi$. Then we obtain the formula (27) for the elevation at the limiting plane. So far good. But the first line in (120) changes sign with ϕ. That is bad, but may be righted by the prefix $\mp$ instead of $-$, according as ϕ is $\pm$. Assuming that this formula (120) is correct from $\phi=0$ to $\frac{1}{2}\pi$, it is eminently suitable for showing the meaning without lengthy calculations. The exponential factor is very significant. The function $t\epsilon^{-y\cos\phi}$ increases with the time to a maximum and then falls to zero, provided ϕ is less than $\frac{1}{2}\pi$. But when $\phi=\frac{1}{2}\pi$, the function increases for ever without limit. That is to say, at any point under the limiting plane, the oscillation set up by the initial condensed hump at the origin (or by a diffused hump at a lower level) increases in amplitude up to a certain amount, and then subsides to rest. But at the limiting level, the maximum will never be reached, so no subsidence is shown. This behaviour agrees with that of the original convergent solution (13). The limiting surface is therefore a surface of physical absurdity. We must go deeper, and have a diffused hump for physical intelligibility.

Now, on treating the horizontal displacement in a similar manner, we come of course to a formula of the same type as (120), but with differences. It has the same value for $\phi=\frac{1}{2}\pi$ and $-\frac{1}{2}\pi$, instead of the opposite, like the vertical displacement. Moreover, it has a finite value for $\phi=0$, whereas it should be zero, by symmetry. Taking D_z and D_x together, then, the interior state indicated by (120) and its companion, taken unchanged on both sides of the origin, indicate an initial state not alike on the two sides, but opposite. So, if D_z is correctly given on the positive side, as seems from its values at $\phi=0$ and $\frac{1}{2}\pi$, there is some imperfection in the completer treatment to be removed, perhaps later. The trouble is to extend safely generalised series from real to complex quantities.

The Solution of Definite Integrals by Differential Transformation.

§ 526. The last article contains several examples of the use of operational methods in the solution of definite integrals. They are however, of a special class, and in the following the consideration is more varied. The treatment might be made to cover the greater part of the field of real definite integrals. Only a small part of it, however, will be gone over, mostly of physical interest. My usual practice is to avoid definite integrals and obtain functional solutions direct from the operational solutions derived from the differential equations. Occasionally, however, I have applied the same processes to definite integrals. Stated roughly, the general idea in solving them is to convert them to differential form, and then apply the operational processes.

There are no particular rules for solving or functionising definite integrals, like those for differentiating functions. The work is largely experimental, though, of course, rigorous mathematicians conceal this as well as they can when they write treatises, and pretend to be omniscient by avoiding their failures, and also the many interesting things they cannot explain. In the first place, if we know that a given function has another function for differentiant, then conversely we know the indefinite integral or integrant of the second function. Moreover, it is generally possible to put the differentiant in the form of a power series, and then the integrant is also obtainable in a power series. But this is not always a satisfactory way of solving a definite integral, for the integrant may require interpretation at the limits of the integral. This is often practically impossible, for the power series expressing the integrant takes the form of a series of infinities at one or both limits. There are crowds of definite integrals of this sort. Their solutions are usually obtained indirectly, say by working out the problem concerned in some other way. Then a stock of known definite integrals can be accumulated. After that, there are various devices for deriving the solutions of new integrals from old ones. One is by changing the variable of the integration. This makes wonderful transformations. Another is the use of Fourier's theorem, whereby we can, given one integral, immediately derive another. The corresponding Bessel theorem does the same work through a wider range. Amongst minor devices may be mentioned the differentiation of an integral with respect to constants contained in it.

Now these processes have a valuable auxiliary in the application of structural integration as carried out in vol. 2 of this work and in previous work. Roughly stated, it consists in the conversion of functions of variables to functions of their differentiators, and also conversely, for one aim is to have reversible processes, free from indefiniteness, and this is usually secured. For example, if x_1 is the differentiator of x, x_2 of x_1, and so on, we may transform $f(x)$ to $f_1(x_1)$, and this again to $f_2(x_2)$, and so on, or the other way. Sometimes an infinite series of real and interpretable functions can arise this way. Or maybe they become divergent and require special interpretation. Usually only one step is required, or perhaps one step down and another up. But sometimes two steps in one direction are useful in definite integrals.

In practical work the methods to be employed with advantage vary with circumstances, and judicious selection is needed. For one method may be lengthy and another method short, in a certain case, and exactly the reverse in another case. In the present article, naturally, the treatment is almost entirely evolutionary and involutionary, irrespective of shortness. In one respect it is more comfortable than all the other ways, for it can usually be carried out by itself, without that distressing hunting up of known results in books. It is also good when a definite integral is of a kind that you cannot see how to reduce to known results or to develop therefrom. Also in getting convergent results out of divergent. Lastly, it gives some insight into the differential mechanism of functions. In the following is a collection of examples, beginning with easy ones.

(1). Let x and t be two real variables, positive, ranging from 0 to ∞, and let x_1 and t_1 be their differentiators, or $x_1 = d/dx$, $t_1 = d/dt$. This notation is, for the present article, convenient in easing the memory of a lot of work. You see instantly what variable a differentiator belongs to. Then, in accordance with the principles explained in vol. 2, which must be taken as understood, at least in their elementary working, $t_1 1$ means a unit impulse at $t=0$, and $\epsilon^{-t_1 x} t_1$ means a unit impulse at $t=x$. Similarly $x_1 1$ is a unit impulse at $x=0$, and $\epsilon^{-x_1 t} x_1$ a unit impulse at $x=t$. In the first case it is a function of t wholly condensed at $t=x$. In the second case it is a function of x wholly condensed at $x=t$. The condensation and coincidence makes them quantitatively identical. They ought, therefore, to have the same generator. They have, for they are both evolved from $t_1 x_1/(t_1+x_1)$. Thus,

$$\frac{t_1 x_1}{x_1+t_1}=\frac{x_1}{1+x_1/t_1}=\epsilon^{-x_1 t} x_1, \tag{1}$$

$$\frac{t_1 x_1}{x_1+t_1}=\frac{t_1}{1+t_1/x_1}=\epsilon^{-t_1 x} t_1. \tag{2}$$

The generator is symmetrical with respect to x_1 and t_1. The two results are not, but they are equivalent. It will be remembered that a unit operand is always understood, meaning the function whose value is $+1$ for $+$ values only of the variable.

From the above it follows that

$$f(t)=\int_0^\infty f(x)(\epsilon^{-x_1 t} x_1)\,dx, \tag{3}$$

because the function integrated is an impulse of size $f(x)$ at the point $x=t$.

Substituting the other form, we also have

$$f(t)=\int_0^\infty f(x)(\epsilon^{-t_1 x} t_1)\,dx. \tag{4}$$

Now it is this second form that is useful in definite integrals, though not by itself so obvious as the first form.

If the limits are a and b, the result is $f(t)$ only if the value t lies between a and b. The result is zero otherwise. But $f(t)$ may be defined to be existent only between a and b; then the full limits 0 and ∞ may be used, and this is often convenient.

If we transform $f(t)$ to $f_1(t_1)$, then (4) becomes

$$f_1(t_1)=\int_0^\infty f(x)\,t_1\epsilon^{-t_1 x}\,dx. \tag{5}$$

This is a real quantitative definite integral and its solution. That is, t_1 may be a positive variable.

(2). To show the inner working when the limits are a and b, let $f(x)=1$, then (4) becomes

$$\int_a^b \epsilon^{-x t_1} t_1\,dx=(\epsilon^{-a t_1}-\epsilon^{-b t_1})1, \tag{6}$$

by integration in the usual way. It is true when t_1 is a $+$ quantity, and also when t_1 is a differentiator. In the latter case the result tells us explicitly that the value is 1 when t lies between a and b, and zero else. For $\epsilon^{-at_1}1$ begins at $t=a$, and then comes $-\epsilon^{-bt_1}1$ later and cuts it off.

Again, $f(x)=x$. Then

$$\int_a^b x\epsilon^{-xt_1}t_1 dx = -\left[\left(x+\frac{1}{t_1}\right)\epsilon^{-xt_1}\right]_a^b = \epsilon^{-at_1}(a+t) - \epsilon^{-bt_1}(b+t). \quad (7)$$

That is, the value of $f(t)$ is zero from $t=0$ to a, then t from a to b, and then zero again.

Similarly as regards $f(x)=x^2$, &c., but there is no need to elabo rate these details.

(3). There is no necessary confinement to single integrals. For example,

$$f_1(x_1, y_1, z_1) = f(x, y, z) = \iiint f(l, m, n)\epsilon^{-(x_1 l + y_1 m + z_1 n)} x_1 y_1 z_1 \, dl \, dm \, dn \quad (8)$$

is a triple integral on a triply impulsive foundation, which may be interpreted functionally when x_1, y_1, z_1 are quantities.

(4). If $f(x) = x^n/\underline{|n}$, then

$$\int_0^\infty \frac{x^n}{\underline{|n}} t_1 \epsilon^{-t_1 x} dx = t_1^{-n} = \frac{t^n}{\underline{|n}}. \quad (9)$$

When t_1 is a $+$constant, this is Euler's most valuable fundamental integral, which is always turning up. We also see its impulsive meaning, for the operator $t_1\epsilon^{-t_1 x}$ spots a single value of the function $x^n/\underline{|n}$. Of all definite integrals I admire it the most, because of its leading to so many others so easily; not only ordinary integrals, but those of the Fourier and Bessel theorems, and elliptic functions, and all sorts of things, like generalised differentiation and divergent series. But it must usually be interpreted in accordance with the impulsive integral (4), or else (5), to show its power.

(5). Another way of exhibiting this integral (4) is to divide by t_1. Then

$$\frac{1}{t_1} f(t) = \int_0^\infty \epsilon^{-t_1 x} f(x) dx = \int_0^t f(x) dx. \quad (10)$$

The interpretation now is that $\epsilon^{-t_1 x}$ is zero before and 1 after $t=x$. That is, it is zero when $x>t$. So it is the same as making the limits be 0 and t. But in Eulerian applications, the full limits are wanted when we consider the function of t_1; *e.g.*, $f(x) = x^n/\underline{|n}$,

$$\int_0^\infty \epsilon^{-t_1 x} \frac{x^n}{\underline{|n}} dx = \frac{1}{t_1^{n+1}} = \frac{t^{n+1}}{\underline{|n+1}} \quad (11)$$

(6). If in the elementary case $n=0$ of the last, we turn t_1 to t_1^{-1}, we get

$$\int_0^\infty \epsilon^{-\frac{x}{t_1}} dx = t_1, \quad (12)$$

when t_1 is a + constant. But if it is d/dt, then it means that the *result* is impulsive. In terms of t,

$$\epsilon^{-\frac{x}{t_1}} = 1 - \frac{x}{\underline{|1}\, t_1} + \dots = 1 - \frac{xt}{\underline{|1}\,\underline{|1}} + \dots = J_0 2\sqrt{xt}; \tag{13}$$

so we obtain

$$\int_0^\infty J_0 2\sqrt{xt}\, dx = t_1. \tag{14}$$

That is, regarded as a function of t, the integral is zero save when $t=0$, then its total is 1. Here we have jumped to the zeroth Bessel function.

(7). A further short step leads to the Bessel double integral. For let

$$u = \int_0^\infty J_0(2\sqrt{yt}) J_0(2\sqrt{yx})\, dx. \tag{15}$$

Then, by (13),

$$u = \int_0^\infty \epsilon^{-\frac{y}{t_1}} \epsilon^{-\frac{y}{x_1}} dy = \frac{t_1 x_1}{t_1 + x_1}, \tag{16}$$

the integration being done by Euler's integral. Comparing with equations (1), (2), we see that u is an impulsive function of t (or of x) condensed at x (or at t). So

$$f(t) = \int_0^\infty f(x)\, dx \int_0^\infty J_0(2\sqrt{yt}) J_0(2\sqrt{yx})\, dy, \tag{17}$$

which is the Bessel expansion theorem for the zeroth function. It comes out in a singularly simple manner.

(8). The corresponding theorem for J_m is much more difficult, though essentially similar. Thus, the series definition makes

$$J_m(2\sqrt{yt}) = \frac{(yt)^{\frac{1}{2}m}}{\underline{|0}\,\underline{|m}} + \frac{(yt)^{\frac{1}{2}m+1}}{\underline{|1}\,\underline{|m+1}} + \dots = \left(\frac{y}{t}\right)^{\frac{1}{2}m} \frac{1}{t_1^{\,m}} \epsilon^{-\frac{y}{t_1}}, \tag{18}$$

by the usual transformation $t^n/\underline{|n} = 1/t_1^{\,n}$. So,

$$u = \int J_m(2\sqrt{yt}) J_m(2\sqrt{xy})\, dy = \frac{1}{(tx)^{\frac{1}{2}m}} \frac{1}{(t_1 x_1)^m} \int y^m \epsilon^{-y\left(\frac{1}{t_1} + \frac{1}{x_1}\right)} dy. \tag{19}$$

To this apply Euler's integral, then

$$u \; \frac{\underline{|m}}{(tx)^{\frac{1}{2}m}} \frac{t_1 x_1}{(t_1 + x_1)^{m+1}}. \tag{20}$$

This looks hard, but is readily simplified by the shift transformation

$$F(t_1) = F(t_1) \epsilon^{-tx_1} \frac{t_1}{t_1 - x_1} = \epsilon^{-tx_1} F(t_1 - x_1) \frac{t_1}{t_1 - x_1}, \tag{21}$$

a transformation of frequent utility. Applied to (20), it makes

$$u = \frac{\underline{|m}}{(tx)^{\frac{1}{2}m}} \epsilon^{-tx_1} \frac{x_1}{t_1^{\,m}} = \left(\frac{t}{x}\right)^{\frac{m}{2}} \epsilon^{-tx_1} x_1; \tag{22}$$

or, since x and t may be interchanged throughout,

$$u = \left(\frac{x}{t}\right)^{\frac{m}{2}} \epsilon^{-xt_1} t_1. \tag{23}$$

One means a unit impulsive function of x at $x=t$, the other of t at $t=x$, since the factor $(x/t)^{\frac{1}{2}m}$ is 1 at $x=t$. So

$$f(t)=\int f(x)u\,dx=\int f(x)dx\int \mathrm{J}_m(2\sqrt{xy})\mathrm{J}_m(2\sqrt{yt})dy. \tag{24}$$

In practical use with definite integrals, it is not necessary to use the double integral and it saves useless work not to. For if

$$\mathrm{F}(y)=\int f(x)\mathrm{J}_m(2\sqrt{xy})dx. \tag{25}$$

$$\therefore\quad f(t)=\int \mathrm{F}(y)\mathrm{J}_m(2\sqrt{yt})dy, \tag{26}$$

by using (24). In (26) turn t to y and y to x. It becomes then the same as (25) with F and f interchanged. So, given an integral of the form (25), we can immediately derive another by interchanging F and f.

If we put $y=\frac{1}{2}\mathrm{Y}^2$, $x=\frac{1}{2}\mathrm{X}^2$, $t=\frac{1}{2}\mathrm{T}^2$,

then
$$u=\int \mathrm{J}_m(\mathrm{XY})\mathrm{J}\ (\mathrm{YT})\mathrm{Y}d\mathrm{Y}; \tag{27}$$

and since
$$\int u\,dx=1=\int u\mathrm{X}d\mathrm{X}, \tag{28}$$

it is now uX that is the impulsive function of X, making

$$\mathrm{F}(\mathrm{T})=\int \mathrm{F}(\mathrm{X})\mathrm{X}d\mathrm{X}\int \mathrm{J}_m(\mathrm{YX})\mathrm{J}_m(\mathrm{YT})\mathrm{Y}d\mathrm{Y}, \tag{29}$$

which is the usual form, and may be more convenient sometimes, though less easily worked operationally.

(9). The reduction of (29) to Fourier's form is done by $m=-\frac{1}{2}$ and $\frac{1}{2}$.

Thus, with $m=-\frac{1}{2}$, say $v=u\mathrm{X}$, then

$$v=\int \mathrm{X}\left(\frac{2}{\pi\mathrm{XY}}\right)\cos\mathrm{XY}\left(\frac{2}{\pi\mathrm{YT}}\right)^{\frac{1}{2}}\cos\mathrm{YT}\,.\,\mathrm{Y}d\mathrm{Y}$$
$$=2\int \cos\mathrm{XY}\,.\cos\mathrm{YT}\,d\mathrm{Y}. \tag{30}$$

This is the cosine unit impulse. The sine unit impulse is got by $m=\frac{1}{2}$ in a similar manner. Compare with vol. 2, p. 99.

(10). This is a known integral typical of many.

$$\int \epsilon^{-xt_1}\mathrm{J}_0(bx)dx=\frac{1}{t_1}\mathrm{J}_0(bt)=\frac{1}{(t_1^2+b^2)^{\frac{1}{2}}}. \tag{31}$$

The interpretation is double; t_1 as a + constant, or a differentiator. The binomial theorem converts the third form to the second.

If we turn b^2 to $-b^2$, then

$$\int \epsilon^{-xt_1}\mathrm{I}_0(bx)dx=\frac{1}{t_1}\mathrm{I}_0(bt)=\frac{1}{(t_1^2-b^2)^{\frac{1}{2}}}. \tag{32}$$

Here is a commonly occurring property. In (31), t_1 may have any value from 0 to ∞. But in (32) $t_1>b$. The value becomes ∞ when t_1 is reduced to b, and remains ∞ ever after; that is down to $t_1=0$. But if t_1 is a differentiator this limitation does not exist.

(11). It is similar in

$$\int t_1\epsilon^{-xt_1}\cos mx\,dx = \cos mt = \frac{1}{1+m^2/t_1^2}. \tag{33}$$

$$\int t_1\epsilon^{-xt_1}\cosh mx\,dx = \cosh mt = \frac{1}{1-m^2/t_1^2}. \tag{34}$$

In the first t_1 as a quantity may have any + value, in the second it must be $>m$.

(12). By using (14), we get

$$\epsilon^{-\frac{m}{t_1}} = J_0(2\sqrt{mt}) = \int (J_0 2\sqrt{mx})\epsilon^{-xt_1}t_1 \,.\, dx, \tag{35}$$

where notice the change in position of t_1 in the two exponentials.

(13). In elastic waves, we have this integral.

$$K_0(t_1 r)t_1 = \frac{2}{\pi(t^2-r^2)^{\frac{1}{2}}} = \frac{2}{\pi}\int_r^\infty \frac{\epsilon^{-xt_1}t_1}{(x^2-r^2)^{\frac{1}{2}}}dx. \tag{36}$$

We require $t>r$ in the middle, and $x>r$ in the integral, making the effective limits be r to ∞. The unreal parts of radicals are always omitted in real physical applications. Or, in a better form, the function does not exist unless it is real.

(14). Turn t_1 to $1/t_1$ in the last. Then

$$K_0\left(\frac{r}{t_1}\right) = \frac{2}{\pi}\int_r^\infty \frac{J_0(2\sqrt{xt})}{(x^2-r^2)^{\frac{1}{2}}}\,dx, \tag{37}$$

by using (13). Now to put the left side in terms of t_1, use the divergent formula, making

$$\epsilon^{-\frac{r}{t_1}}\left(\frac{2t_1}{\pi r}\right)^{\frac{1}{2}}\left(1 - \frac{t_1}{\underline{|1}(8r)} + \frac{1^2 3^2 t_1^2}{\underline{|2}(8r)^2} - \dots\right). \tag{38}$$

Compare with (18), with $m = -\frac{1}{2}, -1\frac{1}{2}$, &c., we see that (38) it can be turned to a function of t, term by term. It is not necessary to elaborate this, as there is an easier way, coming later.

(15). This case occurs in the telegraph theory.

$$\frac{t_1^{\frac{1}{2}}}{(t_1+2\sigma)^{\frac{1}{2}}} = \epsilon^{-\sigma t}I_0(\sigma t) = \int I_0(\sigma x)\epsilon^{-x(\sigma+t_1)}t_1 dx. \tag{39}$$

It may readily be harmonised with (31) above.

(16). The solution of Weber's integral

$$\int x^m J_0(xy)\,dx = \frac{2^m}{y^{m+1}}\,\frac{\underline{|\tfrac{1}{2}(m-1)}}{\underline{|-\tfrac{1}{2}(m+1)}} \tag{40}$$

is immediate. Thus,

$$\int \frac{x^n}{\underline{|n}} J_0 2\sqrt{xy}\,dx = \int \frac{x^n}{\underline{|n}}\epsilon^{-\frac{x}{y_1}}\,dx = y_1^{n+1} = \frac{y^{-(n+1)}}{\underline{|-(n+1)}}, \tag{41}$$

by Euler's integral.

(17). A more advanced case is

$$\int \epsilon^{-ax} J_m 2\sqrt{tx}\,(tx)^{\frac{1}{2}m}\,dx = \int \epsilon^{-ax}\left(\frac{x}{t_1}\right)^m \epsilon^{-\frac{x}{t_1}}\,dx = \frac{\underline{|m}\,t_1}{(1+at_1)^{m+1}}, \tag{42}$$

by Euler's integral. To put in terms of t, put on the prefactor $\epsilon^{-t/a}$ by the transformation (21). The result is

$$\frac{\underline{|m}\,\epsilon^{-t/a}}{a^{m+1}}\frac{1}{t_1{}^m} = \frac{t^m}{a^{m+1}}\epsilon^{-t/a}, \tag{43}$$

agreeing with another of Weber's formulæ.

(18). A still more advanced case is

$$\int \frac{x^n}{\underline{|n}} J_0 2\sqrt{xt}\; J_0 2\sqrt{xy}\,dx = \int \frac{x^n}{\underline{|n}} \epsilon^{-x\left(\frac{1}{t_1}+\frac{1}{y_1}\right)}dx = \left(\frac{t_1 y_1}{t_1+y_1}\right)^{n+1}, \tag{44}$$

by Euler's integral. Here we have several solutions, viz., in terms of t_1 and y_1, or by the binomial theorem, in terms of t and y_1, or y_1 and t, or finally t and y. The first is as in (44). The second comes by expanding in powers of t_1^{-1}. Thus

$$= y_1^{n+1}\left\{1 - \frac{n+1}{\underline{|1}}\frac{y_1}{t_1} + \frac{(n+1)(n+2)}{\underline{|2}}\frac{y_1^2}{t_1^2} + \dots\right\}, \tag{45}$$

So we obtain

$$\int \frac{x^n}{\underline{|n}} \epsilon^{-\frac{x}{y_1}} J_0 2\sqrt{xt}\,dx = y_1^{n+1}\left\{1 - \frac{n+1}{\underline{|1}}\frac{y_1 t}{\underline{|1}} + \frac{(n+1)(n+2)}{\underline{|2}}\frac{y_1^2 t^2}{\underline{|2}} + \dots\right\}. \tag{46}$$

The third form is got by turning t to y and y_1 to t_1 on both sides of (46). The fourth form is got by evolution, putting both sides of (46) in terms of y, making

$$\frac{1}{y^{n+1}\underline{|-n-1}}\left\{1 + \frac{(n+1)^2}{(\underline{|1})^2}\frac{t}{y} + \frac{(n+1)^2(n+2)^2}{(\underline{|2})^2}\frac{t^2}{y^2} + \dots\right\}, \tag{47}$$

which is a part of the solution of the integral containing the two Bessel functions. The other part, on account of the interchangeability of t and y, arises by interchanging t and y in (47). The meaning is that the two together make the complete convergent solution, because, for any value of y, (47) is convergent when t is less than y, and then the other form is convergent when t is greater than y. Of course, the value of n need not be so great as to make the definite integral be divergent. Inspection of the integral (44) shows that $n = \frac{1}{2}$ is the limiting value. But this matter is not under discussion.

If $y_1 = \infty$ on the left side of (46), it reduces to the form (41). But the right side does not, as it assumes the form of a succession of infinities. This curious behaviour often occurs. The remedy, if it is desired to evaluate (41) in this way, is to interchange t_1 and y_1 in

(45); that is, to seek the solution in descending powers of y_1; we then have

$$t_1^{n+1}\left\{1-\frac{n+1}{\lfloor 1}\frac{t_1}{y_1}+\dots\right\}=\frac{t^{-(n+1)}}{\lfloor -n-1}\left\{1+\frac{(n+1)^2}{ty_1}+\frac{(n+1)^2(n+2)^2}{t^2y_1^2\lfloor 2}+\right\}, \quad (48)$$

and now we see that $y_1=\infty$ reduces correctly to the proper value.

We must not conclude that (48) is fully equivalent to (46) for any value of n; that requires special examination. They are numerically equivalent when $n=-\frac{1}{2}$, as will be noticed later.

If $n=-1$, the value shows an impulsive function of x condensed at $x=0$. The original integral (44) also shows this when combined with (16) above. When $n=-\frac{1}{2}$, the two forms of solution taken together, (47) and its companion, represent a wave formula expressed by a complete first elliptic function, which will turn up in another way.

(19). The integral

$$\int J_0(2n\sqrt{x})\cos nt\,dn=\int \epsilon^{-n^2/x_1}\left(1-\frac{n^2t^2}{\lfloor 2}+\frac{n^4t^4}{\lfloor 4}-\right)\frac{dn^2}{2\sqrt{n^2}} \quad (49)$$

is solved in a series by Euler's integral, with variable n^2, making

$$\frac{1}{2}\sqrt{\pi}\left(x_1^{\frac{1}{2}}-\frac{1}{2}\frac{t^2}{\lfloor 2}x_1^{1\frac{1}{2}}+\frac{1.3}{2.2}\frac{t^4}{\lfloor 4}x_1^{2\frac{1}{2}}+\dots\right)=\frac{\sqrt{\pi x_1}}{2}\epsilon^{-\frac{t^2x_1}{4}} \quad (50)$$

$$=\frac{1}{2\sqrt{x}}\left(1+\frac{1}{2}\frac{t^2}{4x}+\frac{1.3}{2.4}\frac{t^4}{(4x)^2}+\dots\right)=\frac{1}{(4x-t^2)^{\frac{1}{2}}}. \quad (51)$$

Here we solve two integrals; viz., in terms of x_1 and t, considered as quantities, and x and t. In the last, $4x>t^2$, else the value is zero.

If we turn x_1 to $1/x_1$ in (50), we get

$$\int \epsilon^{-n^2x_1}\cos nt\,dn=\frac{1}{t}\sin nt, \quad (52)$$

which is a case of (10) above.

(20). A very interesting integral is

$$v=\int J_0(2\sqrt{xn})J_0(2\sqrt{yn})J_0(2\sqrt{zn})dn, \quad (53)$$

involving three variables (positive), x, y, z. It may be at once turned to

$$v=\int \epsilon^{-\frac{n}{x_1}}J_0(2\sqrt{yn})J_0(2\sqrt{zn})dn=\int \epsilon^{-n\left(\frac{1}{x_1}+\frac{1}{y_1}\right)}J_0(2\sqrt{zn})dn$$

$$=\int \epsilon^{-n\left(\frac{1}{x_1}+\frac{1}{y_1}+\frac{1}{z_1}\right)}dn=\frac{1}{\frac{1}{x_1}+\frac{1}{y_1}+\frac{1}{z_1}}, \quad (54)$$

the final integration being done by Euler's integral. According as x_1, y_1, z_1 are constants or differentiators, we get several results.

First treat x_1 as a constant, then

$$v=\frac{x_1y_1z_1}{(y_1+x_1)(z_1+x_1)-x_1^2}\left(\epsilon^{-x_1(y+z)}\frac{y_1}{y_1-x_1}\frac{z_1}{z_1-x_1}\right), \tag{55}$$

where the (...) is equivalent to unit operands for y_1 and z_1. Then shift the exponential to the left, and we get

$$\epsilon^{-x_1(y+z)}\frac{x_1y_1z_1}{y_1z_1-x_1^2}=x_1\epsilon^{-x_1(y+z)}\left(1+\frac{x_1^2}{y_1z_1}+\frac{x_1^4}{y_1^2z_1^2}+\ldots\right). \tag{56}$$

So, integrating with respect to y and z, we come to

$$\int\epsilon^{-\frac{n}{x_1}}\,J_0(2\sqrt{yn})\,J_0(2\sqrt{zn})dn=x_1\epsilon^{-x_1(y+z)}\,I_0(2x_1\sqrt{yz}), \tag{57}$$

which is, x_1 being a constant, one of Weber's integrals.

Lastly, to integrate to x, we may use the result (51). Thus,

$$v=\epsilon^{-x_1(y+z)}\,I_0(2x_1\sqrt{yz})\,\frac{1}{\pi}\int\cos nx\,dn \tag{58}$$

$$=\frac{1}{\pi}\epsilon^{-x_1(y+z)}\int J_0(2x_1\sqrt{yz})\cos nx\,dn$$

$$=\frac{1}{\pi}\epsilon^{-x_1(y+z)}\frac{1}{(4yz-x^2)^{\frac{1}{2}}}=\frac{1}{\pi}\frac{1}{\{4xy-(z-x-y)^2\}^{\frac{1}{2}}}. \tag{59}$$

This solves the original integral containing x, y, z.

But without using (51) we must use the divergent form of I_0, which is specially adapted for operational work of this kind. Then (57) is the same as

$$v=x_1\epsilon^{-x_1(y+z)}\frac{\epsilon^{2x_1\sqrt{yz}}}{(4\pi x_1\sqrt{yz})^{\frac{1}{2}}}\left(1+\frac{1^2}{16x_1\sqrt{yz}}+\frac{1^23^2}{\lfloor 2(16x_1\sqrt{yz})^2}+\ldots\right)$$

$$=\frac{\epsilon^{-x_1(\sqrt{y}-\sqrt{z})^2}}{(4\pi\sqrt{yz})^{\frac{1}{2}}}x_1^{\frac{1}{2}}\left(1+\frac{1^2}{16x_1\sqrt{yz}}+\ldots\right). \tag{60}$$

This is functionised to x by the usual way, making

$$v=\frac{\epsilon^{-Dx_1}}{(4\pi\sqrt{yz})^{\frac{1}{2}}}\frac{1}{(\pi x)^{\frac{1}{2}}}\left(1+\frac{x}{8\sqrt{yz}}+\frac{1.3}{\lfloor 2(8\sqrt{yz})^2}+\ldots\right)$$

$$=\;.\;.\;.\;.\;.\;.\;.\;\frac{1}{\left(1-\frac{x}{4\sqrt{yz}}\right)^{\frac{1}{2}}}$$

$$=\frac{1}{\pi}\frac{1}{(x-D)^{\frac{1}{2}}(S-x)^{\frac{1}{2}}}=\frac{1}{\pi}\frac{1}{(2xy+2yz+2xz-x^2-y^2-z^2)^{\frac{1}{2}}}, \tag{61}$$

if $D=(\sqrt{y}-\sqrt{z})^2$, $S=(\sqrt{y}+\sqrt{z})^2$. The range of x is therefore from D to S. Similarly regarding the range of y in terms of x and z, and of z in terms of x and y. When symmetrically written, these ranges do not show themselves so easily. The function v is ∞ on the boundary of the region, finite inside and zero outside.

(21). A little more advanced, though very similar, is

$$v=\int n^r \mathrm{J}_m(2\sqrt{nx})\mathrm{J}_m(2\sqrt{ny})\mathrm{J}_m(2\sqrt{nz})dn \tag{62}$$

$$=\int n^r\left(\frac{n}{x}\right)^{\frac{1}{2}m}\left(\frac{n}{y}\right)^{\frac{1}{2}m}\left(\frac{n}{z}\right)^{\frac{1}{2}m}\frac{1}{(x_1y_1z_1)^m}\epsilon^{-n\left(\frac{1}{x_1}+\frac{1}{y}+\frac{1}{z_1}\right)}dn$$

$$=\frac{1}{(xyz)^{\frac{1}{2}m}}\frac{1}{(x_1y_1z_1)^m}\frac{\underline{|r+\frac{3}{2}m}}{\left(\frac{1}{x_1}+\frac{1}{y_1}+\frac{1}{z_1}\right)^{r+1+\frac{3}{2}m}}$$

$$=\frac{\underline{|r+1\frac{1}{2}m}}{(xyz)^{\frac{1}{2}m}}\epsilon^{-x_1(y+z)}\frac{x_1^{\,r+1+\frac{1}{2}m}(y_1-x_1)^{r+\frac{1}{2}m}(z_1-x_1)^{r+\frac{1}{2}m}y_1z_1}{(y_1z_1-x_1^2)^{r+1+\frac{3}{2}m}}, \tag{63}$$

by the shift formula (21). Here we see there is no apparent simplification, except by giving r the value $-\frac{1}{2}m$. Then

$$v=\frac{\underline{|m}}{(xyz)^{\frac{1}{2}m}}\epsilon^{-x_1(y+z)}\frac{x_1}{\left(1-\frac{x_1^{\,2}}{y_1z_1}\right)^{1+m}(y_1z_1)^m}$$

$$=\frac{\underline{|m}}{(xyz)^{\frac{1}{2}m}}\epsilon^{-x_1(y+z)}\frac{x_1}{(y_1z_1)^m}\left(1+(1+m)\frac{x_1^{\,2}}{y_1z_1}+\frac{1}{\underline{|2}}(1+m)(2+m)\frac{x_1^{\,4}}{y_1^{\,2}z_1^{\,2}}+\ldots\right)$$

$$=\frac{\epsilon^{-x_1(y+z)}}{(xyz)^{\frac{1}{2}m}}x_1\left(\frac{(y_1z_1)^m}{\underline{|0}\,\underline{|m}}+x_1^{\,2}\frac{(y_1z_1)^{m+1}}{\underline{|1}\,\underline{|m+1}}+\ldots\right)$$

$$=\frac{\epsilon^{-x_1(y+z)}}{x^{\frac{1}{2}m}}\frac{x_1}{x_1^{\,m}}\mathrm{I}_m(2x_1\sqrt{yz}). \tag{64}$$

This comes out in the convergent form, and solves the integral when x_1 is constant. To turn to a function of x, use the divergent form, then

$$v=\frac{\epsilon^{-x_1(y+z)}}{x^{\frac{1}{2}m}}x_1^{\,1-m}\frac{\epsilon^{2x\sqrt{yz}}}{(4\pi x_1\sqrt{yz})^{\frac{1}{2}}}\left\{1+\frac{1^2-4m^2}{16x_1\sqrt{yz}}+\ldots\right\}$$

$$=\frac{\epsilon^{-x_1\mathrm{D}}}{x^{\frac{1}{2}m}(4\pi\sqrt{yz})^{\frac{1}{2}}}\left\{\frac{x^{m-\frac{1}{2}}}{\underline{|m-\frac{1}{2}}}+\frac{(1^2-4m^2)x^{m+\frac{1}{2}}}{16\sqrt{yz}\,\underline{|m+\frac{1}{2}}}+\ldots\right\}$$

$$=\frac{\epsilon^{-x_1\mathrm{D}}}{x^{\frac{1}{2}m}(4\pi\sqrt{yz})^{\frac{1}{2}}}\frac{x^{m-\frac{1}{2}}}{\underline{|m-\frac{1}{2}}}\left(1-\frac{x}{4\sqrt{yz}}\right)^{m-\frac{1}{2}}$$

$$=\frac{1}{4^m}\frac{(\mathrm{S}-x)^{m-\frac{1}{2}}(x-\mathrm{D})^{m-\frac{1}{2}}}{(xyz)^{\frac{1}{2}m}\sqrt{\pi}\,\underline{|m-\frac{1}{2}}}, \tag{65}$$

where S and D have the same meaning as before. The range of x, y, z is also the same. This agrees with Sonine's integral (Gray and Matthew, "Bessel Functions," p. 239. Ex. 52). It is very complicated in the work, though I do not know whether it is any easier by rigorous methods. I have only worked it out to confirm Sonine's result. The cases $m=\frac{1}{2}$, or $-\frac{1}{2}$, &c., can be more simply done separately, since then I_m becomes a finite formula with an exponential factor.

(22). The integral

$$v=\int z^r\,J_m(2\sqrt{zx})J_n(2\sqrt{zy})dz \tag{66}$$

is comprehensive with five variables r, x, y, m and n, though only x and y will be differentially used. It is the same as

$$v=\frac{1}{x^{\frac{1}{2}m}y^{\frac{1}{2}n}}\,\frac{1}{x_1{}^m y_1{}^n}\int z^s\,\epsilon^{-z\left(\frac{1}{x_1}+\frac{1}{y_1}\right)}dz, \tag{67}$$

where $s=r+\frac{1}{2}m+\frac{1}{2}n$. Or, by Euler's integral,

$$v=\frac{\lfloor s}{x^{\frac{1}{2}m}y^{\frac{1}{2}n}}\,\frac{1}{x_1{}^m y_1{}^n}\left(\frac{x_1y_1}{x_1+y_1}\right)^{s+1}. \tag{68}$$

Here we see that the use of the binomial theorem, and then evolution, gives two series, one in powers of x/y, the other in powers of y/x. They are

$$v_1=\frac{\lfloor s}{\lfloor 0}\,\frac{x^{\frac{1}{2}m}}{\lfloor m}\,\frac{y^{\frac{1}{2}n-1-s}}{\lfloor n-1-s}-\frac{\lfloor s+1}{\lfloor 1}\,\frac{x^{\frac{1}{2}m+1}}{\lfloor m+1}\,\frac{y^{\frac{1}{2}n-2-s}}{\lfloor n-2-s}+\dots, \tag{69}$$

$$v_2=\frac{\lfloor s}{\lfloor 0}\,\frac{y^{\frac{1}{2}n}}{\lfloor n}\,\frac{x^{\frac{1}{2}m-1-s}}{\lfloor m-1-s}-\frac{\lfloor s+1}{\lfloor 1}\,\frac{y^{\frac{1}{2}n+1}}{\lfloor n+1}\,\frac{x^{\frac{1}{2}m-2-s}}{\lfloor m-2-s}+\dots. \tag{70}$$

Only two terms are needed to show the full construction. To turn v_1 to v_2, interchange x and y, and also m and n. If v_1 is convergent when $x<y$, and v_2 is convergent when $y<x$, then the two series together make up the complete convergent solution.

A complete discussion would be out of place, but there are several cases of simplification. If $m=0$, $n=0$, $r=-\frac{1}{2}$, then

$$v=\int z^{-\frac{1}{2}}J_0(2\sqrt{zx})J_0(2\sqrt{zy})dz=\pi^{\frac{1}{2}}\left(\frac{x_1y_1}{x_1+y_1}\right)^{\frac{1}{2}} \tag{71}$$

and

$$v_1=\frac{1}{y^{\frac{1}{2}}}\left(1+\frac{1^2}{2^2}\frac{x}{y}+\frac{1^23^2}{2^24^2}\frac{x^2}{y^2}+\dots\right), \tag{72}$$

$$v_2=\frac{1}{x^{\frac{1}{2}}}\left(1+\frac{1^2}{2^2}\frac{y}{x}+\frac{1^23^2}{2^24^2}\frac{y^2}{x^2}+\dots\right). \tag{73}$$

The relation $s=n$ makes $v_1=0$, and $s=m$ makes $v_2=0$. The region of the integral (66) is then confined to one or other side of the point $x=y$. But $s=m=n$ makes both vanish, unless $x=y$, a case of condensation. Also $n-1-s=$negative integer makes $v_1=0$, and $m-1-s=$negative integer makes $v_2=0$. But in general both are finite. Sometimes one of the two series becomes a stopping series, or even one term only. Thus, $s=n-1$, or $r=\frac{1}{2}(n-m)-1$, makes

$$v_1=\frac{\lfloor n-1}{\lfloor m}\,\frac{x^{\frac{1}{2}m}}{y^{\frac{1}{2}n}},\quad (x<y), \tag{74}$$

whilst the other remains an infinite series, namely,

$$v_2=\frac{x^{\frac{1}{2}m-n}y^{\frac{1}{2}n}}{\lfloor m-n}\left(\frac{1}{n}-\frac{m-n}{n+1}\,\frac{y}{x}+\frac{(m-n)(m-n-1)}{\lfloor 2(n+2)}\,\frac{y^2}{x^2}-\dots\right). \tag{75}$$

There is continuity at the junction.

Similarly $s=m-1$ reduces v_2 to one term, v_1 being an infinite series. Again, let $s=m$, or $r=\frac{1}{2}(m-n)$, then $v_2=0$, and

$$v_1=\frac{x^{\frac{1}{2}m}}{y^{\frac{1}{2}n}}\frac{(y-x)^{n-m-1}}{\lfloor n-m-1}. \quad (x<y). \tag{76}$$

This confirms Sonine ("G. and M.," p. 239, Ex. 51).

We may also turn (68) to

$$v=\frac{\lfloor s}{x^{\frac{1}{2}m}y^{\frac{1}{2}n}}\frac{(x_1y_1)^s}{x_1^{\,m}y_1^{\,n}}\epsilon^{-y_1x}\frac{y_1}{x_1^{\,s}}, \tag{77}$$

and to another form by interchanging x, y, and x_1, y_1, and m, n.

In the special case $m=n$, of course, there is a further simplification. When $r=0$, also, we come back to the Bessel impulsive function for J_m.

(23). The next integral is done by Euler's integral term by term.

$$\int\epsilon^{-ax}\mathrm{I}_n(bx)dx=\int\epsilon^{-ax}\left\{\frac{(\frac{1}{2}bx)^n}{\lfloor 0\lfloor n}+\frac{(\frac{1}{2}bx)^{n+2}}{\lfloor 1\lfloor n+1}+\dots\right\}dx$$

$$=\frac{1}{a}\left\{\left(\frac{b}{2a}\right)^n+\left(\frac{b}{2a}\right)^{n+2}\frac{\lfloor n+2}{\lfloor 1\lfloor n+1}+\left(\frac{b}{2a}\right)^{n+4}\frac{\lfloor n+4}{\lfloor 2\lfloor n+2}+\dots\right\} \tag{78}$$

$$=\frac{(b/a)^{n+1}}{b(1-b^2/a^2)^{\frac{1}{2}}\{1+\sqrt{1-b^2/a^2}\}^n}. \tag{79}$$

Similarly, by alternating signs in the series, we get

$$\int\epsilon^{-ax}\mathrm{J}_n(bx)dx=\frac{(b/a)^{n+1}}{b(1+b^2/a^2)^{\frac{1}{2}}\{1+\sqrt{1+b^2/a^2}\}^n}. \tag{80}$$

In (79), $b<a$, else the integral is ∞. But in (80), b/a may have any value. It may well be asked here how the transition from (78) to (79) is done. No one could be expected to recognise that the series was the expansion of the function in (79), it is such a complicated function, unless one had a surprising memory, or by some accident. I did it in this roundabout way. In (78) we do not require any series in powers of a/b. But in (80) we do, or may and can get it thus. Put $b^2=4t$, then

$$\int\epsilon^{-ax}\mathrm{J}_n(bx)dx=\int\epsilon^{-a\sqrt{x}}\left(\frac{x}{t}\right)^{\frac{n}{2}}t_1^{-n}\epsilon^{-\frac{x}{t_1}}\frac{dx}{2\sqrt{x}}$$

$$=\tfrac{1}{2}t^{-\frac{1}{2}n}t_1^{-n}\int\left(1-ax^{\frac{1}{2}}+\frac{a^2x}{\lfloor 2}-\dots\right)x^{\frac{1}{2}(n-1)}\epsilon^{-\frac{x}{t_1}}dx$$

$$=\tfrac{1}{2}\left\{t^{-\frac{1}{2}}-at^{-1}\frac{n}{2}+\frac{a^2t^{-1\frac{1}{2}}}{\lfloor 2}\left(\frac{n}{2}+\frac{1}{2}\right)\left(\frac{n}{2}-\frac{1}{2}\right)-\dots\right\},$$

by Euler's integral. Or, in terms of b,

$$=\frac{1}{b}\left\{1+\frac{n^2-1^2}{\lfloor 2}\frac{a^2}{b^2}+\frac{(n^2-1^2)(n^2-3^2)}{\lfloor 4}\frac{a^4}{b^4}+\dots\right\}$$

$$-\frac{an}{b^2}\left\{1+\frac{n^2-2^2}{\lfloor 3}\frac{a^2}{b^2}+\frac{(n^2-2^2)(n^2-4^2)}{\lfloor 5}\frac{a^4}{b^4}+\dots\right\}. \tag{81}$$

This very curious series seemed familiar, so I hunted it up, and found in Todhunter's Diff. Calc. the following,

$$\epsilon^{a \sin^{-1} x} = 1 + ax + \frac{a^2x^2}{\lfloor 2} + \frac{a(a^2+1)x^3}{\lfloor 3} + \frac{a^2(a^2+2^2)x}{\lfloor 4} + \dots \qquad (82)$$

Differentiate to x, and divide by a; then put $a = ni$, and $x = ia/b$. The result is

$$\frac{\epsilon^{-n \operatorname{shin}^{-1}(a/b)}}{(1+a^2/b^2)^{\frac{1}{2}}} = \text{series (81)} \times b,$$

$$= (1+a^2/b^2)^{-\frac{1}{2}} \{a/b + \sqrt{1+a^2/b^2}\}^{-n} \qquad (83)$$

This proves (79) and (80). Using (79), a fresh integral can be derived, by differentiating to n. Thus,

$$-\frac{2}{\pi}\frac{d}{dn_0}\int \epsilon^{-ax} \mathrm{I}_n(bx)\,dx = \int \epsilon^{-ax} \mathrm{K}_0(bx)\,dx$$

$$= \frac{2}{\pi a}\frac{\log \{a/b + \sqrt{a^2/b^2 - 1}\}}{(1-b^2/a^2)^{\frac{1}{2}}}. \qquad (84)$$

This is one of Basset's integrals. From it, by putting $b = ci$, and supposing c to be real positive, arise two more,

$$\int \epsilon^{-ax} \mathrm{J}_0(cx)\,dx = \frac{1}{\sqrt{a^2+c^2}} \qquad (85)$$

$$\int \epsilon^{-ax} \mathrm{G}_0(cx)\,dx = \frac{2}{\pi\sqrt{a^2+c^2}} \log\left(\frac{a}{c} + \sqrt{1+\frac{a^2}{c^2}}\right). \qquad (86)$$

The first of these occurred before. The notation is that of vol. 2; namely, $\mathrm{K}_0(ci) = \mathrm{G}_0(c) - i\mathrm{J}_0(c)$, the proper notation to harmonise with $\epsilon^{-ci} = \cos c - i \sin c$, as may be seen in any simply periodic applications of the K_0 function, especially progressive waves.

An application of (84) is to find the function of a_1 which generates $\mathrm{K}_0(ab)$. For, by (4),

$$\mathrm{K}_0(ab) = \int a_1 \epsilon^{-a_1 x} \mathrm{K}_0(bx)\,dx = -\frac{2}{\pi}\int \frac{d}{dn_0} a_1 \epsilon^{-a_1 x} \mathrm{I}_n(bx)\,dx$$

$$= -\frac{2}{\pi} a_1 \frac{\log \{a_1/b + \sqrt{a_1^2/b^2 - 1}\}}{(a_1^2 - b^2)^{\frac{1}{2}}}. \qquad (87)$$

This may be expanded to

$$-\frac{2}{\pi}\left\{\frac{\log(b/2a_1)}{(1-b^2/a_1^2)^{\frac{1}{2}}} + \left(\frac{b}{2a_1}\right)^2 + \frac{7}{\lfloor 2}\left(\frac{b}{2a_1}\right)^4 + \frac{74}{\lfloor 3}\left(\frac{b}{2a_1}\right)^6 + \dots\right\}. \qquad (88)$$

This generates the convergent formula for $\mathrm{K}_0(ab)$. The treatment of the log term need not be given here, as it will come in a later example. The coefficients 1, 7, 74, etc., may be found by doing the differentiation to n upon the series (78), instead of upon (79). It produces (88), on remembering that the original a is replaced by a_1.

(24). Now some examples of two steps of involution. We have

$$\int f(x)\epsilon^{-xt_1}\,dx = \int \frac{f(x)\,dx}{1+x/t_2} = \frac{1}{t_1}f(t) = f_1(t_1) = f_2(t_2), \qquad (89)$$

if t_2 is d/dt_1, and t_1 is d/dt. Similarly $f_2(t_2)$ generates $f_1(t_1)$.

Say $f(x)=x^n/\underline{|n}$. Then

$$\int\frac{x^n dx}{\underline{|n}(1+x/t_2)}=\frac{1}{t_1}\frac{t^n}{\underline{|n}}=\frac{1}{t_1^{n+1}}=t_2^{n+1}\underline{|-n-1}. \tag{90}$$

When n goes from -1 to 0, the function of t_2 goes from 1, expressing the integral of the impulsive function $x^{-1}/\underline{|-1}$, or $x_1 1$, to ∞. Between the two it is a continuous finite function. Thus, $n=-\frac{1}{2}$ makes

$$\int\frac{dx}{\sqrt{x}(1+x/t_2)}=\pi\sqrt{t_2}=\int\frac{2dy}{1+y^2/t_2}, \tag{91}$$

the last form being a corroboration by an elementary integral, using $x=y^2$.

(25). More advanced and varied is the following, with two variables s and t.

$$\int\epsilon^{-xt_1}\epsilon^{-\frac{x}{s_1}}dx=\frac{1}{t_1+\frac{1}{s_1}}=\frac{1}{t_1}\epsilon^{-\frac{t}{s_1}}=\frac{1}{t_1}\epsilon^{-\frac{s}{t_1}}=\frac{1}{t_1}\mathrm{J}_0(2\sqrt{st})=\left(\frac{t}{s}\right)^{\frac{1}{2}}\mathrm{J}_1(2\sqrt{st})$$

$$=\int\frac{\epsilon^{-\frac{x}{s_1}}}{1+\frac{x}{t_2}}dx=\int\frac{\mathrm{J}_0(2\sqrt{sx})}{1+\frac{x}{t_2}}dx=\int\epsilon^{-xt_1}\mathrm{J}_0(2\sqrt{sx})\,dx. \tag{92}$$

All this is done at sight, through the previous relating to the treatment of ϵ^{-xt_1} and ϵ^{-x/t_1}. We obtain the solution of a quantitative definite integral by pairing together one of the integrals and the particular solution that contains the same symbols. But some are not done, for we have s_1 and t_2 in one integral, and s and t_2 in another, unsolved explicitly. Do these.

$$\frac{s_1}{1+s_1t_1}=s_1(1-s_1t_1+s_1^2t_1^2-\ldots)=s_1(1-\frac{s_1}{t_2}+\frac{s_1^2\underline{|2}}{t_2^2}-\ldots). \tag{93}$$

Here we have the s_1, t_2 solution in a divergent form. But it can be turned to convergent form by vol. 2, p. 452, eq. (93), which makes

$$t_2\left(\frac{t_2}{s_1}+\frac{1+\frac{1}{2}}{\underline{|2}}\frac{t_2^2}{s_1^2}+\frac{1+\frac{1}{2}+\frac{1}{3}}{\underline{|3}}+\ldots\right)-t_2\epsilon^{t_2/s_1}\left(\log\frac{t_2}{s_1}+\mathrm{C}\right), \tag{94}$$

where $\mathrm{C}=0{\cdot}5772$.

Next, to obtain the s, t_2 solution, note that

$$\log\frac{t_2}{s_1}=\frac{d}{dn_0}\left(\frac{t_2}{s_1}\right)^n=\frac{d}{dn_0}\frac{(st_2)^n}{\underline{|n}}. \tag{95}$$

Applying this to (94), we get

$$t_2\left(st_2+(1+\tfrac{1}{2})\frac{s^2t_2^2}{\underline{|2}\underline{|2}}+\ldots\right)-\mathrm{C}t_2\,\mathrm{I}_0(2\sqrt{st_2})$$
$$-t_2\frac{d}{dn_0}\left\{\left(\frac{t_2}{s_1}\right)^n+\frac{1}{\underline{|1}}\left(\frac{t_2}{s_1}\right)^{n+1}+\ldots\right\}, \tag{96}$$

where the third part is

$$-t_2\frac{d}{dn_0}\left\{\frac{(st_2)^n}{\lfloor 0 \lfloor n}+\frac{(st_2)^{n+1}}{\lfloor 1 \lfloor n+1}+\frac{(st_2)^{n+2}}{\lfloor 2 \lfloor n+2}+\dots\right\}$$

$$=t_2\log st_2\,\mathrm{I}_0(2\sqrt{st_2})-\mathrm{C}t_2\,\mathrm{I}_0(2\sqrt{st_2})+st_2+\frac{(st_2)^2}{\lfloor 2 \lfloor 2}(1+\tfrac{1}{2})+\dots \qquad (97)$$

Inserting this in (96), the final result is

$$t_2\pi\,\mathrm{K}_0(2\sqrt{st_2})=\int\frac{\mathrm{J}_0(2\sqrt{sx})}{1+x/t_2}dx. \qquad (98)$$

It means the same as

$$\mathrm{K}_0(ab)=\frac{2}{\pi}\int\frac{\mathrm{J}_0(ax)x\,dx}{b^2+x^2}, \qquad (99)$$

which is (with $b=1$), one of Basset's integrals. Observe that a and b may be interchanged. Also, it follows that

$$\frac{1}{a+b}=\pi\int\mathrm{K}_0(2\sqrt{xb})\,\mathrm{J}_0(2\sqrt{xa})\,dx, \qquad (100)$$

by the interchangeability of functions in (98), with the substitutions $s=a$, $t_2=b$. This can also be verified. For the integral in (100) is

$$=\pi\int\epsilon^{-\frac{x}{a_1}}\mathrm{K}_0(2\sqrt{xb})dx=-2\frac{d}{dn_0}\int\epsilon^{-\frac{x}{a_1}}\left(\frac{x}{b}\right)^{\frac{n}{2}}b_1^{-n}\,\epsilon^{\frac{x}{b_1}}\,dx$$

$$=-2\frac{d}{dn_0}b^{-\frac{n}{2}}\lfloor\tfrac{1}{2}n\frac{a_1^{1+\frac{1}{2}n}}{\left(1-\frac{a_1}{b_1}\right)^{1+\frac{1}{2}n}}\frac{1}{b_1^{\,n}}$$

$$=-2\frac{d}{dn_0}\lfloor\tfrac{1}{2}n\left\{\frac{x^{-(1+\frac{1}{2}n)}}{\lfloor -1-\frac{1}{2}n}\frac{b^{\frac{1}{2}n}}{\lfloor n}+\frac{x^{-(2+\frac{1}{2}n)}}{\lfloor -2-\frac{1}{2}n}\frac{b^{\frac{1}{2}n+1}}{\lfloor n+1}\frac{1+\frac{1}{2}n}{\lfloor 1}+\dots\right\}. \qquad (101)$$

This looks bad, but note that all terms vanish on account of the denominators $\lfloor -1-\frac{1}{2}n$, &c., except the terms containing the derivatives of these denominators. So we come to

$$\frac{1}{a}+\frac{b}{a^2}g'(-2)+\frac{b^2}{a^3}\frac{g'(-3)}{\lfloor 2}+\dots=\frac{1}{a}\left(1-\frac{b}{a}+\frac{b^2}{a^2}-\dots\right)=\frac{1}{a+b}, \qquad (102)$$

as required.

After equation (98), up to (102) is a confirmatory interpolation. Returning to (98), I have also put the integral in terms of t and s_2. It is done thus. Go back to (93), the middle form, and put in terms of s_2, t_1. It makes

$$\frac{1}{s_2}\left(1-\frac{\lfloor 2\,t_1}{s_2}+\frac{\lfloor 3\,t_1^{\,2}}{s_2^{\,2}}-\dots\right)=\frac{1}{t_1}\left\{1-\left(1-\frac{t_1}{s_2}+\frac{\lfloor 2\,t_1^{\,2}}{s_2^{\,2}}-\dots\right)\right\}$$

$$=\frac{1}{t_1}\left\{1-\frac{s_2}{t_1}\left(\frac{s_2}{t_1}+\frac{1+\frac{1}{2}}{\lfloor 2}\frac{s_2^{\,2}}{t_1^{\,2}}+\dots\right)-\epsilon^{s_2/t_1}\left(\log\frac{s_2}{t_1}+\mathrm{C}\right)\right. ; \qquad (103)$$

by the same transformation as from (93) to (94);

$$=t-t\left(\frac{s_2^{\,2}t^2}{\lfloor 1\lfloor 3}+\frac{s_2^{\,3}t^3}{\lfloor 2\lfloor 4}(1+\tfrac{1}{2})+\dots\right)+\frac{1}{t_1}\frac{s_2}{t_1}\epsilon^{s_2/t_1}\left(\log\frac{s_2}{t_1}+\mathrm{C}\right); \qquad (104)$$

where the last part, in terms of t_1, becomes

$$Ct\,\mathrm{I}_2(2\sqrt{s_2t})+\frac{1}{s_2}\frac{d}{dn_0}\left\{\frac{(s_2t)^{n+2}}{\lfloor 1\lfloor n+2}+\frac{(s_2t)^{n+3}}{\lfloor 2\lfloor n+3}+\dots\right\}. \tag{105}$$

Carrying this out in the same way as before, the final result is

$$u=t\left(\frac{1}{s_2t}-\pi\mathrm{K}_2(2\sqrt{s_2t})\right), \tag{106}$$

using the notation of vol. 2, p. 250, eq. (48), for K_2.

Looking through (92), there is no corresponding integral, with x as the variable of integration. But we have the form

$$u=\frac{1}{t_1}\epsilon^{-\frac{t}{s_1}} \quad \therefore \ t_1u=\epsilon^{-\frac{t}{s_1}}=-\pi t_1t\,\mathrm{K}_2(2\sqrt{s_2t})\,; \tag{107}$$

so we come to

$$u=\int\epsilon^{-xt_1}\epsilon^{-\frac{x}{s_1}}dx=-\int\epsilon^{-xt_1}\pi x_1x\,\mathrm{K}_2(2\sqrt{s_2x})\,dx, \tag{108}$$

which is in terms of s_2 and t, because ϵ^{-xt_1} is a known function of t.

The above is an indirect way of finding what ϵ^{-x/s_1} should mean in terms of s_2. I have not found a direct way yet. It looks indeterminate directly. For

$$\epsilon^{-\frac{x}{s_1}}=1-\frac{x}{s_1}+\frac{x^2}{\lfloor 2 s_1^2}-\dots=1-x\left(1+\frac{s_2x}{\lfloor 1\lfloor 2}+\frac{s_2^2x^2}{\lfloor 2\lfloor 3}+\dots\right)\frac{1}{s_1}. \tag{109}$$

Now

$$(\log s_2)1=\left(\frac{d}{dn_0}s_2^n\right)1=\frac{d}{dn_0}s_1^{-n}\lfloor -n=-(\log s_1+\mathrm{C}), \tag{110}$$

and therefore

$$(s_2\log s_2)1=-\frac{1}{s_1}. \tag{111}$$

These results (110), (111) will be found to work quite well in general. But if we substitute for $1/s_1$ in (109) the form $s_2 \log s_2{}^{-1}$ of (111), it makes only a part of the required function. We get

$$1-\left(\frac{x}{s_2}\right)^{\frac{1}{2}}\mathrm{I}_1(2\sqrt{s_2x})\frac{1}{s_1}.$$

There are really additional terms in the s_2 function, like $(\mathrm{A}s_2+\mathrm{B}s_2^2+\dots)1$, which on conversion to s_1 form, make zero. It is an object in the development of operational mathematics to do away with indeterminateness of this sort, for in physical applications there is really no indeterminateness. I have been generally successful in this respect. But here we have found a function of s_2 which generates the given function of s_1, but it is not the proper generator, one to harmonise with the other related functions. I think the remedy is to be found by regarding ϵ^{-x/s_1} as the limiting form of a more general function.

(26). The Eulerian integral is also useful in a different class of integrals, of which the most elementary is

$$u=\frac{2}{\pi}\int_0^{\sqrt{a}}\frac{dx}{(a-x^2)^{\frac{1}{2}}}=1. \tag{112}$$

It is usually solved by recognising the indefinite integral to be an inverse sine function. But also thus,

$$u=\frac{2}{\pi}\int\epsilon^{-x^2a_1}(\pi a_1)^{\frac{1}{2}}\frac{dx^2}{2\sqrt{x^2}}=1 \tag{113}$$

by Euler's integral. The limits are here 0 and ∞. So they are in (112), considered as a real integral.

Put t^2 for a in (112), and change x^2 to $r^2=x^2+y^2$. Then, by (36), we get this integral,

$$\int_0^\infty t_1\mathrm{K}_0(t_1r)dy=\frac{2}{\pi}\int\frac{dy}{(t^2-r^2)^{\frac{1}{2}}}=\epsilon^{-t_1x}1; \tag{113A}$$

that is, the value is 1 when $t>x$, and zero when $t<x$. But if t_1 is a + quantity, then the third form evaluates the first integral.

A similar one to (112) is

$$u=\int_0^a\frac{dx}{(a-x)^{\frac{1}{2}}}=\int\epsilon^{-xa_1}(\pi a_1)^{\frac{1}{2}}dx=\left(\frac{\pi}{a_1}\right)^{\frac{1}{2}}=2\sqrt{a}. \tag{114}$$

Now take a very complicated one. Say

$$u=\int\frac{x^ndx}{(a-x^m)^l}=\int\epsilon^{-x^ma_1}a^l_1\,\underline{|-l}\,(x^m)^{\frac{n}{m}}\frac{dx^m}{m(x^m)^{\frac{m-1}{m}}}. \tag{115}$$

Here we make x^m the variable of integration, and m is to be positive, so that the limits are 0 to $a^{1/m}$ in the first, and 0 to ∞ in the second integral. Then Euler's integral makes

$$u=\frac{\underline{|-l}\;a^{\frac{n+1}{m}-l}}{m\;\underline{\left|\frac{n+1}{m}-l\right.}}\;\underline{\left|\frac{n+1}{m}-1\right.}. \tag{116}$$

(27). Similarly, with two radicals, this is interesting.

$$u=\int\frac{dx}{(a-x)^{\frac{1}{2}}(b-x)^{\frac{1}{2}}}=\int\epsilon^{-a_1x}\epsilon^{-b_1x}\pi\sqrt{a_1b_1}\,dx=\pi\frac{\sqrt{a_1b_1}}{a_1+b_1}. \tag{117}$$

By symmetry there are two forms of a, b solution, according as $a>b$ or $b>a$, it being supposed that the integral is real and complete, so that x goes from 0 to the smaller of a and b in the first integral, and from 0 to ∞ in the second. Then

$$u=\pi\left(\frac{b_1}{a_1}\right)^{\frac{1}{2}}\left(1-\frac{b_1}{a_1}+\frac{b_1^2}{a_1^2}-\ldots\right)=2\sqrt{\frac{a}{b}}\left(1+\frac{1}{3}\frac{a}{b}+\frac{1}{5}\frac{a^2}{b^2}+\ldots\right) \tag{118}$$

$$=\sqrt{\frac{b}{a}}\log\frac{b+a}{b-a}, \tag{119}$$

when $a<b$. This function can be evolved any number of times with convergent results. Thus

$$\frac{\sqrt{a_1 b_1}}{a_1+b_1}=\frac{2}{\pi}\left(\frac{a}{b}\right)^{\frac{1}{2}}\left(1+\frac{1}{3}\frac{a}{b}+\frac{1}{5}\frac{a^2}{b^2}+\dots\right)$$

$$=\left(\frac{2}{\pi}\right)^2\left(\frac{b_{-1}}{a_{-1}}\right)^{\frac{1}{2}}\left(1-\frac{1}{3^2}\frac{b_{-1}}{a_{-1}}+\frac{1}{5^2}\frac{b^2{}_{-1}}{a^2{}_{-1}}-\dots\right), \qquad (120)$$

$$=\left(\frac{2}{\pi}\right)^3\left(\frac{a_{-2}}{b_{-2}}\right)^{\frac{1}{2}}\left(1+\frac{1}{3^3}\frac{a_{-2}}{b_{-2}}+\frac{1}{5^3}\frac{a^2{}_{-2}}{b^2{}_{-2}}+\dots\right); \qquad (121)$$

and so on to any extent. The other way, the way of involution, leads to a similar set of divergent series.

(28). Again with two radicals,

$$\int\frac{dx}{(a-x^2)^{\frac{1}{2}}(b-x^2)^{\frac{1}{2}}}=\int\epsilon^{-(a_1+b_1)x^2}\pi\sqrt{a_1 b_1}\frac{dx^2}{2\sqrt{x^2}}=\frac{\pi^{1\frac{1}{2}}}{2}\left(\frac{a_1 b_1}{a_1+b_1}\right)^{\frac{1}{2}}. \qquad (122)$$

This is the first elliptic integral function of a, b, as in (72), (73) above, where it is fully expanded. The limits of the first integral are from 0 to the smaller of $a^{\frac{1}{2}}$ and $b^{\frac{1}{2}}$. After that it has no existence. Then, when x passes the value of the greater of $a^{\frac{1}{2}}$ and $b^{\frac{1}{2}}$ it becomes real again. This outer integral is entirely excluded from the second form of the integral by its manner of construction. It was the same in the last example. How shall we modify the construction of the second integral above in order to exhibit the outer integral? The change of variable $x^2=y^{-1}$ will do it. If b is the greater, then the first integral from $b^{\frac{1}{2}}$ to ∞ becomes

$$\int_0^{\frac{1}{b}}\frac{dy}{2y^{\frac{1}{2}}\sqrt{ab}\left(\frac{1}{a}-y\right)^{\frac{1}{2}}\left(\frac{1}{b}-y\right)^{\frac{1}{2}}}=\frac{\pi^{1\frac{1}{2}}}{2\sqrt{ab}}\left\{\frac{\left(\frac{1}{a}\right)_1\left(\frac{1}{b}\right)_1}{\left(\frac{1}{a}\right)_1+\left(\frac{1}{b}\right)_1}\right\}^{\frac{1}{2}}. \qquad (123)$$

So the new result is got from the old (in terms of a, b) by turning a and b to their reciprocals, and dividing by $\sqrt{ab}$. It will be found by inspection of (72) that the new result is the same as the old, or the outer integral has the same value as the inner.

The middle integral, realised by altering the sign of one of the quantities under radical sign, and putting $x^2=y$, makes

$$\int_a^b\frac{dy}{2y^{\frac{1}{2}}(b-y)^{\frac{1}{2}}(y-a)^{\frac{1}{2}}}=\int_0^{b-a}\frac{dz}{2z^{\frac{1}{2}}(b-z)^{\frac{1}{2}}\{(b-a)-z\}^{\frac{1}{2}}}, \qquad (124)$$

where $b-y=z$. So this integral now comes under the same rule, producing

$$\frac{\pi^{1\frac{1}{2}}}{2}\left\{\frac{b_1(b-a)_1}{b_1+(b-a)_1}\right\}^{\frac{1}{2}}; \qquad (125)$$

and, fully developed, is got by turning a to $b-a$ in the original solution of (122). Or, to suit all three integrals,

$$\left(\frac{a_1 b_1}{a_1+b_1}\right)^{\frac{1}{2}}=\frac{1}{(\pi b)^{\frac{1}{2}}}\left\{1+\frac{1^2}{2^2}\frac{a}{b}+\frac{1^2 3^2}{2^2 4^2}\frac{a^2}{b^2}+\dots\right\}, \qquad (126)$$

which suits (122), and then with the changes in a and b specified to suit (123) and (124).

(29). Some remarks on the solution in (126) are here inserted. If it is given that

$$v=\{a_1b_1/(a_1+b_1)\}^{\frac{1}{2}}, \tag{127}$$

the complete convergent solution in terms of a and b, both ranging from 0 to ∞, is contained in

$$v_1=\frac{1}{(\pi b)^{\frac{1}{2}}}\left\{1+\frac{1^2}{2^2}\frac{a}{b}+\dots\right\},\quad v_2=\frac{1}{(\pi a)^{\frac{1}{2}}}\left\{1+\frac{1^2}{2^2}\frac{b}{a}+\dots\right\}, \tag{128}$$

and they are got by the binomial theorem first, and then $a_1^{-n}=a^n/\underline{|n}$, $b_1^{-n}=b^n/\underline{|n}$. First the expansion.

$$v_1=\frac{b_1^{\frac{1}{2}}}{(1+b_1/a_1)^{\frac{1}{2}}}=b_1^{\frac{1}{2}}\left(1-\frac{1}{2}\frac{b_1}{a_1}+\frac{1.3}{2.4}\frac{b_1^2}{a_1^2}-\dots\right), \tag{129}$$

$$v_2=\frac{a_1^{\frac{1}{2}}}{(1+a_1/b_1)^{\frac{1}{2}}}=a_1^{\frac{1}{2}}\left(1-\frac{1}{2}\frac{a_1}{b_1}+\frac{1.3}{2.4}\frac{a_1^2}{b_1^2}-\dots\right). \tag{130}$$

Now carry out not both, but only one of the integrations, say to b.

$$\therefore\quad v_1=\frac{1}{(\pi b)^{\frac{1}{2}}}\left\{1+\frac{1^2}{4ba_1}+\frac{1^2 3^2}{\underline{|2}(4ba_1)^2}+\dots\right\} \tag{131}$$

$$v_2=\sqrt{a_1}\left\{1-\frac{ba_1}{2}+\frac{3}{4}\left(\frac{ba_1}{2}\right)^2-\dots\right\}. \tag{132}$$

These, then, are two forms of solution of (127) when a_1 is a constant. They are the same as

$$v_1=a_1^{\frac{1}{2}}\tfrac{1}{2}\mathrm{H}_0(\tfrac{1}{2}ba_1)\,.\,\epsilon^{-\frac{1}{2}ba_1},\qquad v_2=a_1^{\frac{1}{2}}\mathrm{I}_0(\tfrac{1}{2}ba_1)\,.\,\epsilon^{-\frac{1}{2}ba_1}. \tag{133}$$

Are they supplementary, like the two in (128)? No, because v_2 is convergent all the way from $b=0$ to ∞, whilst v_1 is divergent all the way. They do not exist separately, like (128). They are equivalent forms of one solution, a numerical equivalence which is only limited by the size of the smallest term in the divergent series. This equivalence occurs in the telegraph theory, considered in vol. 2 (p. 292). Then comes the second integration. If the series in (131) is turned to a function of a, it makes the a/b series, whilst the series in (132) similarly makes the b/a series. If we keep to their convergent regions, they supplement one another, and both are required. But each of these formulæ has its divergent region as well. It asks itself then, is the divergent part of v_1 $(a > b)$ equivalent to the convergent part of v_2 $(a > b)$? Numerical calculation shows that they are not equivalent in the same way as (128) above. Then what do the divergent formulæ mean? What is the relation between v_1 and v_2 in (128), when a/b is given a series of values a little over unity, with distinct initial convergence of v_1? This question will be considered a little later on.

(30). The v formula becomes preposterously slowly convergent as a/b goes from 0·9 up to 1, where the infinite value occurs. It is therefore desirable to transform the series to a more rapidly convergent one. This can be done. Thus, try by introducing the factor ϵ^{-ab_1}. The shift transformation turns v to

$$v=\epsilon^{-ab_1}\left(\frac{(a_1-b_1)b_1}{a_1}\right)^{\frac{1}{2}}\frac{a_1}{a_1-b_1}=\epsilon^{-ab_1}\left(\frac{a_1b_1}{a_1-b_1}\right)^{\frac{1}{2}}$$

$$=\epsilon^{-ab_1}b_1^{\frac{1}{2}}\left(1+\tfrac{1}{2}ab_1+\frac{1.3}{2.4}\frac{a^2b_1^2}{\lfloor 2}+\dots\right),$$

$$\therefore\quad v_1=\frac{1}{\{\pi(b-a)\}^{\frac{1}{2}}}\left\{1-\frac{1^2}{2^2}\frac{a}{b-a}+\frac{1^23^2}{2^24^2}\frac{a^2}{(b-a)^2}-\dots\right\}, \tag{135}$$

by the binomial theorem in terms of b_1/a_1 first, then conversion to a function of a and b_1, and finally of a and b. Here $a<b$. Interchange a and b to get the v_2 form. Now this differs from the standard v_1 in the substitution of $b-a$ for b, and in alternating the signs. The result is to produce a series which is convergent up to $a=\frac{1}{2}b$, and is then alternatingly divergent up to $a=b$. Moreover, $a/(b-a)$ is greater than a/b. So there is no advantage in this variation.

(31). If, for brevity, we write

$$F(x)=1+\frac{1^2}{2^2}x+\frac{1^23^2}{2^24^2}x^2+\dots, \tag{135A}$$

the function thus defined can be calculated convergently from $x=+1$ to $x=-1$. The portion from $x=0$ to $+1$ is that concerned in the first complete elliptic integral. Below $x=-1$ the series becomes alternatingly divergent, and can still be calculated for some distance. But how shall we find the complete curve down to $x=-\infty$? Equation (135) above supplies the answer. For it proves that

$$\sqrt{1-x}\,F(x)=F\left(\frac{x}{x-1}\right). \tag{135B}$$

Here let x range from 0 to 1. Then $x(x-1)^{-1}=z$, say, will range from 0 to $-\infty$. Or, if x ranges from 0 to $-\infty$, then z will range from 0 to 1. For example,

$$\sqrt{\tfrac{3}{4}}\,F(\tfrac{1}{4})=F(-\tfrac{1}{3}),\quad \sqrt{\tfrac{1}{2}}\,F(\tfrac{1}{2})=F(-1),\quad \sqrt{\tfrac{1}{4}}\,F(\tfrac{3}{4})=F(-3).$$

So the equivalence (135B) allows us to construct the complete curve of $F(x)$ from $x=1$ to $-\infty$ in terms of the known portion from $x=1$ to 0. This curve rises from 0 at $x=-\infty$, through 1 at $x=0$, to ∞ at $x=1$.

Now pass to the other side of the ∞. $F(x)$ becomes directly divergent, but possesses initial convergence from $x=1$ up to $x=4$. The curve is definitely determinable if the value is estimated by the initially convergent part in the same way as is true in the numerous divergent series that occur in the treatment of practical physical problems. At the one end, near $x=1$, the value can be closely calculated. Also fairly well up to $x=2$. After that, roughly. At the limit $x=4$, we get $F(4)=1+1+\ldots$, which is, to a first approximation, 1. How near this is will appear later. But equation (135B) above, with the radical taken real, making it $\sqrt{x-1}$, allows us to utilise the good part of the curve, that is from $x=1$ to $x=2$, for the plotting of all the rest, up to $x=\infty$. For example, $2F(5)=F(1\frac{1}{4})$, and so on. I have carried out this work, and a nice smooth curve results.

Now what relation does it bear to the curve on the other side? To all appearance, it is simply a copy, turned the other way, but magnified. The ratio is about 1·31. That is,

$$\frac{F(1+y)}{F(1-y)}=1\cdot31, \quad \text{when } y \text{ is } +. \tag{135c}$$

This determination is empirical, and it would be better to have an analytical investigation. I will not give lengthy details of the calculations, but only some remarks thereon. In the initially convergent part of the divergent $F(x)$, the ratio of the l.c.t. to the total varies from 0 at $x=1$ to 1 at $x=4$, or the error might vary from 0 to 100 per cent. If we took values of x at random between $x=1$ and 4, we might expect an error of 50 per cent. of the l.c.t. Even if we keep entirely to the good part of the curve, from $x=1$ to 2, the curve will tend to get a little staggery on approaching $x=2$. But there is a way of largely reducing the size of the error. It is (as in vol. 2, p. 484) to select values of x which produce a pair of equal bottom terms. Then stop counting in the middle, that is, count the first l.c.t. only. Another way is to choose x so that the term before the l.c.t. equals the term after it. Then stop at the middle of the l.c.t., of course. The first way is more convenient. By its means the error is usually reduced from a large to a small fraction of the l.c.t. Then a very smooth curve results. (In some series the error is made in this way a quite minute fraction of the l.c.t.) In the present case the error is not very large even at the extreme limit, $x=4$. The rule makes 1, instead of about 0·9. This 10 per cent. error falls rapidly as x is reduced.

As we do not eliminate the error by using equal bottom terms, so the value of the ratio of $F(1+y)$ to $F(1-y)$ would not come out quite constant, even if we knew that the true value was really a constant. That is, however, just what has to be found. In fact, as y goes from 0·7777 down to 0·07015, the ratio falls from 1·358 to 1·316. But in the first case the l.c.t. was large, in the second relatively small, about $2\frac{1}{4}$ per cent. of the total of the series; so the fall from 1·35 to 1·31 should rather be attributed to the imperfection of the rule of calculation by equal bottom terms than to the ratio being really a function of x, although that may not be impossible.

(32). Now use the prefactor $\epsilon^{-\frac{1}{2}ab_1}$ in (127). Then

$$v=\epsilon^{-\frac{1}{2}ab_1}\left\{\frac{(a_1-\frac{1}{2}b_1)b_1}{a_1+\frac{1}{2}b_1}\right\}^{\frac{1}{2}}\frac{a_1}{a_1-\frac{1}{2}b_1}=\epsilon^{-\frac{1}{2}ab_1}\frac{a_1b_1^{\frac{1}{2}}}{(a_1^2-\frac{1}{4}b_1^2)^{\frac{1}{2}}}. \tag{136}$$

Expand in powers of b_1/a_1. Then

$$v_1=\epsilon^{-\frac{1}{2}ab_1}b_1^{\frac{1}{2}}\left\{1+\frac{1}{2}\frac{b_1^2}{4a_1^2}+\frac{1.3}{2.4}\left(\frac{b_1^2}{4a_1^2}\right)^2+\ldots\right\}$$

$$=\epsilon^{-\frac{1}{2}ab_1}b_1^{\frac{1}{2}}\left\{1+\frac{(ab_1)^2}{\underline{|2}.2^2.2}+\frac{(ab_1)^4 1.3}{\underline{|4}.2^4.2.4}+\ldots\right\}=\ldots\{\mathrm{I}_0(\tfrac{1}{2}ab_1)\}$$

$$=\frac{1}{\{\pi(b-\frac{1}{2}a)\}^{\frac{1}{2}}}\left\{1+\left(\frac{a}{b-\frac{1}{2}a}\right)^2\frac{1.3}{2^5}+\left(\frac{a}{b-\frac{1}{2}a}\right)^4\frac{1.3.5.7}{2^{12}(\underline{|2})^2}+\ldots\right\}. \tag{137}$$

Or, more conveniently, let $a/4(2b-a)=c$, then

$$v_1=\left\{\frac{2}{\pi(2b-a)}\right\}^{\frac{1}{2}}\left\{1+1.3\,c^2+\frac{1.3.5.7}{(\underline{|2})^2}c^4+\frac{1.3.\ \ldots 9.11}{(\underline{|3})^2}c^6+\right\} \tag{138}$$

This is convergent up to $a=b$, and is a capital improvement. It converges doubly as fast as the standard v_1 because it goes by squares instead of first powers, and is additionally convergent by smaller terms. In a numerically tested case, where four terms of the new series give v_1 correct to 3 figures, eight terms of the old series only make it right to 2 figures, and several more terms are wanted.

(33). The usual potential of a steady fixed point source g being $g/4\pi\mathrm{R}$ at distance R, when g varies in intensity with the time, the potential will be the same $g/4\pi\mathrm{R}$ only if propagation is instantaneous using the momentary value of g. But if the propagation is done isotropically at speed v, then the potential is

$$\mathrm{V}=\frac{\epsilon^{-q\mathrm{R}}}{4\pi\mathrm{R}}g, \quad \text{where} \quad q=d/d(vt); \tag{139}$$

i.e., the value of g is taken to be the value at the source at the moment $t-\mathrm{R}/v$, in order that V shall be the value of the potential at the end of R at the moment t. It is desirable to have a physical foundation. The characteristic $\nabla^2\mathrm{V}=q^2\mathrm{V}$, when referred to the single space co-ordinate R, where $\mathrm{R}^2=x^2+y^2+z^2$, is

$$\frac{d^2}{d\mathrm{R}^2}(\mathrm{RV})=q^2(\mathrm{RV}), \tag{140}$$

with two obvious solutions. Selecting the one for outward waves only, it is the above (139), g meaning any function of t. To find the meaning of g more closely, we have

$$4\pi\mathrm{R}^2(-d\mathrm{V}/d\mathrm{R})=\mathrm{R}^2(q/\mathrm{R}+1/\mathrm{R}^2)g=(1+q\mathrm{R})g. \tag{141}$$

If $\mathrm{R}=0$, the value is g. That is, the surface integral of $-d\mathrm{V}/d\mathrm{R}$ over an infinitely small sphere centred at the origin is g. If, then, $-\nabla\mathrm{V}$ is the density of a flux (surface density) of fluid of unit standard density, g is the rate of generation of fluid at the origin, or the amount added per second. It remains in the form of fluid condensation, so that the time integral of g is always equal to the volume integral of the condensation. Also, the condensation s itself is given by $sv^2 = \dot{V}$; or $sv = qV$. The condensation only exists at the end of R, at the moment t, when V is increasing, and therefore only if g is increasing at the previous moment $t - R/v$. This is a very brief statement of the acoustic analogy. But in sound waves, according to these equations, the quantity s must be very small. In the general theory, however, there is no limitation of this kind, and g may be any $f(t)$, and s have any value. I find, however, the general ideas of the acoustic analogy to be useful even then, and it supplies us with a terminology.

Observe that (apart from the analogy), the above supplies us with the simplest proof that all disturbances, subject to $\nabla^2 V = q^2 V$, are propagated at speed v through the medium. For what is true for one point source is true for another, and so on to any distribution of sources. Moreover, an initial state without sources is brought under the same rule by producing the initial state impulsively. If $g = vqQ$, it means the impulsive introduction of the amount Q of condensation at the origin. Therefore

$$V = \Sigma \frac{\epsilon^{-qR} vq\sigma}{4\pi R} \tag{142}$$

is the potential at the moment t due to the initial state of condensation σ (vol. density), the summation being a volume integral over all space. This is an elementary case of the pan-potential, to be considered later in connection with electromagnetics.

The solution

$$V = Q \frac{\epsilon^{-qR} qv}{4\pi R} \tag{143}$$

is purely impulsive, remembering that $q = d/d(vt)$. It is simply confined to a spherical shell of no depth, of radius $R = vt$, with condensation on its outside, and rarefaction on its inside, both increasing in total amount uniformly with the time, the amount of the condensation always exceeding that of the rarefaction by Q.

But if the source is steady, of intensity g, beginning $t=0$, then V exists from $R=0$ up to $R=vt$, and on the boundary of this region is the condensational wave without any rarefaction anywhere, and gt is the total condensation. (*See* vol. 2, p. 500, for illustration.)

(34). In the case of an impulsive line source, of linear density σ,

$$V = \int \frac{vq\,\epsilon^{-qR}\,\sigma}{4\pi R} ds. \tag{144}$$

Supposing σ is the same for every element, this is the same as

$$V = \frac{\sigma v}{4\pi} \int \frac{q\epsilon^{-qR}}{R} \frac{ds}{dR} dR. \tag{145}$$

In case of a circle, we get a remarkably simple result. Let r, θ, and z be columnar co-ordinates of the point at end of R where V is wanted, so that

$$R^2 = a^2 + r^2 - 2ar\cos\theta + z^2, \tag{146}$$

$$R\,dR = ar\sin\theta\,d\theta. \tag{147}$$

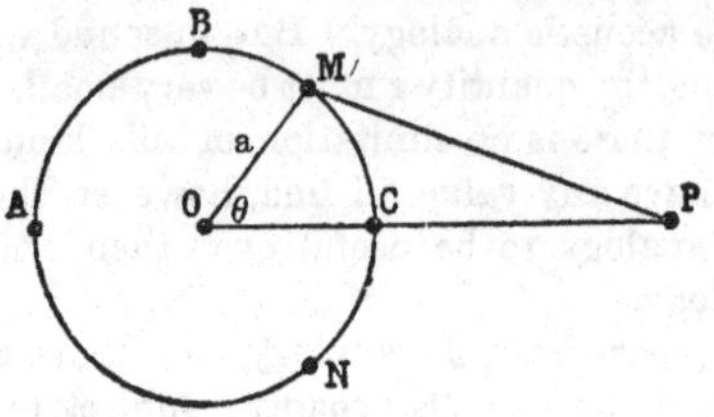

In the figure P is the projection of the R point on the plane of the circle. At the moment represented, the elements at M and N are those alone concerned, and

$$V = \frac{\sigma v}{2\pi}\int_0^\pi \frac{q\epsilon^{-qR}}{r\sin\theta}\,dR = \frac{\sigma v}{2\pi r\sin\theta} \tag{148}$$

The value of θ in the result belongs to $R = vt$, found by (146), so that

$$V = \frac{\sigma v}{2\pi r\sqrt{1 - \left(\frac{v^2t^2 - a^2 - r^2 - z^2}{2ar}\right)^2}}, \tag{149}$$

in terms of t, r and z. The volume occupied by V is entirely bounded by the reality of the radical. It begins as a circle, which expands as a tore, or anchor ring; but as soon as it has swollen so much that it crosses O inwardly, the overlapping region is excluded. Later on the region becomes a shell of depth varying from 0 on the axis (both sides of O) to $2a$ in the plane of the circle. Outside this region $V=0$. Inside the region it is always finite, but increases to ∞ on the boundary. (Compare with vol. 1, p. 114.) In the middle of the shell the value of V is $\sigma v/2\pi r$.

For other shapes of line sources, if $R^{-1}ds/dR$ is simply expressible, useful solutions are also obtainable.

In case of an impulsive surface source, density σ,

$$V = \frac{\sigma v}{4\pi}\iint \frac{q\epsilon^{-qR}}{R}\frac{dS}{dR}\,dR, \tag{150}$$

where dS is element of surface. If a plane,

$$V = \sigma v\int \frac{q\epsilon^{-qR}}{4\pi R}\frac{d\pi r^2}{dR}\,dR = \frac{\pi}{4\pi R}\frac{dr^2}{dR}\sigma v. \tag{151}$$

Here $\qquad R^2 = a^2 + r^2, \quad \frac{dr}{dR} = \frac{R}{r}, \quad$ and $\quad V = \frac{1}{2}\sigma v. \qquad (152)$

It is the same on the other side of the plane. So the region of V is a slab of depth $2a$, within which V is constant. This serves to illustrate the artificial nature of V. The consideration of condensation is more natural. It starts as a plane sheet of density σ, which splits into two equal ones of half density, which then travel right and left at speed v. In practical work there is no recommendation in general of this method for constructing the solutions for surface

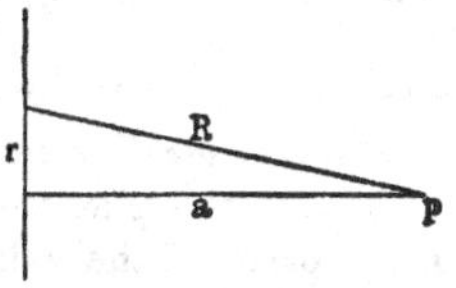

sources, because so much work will be required in integration. It is generally much simpler to make use of the functions known to belong to the surface in question, sphere, &c, which themselves express the result of the integration.

(35). Having got the solution (149) for a circular source, we can make it the element for a cylindrical source, regarding the cylinder as an assemblage of circles. Let $\rho=$ amount impulsively generated per unit length of cylinder of radius a, so that the surface density of source is $vq\rho/2\pi a$; then

$$V=\frac{2\rho v}{4\pi^2 ar}\int\frac{dz}{\sqrt{1-\left(\frac{v^2t^2-a^2-r^2-z^2}{2ar}\right)^2}} \tag{153}$$

$$=\frac{\rho v}{\pi^2}\int\frac{dz}{\sqrt{A-z^2}\,\sqrt{z^2-B}}; \tag{154}$$

if $A=v^2t^2-(a-r)^2$, $B=v^2t^2-(a+r)^2$. Or, since $zdz=ar\sin\theta\,d\theta$, by (146),

$$V=\frac{\rho v}{\pi^2}\int_0^{\frac{\pi}{2}}\frac{d\phi}{(A-C\sin^2\phi)^{\frac{1}{2}}} \tag{155}$$

if $\phi=\frac{1}{2}\theta$, $C=4ar$. In all these formulæ the function does not exist when unreal, but in the form (155), ϕ must not exceed $\frac{1}{2}\pi$. When $A<C$, the upper limit is that value of ϕ which reduces the denominator to zero. Here we have the first elliptic function again. But regarding (154), a note is necessary. A is always +, but B is not.

One way of solution is

$$V=\frac{\rho v}{\pi^2}\int\frac{dz}{(A-z^2)^{\frac{1}{2}}(z^2-\overline{A-C})^{\frac{1}{2}}}=\frac{\rho v}{\pi^2}\int\epsilon^{-z^2A_1}\frac{dz}{A^{\frac{1}{2}}(C-A)^{\frac{1}{2}}}. \tag{156}$$

If $z=0$, the function of A and C is (to a constant) the value of V on the plane of the circle, due to a circular source. Then $C-A$ is +, and we have

$$\frac{1}{A^{\frac{1}{2}}(C-A)^{\frac{1}{2}}}=\frac{1}{(AC)^{\frac{1}{2}}}\left(1+\frac{1}{2}\frac{A}{C}+\frac{1.3}{2.4}\frac{A^2}{C^2}+\ \right)=\frac{\pi(A_1C_1)^{\frac{1}{2}}}{(1+C_1/A_1)^{\frac{1}{2}}}, \tag{157}$$

so
$$V=\frac{\rho v}{\pi}\int \epsilon^{-z^2A_1}\frac{dz^2}{2\sqrt{z^2}}\frac{A_1C_1^{\frac{1}{2}}}{(A_1+C_1)^{\frac{1}{2}}}=\frac{\rho v}{2\pi^2}\left(\frac{A_1C_1}{A_1+C_1}\right)^{\frac{1}{2}}, \tag{158}$$

by Euler's integral. Or in full, by (128),

$$V_2=\frac{\rho v}{2\pi\{v^2t^2-(a-r)^2\}^{\frac{1}{2}}}\left\{1+\frac{1^2}{2^2}\frac{4ar}{v^2t^2-(a-r)^2}+\dots\right\}, \tag{159}$$

$$V_1=\frac{\rho v}{2\pi\sqrt{4ar}}\left\{1+\frac{1^2}{2^2}\frac{v^2t^2-(a-r)^2}{4ar}+\dots\right\}. \tag{160}$$

Or, we may use (154) in the sense of (124), with this difference, that the limits are 0 and $\sqrt{A}$ when B is negative, instead of $\sqrt{B}$ and $\sqrt{A}$. This occurs in the early stage of the phenomenon, before disturbance from the more distant parts of the cylinder has reached the point of observation.

That this extension of (124) is allowable, the following shows. Let $b-x=y$, we get

$$\int_0^b\frac{dx}{x^{\frac{1}{2}}(x-a)^{\frac{1}{2}}(b-x)^{\frac{1}{2}}}=\int_0^b\frac{dy}{y^{\frac{1}{2}}(b-y)^{\frac{1}{2}}(b-a-y)^{\frac{1}{2}}}=\pi^{1\frac{1}{2}}\left\{\frac{b_1(b-a)_1}{b_1+(b-a)_1}\right\}^{\frac{1}{2}}, \tag{161}$$

by Euler's integral, provided b and $b-a$ are both positive, which allows a to be negative. It is the same formula as when a is +, only then with a far lower limit. The following diagram will explain the practical meaning of V_1 and V_2.

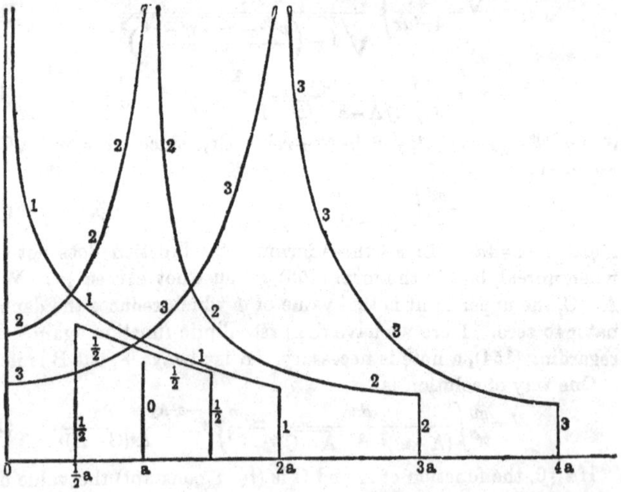

Along the horizontal axis is measured the distance r from the axis, and the vertical ordinate measures V at different times. When $t=0$, V is merely the vertical line 0 at $r=a$, where the source momentarily exists. This spreads out both ways, attenuating outward and condensing inwards, due to geometrical expansion and contraction. When $vt=\frac{1}{2}a$, the curve $\frac{1}{2}$ shows the distribution of V. It has increased slightly at the place of the source; at the wave front it varies as $\sqrt{1/r}$ both ways. When $vt=a$, the curve 1 shows V, with an infinite momentary value. So far we have only one curve and it is given by the formula V_1, equation (160). After this, the second curve V_2 (equation (159)) comes into existence and moves outward. At the moment $vt=2a$, the curve 2 on the left side shows V_2, and on the right side V_1. A later stage is shown by the curve 3, belonging to $vt=3a$. For very large values of the time, V consists of the outer shell of depth $2a$, in which V rises from nearly 0 up to ∞, followed by a rapid drop to small value behind the shell. (The curves are not drawn to scale, except for principal values.)

The infinite value at the back of the shell can be abolished by spreading out the source. Instead of a surface distribution of the source, let the same amount be spread throughout the space between $r=a(1+n)$ and $a(1-n)$, where n is as small a fraction as you like, though not zero. The result will be to round off the sharp infinity, making it become a finite maximum. Apparently then, the two formulæ of the above pointed curve should be capable of representation by a single series. Against this, however, it might be objected that this curve at the wave front will no longer be vertical but will make a rapid continuous rise instead. The single formula will turn up shortly.

Since the condensation s is given by $sv^2=\dot{V}$, its most important part is at the wave fronts. In the first stage we have two waves of condensation, with a weak condensation between them. In the second stage at a given point there is first the wave of surface condensation at the wave front, followed by distributed volume condensation in the shell, increasing continuously to ∞ at the back of the shell, then dropping to $-\infty$ (rarefaction) which goes continuously to a small negative value.

The total condensation (reckoning the later rarefaction negatively) is ρ per unit length of cylinder. And since $sv^2=\dot{V}$, whilst $\Sigma sv^2=\rho v^2$, it follows that $\Sigma V=\rho v^2 t$, or

$$\int 2\pi r V dr=\rho v^2 t,$$

or the area of the curve of rV varies as t.

At the initial moment, the total condensation in each of the two waves is $\frac{1}{2}\rho$. But since V varies at $r^{-\frac{1}{2}}$ at the wave front, the total condensation at the wave fronts varies as $r^{\frac{1}{2}}$. So it increases indefinitely, and always as $r^{\frac{1}{2}}$, at the outer wave front. But at the inner one it falls continuously, and is zero at the moment $t=a/v$. At the same moment the total condensation is $\rho\sqrt{\frac{1}{2}}$ at the outer wave front. Therefore the volume condensation in the wave 1 is $\rho(1-\sqrt{\frac{1}{2}})$. It is the velocity $-dV/dr$ being always outward in this stage (in the extended acoustical analogy) combined with the cylindrical expansion and contraction, that accounts for the redistribution of condensation. (In a spherical outward surface wave of condensation, the total would increase as r, the radius.) After this, the condensation, surface and volume, in the shell that is formed, goes on increasing, but it is followed by the rarefaction, the amount of which also increases continuously, so as to keep the total condensation constant. In the surface wave at the outer wave front the velocity is outward; but it is inward in the inner one. In both the velocity is proportional to the condensation. So the total flux of fluid inward in the inner surface wave varies as $r^{\frac{1}{2}}$ and vanishes the moment it reaches the axis. Also, there is no source of fluid at the axis. So the infinite value of V which is propagated outward immediately after does not correspond to a surface wave like the outward one. The inward one is gone for good, and is replaced by a continuous wave of volume rarefaction.

If the source on the surface of the cylinder is not impulsive, but steadily continuous, generating or introducing fluid at a uniform rate, commencing to do so when $t=0$, then the V curves in the above represent the condensation itself. There are now no surface waves of condensation; it is now entirely volume condensation. This is a simplification. On the other hand, the velocity is troublesome to obtain, requiring a time integration.

(36). Electromagnetic solutions represented by the V curves may be either in terms of E or of H, by the now well-known method of translation of electric and magnetic relations. Choosing H, the lines of magnetic force (and induction) in a uniform isotropic dielectric (the ether, most simply) must be parallel to the axis of the cylinder. Then, remembering that the volume integral of V varies as t, we see that the source at the surface of the cylinder must generate magnetic induction at a uniform rate. The V curves then show how the "lines of force" spread over the field, at first both ways, later entirely outward. Here again there are no surface waves of condensed H. The wave fronts merely represent discontinuities in H. What is infinite there is the electric current, but this infiniteness has no energetic significance. Nor is the travelling ∞ of any consequence. There is nothing in it.

(37). Another solution derivable from the tore solution. Find the ultimate steady V due to a steady source on the *circle* of radius a. This involves a time integration of the tore solution. Thus, if Q = total round circle of fluid introduced per second,

$$V=\frac{Qv}{2\pi^2}\int\frac{dt}{\{v^2t^2-(a-r)^2-z^2\}^{\frac{1}{2}}\{z^2+(a+r)^2-v^2t^2\}^{\frac{1}{2}}} \qquad (162)$$

$$=\frac{Q}{2\pi^2}\int\frac{d(vt)}{(Z-v^2t^2)^{\frac{1}{2}}(v^2t^2-Y)^{\frac{1}{2}}}, \qquad (163)$$

where Y and Z define themselves by comparison. Comparison with (154) shows the similarity, with a notable difference. Y is always +, like Z, and the limits of reality are from $vt=\mathrm{Y}^{\frac{1}{2}}$ to $\mathrm{Z}^{\frac{1}{2}}$, so the result is

$$\mathrm{V}=\frac{\mathrm{Q}}{4\pi\{z^2+(a+r)^2\}^{\frac{1}{2}}}\left\{1+\frac{1^2}{2^2}\frac{4ar}{z^2+(a+r)^2}+\dots\right\}. \tag{164}$$

There is only one form of solution, because $4ar$ is smaller than the denominator.

But suppose we want to show the way of establishment of V beginning $t=0$. Then the limits for vt in the above must be from $\mathrm{Y}^{\frac{1}{2}}$ to vt during the passage of the shell past a given point, finishing with the full limits $\mathrm{Y}^{\frac{1}{2}}$ and $\mathrm{Z}^{\frac{1}{2}}$ as soon as the shell is clear away. This incomplete elliptic function is very disagreeable, and I shall give no formulæ of development. One way is by

$$\mathrm{V}=\frac{\mathrm{Q}}{2\pi^2}\frac{1}{\mathrm{T}_1}\epsilon^{-\mathrm{TZ}_1}\frac{1}{\mathrm{Z}^{\frac{1}{2}}(4ar-\mathrm{Z})^{\frac{1}{2}}}, \tag{165}$$

when $\mathrm{T}=v^2t^2$, and others may be found if this is not convenient.

Equation (164) may be confirmed by the direct way for a steady state, or $\mathrm{V}=\Sigma\rho/4\pi\mathrm{R}$. This makes

$$\mathrm{V}=2\int_0^{\pi}\frac{\frac{\mathrm{Q}}{2\pi a}a\,d\theta}{4\pi\{a^2+r^2+z^2-2ar\cos\theta\}^{\frac{1}{2}}}=\frac{\mathrm{Q}}{2\pi^2}\int_0^{\frac{\pi}{2}}\frac{d\phi}{(\mathrm{Z}-4ar\sin^2\phi)^{\frac{1}{2}}}, \tag{166}$$

meaning the same as (164), as can be seen by comparison with (155). The formula (164) is also that for the gravitational potential of a circular line of mass Q. Moreover, if we draw the equipotential lines in a plane through the axis, they will also represent the lines of magnetic force of a steady circular electric current. This is true even in the variable period provided the shell is fully formed, and we keep behind it—that is, nearer the centre of the circle than its inner boundary.

(38). The above cylindrical source may be also regarded as an assemblage of straight line sources, parallel to the axis. First, as regards V for a single straight line source. This is easy. Let σ be the linear density (impulsive). Then

$$\mathrm{V}=\sigma v\int_{-\infty}^{\infty}\frac{q\epsilon^{-q\mathrm{R}}}{4\pi\mathrm{R}}dz=\frac{\sigma v}{2\pi}\int_0^{\infty}\frac{q\epsilon^{-q\mathrm{R}}}{\sqrt{\mathrm{R}^2-r^2}}d\mathrm{R}=\frac{\sigma v}{2\pi(v^2t^2-r^2)^{\frac{1}{2}}}, \tag{167}$$

because $q\epsilon^{-q\mathrm{R}}$ is impulsive at $\mathrm{R}=vt$.

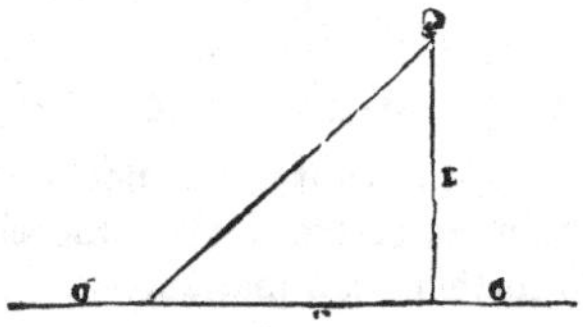

This avoids the $K_0(qr)$ function altogether, which, however, is necessary in a complete treatment. The point P at moment t is only receiving V from two elements on the axis, and the slope is allowed for. The wave front is at $r=vt$, a cylinder. But all points inside it have finite V; for although the source is existent only at the moment $t=0$, it takes time for disturbance from more distant parts of the source to reach P.

In general, if g is the linear intensity of the line source, a function of the time, then

$$V=\tfrac{1}{4}K_0(qr)g. \tag{168}$$

To prove this, first note that $K_0'qr)$ is the proper operator for an outward wave (vol. 2, p. 240), and then prove by the formula for K_0 that

$$2\pi r(-dV/dr)=g, \quad \text{when} \quad r=0, \tag{169}$$

showing that the amount of fluid g is introduced and passed outward per second from the straight line per unit of its length.

With an impulsive source, $g=vq\sigma$. Therefore, by (167),

$$K_0(qr)q=\frac{2}{\pi}\int_0^\infty \frac{q\epsilon^{-qR}}{R}dz=\frac{2}{\pi}\int_r^\infty \frac{q\epsilon^{-qR}dR}{(R^2-r^2)^{\frac{1}{2}}}=\frac{2}{\pi(v^2t^2-r^2)^{\frac{1}{2}}}. \tag{170}$$

These are true whether q is a +constant, or the differentiator $d/d(vt)$. Compare with (167), and note how even the imperfect acoustical analogy assists the understanding of the meaning of the definite integrals in one way, whilst the impulsive function $q\epsilon^{-qR}$ does so in another.

(39). Passing now to the cylindrical source of radius a, regarded as an assemblage of straight line sources, there is one place where the solution can be at once recognised—viz., the axis. For it is equidistant from the line sources all round, all of them behaving according to (170), with $r=a$, so

$$V_0=\frac{Qv}{2\pi(v^2t^2-a^2)^{\frac{1}{2}}} \tag{171}$$

is the potential on the axis, but only from the moment $t=a/v$.

Next, if $\qquad R^2=a^2+r^2-2ar\cos\theta,$

then

$$V=\frac{Qv}{(2\pi)^2}\int_0^{2\pi}\frac{d\theta}{(v^2t^2-R^2)^{\frac{1}{2}}}.$$

(*See* figure, p. 260, and keep in the plane of the diagram.) Or, with $\theta=2\phi$,

$$V=\frac{Qv}{\pi^2}\int_0^{\frac{1}{2}\pi}\frac{d\phi}{\{v^2t^2-(r-a)^2-4ar\sin^2\phi\}^{\frac{1}{2}}}, \tag{172}$$

where the limits are those of reality, but not exceeding $\frac{1}{2}\pi$ at the top. This is (155) again, as obtained from the circular source, integrated, so needs no further examination.

(40). Returning to (171), we can derive V at r from V_0 by

$$V_2 = I_0(qr)V_0 = I_0(qr)\frac{Qv}{2\pi(v^2t^2-a^2)^{\frac{1}{2}}} = \frac{Qv}{2\pi}I_0(qr)I_0(qa)\frac{1}{vt}, \tag{173}$$

and the region of validity is from $r=0$ up to $r=vt-a$, that is, the back of the shell, because V_0 only begins the moment the back of the shell begins to exist. So the V_2 solution is expressible in a series consisting of the simple solution for a line source, and its second, fourth, &c., differentiants. This is obvious work.

Another way of expansion. Let $\Delta_1 = d/d(\frac{1}{4}r^2)$, $\Delta_2 = d/d(\frac{1}{4}a^2)$, then, by (13),

$$V_2 = \frac{Qv}{2\pi}\epsilon^{q^2\left(\frac{1}{\Delta_1}+\frac{1}{\Delta_2}\right)}\frac{1}{vt}$$

$$= \frac{Qv}{2\pi vt}\left\{1+\left(\frac{r^2}{4}+\frac{a^2}{4}\right)\frac{\lfloor 2}{v^2t^2}+\left(\frac{(\frac{1}{4}r^2)^2}{\lfloor 2}+2\frac{r^2}{4}\frac{a^2}{4}+\frac{(\frac{1}{4}a^2)^2}{\lfloor 2}\right)\frac{\lfloor 4}{\lfloor 2\, v^4t^4}+\dots\right\}, \tag{174}$$

which is got by exponential expansion first, and then obvious integrations through Δ_1 and Δ_2.

(41). Now as regards the V_1 formula. Consider the state of things before $t=a/v$. There is an inward wave of the type $H_0(qr)A$, and an outward wave of the type $K_0(qr)B$ (vol. 2, p. 240). To find A and B we have the sole conditions that V is continuous, and that $-dV/dr$ is discontinuous by the amount of $vqQ/2\pi a$ at the surface $r=a$. It is only initially discontinuous, but that is accidental. These definitely determine A and B, and make

$$V_1 = \tfrac{1}{8}Qv\,H_0(qa)K_0(qr)q, \quad \text{(out)} \tag{175}$$

$$v_1 = \tfrac{1}{8}Qv\,H_0(qr)K_0(qa)q. \quad \text{(in)} \tag{176}$$

These have been got without any troublesome integration. Use the divergent expressions for H_0 and K_0. Then

$$V_1 = \tfrac{1}{4}Qv\frac{\epsilon^{-q(r-a)}}{\pi q\sqrt{ar}}\left\{1+\frac{1^2}{8qa}+\frac{1^23^2}{\lfloor 2(8qa)^2}+\dots\right\}\left\{1-\frac{1^2}{8qr}+\frac{1^23^2}{\lfloor 2(8qr)^2}-\dots\right\} \tag{177}$$

$$= \frac{Qv\epsilon^{-\dots}}{4\pi\sqrt{ar}}\left\{1+\frac{1}{(8q)}\left(\frac{1}{a}-\frac{1}{r}\right)+\frac{1}{(8q)^2}\left(\frac{1^23^2}{\lfloor 2\,a^2}-\frac{1}{ar}+\frac{1^23^2}{\lfloor 2\,r^2}\right)+\dots\right\}$$

$$= \frac{Qv}{4\pi\sqrt{ar}}\left\{1+\frac{vt_0}{8}\left(\frac{1}{a}-\frac{1}{r}\right)+\frac{(vt_0)^2}{\lfloor 2\,8^2}\left\{\frac{1^23^2}{\lfloor 2\,a^2}-\frac{1}{ar}+\frac{1^23^2}{\lfloor 2\,r^2}\right\}+\dots\right\}. \tag{178}$$

The other one, v_1, is obtained by interchanging a and r, not forgetting to do it in t_0. The value of t_0 is given by $vt_0 = vt-(r-a)$. They hold good between $r=a-vt$ and $a+vt$ until $vt=a$; that is, for example, for the V curves in the figure on p. 260 which lie between the first 0 curve and the curve 1,1,1. But both the V_1 of (178) and its companion v_1 are represented by (160) above. That is, (160) is equivalent to (178) when $r>a$, and to its companion v_1 when $r<a$.

After $vt=a$, the shell being fully formed and moving outward, the range of v_1 gets smaller, and that of V_1 larger, these ranges together making the complete shell. When $vt=2a$, the range of v_1 disappears altogether, because the shell is wholly outside $r=a$. From this moment onward the V in the whole shell is represented by V_1, either by (178) or by (160). What follows behind it is obtainable by a different way of considering the discontinuity at the source. In equations (175), (176) the operators are explicitly appropriate to inward and outward waves; but should the cylindrical source at $r=a$ be receiving a reflected wave from the axis as well as sending a wave in, we should substitute $2I_0$ for H_0 in those formulæ, making

$$V_2=\tfrac{1}{4}I_0(qa)K_0(qr)qQv, \qquad \text{(out)} \quad (r>a) \quad (179)$$

$$v_2=\tfrac{1}{4}I_0(qr)K_0(qa)qQv, \qquad \text{(in)} \quad (r<a) \quad (180)$$

where vqQ is the special value of the source $2\pi ag$, per unit length of cylinder, regarded as a function of the time. In general this is continuous; though, as before said, it is accidentally merely impulsive. There is continuity of V_2 and v_2 at $r=a$, and

$$2\pi a\left(\frac{dv_2}{dr}-\frac{dV_2}{dr}\right)=Qqv \qquad (181)$$

at $r=a$, so the differential accuracy of (179), (180) is established. The distinction is that H_0 and I_0 operate during different periods of time; H_0 is only valid for an inward wave, and ceases to be valid when the disturbance it sends inward comes back to the source from the axis. If we now carry out the operations in (179), (180), we come to the same results as before. For we get

$$V_2=I_0(qa)\frac{Qv}{2\pi(v^2t^2-r^2)^{\frac{1}{2}}}, \qquad v_2=I_0(qr)\frac{Qv}{2\pi(v^2t^2-a^2)^{\frac{1}{2}}}. \qquad (182)$$

But these are identical, in the present case of a momentary source, because

$$I_0(qa)\frac{1}{vt}=\frac{1}{(v^2t^2-a^2)^{\frac{1}{2}}}, \qquad (183)$$

and similarly with r substituted for a. It follows that both of (182) lead to the same formula (173) or (174), which moreover, we found to be equivalent to (159). That is, the present V_2 and v_2 mean the same, and the same as V_2 of the previous investigation, valid from the origin up to the back of the shell.

(42). Now if we construct the formula for V out of the line source formula (168) by a circular integration, it makes

$$V=2\int_0^{\pi}\tfrac{1}{4}K_0(qR)ga\,d\theta. \qquad (184)$$

In this, we have

$$R^2=r^2+a^2-2ar\cos\theta,$$

$$\frac{d\theta}{dR}=\frac{R}{ar\sin\theta}=\frac{2R}{\sqrt{(a+r)^2-R^2}\sqrt{R^2-(a-r)^2}}; \tag{185}$$

so

$$V=\int\frac{K_0(qR)RdR\,.\,ag}{\sqrt{(a+r)^2-R^2}\sqrt{R^2-(a-r)^2}}, \tag{186}$$

where the limits are those of reality. Comparing with (179), (180), and noting that $Qqv=2\pi ag$, the result makes

$$\int\frac{K_0(qR)RdR}{\sqrt{(a+r)^2-R^2}\sqrt{R^2-(a-r)^2}}=\frac{\pi}{2}I_0(qr)K_0(qa),\quad (r<a)$$

$$\text{or}\quad\frac{\pi}{2}I_0(qa)K_0(qr),\quad (r>a). \tag{187}$$

Here we have obtained the solution of a rather formidable definite integral, when q is a $+$ constant. There are two results, because $(a-r)^2$ in the integral is essentially positive, whereas $a-r$ may be positive or negative. Or, say $y^2=(a+r)^2$, $x^2=(a-r)^2$, then

$$\int\frac{K_0(qR)RdR}{\sqrt{(y^2-R^2)(R^2-x^2)}}=\frac{\pi}{2}I_0\{\tfrac{1}{2}q(y-x)\}K_0\{\tfrac{1}{2}q(y+x)\}, \tag{188}$$

provided x is understood to be always positive, whether $a>$ or $<r$. The limits are from x to y.

From (188), by $q=si$, if this substitution is valid, we obtain two other integrals, by using $G_0(ai)=G_0(a)-iJ_0(a)$, and $I_0(ai)=J_0(a)$. They are

$$\int_x^y\frac{G_0(sR)RdR}{\sqrt{(y^2-R^2)(R^2-x^2)}}=\frac{\pi}{2}J_0\{\tfrac{1}{2}s(y-x)\}G_0\{\tfrac{1}{2}s(y+x)\}, \tag{189}$$

$$\int_x^y\frac{J_0(sR)RdR}{\sqrt{(y^2-R^2)(R^2-x^2)}}=\frac{\pi}{2}J_0\{\tfrac{1}{2}s(y-x)\}J_0\{\tfrac{1}{2}s(y+x)\}. \tag{190}$$

The transformation $q=si$ sometimes fails in definite integrals, but we can obtain a corroboration of one of these by the property of interchangeability of functions mentioned before, F and f in (25). Applying this to (190), the result is equivalent to Sonine's integral containing the three J_0 functions, (53) and (61) above. This is satisfactory. But the interchange property does not apply to the other integral, so it remains without corroboration.

(43). Now integrate (188) to q, using the result $\int K_0(qR)dq=R^{-1}$, to be proved separately later. We get

$$\int\frac{dR}{\sqrt{(y^2-R^2)(R^2-x^2)}}=\frac{\pi}{2}\int I_0\{\tfrac{1}{2}q(y-x)\}K_0\{\tfrac{1}{2}q(y+x)\}\,dq. \tag{191}$$

But the integral on the left side is known, as in (122), (124) to (126). So we get

$$\int I_0\{\tfrac{1}{2}q(y-x)\}K_0\{\tfrac{1}{2}q(y+x)\}\,dq=\frac{1}{y}\left(1+\frac{1^2}{2^2}\frac{y^2-x^2}{y^2}+\dots\right); \tag{192}$$

or, in terms of r and a,

$$\int I_0(qa)K_0(qr)dq = \frac{1}{r+a}\left(1+\frac{1^2}{2^2}\frac{4ra}{(r+a)^2}+\ldots\right) \qquad (193)$$

if $r>a$, but with r and a interchanged in the integral if $r<a$. That is, the larger of r and a is to go to the K_0 function. As for the right side, the interchange of r and a makes no difference. It may readily be suspected from this property of a change on one side without any on the other, that the right member is transformable to two other forms, one to suit $r>a$, the other $r<a$. And in fact if we suppose $r>a$, every term in the series on the right may be expanded convergently in rising powers of a/r. On rearrangement of the whole in powers of a/r, the coefficients of all odd powers of a vanish, and the even powers make

$$\int I_0(qa)K_0(qr)dq = \frac{1}{r}\left(1+\frac{1^2}{2^2}\frac{a^2}{r^2}+\ldots\right), \qquad (194)$$

when $r>a$. We must now interchange r and a on both sides at once when $a>r$. We have here a practical example of the old theorem

$$F(x)(1+\sqrt{x}) = F\left\{\frac{4\sqrt{x}}{(1+\sqrt{x})^2}\right\}, \qquad (195)$$

using the notation of (135A) above. The use of this theorem mentioned in books is to allow of the calculation of $F(x)$ for large values of x, where the series is very slowly convergent, in terms of the values of $F(x)$ for smaller values. For example

$$F(0{\cdot}25) = \tfrac{2}{3}F(0{\cdot}8888), \quad F(0{\cdot}81) = \frac{1}{1{\cdot}9}F\left(\frac{360}{361}\right).$$

Of course, the calculation of $F(360/361)$ by itself would be very lengthy.

But in the present circumstances there is another use for the theorem. It was suggested before (p. 261) that the curve of V on p. 260, should be expressible by one formula, in spite of the infinity. Now it has turned up naturally. We have

$$\pi^{\frac{1}{2}}\left(\frac{m_1n_1}{m_1+n_1}\right)^{\frac{1}{2}} = \frac{1}{\sqrt{m}}F\left(\frac{n}{m}\right) \quad \text{or} \quad \frac{1}{\sqrt{n}}F\left(\frac{m}{n}\right)$$

$$= \frac{1}{\sqrt{m}+\sqrt{n}}F\left(\frac{4\sqrt{mn}}{(\sqrt{m}+\sqrt{n})^2}\right), \qquad (196)$$

and V_1, V_2 of (159), (160) are both contained in

$$V = \frac{\rho v/2\pi}{\sqrt{4ar}+\sqrt{v^2t^2-(a-r)^2}}F\left(\frac{4\sqrt{4ar}\sqrt{v^2t^2-(a-r)^2}}{(\sqrt{4ar}+\sqrt{v^2t^2-(a-r)^2})^2}\right), \qquad (197)$$

for all possible values of t, a and r. That is, $v^2t^2-(a-r)^2$ begins with the value 0. But the two separate series are simpler and are usually more convenient.

(44). If we give negative values to t, and let t increase from a large negative value up to zero, we have, provided we reverse the velocity everywhere, to contemplate the singular spectacle of the march of the function V taking place exactly in the reverse manner. The infinity moves in to the axis, and disappears, after which all the condensation proceeds towards and condenses itself at the surface of the cylinder $r=a$, at the moment $t=0$. If at this moment we introduce a momentary negative source, or sink, of the right amount, there is an end of the automatic dynamics. But if no sink be introduced, the arrived condensation will itself act as a positive source, and the later march of V will be precisely as before described, with t positive.

(45). If in (194) we change the sign of a^2, say $=-b^2$, then we get

$$\int_0^\infty \mathrm{J}_0(qb)\mathrm{K}_0(qr)dq=\frac{1}{r}\left(1-\frac{1^2}{2^2}\frac{b^2}{r^2}+\dots\right). \tag{198}$$

This is not the standard form of series, since the signs alternate. But see (135B). It converts our result to

$$\int_0^\infty \mathrm{J}_0(qb)\mathrm{K}_0(qr)dq=\frac{1}{\sqrt{r^2+b^2}}\left(1+\frac{1^2}{2^2}\frac{b^2}{r^2+b^2}+\dots\right), \tag{199}$$

which is in the standard form. It is one of Basset's integrals. I always rejoice in coming to the known for confirmation, even though compelled to return to my own ways of work to make further progress.

(46). Some of the K_0 integrals are very interesting, and it is as well to construct them systematically. This can be done by first finding $\int x^m\mathrm{K}_0(qx)dx$. Thus, if $t^2=\mathrm{T}$,

$$\int_0^\infty x^m\mathrm{K}_0(t_1x)dx=\frac{2}{\pi t_1}\int_0^t\frac{x^m}{(t^2-x^2)^{\frac{1}{2}}}dx=\frac{2}{\pi t_1}\int(\mathrm{T}_1\pi)^{\frac{1}{2}}\epsilon^{-\mathrm{T}_1x^2}(x^2)^{\frac{1}{2}m}\frac{dx^2}{2\sqrt{x^2}}$$

$$=\frac{1}{t_1\sqrt{\pi}}\frac{\underline{\lfloor \tfrac{1}{2}m-\tfrac{1}{2}}}{\mathrm{T}_1^{\frac{1}{2}m}}=\frac{t^m}{t_1\sqrt{\pi}}\frac{\underline{\lfloor \tfrac{1}{2}m-\tfrac{1}{2}}}{\underline{\lfloor \tfrac{1}{2}m}}=\frac{\underline{\lfloor m}\,\underline{\lfloor \tfrac{1}{2}m-\tfrac{1}{2}}}{\underline{\lfloor \tfrac{1}{2}m}\,t_1^{m+1}\sqrt{\pi}}. \tag{201}$$

This is true from $m=-1$ to ∞. A more convenient form arises by using

$$\sqrt{\pi}\,\underline{\lfloor m}=\underline{\lfloor \tfrac{1}{2}m}\,\underline{\lfloor \tfrac{1}{2}m-\tfrac{1}{2}}\,2^m, \tag{202}$$

which identity is a special case of (188), in vol. 2, p. 472. So we have

$$\int_0^\infty x^m\mathrm{K}_0(t_1x)dx=\frac{2^m}{\pi t_1^{m+1}}(\underline{\lfloor \tfrac{1}{2}m-\tfrac{1}{2}})^2. \tag{203}$$

It is easier to refer to special values, and to even and odd m's separately. Thus

$$m=0,\quad 2,\quad 4,\quad 6,\quad 8,$$

$$\frac{1}{t_1},\quad \frac{1^2}{t_1^3},\quad \frac{1^23^2}{t_1^5},\quad \frac{1^23^25^2}{t_1^7},\quad \frac{1^23^25^27^2}{t_1^9}, \tag{204}$$

for even values of m, and

$$m=1,\qquad 3,\qquad 5,\qquad 7,$$

$$\frac{2}{\pi t_1^{\,2}},\quad \frac{2}{\pi t_1^{\,4}}2^2,\quad \frac{2}{\pi t_1^{\,6}}2^2 4^2,\quad \frac{2}{\pi t_1^{\,8}}2^2 4^2 6^2, \tag{205}$$

for odd values. We can now write out the solutions of various integrals involving K_0 by reference to (204), (205).

(47). Begin with one that is infinite.

$$\int_0^\infty (1+a^2x^2+a^4x^4+\dots)K_0(qx)dx=\frac{1}{q}\left(1+\frac{a^2}{q^2}+1^2 3^2\frac{a^4}{q^4}+\dots\right). \tag{206}$$

The sum of any set of terms on one side is, by the manner of construction, equal to the sum of the corresponding set on the other side. If, then, the number in the set is infinite, the integral is infinite in value, as is easy to see from the form of the series. We may, however, even then, derive a finite integral. For, if $q=t_1$, we get

$$\frac{2}{\pi}\int_0^t \frac{dx}{(t^2-x^2)^{\frac{1}{2}}(1-a^2x^2)}=\frac{1}{(1-a^2t^2)^{\frac{1}{2}}}. \tag{207}$$

We require $t<a^{-1}$.

(48). Change sign of a^2. Put $a^2=-b^2$. Then instead of (206) we have

$$\int_0^\infty \frac{K_0(qx)dx}{1+b^2x^2}=\frac{1}{q}\left\{1-\frac{b^2}{q^2}+\frac{1^2 3^2 b^4}{q^4}-\dots\right\} \tag{208}$$

$$=\frac{\pi}{2b}\left\{G_0\left(\frac{q}{b}\right)+\frac{2}{\pi}\left(\frac{q}{b}-\frac{q^3}{1^2 3^2 b^3}+\frac{q^5}{1^2 3^2 5^2 b^5}-\right)\right\}. \tag{209}$$

The integral is obviously finite. The immediate result is alternatingly divergent. It is converted to a purely convergent form in (209) by the formula (115), vol. 2, p. 455.

In passing, note that the complete series on the right side of (206) is also, when calculated numerically by its initial convergence, equivalent to a convergent function, as in vol. 2, p. 441, eq. (31). It may be possible to modify the interpretation or expression of the integral to suit this.

(49). For variety and interesting views, change the order of work. If $a_0<t$, then

$$\frac{2}{\pi}\int_0^{a_0}\frac{dx}{(t^2-x^2)^{\frac{1}{2}}}=\frac{2}{\pi}\sin^{-1}\frac{a_0}{t}=\frac{2}{\pi}\left\{\frac{a_0}{t}+\frac{1}{3}\frac{1}{2}\frac{a_0^{\,3}}{t^3}+\frac{1}{5}\frac{1.3}{2.4}\frac{a_0^{\,5}}{t^5}+\dots\right\}. \tag{210}$$

This is elementary. Change the upper limit to t without altering the value, as in (10);

$$=\frac{2}{\pi}\int_0^t\frac{\epsilon^{-ax}}{(t^2-x^2)^{\frac{1}{2}}}dx=\frac{2}{\pi}\left(\frac{1}{at}+\frac{1^2}{a^3t^3}+\frac{1^2 3^2}{a^5t^5}+\dots\right) \tag{211}$$

$$=I_0(at)-\frac{2}{\pi}\left(at+\frac{a^3t^3}{1^2 3^2}+\frac{a^5t^5}{1^2 3^2 5^2}+\dots\right). \tag{212}$$

Here the right member of (211) comes from (210) by converting it to a function of a. It is divergent, but is converted to the convergent form (212) by eq. (31), vol. 2, p. 441. At the same time this convergent form is obtainable from the integral (211) by direct integration.

Next pass from a to a_1, and then again from t to $t_1=q$, and finally from a_1 to a.

$$=\frac{2}{\pi}\int_0^t \frac{dx}{(1+x/a_1)(t^2-x^2)^{\frac{1}{2}}}=\frac{1}{(1-t^2/a_1^2)^{\frac{1}{2}}}-\frac{2}{\pi}\left(\frac{t}{a_1}+\frac{2}{3}\frac{t^3}{a_1^3}+\frac{2.4}{3.5}\frac{t^5}{a_1^5}+\dots\right), \tag{213}$$

$$=\int_0^\infty \frac{q\mathrm{K}_0(qx)dx}{1+x/a_1}=\left(1+\frac{1}{a_1^2q^2}+\frac{1^23^2}{a_1^4q^4}+\dots\right)$$
$$-\frac{2}{\pi}\left(\frac{1}{a_1q}+\frac{2^2}{a_1^3q^3}+\frac{2^24^2}{a_1^5q^5}+\dots\right), \tag{214}$$

$$=\int_0^\infty \epsilon^{-ax}\mathrm{K}_0(qx)q\,dx=\frac{1}{(1-a^2/q^2)^{\frac{1}{2}}}-\frac{2}{\pi}\left(\frac{a}{q}+\frac{2}{3}\frac{a^3}{q^3}+\frac{2.4}{3.5}\frac{a^5}{q^5}+\dots\right). \tag{215}$$

Here (213) comes from (212), (214) from (213), and (215) from (214), and all is done at sight, without troublesome references. In the end we come to two K_0 integrals, both of which may be directly constructed from (204), (205). The whole process is reversible, so that we may work back step by step to the elementary form (210). Another form of (215) is (see also p. 247),

$$=\frac{2}{\pi}\frac{q}{a}\frac{\log\{a/q+\sqrt{a^2/q^2-1}\}}{(1-q^2/a^2)^{\frac{1}{2}}}=\frac{q}{a}\frac{2}{\pi}\left\{1+\tfrac{1}{3}\left(1-\frac{q^2}{a^2}\right)+\tfrac{1}{5}\left(1-\frac{q^2}{a^2}\right)^2+\dots\right\}. \tag{216}$$

Observe the series on the right sides of (213), (215). They are of the same form and are convertible by changing t to a, and a_1 to t_1. Both are the difference of two convergent series. But in (214) the result is the difference of two divergent series. One of these, the first, is known convergently. See (211), (212) for this equivalence. The other has not previously turned up. I will give the method of convergent evaluation separately later on.

(50). Another integral coming from (204) directly is

$$\int_0^\infty \epsilon^{-a^2x^2}\mathrm{K}_0(qx)dx=\frac{1}{q}\left(1-\frac{a^2}{q^2}+\frac{1^23^2}{\lfloor 2}\frac{a^4}{q^4}-\frac{1^23^25^2}{\lfloor 3}\frac{a^6}{q^6}+\dots\right) \tag{217}$$

$$=\frac{\sqrt{\pi}}{4a}\epsilon^{q^2/8a^2}\mathrm{K}_0\left(\frac{q^2}{8a^2}\right). \tag{218}$$

The immediate result is divergent, but it is a well-known formula, of the convergent form (218). Put $q=si$ in (218); then we obtain

$$\int \epsilon^{-a^2x^2}\mathrm{G}_0(sx)dx=\frac{\sqrt{\pi}}{4a}\epsilon^{-s^2/8a^2}\mathrm{K}_0\left(\frac{s^2}{8a^2}\right), \tag{219}$$

$$\int \epsilon^{-a^2x^2}\mathrm{J}_0(sx)dx=\frac{\sqrt{\pi}}{2a}\epsilon^{-s^2/8a^2}\mathrm{I}_0\left(\frac{s^2}{8a^2}\right). \tag{220}$$

These are ascribed to Basset in Gray and Matthews "Bessel Functions," p. 227, example 8, with a difference, however. In that place, the formula corresponding to (219) is discrepant with the one corresponding to (218) by the factor $2/\pi$.

(51). To clear this up, and get new results, let

$$v=\int_0^\infty \epsilon^{-a^2x^2}\mathrm{J}_n(cx)dx=\int \epsilon^{-by}\mathrm{J}_n\{2\sqrt{ey}\}\frac{dy}{2\sqrt{y}}$$

$$=\tfrac{1}{2}\left(\frac{1}{e}\right)^{\frac{1}{2}n}\frac{1}{e_1{}^n}\frac{\underline{|\tfrac{1}{2}n-\tfrac{1}{2}}}{(b+1/e_1)^{\frac{1}{2}n+\frac{1}{2}}}, \tag{221}$$

where $x^2=y$, $a^2=b$, $4e=c^2$, and the integration is done as in the previous work concerning J_n by putting it in exponential form, and then using Euler's integral. We have now to turn this result to a function of e. There are two ways (by the binomial theorem), in rising and in descending powers. The convergent one will be right; the other will take its chance. The descending series comes to

$$v_2=\frac{1}{c}\left\{1-(n^2-1)\frac{a^2}{c^2}+\frac{1}{\underline{|2}}(n^2-1^2)(n^2-3^2)\frac{a^4}{c^4}-\dots\right\}$$

$$=\frac{\sqrt{\pi}}{4a}\mathrm{H}_{\frac{1}{2}n}\left(\frac{c^2}{8a^2}\right)\epsilon^{-c^2/8a^2}. \tag{222}$$

This is the divergent form. In the case $n=0$, it agrees with (220).

The other way of expanding (221) leads to

$$v_1=\frac{1}{2a}\left(\frac{c}{2a}\right)^n\left[\frac{\underline{|\tfrac{1}{2}n-\tfrac{1}{2}}}{\underline{|0}\,\underline{|n}}-\frac{\underline{|\tfrac{1}{2}n+\tfrac{1}{2}}}{\underline{|1}\,\underline{|n+1}}\left(\frac{c}{2a}\right)^2+\frac{\underline{|\tfrac{1}{2}n+1\tfrac{1}{2}}}{\underline{|2}\,\underline{|n+2}}\left(\frac{c}{2a}\right)^4-\dots\right]. \tag{223}$$

This is convergent, but for comparison with (222), we should introduce the prefactor $\epsilon^{-c^2/8a^2}$, or $\epsilon^{-e/2b}$. This is best done on (221). It makes

$$v_1=\tfrac{1}{2}e^{-\frac{1}{2}n}\epsilon^{-e/2b}\,\underline{|\tfrac{1}{2}n-\tfrac{1}{2}}\,e_1{}^{-n}\left(1-1/4b^2e_1{}^2\right)^{-\frac{1}{2}n-\frac{1}{2}}. \tag{224}$$

On expansion, this produces

$$v_1=\frac{\sqrt{\pi}}{2a}\epsilon^{-c^2/8a^2}\mathrm{I}_{\frac{1}{2}n}\left(\frac{c^2}{8a^2}\right). \tag{225}$$

Comparing with (222), they agree when $\frac{1}{2}n$ is integral, not otherwise.

Now, if we change J_n to I_n in the original integral in (221), the only effect is to change the sign of e_1 in the exponential form first, and later on in the radical. Then, on introducing the prefactor $\epsilon^{+e/2b}$, the result is that everything is the same as in (224) save in this changed factor. So we come to

$$\int_0^\infty \epsilon^{-a^2x^2}\mathrm{I}_n(qx)dx=\frac{\sqrt{\pi}}{2a}\epsilon^{q^2/8a^2}\mathrm{I}_{\frac{1}{2}n}\left(\frac{q^2}{8a^2}\right). \tag{226}$$

These results entirely confirm the preceding. For, by differentiating (226) to n, and putting $n=0$ after, (226) leads to (218); and by performing the same operation upon (225), the result is (220).

(52). The v_2 result, converted to convergent form, would need the use of $H_{\frac{1}{2}n} = I_{\frac{1}{2}n} + I_{-\frac{1}{2}n}$, for numerical equivalence, in disagreement with the convergent result v_1. In connection with this, it is to be remarked that in doing Sonine's integral on p. 244, the use of $\frac{1}{2}H_n$ instead of I_n, when worked differentially, led to Sonine's result. That is, the numerical conflicts with the operational equivalence when n is fractional. Things of this sort are not without importance in the complete laws of divergent series.

If we employ the formula

$$\int J_n(sx) x^m dx = \frac{2^m}{s^{m+1}} \frac{\lfloor\underline{\tfrac{1}{2}n + \tfrac{1}{2}m - \tfrac{1}{2}}}{\lfloor\underline{\tfrac{1}{2}n - \tfrac{1}{2}m - \tfrac{1}{2}}}. \tag{227}$$

to evaluate (221), by expanding the exponential function in rising powers of a^2, the result is the sometimes correct divergent formula (222). But if we expand the J_n function in rising powers of c, the result is the fully correct formula (225), when done by Euler's integral.

And, by q or $t_1 = si$ in (203), we get not only the special case $n = 0$ of (227) just written, but also

$$\int G_0(sx) x^m dx = -\frac{\sin \frac{1}{2}m\pi}{s^{m+1}} \frac{\lfloor\underline{m}\ \lfloor\underline{\tfrac{1}{2}m - \tfrac{1}{2}}}{\lfloor\underline{\tfrac{1}{2}m}\ \sqrt{\pi}}, \tag{228}$$

which takes the value zero when m is 0 or an even number. If then, we should try to evaluate the integral (219) by expanding the exponential in powers of a^2, and then applying (228), the result would be zero. But if we expand G_0 in powers of s (the logarithmic formula), and then apply Euler's integral, we confirm the result in (219). As follows.

(53). Verification of (219). Put $G_0(sx)$ in logarithmic form, with $x^2 = y$, and $\lambda = \log \frac{1}{2}sx$. Then the integral is

$$-\frac{2}{\pi}\int \epsilon^{-a^2 y}\left\{(\lambda + G_0) - \frac{\frac{1}{4}s^2 y}{\lfloor\underline{1}\,\lfloor\underline{1}}(\lambda + G_1) + \frac{(\frac{1}{4}s^2 y)^2}{\lfloor\underline{2}\,\lfloor\underline{2}}(\lambda + G_2) - \dots\right\}\frac{dy}{2\sqrt{y}}, \tag{229}$$

where $G_n = g'(n)/g(n)$. Do this by Euler's integral. The result is

$$\frac{1}{2a\sqrt{\pi}}\left\{(\lambda_1 + 2G_0 - G_{-\frac{1}{2}}) - \frac{\frac{1}{2}(s^2/4a^2)}{\lfloor\underline{1}\,\lfloor\underline{1}}(2G_1 - G_{\frac{1}{2}} + \lambda_1) + \frac{\frac{1.3}{2.2}(s^2/4a^2)^2}{\lfloor\underline{2}\,\lfloor\underline{2}}(2G_2 - G_{1\frac{1}{2}} + \lambda_1) - \right\}, \tag{230}$$

where $\lambda_1 = \log(s/2a)$. Using the values of the G functions, this makes

$$-\frac{1}{2a\sqrt{\pi}}\left\{(G_0 + \lambda_0) - \frac{s^2/8a^2}{\lfloor\underline{1}\,\lfloor\underline{1}}(G_0 + \lambda_0) + \frac{1.3(s^2/8a^2)^2}{\lfloor\underline{2}\,\lfloor\underline{2}}(G_0 + \lambda_0 - \tfrac{1}{3}) - \frac{1.3.5(s^2/8a^2)^3}{\lfloor\underline{3}\,\lfloor\underline{3}}(G_0 + \lambda_0 - \tfrac{3}{5}) + \right\}, \tag{231}$$

where $\lambda_0 = \log(s^2/16a^2)$. This is also what results from (219) by multiplying together the expansions of the exponential function and of the K_0 function.

(54). Here are the necessary formulæ relating to G_n. By (202),

$$g(m) = \frac{\sqrt{\pi}}{2^m}\, g\left(\frac{m}{2}\right) g\left(\frac{m}{2}-\frac{1}{2}\right). \tag{232}$$

Logarize and differentiate to m. Then

$$G(m) = \tfrac{1}{2}G(\tfrac{1}{2}m) + \tfrac{1}{2}G(\tfrac{1}{2}m - \tfrac{1}{2}) - \log 2. \tag{233}$$

Put $m=0$. Then follows

$$G_0 + \log 4 = G_{-\frac{1}{2}} = 0{\cdot}5772 + 1{\cdot}3863 = 1{\cdot}9635. \tag{234}$$

All the rest follow from G_0 and $G_{-\frac{1}{2}}$, which are known, through

$$G(n) + n^{-1} = G(n-1), \tag{235}$$

which follows from $ng(n) = g(n-1)$, the fundamental property.

(55). Closely connected with this matter is the behaviour of (198) on conversion of $K_0(qr)$ to $G_0(qs) - iJ_0(qs)$ by the change $r = si$. We get

$$\int J_0(qb)G_0(qs)\,dq = 0, \qquad \int J_0(qb)J_0(qs)\,dq = \frac{1}{s}\left(1 + \frac{1^2}{2^2}\frac{b^2}{s^2} + \dots\right). \tag{236}$$

The second of these is right, but as regards the first, we have to ask whether there is any restriction upon the value of s. We have $b<s$ in the second one, therefore presumably also in the first. But if so, it still remains to find the value when $b>s$. To do this, go back to the operational treatment. Let

$$u = \int x^{-\frac{1}{2}} J_0(2\sqrt{xa})J_n(2\sqrt{xb})\,dx = b^{-\frac{1}{2}n}\,\frac{\lfloor \tfrac{1}{2}n - \tfrac{1}{2}}{b_1^{\,n}}\left(\frac{a_1 b_1}{a_1+b_1}\right)^{\frac{1}{2}n+\frac{1}{2}}. \tag{237}$$

It makes two series

$$u_2 = \frac{1}{\sqrt{b}}\left\{1 - \frac{n^2-1}{4}\frac{a}{b}\left(1 - \frac{n^2-3^2}{4.2^2}\frac{a}{b}\left(1 - \frac{n^2-5^2}{4.3^2}\frac{a}{b}\right(\dots,\right.\right.\right. \tag{238}$$

$$u_1 = \left(\frac{b}{a}\right)^{\frac{1}{2}n}\frac{1}{\sqrt{a}}\left\{\frac{\lfloor \tfrac{1}{2}n - \tfrac{1}{2}}{\lfloor -\tfrac{1}{2} - \tfrac{1}{2}n}\,\frac{1}{\lfloor 0 \,\lfloor n} - \frac{\lfloor \tfrac{1}{2}n + \tfrac{1}{2}}{\lfloor -1\tfrac{1}{2} - \tfrac{1}{2}n}\,\frac{b/a}{\lfloor 1 \,\lfloor n+1} + \dots\right\}. \tag{239}$$

Here u_1 and u_2 are both convergent, one from $a=0$ to b, the other from $a=b$ to ∞, so the formulæ are complementary.

Now $G_0(y) = -(2/\pi)(d/dn_0)J_n(y)$. So we evaluate thus,

$$\int x^{-\frac{1}{2}} J_0(2\sqrt{xa})G_0(2\sqrt{xb})\,dx = -\frac{2}{\pi}\frac{d}{dn_0}(u_1 \text{ and } u_2). \tag{240}$$

But du_2/dn_0 is evidently zero. That is, when $a<b$. So we confirm the vanishing of the first of (236) when $b<s$. Also, du_1/dn_0 is not zero. So the first integral in (236) is not zero when $b>s$, and its expression is obtainable from u_1 through (240).

(56). Now examine whether

$$u=\int \mathrm{J}_0(2\sqrt{ax})\,\mathrm{J}_n(2\sqrt{bx})dx=\frac{\lfloor\underline{\tfrac{1}{2}n}}{b^{\frac{1}{2}n}}\frac{1}{b_1{}^n}\left(\frac{a_1b_1}{a_1+b_1}\right)^{\frac{1}{2}n+1} \tag{241}$$

vanishes on either side of $a=b$ when the result is differentiated to n_0. The two formulæ are

$$u_1=\frac{n}{2b}\left\{1-\left(\frac{n^2}{4}-1\right)\frac{a}{b}+\frac{(\frac{1}{4}n^2-1^2)(\frac{1}{4}n^2-2^2)}{\lfloor\underline{2}\lfloor\underline{2}}\frac{a^2}{b^2}-\dots\right\}, \tag{242}$$

$$u_2=\frac{b^{\frac{1}{2}n}}{a^{\frac{1}{2}n+1}}\left\{\frac{\lfloor\underline{\tfrac{1}{2}n}}{\lfloor\underline{n}\lfloor\underline{-\tfrac{1}{2}n-1}}-\frac{\lfloor\underline{\tfrac{1}{2}n+1}\;b/a}{\lfloor\underline{n+1}\lfloor\underline{-\tfrac{1}{2}n-2}\lfloor\underline{1}}+\frac{\lfloor\underline{\tfrac{1}{2}n+2}\;b^2/a^2}{\lfloor\underline{n+2}\lfloor\underline{-\tfrac{1}{2}n-3}\lfloor\underline{2}}-\dots\right\}; \tag{243}$$

and, by differentiation,

$$\frac{du_1}{dn_0}=\frac{1}{2b}\left(1+\frac{a}{b}+\frac{a^2}{b^2}+\dots\right)=\frac{1}{2(b-a)},\quad \text{if } a<b, \tag{244}$$

$$\frac{du_2}{dn_0}=-\frac{1}{2a}\left(1+\frac{b}{a}+\frac{b^2}{a^2}+\dots\right)=\frac{1}{2(b-a)},\quad \text{if } a>b. \tag{245}$$

The result is given by the same formula in both cases, but the value reverses itself. It follows that

$$\int \mathrm{J}_0(2\sqrt{ax})\mathrm{G}_0(2\sqrt{bx})dx=\frac{1}{\pi(a-b)},\quad \int \mathrm{J}_0(ry)\mathrm{G}_0(sy)y\,dy=\frac{2}{\pi(r^2-s^2)}. \tag{246}$$

Finally, put $q=si$, then, remembering that $\int \mathrm{J}_0(ry)\mathrm{J}_0(sy)y\,dy=0$, we come to

$$\int \mathrm{J}_0(ry)\mathrm{K}_0(qy)y\,dy=\frac{2}{\pi(r^2+q^2)}, \tag{247}$$

which is true whether r is $<$ or $>q$.

(57). The formula (203) leads directly to

$$\int \mathrm{I}_n(ax)\mathrm{K}_0(bx)\,(\tfrac{1}{2}ax)^m dx=\frac{1}{\pi b}\left(\frac{a}{b}\right)^{n+m}\left[\frac{(\lfloor\underline{\tfrac{1}{2}(m+n)-\tfrac{1}{2}})^2}{\lfloor\underline{0}\lfloor\underline{n}}+\frac{a^2}{b^2}\frac{(\lfloor\underline{\tfrac{1}{2}(m+n)+\tfrac{1}{2}})^2}{\lfloor\underline{1}\lfloor\underline{n+1}}+\dots\right], \tag{248}$$

which includes some previous formulæ. To get simpler ones, say $m=1$. Then,

$$\int \mathrm{I}_n(ax)\mathrm{K}_0(bx)x\,dx=\frac{2}{\pi ab}\left(\frac{a}{b}\right)^{n+1}\left[\frac{(\lfloor\underline{\tfrac{1}{2}n})^2}{\lfloor\underline{0}\lfloor\underline{n}}+\frac{a^2}{b^2}\frac{(\lfloor\underline{\tfrac{1}{2}n+1})^2}{\lfloor\underline{1}\lfloor\underline{n+1}}+\dots\right]. \tag{249}$$

It is necessary that $a<b$, because $\mathrm{I}_n(ax)\propto\epsilon^{ax}/x^{\frac{1}{2}}$, and $\mathrm{K}_0(bx)\propto\epsilon^{-bx}/x^{\frac{1}{2}}$, when x is very large.

Differentiate (249) to n. We have

$$\frac{d}{dn_0}\frac{(\lfloor\underline{\tfrac{1}{2}n+s})^2}{\lfloor\underline{s}\lfloor\underline{n+s}}=\frac{(\lfloor\underline{s})^2}{\lfloor\underline{s}\lfloor\underline{s}}(\mathrm{G}(s)-\mathrm{G}(s))=0. \tag{250}$$

Apply this to (249). It makes

$$\int K_0(ax)K_0(bx)x\,dx = \frac{4}{\pi^2}\frac{\log(b/a)}{b^2-a^2}: \tag{251}$$

from which, by $a=ri$, are derived

$$\int G_0(rx)K_0(bx)x\,dx = \frac{4}{\pi^2}\frac{\log(b/r)}{b^2+r^2}, \tag{252}$$

and another one, which is represented by (247) above. Also, by $b=si$,

$$\int G_0(rx)G_0(sx)\,x\,dx = \frac{4}{\pi^2}\frac{\log(s/r)}{s^2-r^2}, \tag{253}$$

and another one, equivalent to (246).

Compare (253) with (251). They make

$$\int K_0(ax)K_0(bx)x\,dx = \int G_0(ax)G_0(bx)x\,dx. \tag{254}$$

When $a=b$, the common value is $2/\pi^2a^2$.

(58). If, in (248), we make $m=0$, and then differentiate to n, and finally put $n=0$, we come to

$$\int K_0(ax)K_0(bx)dx = -\frac{2}{b\pi}\left[(\lambda+G_0-G_{-\frac{1}{2}}) + \frac{1^2}{2^2}\frac{a^2/b^2}{\lfloor 1\lfloor 1}(\lambda+G_1-G_{\frac{1}{2}})+\dots\right], \tag{255}$$

where $\lambda=\log(a/b)$. This function bears a somewhat similar relation to the first elliptic function as K_0 does to I_0. In full, by the values of G_n already given, the result is

$$= \frac{2}{b\pi}(\beta-\gamma-\lambda)\left(1+\frac{1^2}{2^2}\frac{a^2}{b^2}+\dots\right) - \left[\frac{1^2}{2^2}\frac{a^2}{b^2}+\frac{1^23^2}{2^24^2}\frac{a^4}{b^4}\left(1+\frac{1}{2.3}\right)\right.$$
$$\left.+\frac{1^23^25^2}{2^24^26^2}\frac{a^6}{b^6}\left(1+\frac{1}{2.3}+\frac{1}{3.5}\right)+\dots\right], \tag{256}$$

where $\gamma=G_0=0{\cdot}5772$, $\beta=G_{-\frac{1}{2}}=1{\cdot}9635$. This form suits $a<b$. If $a>b$, then a and b must be interchanged.

(59). By direct use of (204) we get

$$\int K_0(qx)\left(1+\frac{x^2}{a^2}+\frac{x^4}{a^4}\frac{1}{1^23^2}+\dots\right)dx = \frac{1}{q}\left(1+\frac{1}{a^2q^2}+\frac{1}{a^4q^4}+\dots\right)$$
$$= \frac{1}{q(1-1/a^2q^2)}. \tag{257}$$

Therefore $q>a^{-1}$ for finiteness. This is evident from the result. But to see the reason we should evaluate the function integrated for $x=\infty$. By the equivalence (31), vol. 2, p. 441, the function is proportional to $(\epsilon^{-qx}x^{-\frac{1}{2}})\times(x^{\frac{1}{2}}\epsilon^{x/a})$ when x is ∞, or to $\epsilon^{-x(q-1/a)}$. So the integral is finite only when $q>1/a$. It becomes infinite when $q=1/a$, and stays infinite later.

But if we alternate the signs, we have

$$\int K_0(qx)\left(1-\frac{x^2}{a^2}+\frac{x^4}{a^4}\frac{1}{1^23^2}-\dots\right)dx = \frac{1}{q(1+1/a^2q^2)}, \tag{258}$$

and now q^2a^2 may have any value, the integral being always finite. From (257) follows

$$\frac{2}{\pi}\int_0^t \frac{1+x^2/a^2+x^4/a^4 1^2 3^2+\dots}{(t^2-x^2)^{\frac{1}{2}}}dx=\cosh\frac{t}{a} \tag{259}$$

$$=\frac{2}{\pi}\int_0^t \frac{1+\frac{x^2a_0^2}{\lfloor 2}+\frac{x^4a_0^4}{\lfloor 4\, 1^2 3^2}+\dots}{(t^2-x^2)^{\frac{1}{2}}}dx=1+\frac{a_0^2t^2}{\lfloor 2\lfloor 2}+\frac{a_0^4t^4}{\lfloor 4\lfloor 4}+\dots \tag{260}$$

(60). The following is similar, with differences. By (203),

$$q\int \mathrm{K}_0(qx)\left(1+\frac{x^2}{a^2}+\frac{x^4}{a^4 1.3}+\frac{x^6}{a^6 1.3.5}+\dots\right)dx$$
$$=1+\frac{1}{a^2q^2}+\frac{1.3}{a^4q^4}+\frac{1.3.5}{a^6q^6}+\dots \tag{261}$$

The value is infinite for any positive values of a and q, in contrast with (257). The function to be integrated is equivalent to

$$q\mathrm{K}_0(qx)\left(\frac{\pi}{2}\right)^{\frac{1}{2}}\frac{x}{a}\left[\epsilon^{x^2/2a^2}-\frac{(x^2/2a^2)^{-1\frac{1}{2}}}{\lfloor -1\frac{1}{2}}-\frac{(x^2/2a^2)^{-2\frac{1}{2}}}{\lfloor -2\frac{1}{2}}-\dots\right]$$

by the generalised exponential function, vol. 2, p. 439, eq. (24). So, at $x=\infty$ it is proportional to $x^{\frac{1}{2}}\epsilon^{-qx}\times\epsilon^{x^2/2a^2}$, that is, ∞.

But if we alternate the signs, the integral is finite. To see this, use the generalised $\epsilon^{-x}\cos r\pi$ formula (140), vol. 2, p. 466. Then, instead of (261) we have a function which vanishes at ∞. So

$$q\int \mathrm{K}_0(qx)\left(1-\frac{x^2}{a^2}+\frac{x^4}{a^4.1.3}-\dots\right)dx=1-\frac{1}{a^2q^2}+\frac{1.3}{a^4q^4}-\frac{1.3.5}{a^6q^6}+\dots \tag{262}$$

$$=aq\left(\frac{\pi}{2}\right)^{\frac{1}{2}}\epsilon^{a^2q^2/2}-a^2q^2\left(1+\frac{a^2q^2}{1.3}+\frac{a^4q^4}{1.3.5}+\dots\right), \tag{263}$$

where the second form of result, which is convergent, is obtained from the first form by the use of the generalised ϵ^x formula, vol. 2, p. 439. The form (263) is for small values of aq, and (262) for large values.

It is to be carefully noted that we cannot in general make such changes in a function which is to be integrated from 0 to ∞, and all terms counted; the result will obviously differ from the true result even when the substituted function is equivalent to the original.

From (261) is derived

$$\frac{2}{\pi}\int_0^t \frac{1+x^2/a^2+x^4/1.3a^4+\dots}{(t^2-x^2)}dx=\epsilon^{t^2/2a^2} \tag{264}$$

In this, put $a^2=\mathrm{A}_1$, turn both sides to functions of A, and put $2\mathrm{A}=c^2$; then

$$\frac{2}{\pi}\int_0^t \frac{\cosh cx}{(t^2-x^2)^{\frac{1}{2}}}dx=\mathrm{I}_0(ct). \tag{265}$$

Now, convert back from t to t_1 or q. Then

$$\int_0^\infty q\mathrm{K}_0(qx)\,.\,\cosh cx\,.\,dx = \frac{q}{(q^2-c^2)^{\frac{1}{2}}}. \tag{266}$$

Lastly, confirm this by (203).

(61). Sometimes the generalised ϵ^{-x} formula comes in to obviate an apparent failure in an integrational process. Thus, if $s^2=4r$, $x^2=y$,

$$\int_0^\infty \mathrm{J}_0(sx)\sin tx\,dx = \int \epsilon^{-y/r_1}\sin t\sqrt{y}\,\frac{dy}{2\sqrt{y}}. \tag{267}$$

Expand in powers of t and evaluate by Euler's integral. The result is

$$\frac{tr_1}{2}\left(1-\frac{t^2r_1}{\lfloor 3}+\frac{t^4r_1^{\,2}\lfloor 2}{\lfloor 5}-\dots\right)=\frac{\sqrt{\pi r_1}}{2}\left\{\frac{(\frac{1}{4}t^2r_1)^{\frac{1}{2}}}{\lfloor\frac{1}{2}}-\frac{(\frac{1}{4}t^2r_1)^{1\frac{1}{2}}}{\lfloor 1\frac{1}{2}}+\dots\right\}. \tag{268}$$

Being in rising integral powers of r_1, how can we turn it to a function of r? Observe that the result represents a portion of the $\epsilon^{-x}\cos\frac{1}{2}\pi$ expansion, vol. 2, p. 466. We can therefore substitute the negative of the other portion, since the total is zero. This brings us to

$$\frac{\sqrt{\pi r_1}}{2}\left\{\frac{(\frac{1}{4}t^2r_1)^{-\frac{1}{2}}}{\lfloor-\frac{1}{2}}-\frac{(\,.\,)^{-1\frac{1}{2}}}{\lfloor-1\frac{1}{2}}+\dots\right\}$$

$$=\frac{1}{t}\left\{1+\frac{2r}{t^2}+\frac{1.3}{\lfloor 2}\left(\frac{2r}{t}\right)^2+\frac{1.3.5}{\lfloor 3}\left(\frac{2r}{t}\right)^3+\dots\right\}=\frac{1}{(t^2-s^2)^{\frac{1}{2}}}; \tag{269}$$

the known result, though zero when $t<s$. So we have a batch of connected results.

$$\int_0^\infty \mathrm{J}_0(sx)\sin tx\,dx=\frac{1}{(t^2-s^2)^{\frac{1}{2}}}=\mathrm{I}_0(t_1s)\frac{1}{t}=\frac{\pi}{2}t_1\mathrm{K}_0(t_1s) \tag{270}$$

$$=\int_0^\infty\frac{\sin tx\,dx}{(1+x^2/s_1^{\,2})^{\frac{1}{2}}}=\int_0^\infty\frac{\mathrm{J}_0(sx).x/t_1}{1+x^2/t_1^{\,2}}dx \tag{271}$$

$$=\tfrac{1}{2}\int_0^\infty \mathrm{K}_0(t_1s)\cos xt\,dx=\int_0^\infty \mathrm{G}_0(xs)\cos xt\,dx; \tag{272}$$

most of which turn up in the theory of cylindrical elastic waves. The way of deriving these should be clear enough to one who has studied parts of the preceding of a similar nature.

On the other hand, if we turn sin to cos in (267), working the same way, the result corresponding to (268) is $\frac{1}{2}\sqrt{\pi r_1}\,\epsilon^{-t^2r_1/4}$, which does not present any difficulty in turning to a function of r. We get

$$\int_0^\infty \mathrm{J}_0(sx)\cos tx\,dx=\frac{1}{(s^2-t^2)^{\frac{1}{2}}}, \tag{273}$$

and zero if $s<t$. Some of the related integrals are

$$=\mathrm{I}_0(s_1t)\frac{1}{s}=\frac{\pi}{2}s_1\mathrm{K}_0(s_1t)=\int_0^\infty\frac{\cos tx\,dx}{(1+x^2/s_1^{\,2})^{\frac{1}{2}}}=\int_0^\infty\frac{\mathrm{J}_0(sx)}{1+x^2/t_1^{\,2}}dx. \tag{274}$$

(62). If T and C are both positive,

$$\frac{2}{\pi}\int_0^{\sqrt{T}} \frac{x\,dx}{(T-x^2)^{\frac{1}{2}}(C+x^2)_{\frac{1}{2}}} = \frac{1}{\pi}\int_0^{\infty} \epsilon^{-x^2(T_1-C_1)}\, \pi\sqrt{T_1C_1}\, dx^2 = \frac{\sqrt{T_1C_1}}{T_1-C_1} \quad (275)$$

$$\frac{2}{\pi}\left(1-\frac{1}{3}\frac{T}{C}+\frac{1}{5}\frac{T^2}{C^2}-\dots\right)\left(\frac{T}{C}\right)^{\frac{1}{2}} = \frac{2}{\pi}\tan^{-1}\left(\frac{T}{C}\right)^{\frac{1}{2}}. \quad (276)$$

This applies when $T<C$, or $T_1>C_1$. For, if $T_1<C_1$, Euler's integral does not apply. But since $\tan^{-1}x+\tan^{-1}x^{-1}=\frac{1}{2}\pi$, the other form is evidently $1-(2/\pi)\tan^{-1}(C/T)$. Here we may note that the use of Euler's integral, as if it were valid when $T_1<C_1$, leads to the second form, all except the additional term 1, which is very curious. Moreover, note that the two $\tan^{-1}$ solutions are not only trigonometrically equivalent, but are also equivalent in the series form, when the argument is real positive.

Another way is to expand $(C+x^2)^{-\frac{1}{2}}$ in rising powers of x^2 by the binomial theorem, and use only $\epsilon^{-x^2T_1}$ in combination with it. The result will be (276), without raising the question of the relative size of T_1 and C_1. The binomial expansion is valid irrespective of its convergency because C is positive.

Now put $T=t^2$, $C=1/c_1^2$. Then, by the above,

$$(A)=\frac{2}{\pi}\int_0^t \frac{x\,dx}{(t^2-x^2)^{\frac{1}{2}}(x^2+1/c_1^2)^{\frac{1}{2}}} = (A_1) = \frac{2}{\pi}\tan^{-1}c_1t$$

$$= (A_2) = 1-\frac{2}{\pi}\tan^{-1}\frac{1}{c_1t}. \quad (277)$$

Turn to functions of t_1. Then we have

$$(B)=\int_0^{\infty}\frac{t_1K_0(t_1x)x\,dx}{(x^2+1/c_1^2)^{\frac{1}{2}}} = (B_1) = \frac{2}{\pi}\left\{\frac{c_1}{t_1} - \lfloor 2\,\frac{c_1^3}{t_1^3} + \lfloor 4\,\frac{c_1^5}{t_1^5} - \dots\right\}. \quad (278)$$

Here (B) is got from (A) by the usual process $1/c_1^n = c^n/\lfloor n$, which is multiplex integration, or multegration, or gammatic integration, or gamtegration, or whatever else it may be termed. Also (B_1) comes direct from (A_1) by the same process. That it is reversed does not matter. And (B_1) also comes direct from (B) by (203) above, when the binomial theorem is applied to the denominator. But (B_1), which is the inverted cosine series multiplied by $(2/\pi)(c_1/t_1)$, being divergent, necessitates another form, say (B_2) to meet the case of large values of c_1/t_1. Here is the transformation.

$$(B_1)=\frac{2}{\pi}\left\{\frac{c_1}{t_1}g'(-1)-\frac{c_1^3}{t_1^3}g'(-3)+\dots\right\}$$

$$=\frac{2}{\pi}\frac{d}{dn_0}\left\{\frac{(t_1/c_1)^{n-1}}{\lfloor n-1}-\frac{(t_1/c_1)^{n-3}}{\lfloor n-3}+\dots\right\} \quad (279)$$

$$=\frac{2}{\pi}\frac{d}{dn_0}\left\{-\sin\left(\frac{t_1}{c_1}-\frac{n\pi}{2}\right)+\frac{(t_1/c_1)^{n+1}}{\lfloor n+1}-\frac{(t_1/c_1)^{n+3}}{\lfloor n+3}+\dots\right\} \quad (280)$$

$$=(B_2)=\cos\frac{t_1}{c_1}+\frac{2}{\pi}\left[\frac{t_1}{c_1}\left(\log\frac{t_1}{c_1}+G_1\right)-\frac{(t_1/c_1)^3}{\lfloor 3}\left(\log\frac{t_1}{c_1}+G_3\right)+\dots\right]. \quad (281)$$

Here we have the convergent function of which (B_1) is the divergent equivalent. The transformation from (279) to (280) is done by the generalised sine series, eq. (20), p. 216, vol. 3, or (17), p. 198. (B_1) and (B_2) are only equivalent when c_1/t_1 is positive.

Now go back to (B), and turn it to a function of c. It makes, by multegration,

$$(C)=\int_0^\infty J_0\left(\frac{c}{x}\right)K_0(t_1x)t_1dx=(C_2)=1-\frac{t_1^2c^2}{\underline{|2}\,\underline{|2}}+\frac{t_1^4c^4}{\underline{|4}\,\underline{|4}}-\dots$$
$$+\frac{2}{\pi}\left\{\frac{t_1c}{\underline{|1}\,\underline{|1}}(\log t_1c+2G_1)-\frac{t_1^3c^3}{\underline{|3}\,\underline{|3}}(\log t_1c+2G_3)+\dots\right\}. \quad (282)$$

Here (C) comes from (B) and (C_2) from (B_2). The working formula for the log is

$$\frac{1}{c_1^m}\log\frac{1}{c_1}=\frac{c^m}{\underline{|m}}(\log c+G(m)), \quad (283)$$

as on p. 220, eq. (40), vol. 3, slightly varied. This is done by substituting $(d/dn)c_1^n$ for $\log c_1$, then turning $c_1^{-m}(d/dn)c_1^{-n}$ to a function of c, and finally putting $n=0$. But its use may be avoided altogether by first turning (280) to a function of c; then, on carrying out d/dn_0, the result (C_2) follows.

Lastly, turn (C) to a function of t. We get

$$(D)=\frac{2}{\pi}\int_0^t\frac{J_0(c/x)}{(t^2-x^2)^{\frac{1}{2}}}dx=(D_2)=1-\frac{2}{\pi}\left\{\frac{c}{t}-\frac{1}{3\,\underline{|3}}\frac{c^3}{t^3}+\frac{1}{5\,\underline{|5}}\frac{c^5}{t^5}-\dots\right\}. \quad (284)$$

Here (D_2) comes from (C_2) by the use of (283). And finally, to finish the round, (D_2) may be directly converted to (A_2), or conversely.

Now some remarks upon the above. The integral (C) is obviously convergent, but if we expand $J_0(c/x)$ in the usual series $1-c^2/4x^2+\dots$, and then attempt to apply (203), with negative integral values of m, we meet the difficulty that these integrals are all infinite, except for the first term $(m=0)$. Nevertheless, the use of the formula (203), with $m=-2, -4, -6$, &c., does actually lead to the part of (C_2) in the first line of (282). So there is something to be got from the invalid use of infinite integrals, for examination, suggestion, and supplementation. This may be compared with (275) above, when $T_1<C_1$. As already noticed, a part of the proper result comes out right, although the integral is infinite.

Next, as regards (B). If we put $t_1=si$, and suppose s real positive, we must use the convergent formula (B_2) to obtain fully equivalent results. Thus

$$(E)=\int_0^\infty\frac{sxG_0(sx)dx}{(x^2+1/c_1^2)^{\frac{1}{2}}}$$
$$=(E_1)=\frac{2}{\pi}\left\{\frac{s}{c_1}\left(\log\frac{s}{c_1}+G_1\right)+\frac{s^3}{\underline{|3}\,c_1^3}\left(\log\frac{s}{c_1}+G_3\right)+\dots\right\}, \quad (285)$$

$$(F)=\int\frac{sxJ_0(sx)dx}{(x^2+1/c_1^2)^{\frac{1}{2}}}=\epsilon^{-s/c_1}. \quad (286)$$

But if we use $t_1=si$ in the divergent form (B_1), we get

$$(E)=\int_0^\infty \frac{G_0(sx)sx\,dx}{(x^2+1/c_1^2)^{\frac{1}{2}}}=(E_2)=-\frac{2}{\pi}\left\{\frac{c_1}{s}+\lfloor 2\left(\frac{c_1}{s}\right)^3+\lfloor 4\left(\frac{c_1}{s}\right)^5+\dots\right\}, \tag{287}$$

whilst the companion formula, corresponding to (F) is missing. Nevertheless, the two formulæ (E_1) and (E_2) are equivalent, remembering now that c_1/s is real positive. For

$$(E_2)=-\frac{2}{\pi}\left\{\frac{c_1}{s}g'(-1)+\left(\frac{c_1}{s}\right)^3 g'(-3)+\dots\right\} \tag{288}$$

$$=\frac{2}{\pi}\frac{d}{dn_0}\left\{\frac{\epsilon^{s/c_1}-\epsilon^{-s/c_1}\cos n\pi}{2}-\frac{(s/c_1)^{n-1}}{\lfloor n-1}-\frac{(s/c_1)^{n-3}}{\lfloor n-3}-\dots\right\} \tag{289}$$

$$=\frac{2}{\pi}\frac{d}{dn_0}\left\{\frac{(s/c_1)^{n+1}}{\lfloor n+1}+\frac{(s/c_1)^{n+3}}{\lfloor n+3}+\dots\right\}=(E_1). \tag{290}$$

Here the transition from (288) to (290) is done by means of my *two* generalised formulæ for ϵ^x and $\epsilon^{-x}\cos n\pi$ (vol. 2, p. 439, eq. (24), and p. 466, eq. (140)); that is,

$$\epsilon^x=\dots+\frac{x^{n-1}}{\lfloor n-1}+\frac{x^n}{\lfloor n}+\frac{x^{n+1}}{\lfloor n+1}+\dots, \tag{291}$$

$$\epsilon^{-x}\cos n\pi=\dots-\frac{x^{n-1}}{\lfloor n-1}+\frac{x^n}{\lfloor n}-\frac{x^{n+1}}{\lfloor n+1}+\dots; \tag{292}$$

making $$\frac{\epsilon^x-\epsilon^{-x}\cos n\pi}{2}=\dots+\frac{x^{n-1}}{\lfloor n-1}+\frac{x^{n+1}}{\lfloor n+1}+\dots, \tag{293}$$

where x must be positive. It is remarkable that on differentiation to n_0, the exponential functions go out altogether. Contrast with the generalised formula involved in (279), (280). On differentiation to n_0 the circular function (corresponding to the exponentials) does not go out.

It will be interesting to see how (E) works out when done by Weber's integral (227). That is, by the binomial theorem, and then Weber, we come to

$$\int_0^\infty \frac{sxJ_n(sx)dx}{(x^2+1/c_1^2)^{\frac{1}{2}}}=c_1 s\left(\frac{2}{s^2}\frac{\lfloor\frac{1}{2}n}{\lfloor\frac{1}{2}n-1}-\tfrac{1}{2}c_1^2\frac{2^3}{s^4}\frac{\lfloor\frac{1}{2}n+1}{\lfloor\frac{1}{2}n-2}+\dots\right). \tag{294}$$

On this perform the operation $-(2/\pi)d/dn_0$. The result is the divergent form (E_2) above, and is quite correct, though it needs to be supplemented by the convergent (E_1) for calculations throughout the entire range of the argument.

The auxiliary integral (F) leads to

$$\int J_0(c/x)J_0(sx)s\,dx=J_0(2\sqrt{sc}), \tag{295}$$

and to various others. But that is of no consequence. The present object has been to show explicitly the application of my generalised formulæ to definite integrals, besides their multegrational transformation. In the following is a further step.

(63). In vol. 2, p. 452 and p. 473, I obtained incidentally the convergent functions equivalent to the inverted ϵ^x and ϵ^{-x} functions, that is, inverted term by term. There is an application of one of these on p. 37, vol. 3. Also, the inverted sine and cosine functions, and shin and cosh, have been considered in the present article. I have now to convert the inverted I_0 function to a convergent form. It turned up in eq. (214) above as a part of the solution of a finite definite integral, the other part being known. We have to find the value of

$$Y=\frac{1}{y}+\frac{\lfloor 1 \lfloor 1}{y^2}+\frac{\lfloor 2 \lfloor 2}{y^3}+\frac{\lfloor 3 \lfloor 3}{y^4}+\dots. \qquad (296)$$

The curve of Y is easily drawn for values of y from ∞ down to about 4, and then roughly from 4 down to 1, where the initial convergency ceases. From $y=1$ down to $y=0$ the formula gives no information at all. Y increases from 0 at ∞ up to about 1 at $y=1$, but we have no means of telling what happens after that.

If $y=1$, $Y=1+1+\dots$ This counts roughly for 1. If $y=2$, $Y=\frac{1}{2}(1+\frac{1}{2}+1+\dots)$. This counts for ·75 roughly, but more closely than the last. If $y=4$, $Y=\frac{1}{4}(1+\frac{1}{4}+\frac{1}{4}+\dots)$. This counts for ·3125, more closely. If $y=9$, $Y=\frac{1}{9}(1+\frac{1}{9}(1+\frac{4}{9}(1+1+\dots$ This counts for ·1288 still more closely. After that, the error gets smaller and smaller to any extent. See remarks on p. 255, about manner of calculation when only a small number of terms is counted.

Now Y is the same as

$$Y=\frac{(g'_{-1})^2}{y}+\frac{(g'_{-2})^2}{y^2}+\frac{(g'_{-3})^2}{y^3}+\dots. \qquad (297)$$

This series is not complete, because the similar terms for y^n with n positive, are finite. Now the complete series can be converted to convergent form.

Use eq. (160), vol. 2, p. 469, or

$$\tfrac{1}{2}H_n-\tfrac{1}{2}K_n\sin 2\pi(r+\tfrac{1}{2}n)=\Sigma\frac{y^{\frac{1}{2}n+r}}{\lfloor r \lfloor n+r}. \qquad (298)$$

Here H_n and K_n stand for $H_n(x)$ and $K_n(x)$, with $y=\frac{1}{4}x^2$, being the divergent functions as defined in vol. 2, p. 468, or p 240. In another form,

$$\tfrac{1}{2}H_s-\tfrac{1}{2}K_s\sin t\pi=\dots+\frac{y^{\frac{1}{2}t}}{\lfloor \frac{1}{2}(t+s) \lfloor \frac{1}{2}(t-s)}+\frac{y^{\frac{1}{2}t+1}}{\lfloor \frac{1}{2}(t+s)+1 \lfloor \frac{1}{2}(t-s)+1}+\dots, \qquad (299)$$

the series to be complete both ways. Instead of H_s and K_s we may use their convergent equivalents

$$H_s=\tfrac{1}{2}(I_s+I_{-s}), \qquad K_s=\frac{I_{-s}-I_s}{\sin s\pi}, \qquad (300)$$

as in vol. 2, p. 246. These make

$$\tfrac{1}{2}I_s\left(1+\frac{\sin t\pi}{\sin s\pi}\right)+\tfrac{1}{2}I_{-s}\left(1-\frac{\sin t\pi}{\sin s\pi}\right)=\text{series in (299).} \qquad (301)$$

Now differentiate (299) to t, and then put $t=s$. The result is

$$-\frac{\pi}{2}K_s\cos s\pi=\ldots+\frac{\frac{1}{2}y^{\frac{1}{2}s}}{\underline{|s}\,\underline{|0}}(\lambda+G_s+G_0)+\frac{\frac{1}{2}y^{\frac{1}{2}s+1}}{\underline{|s+1}\,\underline{|1}}(\lambda+G_{s+1}+G_1)+\ldots, \qquad (302)$$

where $\lambda=\log y$. If we perform the same operations upon (301), the result is the same as (302), provided K_s is interpreted as in (300).

Now differentiate (302) to s and then put $s=0$. Then, since $dK_s/ds_0=0$, we get

$$0=\ldots+\frac{\frac{1}{2}y^0}{\underline{|0}\,\underline{|0}}\left((\tfrac{1}{2}\lambda+G_0)(\lambda+2G_0)+G_0'\right)+\ldots. \qquad (303)$$

This result contains the evaluation of $\Sigma\, G_n'y^n/\underline{|n}\,\underline{|n}$ in convergent form. To corroborate, work another way. Differentiate (301) to s and then put $s=t$. We get

$$\frac{dI_t}{dt}+\frac{\pi}{2}\cos t\pi\,.\,K_t=\ldots+\frac{\frac{1}{2}y^{\frac{1}{2}t}}{\underline{|0}\,\underline{|t}}(G_t-G_0)+\ldots. \qquad (304)$$

Differentiate this next to t and then put $t=0$. We get

$$\frac{d^2I_t}{dt_0^2}=\tfrac{1}{2}\sum_{-\infty}^{\infty}\frac{G_n'y^n}{\underline{|n}\,\underline{|n}}, \qquad (305)$$

where, by using the convergent I_t formula,

$$\frac{d^2I_t}{dt_0^2}=\frac{y^0}{\underline{|0}\,\underline{|0}}\left((\tfrac{1}{2}\lambda+G_0)^2+G_0'\right)+\ldots. \qquad (306)$$

By combining (305), (306), the result is (303) again. It is to be observed that whereas I_t is a function of t, H_t is a function of t^2; so dH_t/dt_0 is zero, but not dI_t/dt_0. This does not prevent

$$\frac{d^2I_t}{dt_0^2}=\tfrac{1}{2}\frac{d^2H_t}{dt_0^2} \qquad (307)$$

from being true. So we have three forms which are equivalent, exhibited by (305), (306), (307).

As regards the G_n' function, we have $-G_0'=\pi^2/6=1\cdot6449339$, and the rest follow by

$$-\frac{1}{n^2}+G_n'=G'_{n-1}, \qquad (308)$$

making

$$G_1'=G_0'+1,\quad G_2'=G_0'+1+\tfrac{1}{4},\quad G_3'=G_0'+1+\tfrac{1}{4}+\tfrac{1}{9}, \qquad (309)$$

and so on; whilst, since

$$\frac{G'(n)}{\underline{|n}\,\underline{|n}}=g'(n)g''(n)-g'(n)g'(n), \qquad (310)$$

it follows that when the values of n are negative integers, then

$$\frac{G'(n)}{\underline{|n}\,\underline{|n}}=-g'(n)g'(n),\quad\text{or}\quad G_n'=-G_n^2. \qquad (310\text{A})$$

We have, therefore, A = B = C, where

$$\mathrm{A}=\tfrac{1}{2}\left\{\frac{\mathrm{G}_0'}{|\underline{0}\,|\underline{0}}+\frac{\mathrm{G}_1'}{|\underline{1}\,|\underline{1}}y+\frac{\mathrm{G}_2'}{|\underline{2}\,|\underline{2}}y^2+\ldots-\mathrm{Y}\right\}=\tfrac{1}{2}(\mathrm{X}-\mathrm{Y}), \tag{311}$$

$$\tfrac{1}{2}\frac{d^2\mathrm{H}_t}{dt_0^2}=\mathrm{B}=-\frac{1}{x}\frac{\epsilon^x}{(2\pi x)^{\frac{1}{2}}}\left\{1+\frac{10}{|\underline{2}\,.\,8x}+\frac{259}{|\underline{3}(8x)^2}+\frac{12916}{|\underline{4}(8x)^3}+\ldots\right\} \tag{312}$$

$$\frac{d^2\mathrm{I}_t}{dt_0^2}=\mathrm{C}=\tfrac{1}{2}\lambda\left\{-\frac{\pi}{2}\mathrm{K}_0+\frac{\mathrm{G}_0}{|\underline{0}\,|\underline{0}}+\frac{\mathrm{G}_1}{|\underline{1}\,|\underline{1}}y+\frac{\mathrm{G}_2}{|\underline{2}\,|\underline{2}}y^2+\ldots\right\}$$
$$+\left(\frac{\mathrm{G}_0'}{|\underline{0}\,|\underline{0}}+\frac{\mathrm{G}_1'}{|\underline{1}\,|\underline{1}}y+\frac{\mathrm{G}_2'}{|\underline{2}\,|\underline{2}}y^2+\ldots\right)+(g_0'^2+g_1'^2y+g_2'^2y^2+\ldots), \tag{313}$$

where K_0 may be either of the divergent or the convergent form. So C is wholly convergent, B is divergent, and A is mixed. Y is known in terms of X and C or of X and B.

As these formulæ are elaborate, I have found it desirable to undertake the severe labour of numerically testing the equivalence of A, B, and C in the overlapping region. That is, for values of y large enough to admit of some reasonably close estimate of the value of the divergent Y by itself, and yet not too large for accurate calculation of the convergent C in a reasonable time. For it is the convergent calculations, of X and C, especially of C, that are so lengthy and troublesome, even when y is as low as 4. The work must be done at length, to at least six places of decimals, and then throwing away the end figures. (Four figure logarithms led to large final errors.) I shall not give details of this work, but only quote the results in one case. Say $x=y=4$. Then

$$\mathrm{A}=-3{\cdot}367, \qquad \mathrm{B}=-3{\cdot}368, \qquad \mathrm{C}=-3{\cdot}3623.$$

B is easily done, eight terms, and the ratio of the l.c.t. to the total is 1/353. Two terms only are used in Y, and the ratio of the l.c.t. to the total A is 1/82. That A and B agree to 1/3000 part is no doubt accidental, though assisted by the equality of l.c.t.'s in the Y series. To bring A into full agreement with C, it is necessary to count, not the whole, but only 5/6 of the l.c.t. That is, the outstanding error is 1/5 of the l.c.t., which, considering that only two terms are concerned, cannot be considered bad.

Some of the data.

$$\mathrm{Y}=+{\cdot}3125, \qquad \mathrm{B}=-2{\cdot}72266\times1{\cdot}23614,$$

$$\mathrm{C}=(1{\cdot}847578\times11{\cdot}301921-28{\cdot}737392)\times{\cdot}693147$$
$$-6{\cdot}421530+8{\cdot}504754,$$

$$\mathrm{I}_0=11{\cdot}301921, \qquad \tfrac{1}{2}\lambda={\cdot}693147, \qquad 2\mathrm{G}_0+\tfrac{1}{2}\lambda=1{\cdot}847578,$$

$$\mathrm{X}=-6{\cdot}421530, \qquad \sum_0^\infty g_n'^2y^n=8{\cdot}504754.$$

Also, $\mathrm{K}_0={\cdot}007102$, though it was not explicitly used.

Going back to the integral (214). It is finite, and the convergent meaning of the result has been found. But if we take only the even powers of x in the integral, or only odd powers, making two integrals, each with a divergent result, then an entirely different state of things is created. Each of the integrals is, by the manner of construction and term by term evaluation, infinite, and so are the results, because the evaluation by initial convergence has no application. Taken together, however, we have

$$\int_0^\infty \frac{q\mathrm{K}_0'(qx)}{1+x/p}dx = \frac{pq\pi}{2}\left\{\mathrm{I}_0(pq) - \frac{2}{\pi}\left(pq + \frac{p^3q^3}{1^2 3^2} + \dots\right)\right\}$$
$$+\frac{p}{\pi}\left\{\frac{d^2}{dn_0^2}\mathrm{I}_n(pq) - \tfrac{1}{2}\sum_0^\infty \frac{\mathrm{G}_n'}{\lfloor n \lfloor n}\left(\frac{pq}{2}\right)^{2n}\right\}, \qquad (314)$$

convergently, provided no mistakes are involved. This means that I think the principle of the evaluation is correct, even though the execution may be faulty. The principle is that if the integral is finite, then its divergent evaluation has an equivalent convergent form, when the successive terms in the integral are correctly evaluated.

(64). In this connection the integral and solution (215) is of much interest. If we put $a^2/q^2 = c$, and compare it with the form of solution (216), we see that

$$\frac{\frac{1}{2}\pi}{(1-c)^{\frac{1}{2}}} - \sqrt{c}\left(1 + \tfrac{2}{3}c + \frac{2.4}{3.5}c^2 + \dots\right)$$
$$= \frac{1}{\sqrt{c}}\left\{1 + \tfrac{1}{3}\left(1-\frac{1}{c}\right) + \tfrac{1}{5}\left(1-\frac{1}{c}\right)^2 + \right\}, \qquad (315)$$

provided $c<1$. It is, in fact, then a pure identity. And, since the right member represents $\{\tan^{-1}(1/c-1)^{\frac{1}{2}}\}(1-c)^{-\frac{1}{2}}$, it follows that

$$\sqrt{c}\sqrt{1-c}\left(1 + \tfrac{2}{3}c + \frac{2.4}{3.5}c^2 + \dots\right) = \tan^{-1}\left(\frac{c}{1-c}\right)^{\frac{1}{2}}, \qquad (316)$$

which is, when expanded in rising powers of c, after multiplication by $(1-c)^{\frac{1}{2}}$, an identity for every power of c.

Now the right member of (315) shows continuity of calculation when c passes through 1 to be >1, up to $c=\infty$. We may infer that the left member will also show continuity, in spite of the fact that the two series concerned each pass through ∞ into divergent regions. What happens here is that the point of convergence (in each series), which is at the end of the series when $c<1$, moves towards the beginning when c is made to pass beyond 1. We have now to find whether the estimation of the value of the two series, by their initial convergence, will lead to the same value, by their difference, as the convergent series on the right side.

This stands numerical test very well. Details are too long for insertion; I give some leading results. First show that

$$1 + \tfrac{1}{2}c + \frac{1.3}{2.4}c^2 + \dots = \frac{1}{(1-c)^{\frac{1}{2}}} = \frac{\cdot 953}{(c-1)^{\frac{1}{2}}} \text{ when } c>1. \qquad (317)$$

This is experimental. It is not proved that the ratio of the divergent $(1-x)^{-\frac{1}{2}}$ to $(x-1)^{-\frac{1}{2}}$ is $\cdot953$ for all values of $x>1$, but it is proved that the ratio has this value when x is a little over 1, and that it shows no sensible variation as x is increased. The initial convergency ceases when $x=2$. If the ratio is a function of x, it does not vary much between these limits.

Using this formula, apply it to (315), calculating the second divergent series separately. Then, with $c=1\cdot05$, we get

$$\left(\frac{11}{7}\times 4\cdot4721\times\cdot953\right)-(1\cdot0247\times 5\cdot5695)=6\cdot6972-5\cdot7068=\cdot9904.$$

The true value, by the convergent series, is ·9918. Again $c=1\cdot1$,

$$\left(\frac{11}{7}\times\frac{\cdot953}{\cdot3162}\right)-(1\cdot0488\times 3\cdot5821)=4\cdot7361-3\cdot7568=\cdot9793.$$

The true value is ·9840. Again, $c=1\cdot25$,

$$\left(\frac{11}{7}\times\cdot953\times 2\right)-(1\cdot1180\times 1\cdot8333)=2\cdot9951-2\cdot0495=\cdot9456.$$

The true value is ·962.

The above divergent calculations were made to four figures all through, so that the last figure is useless. There is practical identity in the first case, an error of ½ per cent. in the second, and an error of 1·7 per cent. in the third. Now the ratio of the l.c.t. to the total in the series summing up to 5·5695 was no less than 1/14. How is it the final errors are only from 0 to 1·7 per cent.? The accuracy seems impossible. The answer is, first, the close estimate of the number ·953; next, the choice of x to make two equal bottom terms; thirdly, the slow variation in size of successive terms. For example, the 5·5695 is made up as in the sum at the side of the page. If we draw the curve, as described on p. 484, vol. 2, it will be seen to be nice and smooth in the lower part, so that a close estimate of the minimum point can be made. The last term counted is followed by another of the same size, and then by slowly increasing terms.

1
·7
·588
·5292
·4939
·4714
·4567
·4476
·4423
·4400
5·5695

I do not think there can be any reasonable doubt that the difference of the two divergent series involves continuity of calculation in the way specified, right past the infinite values into the region of direct divergency. It is a fact that the calculations work right, whatever the explanation may be.

It was not necessary to use the equivalence (317), real or apparent. The calculated values of $(1-x)^{-\frac{1}{2}}$ would lead to the same results, without relation to $(x-1)^{-\frac{1}{2}}$. But I had already worked out that relation, and therefore utilised it. An account will follow, D.V.

By rearranging the terms in the two divergent series, their difference becomes a single alternatingly divergent series. The value may be calculated this way also. But there are plenty of directly divergent series which stand alone, and represent by their initial convergence equivalent forms of convergent series. The marked utility of these, as well as of some of the alternatingly divergent ones, is, in physical mathematics, to take the place of convergent solutions which, though theoretically perfect, are practically not amenable to calculation save within a certain range. Conversely, a divergent solution, though easily calculable through a wide range, requires to be supplemented by another form in the remainder of the range. Generally speaking, two forms of solution at least should always be searched for. Mathematicians in general are, I find, exceedingly conservative and prejudiced. Nevertheless, I am confident of a great future for the practical use of divergent series, as well as for the generalized analysis which connects them with the convergent ones, because both these matters are concerned in the operational treatment of physical differential equations.

(65). Two more integrals remain to be considered, in connection with the above. The function of the second complete elliptic integral is related to the first by a single integration. Thus,

$$v=\frac{2}{\pi}\int\frac{d\theta}{(r-s\sin^2\theta)^{\frac{1}{2}}}=\frac{2}{\pi}\int\frac{dx}{(r-x^2)^{\frac{1}{2}}(s-x^2)^{\frac{1}{2}}}=\pi^{\frac{1}{2}}\left(\frac{r_1s_1}{r_1+s_1}\right)^{\frac{1}{2}} \tag{318}$$

summarizes the first complete elliptic integral, and

$$w=\frac{2}{\pi}\int(r-s\sin^2\theta)^{\frac{1}{2}}d\theta=\frac{2}{\pi}\int\left(\frac{r-x^2}{s-x^2}\right)^{\frac{1}{2}}dx=\frac{\pi^{\frac{1}{2}}}{2r_1}\left(\frac{r_1s_1}{r_1+s_1}\right)^{\frac{1}{2}} \tag{319}$$

summarizes the second one. The integration $1/2r_1$ finds the second function from the first. The first one having two forms, so therefore has the second. They are

$$w_1=\frac{1}{2r_1}\frac{\mathrm{F}(r/s)}{\sqrt{s}}=\frac{r}{2\sqrt{s}}\left(1+\tfrac{1}{2}\frac{1^2}{2^2}\frac{r}{s}+\tfrac{1}{3}\frac{1^23^2}{2^24^2}\frac{r^2}{s^2}+\tfrac{1}{4}\frac{1^23^25^2}{2^24^26^2}\frac{r^3}{s^3}+\cdots\right), \tag{320}$$

$$w_2=\frac{1}{2r_1}\frac{\mathrm{F}(s/r)}{\sqrt{r}}=\sqrt{r}\left(1-\frac{1^2}{2^2}\frac{s}{r}-\tfrac{1}{3}\frac{1^23^2}{2^24^2}\frac{s^2}{r^2}-\tfrac{1}{5}\frac{1^23^25^2}{2^24^26^2}\frac{s^3}{r^3}-\cdots\right). \tag{321}$$

There is continuity from w_1 to w_2. Suppose s is constant. Then $2w_1$ is the area bounded by the v curve between $r=0$ and any value of r not exceeding s, whilst $2w_2$ is the area between the same origin and any value of r not less than s. The common value of w_1 and w_2 when $r=s$ is $2r/\pi$. In the integrals containing x, x ranges from 0 to the smaller of $r^{\frac{1}{2}}$ and $s^{\frac{1}{2}}$. In the trigonometrical forms the limits are those of reality from $\theta=0$ upwards when $r<s$, but when the upper limit reaches $\frac{1}{2}\pi$ it stays there.

The more rapidly converging solution for w corresponding to (138), p. 256, is

$$w_2 = \frac{\pi^{\frac{1}{2}}}{2} \epsilon^{-\frac{1}{2}r_1 s} \frac{s_1 r_1^{-\frac{1}{2}}}{(s_1^2 - \frac{1}{4}r_1^2)^{\frac{1}{2}}} \tag{322}$$

$$= (r - \tfrac{1}{2}s)^{\frac{1}{2}} \left\{ 1 - c^2 - \frac{3.5}{\lfloor 2 \lfloor 2} c^4 - \frac{3.5.7.9}{\lfloor 3 \lfloor 3} c^6 - \dots \right\} \tag{323}$$

where $c = s/4(2r - s)$, and $s < r$. The function v on p. 256 must, however, be multiplied by $\pi^{\frac{1}{2}}$ to meet the present case.

Only the second form of w, that is w_2, is concerned in the length of an elliptic quadrant, for if a and b are the semi-axes, b being the smaller, the length is

$$(1 - b^2/a^2)^{\frac{1}{2}} \int_0^a \left\{ \frac{a^2(1 - b^2/a^2)^{-1} - x^2}{a^2 - x^2} \right\}^{\frac{1}{2}} dx$$

$$= \tfrac{1}{2} a\pi \left\{ 1 - \frac{1^2}{2^2}\left(1 - \frac{b^2}{a^2}\right) - \frac{1}{3}\frac{1^2 3^2}{2^2 4^2}\left(\cdot \right)^2 - \right\}, \tag{324}$$

so we have $r = a^2(1 - b^2/a^2)^{-1}$ and $s = a^2$, making $r > s$. But there is no need to confine b to be $< a$. The series remains convergent until $b^2 = 2a^2$. After that, it is alternatingly divergent. The integral (324) may be written

$$(b^2/a^2 - 1)_{\frac{1}{2}} \int_0^a \left\{ \frac{a^2(b^2/a^2 - 1)^{-1} + x^2}{a^2 - x^2} \right\}^{\frac{1}{2}} dx. \tag{325}$$

Here, if $a^2(b^2/a^2 - 1)^{-1} > a^2$, the solution is just the same as (324), but if $< a^2$, we require a fully convergent form in addition. The following shows how to get it.

$$\frac{2}{\pi}\int_0^a \left(\frac{r + x^2}{s - x^2}\right)^{\frac{1}{2}} dx = \frac{\sqrt{\pi}}{2r_1}\left(\frac{r_1 s_1}{s_1 - r_1}\right)^{\frac{1}{2}} = \frac{\sqrt{\pi}}{2r_1}\epsilon^{sr_1}\left(\frac{r_1 s_1}{r_1 + s_1}\right)^{\frac{1}{2}}. \tag{326}$$

Applying this to (325), the result will be found to differ from (324) only in the interchange of a and b. It is obvious that the length of the elliptic quadrant is the same whichever way it is measured. But we get more than that, viz.,

$$\tfrac{1}{2}\pi a \left\{ 1 - \frac{1^2}{2^2}\left(1 - \frac{b^2}{a^2}\right) - \dots \right\} = \tfrac{1}{2}\pi b \left\{ 1 - \frac{1^2}{2^2}\left(1 - \frac{a^2}{b^2}\right) - \dots \right\}, \tag{327}$$

for any values of a and b. The equivalence is remarkable in this way. Either side may be alternatingly divergent, the other being convergent; but there is a region, from $a^2 = 2b^2$ to $b^2 = 2a^2$, in which both sides are convergent at once.

Taking $a = 1$, and $1 - b^2 = x$, and defining E by

$$\mathrm{E}(x) = 1 - \frac{1^2}{2^2}x - \frac{1}{3}\frac{1^2 3^2}{2^2 4^2}x^2 - \frac{1}{5}\frac{1^2 3^2 5^2}{2^2 4^2 6^2}x^3 - \dots, \tag{328}$$

then (327) expresses that

$$\mathrm{E}(x) = \sqrt{1-x}\,\mathrm{E}\left(\frac{x}{x-1}\right). \tag{329}$$

The explanation is that if, keeping the a axis constant, we vary the b axis from 0 to ∞, every value of b can be paired with another value for which the curve has the same shape. This relation (329), which might be made a starting point, being visibly true on examination, is the analogue of (135B) concerning F; and in fact the F relation can be derived from the E relation by differentiation; that is, by carrying out

$$\frac{\mathrm{F}(x^{-1})}{\sqrt{x}} = 2\frac{d}{dx}\left\{\sqrt{x}\,\mathrm{E}(x^{-1})\right\}, \quad \text{or} \quad \mathrm{F}\left(\frac{1}{c^2}\right) = \frac{d}{dc}\left\{c\mathrm{E}\left(\frac{1}{c^2}\right)\right\}, \tag{330}$$

which follow from (318), (319). On the other hand, the relation (195) has no strict parallel in the E's; the nearest approach being

$$(1+\sqrt{x})\mathrm{E}\left\{\frac{4\sqrt{x}}{(1+\sqrt{x})^2}\right\} = 2\mathrm{E}(x) - (1-x)\mathrm{F}(x)$$

$$= 1 + \frac{1}{2^2}x + \frac{1^2 3^2}{2^2 4^2 6^2}x^2 + \frac{1^2 3^2 5^2}{2^2 4^2 6^2 8^2}x^3 + \dots, \tag{331}$$

which is rapidly convergent. This is a pure identity, unlike (329).

(66). The so-called third elliptic integral is merely one of a large number of integrals which reduce to the first kind under special circumstances. I do not know exactly why it should be regarded as a fundamental integral, like the first and second kinds. It is usually evaluated by expressing it in terms of first and second integrals, partly complete and partly incomplete. This is elaborate and complicated, and it is not easy to see from such expressions what the function concerned is really like. But we can readily obtain understandable series formulæ. Thus, let

$$\mathrm{W} = \frac{2}{\pi}\int \frac{dx}{(1+x^2/t_1)(r-x^2)^{\frac{1}{2}}(s-x^2)^{\frac{1}{2}}} = \frac{2}{\pi}\int \frac{\epsilon^{-tx^2}\,dx}{(r-x^2)^{\frac{1}{2}}(s-x^2)^{\frac{1}{2}}}. \tag{332}$$

Here we have two connected integrals, of which the first is called the third elliptic integral. The limits are to be from 0 to the smaller of $r^{\frac{1}{2}}$ and $s^{\frac{1}{2}}$. But we may fix s to be the smaller. The equivalent auxiliary integral is useful in obtaining the solution. By Euler's gamma integral it may be converted to

$$\mathrm{W} = \frac{2}{\pi}\int_0^\infty \epsilon^{-(r_1+s_1+t)x^2}\pi(r_1 s_1)^{\frac{1}{2}}\frac{dx^2}{2\sqrt{x^2}} = \pi^{\frac{1}{2}}\left(\frac{r_1 s_1}{r_1+s_1+t}\right)^{\frac{1}{2}}. \tag{333}$$

This function may now be turned to a function of r, s, t_1. Thus,

$$\mathrm{W} = (\pi r_1 s_1)^{\frac{1}{2}}(r_1+s_1+t)_1^{\frac{1}{2}}\pi^{\frac{1}{2}} = \epsilon^{(r_1+s_1)t_1}(\pi r_1 s_1 t_1)^{\frac{1}{2}}$$

$$= \frac{t_1^{\frac{1}{2}}}{(r+t_1)^{\frac{1}{2}}(s+t_1)^{\frac{1}{2}}} + \mathrm{W}_1 = \mathrm{W}_0 + \mathrm{W}_1. \tag{334}$$

Here W_0 is the explicit function in terms of r, s, t_1. But since $t_1=\infty$ does not reduce it to the first elliptic function, the auxiliary function W_1 is added to make it do so. This auxiliary is implied in W_0 itself, because the first elliptic function is in rising powers of s/r, and W_0 may be expanded in that way together with other terms. So we can find what W_1 is. Thus,

$$W=W_1+\frac{1}{(1+t_1/r)^{\frac{1}{2}}(1+s/t_1)^{\frac{1}{2}}r^{\frac{1}{2}}}$$

$$=W_1+\frac{1}{r^{\frac{1}{2}}}\left[1-\tfrac{1}{2}\left(\frac{t_1}{r}+\frac{s}{t_1}\right)+\left(\frac{1.3}{2.4}\frac{t_1^2}{r^2}+\frac{1}{2}\frac{1}{2}\frac{s}{r}+\frac{1.3}{2.4}\frac{s^2}{t_1^2}\right)-\ldots\right]. \quad (335)$$

Put $t_1=\infty$. To reduce W to the first elliptic function requires that $-W_1=$ sum of all terms involving rising powers of t_1. That is,

$$W_1=-\frac{1}{\sqrt{r}}\frac{t_1}{2r}\left[1-\frac{3}{4}\frac{t_1}{r}\left(1-\frac{1}{2}\frac{s}{r}\right)+\frac{3.5}{4.6}\frac{t_1^2}{r^2}\left(1-\frac{1}{2}\frac{s}{r}+\frac{1.3}{2.4}\frac{s^2}{r^2}\right)-\ldots\right]. \quad (336)$$

The work is now done. The function W_0 wants no reservation about convergency. So $W=W_0+W_1$ is suitable for calculation when $t_1<r$. The functions of s/r that occur represent the sum of first n terms of $(1+s/r)^{-\frac{1}{2}}$.

But when $t_1>r$, we should discard W_1 altogether, and use the rest of the series in (335), that is, the part in descending powers of t_1. This makes

$$W=\frac{1}{\sqrt{r}}\left[\left(1-\frac{1}{2}\frac{s}{t_1}+\frac{1.3}{2.4}\frac{s^2}{t_1^2}-\frac{1.3.5}{2.4.6}\frac{s^3}{t_1^3}+\ldots\right)\right.$$

$$+\frac{1^2}{2^2}\frac{s}{r}\left(1-\frac{3}{4}\frac{s}{t_1}+\frac{3.5}{4.6}\frac{s^2}{t_1^2}-\frac{3.5.7}{4.6.8}\frac{s^3}{t_1^3}+\ldots\right)$$

$$\left.+\frac{1^23^2}{2^24^2}\frac{s^2}{r^2}\left(1-\frac{5}{6}\frac{s}{t_1}+\frac{5.7}{6.8}\frac{s^2}{t_1^2}-\frac{5.7.9}{6.8.10}\frac{s^3}{t_1^3}+\ldots\right)+\ldots\right]. \quad (337)$$

The structure is perfectly clear, viz., the first elliptic function with every term multiplied by a function of t. These functions, moreover, are

$$A_0=(1+s/t_1)^{-\frac{1}{2}},\quad A_1=-\frac{2}{1}\frac{t_1}{s}(A_0-1),\quad A_2=\frac{2.4}{1.3}\frac{t_1^2}{s^2}\left(A_0-1+\frac{1}{2}\frac{s}{t_1}\right), \quad (338)$$

and so on.

Now if, in (337), we turn t_1^{-n} to $t^n/\underline{|n}$, we obtain the solution of the second integral in (332), the auxiliary integral, in rising powers of t. Of course, (337) itself may be regarded as expressing this also. To corroborate, expand the result (333) in rising powers of t, making

$$W=\left(\frac{\pi r_1 s_1}{r_1+s_1}\right)^{\frac{1}{2}}\left\{1-\frac{1}{2}\frac{t}{r_1+s_1}+\frac{1.3}{2.4}\frac{t^2}{(r_1+s_1)^2}-\ldots\right\}. \quad (339)$$

Here we have

$$\frac{(\pi r_1 s_1)^{\frac{1}{2}}}{(r_1+s_1)^{n+\frac{1}{2}}}=(\pi r_1)^{\frac{1}{2}}s_1^{-n}\left\{1-(n+\tfrac{1}{2})\frac{r_1}{s_1}+\frac{(n+\frac{1}{2})(n+1\frac{1}{2})}{\underline{|2}}\frac{r_1^2}{s_1^2}-\ldots\right\}, \quad (340)$$

in which n has to receive the values 0, 1, 2, 3, &c. On turning (339) to a function of r, s, t through (340), the result is (337) above, only with $t^n/\underline{|n}$ instead of t_1^{-n}. So we come round to the third elliptic again, by putting our result in terms of t_1. And then, further, by using (338), we shall come to the form W_0+W_1.

This is enough for the present about the operational treatment of definite integrals, which might go on for ever. The above may help others on the way. But perhaps, like the fishes who were preached to by the saint, "Much edified were they, but preferred the old way." Very well, then there let them stay. More important than the operational treatment of definite integrals is the operational treatment of the definite differential solutions of physical problems. They are themselves definite integrals in a peculiar way, through multiplex integration or differentiation. Moreover, they may be converted to definite integrals of the common kind, in a more or less complicated manner. There is not much use in that, in general. Rather, it is the converse procedure that is useful.

Given the Effect, Find the Cause. The Inversion of Operations.

§ 527. Nothing metaphysical is meant in the use of the words cause and effect. Think rather of a definite mechanical or physical system, wherein by applying force in a certain way to some part, we produce some effect, as motion in a certain way, at some other (or the same) place. Then the applied force may be called the cause, and the resulting motion the effect. Now, in general, the connection between the cause and the effect is not a functional one, when mathematically expressed, but a differential one; and yet, although differential, is quite definite. See vol. 1, p. 391, for an explanation and illustration of this important proposition. There are plenty of examples in vol. 2. If the cause, say a force, is called E, and the effect C, then their relation is $C=YE$, where Y is a definite differential operator, which contains no arbitrariness in a definite physical problem. The class of function C belongs to is usually indeterminate in the sense that although Y is definite, the force E may be usually quite an arbitrary function of the time, so that, when modified by Y to produce the effect C, it may itself turn out arbitrarily, though at the same time there may be exceptional peculiarities in the connector Y which limit the extent of the arbitrariness.

Now the present question is, given the effect, find the cause. Properly interpreted, this involves the same determinateness through the same connections. It is easy, by enlarging the question, to make the answer be indeterminate. We may observe the effect only, and not know where the cause was. But that is not what is implied in $C=YE$. If this be inverted, and turned to $E=Y^{-1}C$, and then, C being given, it is asked what E produced it, it is implied that the cause of the observed effect is at the same place as before, and that the cause and the effect have the same connection. Now one and the same C might be due either to E at the original place or at some other place. Then the structure of the operator Y would itself reveal the possibility.

We better think of a definite practical case to avoid the vagueness and generalities with respect to exceptional peculiarities. Say there is a telegraph circuit from Europe to America, with all its complicated parts, and its terminal arrangements for sending and receiving signals. Say E_1 is an impressed voltage in some part of the European arrangement, and C_2 the current in some part of the American arrangement; then $C_2=YE_1$, and Y can be determined definitely. Or rather, we substitute for the reality an ideal system which professes to imitate practically and sufficiently the fullness of the real one, and Y belongs to the ideal. Evidently when E_1 is given as a function of the time, we can find C_2 as another and corresponding function of the time. We know, then, the answer to the question, given C_2, find E_1, for a particular kind of C_2; and similarly, by working other direct problems, we may obtain other particular answers. But we must go further than that, and determine E_1 from C_2 only, without knowing E_1 first.

An easy case of indeterminateness could arise thus. Let there be a second cable just like the first, and alike the first also in its European terminal arrangements, and let both be connected similarly to the American arrangement. Then E_1 in the original European arrangement and e_1 in the new one, would each, acting alone, produce the same C_2, provided E_1 and e_1 were the same functions of the time. Observation of C_2 only would not settle where it came from, though a practical man would soon find out by observing C_2 and something else as well. The connection would be of the kind $C_2=YE_1+ye_1$; if e_1 is known to be zero, then E_1 becomes known in terms of C_2; and similarly for the other one. Y and y may be identical copies; but since the two cables are in connection with one another, Y is not the same as before, but becomes a function of the constants of the two cables and apparatus.

Let us keep to a single cause and single effect. In vol. 2 will be found some examples of the finding of causes from given effects. In the following are a few more, with a larger scope.

(1). Go back to p. 289, and look at equation (334). Throw away W_1 altogether, and write

$$W_0=\frac{t_1^{\frac{1}{2}}}{(r+t_1)^{\frac{1}{2}}(s+t_1)^{\frac{1}{2}}}. \tag{1}$$

Now say that r and s are constants, and that t_1 is the time differentiator, what physical problem does this equation represent? A multitude; but if we take the telegraph theory of vol. 2, chap. vii., for basis, then the answer becomes definite. As on p. 308, vol. 2, if V_0 is the voltage impressed at $x=0$, and V the resulting voltage at x, then

$$V=\epsilon^{-qx}V_0, \qquad \text{where} \qquad q=v^{-1}(r+t_1)^{\frac{1}{2}}(s+t_1)^{\frac{1}{2}}, \tag{2}$$

if $r=\mathrm{R/L}$, $s=\mathrm{K/S}$, and $v=(\mathrm{LS})^{-\frac{1}{2}}$. So

$$-\frac{dV}{dx}=q\epsilon^{-qx}V_0, \quad \text{and} \quad V_0=-\frac{1}{q}\frac{dV_0}{dx}. \tag{3}$$

If then,
$$-\frac{dV_0}{dx}=et_1^{\frac{1}{2}}/v, \qquad \text{we have} \quad V_0=\frac{e}{qv}t_1^{\frac{1}{2}}. \tag{4}$$

Comparing with (1), which is $W_0=t_1^{\frac{1}{2}}/qv$, we see that W_0 means V/e. So the problem is this. Given that the slope of the potential at the origin is $e/v(\pi t)^{\frac{1}{2}}$, what is the voltage at the same place? It is implied by the use of the operator ϵ^{-qx} that disturbances arise on the left side, so the problem is quite definite.

(2). There are two primary solutions, in rising, and in falling powers of t. The first comes by expanding W_0 in rising powers of $1/t_1$, and makes

$$w_1=\frac{t^{\frac{1}{2}}}{\lfloor\frac{1}{2}}-\frac{r+s}{2}\frac{t^{1\frac{1}{2}}}{\lfloor 1\frac{1}{2}}+\left(\frac{1.3}{2.4}r^2+\frac{1}{2}\frac{1}{2}rs+\frac{1.3}{2.4}s^2\right)\frac{t^{2\frac{1}{2}}}{\lfloor 2\frac{1}{2}}-\dots \tag{5}$$

This suits the beginning part of the curve required. But, remembering the meaning of r and s, it is easy to see that rt, and perhaps st as well, may get unmanageably large, so a second form of solution is wanted. It is got by expanding W_0 in rising powers of t_1, and makes

$$w_2=\frac{1}{(rs)^{\frac{1}{2}}}\left\{\frac{t^{-\frac{1}{2}}}{\lfloor -\frac{1}{2}}-\frac{r^{-1}+s^{-1}}{2}\frac{t^{-1\frac{1}{2}}}{\lfloor -1\frac{1}{2}}+\left(\frac{1.3}{2.4}r^{-2}+\frac{1}{2}\frac{1}{2}r^{-1}s^{-1}+\frac{1.3}{2.4}s^{-2}\right)\frac{t^{-2\frac{1}{2}}}{\lfloor -2\frac{1}{2}}\right\}, \tag{6}$$

which is suitable as soon as rt and st are large enough to produce distinct initial convergence of the series. Beginning with the value 0, W_0 increases at first as $t^{\frac{1}{2}}$, then more slowly, makes a hump, and finally subsides to zero again according to $t^{-\frac{1}{2}}$.

(3). In the distortionless case, $r=s$, and

$$w_1=\frac{2\sqrt{t}}{\sqrt{\pi}}\left\{1-\frac{2rt}{3}+\frac{(2rt)^2}{3.5}-\frac{(2rt)^3}{3.5.7}+\dots\right\}, \tag{7}$$

$$w_2=\frac{1}{r\sqrt{\pi t}}\left\{1+\frac{1}{2rt}+\frac{3}{(2rt)^2}+\frac{3.5}{(2rt)^3}+\dots\right\}. \tag{8}$$

The equivalence of w_1 and w_2 is a case of the generalised $\epsilon^{-x}\cos n\pi$ formula. Both come from

$$W_0=\frac{\sqrt{t_1}}{r+t_1}. \tag{9}$$

(4). This may also be made to represent a simple coil problem. For if the impressed voltage on a coil varies as $t^{-\frac{1}{2}}$, the current will vary as W_0 in (9), fully expanded functionally in (7) and (8).

Or, if the current in a conducting condenser varies as $t^{-\frac{1}{2}}$, the voltage will be proportional to W_0.

(5). Evidently (6) entirely fails when $s=0$, which occurs when there is no leakage. Then (5) remains, in the form

$$w_1=\frac{t^{\frac{1}{2}}}{\lfloor\frac{1}{2}}-\tfrac{1}{2}r\frac{t^{1\frac{1}{2}}}{\lfloor 1\frac{1}{2}}+\frac{1.3}{2.4}r^2\frac{t^{2\frac{1}{2}}}{\lfloor 2\frac{1}{2}}-\ldots=2\left(\frac{t}{\pi}\right)^{\frac{1}{2}}\left\{1-\frac{rt}{\lfloor 1.3}+\frac{r^2t^2}{\lfloor 2.5}-\right\}, \quad (10)$$

and now the question is, what is the second form for large values of t? The reduced form of W_0 is

$$W_0=(r+t_1)^{-\frac{1}{2}}, \quad (11)$$

and expansion in rising powers of t_1 gives the answer $w_2=r^{-\frac{1}{2}}$. This is in fact the value tended to when t is large. There is no leakage now, and that makes a difference.

This result $w_2=r^{-\frac{1}{2}}$ is evidently not a complete equivalent of the convergent solution w_1. In the usual use of a divergent paired with a corresponding convergent series, say (7) and (8) above, the range of practical use of the divergent series may be found by inspection of its initial convergence. This fails with $w_2=r^{-\frac{1}{2}}$, because the series has degenerated to its first term. So it is only by actual comparison with the convergent series that we can find how small rt may be to allow of the use of the second formula. I find that the error, which is considerable when $rt=1$, is reduced to about 1 per cent. when $rt=4$. So from $rt=4$ up to ∞ we may use the short formula, and use the long one only for the piece from $rt=0$ up to $rt=4$.

(6). As to why the series should degenerate, that is rather obscure, but the following will give a partial explanation. Consider the translation operator ϵ^{-xt_1}, which is in constant use in differential transformations, making, for example,

$$\epsilon^{-xt_1}f(t)=f(t-x). \quad (12)$$

This operator shifts the function bodily through the distance x. Applied to unit operand, positive, it makes $\epsilon^{-xt_1}1=(t-x)^0$, that is, 1, when $t>x$, and zero when $t<x$. Yet in powers of t_1 we have

$$\epsilon^{-xt_1}=1-xt_1+\frac{x^2}{\lfloor 2}t_1^{\,2}-\frac{x^3}{\lfloor 3}t_1^{\,3}+\ldots, \quad (13)$$

where (with unit operand) t_1 and its powers are all vanishing quantities. But then there is an infinite number of them, and their separate vanishing does not necessitate the vanishing of the sum. To see this here, put $t^{-n}/\lfloor -n$ for $t_1^{\,n}$; then, remembering that $\lfloor n=n\lfloor n-1$, we may write (13) thus,

$$\epsilon^{-xt_1}=1-\frac{x/t}{\lfloor -1}\left\{1+\frac{x/t}{2}+\frac{(x/t)^2}{3}+\frac{(x/t)^4}{4}+\ldots\right\}. \quad (14)$$

Here the quantity in the { } is finite when $x<t$. The outside factor is zero. So $\epsilon^{-xt_1}1=1$ when $t>x$. But the $\{\cdots\}$ becomes ∞ when $x=t$, and if all terms are counted, is ∞ when $x>t$. So the result is $1-0\times\infty$, that is, indeterminate in this way. We know by the preceding that the $0\times\infty$ means 1.

We may also put it thus. What is the value of the series

$$\frac{(x/t)^{n-1}}{\underline{|n-1}}-\frac{(x/t)^{n-2}}{\underline{|n-2}\,\underline{|2}}+\frac{(x/t)^{n-3}}{\underline{|n-3}\,\underline{|3}}-\dots \tag{15}$$

when $n=0$, when it assumes an indefinite form? The answer is, 1 when $x>t$, and 0 when $x<t$.

(7). The above is an explicit illustration of a useful test to apply to series in powers of a differentiator, to make sure that an infinite series of 0's really means 0 in the total, or to suggest and perhaps prove that it is finite. For, if it is $(t^{-1}/\underline{|-1})\mathrm{F}(t)$, and $\mathrm{F}(t)$ is finite, the result is zero. But if $\mathrm{F}(t)$ is infinite, then the result may be finite, and most likely is.

As another illustration, which works out differently, consider the operator ϵ^{-qx} which occurs in the electrostatic portion of the telegraph theory, vol. 2, p. 49 and after. We have $q=t_1^{\frac{1}{2}}$, and in all the series developments done there, the terms t_1, t_1^2, &c., are taken to be zero. For example,

$$\epsilon^{-t_1}=\left(1+\frac{t_1}{\underline{|2}}+\frac{t_1^2}{\underline{|4}}+\dots\right)-t_1^{\frac{1}{2}}\left(1+\frac{t_1}{\underline{|3}}+\frac{t_1^2}{\underline{|5}}+\dots\right). \tag{16}$$

In turning this to a function of t, the first series degenerates to 1 simply.

In full, this series is

$$1+\frac{t^{-1}}{\underline{|-1}}\left\{\tfrac{1}{2}-\frac{t^{-1}}{2.3.4}+\frac{t^{-2}}{3.4.5.6}-\frac{t^{-3}}{4.5.6.7.8}+\dots\right\}. \tag{17}$$

Here the function in the { } is finite for all positive values of t, so the result is simply 1, and in (16), we count only the second series, together with 1 to represent the first series. Similar treatment applies in various other diffusion problems.

(8). Go back to (11), and see how it behaves when treated in the same way.

It makes

$$w_2=\frac{1}{\sqrt{r}}\left\{1-\frac{(rt)^{-1}}{\underline{|-1}}\left(\tfrac{1}{2}+\frac{1.3}{2.4}\frac{1}{rt}+\frac{1.3.5}{2.4.6}\underline{|2}\frac{1}{r^2t^2}+\frac{1.3.5.7}{2.4.6.8}\underline{|3}\frac{1}{r^3t^3}+\dots\right)\right\}. \tag{18}$$

Here, counting all terms, the result is $r^{-\frac{1}{2}}(1-0\times\infty)$, indefinite. The conclusion is that w_2 is not really reduced to its first term, although, as before seen, it approximates to that result pretty quickly.

Now examine the reduced form of the convergent solution, that is, *the* solution, in the sense of rigorous mathematicians, when $\mathrm{L}=0$, or the self-induction is done away with. The solution for V_0, to make $-dV_0/dx$ vary as $et_1^{\frac{1}{2}}$, takes the form of an oscillating series of infinities. The real solution is given by w_2. It is, $V_0=e/(RS)^{\frac{1}{2}}$, as may be seen by inspection of (4), with the changes needed to get rid of L and v. That is to say, a constant impressed voltage is needed to make the slope of potential vary as $t^{-\frac{1}{2}}$.

The remarkable thing here is that the solution of the electrostatic problem regarded as a reduced case of the electromagnetic, arises by degeneration of a divergent series, whilst the convergent solution becomes useless. This occurs in the reduction of all electromagnetic solutions to electrostatic, when the electromagnetic are expressed in terms of positive powers of R/L, by convergent series. The other way, with L in the numerators, will be noticed presently.

(9). If V_0 is voltage impressed at the beginning of a cable, and V is the result (transverse voltage, or the potential) at distance x, then

$$V=\epsilon^{-qx}V_0, \quad \text{where } q=\sqrt{(R+Lt_1)(K+St_1)}. \tag{19}$$

The inverse problem is given V_0, find V. The anwser is $V_0=\epsilon^{qx}V$; but to allow for changed functional form, we better write it thus,

$$W_0=\epsilon^{qx}W. \tag{20}$$

If, for example, V_0 is constant, beginning $t=0$, then V is the (usually curved) arrival curve of the potential at x. But let W be constant, beginning $t=0$, then W_0 is the impressed voltage that produces the perpendicular arrival curve, or more strictly, perpendicular first, and horizontal later. This solves the problem of how to make square signals. Now W in (20) is the same as V_0 in (19), but W_0 is usually very different in shape from V. Yet all the change in the formula is in the reversal of the sign of x.

The distortionless case is very simple. A constant W_0 suddenly put on will produce a constant W, also suddenly coming on. Here

$$q=(r+t_1)/v, \quad \text{making} \quad W_0=\epsilon^{rx/v}\epsilon^{t_1x/v}W; \tag{21}$$

showing that W_0 must be $\epsilon^{rx/v}$ times as big as W, and start x/v sooner.

But if we do not have the critical relation R/L=K/S, then all we can do without analysis is to find the initial and final values of W_0. The final value is got by putting $t_1=0$ in q. It makes

$$W_0=\epsilon^{x(RK)^{\frac{1}{2}}}W. \tag{22}$$

This is the steady W_0 needed to maintain W steady at x. When there is no leakage, $W_0=W$.

The initial value of W_0 is got by a consideration of the attenuation at a wave front due to constant V_0. This makes the initial W_0 be given by

$$W_0=\epsilon^{(R/2L+K/2S)x}W. \tag{23}$$

So the curve of W_0 falls from the value given by (23) at the moment $t=-x/v$, to the value given by (22), when $t=\infty$. Practically, of course the steady value is nearly reached in a short interval of time. But, practically also, whilst the final value of W_0 is pretty definite, the initial value is not, because it is not possible to realise constancy of the constants R, L, K, S. Increased resistance at the wave front due to the initial stage of imperfect penetration will necessitate a greater initial value of W_0 to counterbalance it. But except at its beginning, the curve of W_0 may fairly approximate to real fact.

Note that if $L=0$, an infinite initial value is wanted. But the inductance never is zero, so the failure, by supposed infiniteness of voltage needed, is not real.

Similar considerations apply to the determination of W_0 to produce a perpendicular arrival curve of current at x. Since

$$-\frac{dV}{dx}=(R+Lt_1)C, \text{ therefore } C=\frac{q}{R+Lt_1}V=\left(\frac{K+St_1}{R+Lt_1}\right)^{\frac{1}{2}}\epsilon^{-qx}V_0. \quad (24)$$

So in the inverse problem

$$W_0=\left(\frac{R+Lt_1}{K+St_1}\right)^{\frac{1}{2}}\epsilon^{qx}C. \quad (25)$$

So W_0/C falls from the value $(L/S)^{\frac{1}{2}}\epsilon^{x(R/2Lv+K/2Sv)}$ to the value $(R/K)^{\frac{1}{2}}\epsilon^{x\sqrt{RK}}$ in order that the curve of C shall be first perfectly steep, and then horizontal later. The final value is to maintain the leakage current, when K is finite. But when K is zero, there is no final steady W_0. The impressed voltage, after its first drop, must increase again without limit to maintain the current constant at x.

Similarly, if we terminate the cable through apparatus, we can determine W_0 finitely to produce definite results in particular ways; a steady current, for instance, through a coil which has a through conducting circuit. But if we ask for the maintenance of a steady current through a non-conducting condenser, then W_0 must ultimately increase without limit.

(10). To make a square arrived signal, of potential or of current, of given duration a. Superpose upon the curve of W_0 calculated for continued constant W or C, the negative of the same, beginning later by the interval a. The resultant makes the curve of W_0, which will make W (or C) jump instantly from 0 up to a certain value, remain at that value for the interval a, and then suddenly drop to the value 0, and remain there. So, provided L is not zero (and it never is), the impressed voltage required to produce any signals of finite size is always finite and calculable.

Keeping to a single square signal, the resultant W_0 will vary notably with the duration of the signal. If very long, then the first part of W_0 may have fallen down greatly towards its final value (assuming that there is one); so the second part will be nearly a copy of the first, reversed in sign. But if the duration is so short that the second part begins before there is a large drop in the first part, then the resultant is first a strong short W_0, followed by a long weak tail of the opposite sign. The initial impulse makes the square signal, the rest wipes it out. Construct a machine to do this properly, and you may see wonders. But I fear that strong impulses are just what the cable people will not have. Then their messages can go dribbling along, as usual.

(11). It is not necessary to say much concerning the way of calculating the shape of the W_0 curve, because it is much the same as the work done in vol. 2, in calculating the effects produced at a distance by voltage impressed at the beginning of a cable. But I gave little detail regarding one way, in power series of x and t, so this method may be given here. It has to be harmonized with the other ways.

If $K=0$, no leakage, then

$$V/V_0=\epsilon^{-qx}=\epsilon^{-yY}, \tag{26}$$

if

$$y=x\sqrt{RS}, \qquad L/R=a, \qquad Y=\sqrt{t_1(1+at_1)}. \tag{27}$$

This expands to

$$\epsilon^{-qx}=\left(1+\frac{y^2Y^2}{\underline{|2}}+\frac{y^4Y^4}{\underline{|4}}+\dots\right)-yY\left(1+\frac{y^3Y^3}{\underline{|3}}+\frac{y^5Y^5}{\underline{|5}}+\dots\right). \tag{28}$$

In accordance with experience in this class of problems, the even series reduces to 1 simply, because Y^2, &c., contain only integral rising powers of t_1. But let us verify this by expanding the even series in powers of a. The constant part has already been found to reduce to 1. The coefficient of a is

$$t_1\left(\frac{t_1y^2}{\underline{|2}}+2\frac{t_1^2y^4}{\underline{|4}}+3\frac{t_1^3y^6}{\underline{|6}}+\dots\right)=\frac{y^2/t^2}{2\underline{|-2}}\left(1-\frac{y^2/t^2}{3}+\frac{y^4/t^4}{4.5}-\frac{y^6/t^6}{5.6.7}+\dots\right). \tag{29}$$

This is $0\times$ finite, or 0. The coefficient of a^2 is

$$t_1^2\left(\frac{t_1^2y^4}{\underline{|4}}+3\frac{t_1^3y^6}{\underline{|6}}+6\frac{t_1^4y^8}{\underline{|8}}+10\frac{t_1^5y^{10}}{\underline{|10}}+\right)$$

$$=\frac{y^4/t^4}{\underline{|4}\,\underline{|-4}}\left(1-\frac{3.4}{5.6}\frac{y^2}{t^2}+\frac{1}{7.8}\frac{y^4}{t^4}-\frac{y^6/t^6}{5.7.8.9}+\right), \tag{30}$$

which is also $0\times$ finite, or 0. So are the rest, more and more so.

Having thus cleared the ground, there is left

$$\epsilon^{-qx}=1-\left(yY+\frac{y^3Y^3}{\underline{|3}}+\frac{y^5Y^5}{\underline{|5}}+\dots\right). \tag{31}$$

The next step is to find Y^n in terms of t. There are two ways, one convergent, the other divergent. First, expand Y^n by the binomial theorem, in rising powers of t_1, and then multegrate. This makes

$$\mathrm{Y}^n = t_1^{\frac{1}{2}n}\left\{1+\tfrac{1}{2}n\,at_1+\frac{\frac{1}{2}n(\frac{1}{2}n-1)}{\lfloor\underline{2}}a^2t_1^{\,2}+\dots\right\} \tag{32}$$

$$=\frac{t^{-\frac{1}{2}n}}{\lfloor\underline{-\tfrac{1}{2}n}}\left\{1-n^2\frac{a}{4t}+\frac{n^2(n^2-2^2)}{\lfloor\underline{2}}\left(\frac{a}{4t}\right)^2-\frac{n^2(n^2-2^2)(n^2-4^2)}{\lfloor\underline{3}}\left(\frac{a}{4t}\right)^3+\dots\right\}. \tag{33}$$

Use this in (31), with $n=1, 3, 5, \dots$. It is desirable to have the commencement explicitly, being the important part. Thus,

$$\begin{aligned}\epsilon^{-qx}=1&-\frac{y}{(\pi t)^{\frac{1}{2}}}\left\{1-\frac{1}{2}\frac{1}{2}\frac{a}{t}-\frac{1.1}{2.4}\frac{1.3}{2.2}\frac{a^2}{t^2}-\frac{1.1.3}{2.4.6}\frac{1.3.5}{2.2.2}\frac{a^3}{t^3}-\dots\right\}\\&-\frac{y^3}{\lfloor\underline{3}\,t^{1\frac{1}{2}}\lfloor\underline{-1\tfrac{1}{2}}}\left\{1-\frac{3}{2}\frac{3}{2}\frac{a}{t}+\frac{3.1}{2.4}\frac{3.5}{2.2}\frac{a^2}{t^2}+\frac{3.1.1}{2.4.6}\frac{3.5.7}{2.2.2}\frac{a^3}{t^3}+\dots\right\}\\&-\frac{y^5}{\lfloor\underline{5}\,t^{2\frac{1}{2}}\lfloor\underline{-2\tfrac{1}{2}}}\left\{1-\frac{5}{2}\frac{5}{2}\frac{a}{t}+\frac{5.3}{2.4}\frac{5.7}{2.2}\frac{a^2}{t^2}-\frac{5.3.1}{2.4.6}\frac{5.7.9}{2.2.2}\frac{a^3}{t^3}-\dots\right\}-\dots\end{aligned} \tag{34}$$

Observe that the terms in the t series all ultimately become of one sign, beginning one term later in passing from one series to the next. Note also that $a=0$ cuts off all but the electrostatic part. The same is done by t large. That is, in the later stage, self-induction ceases to operate sensibly.

The convergent series is got by expanding Y^n in falling powers of t_1 by the binomial theorem, making

$$\mathrm{Y}^n=a^{\frac{1}{2}n}t_1^{\,n}\left\{1+\tfrac{1}{2}n\frac{1}{at_1}+\frac{\frac{1}{2}n(\frac{1}{2}n-1)}{\lfloor\underline{2}}\frac{1}{a^2t_1^{\,2}}+\dots\right\}. \tag{35}$$

Here, on multegrating, some terms at the beginning go out, leaving

$$\mathrm{Y}^n=\frac{y^n}{\lfloor\underline{n}}\frac{1}{2^na^{\frac{1}{2}n}}\frac{1.3.5\dots(n-2)}{2.4.6\dots(n-1)}\left\{1-\frac{n}{n+1}\frac{t}{2a}\left(1-\frac{n+2}{n+2}\frac{t}{4a}\left(1-\frac{n+4}{n+3}\frac{t}{6a}\dots\right\}\right.\right., \tag{36}$$

where, as before, $n=1, 3, 5$, &c. So, writing out the leading part

$$\begin{aligned}\epsilon^{-qx}=1&-\frac{y}{\lfloor\underline{1}.2a^{\frac{1}{2}}}\left\{1-\frac{1}{1.2}\frac{t}{2a}\left(1-\frac{3}{2.3}\frac{t}{2a}\left(1-\frac{5}{3.4}\frac{t}{2a}\left(1-\frac{7}{4.5}\frac{t}{2a}\dots\right\}\right.\right.\right.\\&+\frac{y^3}{\lfloor\underline{3}\,2^3a^{1\frac{1}{2}}}\frac{1}{2}\left\{1-\frac{3}{1.4}\frac{t}{2a}\left(1-\frac{5}{2.5}\frac{t}{2a}\left(1-\frac{7}{3.6}\frac{t}{2a}\left(1-\frac{9}{4.7}\frac{t}{2a}\dots\right\}\right.\right.\right.\\&-\frac{y^5}{\lfloor\underline{5}\,2^5a^{2\frac{1}{2}}}\frac{1.3}{2.4}\left\{1-\frac{5}{1.6}\frac{t}{2a}\left(1-\frac{7}{2.7}\frac{t}{2a}\left(1-\frac{9}{3.8}\frac{t}{2a}\left(1-\frac{11}{4.9}\frac{t}{2a}\dots\right\}\right.\right.\right.\\&+\dots\end{aligned} \tag{37}$$

This shows the structure plainly, and is easier to use than (36). In this form L comes in the denominators, so that we cannot reduce to the electrostatic form directly.

When $t=0$, (37) makes a definite function of y or x. But this is a deception. The formula is only valid from $x=0$ up to $x=vt$. Beyond that, there is nothing. So if we reduce t to 0 we shift the wave front up to the origin. There is left merely the result $V/V_0=1$ at $x=0$, $t=0$, which is correct, of course.

Now in vol. 2, p. 332, eq. (33), is the following form of convergent solution,

$$\epsilon^{-qx}=\epsilon^{-\rho t}\left\{P_0(z)+2wP_1(z)+2\frac{w^2}{\underline{|2}}P_2(z)+2\frac{w^3}{\underline{|3}}P_3(z)+\dots\right\} \qquad (38)$$

$$=\epsilon^{-\rho t}\left\{I_0(z)+2\left(\frac{t-x/v}{t+x/v}\right)^{\frac{1}{2}}I_1(z)+2\left(\frac{t-x/v}{t+x/v}\right)I_2(z)+\dots\right\}, \qquad (39)$$

where $P_n(z)$ is $I_n(z)$ divided by its first term, and $z=\rho(t^2-x^2/v^2)^{\frac{1}{2}}$, and $w=\frac{1}{2}\rho(t-x/v)$, ρ being R/2L, that is $1/2a$. What relation does it bear to (37)? To answer this, expand (38) in powers of x. The coefficients are functions of t. The first function is $\epsilon^{\rho t}\times\epsilon^{-\rho t}=1$, in agreement with (37). The next is $-(\rho x/v)(1+\frac{1}{2}\rho t+\frac{1}{4}\rho^2t^2+\dots)\times\epsilon^{-\rho t}$. Carrying out the multiplication, the result is the y term in (37). Then the x^2 term in (38) comes to 0, in agreement with (37). It follows that the two convergent solutions (37) and (38) are identical, one being a mere rearrangement of the other, when the factor $\epsilon^{-\rho t}$ is allowed for.

The next question is, what relation does the convergent (38) bear to (34)? The answer is less easy. For, if we substitute $\frac{1}{2}H_n(z)$ for $I_n(z)$ in (39), producing

$$\epsilon^{-qx}=\frac{1}{(2\pi z)^{\frac{1}{2}}}\left\{1+\frac{1^2}{8z}+\frac{1^2 3^2}{\underline{|2}(8z)^2}+\dots\right.$$
$$\left.+2\sum_{m=1}^{\infty}\left(\frac{t-x/v}{t+x/v}\right)^{\frac{1}{2}m}\left(1+\frac{1^2-4m^2}{8z}+\dots\right)\right\}, \qquad (40)$$

we have a form equivalent to (38), and therefore to (37) and (34), but which does not reduce identically to (34), because it contains the square root of z. The transformation must take place in a transcendental manner.

But there is another way of passing from (38) to (34), not involving the square root of z. Expand, as before, (38) in powers of x, with coefficients of the form $\epsilon^{-\rho t}f(t)$. Then put $f(t)$ in the form $\sum A_s I_s(\rho t)$, which is an identical substitution. Thirdly, substitute $\frac{1}{2}H_s$ for I_s. Say the coefficient of x^n becomes

$$\epsilon^{-\rho t}f(t)=\tfrac{1}{2}\sum A_s\left(1+\frac{1^2-4s^2}{8\rho t}+\dots\right); \qquad (41)$$

then *this* will be found to be identically the coefficient of x^n in (34). The best way to carry out this process is by means of the $t_1=\frac{1}{2}(a+a^{-1})$ transformation of vol. 2, p. 348, applied to the functions $\epsilon^{\rho t}Y^n$ in terms of t_1, so as to bring them at once to the required form. Then compare with (38). Although the harmonisation is complicated, it is satisfactory to see how it is done.

It also shows how to pass from (37) to (34) through the Bessel functions. But whether (34) is practical for calculation, I have not examined. I prefer (37), &c., to (40).

(12). Now pass to the inverse problem. Use (38). Change the sign of x first. This turns w to $w'=\frac{1}{2}\rho(t+x/v)$, whilst z remains the same. So

$$\epsilon^{qx}=\epsilon^{-\rho t}\{P_0(z)+2w'P_1(z)+2(w'^2/\underline{|2})P_2(z)+\dots\}. \qquad (42)$$

But this begins at the moment $t=-x/v$. So put $t=t'-x/v$, making $w'=\frac{1}{2}\rho t'$, and $z'=\rho\{t'(t'-2x/v)\}^{\frac{1}{2}}$. Then

$$\frac{W_0}{W}=\epsilon^{-\rho t'}\epsilon^{\rho x/v}\{P_0(z')+2(\tfrac{1}{2}\rho t')P_1(z')+2(\tfrac{1}{2}\rho t')^2/\underline{|2}\,.\,P_2(z')+\dots\}. \qquad (43)$$

Now t' starts at the value 0, and W_0 is the impressed voltage at $x=0$ which will produce constant potential W at x, beginning there at the moment $t'=x/v$.

Without calculating the whole curve, there are four points in it which serve very well to exhibit the main features. First, $t'=0$. Then $W_0/W=\epsilon^{\rho x/v}$. Next $t'=\rho x/v$. Then

$$W_0/W=(J_0+2J_1+2J_2+2J_3+\dots)(\rho x/v) \qquad (44)$$

$$=1+\frac{\rho x}{v}-\frac{1}{2}\frac{(\rho x/v)^3}{\underline{|3}}+\frac{1.3}{2.4}\frac{(\rho x/v)^5}{\underline{|5}}-\dots \qquad (45)$$

$$=1+\frac{\rho}{v}\int_0^x J_0\left(\frac{\rho x}{v}\right)dx=1+\{1+(\rho x/v)_1^2\}^{-\frac{1}{2}}. \qquad (46)$$

The immediate result (44) is ornamental. Putting it in a power series (45), its meaning can be seen, and is easily calculable. It can be recognised to be given also by the integral form in (46), and also by the differential form which follows, where $(\rho x/v)_1$ means the differentiator with respect to $\rho x/v$. The last shows that the value tended to when $\rho x/v$ is large, is 2. This is confirmed by the definite integral, because $\int_0^\infty J_0(y)dy=1$.

The third value of t' is $2x/v$, making $w'=\rho x/v$, and $z'=0$. So,

$$\frac{W_0}{W}=\epsilon^{-\rho x/v}\left\{1+2\frac{\rho x}{v}+2\frac{(\rho x/v)^2}{\underline{|2}}+\dots\right\}=2-\epsilon^{-\rho x/v}. \qquad (47)$$

The fourth value of t' is ∞, making $W_0/W=1$. The value of the quantity $\rho x/v$, or $Rx/2Lv$, determines the range of W_0/W. If it is as small as 1, the four values of W_0/W are 2·71, 1·92, 1·63, and 1. If it is as large as 10, the first value is 22,026; the third is nearly 2, and the second is not much more. That is, W_0 drops from a very large to less than twice the final value by the time the wave front due to W_0 has reached the point $2x$, the point of observation of W being at x, when x is large enough to make $\epsilon^{\rho x/v}$ be very large. The remaining drop from 2 to 1 takes place slowly.

(13). When there is leakage, the more developed formula for V/V_0 or ϵ^{-qx} should be used, eq. (57), p. 337, vol. 2. Else it is the same in treatment; that is, first alter the sign of x, and then alter the zero of time, as done above.

But when the formula is wanted for the impressed voltage that will produce a square current signal at x, a further change is required. For

$$\frac{C}{V_0}=\left(\frac{K+St_1}{R+Lt_1}\right)^{\frac{1}{2}}\epsilon^{-qx} \quad \text{makes} \quad \frac{W_0}{C_x}=\left(\frac{R+Lt_1}{K+St_1}\right)^{\frac{1}{2}}\epsilon^{qx}; \tag{48}$$

(vol. 2, p. 308, eq. (105)). So, knowing the formula for C/V_0, to obtain that for W_0/C_x, when W_0 is the impressed voltage that will make C_x square at x, we must, besides changing the sign of x, also interchange R and K, L and S. That is, when the results are, as usually in vol. 2, in terms of the two quantities ρ and σ given by $R/2L \pm K/2S$, ρ is unchanged, but σ must be reversed.

(14). For instance, if there is no leakage,

$$\frac{C}{V_0}=\left(\frac{S}{L}\right)^{\frac{1}{2}}I_0(z)\,.\,\epsilon^{-\rho t} \quad \text{becomes} \quad \frac{W_0}{C_x}=\left(\frac{L}{S}\right)^{\frac{1}{2}}I_0(z)\,.\,\epsilon^{-\rho t} \tag{49}$$

apparently. On examination, it makes nonsense. The reason is that ρ is $=\sigma$ in this case, so a contradiction is involved in keeping ρ the same and reversing σ. The proper way is to start with ρ not $=\sigma$; reverse σ, and then put it There is no contradiction, and the process works properly. We might anticipate some failure in (49) from K being zero whilst R is finite.

(15). The full formula is (53), p. 336, vol. 2, for the direct problem. Or

$$\frac{LvC}{V_0}=\epsilon^{-\rho t}\left[P_0(z)+2\left(\frac{\rho}{\sigma}-1\right)a\{1+a(c-a)+a^2(c-a)^2+\ldots\}\right], \tag{50}$$

where $c=2\rho/\sigma$, and where, on arrangement in powers of a, a^n is to mean

$$a^n=\frac{w^n}{\underline{|n}}P_n(z), \quad w=\tfrac{1}{2}\sigma(t-x/v), \quad z=\sigma(t^2-x^2/v^2)^{\frac{1}{2}}. \tag{51}$$

It is now σ, not ρ, that occurs in w and z. Note that $\rho=\sigma$ makes all terms go out save the first.

Now change the sign of x and of σ. The P's are functions of z^2, so do not change; but w becomes $-\frac{1}{2}\sigma(t+x/v)=-u$, say. Also ρ/σ and c change sign, and the sign of a must be reversed. So,

$$\frac{W_0}{LvC_x}=\epsilon^{-\rho t}\left[P_0(z)+2\left(\frac{\rho}{\sigma}+1\right)a\{1+a(c-a)+a^2(c-a)^2+\ldots\}\right], \tag{52}$$

where

$$a^n=\frac{u^n}{\underline{|n}}P_n(z). \tag{53}$$

This solves the inverse problem, provided t begins at the moment $t=-x/v$. That is, the impressed W_0 makes C_x be zero before and constant after the moment $t=0$. The extra terms are now all retained when we put $\rho=\sigma$. This makes $c=2$, and reduces (52) to

$$\frac{W_0}{LvC_x}=\epsilon^{-\rho t}\left[P_0(z)+4uP_1(z)+8\frac{u^2}{\lfloor 2}P_2(z)+12\frac{u^3}{\lfloor 3}P_3(z)+\ldots\right], \qquad (54)$$

when there is no leakage. We need not always trouble to alter the time zero explicitly. But a check upon the accuracy of the sign reversals is desirable. Say $x=0$. This makes

$$W_0/LvC_0=\epsilon^{-\rho t}\{I_0+4I_1+8I_2+12I_3+\ldots\}(\rho t). \qquad (55)$$

This must be the expansion of

$$\frac{1}{Lv}\left(\frac{R+Lt_1}{St_1}\right)^{\frac{1}{2}}=\left(1+\frac{2\rho}{t_1}\right)^{\frac{1}{2}}=1+\rho t-\tfrac{1}{4}\rho^2t^2+\ldots. \qquad (56)$$

To prove this, use the Expansion theorem of p. 348, vol. 2. Thus,

$$(1+2\rho/t_1)^{\frac{1}{2}}=\epsilon^{-\rho t}\left(\frac{t_1+\rho}{t_1-\rho}\right)^{\frac{1}{2}}\frac{t_1}{t_1-\rho}$$

$$=\epsilon^{-\rho t}\frac{1+a}{1-a}\frac{1+a^2}{(1-a)^2}\frac{1-a^2}{1+a^2}=\epsilon^{-\rho t}\left(\frac{1+a}{1-a}\right)^2. \qquad (57)$$

Expand this result in powers of a, and we get (55), by substituting $I_n(\sigma t)$ for a^n. Besides this corroboration, we see that we may, if we like, derive the solution for C_x from the C_0 solution (57), by altering the meaning of a^n from $I_n(\sigma t)$ to $(u^n/\lfloor n)P_n(z)$.

The four principal values in (54) are first, $t=-x/v$, making $\epsilon^{\rho x/v}$. Second, $t=0$, making

$$(J_0+4J_1+8J_2+12J_3+\ldots)(\rho x/v)$$
$$=1+2(\rho x/v)+\tfrac{3}{4}(\rho x/v)^2-0-5/12.16(\rho x/v)^4+. \qquad (58)$$

Third, $t=x/v$, making

$$\epsilon^{-\rho x/v}\{1+4\rho x/v+4(\rho x/v)^2+4(\rho x/v)^3/\lfloor 2+4(\rho x/v)^4/\lfloor 3+\ldots\}$$
$$=4\rho x/v+\epsilon^{-\rho x/v}. \qquad (59)$$

Fourth, $t=\infty$ makes ∞. The third value is about 40 when $\rho x/v=10$, as in a former case, when the result was 2, in producing constant potential. It is to be constant current now at x. The subsequent rise of W_0 is of no consequence in the calculation relating to a square signal of finite duration, which involves taking a difference.

(16). Another interesting case of (52) is $R=0$, K finite. Here $\sigma=-\rho$, so all the extra terms go out, leaving

$$\frac{W_0}{LvC_x}=\epsilon^{-\rho t}I_0\left[\rho(t^2-x^2/v^2)^{\frac{1}{2}}\right], \qquad (60)$$

where $\rho=K/2S$. The four principal values are

$$\epsilon^{\rho x/v}, \qquad J_0(\rho x/v), \qquad \epsilon^{-\rho x/v}, \qquad \epsilon^{-\rho x/v}/(2\pi\rho t)^{\frac{1}{2}}. \qquad (61)$$

The last is for large t. It tends to zero. The first and third values show the magnitude of the drop in the interval between $t=-x/v$ and $+x/v$. But observe the second value, for $t=0$. It may be positive or negative, or zero, though never greater than +1. Say $\rho x/v = 2\cdot4$, then the four values are about 10, 0, 1/10, 0. Here W_0 drops to zero momentarily, though it does not change sign. But give $\rho x/v$ a very large value, and examine how $\epsilon^{-\rho t}I_0(z)$ behaves as t goes from $-x/v$ to $+x/v$. There is first an immense drop to a small value, followed by oscillations from + to − and back again. There is only a limited number of these oscillations, and when they end W_0 becomes +, and remains +, ending with a long tail, according to the fourth value in (61).

This may seem an extraordinary way of initiating and maintaining a constant current at x. The same phenomenon, in the form of small fluctuations, may be perceived in previous formulæ. Here it is accentuated by the absence of resistance allowing us to examine a simpler formula. It is difficult to give a visible explanation of the necessity of the fluctuations or oscillations after the first big drop, because it is such an involved matter. But as a check upon (60), I have taken it to represent an arbitrarily given impressed voltage at $x=0$, and have worked out the direct problem of finding the current it produces at x. The result is $C=0$ before and constant after $t=0$, as required.

The fall of W_0 towards zero ultimately is easily explained. If we take the time integral of the fourth form in (61), the result varies as $t^{\frac{1}{2}}$, and shows that the time integral of W_0 goes on increasing without limit, though at an exceedingly slow rate. This is to supply current to the very remote parts of the cable (assumed to be infinitely long) that have not been reached by the disturbance. Keeping in the near parts, however, there is constant current at x maintained by an insensible impressed force, tending to zero. The conclusion is that $V=0$, and $C=$constant, having the same value as at x. For there is no resistance in circuit, and when $V=0$, no leakage current.

Then we may ask why C assumes the same value everywhere (in the near parts) ultimately. This is an example of a general property of magnetic induction in plane sheets which was first proved by the distortionless circuit in 1887. Given a line charged and currented in any way, and connected to earth at both ends. All the charge will eventually disappear, by leakage and terminally, but the current will persist if there is no resistance, and redistribute itself so as eventually to make a steady uniform current all along the line, if there is no resistance in circuit anywhere. (See *Elec. Pa.*, vol. 2, p. 144, for particulars.) Now in our present case $C_x =$ constant inexorably. So the rest of the cable (in the near parts) must also have the same current ultimately, as W_0 becomes practically zero. It is done by natural redistribution of current under the circumstance of no resistance. Of course it will be understood that x in the formula is the only place where the value of the current is constant all the time.

To further illustrate the principle concerned, take off the very small residual W_0 altogether. For some time after, no sensible change will occur; but given time enough, the uniform current will all disappear, not by annihilation, but by spreading into the remote parts of the cable, and so attenuating to zero density. When speaking of "current," sheets of magnetic induction should be pictured, not merely their cores.

(17). It will be as well to translate the cable problem into the corresponding one relating to the propagation of plane sheets of magnetic induction in Maxwell's conducting dielectric. Imagine the dielectric to be unbounded one way ($x=\infty$), but terminated at $x=0$ by a perfectly conducting sheet. Let, initially, there be no disturbance. Then, at the moment $t=0$, let a source of magnetic induction of strength f start uniformly all over the plane. That is, f tubes of induction are generated per unit area per second, so that $\int f dt=\int \mu \mathrm{H}\,.\,dx$, on account of the persistence of the induction when once created. Then the equation connecting H at x with f at $x=0$ is

$$\mathrm{H}=\left(\frac{k+ct_1}{g+\mu t_1}\right)^{\frac{1}{2}}\epsilon^{-qx}f. \tag{62}$$

Compare with the first of (48) above. The changes are quite formal, because it is physically the same problem, provided g, or magnetic conductivity, is introduced to imitate R in the cable problem. So, by (50),

$$\frac{\mu v\mathrm{H}}{f}=\epsilon^{-\rho t}\left[\mathrm{P}_0(z)+2\left(\frac{\rho}{\sigma}-1\right)a\{1+a(c-a)+a^2(c-a)^2+\dots\}\right] \tag{63}$$

shows the H at distance x due to f constant, beginning $t=0$, where ρ or $\sigma=g/2\mu \pm k/2c$. In the real Maxwellian dielectric $\rho=-\sigma$, so all the extra terms stay in. Induction is then not only generated at a uniform rate, but persists in the dielectric as well. H begins at x at the moment $t=x/v$, with the value $f\epsilon^{-\rho x/v}$, and then rises up to the final value f later.

Now pass to the inverse problem. (52) shows that

$$\frac{f}{\mu v\mathrm{H}_x}=\epsilon^{-\rho t}\left[\mathrm{P}_0(z)+2\left(\frac{\rho}{\sigma}+1\right)a\{1+a(c-a)+a^2(c-a)^2+\dots\}\right] \tag{64}$$

determines the nature of f needed to make H_x at x begin suddenly and stay constant. This is when there is magnetic conductivity causing destruction of induction after its creation and transmission into the dielectric. Do away with this decay, by $g=0$, or $\rho=-\sigma$, then we come to (60), which makes

$$\frac{f}{\mu v\mathrm{H}_x}=\epsilon^{-\rho t}\mathrm{I}_0\{\sigma(t^2-x^2/v^2)^{\frac{1}{2}}\}, \tag{65}$$

in Maxwell's conducting dielectric. The time integral of f, or the total induction in the dielectric, ultimately varies as $t^{\frac{1}{2}}$. If what remains of f is taken off altogether, then the total induction existent at that moment is conserved. For it cannot pass through the plane boundary, for one thing. It cannot be destroyed by the conductivity of the medium, for another. But it will ultimately attenuate to zero density, by spreading out into the remote parts of the unbounded dielectric.

(18). Yet one thing more. It remains to show how H is established by the determined f in other parts of the medium, say at y, which may have any value from 0 to ∞. It may be convenient now to alter the time zero so that f begins at the moment $t=0$, and H_x when $t=x/v$. Then, by (65),

$$f/\mu v H_x = \epsilon^{-\rho(t-x/v)} I_0\left[\sigma\{(t-x/v)^2-x^2/v^2\}^{\frac{1}{2}}\right]. \tag{66}$$

This f is now to be regarded as a given impressed source, and the H it produces everywhere is wanted. This is a direct problem.

The first step is to expand $I_0(\ldots)$ in the form $\sum A_n I_n(\sigma t)$. Thus

$$I_0(\ldots) = \epsilon^{-xt_1/v} I_0\{\sigma(t^2-x^2/v^2)^{\frac{1}{2}}\} = \epsilon^{-xt_1/v}\,\epsilon^{-x(t_1{}^2-\sigma^2)^{\frac{1}{2}}} I_0(\sigma t). \tag{67}$$

In this, put $t_1=\frac{1}{2}\sigma(a+a^{-1})$, reducing it to $\epsilon^{-\sigma x/va} I_0(\sigma t)$ by one way, or else to the same with a turned to its reciprocal by another. The ambiguity arises from the radical. But a^n and a^{-n} mean the same. with the operand $I_0(\sigma t)$; so both forms expand alike, and make

$$\frac{f}{\mu v H_x} = \epsilon^{-\rho(t-x/v)}\left\{ I_0 - \frac{\sigma x}{v} I_1 + \frac{(\sigma x/v)^2}{\lfloor 2} I_2 - \ldots\right\}(\sigma t). \tag{68}$$

Now this form of result I see represents the series development of a much simpler functional form; and this, again, shows that it conduces to simplicity not to change the time zero so as to make the impressed force begin when $t=0$, but to keep it so that the given *effect* (here H_x) begins then. On this understanding,

$$E_0/H_x = \epsilon^{qx}\mu t_1/q = \mu v\,\epsilon^{-\rho t} P_0(z)_{-x}, \tag{69}$$

because $f=E_0$ numerically. Vectorially, $\mathbf{f}=\text{curl}\,\mathbf{E}_0$. Here z is the regular $\sigma(t^2-x^2/v^2)^{\frac{1}{2}}$, whilst the outside suffix indicates the value to give to x in z. It also indicates the initial value of vt for the function determined. Thus, in (69), t ranges from $-x/v$ to ∞. This notation will be found useful in the following.

The corresponding value of E_y is

$$E_y = \epsilon^{-qy} E_0 = H_x\,\epsilon^{-q(y-x)}\mu t_1/q = \mu v H_x\,\epsilon^{-\rho t} P_0(z)_{y-x}. \tag{70}$$

This is for any point y. The value subsides towards zero, not only at x, but all over as well. On the other hand,

$$H_y = \left(\frac{k+ct_1}{\mu t_1}\right)^{\frac{1}{2}} E_y = \epsilon^{-\rho t}\frac{1-a}{1+a} P_0(z)_{y-x} H_x$$
$$= \epsilon^{-\rho t}(1-2a+2a^2-2a^3+\ldots)P_0(z)_{y-x} H_x. \tag{71}$$

This is H_x steadily at $y=x$; elsewhere it tends to that value in time; subject to the reservation already mentioned.

(19). Now consider the effect of terminal connections, applied to a finite length of cable, beginning with the simple case of terminal short circuits. The reflection of a wave is positive as regards C, and negative as regards V. Let an impressed voltage e be inserted at $x=0$, producing $V_0=e$ there. Then

$$C_1=\left(\frac{K+St_1}{R+Lt_1}\right)^{\frac{1}{2}} 2(\epsilon^{-ql}+\epsilon^{-3ql}+\epsilon^{-5ql}+\dots)V_0=\left(\frac{K+St_1}{R+Lt_1}\right)^{\frac{1}{2}}\frac{2V_0}{\epsilon^{ql}-\epsilon^{-ql}} \quad (72)$$

is the current at $x=l$ due to V_0, being the initial wave followed by its positive reflections, in an infinite series. But if we invert the problem, and write

$$V_0=\left(\frac{R+Lt_1}{K+St_1}\right)^{\frac{1}{2}}\tfrac{1}{2}(\epsilon^{ql}-\epsilon^{-ql})C_1, \quad (73)$$

there are only two waves. This makes the present and other inverse problems much simpler than the direct. But remember that C_1 does not cause V_0 (that would make a distinctly different problem), but is caused by V_0. Suppose C_1 starts when $t=0$, then V_0 begins at the previous moment $t=-l/v$, making the wave $\epsilon^{ql}\dots$; and, at the later moment $t=l/v$, which is the moment the front of the reflection of the first wave reaches the source, a second wave of V_0 begins, given by the formula $\epsilon^{-ql}\dots$. There are no others, because the second impressed force is so graduated and timed as to destroy the effect of all the subsequent reflections which would exist if V_0 impressed continued according to the first formula alone. Both these formulæ have been already worked out, but before noticing them, let us go to the evergreen distortionless circuit for simple information without complication.

Putting $\sigma=0$ makes $q=(t_1+\rho)/v$; so

$$V_0=\tfrac{1}{2}Lv[\epsilon^{\rho l/v}\epsilon^{t_1 l/v}-\epsilon^{-\rho l/v}\epsilon^{-t_1 l/v}]C_1. \quad (74)$$

Here C_1 (as in the preceding also) may be any function of the time, say $f(t)$. Then

$$V_0=\tfrac{1}{2}Lv[\epsilon^{\rho l/v}f(t+l/v)-\epsilon^{-\rho l/v}f(t-l/v)] \quad (75)$$

explicitly. But (74) is equally explicit, in a less conventional way.

The meaning can be seen by letting C_1 be a mere pulse, at the moment $t=0$, and nothing else. Equation (74) shows that two pulses of V_0 are required to produce it, one at the moment $t=-l/v$, magnified $\frac{1}{2}Lv\,\epsilon^{\rho l/v}$ times; the other at the moment $t=l/v$, minified (and reversed) $-\frac{1}{2}Lv\,\epsilon^{-\rho l/v}$ times. To see the significance of the second pulse, consider what the first pulse of V_0 does. It sends a pulse of $V=LvC$ along the line, which attenuates in transit to half the size of the required pulse at the end of the line. The positive reflection of current immediately doubles this, making the required pulse of current at the end be of full size. Then follows the reflected pulse along the line, in which $V = -LvC$. On arriving at the source it is minified by the factor $\epsilon^{-\rho l/v}$. It is negative as regards V and positive as regards C. The usual reflection would then turn it to an ordinary + pulse like the first, to be sent along the line and produce a second pulse at the end, to be followed by an infinite series of smaller pulses. But at the moment $t = l/v$ the second, or negative pulse of V_0 acts. Its effect, by itself, would be an infinite series of pulses. But they are exactly the negatives of the series already described. So, in reality, the second pulse of V_0 stops all further action. There are just two pulses of V_0, two wave pulses in the line, and one doubled pulse of current at its end. It is beautifully simple in operation because it is done with ideally simple machinery. But though not so simple, the action is equally perfect in circuits possessing distortion. The impressed force is of two types, and the second one is constructed so as to cut off the after effects due to the first. If the reader is not familiar with, but wants to understand these things scientifically, and not merely from the analytical, nor yet merely from the practician's point of view, he should read my account (1887) of the properties of the distortionless circuit (*Elec. Pa.*, vol. 2, pp. 119 to 155, and p. 307). It requires no high mathematics to add all sorts of examples in additional illustration.

Equation (74) also shows that if

$$V_0 = \tfrac{1}{2}Lv\,\epsilon^{\rho l/v}C_1, \qquad \text{from } t = -l/v \text{ to } t = 0,$$

is followed by $V_0 = Lv \operatorname{shin}(\rho l/v) \, . \, C_1$, from $t = 0$ onward,
the result is steady C_1 at the end of the line. Here the first V_0 alone, and kept on, would in time make C_1 jump up higher and higher, staircase fashion, to a limiting value. All these later changes are cut off by the alteration in the type of the impressed V_0 which occurs when the reflected wave reaches the source. Note that this reflected wave establishes the steady state as it moves on, by union with the primary wave.

(20). Returning to an ordinary circuit in which the distortion is not destroyed, I will give in brief the full process of developing (73), instead of referring to the results of direct problems worked out in vol. 2, and making changes of sign or other alterations. This will be useful in confirming some of the preceding work, and to show how to work in other cases when it may not be convenient to modify previously worked out solutions.

First put q in terms of t_1, ρ, and σ, making $vq = \{(t_1+\rho)^2 - \sigma^2\}^{\frac{1}{2}}$. Then

$$\frac{V_0}{C_1} = \tfrac{1}{2}Lv\left(\frac{t_1+\rho+\sigma}{t_1+\rho-\sigma}\right)^{\frac{1}{2}}\left\{\epsilon^{\{(t_1+\rho)^2-\sigma^2\}^{\frac{1}{2}}l/v} - \epsilon^{-\text{same}}\right\}1. \tag{76}$$

Now here

$$1 = \epsilon^{-\rho t}\epsilon^{\rho t} = \epsilon^{-\rho t}t_1(t_1-\rho)^{-1}. \tag{77}$$

Substituting in (76) and shifting $\epsilon^{-\rho t}$ to the beginning, makes

$$\frac{V_0}{C_1}=\tfrac{1}{2}Lv\,\epsilon^{-\rho t}\left(\frac{t_1+\sigma}{t_1-\sigma}\right)^{\frac{1}{2}}\frac{t_1}{t_1-\rho}\left\{\epsilon^{\{t_1^2-\sigma^2\}^{\frac{1}{2}}l/v}-\epsilon^{-\text{same}}\right\}1. \tag{78}$$

Now if $u=\frac{1}{2}\sigma(t+x/v)$, $w=\frac{1}{2}\sigma(t-x/v)$, $4uw=z^2=\sigma^2(t^2-x^2/v^2)$, (79)

and a and β are the differentiators with respect to u and w, as in vol. 2, p. 328, then

$$P_0(z)=1+\frac{1}{a\beta}+\frac{1}{a^2\beta^2}+\ldots,\quad a^nP_0(z)=\beta^{-n}P_0(z)=\frac{w^n}{\lfloor n}P_n(z). \tag{80}$$

Also, the identity

$$\frac{1-a^2}{1+a^2}P_0(z)=\left(P_0-2\frac{w^2}{\lfloor 2}P_2+2\frac{w^4}{\lfloor 4}P_4-\ldots\right)(z) \tag{81}$$

reduces when $x=0$ to the identity

$$\frac{1-a^2}{1+a^2}I_0(\sigma t)=1=(I_0-2I_2+2I_4-2I_6+\ldots)(\sigma t). \tag{82}$$

Substitute this for the 1 operand in (78), and make use of

$$\epsilon^{-(t_1^2-\sigma^2)^{\frac{1}{2}}x/v}I_0(\sigma t)=P_0(z), \tag{83}$$

and also put $t_1=\frac{1}{2}\sigma(a+a^{-1})$, because a and a^{-1} are equivalent when the operand is $P_0(z)$. Then (78) becomes, if $c=2\rho/\sigma$,

$$\frac{V_0}{C_1}=\tfrac{1}{2}Lv\,\epsilon^{-\rho t}\frac{1+a}{1-a}\,\frac{1+a^2}{1+a^2-ca}\,\frac{1-a^2}{1+a^2}\{P_0(z)_{-l}-P_0(z)_l\} \tag{84}$$

$$=\tfrac{1}{2}Lv\,\epsilon^{-\rho t}\left(1+\frac{2(\rho/\sigma+1)a}{1-a\;\;c-a)}\right)\{P_0(z)_{-l}-P_0(z)_l\}. \tag{85}$$

This is complete, because on expansion of the a function in rising powers of a, it makes, by (80), a full development in $w^nP_n(z)$ functions. In passing, I remark here that it is not usually desirable to write out such expansions. They can be understood to exist, and it is far simpler to manipulate the a functions which specify them.

In (85) the two z's have the same value, but the associated w factors have not. In the first $w_1=\frac{1}{2}\sigma(t+l/v)$, and in the second, $w_2=\frac{1}{2}\sigma(t-l/v)$. So the two series of P_n functions do not usually destroy one another when coexistent.

(21). But there is a case of great simplification. If $\rho/\sigma=-1$, that is, $R=0$, K finite, it is the case of plane waves in Maxwell's conducting dielectric. We reduce (85) to

$$V_0/C_1=\tfrac{1}{2}Lv[P_0(z)_{-l}-P_0(z)_l]. \tag{86}$$

This says that

$$V_0/C_1=\tfrac{1}{2}Lv\,\epsilon^{-\rho t}J_0\{\rho(l^2/v^2-t^2)^{\frac{1}{2}}\},\qquad\text{from } t=-l/v \text{ to } +l/v. \tag{87}$$

That is, the impressed voltage only lasts for the interval $2l/v$ and then entirely ceases. It ceases the moment the reflected wave from the end reaches the source. It follows that at this moment the steady state is established everywhere.

To confirm this conclusion, find the wave of C anywhere due to V_0. In general,

$$C=\{\epsilon^{-qx}+\epsilon^{-q(2l-x)}+\epsilon^{-q(2l+x)}+\dots\}C_0=\frac{\epsilon^{-qx}+\epsilon^{-q(2l-x)}}{1-\epsilon^{-2ql}}C_0, \qquad (88)$$

being the initial wave and its reflections, all positive. Put $x=l$ to find C_1 in terms of C_0. Then eliminate C_0. The result is

$$C=\tfrac{1}{2}\{\epsilon^{q(l-x)}+\epsilon^{-q(l-x)}\}C_1. \qquad (89)$$

This is also relatively simple. C in the line consists of the primary wave and its reflection. The full development is the same as for ϵ^{-qx}, with two special values given to x. Thus,

$$\frac{C}{C_1}=\tfrac{1}{2}\epsilon^{-\rho t}\frac{1-a}{1+a}\left(1+\frac{2(\rho/\sigma+1)}{1-a(c-a)}\right)\{P_0(z)_{x-l}+P_0(z)_{l-x}\}. \qquad (90)$$

Here $\rho=-\sigma$ reduces the result to

$$\frac{C}{C_1}=\tfrac{1}{2}\epsilon^{-\rho t}\frac{1-a}{1+a}\{P_0(z)_{x-l}+P_0(z)_{l-x}\} \qquad (91)$$

$$=\tfrac{1}{2}\epsilon^{-\rho t}\left\{P_0(z_1)-2w_1P_1(z_1)+2\frac{w_1^2}{\lfloor 2}P_2(z_1)-2\frac{w_1^3}{\lfloor 3}P_3(z_1)+\dots\right\}$$

$$+\tfrac{1}{2}\epsilon^{-\rho t}\left\{P_0(z_2)-2w_2P_1(z_2)+2\frac{w_2^2}{\lfloor 2}P_2(z_2)-2\frac{w_2^3}{\lfloor 3}P_3(z_2)+\dots\right\}, \qquad (92)$$

where

$$w_1=\tfrac{1}{2}\sigma\{t+(l-x)/v\},\quad w_2=\tfrac{1}{2}\sigma\{t-(l-x)/v\},\quad z_1=z_2=\sigma\{t^2-(l-x)^2/v^2\}^{\frac{1}{2}}.$$

Now in vol. 2, p. 334, eq. (43), I showed that

$$\epsilon^{\sigma t}=P_0(z)+\frac{u+w}{\lfloor 1}P_1(z)+\frac{u^2+w^2}{\lfloor 2}P_2(z)+\dots, \qquad (93)$$

and called it an important identity. Applying it to (92), remembering that $\rho=-\sigma$, it proves that $C=C_1$ in the overlap, that is, in the region where the primary wave and its reflection coexist. There is an easier way of showing that this is so, to be come to later; but it is satisfactory to have a confirmation of the accuracy of elaborate formulæ.

From the steadiness of C in the overlap, it follows that $V=0$; and this we can confirm easily. For the V formula is

$$V=\tfrac{1}{2}\left(\frac{R+Lt_1}{K+St_1}\right)^{\frac{1}{2}}\{\epsilon^{q(l-x)}-\epsilon^{-q(l-x)}\}C_1, \qquad (94)$$

which develops to

$$\frac{V}{LvC_1}=\tfrac{1}{2}\epsilon^{-\rho t}\left(1+\frac{2a(\rho/\sigma+1)}{1-a(c-a)}\right)\{P_0(z)_{x-l}-P_0(z)_{l-x}\}. \qquad (95)$$

Here it is visible by inspection that $\rho=-\sigma$ makes $V=0$ in the overlap.

Summing up, we find that when $R=0$ the impressed force

$$V_0=\tfrac{1}{2}LvC_1\,\epsilon^{-\rho t}\,J_0\{\rho(l^2/v^2-t^2)^{\frac{1}{2}}\}, \qquad (96)$$

acting at $x=0$ from $t=-l/v$ to $t=+l/v$, produces the wave

$$V=\tfrac{1}{2}LvC_1\,\epsilon^{-\rho t}\,J_0[\rho\{(l-x)^2/v^2-t^2\}^{\frac{1}{2}}], \qquad (97)$$

from $t=-$ to $+(l-x)/v$, and $C=C_1$, $V=0$, in the overlap. The region occupied by the V wave (97) begins at $x=0$, increases to $x=l$, and then falls back again to the source. At the moment $t=0$, when V occupies the whole line, it may be all of one sign, or else be wavy, according to the size of ρ.

(22). To make an equally simple problem concerning a real cable, in which R is finite, and K zero (practically), let the far end be insulated, making $C_1=0$ there, with + reflection of V and − of C. At $x=0$ put on e in short-circuit producing $e=V_0$, to be found in such a way that V_1 at the far end is steady, beginning when $t=0$. We have

$$V=\frac{\epsilon^{-qx}+\epsilon^{-q(2l-x)}}{1+\epsilon^{-2ql}}V_0, \qquad C=\left(\frac{K+St_1}{R+Lt_1}\right)^{\frac{1}{2}}\frac{\epsilon^{-qx}-\epsilon^{-q(2l-x)}}{1+\epsilon^{-2ql}}V_0. \quad (98)$$

The first of these makes $V_0=\frac{1}{2}(\epsilon^{ql}+\epsilon^{-ql})V_1$, and this, used in (98), produces

$$V=\tfrac{1}{2}(\epsilon^{q(l-x)}+\epsilon^{-q(l-x)})V_1, \quad C=\left(\frac{K+St_1}{R+Lt_1}\right)^{\frac{1}{2}}\tfrac{1}{2}(\epsilon^{q(l-x)}-\epsilon^{-q(l-x)})V_1. \quad (99)$$

The first develops to

$$\frac{V}{V_1}=\tfrac{1}{2}\epsilon^{-\rho t}\left(1+\frac{2a(\rho/\sigma-a)}{1-a(c-a)}\right)\{P_0(z)_{x-l}+P_0(z)_{l-x}\}, \quad (100)$$

and LvC/V_1 is obtained by multiplying by $(1-a)(1+a)^{-1}$, and changing the sign of the second P_0.

Now here it is by $\rho=\sigma$, that is, $K=0$, that we destroy the complication. We reduce to

$$\frac{V}{V_1}=\tfrac{1}{2}\epsilon^{-\rho t}\frac{1+a}{1-a}\left[P_0(z)_{x-l}+P_0(z)_{l-x}\right], \quad (101)$$

$$\frac{LvC}{V_1}=\tfrac{1}{2}\epsilon^{-\rho t}\left[P_0(z)_{x-l}-P_0(z)_{l-x}\right]. \quad (102)$$

It follows now that the current is zero in the overlap. This requires that $-dV/dx=0$ in the overlap; or V is steady, $=V_1$. The identity (93) applied to (101), confirms this conclusion. The steady state of constant V and zero C begins at the insulated end of the line at the moment $t=0$, and travels at speed v to the beginning of the line. When this is reached, the impressed voltage e, which was previously variable, assumes the constant value V_1. After that, nothing more happens. Nor will anything more happen if we remove e, provided we insulate the beginning, which is a common practical procedure.

Having thus charged the cable to uniform potential V in the short period $2l/v$, it may be discharged completely equally quickly by applying the same voltage reversed.

(23). The instantaneous assumption of steady states is not wonderful on a distortionless circuit, because that is made so as to do it naturally. But it is wonderful when the circuit is not balanced for distortion, until one gets used to the idea. It is merely a question of habit; wonder always ceases with sufficient familiarity. Nothing is really more wonderful than anything else, and nobody really understands anything. In the present case the wonder ceases when it comes to be considered that the impressed force itself which produces the electromagnetic wave, is graduated in its strength just in the right way to counterbalance the disturbing effect of the unbalanced distortion on the circuit, as well as the reflective effects. The source itself is distorted, so to speak.

But, so far, I have only proved this property for the two simplest terminal conditions, earth or insulation at the distant end, and then only when either he leakance or the resistance is zero. I shall now, however, show that the same property is true when both these quantities are finite, and have any values; and moreover, when there is terminal apparatus, which may be of several different types, though it cannot be of any type. The effect of the line distortion, and of the terminal apparatus, will all be allowed for in such a way as not only to produce right-angled arrival curves, which is always possible, but also to make the reflected wave, by union with the primary wave, at once produce the steady state of V, or of C, or of both together sometimes. This can be proved without actually developing the form of the primary wave, because the result of its union with the reflected wave may be found by uniting the two waves in differential operational form, instead of in functional form. Of course the development of the primary wave is needed to complete the solution of the problem.

If, after the manner of vol. 2, p. 386, ρ_0 and ρ_1 are the reflection coefficients at $x=0$ and l, the complete V due to an impressed source at $x=0$ is

$$V=\{\epsilon^{-qx}+\rho_1\epsilon^{-q(2l-x)}+\rho_0\rho_1\epsilon^{-q(2l+x)}+\rho_0\rho_1^2\,\epsilon^{-q(4l-x)}+\dots\}f. \quad (103)$$

Here f is not V_0, but is what V_0 would be if the line were infinitely long. This V shows the primary wave and the succeeding reflected waves. Or, condensed,

$$V=\frac{\epsilon^{-qx}+\rho_1\epsilon^{-q(2l-x)}}{1-\rho_0\rho_1\epsilon^{-2ql}}f, \qquad C=\left(\frac{K+St_1}{R+Lt_1}\right)^{\frac{1}{2}}\frac{\epsilon^{-qx}-\rho_1\epsilon^{-q(2l-x)}}{1-\rho_0\rho_1\epsilon^{-2ql}}f. \quad (104)$$

By putting $x=l$, it follows that

$$f=\frac{\epsilon^{ql}-\rho_0\rho_1\epsilon^{-ql}}{1+\rho_1}V_1. \quad (105)$$

Put this f in (104). It is then dismissed for good. We get

$$V=\frac{\epsilon^{q(l-x)}+\rho_1\epsilon^{-q(l-x)}}{1+\rho_1}V_1, \quad C=\left(\frac{K+St_1}{R+Lt_1}\right)^{\frac{1}{2}}\frac{\epsilon^{q(l-x)}-\rho_1\epsilon^{-q(l-x)}}{1+\rho_1}V_1. \quad (106)$$

Here V and C in the line are expressed in terms of the terminal potential. Since, moreover, $V_1=Z_1C_1$, if Z_1 is the resistance operator of the $x=l$ arrangement, we also express V and C in terms of C_1. Either of these may be given as a datum; or, if we please, some other quantity in Z_1 itself. In any case, the fundamental point is that some quantity away from the source (or, if we please, at the source, or another part of the terminal arrangement there) may be regarded as given, and everything else has to be found to match, including the intensity of the source itself. Thus, given V_x, find V_y. With a given situation and kind of source, this also involves V and C being found everywhere, *and* the intensity of the source.

As before, when the source is at $x=0$, there are only two waves, in contrast with the multiplicity in direct problems. One of these, the primary wave, must be found. But the other need not, though it comes from the primary by easy changes. For we may find the V and C in the overlap independently.

Observe that neither e nor ρ_0 occur in (106). This means that for the same V_1, and without any source to the right of $x=0$, V and C everywhere to the right of $x=0$ are always the same. If we alter Z_0, there is a balancing alteration required in e; or if we alter the situation of e in Z_0, we must correspondingly alter its intensity. In fact the determination of e forms a distinct problem.

But ρ_1 must be put in terms of Z_1. If

$$Z=\left(\frac{R+Lt_1}{K+St_1}\right)^{\frac{1}{2}}, \text{ then } \rho_1=\frac{Z_1-Z}{Z_1+Z}, \text{ and } \frac{1}{1+\rho_1}=\tfrac{1}{2}\left(1+\frac{Z}{Z_1}\right). \quad (107)$$

This brings (106) to

$$V=\frac{(Z_1+Z)\epsilon^{q(l-x)}+(Z_1-Z)\epsilon^{-q(l-x)}}{2Z_1}V_1,$$
$$C=q\frac{(Z_1+Z)\epsilon^{q(l-x)}-(Z_1-Z)\epsilon^{-q(l-x)}}{2ZZ_1}V_1. \quad (108)$$

Before $t=(l-x)/v$ only the first exponential is active. But after that, when both are active, we may unite them, making

$$V=\left\{\cosh q(l-x)+\frac{Z}{Z_1}\operatorname{shin} q(l-x)\right\}V_1,$$
$$C=\left\{\frac{q}{Z}\operatorname{shin} q(l-x)+\frac{q}{Z_1}\cosh q(l-x)\right\}V_1. \quad (109)$$

To obtain the ultimate steady state of V and C, put $t_1=0$. This makes

$$q=\sqrt{RK}, \qquad Z=\sqrt{R/K}, \qquad Z_1=R_1, \quad (110)$$

if R_1 is the effective steady resistance of the Z_1 apparatus. Then

$$V=\{\cosh\sqrt{RK}(l-x)+\sqrt{R/K}/R_1 \,.\, \operatorname{shin}\sqrt{RK}(l-x)\}V_1, \quad (111)$$

$$C=\{\sqrt{K/R}\operatorname{shin}\sqrt{RK}(l-x)+1/R_1 \,.\, \cosh\sqrt{RK}(l-x)\}V_1. \quad (112)$$

This process of finding the ultimate state is valid irrespective of whether direct or inverse differential operations are concerned, or whether they are rational or irrational. If there is an ultimate steady state, then $t_1=0$ in (109) will find it. If there is no steady state, the form of the operators will show why, and the physics will also do so. As regards irrationality, Z_1 might be a cable itself.

(24). Now in general it takes an infinite time to reach the steady state. But in these inverse problems the time taken may be either finite or infinite. We may discriminate by examining the forms of the operators in (109). Looking at the V_1 formula, the cosh operator is a rational integral function of t_1. So is the shin operator when divided by q. If, then, Zq/Z_1 or $(R+Lt_1)/Z_1$ is also of the same kind, then $V=(A+Bt_1+Ct_1^2+...)V_1=AV_1$, when V_1 is constant. But this is the steady state. It is therefore assumed suddenly at the moment $t=(l-x)/v$ whenever $(R+Lt_1)/Z_1$ is rational integral. This occurs when Z_1 is a mere resistance, of any size; or a conducting condenser, say, $Z_1=(K_1+S_1t_1)^{-1}$, of any size and any time constant. But if it is a coil, say $Z_1=R_1+L_1t_1$, then it fails, unless the time constant L_1/R_1 is the same as L/R. If Z_1 is a condenser and coil in sequence, it cannot be done. But if the coil and condenser are in parallel, then it can, provided $R/L=R_1/L_1$ as before. In any of these cases (and similar ones) V is given by (111) in the overlap, from the moment $t=(l-x)/v$, being the steady state of V, beginning at $x=l$, and travelling towards $x=0$ at speed v.

But it does not follow that C is also steady in the overlap when V is. Inspection of the C formula in (109) shows that $1/Z_1$ must be rational integral. That is, Z_1 may be a resistance, or a condenser; but not a coil, not even when its time constant is the same as L/R. But the difference made is easily found.

If a coil, everything is rational integral except the term $(Lv/Z_1)V_1$, the steady part of which is $(Lv/R_1)V_1$, and is included in the steady solution. So there is an additional term

$$\left(\frac{Lv}{Z_1}-\frac{Lv}{R_1}\right)V_1=-\frac{Lv}{R_1}\frac{t_1}{t_1+R_1/L_1}V_1=-\frac{LvV_1}{R_1}\epsilon^{-R_1t/L_1}. \qquad (113)$$

That is, in the overlap, C is given by (112), with the extra current (113) added. Though not steady, it is at a given moment constant within its range. At the moment $t=l/v$, and after, this extra current occupies the whole line, besides the steady state of V and C.

The above is an easy example of the finding of the state of affairs in the overlap when there is only a partial assumption of the steady state. It was complete as regards V, partial as regards C. But it may easily be the other way. V may require extra terms, not C. In general, both V and C require extra terms, and it may, even when there are several, be of advantage to exhibit them separately. In any case, the above shows how to find the resultant state in the overlap, without using the primary wave and its reflection developed in terms of x and t. Separate the operator into a rational integral part and a part requiring integrations. The first part, together with the steady part of the second, make the steady state. What is left over constitutes the extra terms.

If Z_1 is a cable of a different type to the original, its resultant effect on the first cable will be fully allowed for. Besides that, we may calculate the state of things in the new cable, since it is acted upon by the impressed voltage V_1. This makes a direct problem. But if the new is of the same type as the old cable, it will act just like a continuation of the old. Then cosh and shin will unite to make just the original $\epsilon^{q(l-x)}$ operator considered before we came to terminal reflection. That is, there is no second wave, and no overlapping to make a steady state, without or with extra terms.

(25). To complete the solution for the main circuit, it remains to express the primary wave in functional form, valid before overlapping begins at any place. To do this, reduce t_1 to $t_1-\rho$ in q and Z_1, and put t_1 in terms of a as before. Let ρ_1 become σ_1 in this way, and Z_1 become $\overline{Z}_1$. The results are

$$\frac{V}{V_1}=\epsilon^{-\rho t}\frac{1}{1+\sigma_1}\,\frac{1-a^2}{1+a^2-ca}P_0(z)_{l-x}, \tag{114}$$

$$\frac{LvC}{V_1}=\epsilon^{-\rho t}\frac{1}{1+\sigma_1}\,\frac{(1-a)^2}{1+a^2-ca}P_0(z)_{l-x}, \tag{115}$$

in which only σ_1 needs specification in terms of a to make them complete.

Do this for a coil, $Z_1=R_1+L_1t_1$.

$$\rho_1=\frac{Z_1-Z}{Z_1+Z},\qquad \sigma_1=\frac{\overline{Z}_1-Lv(1+a)(1-a)^{-1}}{\overline{Z}_1+Lv(1+a)(1-a)^{-1}}, \tag{116}$$

$$\frac{1}{1+\sigma_1}=\frac{\overline{Z}_1+Lv(1+a)(1-a)^{-1}}{2\overline{Z}_1}=\tfrac{1}{2}+\frac{\tfrac{1}{2}Lv(1+a)(1-a)^{-1}}{R_1+L_1\sigma(1+a^2-ca)/2a}. \tag{117}$$

This is for the general case of any R and K, and any R_1 and L_1. But if $K=0$, or $c=2$, and $R/L=R_1/L_1$, it reduces to

$$\frac{1}{1+\sigma_1}=\tfrac{1}{2}+\frac{Lv}{R_1}\,\frac{2a}{1-a^2}. \tag{118}$$

The expansion of the a function then brings the primary V wave to the form $\epsilon^{-\rho t}\Sigma A_n w^n P_n(z)$. Then changing the sign of $l-x$ produces the second wave. Their sum is known; so we have the construction of an identity like (93), though far more complicated. Similarly as regards the C waves.

(26). The above was simplified by the absence of any necessary consideration of e and Z_0, the source and the electrical combination in which it is situated. Z_0 may be any combination, and e may be in any part of it suitable for supplying current to the line. It is useless to formulate any general connection between e and V_0 or V_1. We must specialize in some way. The easiest of all is to put e in between Z_0 and the line. This makes

$$\frac{f}{Z}=\frac{e}{Z_0+Z}, \qquad \text{where} \qquad Z=\left(\frac{R+Lt_1}{K+St_1}\right)^{\frac{1}{2}}. \tag{119}$$

This Z is the resistance operator when the line is infinitely long, and f is the V_0 to match. But, by (104) and (106), the real V_0 is

$$V_0=\frac{1+\rho_1\epsilon^{-2ql}}{1-\rho_0\rho_1\epsilon^{-2ql}}f, \qquad \text{and} \qquad V_0=\frac{\epsilon^{ql}+\rho_1\epsilon^{-ql}}{1+\rho_1}V_1; \tag{120}$$

therefore

$$f=\frac{\epsilon^{ql}-\rho_0\rho_1\epsilon^{-ql}}{1+\rho_1}V_1 \qquad \text{and} \qquad e=\left(1+\frac{Z_0}{Z}\right)\frac{\epsilon^{ql}-\rho_0\rho_1\epsilon^{-ql}}{1+\rho_1}V_1, \tag{121}$$

which gives e in terms of V_1. But here

$$\rho_0=\frac{Z_0-Z}{Z_0+Z}, \quad \text{so} \quad 1+\frac{Z_0}{Z}=\frac{2}{1-\rho_0}, \quad \text{and} \quad e=\frac{2(\epsilon^{ql}-\rho_0\rho_1\epsilon^{-ql})}{(1-\rho_0)(1+\rho_1)}V_1, \tag{122}$$

in terms of ρ_0 and ρ_1.

There is another view of the matter, avoiding f. For

$$C_0=\frac{e}{Z_0+|Z|}=\frac{V_0}{|Z|}, \tag{123}$$

where $|Z|$ is the complete resistance operator of the line, with allowance made for the far end Z_1. It is given by

$$|Z|=\frac{V_0}{C_0}=\left(\frac{R+Lt_1}{K+St_1}\right)^{\frac{1}{2}}\frac{\epsilon^{ql}+\rho_1\epsilon^{-ql}}{\epsilon^{ql}-\rho_1\epsilon^{-ql}}. \tag{124}$$

This way also leads to the same relation between e and V_1.

As regards functionizing e, do as before. Put $t_1=0$ to find the steady value reached; then find the value when both waves are on, from the moment $t=l/v$, and then finally

$$e=\frac{2\epsilon^{ql}}{(1-\rho_0)(1+\rho_1)}V_1$$

finds e from $t=-l/v$ up to $t=l/v$, in the form $\Sigma A_n v^n P_n(z)_{-l}$. Of course the V_0 solution may be utilized, by putting $(1-\rho_0)^{-1}$ in terms of α.

The source e does not necessarily begin to act at the moment $t=-l/v$ when it is not situated between e and Z_0. It will do so when Z_0 is a finite combination of finite coils and condensers, that is, according to the usual theory of such things, which entirely ignores propagation of effects in time by assuming $v=\infty$ in the medium surrounding them. But if Z_0 is a cable, it is clear that e may be shifted back any distance, and will then have to begin to act so much the earlier in order that V and C may be the same at the same time in the line; besides, of course, varying in a different manner.

(27). The following is to illustrate simply the sudden establishment of V and C in a line by the passage of the reflected wave, and also the establishment of a uniform extra current along with the steady state. The last is so curious as to make explicit exhibition in the nature of the primary and reflected waves desirable. The distortionless circuit will show the phenomenon plainly.

But first let Z_1 be a condenser. This case is easier, and has also an interesting peculiarity. Let $Y_1 = K_1 + S_1 t_1$ be the conductance operator of the condenser. Then since $vq = t_1 + \rho$ in the distortionless circuit, equation (108) becomes

$$V = [\epsilon^{(\rho+t_1)(l-x)/v}\tfrac{1}{2}(1+LvY_1) + \epsilon^{-\text{same}}\tfrac{1}{2}(1-LvY_1)]V_1. \qquad (125)$$

Here the primary wave is

$$\epsilon^{\rho(l-x)/v}[\tfrac{1}{2}(1+LvK_1)V_1 + \tfrac{1}{2}LvS_1\epsilon^{t_1(l-x)/v}\,t_1 V_1]. \qquad (126)$$

The first part, extending from $x=0$ up to the wave front, is steady. But the second part represents a pulse at the wave front itself, of such size that when it reaches the terminal condenser the charge it carries is $\frac{1}{2}S_1V_1$. But the condenser acts (momentarily) as a short-circuit, reflecting V negatively and C positively. So, whilst half the proper charge is sent in by the pulse, an equal amount of negative charge simultaneously comes out. The condenser therefore becomes fully charged instantly. At the same time the negative charge that comes out makes a negative pulse at the front of the reflected wave.

Superposing the two waves produces

$$V = \{\cosh \rho(l-x)/v + LvK_1 \text{ shin } \rho(l-x)/v\}V_1 \qquad (127)$$

in the overlap. This is the steady state. But remember the second pulse at the front. The moment it reaches $x=0$ (attenuated), it does nothing more, for the impressed V_0 destroys it. The steady state is now established all over.

(28). The case of a terminal coil with V_1 steady is easier in one way, because there are no pulses, and harder in another, on account of the extra current. We have

$$\frac{V}{V_1} = \epsilon^{\rho(l-x)/v}\,\epsilon^{t_1(l-x)/v}\,\tfrac{1}{2}\left(1+\frac{Lv}{R_1+L_1t_1}\right) + \text{second wave}. \qquad (128)$$

The second wave is got by turning v to $-v$ all through the expression for the first wave. Also $V = LvC$ in the first, and $V = -LvC$ in the second wave.

The final steady state is

$$V/V_1 = \cosh \rho(l-x)/v + Lv/R_1 \text{ shin } \rho(l-x)/v, \qquad (129)$$

from which LvC/V_1 final is got by changing the sign of the second term. But this state of V is not set up immediately by the reflected wave in general. For the primary wave is

$$\frac{V}{V_1} = \epsilon^{\rho(l-x)/v}\tfrac{1}{2}\left\{1+\frac{Lv}{R_1}\left(1-\epsilon^{-R_1\{t+(l-x)/v\}/L_1}\right)\right\}, \qquad (130)$$

in the region from $x=0$ up to $x=vt+l$ at the moment t, before the end is reached, and still existent alone up to $x=l-vt$, after that.

On uniting (130) with the reflected wave, the exponential ϵ^{-R_1t/L_1} does not go out from the result, except when $R/L=R_1/L_1$. In this case, V in the overlap is the steady V of (129).

Similarly as regards C, with the difference that even when the time constants are equal, the exponential $\epsilon^{-\rho t}$ remains in the overlap, where C is now the steady C *plus* the extra current $-(V_1/R_1)\epsilon^{-\rho t}$. This extra current in the overlap does not interfere with the steady state of V and C (it is merely to be regarded as superimposed). Nor is there any real discontinuity at the front of the return wave. The extra current is necessitated by the datum V_1=constant from the moment $t=0$. It can only produce C_1 in the coil in the gradual manner $C_1=(V_1/R_1)(1-\epsilon^{-R_1t/L_1})$. The "extra current" is the negative part of this, and continues. So it shows completely in the overlap, and is also present (halved) in the two waves, and is fully allowed for in the variable strength of the source. After the moment $t=l/v$, the source becomes steady, and the extra current exists through the whole circuit.

But if it is given that C_1, the coil current, is to be steady, then there is a pulse wave sent out to energise the coil instantly, and a return pulse along the line which is destroyed by the source of energy. In fact, the V formula is now

$$V=\tfrac{1}{2}\epsilon^{(\rho+t_1)(l-x)/v}\{R_1+L_1t_1+Lv\}\,C_1+(\text{same with}-v), \qquad (131)$$

showing the pulse wave, and also that there is now no extra current. So both V and C are steady in the overlap.

(29). In the above I have taken V_1=constant, or C_1=constant, after $t=0$, to be the given datum, because it is most useful in general. By a time differentiation the datum becomes impulsive, so the solution for that case is known immediately. Moreover, a succession of pulses makes a continuous function in the limit, so we can write definite integrals to express V and C in terms of the datum expressed as any function of the time. But such integrals are only ornamental as a rule, and not worth writing out.

The practical way is the operational. Say V_1 is given as a function of the time, beginning $t=0$. Put it in terms of t_1; then the operational solution for V is made entirely operational with operand 1, and can be developed in the usual way. Or, put $V_1\epsilon^{\rho t}$ in terms of t_1, and the result in terms of a, after the manner of vol. 2, p. 347. The developed solutions obtained in this way, to find V and C to make V_1 or C_1 vary as a definite function of the time, are sometimes much simpler than those in which V_1 or C_1 are constant. The operational solutions themselves (in terms of t_1) will suggest particular forms of V_1 or C_1 which will simplify the results. For example, if in the problem of which (114), (115) is the solution for V_1=constant, we alter things so that $V_1\epsilon^{\rho t}$ is constant, the effect is the same as making $c=0$ in those formulæ, thus simplifying the expansions greatly. But the treatment of such modifications is, after the preceding, too obvious to require separate development.

(30). Cylindrical electromagnetic waves present some interesting features in connection with the relation between a source of disturbance and its consequences, or equivalent pre-existing states. By the selection of relatively simple solutions for examination, a good deal more can be learnt than by the construction of very complicated ones.

Consider first the fundamental wave solution

$$\mathrm{H}=\frac{f}{2\pi\mu v(v^2t^2-r^2)^{\frac{1}{2}}}, \qquad \mathrm{E}=\frac{fvt/r}{2\pi(v^2t^2-r^2)^{\frac{1}{2}}}. \tag{132}$$

Here H is defined to be the magnetic force and E the electric force at time t and at distance r from a straight axis, on the understanding that H is parallel to the axis, and E circular round it. Test that they satisfy

$$-\frac{d\mathrm{H}}{dr}=c\dot{\mathrm{E}}, \qquad -\frac{1}{r}\frac{d}{dr}r\mathrm{E}=\mu\dot{\mathrm{H}}, \tag{133}$$

which are the reduced forms of the two circuital laws to suit the circumstances, if $\mu cv^2=1$. The wave front is at $r=vt$, and since E and H have the same sign when t is +, the wave motion is outward. It is easy to show that

$$\int_0^{vt} 2\pi r\mu\mathrm{H}dr=ft. \tag{134}$$

This is the cardinal property of the wave. It means that the total induction is increasing at a uniform rate, and that to complete the system there must be a source of induction at the axis, generating the amount f per second steadily, and starting at the moment $t=0$. The ∞ value at the wave front is of no consequence, because it is empty. See p. 39 for a diagram of the H wave going out. The differential form is

$$\mathrm{H}=\frac{1}{4\mu v}\mathrm{K}_0(qr)qf, \qquad \text{if} \qquad q=d/d(vt); \tag{135}$$

and in fact (132) is derived from (135) by multegration. Compare with (170), p. 264.

The straight line source f may be regarded as a condensation of curl **e**, if **e** is impressed electric force. It may also be regarded as an impressed linear magnetic current of strengh $-f$, or a current of magnetons, if allowance is made for the radial magnetic force. But if we have **E** straight, and **H** circular, and write

$$\mathbf{E}=-\frac{\mu v\mathrm{C}}{2\pi(v^2t^2-r^2)^{\frac{1}{2}}}, \qquad \mathrm{H}=\frac{(vt/r)\mathrm{C}}{2\pi(v^2t^2-r^2)^{\frac{1}{2}}}, \tag{135A}$$

then the linear "source" C generates displacement cE negatively, at the rate $-$C. It is an impressed electric current of strength C, or a current of electrons (with proper allowance made for the outward electric force, if necessary), or it may be **C** = curl **h**, condensed, if **h** is impressed magnetic force.

(31). Now make t negative in (132). H is the same and E is reversed, compared with the values for equally large positive t. The wave is therefore (if it is possible) going in to the axis, and the total induction is decreasing at uniform rate, or is being swallowed up at the axis. This looks like a negative axial source, of strength $-f$. But if there were such a source, beginning at the moment $t=s$, it would produce an outward wave like (132), with f turned to $-f$ and t to $t-s$. Even if we make $s=-\infty$, and reduce this new H wave to nothing (within range), the source $-f$ will still be associated with it, and the supposed inward wave (132) is left without a source at the axis. This incoherency shows that (132) is not a possible wave of H and E when t is negative. Besides that, E in the new wave does not tend to vanish, like H.

But by adding to the inward wave, which is impossible by itself, an outward wave whose source at the axis cancels the sink of the inward wave, we come to a possible combination, without any source anywhere within range, though there may be or have been at some previous time sources at a distance. The pair of waves makes a perfectly free combination. Say that

$$\mathrm{H}=\frac{f_1}{2\pi\mu v\{v^2(t-s_1)^2-r^2\}^{\frac{1}{2}}}+\frac{f_2}{2\pi\mu v\{v^2(t-s_2)^2-r^2\}^{\frac{1}{2}}}, \tag{136}$$

with corresponding E's, after the manner of (132). These may mean that there are two outward waves; one due to f_1 beginning when $t=s_1$, the other due to f_2, beginning when $t=s_2$. But if $f_1=-f_2$, there is no axial source when both are on at once, or any source within range. Then the question is, how far back in time can (136) exist? Examination shows that the time may go back to any extent, *provided* we reverse the sign of H_1 or H_2 when either wave disappears at the axis, without reversing the corresponding E. The leading idea to be kept in mind is that the total induction must be kept constant.

Begin with a pair of inward waves. H must be + in one and $-$ in the other. The total induction is $f(s_2-s_1)$, positive when f is positive and $s_2>s_1$. This is the first stage, when $t-s_1$ and $t-s_2$ are both negative. But when the wave 2 reaches the axis and disappears, if it remained non-existent thenafter, the inward wave 1 would be impossible. So let the change at the axis consist in the continuance of the wave 2 as an outward wave, with E the same and H reversed. The second figure shows the second stage, an inward wave 1 and outward 2, both with positive H, and such that the wave 2 gains induction as fast as the wave 1 loses it. At the moment $t=\frac{1}{2}(s_1+s_2)$, there is a momentary coincidence, with zero E and doubled H.

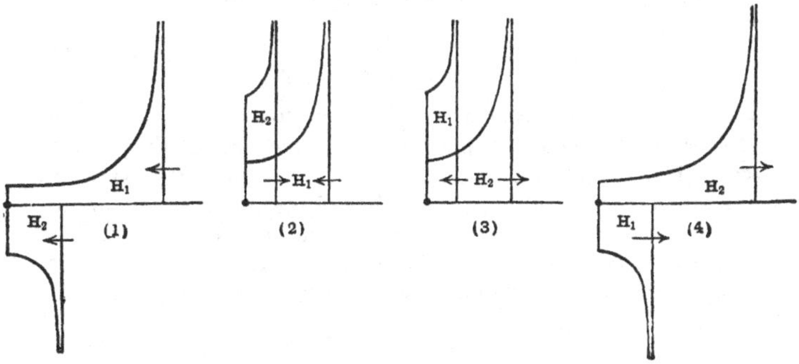

After that the waves separate, and we have the third stage. Then the inward wave 1 reaches and disappears at the axis, to emerge as an outward wave of negative H. This is the last stage, for we now have a pair of waves going out with nothing to stop them.

Incidentally, we solve this problem. Given the initial state at the moment $t=\frac{1}{2}(s_1+s_2)$,

$$H=\frac{2f}{2\pi\mu v\{v^2(\frac{1}{2}s_2-\frac{1}{2}s_1)^2-r^2\}^{\frac{1}{2}}}, \tag{137}$$

without any E. What happens, if there is no source? The answer is, the generation of a pair of electromagnetic waves like (136), with similar H's and opposite E's. The first goes out, the second goes in to the axis and comes out again with H reversed.

We also solve this problem. Given the initial state (132) of E and H, at the moment $t=s$, but without any source. What happens? The answer is

$$H=\frac{f}{2\pi\mu v(v^2t^2-r^2)^{\frac{1}{2}}}-\frac{f}{2\pi\mu v\{v^2(t-s)^2-r^2\}^{\frac{1}{2}}}, \tag{138}$$

from $t=s$ onward. A new outward wave of negative H instantly begins at the axis.

All conditions of continuity, &c., are satisfied by the above, and yet there may seem to be something mysterious about the reflected waves from the axis, in the reversal of H. But it is not the true reflection that occurs when a proper solitary wave reaches the axis, without any source, as will be seen later. It is a property not of one wave, but of a pair of waves, of a particular type, and is necessitated by the condition of absence of a source of induction.

(32). If the two waves in the first diagram differ only slightly in phase, so that their fronts (or rather backs) are close together, the resultant is a thin slab of E and H to represent the remaining solitary portion of wave 1. and a continuous negative distribution for the overlap. Head and tail, so to speak, only the tail goes in first, and is followed by the head. In the limit, with magnification, we come to the wave

$$\mathrm{H}=f_0\frac{d}{dt}\frac{1}{2\pi\mu v(v^2t^2-r^2)^{\frac{1}{2}}}. \tag{139}$$

This is, when t is $+$, the wave of H due to an impulsive source generating the amount of induction f_0 at the axis at the moment $t=0$. But it may also be interpreted without any axial source as first an inward wave of H, going in tail first, followed by an outward wave of the same type, going out head first. The details of the transformation from the inward to the outward wave are best followed in the previous manner, considering the heads to be of

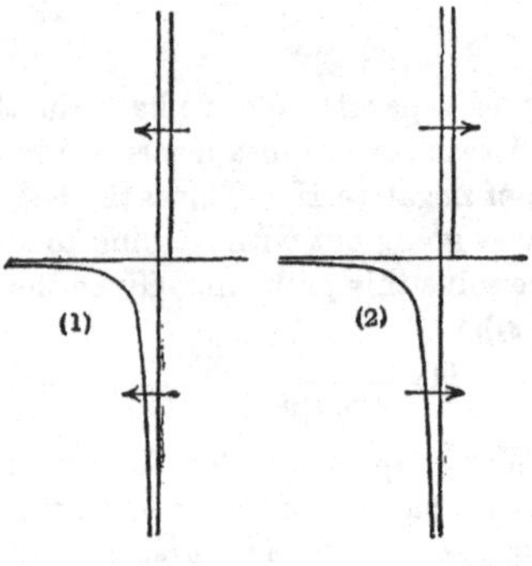

small finite depth. It is to be noticed that the amount of induction in the head when fully impulsive is $\infty+f_0$, in the tail $-\infty$; resultant $f_0=$constant always. The $-\infty$ is of course mostly close to the head, so that when $t=0$ and the tail of the inward wave disappears altogether, there is a momentary concentration of the amount f_0 of induction at the axis.

By making f_0 infinitesimal, say $=fdst$, and then packing together impulses of this sort to make a continuous function, we may use (139) to construct an integral showing H due to a continuous variable source at the axis. Say H$=$Z, then

$$\mathrm{Z}=\int_{-\infty}^{t-r/v}\frac{f(s)ds}{2\pi\mu v\{v^2(t-s)^2-r^2\}^{\frac{1}{2}}}=\frac{1}{4\mu v^2}\mathrm{K}_0(qr)f(t), \tag{140}$$

where the lower limit is $-\infty$ to include the action of the source in past time. Here $f(t)$ is the rate of generation of induction at the axis. From this integral several of those in the last article may be derived. The differential form of solution is also exhibited in (140).

(33). Returning to a steady source, if it is not condensed at the axis, but is distributed over the surface of the cylinder $r=a$, we must first have an inward wave to the axis, as well as an outward wave, before we come to the generalisation of (132), which is the elliptic wave of two branches, as in the figure on p. 260, which suits the case $f=$constant from $t=0$, provided the ordinate of the curve represents H. See also remarks on p. 262. By reducing a from finiteness to zero, and so condensing f at the axis, the outer branch of the elliptic curve goes out of existence, and the inner branch reduces to (132). The solution (173), p. 265, shows how this takes place, showing that (132) is the leading term of the elliptic formula.

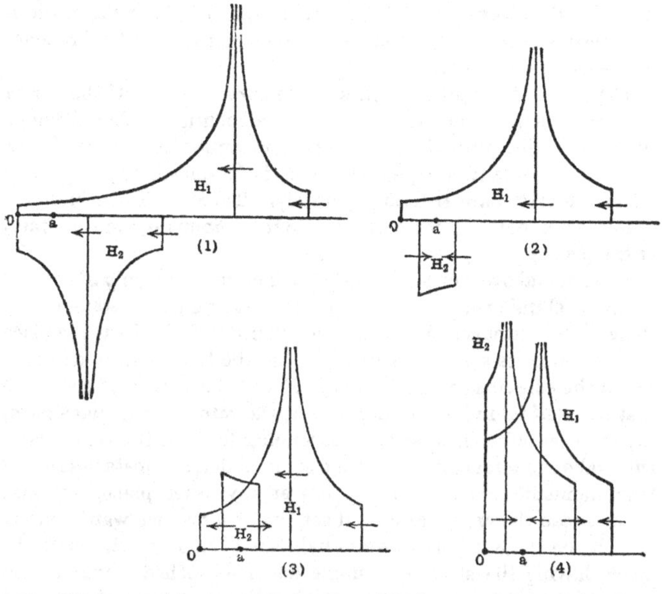

To dispense with a source in the elliptic case, we require two solutions of this sort combined, one for source f, the other for source $-f$. The total induction persists in amount. Start with a pair of inward going double-branched elliptic waves, one +, the other −. When the ∞ value in wave 2 reaches the axis, its H is not reversed, but keeps its sign negative and immediately becomes finite at the return wave front. At the axis itself the drop is to zero, and it remains zero, there being a vacant space between the axis and the return wave front. At a certain moment the wave 2 shrinks to nothing. This is at the place $r=a$. At this moment there is left simply the wave 1 going in. It is now necessary for wave 2 to reappear, positively, at $r=a$. If not, then wave 1 could not exist. Here we have a better understanding of the apparent reflection of H at the axis with reversal of sign which occurred before. It was merely because the source (real or fictitious) was there. At present it is at $r=a$, and it is there that the wave which was swallowed up comes out again reversed. The later history of this renovated wave 2 is as described on p. 261 for V, viz., a double wave first, in and out, to set up the outer branch of the elliptic curve, and then the passage out of the inner branch behind it.

But there is also wave 1 to be considered. Its ∞ goes in to the axis, and disappears, and then a little later the whole of wave 1 is swallowed up at the place $r=a$, to immediately reappear reversed (in order that wave 2 may be maintained). The later history is as described for wave 2; the last stage of all being two double-branched waves going out indefinitely.

The above description applies to the curves of H. If the curves of E are under consideration, it should be mentioned that although there is no discontinuity in E anywhere away from wave fronts when the resultant of a pair of waves is in question, yet there is when the waves are treated separately. There is discontinuity in E to the amount f in one wave, and to the amount $-f$ in the other, at the place $r=a$.

From the above can be seen what will occur if one only of the two complete elliptic waves is given (H and E) as an initial state, in any stage of its existence. A second one will instantly begin at the place $r=a$, of opposite sign to the original, and the later history will consist of the combination of the two. For if the first is of such size that by itself its induction increases at the rate f at the place $r=a$, then the second one, of size corresponding to $-f$ at the same place, will make the total induction in the combination be constant at every later moment. So it would if put at any other place, but that would mean the existence of real sources, whereas we want none at all. If, on the other hand, an initial state of H only, without E, be given, having the shape of a single double-branched wave in any stage of its existence, it will at once split into a pair of waves of half size, both similar as regards H, but opposite as regards E. These waves will then separate, and behave as above described.

Also, from the above, proceeding to the limit, we come to the generalisation of (139) to suit the case of the impulsive source f_0 being distributed at $r=a$. And this leads naturally to the integral corresponding to (140), for a continuous source. But it is complicated, and it will be of more interest to pass to another simple type of cylindrical wave.

(34). The wave (132) was $H=(4\mu v)^{-1}K_0(qr)qf$, with f constant, and the interpretation of f is a source of μH at the axis. But now turn $K_0(qr)$ to $H_0(qr)$, and f to g. The result is

$$H=\frac{g}{2\pi\mu v(r^2-v^2t^2)^{\frac{1}{2}}},\qquad E=\frac{gvt/r}{2\pi(r^2-v^2t^2)^{\frac{1}{2}}}. \tag{141}$$

Here E and H are of the same sign when t is + and of opposite sign when t is −. There is no source at the axis or anywhere within range. So the solution is valid (if desired) for all time, from $t=-$ to $+\infty$. Before $t=0$, we have the solitary H wave 1 going in to the axis. After $t=0$, precisely the same wave 2 of H going out

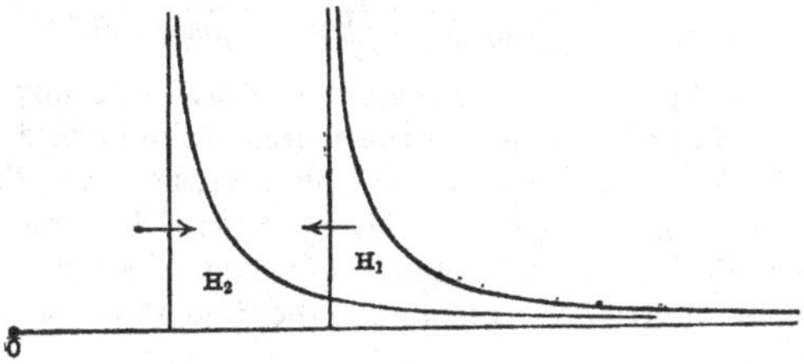

again, with E reversed. The wave front of 1 is at $r=-vt$, and of 2 at $r=vt$. The inner space is vacant.

The total induction between the front and distance r is

$$\int_{vt}^{r} 2\pi r\mu H dr=\frac{g}{v}(r^2-v^2t^2)^{\frac{1}{2}}. \tag{142}$$

This is ∞ if $r=\infty$. But we may have a source at a finite distance, producing the inward wave, and this will make the result up to the source be finite.

At the moment $t=0$, $H=g/2\pi r\mu v$, and $E=0$. This may be regarded as a given initial state of magnetic force only, and (141), with t positive, will show what results, the outward solitary wave 2 in the diagram.

The time-integral of H at r due to the inward and outward waves is

$$\int_{-r/v}^{r/v}\frac{g\,dt}{2\pi\mu v(r^2-v^2t^2)^{\frac{1}{2}}}=\frac{cg}{2\pi}\left[\sin^{-1}\frac{vt}{r}\right]_{-r/v}^{r/v}=\tfrac{1}{2}cg. \tag{143}$$

This is constant at all distances. It is the cardinal property of the wave under discussion. At the axis itself, where H is momentary, (143) shows that it is also impulsive. That is, $H=I_0(qr)\frac{1}{2}cvqg$ gives the external H at later times in terms of the value at the origin. But only up to the distance vt, because of the finite speed of propagation; and here the value is simply zero, as before seen, in the vacant space. In fact the full development is

$$H=\frac{g}{2\mu v}\left\{1+\frac{1}{2}\frac{r}{vt}+\frac{1.3}{2.4}\frac{r^2}{v^2t^2}+\dots\right\}\frac{(vt)^{-1}}{\underline{|-1}}, \tag{144}$$

which is zero before $r=vt$ and indeterminate beyond. The datum here is simply that H at the axis is impulsive at the moment $t=0$. Now there is no source there, so the given state must have arisen from an inward wave. This is the explanation why the substitution of the operator $\frac{1}{2}\mathrm{H}_0(qr)$ for $\mathrm{I}_0(qr)$, applied to the same datum, gives us the inward wave which produces the impulsive state at the axis. Thus,

$$\mathrm{H}=\frac{g}{2\mu v}\frac{\epsilon^{qr}q}{(2\pi qr)^{\frac{1}{2}}}\left\{1+\frac{1^2}{\lfloor 1(8qr)}+\frac{1^2 3^2}{\lfloor 2(8qr)^2}+\cdots\right\}$$

$$=\frac{g}{2\pi\mu v}\epsilon^{qr}\frac{1}{\sqrt{2rvt}}\frac{1}{(1-vt/2r)^{\frac{1}{2}}}=\frac{g}{2\pi\mu v(r^2-v^2t^2)^{\frac{1}{2}}}, \tag{145}$$

as in (141). The process is valid for negative t to any distance, provided we do not pass beyond the source of the inward wave.

We may take advantage of the impulsiveness of H at the axis to construct a definite integral exhibiting H at r, t in terms of the axial value when the sources are out of range. Say that $h(t)$ is the axial value of H given as a function of the time, then the integral is

$$\mathrm{H}_{r,t}=\int_{t-r/v}^{t+r/v}\frac{vh(s)ds}{\pi\{r^2-v^2(t-s)^2\}^{\frac{1}{2}}}=\mathrm{I}_0(qr)h(t). \tag{146}$$

From this may be derived some of the integrals in the last article.

(35). Closely connected with the above problem is another. Given that $\mathrm{H}=h$, at the axis, constant, beginning when $t=0$. Find the inward wave that produces it, and the result after $t=0$, on the understanding that the source is out of range. Here $\mathrm{H}=\mathrm{I}_0(qr)h$ tells us at once that $\mathrm{H}=h$ later up to the distance $r=vt$. That is, the constant value spreads outward. This is due to the superposition of the inward wave producing h and a reflected wave. But beyond $r=vt$ we get no information this way. It is the seat of the inward wave only, which we find thus:

$$\mathrm{H}=\tfrac{1}{2}\mathrm{H}_0(qr)h=\frac{h\epsilon^{qr}}{(2\pi qr)^{\frac{1}{2}}}\left\{1+\frac{1^2}{8qr}+\frac{1^2 3^2}{\lfloor 2(8qr)^2}+\cdots\right\}$$

$$=\frac{2h}{\pi}\epsilon^{qr}\left(\frac{vt}{2r}\right)^{\frac{1}{2}}\left\{1+\frac{1}{2}\frac{1}{3}\frac{vt}{2r}+\frac{1.3}{2.4}\frac{1}{5}\left(\frac{vt}{2r}\right)^2+\cdots\right\}=\frac{2h}{\pi}\epsilon^{qr}\sin^{-1}\left(\frac{vt}{2r}\right)^{\frac{1}{2}}. \tag{147}$$

This gives finally, by extension,

$$\mathrm{H}=\tfrac{1}{2}h\left(1+\frac{2}{\pi}\sin^{-1}\frac{vt}{r}\right),\qquad \mathrm{E}=-\frac{\mu vh}{\pi}\left(1-\frac{v^2t^2}{r^2}\right)^{\frac{1}{2}}, \tag{148}$$

which are valid for all time if we interpret the inverse sine to mean $\frac{1}{2}\pi$ when $vt>r$ as well as when $=r$, and also $=-\frac{1}{2}\pi$ when $vt<-r$ as well as when $=-r$; and in a similar manner consider E to be zero outside the limits of reality as well as at the limits.

The diagram shows the H wave going in (t negative). It is zero at the front, though the curve is perpendicular. At a distance, H is nearly $\frac{1}{2}h$, and when the front reaches the axis, the value is the full $\frac{1}{2}h$ everywhere, and the simultaneous value of $E = -\mu vh/\pi$ everywhere. Remember that **H** is straight, and **E** circular, so that this is a singular momentary state. The field of **H** is uniform, but cannot stay so, on account of the circular **E**.

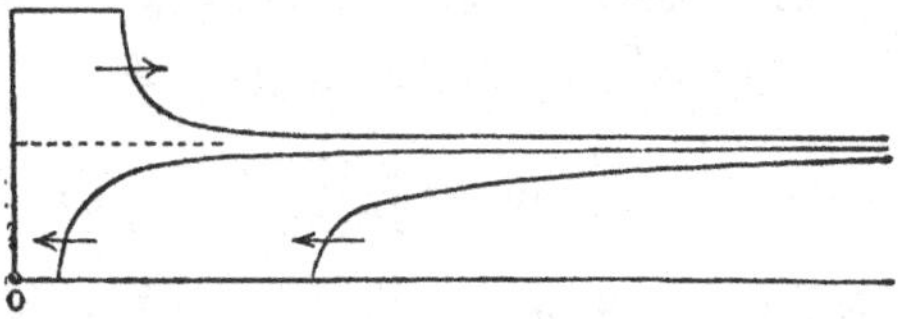

Then H doubles itself to h at the axis, and the region of doubled H spreads outward at speed v. Beyond the distance $r=vt$, H goes on increasing according to the same formula as when t was negative. The curve was convex upwards before $t=0$, and is convex downward after. In the region of $H=h$ there is no E. Elsewhere E is negative; so the flux of energy is inward all along where there is E. The final result is $H=h$, $E=0$ to any distance, or up to the source.

This final state is what would be produced in its interior by a cylindrical current sheet at any distance. So the source may be a current in a cylindrical coil. But that is not the only way. It depends upon the kind of source what the connection is with the effect produced. And since the source produces the inward wave in the first place, and the rest by reflection, we might consider the specification of E only or of H only at any distance to be the effective source, provided the real source is on the further side. Say E_a is the electric force at $r=a$. Its connections with the interior state E, H, are

$$\mu v H = -\frac{I_0(qr)}{I_1(qa)}E_a, \qquad E = \frac{I_1(qr)}{I_1(qa)}E_a. \tag{149}$$

This is settled by the equations (133) above, and is true always if there is no source between $r=0$ and a. Then the inverse problem is, given $H=h$ at the axis, find the E_a that produced it. The answer is

$$E_a = \mu v\, I_1(qa)h, \quad \text{or} \quad E_a = \tfrac{1}{2}\mu v\, H_1(qa)h. \tag{150}$$

Work this out in the same way as (147) was done. The result is the formula for E in (148), with $r=a$.

(36). Now it is possible to choose the source to be of the kind f already used, and to have f at the place of and at every moment proportional to E_a itself as above found. This may be seen partially by comparison with telegraph cable problems. If we want V_0 at the beginning of a cable to represent the impressed voltage itself, we do this by the use of a battery of voltage $e=V_0$ and zero resistance, put on between the beginning of the cable and earth. To imitate this now, we require to place immediately outside the cylindrical source f, a perfectly conducting cylinder to stop the outward wave from f. To prove this definitely, let the source f be at $r=a$, and the perfect reflector at $r=b$, outside. Let H_1, E_1 and H_2, E_2 be the results inside and outside the source. They are given by

$$\begin{aligned} H_1 &= H_0(qr)\left\{K_0(qa)+\frac{K_1(qb)}{H_1(qb)}H_0(qa)\right\}\frac{qf}{8\mu v}, \\ E_1 &= -H_1(qr)\left\{K_0(qa)+\frac{K_1(qb)}{H_1(qb)}H_0(qa)\right\}\frac{qf}{8}, \end{aligned} \qquad (151)$$

$$\begin{aligned} H_2 &= H_0(qa)\left\{K_0(qr)+\frac{K_1(qb)}{H_1(qb)}H_0(qr)\right\}\frac{qf}{8\mu v}, \\ E_2 &= H_0(qa)\left\{K_1(qr)-\frac{K_1(qb)}{H_1(qb)}H_1(qr)\right\}\frac{qf}{8}. \end{aligned} \qquad (152)$$

Though they look difficult, the proof is very easy. For, by inspection, $H_1=H_2$ at $r=a$, and $E_2=0$ at $r=b$. Also, at $r=a$,

$$E_2-E_1=\tfrac{1}{8}qf(H_{1a}K_{0a}+H_{0a}K_{1a})=(qf/8)(4/\pi qa)=f/2\pi a, \qquad (153)$$

by the conjugate property (6), p. 241, vol. 2. Here we have a perfect reflector at b, and a source f at a generating induction at the rate f, and producing the discontinuity in E at a. We may consider $f/2\pi a$ to be $|\,\mathrm{curl}\,\mathbf{e}\,|$, if $\mathbf{e}$ is impressed electric force. Or it might be an impressed magnetic current, as a rotating cylindrical sheet of magnetons.

Now put the reflector close behind the source. Make $b=a$; then by a second use of the same conjugate property the inner E and H are reduced to

$$E=-\frac{H_{1r}}{H_{1a}}\frac{f}{2\pi a}, \qquad H=\frac{H_{0r}}{H_{1a}}\frac{f}{2\pi\mu va}, \qquad (154)$$

which mean exactly the same as (149), because $E_a=-f/2\pi a$.

With this particular kind of source then, and the perfect screen outside it to assist, we know exactly how it must vary in order to make H at the axis jump from 0 to the value h, and remain at that value. And, in fact, to establish the whole interior state up to $r=a$ in the time $2a/v$, in the manner shown by equations (148). The source begins at the moment $t=-a/v$ with the value 0, reaches full value when $t=0$, and then falls in the same way to the value 0 when $t=a/v$, and remains 0 after. The induction being fully established and enclosed within a screen, requires no further source to maintain it. Notice also, that the interior state is set up in a smooth continuous manner everywhere, except at the axis, where there is a solitary jump.

(37). But if the screen does not exist, then f must of course vary in a different way to produce the required jump at the axis. We may utilise (151), (152). Put $b=\infty$ to do away with the screen.

Then the internal E and H are

$$H = H_{0r}K_{0a}\, qf/8\mu v, \qquad E = -H_{1r}K_{0r}\, qf/8; \tag{155}$$

and now, given $H = h$ at the axis beginning when $t = 0$, we have

$$h = K_{0a}\, qf/4\mu v, \qquad f = (qK_{0a})^{-1}\, 4\mu v h. \tag{156}$$

Remark here that whereas previously f was found by H_{1a} acting directly on h, yet now it is by qK_{0a} acting inversely. The change it makes is very great, and can be foreseen. For a constant f would make h drop from ∞ to 0 at the axis, as before found. It follows that for constant h we want f to rise from 0 to ∞. It is the want of a screen that is the cause of the unlimited rise. To find a formula for f, long division makes

$$f = 4\mu v h\, \epsilon^{qa}\left(\frac{\pi a}{2q}\right)^{\frac{1}{2}}\left[1 + \frac{1}{8qa} - \frac{7}{2(8qa)^2} + \frac{59}{2(8qa)^3} - \frac{3013}{8(8qa)^4} + \frac{50735}{8(8qa)^5} - \frac{2110163}{16(8qa)^6} + \dots\right]. \tag{157}$$

By multegration this makes (leaving out ϵ^{qa}, so as to make f begin when $t = 0$),

$$f = 4\mu v h\, (2vta)^{\frac{1}{2}}\left[1 + \frac{vt/4a}{3} - \frac{7}{30}\left(\frac{vt}{4a}\right)^2 + \frac{59}{210}\left(\frac{vt}{4a}\right)^3 - \frac{3013}{8.3.5.7.9}\left(\frac{vt}{4a}\right)^4 + \frac{50735}{8.3.5.7.9.11}\left(\frac{vt}{4a}\right)^5 - \frac{2110163}{16.3.5\dots 11.13}\left(\frac{vt}{4a}\right)^6 + \dots\right]. \tag{158}$$

This begins with the value zero, as expected, and increases continuously, so far as can be seen. It is calculable from $t = 0$ up to about $8a/v$, very closely at first, but loosely at the end. It is an oscillating divergent series, and the point of convergence is at the end of the sixth term, when $vt = 4a$, and then moves in to the first term, when $vt = 12a$. So from about $vt = 8a$ upwards, an equivalent series of a different type is required.

(38). It is desirable to have a different sort of source, and perhaps the best of all is impressed electric current. Say we have an electrified dielectric cylinder, and rotate it by turning a handle. This will produce an electric current sheet, and as its strength will vary strictly as the speed of rotation, it may be varied conveniently. The immediate problem now is how fast to turn the handle to make H at the axis jump from 0 to h and then remain steady. Let C be the impressed current at $r = a$, then the equations of E and H are

$$\text{(in)} \quad H_1 = \tfrac{1}{4}\pi qa\, H_{0r}K_{1a}C, \qquad E_1 = -\tfrac{1}{4}\mu v\pi qa\, H_{1r}K_{1a}C, \tag{159}$$

$$\text{(out)} \quad H_2 = -\tfrac{1}{4}\pi qa\, K_{0r}H_{1a}C, \qquad E_2 = -\tfrac{1}{4}\mu v\pi qa\, K_{1r}H_{1a}C, \tag{160}$$

because they make $E_1 = E_2$, and $H_1 - H_2 = C$, at $r = a$. At the axis, let H be $h\epsilon^{-qa}$, then

$$\tfrac{1}{2}\pi qa\, K_{1a}C = h\epsilon^{-qa}, \qquad \text{and} \qquad C = \frac{2h\epsilon^{-qa}}{\pi qa K_{1a}}. \tag{161}$$

By division this makes

$$C=h\left(\frac{2}{\pi qa}\right)^{\frac{1}{2}}\left[1-\frac{3}{8qa}+\frac{33}{2(8qa)^2}-\frac{249}{2(8qa)^3}+\frac{9963}{8(8qa)^4}-\frac{131229}{8(8qa)^5}+\dots\right] \quad (162)$$

$$=\frac{2h}{\pi}\left(\frac{2vt}{a}\right)^{\frac{1}{2}}\left[1-\frac{vt}{4a}+\frac{11}{10}\left(\frac{vt}{4a}\right)^2-\frac{83}{70}\left(\frac{vt}{4a}\right)^3+\frac{369}{280}\left(\frac{vt}{4a}\right)^4-\frac{2083}{1320}\left(\frac{vt}{4a}\right)^5+\dots\right]. \quad (163)$$

The current rises quickly to about 73 p.c. of full value in the first interval from $t=0$ to a/v, then to 86 p.c. when $t=2a/v$, after which the progress to the full value is very slow. Up to $vt=2a$, C is easily found; up to $3a$ a few more terms are needed; but after $vt=4a$, when the initial convergency ceases, nothing can be done. So from about $vt=3a$ to ∞, another series is wanted. But it is only for the dregs, so does not matter, though it would be interesting to know how to find the result. It will be understood here that the current is producing H outside as well as inside, and only when the external H has entirely gone off can the current assume its steady value, $C=h$.

(39). To abolish this complication, introduce a screen. Thus, let

$$\text{(in)} \quad \begin{aligned} H_1&=\tfrac{1}{4}\pi qa(K_{1a}-yH_{1a})H_{0r}C, \\ E_1&=-\tfrac{1}{4}\mu v\pi qa(K_{1a}-yH_{1a})H_{1r}C, \end{aligned} \quad (164)$$

$$\text{(out)} \quad \begin{aligned} H_2&=-\tfrac{1}{4}\pi qaH_{1a}(K_{0r}+yH_{0r})C, \\ E_2&=-\tfrac{1}{4}\mu v\pi qa(K_{1r}-yH_{1r})H_{1a}C. \end{aligned} \quad (165)$$

These make $E_1=E_2$, and $H_1-H_2=C$ at $r=a$, whatever y may be. Some external condition will find y. If no screen or other change, then $y=0$. But if there is a boundary at b, then y is to be found by a condition there. Say, first, $E_2=0$. This makes $y=K_{1b}/H_{1b}$. But this reduces E and H. In the limit, when $b=a$, the internal E and H become zero. That is, the current C cannot produce any E or H when it is enclosed within a close-fitting perfect electric conductor, a striking result which can perhaps be imperfectly imitated practically.

This kind of screen being no good, use the other kind, a perfect magnetic conductor. Then $H_2=0$ at b. This makes $y=-K_{0b}/H_{0b}$, and in the limit, when $b=a$, makes the internal state be

$$H=\frac{H_{0r}}{H_{0a}}C, \qquad E=-\mu v\frac{H_{1r}}{H_{0a}}C. \quad (166)$$

The outer E and H do not count now. Compare with (159), and again see that the effect of the screen is to substitute an inverse for a direct operation.

Now if we ask how C must vary so that the axial H shall be constant, $=h$, from the moment $t=0$, we have

$$h=\frac{C}{\frac{1}{2}H_{0a}}, \qquad C=\tfrac{1}{2}H_{0a}h. \quad (167)$$

This has already been worked out, in effect, and makes

$$C=\tfrac{1}{2}h\left\{1+\frac{2}{\pi}\sin^{-1}\frac{vt}{a}\right\} \tag{168}$$

beginning when $t=-a/v$. It is, in fact, the simple case (148) already treated, with the new kind of source, and new reflector. C rises smoothly from 0 to h in the period $2a/v$, and remains at the value h.

The impressed activity is $-E_aC$ per unit surface of the current sheet, and since E_a is negative, the activity is positive. Its result is the internal magnetic energy. If, now, C is guided back to zero in a similar way, (say by the superposition of the negative of (168) upon the steady h), then E_a is positive all the way, and the magnetic energy is given out again.

But if this current C be convection current due to the rotation of an electrified dielectric cylinder, the existence of the screen of the kind chosen makes it imperative for us to supplement the (say) positive electrification of the current sheet by compensating negative electrification inside the sheet. Its distribution is immaterial, but it must be kept still. Then the movement of the electrified sheet will set up precisely the same waves inside as described, which will be simply superposed upon the stationary displacement due to the positive and negative electrifications at rest.

Theory of Electric Telegraphy.

["Ency. Brit." Tenth edition. Reprinted by permission of the proprietors of *The Times*. Written June, 1902.]

§ 528. The theory of the electric telegraph enunciated by W. Thomson (Lord Kelvin) in 1855, which has been of importance in Atlantic and cable telegraphy generally, has now become merged in a wider theory, of which it forms an extreme case. This wider theory is founded upon Maxwell's theory of electromagnetic radiation, and may be accordingly termed the radiational theory. The latest type of electric telegraphy, usually termed "wireless," which has been developed from Hertz's experimental researches, also comes under the wider theory. The principal difficulty is to explain the various functions of wires. The radiational part, which is fundamental, is comparatively easy.

Electric telegraphy is done by means of electromagnetic waves through the ether differing in no essential respect from radiation from the Sun. This involves a flux of energy through the ether, travelling at the enormous speed of 30 earth quadrants, or 3×10^{10} cm. per second, which speed will here be denoted by v. The energy arriving at the Earth's surface may be measured by its heating effect. But in transit it is measured by the product of the density of the energy into its speed, that is, by $2Tv$, if $2T$ is the density. In Maxwell's theory of light it is half electric energy, say $U=\frac{1}{2}cE^2$, half magnetic energy, $T=\frac{1}{2}\mu H^2$. Here E and H are the intensities of the electric and magnetic forces, and μ, c are etherial constants. It is a general property of elastic waves that, in order to be transmitted without reflection in transit, the two energies concerned should be equal. In this case, therefore, $\mu H^2=cE^2$ and $E=(\mu/c)^{\frac{1}{2}}H$. The ratio $(\mu/c)^{\frac{1}{2}}$ is the same as μv, as may be proved by adjusting the vectors **E** and **H** so that the two circuital laws of electromagnetics shall be simultaneously obeyed. To secure this result, E must equal μvH, and **E** and **H** must be perpendicular to one another and to the direction of propagation. If a watch indicates three o'clock, and the big hand stands for **E** and the small hand for **H**, then the flux of energy is straight through the watch from front to back. These properties hold good in all parts of plane waves. It is convenient to think of a slab wave. Let the ether between two parallel planes be the seat of uniform **E** and uniform **H**, related as above, and in the plane of the slab. This being the state of things at a given moment, electromagnetics asserts that the slab of **E** and **H** will move through stationary ether perpendicular to itself at speed v, unchanged, and carrying all its properties with it. It is not known what μ and c are separately, but $\mu cv^2=1$ connects them. Also the product μv is known in terms of resistance, being an impedance equivalent to 30 ohms; and cv is the equivalent conductance. The electric displacement is $\mathbf{D}=c\mathbf{E}$, and the magnetic induction is $\mathbf{B}=\mu\mathbf{H}$.

A little later the concrete forms of these equations will be wanted. If a condenser of permittance S has a charge, Q, under the voltage V, then $Q=SV$. This is the concrete form of $D=cE$ for a unit cube; for Q is the total displacement through the condenser, and V the line integral of E, which is, therefore, the voltivity, whilst c, corresponding to S, is the permittivity. The energy of the condenser is $\frac{1}{2}SV^2$. This is the space integral of $\frac{1}{2}cE^2$. Similarly, if P is the total induction through a coil of inductance L under the gaussage C, then $P=LC$. This is the concrete form of $B=\mu H$. So μ is the inductivity, and H the gaussivity, C being the line integral of **H**. The energy of the coil is $\frac{1}{2}LC^2$, which is the space integral of $\frac{1}{2}\mu H^2$. It is important to bear in mind these relations of the concrete and the specific, for they appear frequently in telegraphic theory.

There may be any number of slabs of the above kind, separated, or in contact, of any depths and any strengths. But if $E=\mu v H$ in all, they will all behave alike, independently of one another. This is expressed by

$$E=\mu vH=f(t-x/v), \tag{1}$$

where $f(t)$ expresses the state at the plane $x=0$. If **E** and **H** do not change in direction from one slab to another, the radiation is plane polarised; the plane of polarisation is the plane of **H**. But the direction of **E** (and with it that of **H**) may vary from slab to slab in any way we like. Then the radiation is circularly, or elliptically, or heterogeneously polarised, the last being, in the average, equivalent to no polarisation. In all cases the behaviour of individual slabs is the same. The disturbance is vibratory in solar radiation, but in telegraphy this is not necessary. We may have waves in which **E** and **H** do not change at all. It is the progression through the ether that is the wave, not accidental vibration. There is, it may be noted, no change from electric to magnetic energy anywhere in this sort of radiation.

We can make an immediate application to telegraphy. The slab was above considered to extend indefinitely. But let the tubes of displacement be confined between a pair of parallel perfectly conducting plates upon which they terminate perpendicularly. The tubes of induction are unchanged. One plate is positively, the other negatively electrified to surface density D. In virtue of the perfect conductivity, this terminated plane slab will behave exactly like the complete one; that is, we may send any radiation we like along between the plates, unchanged in transit. This is a rudimentary case of distortionless telegraphy along a pair of conductors. But the plates need not be infinitely extended, nor need they be flat. They may become a pair of cylinders of any form of section. The three most important cases are: (1) a pair of parallel wires; (2) a wire suspended parallel to a flat conductor, the earth; (3) a wire inside a cylinder surrounding it. In all cases the displacement goes from one conductor to the other through the ether between them, leaving one and reaching the other perpendicularly, and its distribution is to be calculated as in two-dimensional electrostatics. In all cases the lines of **H** cut the lines of **E** perpendicularly. They therefore make complete circuits round one or other conductor. In all cases $E = \mu v H$.

If we consider only a unit tube of flux of energy (a beam) there is no difference from the radiational case first considered. The complications due to the varied distributions of **E** and **H** in their plane are avoided by expressing results in terms of the concrete V and C. The line integral of **E** in any wave plane from one conductor to the other is V, the line integral of **H** in a circuit of magnetic force is C. Then we have

$$V = LvC = F(t - x/v); \tag{2}$$

$$C = SvV; \qquad LSv^2 = 1; \qquad \tfrac{1}{2}SV^2 = \tfrac{1}{2}LC^2. \tag{3}$$

The additional L and S are the inductance and permittance per unit depth of the wave, or per unit length of the conductors.

The use of conductors will now be seen partly. They serve to *guide* a wave along from place to place without loss, and with a limited amount of energy. Without them, the complete plane wave would be required. If not complete it will spread out over all space. This is, in fact, what occurs with solar radiation. Although the small curvature makes the waves be practically plane, yet they are really expanding spherical waves. The energy is conserved, and therefore its density varies inversely as the square of the distance from the Sun's centre, whilst E and H vary inversely as the distance. This is the attenuation.

The transition from plane to spherical waves is shown by the theory of waves along conical conductors. Let there be two coaxial conical conductors with a common apex, O. Let the semi-vertical angles be θ_1 and θ_2. The displacement is to go from the inner cone to the outer symmetrically in spherical sheets centred at O. The induction, as usual, is to be perpendicular to the displacement, therefore in circles round the inner cone. As before, make $E=\mu v H$. Then the wave will run along between the cones at speed v, attenuating in transit like solar radiation. We may vary the angles of the two cones as we please. An important case is got by making one angle $\theta_1=0$, the other $\theta_2=\frac{1}{2}\pi$; this gives the case of a flat plane with a vertical wire projecting from it at O, and the waves are hemispherical. The displacement goes from the wire to the plane in quadrantal sectors. This gives "wireless" telegraphy, like that of Lodge and Marconi, though more perfect, having ideal conditions. Signals made at O will be sent out in hemispherical waves quite undistorted, though attenuating as they expand. They are guided by the wire and by the plane representing the surface of the sea. If the wire is of finite length reflected pulses are generated; the waves sent out then become vibratory, of a frequency and wave length determined by the length of the wire. If the angle of one cone is $\theta_1=0$, and of the other $\theta_2=\pi$, we get complete spherical waves upon a wire, generated at O. If the wire is of finite length the terminal reflections make it become a sort of Hertzian vibrator.

Comparing the case of the cylindrical wires with the so-called "wireless" case, though there is little difference in theory, there is great difference in practice. Using wires, we can send radiation anywhere we like in small quantities without loss. One wire and earth is enough, but two parallel wires are preferable, to avoid certain interferences. Were it not for the resistance of the wires, and a little disturbance and loss in turning corners (for the wires need not be straight), telegraphy with wires would be perfect, by any path to any distance. But in "wireless telegraphy," though no expensive connecting wires are required, which gives a remarkable freedom in certain ways, there is enormous loss, and enormous power is required to send workable signals across the Atlantic, since they are being sent simultaneously everywhere. But for this loss by expansion and by resistance, and possible unsettled interferences, there is no reason to limit the distance. The course of a wave round the earth can be easily followed graphically.

When a wave sent along wires comes to a sharp bend in the circuit, a new wave is generated at the bend. This, combined with the old wave, forms the wave after passing the bend. There is a rapid accommodation of the wave round the wire to the new direction. But if the bending is continnous, instead of abrupt, the accommodation goes on continuously also. The reason is that the electrification cannot leave the wires, so the wave in close proximity must accommodate itself to them. A part of the wave, however, really does go off into space with some loss of energy at a sharp corner by its own natural tendency to keep going, but the wire serves to guide the disturbance round the corner as a whole, by holding on to the tubes of displacement by their ends. This guidance is obviously a most important property of wires.

There is something similar in "wireless" telegraphy. Sea water, though transparent to light, has quite enough conductivity to make it behave as a conductor for Hertzian waves, and the same is true in a more imperfect manner of the earth. Hence the waves accommodate themselves to the surface of the sea in the same way as waves follow wires. The irregularities make confusion, no doubt, but the main waves are pulled round by the curvature of the earth, and do not jump off. There is another consideration. There may possibly be a sufficiently conducting layer in the upper air. If so, the waves will, so to speak, catch on to it more or less. Then the guidance will be by the sea on one side and the upper layer on the other. But obstructions, on land especially, may not be conducting enough to make waves go round them fairly. The waves will go partly through them.

The effects of the resistance of the guides are very complicated in general, and only elementary cases can be considered here. Considering the transmission of plane waves in the ether bounded by parallel cylinders; first imagine the ether to be electrically conducting. Then

$$-\frac{dH}{dx} = kE + c\frac{dE}{dt}, \quad \text{or} \quad -\frac{dC}{dx} = KV + S\frac{dV}{dt} \tag{4}$$

expresses the first circuital law, and

$$-\frac{dE}{dx} = \mu\frac{dH}{dt}, \quad \text{or} \quad -\frac{dV}{dx} = L\frac{dC}{dt} \tag{5}$$

expresses the second. The additional quantities here are k the conductivity and K the conductance per unit length along the wires. The effect is to waste energy at the rate kE^2 per unit volume, or KV^2 per unit length. It is wasted in heating the medium or matter in it, according to Joule's law. Two other effects occur, viz., attenuation of the wave in transit, and distortion, or change of shape, due to reflection in transit. The attenuation cannot be prevented, but the distortion can. For let the medium be conducting magnetically, so that gH^2 is the waste per unit volume, and RC^2 the corresponding waste per unit length. This g is analogous to k, and R to K. Then instead of (5) above, we shall have

$$-\frac{dE}{dx}=gH+\mu\frac{dH}{dt}, \quad \text{or} \quad -\frac{dV}{dx}=RC+L\frac{dC}{dt}. \tag{6}$$

These are of the same form as (4). So, if there be no electric conductance, but only the new magnetic conductance, the wave of H will be distorted in the same way as that of E was before, and the wave of E in the same way as that of H was before, and there will be attenuation similarly. But if the two conductivities coexist, though the attenuations are additive the distortions are combative, and may therefore destroy one another. This occurs when R/L = K/S, or $g/\mu = k/c$. The solution expressing the transmission of a plane wave is now

$$E=\mu vH=\epsilon^{-gt/\mu}\, f(t-x/v), \tag{7}$$

or

$$V=LvC=\epsilon^{-Rt/L}\, F(t-x/v). \tag{8}$$

The meaning is that signals are transmitted absolutely without distortion, every slab independently of the rest, but with attenuation in transit according to the time factor $\epsilon^{-Rt/L}$.

This would be very curious, even if it could only be imagined to be done by means of the imaginary magnetic conductivity. What is even more remarkable, however, is that it can be closely imitated by means of the real electric resistance, not of the medium outside the wires, but of the wires themselves. Abolish g altogether, but keep in R. It is to mean the steady resistance of the conducting guides per unit length, previously taken as zero. Then

$$-\frac{dC}{dx}=KV+S\frac{dV}{dt}, \quad \text{and} \quad -\frac{dV}{dx}=RC+L\frac{dC}{dt} \tag{9}$$

are still the proper equations under certain circumstances, to be mentioned later. Thus equation (8) is still the result, *i.e.*, distortionless propagation along wires. The waste RC^2 is now in the wires, instead of outside. It equals the other waste KV^2. It follows that any ordinary telegraph circuit may be made approximately distortionless by adding a certain amount of leakance, or leakage conductance. For it has L, S, and R already, and a little K. Increase K until K/S = R/L. Then the distortion, which may be excessive at first, will gradually disappear, and the signals will be restored to their proper shape, but at the expense of increased attenuation. If K be increased further, distortion will come on again, of the other kind; for at first R was in excess, now it is K.

For example, if R is 1 ohm per kilom., and $Lv=600$ ohms, then

$$\frac{1}{600}=\frac{R}{Lv}=\frac{K}{Sv}=KLv, \quad \text{so} \quad K^{-1}=360{,}000 \text{ ohms per kilom.}$$

This is the insulation resistance required. Also, the attenuation in the distance x is $\epsilon^{-Rx/Lv}$ or $\epsilon^{-x/600}$; that is, from 1 to ϵ^{-1} in 600 kilom., to ϵ^{-2} in 1,200 kilom., and so on.

To understand the reason of the disappearance of the distortion: concentrate the resistance of the wire in detached lumps, with no resistance between them. Let each resistance be r. Similarly concentrate the leakance, each leak being k. Then, between the r's and k's there is natural unattenuated distortionless propagation; so we have only to examine what happens to a slab wave in passing by one of the r's or one of the k's to see the likeness and difference of their effects on V and C. First let a positive wave be passing r. Let V_1, V_2, V_3 be corresponding elements in the incident, reflected and transmitted waves. Then the conditions are

$$V_1+V_2=V_3+rC_3, \qquad C_1+C_2=C_3, \tag{10}$$

and, since $V_1=LvC_1$, $V_2=-LvC_2$, $V_3=LvC_3$, we have the results

$$\frac{V_3}{V_1}=\frac{1}{1+r/2Lv}, \qquad V_1=V_2+V_3. \tag{11}$$

The second of these equations shows that the electrification (and displacement) is conserved. The first shows the ratio of the transmitted to the incident wave. The incident element, on arriving at r, divides into two, both of the same sign as regards V; one V_3 goes forward, the other V_2 backward, increasing the electrification behind. Now suppose a slab wave passes by n resistances in succession in the distance x, such that $nr=Rx$, then the attenuation produced in the distance x is the nth power of V_3/V_1, and in the limit, when n is made ∞, it becomes $\epsilon^{-Rx/2Lv}$. This is when there is no leakage.

Next consider the effect of a single leak. The conditions are

$$C_1+C_2=C_3+kV_3, \qquad V_1+V_2=V_3, \tag{12}$$

and the results are

$$\frac{V_3}{V_1}=\frac{1}{1+k/2Sv}, \qquad C_1=C_2+C_3. \tag{13}$$

Here the second result shows that the induction is conserved, instead of the displacement. A part of C_1 is thrown back and increases C behind the leak. The complete attenuation in the distance x, by similar reasoning to the above, is $\epsilon^{-Kx/2Sv}$, when there is leakage without resistance in the wires.

To compare the two cases; in the first, part of V_1 is reflected positively and part of C_1 negatively, in passing a resistance; and in the second, part of V_1 is reflected negatively, and part of C_1 positively. The effects are opposite. So if the resistance and the leak coexist, there is partial cancellation of the reflection. This compensation becomes perfect when r and k are infinitely small and in the proper ratio. Then there is no reflection, though increased attenuation. So with R and K uniformly distributed there is no reflection in transit anywhere, provided $R/L = K/S$. The attenuation in the distance x is now $\epsilon^{-Rx/Lv}$.

If the circuit stops anywhere, what happens to the wave depends upon the electrical conditions at the terminus. If it is a short-circuit, then the resultant $V=0$ is imposed. This causes complete positive reflection of incident C, and negative of V. If insulated, then the resultant is $C=0$, and there is positive reflection of incident V and negative of C. A remarkable case is that of a terminal resistance, say, R_1. Then

$$V_1 = LvC_1, \qquad V_2 = -LvC_2, \qquad V_1 + V_2 = R_1(C_1 + C_2) \tag{14}$$

are the conditions. So the reflected wave is given by

$$\frac{V_2}{V_1} = \frac{R_1 - Lv}{R_1 + Lv}. \tag{15}$$

V is reflected positively when $R_1 > Lv$, and negatively when $< Lv$. If $R_1 = Lv$, there is no reflection. The energy of the wave is absorbed by the resistance. So we attain not only perfect transmission, but also perfect reception of signals.

The above method of treating resistance and leakance by isolated resistances and leaks may be applied to find out what in general happens when there is either no leakage or else no resistance in a continuous circuit. Suppose, for example, there is no leakage. Given a charge anywhere, initially without current, how will it behave? If there were no resistance, it would immediately split into two; one with positive magnetic force would move to the right, the other with negative magnetic force would move to the left, both at speed v. Now this is also exactly what happens in a resisting circuit at the first moment, namely, the generation of a pair of oppositely travelling pure electromagnetic slab waves. But this state does not continue, because the resistance causes a continuous partial reflection of the two waves. This may be studied in detail by easy numerical operations when we concentrate the resistance of the circuit at a number of equidistant points. The result is as follows: Let V_0 be the original state of V of depth a. Then at a certain time t the distribution consists first of two heads at distance $2vt$ apart. These are two pure electromagnetic waves travelling at speed v. They are the attenuated remains of the initial waves $\frac{1}{2}V_0 = V = LvC$, and $\frac{1}{2}V_0 = V = -LvC$. Between them is the rest of V_0, cast behind by reflection as the two waves progress. In time the heads attenuate to insignificance, and there is left the widespread diffused state of V, having a maximum at the centre. It then tends to obey the Fourier law of diffusion.

If, on the other hand, there is no resistance, then the effect of the leakance alters the nature of the disturbance between the heads. V is positive in the heads, which are nearly as before, but negative in the intervening space. Now, in the former case the charge or integral of SV was conserved. Suppose Q was the original charge. Then $\frac{1}{2}Q\epsilon^{-Rt/2L}$ is the amount in each head, and the rest of Q is diffused between them. But in the second case the charge in each head is $\frac{1}{2}Q\epsilon^{-Kt/2S}$, and the total charge at the same moment is $Q\epsilon^{-Kt/S}$. So we know the amount of negative charge between the two heads.

If the initial state is a pure slab wave $V_0 = LvC_0$, it goes on moving as a pure wave, but attenuating and casting a tail behind it, the tip of which travels backwards at speed v. The tail of V is everywhere positive, and the tail C is everywhere negative at first, and until the head has attenuated very considerably, when by diffusion in the tail itself and attenuation it becomes positive, first near the head and later all the way to the middle. The ultimate result is a widely diffused state of V symmetrical with respect to the origin, where V is a maximum and C zero, whilst elsewhere C is all positive on the right side and negative on the left, in accordance with the diffusion law.

It is now necessary to consider what R, the resistance, means in the above theory. The remarkable property of guidance has been noticed, and also that R in the wires acts oppositely to K outside in reflective power, so that they may neutralize. But there is another remarkable property of wires to be mentioned. They are conductors truly, but it is for conduction or guidance of the wave outside them, in effect, in all the above. Do they conduct the wave similarly inside? Not at all. So far from doing that, they obstruct it. If a plane wave strikes flush against a perfect conductor it is wholly reflected, and with any good conductor it is mostly reflected. When a wave is sent along a wire, a purely superficial conduction is first of all concerned. The boundary of the wire is exposed to magnetic force tangentially. Now, suppose the wire is really a conducting dielectric, like the medium outside it in cases before considered. Then the magnetic force impressed upon its boundary will send an electromagnetic wave inwards. This wave will be nearly of the same type as the plane wave in the medium outside, when there is electric conductivity there, if the conductor has a plane boundary. If it is a wire, the wave will be cylindrical, with complications accordingly. But there is no difference in principle. So the transit of the plane wave along the wire outside it causes the transmission perpendicularly into the wire of another wave, and, just as the former suffers positive reflection of H and negative of E in transit, so does the secondary wave which, so to speak, leaks into the wire and converges to its axis.

But though there is no difference in principle, there is a very great difference in practice. The external conductivity is usually very small, and a wave can go a long way with little change. But the internal conductivity is relatively very large indeed. Hence, if an impulsive wave enters the wire at its boundary, its "head," in the sense used above, attenuates to nothing almost at once, and the wave becomes of the diffusive kind. The magnetic induction therefore diffuses into the wire according to Fourier's law of diffusion. The time interval needed to effect any operation varies as the product μka^2, where μ is the inductivity, k the conductivity and a the radius of the wire. This diffusion is nothing like the elastic wave outside from which it results. It is to be regarded, in the main, as a purely local phenomenon. The electric force in this transverse wave is axial, or parallel to the axis of the wire. Energy is wasted in the wire at the rate $k\mathrm{E}^2$ per unit vol. The transverse wave has no direct action upon the main wave outside the wire from which it is derived, but its indirect action may in time become very great, namely, the distortion and attenuation of the main wave before considered.

Metals vary considerably in the facility with which they are penetrated. Iron and copper are typical examples. Although k is six times as great in copper as in iron, yet μ is 100 or 200 times as great (for small H) in iron as in copper. So iron is much more difficult to penetrate than copper; and it takes a far longer time to set up the state of steady current in iron if the wires are of the same size. When it is reached the steady current in iron is only one-sixth of that in copper, if the impressed voltage be the same. It follows that under the action of an alternating voltage the wave train entering the iron will attenuate much faster than that entering the copper. In either case, increasing the frequency of alternation removes the current from the interior towards the boundary, shortening the wave length. The tendency is towards skin conduction. The attenuation is so rapid in going inward that only one wave length in the wire itself need be considered. This may be confined within a mere skin. Then the interior might be all scooped out without making any sensible difference. With Hertzian waves a mere film may be sufficient. These waves (outside) may be of the order of a few metres long, and the derived waves inside of microscopic length. With telephonic waves, on the other hand, there is very good penetration in the usual copper wires, except at the higher frequencies. On the other hand, if the wires are iron there is bad penetration.

Wires have really no definite resistances as a whole when waves are sent along them. The resistance per unit length will vary according to the amount of penetration at the place. The least resistance is when there is complete penetration. It is then the steady resistance measured by the Christie balance. But if the impressed alternations are simply periodic, and the resultant alternating state is allowed to be assumed, then a definite meaning can be given to the "effective resistance" (Rayleigh). If it is called R' it will be such that $R'\overline{C}^2$ is the average rate of generation of heat, $\overline{C}^2$ meaning the mean square of the current. The more the current is concentrated, the greater the resistance. Thus with Hertzian waves the resistance is largely multiplied (compared with the steady resistance), for it is the resistance of a thin skin. It is also easy to multiply the resistance three times or so with telephonic currents in iron wires, but in copper wires there is little increase.

In skin conduction the resistance depends upon the area of the surface of the conductor. The amount per unit area is the same for a round wire as for a plane boundary when the surface state is the same. But, as before mentioned, the displacement between a pair of parallel cylinders distributes itself (in electromagnetic waves) according to two-dimensional electrostatics, and the magnetic force lines are perpendicular to the displacement. It follows that in such very rapid alternations as lead to skin conduction, the skin current concentrates itself where the displacement is strongest, usually the parts of the two conductors which are nearest together.

The penetration of the magnetic force also increases the inductance L. This increase is a trifling matter in copper; but in iron, since it becomes magnetised, it is a large effect with good penetration, though insignificant when there is skin conduction.

The way the current "rises" in a wire under impressed surface action may be visualised by comparing it with water in a pipe, originally at rest, subjected to surface traction upon its boundary acting longitudinally. The traction will at first only drag along the outermost layer of water. But if the traction continues to act steadily, viscosity will gradually cause similar motion to penetrate inwards, and to any depth. On the other hand, if the traction alternates in direction, not too slowly, only the outer layer will be sensibly moved to and fro.

It may become possible to convert this analogy into something more than an analogy by regarding the surface traction as that exerted by the longitudinal electric force on electrification in the wire, and supposing that the interior electrification (half positive, half negative) is somehow controlled in its average motion by the diffusion law. It has always been difficult to reconcile metallic conduction with electrolytic. There is a slow drift of ions in electrolysis carrying electric charges. But there is no sign of ionic electrolysis in metallic conductors. Again, regarding metals as polarizable, Maxwell's idea that conduction consisted in a breakdown of the electric displacement does not go far in explaining the nature of the intermolecular discharge. Yet the possibility of a convective explanation of metallic conduction in harmony with Maxwell's theory became obvious when it was established that a moving charge in that theory was magnetically the same thing as a "current element," both in itself and under external magnetic force. Only quite lately, however, has it been possible to carry out this notion even tentatively. This has come about by the experimental researches which appear to establish the individuality of electrons of astonishing smallness and mobility. It is now believed by many that the conduction current inside a wire consists of a slow drift of electrons. Naturally, in the present state of ignorance about atoms and molecules, the theory is in an experimental stage. But it does not come into the telegraphic theory sensibly, since the electronic drift is a local phenomenon. It is stationary compared with the wave outside a wire. It may be noted that it is not necessary to consider the electronic drift to be the cause of the electromagnetic wave which has the drift for an after effect.

The resistance of the sea to Hertzian waves in "wireless telegraphy" comes under the same principles, and may be calculated by the same formula as the skin resistance of wires. But this is only a rough beginning, for a reason to be seen presently. Going back to equations (9), consider that R originally meant magnetic conductance outside the wires. The theory was then an exact one; but when R means the resistance of the wires, the theory cannot be exact, because R is not in the right place. It is outside the wave. The theory is now only true, as regards R, for long waves. A long wave length may be considered to be one which is a large multiple of the distance apart of the wires, though there is really no distinct limitation. Telegraphic and telephonic waves are long, of course. So may Hertzian waves be, though on the other hand, as these waves may be made only a few centimetres in length, such waves might require serious correction in the theory. Even when they are many metres long and the wires only a few centimetres apart, although there may be little distortion in transit, the estimation of attenuation may be considerably upset. As for light waves, obviously they are very much too short for the theory. The application of these remarks to waves over the sea is obvious. The waves in the air are too short for the influence of the resistance of the sea to operate on them in so simple a way as in the wire theory. The air itself may also be feebly conductive generally or locally.

Understanding, then, confinement to long waves, equations (9) will apply when the circumstances allow R to be considered fairly constant. This may happen when the conductors are thin sheets of metal permitting very rapid penetration. It may also happen with wires, as when the frequency of the waves is not so great as to increase the resistance greatly; for example, in telephony along copper wires of low resistance. But a considerable extension of the use of equations (9) may be made. Let it be a simply periodic train of waves that is in question, and let R stand for the effective resistance, which becomes definite at a definite frequency. Also make the corresponding change in L, if desired. Then it is a consequence of the general theory of wires that equations (9) are still true under the circumstances. The fundamental solution is

$$V = V_0 \, \varepsilon^{-Px} \sin (nt - Qx), \qquad (17)$$

due to $V_0 \sin nt$ impressed at $x=0$, where P and Q are given by

$$P + Qi = \sqrt{(R+Lni)(K+Sni)}\,; \qquad (18)$$

that is

$$P \text{ or } Q = (\tfrac{1}{2})^{\frac{1}{2}}\{(R^2+L^2n^2)^{\frac{1}{2}}(K^2+S^2n^2)^{\frac{1}{2}} \pm (RK - LSn^2)\}^{\frac{1}{2}}. \qquad (19)$$

Valuable information may be obtained from this equation, by varying the frequency and other constants.

There is an extreme case where R and S are paramount. Then

$$P = Q = (\tfrac{1}{2}RSn)^{\frac{1}{2}}. \qquad (20)$$

This applies to signals received at the end of a long submarine cable, slowly worked. They start as electromagnetic waves, and degenerate to diffusive waves. This is the reason why Lord Kelvin's (W. Thomson's) theory of 1855 applies pretty closely to the reception of signals.

In contrast with this, $R/L = K/S$ produces distortionless transmission, with $P = R/Lv$, $Q = n/v$, if v is defined by $LSv^2 = 1$. Again, if R/Ln and K/Sn are both small, we get approximately.

$$P = \frac{R}{2Lv} + \frac{K}{2Sv}, \qquad Q = \frac{n}{v}. \qquad (21)$$

This is practically very important in telephony. We can have a near approach to distortionless transmission. It is easy to make K/Sn small, because it is naturally small. We may in fact put $K=0$ in effect. So if R/Ln is small, there is little distortion in transit. Observe, too, that the attenuation in the distance x is only $\varepsilon^{-Rx/2Lv}$ instead of $\varepsilon^{-Rx/Lv}$, which is important. The proper leakage to remove the distortion entirely would cure the evil also when R/Ln is big; any value, in fact. Now in telephony we are not concerned with very low frequencies, but only with those beginning with the lowest tone of the human voice. If R/Ln is small for this, it is small for all higher tones, therefore first-rate telephony. Even if R/Ln is 1 for the lowest tone, it will be $\frac{1}{2}$, $\frac{1}{4}$, $\frac{1}{8}$, &c., for the octaves. Then P and Q in (17) will each not vary much in the upper tones, and there can be good transmission.

There are other ways of approximating to the distortionless state than by leakage. In the first example given, no leakage would be wanted, because there would be so little distortion in thousands of miles without the leakage. Other examples relating to telephony were given. Considering the matter more generally, let there be a circuit in which R/L is far greater than K/S, K being small and to be kept small, preferably, to avoid attenuation by leakage. Then we may tend to equalize R/L and K/S in several ways. By reducing S. This is well known. It means with air wires, separating them as much as possible. With cables it means reducing the "spec. ind. capacity" principally, by air or paper insulation. But this reduction of S is not nearly enough. Another way is to reduce R. This is well known and is very important. It can be carried a long way. Cable companies have been much too economical in the matter of copper, in the past especially.

A third way is to increase L. Inductance causes impedance in general. But this way is proper, and is powerful. If we can increase the L of a circuit, without any change at all in R, S, we can approximate to distortionless transmission as much as we please, and with as little attenuation as we please. That is, the effect of increasing L is advantageous in a double sense, for it lessens the attenuation and the distortion simultaneously. To illustrate, if it were possible for L to be zero on say 100 miles of cable of Atlantic type (which it is not) telephony would be very bad indeed or impossible, because there would be such a wide difference in the attenuation produced on tones of different frequencies. The formula (17) above may be used, with $L=0$, $K=0$. But even if L is increased only from 0 to the small value 2 (cm. per cm., B.A. mag. units), the formula will show a remarkable improvement; and if L is increased to 10, the results are excellent. What happens is simply this, that self-induction imparts momentum to the waves, and that carries them on. The formula will show that further increase in L will carry waves across the Atlantic with little loss. But the L needed is exceedingly great. Any increase of L does good, provided it be unaccompanied by increased resistance or other evils to destroy the benefit. It is therefore very important to know how to increase L, and to find out what can be made of the principle concerned in practice. Now there is some practice in it already. Telephone wires in air, for long distance telephony, should be, and are in some places, put up in such a way as to make L as large as possible, although some electricians have maintained that L should be as small as possible. The difference between $L=2$ and $L=20$ or 30, which is produced by separating wires, is enormous, and largely multiplies the distance over which telephonic waves can be carried.

But it is the improvement of signalling through cables that is most important, if it can be done. The most effective way of obtaining self-induction is by the use of iron, suitably arranged and divided; not iron for the guides, but outside or inside them. There is plenty of iron put round cables now, but it is for protection, and is not at all suitably arranged for the purpose required. It is not unlikely that a new type of cable can be designed in which iron can be brought into use to increase the self-induction largely and in a uniform manner.

The writer invented a way of carrying out the principle other than uniformly, and recommended it for trial; viz., by the insertion of inductance coils in the main circuit at regular intervals, say one per mile, according to circumstances. The time constants of these coils should be as large as possible, so that the average inductance may be largely increased without a large increase in the average resistance per mile. The action of the coils is similar to that of small detached loads placed upon a string to give it inertia, and enable it to carry waves farther. If we wish to imitate a curve containing bends of various sizes by means of dots, it is obviously necessary to have several dots in every bend. So, in the substitution of detached inductances for uniform inductance, we need to have several coils in every bend of the electrical wave. It is easy to say that so many coils will be wanted in such a length, in telephony, for example, if we can fix upon the shortest wave length, by using the known formula for the wave length. One could do that without being considered to make a new discovery. But this does not answer the question how few the coils may be for telephony through a cable. Theoretically we want an infinite number, *i.e.*, uniform inductance. We must sacrifice something. The shortest telephonic wave length necessary to be considered is a doubtful element, and there are unmentioned interferences to be allowed for. Hence experiment alone can decide how few the coils need be for telephony through a distance far exceeding that possible without the coils Nothing particular has been done in Great Britain to carry out the writer's invention; but in America some progress has been made by Dr. Pupin, who has described an experiment supporting its practicability; the length telephoned through was increased five times by inserting the coils.

The above radiational theory of telegraphy founded upon Maxwell's theory of light has been proved in all essential points and in various details by important experiments of Hughes, Lodge, Hertz, and others, to say nothing of long distance telephony in America, or the more recent spherical telegraphy. But quite different views have been maintained by the British official electricians, under the guidance of Mr. (now Sir) W. H. Preece. It is only fair to put on record here his leading articles of faith. The information is derived from his papers. Self-induction is harmful to telephony, and is to be avoided and minimised. Nevertheless, experiments upon the Post Office wires showed that though self-induction is sensible in iron wire circuits, it is quite insensible in copper circuits. That is why copper wire is so successful, and iron wires a failure. Also the capacity of circuits (their permittance) is reckoned in a peculiar manner. What controls telephony is "the KR law." If K is the total capacity of a circuit in microfarads, and R its total resistance in ohms, the product KR fixes over what distance telephony can be carried on. It must not exceed 2,000 because of the KR law, the speed of the current varying as the square of the length of the line. Finally, Sir W. H. Preece has a special invention aiming at Atlantic telephony. Imagine a round wire to be slit in two, and the two halves to be slightly separated for insulation, say by a piece of paper. The two halves form the circuit, and it was supposed by the inventor that by close approximation of the flat surfaces the electrostatic and electromagnetic inductions would neutralize one another.

For additional information the reader is referred to the following works: LORD KELVIN. *Mathematical and Physical Papers*, vol. 2. Cambridge, 1884.—CLERK-MAXWELL. *Treatise on Electricity and Magnetism*, vol. 2. Oxford, 1881.—J. J. THOMSON. *Recent Researches in Electricity and Magnetism.*—OLIVER HEAVISIDE. *Electrical Papers*, London, 1892, and *Electromagnetic Theory*, London, vol 1., 1893, vol. 2., 1899.

Some Plane and Cylindrical Waves

§ 529 In § 527 was discussed the general problem of the inversion of operations for plane waves in a conducting dielectric, particularly as regards the telegraphic application. Also the same for some cylindrical waves. Comparing the plane with the cylindrical cases, it may be seen that, whilst the ideas in force are the same, the execution is much more difficult in the cylindrical cases. This arises mathematically from the fact that the operators ϵ^{qx} and ϵ^{-qx} for inward and outward plane waves are reciprocal, and this leads to simple algebraic manipulation. On the other hand, the inward and outward cylindrical operators $H_0(qr)$ and $K_0(qr)$ are not reciprocal, and this leads to more complicated work. At the same time, the property which revealed itself in plane waves, when terminal conditions were imposed, of the setting up of the steady state by a variable source in a very short interval of time, was found to be repeated in the cylindrical cases treated.

(1). But now consider a case of shock. Say there is a cylindrical sheet of circular current C for source, as on p. 330, at the distance $r=a$ from the axis, and just outside it a perfect magnetic reflector, making $H=0$ outside the source, so that only the internal state of H is in question. The equations are (166), p. 330. Let now C be zero before and constant after $t=0$. Find H inside; in particular, its value h at the axis. This is a case of shock, because of the sudden starting of the source. Initially, $H=C$ at $r=a$, and zero elsewhere. The first stage is the journey of the wave front to the axis. At the front itself, $H=C(a/r)^{\frac{1}{2}}$, so it mounts up to ∞ momentarily at the axis. In the second stage, the front travels back to the reflector. By previous experience, the ∞ persists, and then the reflector turns it to $-\infty$. So the value of H at the wave front is $-\infty$ in the third stage. But now we may be guided by the behaviour of the inward waves on p. 323. The ∞ value going to the axis became finite, without change of sign. So we may expect H at the front to be finite in the fourth stage, and therefore, by reflection, finite also in the fifth. This brings us to the axis again, and the creation of an ∞ value, which persists in the sixth stage, and is reversed in the seventh. Then we come to the axis again, with resumption of finiteness of H at wave front. And so on, an eternal series of creation, persistence, reversal, and destruction of an infinite value at the wave front. If this is partly speculative, it can be confirmed or refuted.

A formula for the first wave can be readily found. Put $r=0$ in (167), p. 330, then the axial value is

$$h=\frac{2C}{H_0(qa)}=C\epsilon^{-qa}(2\pi qa)^{\frac{1}{2}}\left[1-\frac{1}{8qa}-\frac{7}{2(8qa)^2}-\frac{59}{2(8qa)^3}-\dots\right]. \quad (1)$$

This series is got by long division. The numbers are as in (157), p. 329, only now with all signs negative after the first. Leave out the factor ϵ^{-qa}, so as to make h begin at the moment $t=0$. Then multegration makes

$$h=C\left(\frac{2a}{vt}\right)^{\frac{1}{2}}\left[1-\frac{1}{2}\left(\frac{vt}{2a}\right)-\frac{7}{24}\left(\frac{vt}{2a}\right)^2-\frac{59}{240}\left(\cdot\right)^3-\frac{3013}{240.56}\left(\cdot\right)^4\right.$$
$$\left.-\frac{50735}{240.56.72}\left(\cdot\right)^5-\frac{2110163}{240.56.72.88}\left(\cdot\right)^6-\dots\right]. \quad (2)$$

It is easy to see that $h=\infty$ at first moment, and that it then falls rapidly, and becomes negative. The coefficients, too, tend towards unity, and this indicates $h=-\infty$ when $vt=2a$. In fact, the curve is something like the cotangent curve, going from + to $-\infty$ as vt goes from 0 to $2a$. But the zero is not in the middle, but at about $1{\cdot}5a$. The + part of the curve is calculable without difficulty, but the negative part requires more and more terms as the negativity comes on, so that a transformation to another series is needed if the latter part of h is to be as easily calculable as the early part. But we see that whereas the first $+\infty$ comes on with a shock, the second $-\infty$, which occurs on the arrival of the reflected wave from the reflector, does not, but is gradually led up to. The same may be inferred to happen with all the later infinite values.

But this is only the first stage in the extraordinarily complicated process of setting up the internal magnetic field. The surging does not decay, because there is no resistance. But there is change of type as the time goes on, so it may be there is an effective simplification in the long run. To determine the later history requires the performance of more elaborate operations.

(2). The alternative method, using the normal functions, leads to a very compact resultant solution. Use the Normal Expansion Theorem, vol. 2, p. 127, eq. (1). It gives

$$\mathrm{H}=\mathrm{C}\frac{\mathrm{I}_0(qr)}{\mathrm{I}_0(qa)}=\mathrm{C}+\mathrm{C}\Sigma\frac{\mathrm{I}_0(qr)\epsilon^{pt}}{p\frac{d}{dp}\mathrm{I}_0(qa)}, \tag{3}$$

if $p=d/dt=qv$, and the summation ranges over the roots of $\mathrm{I}_0(qa)=0$, or $\mathrm{J}_0(sa)=0$, if $q^2=-s^2$. The values of sa are well-known, 2·40, &c. In terms of s, (3) becomes

$$\mathrm{H}=\mathrm{C}\left[1-2\Sigma\frac{\mathrm{J}_0(sr)\cos svt}{sa\,\mathrm{J}_1(sa)}\right]. \tag{4}$$

This is comprehensive. But the numerical calculation for the different stages will be found to be highly complicated. One stage is not enough, for they are all different.

A distortionless extension is obvious. Introduce electric conductivity k, and magnetic conductivity g, balanced so that $k/c=g/\mu=\rho$. This brings in the subsidence factor $\epsilon^{-\rho t}$. If we now make the source be, not C = constant, but $\mathrm{C}\epsilon^{-\rho t}$, then the complete solution becomes

$$\mathrm{H}=\mathrm{C}\epsilon^{-\rho t}[\text{same as in (4)}]. \tag{5}$$

Here there is subsidence to zero, but only because the impressed current subsides. If it does not, but is steady, then the surging will subside, leaving behind a steady state. A different formula results. We have $qv=p+\rho$ now; so the Normal Expansion Theorem makes

$$\mathrm{H}=\frac{\mathrm{I}_0(qr)}{\mathrm{I}_0(qa)}\mathrm{C}=\mathrm{C}\frac{\mathrm{I}_0(\rho r/v)}{\mathrm{I}_0(\rho a/v)}+\mathrm{C}\Sigma\frac{\mathrm{I}_0(qr)\epsilon^{pt}}{p\frac{d}{dp}\mathrm{I}_0(qa)}, \tag{6}$$

and if $q=si$, we have the same values of sa as before, and finally

$$\mathrm{H}=\mathrm{C}\left[\frac{\mathrm{I}_0(\rho r/v)}{\mathrm{I}_0(\rho a/v)}-2\epsilon^{-\rho t}\Sigma\frac{\mathrm{J}_0(sr)\cos svt}{sa\,\mathrm{J}_1(sa)}\frac{1+(\rho/sv)\tan svt}{1+(\rho/sv)^2}\right] \tag{7}$$

Here the final H is smaller at the axis than at the boundary, but the result will be sensibly H=C all over if ρ is very small; and however small it may be, if not actually zero, the surging will disappear in time.

(3). It is possible to recognise the presence of the $\pm\infty$ travelling to and fro from the formula (4), because we always know the situation of the wave front, and can apply a test. Use the convergent formulæ for J_0 and J_1 for the earlier values of sa, and the divergent formulæ for the later. Then the mth root is $(sa)_m = m\pi - \frac{1}{4}\pi$ when m is large, and also

$$J_0(sr) = \left(\frac{2}{\pi sr}\right)^{\frac{1}{2}} \cos(sr - \tfrac{1}{4}\pi), \qquad J_1(sa) = \left(\frac{2}{\pi sa}\right)^{\frac{1}{2}} \cos(sa - \tfrac{3}{4}\pi), \quad (8)$$

(vol. 2, p. 255, eq. (66),). These reduce (7) to

$$H = C\left[1 + X + 2\left(\frac{a}{r}\right)^{\frac{1}{2}} \Sigma \frac{\cos(sr - \frac{1}{4}\pi)\cos svt}{sa \cos m\pi}\right], \qquad (9)$$

where X stands for the sum of the earlier terms. It does not count in the present argument, which concerns the value of the later series. Let the function after Σ be called Y. Choose any point r between 0 and a. The wave front reaches r the first time when $vt = a - r$, and we know that H is finite, so Σ Y must be finite. In this case $Y = (\cos 2sr)/2sa$. But if we make $vt = a + r$, then Y reduces to

$$Y = (1 + \sin 2sr)/2sa. \qquad (10)$$

This makes $\Sigma Y = \infty$, because $m^{-1} + (m+1)^{-1} + (m+2)^{-1} + \ldots = \infty$. This shows the first ∞ value of H, in the second stage. In the third stage, put $vt = 3a - r$. There it will be found that Y reduces to the negative of the value in (10). So H is $-\infty$. In the fourth stage, put $vt = 3a + r$, then $Y = -(\cos 2sr)/2sa$. So H is finite. It is not necessary to go further, because if in the general formula for Y, we increase vt to $vt + 4a$, its value is reversed. That is, Y is periodic, in the period $8a/v$, and a complete cycle of values in Σ Y, and therefore in H, is (1), finite; (2), $+\infty$; (3), $-\infty$; (4), finite; (5), finite; (6), $-\infty$; (7), $+\infty$; (8), finite. But the X part, though always finite, is not periodic.

(4). If now we make the source C be longitudinal, or parallel to the axis of the cylinder, instead of circular round it, **H** becomes circular, and **E** longitudinal inside. The equations are

$$E_1 = -\tfrac{1}{4}\pi qa H_{0r}(K_{0a} + yH_{0a})\mu v C, \quad H_1 = -\tfrac{1}{4}\pi qa H_{1r}(K_{0a} + yH_{0a})C, \quad (11)$$
$$E_2 = -\tfrac{1}{4}\pi qa H_{0a}(K_{0r} + yH_{0r})\mu v C, \quad H_2 = +\tfrac{1}{4}\pi qa H_{0a}(K_{1r} - yH_{1r})C, \quad (12)$$

if E_1, E_2 are the in and out values of E, and similarly as regards H. To prove, test that $E_2 = E_1$, and $H_2 - H_1 = C$ at $r = a$. If we specify y by the condition $E_2 = 0$ at b, outside a, and then shift this electric screen up to just outside a, the result internally is

$E_1=0=H_1$; or there is no disturbance at all. But make $H_2=0$ at b, and then $b=a$. This is a magnetic screen outside the source. The result is, internally,

$$E=-(H_{0r}/H_{1a})\mu v C, \qquad H=-(H_{1r}/H_{1a})C. \tag{13}$$

This case differs remarkably from the last. In both cases the source C per unit area generates displacement **D** at the rate $-$C; that is, the displacement is parallel to **C**, but oppositely directed. But when **D** was circular there was vectorial cancellation, so the steady creation of **D** did not cause unlimited increase in its density. But in the present case, **D** being longitudinal, cannot be destroyed vectorially. Moreover, it is shut in by the screen. So D must mount up without limit, so long as the source C is kept on. The amount of displacement along the cylinder $r=a$ is at every moment the time integral of C. So the cylinder is like an electric condenser of unlimited permittance.

If C is steady from the moment $t=0$, there is also surging, as well as the continuous increase in the total displacement. As before, we can get a formula to show the first wave, with the creation of a $-\infty$ value at the axis, passing to $+\infty$ when the first reflected wave arrives. Say $E=e$ at the axis, then by (13),

$$e=-\mu v C\epsilon^{-qa}(2\pi qa)^{\frac{1}{2}}\Big[1-\frac{3}{8qa}-\frac{33}{2(8qa)^2}-\frac{249}{2(\ \cdot\ )^3}-\frac{9963}{8(\ \cdot\ )^4}-\frac{131229}{8(\ \cdot\ )^5}-\Big]. \tag{14}$$

Leaving out the exponential factor, multegration turns this to

$$e=-\mu v C\Big(\frac{2a}{vt}\Big)^{\frac{1}{2}}\Big[1-\frac{3}{2}\frac{vt}{2a}-\frac{11}{8}\Big(\cdot\Big)^2-\frac{83}{80}\Big(\cdot\Big)^3-\frac{3321}{4480}\Big(\cdot\Big)^4-\frac{43743}{80640}\Big(\cdot\Big)^5-\Big], \tag{15}$$

showing fall from $-\infty$ when $t=0$, to (presumably) $+\infty$ when $vt=2a$, although the later part requires another type of formula for convenient calculation.

The solution in normal series is

$$e=-\mu v C\Sigma\frac{2J_0(sr)\sin svt}{sa(d/dsa)J_1(sa)}-\mu v C\frac{2vt}{a}, \tag{16}$$

the summation ranging over the roots of $J_1(sa)=0$, not counting $sa=0$, which is allowed for by the extra term increasing with the time. This was done by the Normal Expansion Theorem, but is included in the following relating to the distortionless frictional case. Introduce the two conductivities, making also $k/c=g/\mu$, and $p=qv-\rho$. There is now a steady state tended to, got by making $p=0$, and

$$e=-\mu v C\frac{I_0(qr)}{I_1(qa)}=-\mu v C\Big[\frac{I_0(\rho r/v)}{I_1(\rho a/v)}+\epsilon^{-\rho t}\Sigma\frac{I_0(qr)\epsilon^{qvt}}{\Big(q-\frac{\rho}{v}\Big)\frac{d}{dq}I_1(qa)}\Big], \tag{17}$$

because $p(d/dp) = (q - \rho/v)(d/dq)$. The summation must include all the roots of $I_1(qa) = 0$, or $J_1(sa) = 0$. The zero root makes a solitary term, whilst the rest, equally positive and negative, pair together. The result is, in terms of s,

$$e = -\mu v C \left[\frac{I_0(\rho r/v)}{I_1(\rho a/v)} - \frac{2\epsilon^{-\rho t}}{\rho a/v} + 2\epsilon^{-\rho t} \Sigma \frac{J_0(sr)[s \sin - \rho/v \cos] svt}{a(s^2 + \rho^2/v^2)[J_0(sa) - (sa)^{-1} J_1(sa)]} \right]. \quad (18)$$

In this case all the surging vanishes in time. But in the reduced case of no conductivity, the two outside terms in (18) combine, by putting $\rho = 0$, to make the single outside term in (16), showing unlimited accumulation.

(5). There is, of course, no unlimited increase when the source is simply periodic, say $C = C_0 \sin nt$. When the circumstances allow the assumption of a simply periodic state, then we have a stationary vibration inside the source. This is to be got by $p = ni$ in the operational solutions, remembering, however, to use $2I_0$ instead of H_0, $2I_1$ instead of H_1, because H_0 and H_1 belong to solitary inward waves. For example, if there is no resistance, $p = ni$ in (17) makes

$$e = +\mu v [J_0(nr/v)/J_1(na/v)] iC, \quad (19)$$

where iC means $d/d(nt) \cdot C_0 \sin nt = C_0 \cos nt$. But this is not the ultimate result of starting C at any moment, unless we suppose there is a small amount of friction somewhere, not sufficient to sensibly alter the form of (19), and yet sufficient to allow the state to be assumed in a finite time.

External waste acts similarly to resistance. To obtain this waste, take away the screen outside the source, letting it send out disturbance both ways; the result is a stationary vibration inside, and an outward progressive wave outside the source. It is worth while examining the waste formula, in order to see the difference made by changing the direction of the current from being circular round the axis to longitudinal, or parallel to it.

When the current is circular, the equations are

(in) $$H_1 = \tfrac{1}{2}\pi qa I_{0r} K_{1a} C = \tfrac{1}{2}\pi sa J_{0r}(G_{0a} - iJ_{0a}) iC, \quad (20)$$

(out) $$H_2 = -\tfrac{1}{2}\pi qa I_{1a} K_{0r} C = \tfrac{1}{2}\pi sa J_{1a}(G_{0r} - iJ_{0r}) C, \quad (21)$$

by putting $q = si = in/v$, and J_{0r} meaning $J_0(sr)$. The corresponding E's are got by $E = -(cp)^{-1} dH/dr$. Only the outer one is wanted, the inner being a standing vibration, like H_1 in (20). We have

$$E_2 = -\tfrac{1}{2}\mu v \pi qa I_{1a} K_{1r} C = -\tfrac{1}{2}\mu v \pi sa J_{1a}(G_{1r} - iJ_{1r}) iC, \quad (22)$$

and therefore, by (21),

$$E_2 H_2 = -\mu v(\tfrac{1}{2}\pi sa J_{1a} C_0)^2 [G_{1r} \cos + J_{1r} \sin] nt \cdot [G_{0r} \sin - J_{0r} \cos] nt$$
$$= -\mu v(\tfrac{1}{2}\pi sa J_{1a} C_0)^2 [(G_{0r} G_{1r} - J_{0r} J_{1r}) \sin \cos + J_{1r} G_{0r} \sin^2 - J_{0r} G_{1r} \cos^2] nt, \quad (23)$$

of which the mean value is

$$\overline{E_2H_2} = -\tfrac{1}{2}\mu v(\tfrac{1}{2}\pi sa J_{1a} C_0)^2 [J_{1r}G_{0r} - J_{0r}G_{1r}] = \tfrac{1}{4}\mu v \pi sa (J_{1a}C_0)^2. \qquad (24)$$

But when the current is longitudinal, the equations are

$$E_1 = -\tfrac{1}{2}\mu v \pi qa I_{0r} K_{0a} C, \qquad H_1 = -\tfrac{1}{2}\pi qa I_{1r} K_{0a} C, \qquad (25)$$

$$E_2 = -\tfrac{1}{2}\mu v \pi qa I_{0a} K_{0r} C, \qquad H_2 = +\tfrac{1}{2}\pi qa I_{0a} K_{1r} C, \qquad (26)$$

and $q = si$ makes the external state be

$$E_2 = -\tfrac{1}{2}\mu v \pi sa J_{0a}(G_{0r} - iJ_{0r}) iC, \quad H_2 = -\tfrac{1}{2}\pi sa J_{0a}(G_{1r} - iJ_{1r})C, \quad (27)$$

by (65) vol. 2, p. 255; therefore

$$E_2H_2 = -\mu v(\tfrac{1}{2}\pi sa J_{0a} C_0)^2 [G_{0r} \sin - J_{0r} \cos] nt \,.\, [G_{1r} \cos + J_{1r} \sin] nt, \quad (28)$$

of which the mean value reduces to

$$\overline{E_2H_2} = \tfrac{1}{4}\mu v \pi sa (J_{0a}C_0)^2. \qquad (29)$$

Comparing with (24), the only difference is in the change from J_{1a} to J_{0a}. In both, C is the current per unit area of the sheet $r = a$, but the series of frequencies for which the waste is zero or maximal are different. If, however, sa is large, or the wave length is a small fraction of the radius a, the two wastes become the same in corresponding phases, because, when sa is large,

$$J_{0a} = (2/\pi sa)^{\frac{1}{2}} \cos(sa - \tfrac{1}{4}\pi), \qquad J_{1a} = (2/\pi sa)^{\frac{1}{2}} \cos(sa - \tfrac{3}{4}\pi). \quad (30)$$

Taking the circular function to be 1 in either case, the result is

$$\overline{E_2H_2} = \tfrac{1}{2}\mu v C_0^2. \qquad (31)$$

The reason can be easily seen, for $\frac{1}{2}C_0^2$ is the mean square of current, and μv is the impedance outward from unit area of the surface, which, on account of the short wave length, is practically plane in the present argument.

But it is very different when the wave length is a large multiple of the radius a, which requires sa to be small. Then $J_{0a} = 1$, and $J_{1a} = \frac{1}{2}sa$. So the waste varies as the first power of the frequency when the current is longitudinal, and as the cube of the frequency when the current is circular; a remarkable difference in the radiative power.

(6). On p. 330 occurred one case, and on p. 350 another, where a perfect electric conductor stopped a current sheet from sending out any disturbance. The screen was outside the source. The same happens when it is inside, and for the same reason. As a screen it prevents the passage of waves through it, and as a reflector it sends out what it receives from the source with the electric force reversed, so that on the free side of the source the direct and reflected waves annul one another both as regards E and H. To exhibit this formally in the same notation, let the reflector be at $r = b$, inside the source C at $r = a$. Let E_1, H_1 be the inner and E_2, H_2 the outer results. Then, when C is circular, the equations are

$$H_1 = \tfrac{1}{4}\pi qa(H_{0r} + yK_{0r})K_{1a}C. \qquad E_1 = -\tfrac{1}{4}\mu v \pi qa K_{1a}(H_{1r} - yK_{1r})C, \quad (32)$$

$$H_2 = -\tfrac{1}{4}\pi qa K_{0r}(H_{1a} - yK_{1a})C, \quad E_2 = -\tfrac{1}{4}\mu v\pi qa K_{1r}(H_{1a} - yK_{1a})C; \qquad (33)$$

to prove which, note that $E_1 = E_2$, and $H_1 - H_2 = C$ at a, for any y; and that $y = H_{1b}/K_{1b}$ makes $E_1 = 0$ at b. These may be fully elaborated, to show the waste, for example. The point in present question, however, is that if b is made $= a$, then the outer field of E and H disappears. The inner, between a and b, is reduced simply to $H_1 = C$ in a vanishingly thin layer, and $E_1 = 0$.

Similarly, when the current is longitudinal, the equations are

$$E_1 = -\tfrac{1}{4}\mu v\pi qa(H_{0r} + yK_{0r})K_{0a}C, \quad H_1 = \tfrac{1}{4}\pi qa(H_{1r} - yK_{1r})K_{0a}C, \qquad (34)$$

$$E_2 = -\tfrac{1}{4}\mu v\pi qa(H_{0a} + yK_{0a})K_{0r}C, \quad H_2 = -\tfrac{1}{4}\pi qa(H_{0a} + yK_{0a})K_{1r}C, \qquad (35)$$

because $E_1 = E_2$, and $H_1 - H_2 = C$, at $r = a$, for any y. The condition $E_1 = 0$ at b makes $y = -H_{0b}/K_{0b}$; and then increasing b to the value a makes E_2 and H_2 vanish, and also E_1, whilst H_1 becomes simply C.

This impotency of an impressed current sheet to produce any electromagnetic effects when bounded upon either side by a perfect electric conductor has no necessary limitation to a cylindrical sheet. That is quite accidental here. It is interesting in connection with the theory of conduction in wires, in illustration of the difference pointed out on p. 12, between a surface conduction current and a surface convection current. Thus, if a long wire has on its surface, or more distinctly, just outside it, a sleeve of impressed electric current, say C per unit area, acting longitudinally, there will be no magnetic force produced externally save what is due to imperfect conductivity of the wire, and that will tend to vanish with increasing conductivity. The correction due to finite length of sleeve does not enter the present argument. Now, one way of interpreting C is by σu, a convection current, σ being surface density of electrification and u its speed. It is then a sleeve of electrification that is slipped along the wire. There will be no external H produced. Of the two electromagnetic waves sent out by the source, the inward one is turned by total reflection to an outward wave which cancels the primary outward wave. If the compensating opposite electrification exists in close contiguity to the moving sleeve, say in the form of an inner or outer stationary sleeve, there is obviously no external E either, not merely the tangential E of the electromagnetic wave, but of any sort. But if the compensating electrification is outside and right away from the moving sleeve, then of course there is E between them, in the space now made external. But this does not affect the main matter, which is that the motion of the electrified sleeve does not produce any external H when the conductivity of the wire is perfect.

If we construct a distortionless circuit so well that it will practically conduct a single pulse without any back effects, then we may say that electrification travels along the surface of the wire at speed

v in company with the pulse. But it is not a convection current in the proper sense, a convection of the *same* electricity through the ether at speed v. There is a great physical difference. The electrification is a surface state produced by and travelling with the wave along the wire. This is not at all the same thing as moving a sleeve of electrons (always the same) along the surface at speed v. The electrons set moving by the external pulse have their own proper motions, no doubt. But the discovery of the electron does not, in my opinion, furnish any reason for retrograde ideas in electromagnetics, even though the precise nature of metallic conduction remains somewhat vague, in the absence of a knowledge of facts relating to atoms and electrons which would determine it definitely.

(7). Without using the idea of a perfect conductor, we may show how to imitate its theoretical action in causing total reflection, by using a moving sheet of electrification. Consider first a plane sheet of impressed electric current, C per unit area, which may be σu, a convection current, if we make proper allowance for the stationary electric field of σ. What this C does is to generate displacement **D** at the rate $-$C, half above and half below the sheet. Along with this **D**, which is tangential, is magnetic force, according to $\mathrm{E} = \pm\mu v\mathrm{H}$, above and below the sheet, of strength such that $2\mathrm{H} = \mathrm{C}$. So by varying C arbitrarily, we can send out arbitrary plane radiation from the sheet both ways.

But let arbitrary radiation, specified by E, H, be falling flush upon the sheet from above. If the sheet is kept at rest, the radiation will go right through unaltered. But move the sheet so that its speed at every moment is such as to generate $-$E above it, the exact negative of the arriving +E. The remarkable result follows that on the lower side the radiation of the sheet is the negative of the arriving radiation, both as regards E and H, so that there is no disturbance below the sheet. On the upper side the radiation of the sheet is the same as the arriving radiation as regards H, but exactly opposite as regards E. That is, every elementary slab of E, H which arrives at the sheet is sent back again with E reversed and H the same. This, superposed on the arriving radiation, shows the resultant effect on the upper side. Or, we have effectively the boundary condition $\mathrm{E} = 0$. The moving sheet of electrification is therefore equivalent to the surface of a perfect conductor. Its speed varies at every moment as the intensity of the incident E or H.

(8). It is worth while looking to see how small the speed u comes out on certain assumptions. If, as usual, 10^{-8} is the atomic range, there are 10^{16} atoms in a single layer per unit surface of a conductor. Also, 10^{-20} is the electronic charge. Let there be a small number of m layers effectively concerned, and let n electrons per atom be effectively concerned; then $nm10^{-4}=\sigma$ in B.A. magnetic units, and $4\pi\sigma u=2H$ becomes

$$4\pi u=(2H/mn)10^4 \text{ cm. per sec.} \tag{36}$$

The 4π is introduced to meet the needs of the B.A. units, but I will say nothing opprobrious about it, because Lord Rayleigh does not like it. If $H=mn$, u is about 1 kilom. per minute, which is ordinary motor-car speed away from the trap. But u may be far less than that, or it may be greater. The data do not exist for application to real conductors. The above only shows how strong a certain convection current in a layer in the ether must be to reflect radiation.

(9). Since all solids are elastic solids, and mechanism and machinery are made of various solids fitted together, it is obvious that the proper theory of their motions would require the consideration of the elastic yielding throughout, and would be immensely complicated. There would be no instantaneous transmission of motion along any one part, but transmission in time instead. All sorts of shocks and very rapid vibratory motions would require consideration, which is avoided and ignored in general by the substitution of unyielding solids, furnishing a useful and necessary, but very artificial system of statics and dynamics for practical use. It is just the same in electromagnetics, when we do away with, in certain places, the elastic yielding implied in the equation of electric displacement $D=cE$, by making $c=0$, and thereby introducing instantaneous actions, analogous to those of practical mechanics. Some of the mechanical problems become impossibly complicated on allowance for elastic yielding. Others are quite easily treated. For example, think of an infinitely long shaft supported upon numerous (frictionless) bearings to keep it from bending, subjected to a torque at its beginning. If rigid, no finite torque could turn it, owing to the infinite mass. But in reality, a steady applied torque will turn it at constant speed proportional to the torque at its beginning, twisting it, and the twist will travel along the shaft at constant speed. Any amount of turning the same way is permissible, because the strain does not accumulate, but travels out. In fact, the theory is like that of a distortionless circuit without resistance, and is quite elementary. Similarly, the reflections and vibratory effects when the shaft is of finite length make problems like those of a distortionless circuit. Similar remarks apply to the transmission of longitudinal effects, about which Prof. Perry has written. But the bending effects are quite different, and a few remarks upon them will be given separately. What follows now concerns the torsion of two rods in sequence of very different sections, a thick and a thin one, and the practical substitution of a rigid rod for a thick one. But it will be done in terms of the distortionless circuit, for convenience of continuity.

(10). Let there be a distortionless line extending from $x=0$ to l with the constants L_1 and S_1, and another from $x=l$ to ∞ with the constants L and S.

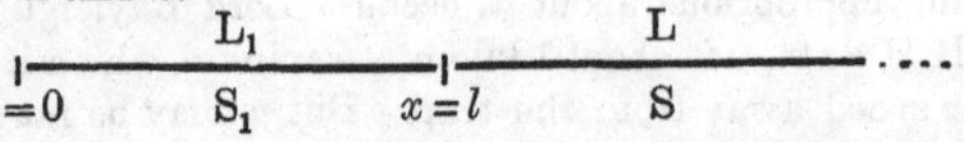

It may be that the effect of S_1 is so trifling that we say $S_1=0$ practically, reducing the first line to an inductance of amount $L_1 l$. The effects of resistance and leakage are not in question. They are zero here. Then if V_0 is impressed at $x=0$, the result is

$$C_1=\frac{V_0}{Lv+L_1 lp}, \tag{37}$$

because the further distortionless circuit behaves as a resistance Lv. Here C_1 is the current in the first line, not only at its beginning, but all along, up to $x=l$. Then, in the rest, we shall have

$$V_x=\epsilon^{-q(x-l)}V_1=LvC_x, \qquad V_1=LvC_1, \tag{38}$$

where V_1 and C_1 are the values at $x=l$. The effect of inserting the first line is merely to produce inertial retardation. Say V_0 is constant, then

$$V_1=LvC_1=V_0[1-\epsilon^{-Lvt/L_1 l}]. \tag{39}$$

When this initial delay is overcome, the effects in the second line are the same as if the first line did not exist.

Now see how this elementary practical solution works down from the true one, with S_1 finite from $x=0$ to l. We want the reflection coefficient for V at the junction. Say that w_1, w_2, w_3 are corresponding values in an incident, reflected, and transmitted slab wave at the junction. Then

$$w_1+w_2=w_3=L_1 v_1 c_3=L_1 v_1(w_1-w_2)/Lv \tag{38A}$$

are the conditions of persistence of voltage and current. So

$$\rho=\frac{w_2}{w_1}=\frac{Lv-L_1 v_1}{Lv+L_1 v_1}, \qquad \sigma=\frac{w_3}{w_1}=\frac{2Lv}{Lv+L_1 v_1}, \tag{39A}$$

if ρ is the reflection factor and σ the transmission factor for a voltage wave. It now follows that

$$V=\frac{\epsilon^{-q_1 x}+\rho\epsilon^{-q_1(2l-x)}}{1+\rho\epsilon^{-2q_1 l}}V_0=[\epsilon^{-q_1 x}+\rho\epsilon^{-q_1(2l-x)}-\rho\epsilon^{-q_1(2l+x)}+\dots]V_0 \tag{40}$$

$$C=\frac{\epsilon^{-q_1 x}-\rho\epsilon^{-q_1(2l-x)}}{1+\rho\epsilon^{-2q_1 l}}\frac{V_0}{L_1 v_1}=[\epsilon^{-q_1 x}-\rho\epsilon^{-q_1(2l-x)}-\rho\epsilon^{-q_1(2l+x)}+\dots]\frac{V_0}{L_1 v_1} \tag{41}$$

are the V and C at x in the first line due to V_0 impressed at $x=0$. The reflection coefficient at $x=0$ is -1 for voltage. The expansions in series show the full history of events, the exponentials being merely operators to fix the moment of starting of the corresponding terms, having the value 0 before and 1 after particular moments.

Put $p=0$, or $q_1=0$, in the condensed solutions to find the ultimate results when V_0 is steady. They are $V=V_0$, and $C=V_0/Lv$. That is, the actions in the second line are just the same as if the first did not exist, after the preliminary irregularity, although the constants of the first line are different.

Now observe the effect upon ρ and σ as S_1 is varied. The transmission factor for voltage goes from 0 to 2 as Lv/L_1v_1 goes from 0 to ∞, and the transmission factor for current goes from 2 to 0 under the same circumstances. But the reflection factor for voltage goes from -1 to $+1$, and that for current from $+1$ to -1 under the same circumstances. In particular, as S_1 is reduced to 0, ρ is reduced to -1. This is the same as for a short-circuit. But there is deception here, for if it were really a short-circuit, nothing would pass into the second line. In fact v_1 in the first line tends to ∞ as S_1 tends to 0, and ρ to -1. So we have a packing together in a moment of innumerable small effects. The whole series requires to be properly allowed for. Since $q_1=p/v_1$, it tends to zero, and the exponential ϵ^{-q_1x} can be replaced by $1-q_1x$, and similarly for all the other exponentials, in proceeding to the limit. Do this in the condensed forms of the solutions. They then reduce simply to

$$V=\left(1-\frac{L_1px}{Lv+L_1pl}\right)V_0, \qquad C=\frac{V_0}{Lv+L_1pl}. \tag{42}$$

These are the reduced solutions when $S_1=0$, for any V_0. When V_0 is constant,

$$V=V_0\left(1-\frac{x}{l}\epsilon^{-Lvt/L_1l}\right), \qquad C=\frac{V_0}{Lv}\left(1-\epsilon^{-Lvt/L_1l}\right). \tag{43}$$

The C is as we got before. In addition, we have the V solution, and see that the impression of V_0 instantly produces $V=V_0(1-x/l)$, which is zero at the end of the first line, and then rises in time to $V=V_0$.

(11). Although this reduction from staircase functions to continuous functions is mathematically inevitable by the disappearance of all the terms which have S_1 as a factor, yet it is in some respects more satisfactory to view the matter differently. Keep v_1 constant, whilst increasing L_1 and reducing S_1. But let S_1 be reduced only to a small value, not zero. Then ρ is a trifle less than -1. Now see the series formula (41) for C. Since L_1 is large, V_0/L_1v_1 is very small. Moreover, all terms have the same sign, and they decrease in magnitude slowly, in pairs. The resulting function of the time is not the continuous curve (43), but is what that curve becomes by having an enormous number of little steps cut in it, of size and position indicated by the series formula. The steps are not the same for all values of x, but the practical difference from place to place is trifling when median curves are drawn through the different stepped curves.

But it is strikingly different with the V series. The steps are of full size V_0 at first, and only fall to insignificance in time. Thus V at any spot jumps from 0 to V_0, stays there for a time; then drops to nearly 0, and stays there for a time; then jumps up to V_0 again, and after a time drops to nearly 0, but a little greater than before; and so on, over and over again. The moments of these jumps up and down, and the durations, vary from place to place. Thus the jumps are large in the variable period. How is this to be harmonised with the practical formula for V? A close examination in detail will show that what the practical V of the rigid theory represents is the mean value (with respect to the time) of its widely varying values in the real yielding theory. This mean value is not the same in different places. This property of the substitution of a mean value in reducing from one theory to another is very convenient in eliminating mathematical complication. At the same time it is well to remember in physics that the simplified results are not always like the real, but are conventional substitutions for the same, to simplify work.

(12). The other way of making the speed in the first line be infinite is by $L_1=0$, with finite S_1. The reduced equations are now

$$V=e, \qquad C=[(Lv)^{-1}+S_1(l-x)p]\,e, \tag{44}$$

in the first line. The interpretation is easy. The first line makes a condenser, all at one potential, with the result of making e act instantly in full strength upon the second line. So far as the actions in the second line go, the interposition of the first line makes no difference. But the first line has to be charged itself, and this is represented in the equation of C. The arrangement is equivalent to a condenser $S_1 l$ and the second line put in parallel.

But if we keep v_1 finite, making S_1 large and L_1 small, we have the representation in (40), (41) of the proper and very complicated oscillations in the first line. The constant $e=V_1$ impressed upon the second line in (44) is represented by big jumps between the values $V_1=0$ and $2e$ at the commencement, but of regularly decreasing range, settling down to the constant value $V_1=e$. This also represents how LvC_1 behaves at the beginning of the second line. But at other parts of the first line the details are quite different for V and for C.

In the torsion illustration the first line should be easily twistable, the second one relatively hard to twist. By the use of a rod of india-rubber to communicate rotation to a straight metal rod, it becomes less difficult to see the general nature of the twisting and untwisting effects in the very yielding connecting link.

(13). If there is a resistance R_0 at the beginning, and a resistance R_1 at the end of a distortionless circuit, the reflection coefficients for a voltage wave are

$$\rho_0 = (R_0 - Lv)(R_0 + Lv)^{-1}, \qquad \rho_1 = (R_1 - Lv)(R_1 + Lv)^{-1}, \tag{45}$$

and the full solution due to e impressed at $x=0$ is

$$V = \frac{Lv}{R_0+Lv}\,\frac{\epsilon^{-qx}+\rho_1\epsilon^{-q(2l-x)}}{1-\rho_0\rho_1\epsilon^{-2ql}}e \qquad C = \frac{1}{R_0+Lv}\,\frac{\epsilon^{-qx}-\rho_1\epsilon^{-q(2l-x)}}{1-\rho_0\rho_1\epsilon^{-2ql}}e. \tag{46}$$

The numerator parts represent the primary wave and the first reflected wave, whilst all the following waves are obtained by expanding the denominator by long division. Here $q=(p+\rho)/v$, so the whole history is known in functional detail by inspecting the series. If $\rho_1=0$, there is just one wave, the primary wave, because it is wholly absorbed by R_1. If ρ_1 is not zero, there is a reflected wave. But if $\rho_0=0$, it is absorbed at $x=0$, so there is nothing more. If ρ_0 and ρ_1 are both finite, there is an infinite series of waves, leading finally to steady states of V and C, which are expressed by (46) when we give to q the reduced value ρ/v, ρ being R/L or K/S.

There is nothing special about the above, it being an easy case of the distortionless circuit which I gave in my treatment in 1887. What is of further interest is to observe that the reflection factors ρ_0 and ρ_1, which are included between the limits -1 and $+1$ when the terminal resistances are positive, may have any real values from $-\infty$ to $+\infty$ given to them without at all interfering with real electromagnetic interpretation of the results. For convenience, however, R_0 may be retained as a positive resistance. Then we may impose at $x=l$ the condition $V_1=R_1C_1$, without restricting R_1 to be positive, and then follow up by the wave series the whole history of the effects due to e, and come finally (usually) to a steady state. If, temporarily, $\rho=0$, or the line itself has no resistance and no leakance, then the final current will be $C=e/(R_0+R_1)$, which may be of any size, positive or negative. But if $R_0=-R_1$, the circuit has no resistance in the whole. We can then say that C will tend to ∞. But will it be $+\infty$ or $-\infty$? That depends on circumstances. The condition $R_0+R_1=0$ makes $\rho_0\rho_1=1$, and the current waves are given by

$$C = \frac{e}{R_0+Lv}\left[\epsilon^{-qx} - \rho_1\epsilon^{-q(2l-x)} + \epsilon^{-q(2l+x)} - \rho_1\epsilon^{-q(4l-x)} + \dots\right], \tag{47}$$

where

$$\rho_1 = (R_0+Lv)(R_0-Lv)^{-1}. \tag{48}$$

Here it is all finite and interpretable in finite time, save when $R_0=Lv$, making $\rho_1=\pm\infty$. It depends on the sign of ρ_1 whether C mounts up positively or negatively. When $R_0>Lv$, ρ_1 is $+$, and

>1, so the reflection of C at $x=l$ is negative, and C tends towards $-\infty$. But if $R_0<Lv$, ρ_1 is $-$, and C tends towards $+\infty$. The first wave is always finite; it is the sign of the second wave which determines the later tendency.

There are corresponding peculiarities when R_0 is negative. It is necessary for ρ_0 and ρ_1 to have both the same sign in the case of no total resistance.

(14). The solutions (46) are remarkable in another way. If there is only one wave, or only two waves, the differential denominator reduces to 1. Then the Fourier normal analysis completely fails. Except in these cases, we may apply the Normal Expansion Theorem, vol. 2, p. 127. It gives, immediately, since $qv=p+\rho$,

$$C=C_0+\frac{e}{R_0+Lv}\sum\frac{\epsilon^{-qx}-\rho_1\epsilon^{-q(2l-x)}}{2pl/v}\epsilon^{pt}, \tag{49}$$

where C_0 is the steady final current, and the summation ranges over the roots of

$$\epsilon^{2ql}=\rho_0\rho_1, \quad \text{or} \quad ql=\lambda+n\pi i, \tag{50}$$

where $\lambda=\log\sqrt{\rho_0\rho_1}$ when $\rho_0\rho_1$ is $+$, or else $-\log\sqrt{-\rho_0\rho_1}$ when $\rho_0\rho_1$ is $-$, if n receives all integral values, positive and negative, and zero. This makes

$$C=C_0+\frac{e\epsilon^{-\rho t}}{R_0+Lv}\sum_{-\infty}^{\infty}\frac{\epsilon^{q(vt-x)}-\rho_0^{-1}\epsilon^{q(vt+x)}}{2(ql-\rho l/v)}. \tag{51}$$

It is entirely real, because the $n=0$ term makes a solitary real term, and the term for $n=+m$ pairs with the term for $n=-m$ to make a real compound term. It is, however, very complicated in circular functions, and need not be written out.

Either $\rho_0=0$ or $\rho_1=0$ makes $\lambda=\infty$, and the series is uninterpretable. But $\rho_0\rho_1=1$ makes $\lambda=0$. This is when $R_0+R_1=0$. Then

$$C=C_0+\frac{e\epsilon^{-\rho t}}{2(R_0+Lv)}\sum_{-\infty}^{\infty}\frac{\epsilon^{(vt-x)n\pi i/l}-\rho_1\epsilon^{(vt+x)n\pi i/l}}{n\pi i-\rho l/v}. \tag{52}$$

If, further, we wish to abolish the resistance and leakance of the line, by $\rho=0$, we should take out the $n=0$ term and unite it with C_0, and then put $\rho=0$. This makes

$$C=\frac{eR_0(1-x/l)-Lv^2t/l}{R_0^2-L^2v^2}+\frac{e}{2(R_0+Lv)}\sum\frac{\epsilon^{(vt-x)n\pi i/l}-\rho_1\epsilon^{(vt+x)n\pi i/l}}{n\pi i}, \tag{53}$$

without any zeroth term in the summation. Here the outside part shows the rise of C in time to $-\infty$ or $+\infty$ according as $R_0>$ or $<Lv$. What the outside part represents is the limit of

$$\frac{e}{R_0+Lv}\left[\frac{\epsilon^{\rho(l-x)/v}-\rho_1\epsilon^{-\rho(l-x)/v}}{\epsilon^{\rho l/v}-\epsilon^{-\rho l/v}}-\frac{(1-\rho_1)\epsilon^{-\rho t}}{2\rho l/v}\right], \tag{54}$$

when $\rho=0$, and ρ_1 is defined by (48).

It may be said, and with reason, that the method of normal functions, even when assisted by the Normal Expansion Theorem, which cuts out a lot of needless work, is not suited for the investigation of wave problems; that the results are too complicated, and the final interpretation very difficult or impossible. Yet it remains true that the method of normal functions is sometimes the only known way, owing to the wave method not having been carried out. This remark does not apply in the present case, of course, or to the far more elaborate results relating to unbalanced circuits considered in vol. 2. But there is plenty of room for extension of the wave method in mathematical physics. In the above it is particularly easy, and though this is of principal importance physically, it also serves the purpose of interpreting the normal series. I wonder how long it will take before writers of books about Fourier series and other normal series will come to see the importance of doing the work in a way suited to practical physics. At present they seem to be too much occupied with questions of convergency, and with logical difficulties to which there is no end, because they can be manufactured interminably. But there are much more important matters than these; for instance, the promotion of natural knowledge.

(15). On p. 357 we saw that when plane waves are traversing a non-conducting dielectric, say the ether, the effect of doing away with the elastic yielding in a portion of finite depth is to produce merely inertial retardation in the transmission of H from one side to the other, although this comes about, when the permittivity is reduced continuously to zero, by the dense compression, finally to instantaneity, of an infinite series of to and fro waves. If, however, the slab of zero permittivity is of unlimited depth, then we do away with the reflection from the distant boundary. There is left only the initial wave to deal with. But this is reduced to zero along with c. So there is no H at all transmitted. From another point of view, the inertia is made ∞ by the rigid connection. So the medium of finite μ and zero c, with a plane boundary on one side and unlimited on the other, acts as a perfect reflector to incident radiation, producing the condition $H=0$ at the boundary. But E passes through, though this has no energetic significance.

We may expect something similar with cylindrical waves, though it is needful to be careful, because they are very peculiar sometimes. Say that such waves are emitted from a wire, due to impressed force in it. If we abolish c in a finite portion of the medium outside it, say from $r=a$ to b, there should be free transmission of H through this portion to the outer ether, only modified by inertial retardation. But if $c=0$ all the way outside the wire, then there can be no H at all. From the wave point of view, the initial wave is made zero, and there is no infinite series of instantaneous reflections to counteract that. From the other point of view, the inductance becomes ∞. That is, uniform longitudinal e in an infinitely long straight wire can produce no current in it when the elastic yielding outside it is done away with. This is like an extreme case of Helmholtz's coil theory. The cylindrical equations are not wanted to prove the above, but there are other results which cannot be foreseen in this way.

(16). If there is longitudinal impressed electric force $e=f(z, t)$ acting uniformly over the section of a wire of radius a, with constants μ_1, c_1, k_1, surrounded by a medium of constants μ_2, c_2, k_2, then **H** is circular, but **E** in general has an outward or radial component F as well as a longitudinal component E. By inspection of small areas bounded by lines parallel to **z**, **r**, and the line perpendicular to both, the circuital equations are reduced to

$$\frac{1}{r}\frac{d}{dr}r\mathrm{H}=(k+cp)\mathrm{E}, \qquad -\frac{d\mathrm{H}}{dz}=(k+cp)\mathrm{F}, \qquad \frac{d\mathrm{E}}{dr}-\frac{d\mathrm{F}}{dz}=\mu p\mathrm{H}, \tag{15}$$

in either medium, with the proper values of μ, c and k. So

$$\frac{1}{r}\frac{d}{dr}r\frac{d\mathrm{E}}{dr}=q^2\mathrm{E}, \qquad \text{where} \quad q^2=\mu kp+\frac{p^2}{v^2}-\frac{d^2}{dz^2}. \tag{16}$$

The E/H operators may therefore be of Besselian type,

$$\frac{\mathrm{E}_1}{\mathrm{H}_1}=\frac{q_1}{k_1+c_1p}\frac{\mathrm{I}_{0r}}{\mathrm{I}_{1}r}, \qquad \frac{\mathrm{E}_2}{\mathrm{H}_2}=-\frac{q_2}{k_2+c_2p}\frac{\mathrm{K}_{0r}+y\mathrm{I}_{0r}}{\mathrm{K}_{1r}-y\mathrm{I}_{1r}}. \tag{17}$$

No y is put in the wire operator, it being assumed that there is no internal boundary. The y in the outer operator must be found by some external condition, say at $r=b$.

The conditions imposed by the source of disturbance (curl **e**) on the surface of the wire make $\mathrm{H}_1=\mathrm{H}_2$, and $\mathrm{E}_1-\mathrm{E}_2=e$, at $r=a$. Applying these to (17), the results are

$$\mathrm{E}_1=\frac{\dfrac{q_1}{k_1+c_1p}\dfrac{\mathrm{I}_{0r}}{\mathrm{I}_{1a}}e}{\mathrm{Z}_1+\mathrm{Z}_2}, \qquad \mathrm{H}_1=\frac{\dfrac{\mathrm{I}_{1r}}{\mathrm{I}_{1a}}e}{\mathrm{Z}_1+\mathrm{Z}_2}, \qquad \mathrm{F}_1=\frac{-\dfrac{\mathrm{I}_{1r}}{\mathrm{I}_{1a}}\dfrac{de}{dz}}{(k_1+c_1p)(\mathrm{Z}_1+\mathrm{Z}_2)}, \tag{18}$$

$$\mathrm{E}_2=\frac{\dfrac{-q_2}{k_2+c_2p}\dfrac{\mathrm{K}_{0r}+y\mathrm{I}_{0r}}{\mathrm{K}_{1a}-y\mathrm{I}_{1a}}e}{\mathrm{Z}_1+\mathrm{Z}_2}, \qquad \mathrm{H}_2=\frac{\dfrac{\mathrm{K}_{1r}-y\mathrm{I}_{1r}}{\mathrm{K}_{1a}-y\mathrm{I}_{1a}}e}{\mathrm{Z}_1+\mathrm{Z}_2}, \qquad \mathrm{F}_2=\frac{-\dfrac{\mathrm{K}_{1r}-y\mathrm{I}_{1r}}{\mathrm{K}_{1a}-y\mathrm{I}_{1a}}\dfrac{de}{dz}}{(k_2+c_2p)(\mathrm{Z}_1+\mathrm{Z}_2)}, \tag{19}$$

$$\mathrm{Z}_1=\frac{q_1}{k_1+c_1p}\frac{\mathrm{I}_{0a}}{\mathrm{I}_{1a}}, \qquad \mathrm{Z}_2=\frac{q_2}{k_2+c_2p}\frac{\mathrm{K}_{0a}+y\mathrm{I}_{0a}}{\mathrm{K}_{1a}-y\mathrm{I}_{1a}}. \tag{20}$$

In these use q_1 for the wire and q_2 for the outer medium. These are the complete results for reference, from which special results may be got in the manner of my papers on cylindrical waves (El Pa., vol. 2, p. 443). If equations of this sort, which are really differential or operational solutions, are apparently unintelligible, they have the advantage of being very readily converted to functional forms in particular cases.

The meaning of Z_1 and Z_2 may be seen thus. Let C be the total current in the wire. It is

$$C = 2\pi a H_a = 2\pi a(Z_1 + Z_2)^{-1} e. \tag{21}$$

So $Z_1/2\pi a$ is the resistance operator of the wire per unit length, and $Z_2/2\pi a$ that of the surrounding medium. Z_2 is indefinite as regards y. If there is no outer boundary, $y=0$. The form of y may, however, be determined to suit the case of any number of coaxial different media, by simply equating the E/H operator on one side of a boundary to that on the other side, since there is continuity in both E and H. If, for instance, at $r=b$, the medium changes its constants to μ_3, k_3, c_3, then by (17) we have the condition

$$-\frac{q_2}{k_2+c_2p}\frac{K_{0b}+yI_{0b}}{K_{1b}-yI_{1b}} = -\frac{q_3}{k_3+c_3p}\frac{K_{0b}+y_3I_{0b}}{K_{1b}-y_3I_{1b}}, \tag{22}$$

using q_2 on the left and q_3 on the right side. This finds y in terms of the third medium, in which y_3 is to be found by a further boundary condition, if there is another boundary, or by $y_3=0$ if there is not. In the last case, of a wire, with only two differing media outside it, the value of y is

$$y = -\left(\frac{q_2}{k_2+c_2p}\frac{K_{0b2}}{K_{0b3}} - \frac{q_3}{k_3+c_3p}\frac{K_{1b2}}{K_{1b3}}\right)\left(\frac{q_2}{k_2+c_2p}\frac{I_{0b2}}{K_{0b3}} + \frac{q_3}{k_3+c_3p}\frac{I_{1b2}}{K_{1b3}}\right)^{-1} \tag{23}$$

where the third suffix refers to q_2 or q_3 as the case may be.

(17). Now to see the effect of abolishing the elastic yielding in the medium just outside the wire, say from $r=a$ up to b, put $k_2=0$, $c_2=0$ in (23), and let q_2 be finite. Then y is reduced to $y=-K_{0b2}/I_{0b2}$, and Z_2 becomes ∞, by (20). So, by (18) and (19), E_1, F_1, H_1 are zero. Also H_2. But

$$E_2 = -\frac{K_{0r}+yI_{0r}}{K_{0a}+yI_{0a}}e, \qquad F_2 = -\frac{K_{1r}-yI_{1r}}{K_{0a}+yI_{0a}}\frac{1}{q_2}\frac{de}{dz}. \tag{24}$$

So E_2 falls from the value $-e$ at $r=a$ to 0 at $r=b$, whilst F_2 is finite throughout, including $r=b$. It follows by continuity of E and H at b that there is no E_2 or F_2 or H in the outer medium. The intermediate medium is apparently impermeable to H, although μ_2 is finite, and the electric force stops dead at the outer boundary. To understand, we must remember the circumstance that q_2 is finite, although c_2 and k_2 are zero. This makes $q^2_2 = -d^2/dz^2$. So, to be explicit, if the type of e is $e=f(t)\cos mz$, we have $q=m$, and the sole result of this distribution of e, when m is finite, is the electric field (24) outside the wire, without energy. It accompanies the changes in e with the time instantly, and is the reaction against e under circumstances which do not permit magnetic force being produced. But still the tube of zero permittivity is of finite depth. On the other hand it is infinitely long, and is compelled by $c_2=0$ to be everywhere in the same phase as regards H at the same distance from the axis in passing along z. It follows that to have any H, or current in the wire, the total e in the wire must be infinite, or the mean value must be finite. But the mean value of the chosen e is zero. So there is no current in the wire, even though the tube of no elastic yielding is of finite depth.

But it is different when $m=0$, say $e=f(t)$. There is now current in the wire, and H in the wire, tube, and outer medium. The expressions are to be found by the consideration that q_2 is zero, as well as k_2 and c_2. The operator $q_2/(k_2+c_2p)$ is now the same as $\mu_2 p/q_2$, and the use of the convergent formulæ for the Bessel operators will reduce Z_2 to

$$Z_2=\mu_2 pa \log\frac{b}{a}+\frac{a}{b}\frac{\mu_3 p}{q_3}\frac{K_{0b_3}}{K_{1b_3}}. \tag{25}$$

That is, Z_2 has split into two parts. Dividing by $2\pi a$, it can be seen that the first part of the result is the same as $L_2 p$, where L_2 is the inductance (without yielding) of the tube per unit length of wire, and the second part is the resistance operator of the outer medium per unit length. Now, of course, electromagnetic waves are sent through to the outer medium.

It comes to this. Given any distribution of e along the wire. Only its mean value is effective in producing H, and sending electromagnetic waves through. All the rest is ineffective, and is statically balanced. But if the outer medium does not exist, or the unyielding medium exists everywhere outside the wire, then there can be no H, and no current in the wire, even from the finite mean value of e.

(18). Now do away with the unyielding tube altogether. This means $y=0$ in equations (18), (19), (20), and a simplification. Another one is produced by supposing the constants of the wire and of the single outer medium are the same. We then have a uniform conducting medium, in a part of which, the wire, e acts. Boundary reflections are done away with too, viz., at $r=a$, and the conjugate property comes into play, and reduces the solutions to

$$E_1=\tfrac{1}{2}\pi qa I_{0r}K_{1a}e, \qquad H_1=(k+cp)\tfrac{1}{2}\pi a I_{1r}K_{1a}e,$$
$$E_2=-\tfrac{1}{2}\pi qa K_{0r}I_{1a}e, \qquad H_2=-(k+cp)\tfrac{1}{2}\pi a K_{1r}I_{1a}e, \tag{26}$$

$$F_1=-\tfrac{1}{2}\pi a I_{1r}K_{1a}de/dz,$$
$$F_2=-\tfrac{1}{2}\pi a K_{1r}I_{1a}de/dz, \qquad q^2=\mu kp+\frac{p^2}{v^2}-\frac{d^2}{dz^2}. \tag{27}$$

When e is of the type $e_0 \cos mz \cos nt$, the only trouble in development is due to the presence of k. There are two wave trains proceeding from the source at $r=a$, inward and outward, and then secondarily outward from the axis. But the practical significance will largely depend upon the size of k, as well as the frequency and wave length. The waves may at one extreme closely resemble diffusion waves, and at the other elastic inertial waves.

If $p=0$, the existence of H depends on that of k. With $k=0$, there is no steady H, and the stored energy due to steady e is entirely electric. Say $e=e_0 \cos mz$, then $q=m$ in the E and F formulæ makes (26), (27) express the steady result. But if $m=0$, and $e=e_0$ all along, then the radial F disappears, and also E_2. There is nothing left but $E_1=e$ in the region of e. To find the electromagnetic waves leading up to this final state. Use (26) with $k=0$, and $q=p/v$; and suppose e starts when $t=0$. Comparing with (36), on p. 262, we see that the results here for E can be derived from those there for H by differentiation with respect to a and a time integration. Thus, comparing (179), (180), p. 266, with (159), (160), p. 260, and then with (26) above, we see that the present solutions are

$$qE = -ea\frac{d}{da}\left[\frac{1}{\sqrt{4ar}}F\left(\frac{v^2t^2-(a-r)^2}{4ar}\right)\right]$$

or else,
$$qE = -ea\frac{d}{da}\left[\frac{1}{\sqrt{v^2t^2-(a-r)^2}}F\left(\frac{4ar}{v^2t^2-(a-r)^2}\right)\right], \qquad (28)$$

where F is as on p. 254. The other way of solution, using the divergent operators, was done in *El. Pa.*, vol. 2, p. 462. Or we may use the solutions (177), (178), p. 265, and then differentiate, as in (28) above.

(19). If the conductivities k_1 and k_2 are unequal, then we return to (18), (19), of course. The results are finite in the wire and outside for E, F, and H, for the steady $e=e_0 \cos mz$. But if we abolish the external conductivity, and have a real wire surrounded by ether, then $q_2/k_2=\infty$; the magnetic force is zero all over, also E_1 and F_1 in the wire. There is nothing left but

$$E_2 = -\frac{K_{0r}}{K_{0a}}e, \qquad F_2 = -\frac{K_{1r}}{K_{0a}}\frac{1}{m}\frac{de}{dz}, \qquad (29)$$

in which $q=m$. The wire is, of course, electrified, and this external field may be regarded as due to the electrification and to the internal impressed force, though, of course, the impressed force has caused the electrification. If, further, $m=0$, then $F_2=0$, and $E_2=-e$. This is a remarkable extreme result. Although there is finite impressed force associated with finite resistance in any finite length, the final result which is tended to is zero current, though it is never quite reached, of course. Here there is elastic yielding, but the magnetic effect passes away.

(20). If the wire has no resistance, then it conducts plane waves of radial **F** and circular **H**, connected by $F=\mu vH$, in the distortionless manner. But when the wire has finite resistance, there is a continuous reflection in transit going on, causing distortion. How shall we abolish this distortion, using only the means allowed above? Very simply, by making the impressed force travel with the outer wave. Thus, let $q_2=0$, and $k_2=0$. Then $d^2/dz^2=p^2/v_2^2$. To be definite, let $e=e_0 f(t-z/v_2)$. Then

$$E_2=0, \qquad H_2=\frac{a}{r}H_a, \qquad F_2=\mu_2 v_2 H_2, \tag{30}$$

represent the outer state, and

$$E_1=\frac{I_{0r}}{I_{0a}}e, \qquad H_1=\frac{k_1+c_1 p}{q_1}\frac{I_{1r}}{I_{0a}}e, \qquad F_1=-\frac{1}{q_1}\frac{I_{1r}}{I_{0a}}\frac{de}{dz}, \tag{31}$$

represent the state in the wire. In the wire formulæ q_1 is to be used, not the vanishing q_2. Suppose H_a determined by (31), then (30) shows distortionless transmission of plane waves in spite of the resistance of the wire. Here the internal q_1 is given by

$$q_1^2=\mu_1 k_1 p+p^2/v_1^2-p^2/v_2^2, \tag{32}$$

so, by varying μ_1, k_1, c_1 in the wire, we have various sorts of results in the wire. A remarkable case occurs when $\mu_1=\mu_2$ and $c_1=c_2$, making the speed of propagation the same in the wire as outside it. Then $v_1=v_2$, and q_1^2 reduces to $\mu_1 k_1 p$.

Since the constants μ and c are the same in and out, the suffixes may be omitted from them. Then we find the whole system thus. Divide H_1 into the two parts H_{1k} and H_{1c}, depending on k and c respectively. Then

$$E_1=\frac{I_{0r}}{I_{0a}}e, \qquad H_{1k}=\frac{k}{q}\frac{I_{1r}}{I_{0a}}e, \qquad \frac{F_1}{\mu v}=H_{1c}=\frac{cp}{k}H_{1k}, \tag{33}$$

$$E_2=0, \qquad \frac{F_2}{\mu v}=H_2=H_a\frac{a}{r}, \qquad q^2=\mu k p. \tag{34}$$

If we find E_1 and H_{1k}, all the rest follows. The other part of H_1 pairs with F_1 to make an electromagnetic wave travelling along the wire inside, in company with another one outside whose size depends on the boundary value H_a of both H_1 and H_2. But the density of electrification is not to be found in the usual way, by $\sigma=cF_2$, but by

$$\sigma=c(F_2-F_1)=c\mu v(H_a-H_{1c})=H_{1k}/v. \tag{35}$$

It is remarkable that E_1 and H_{1k} are to be calculated in the diffusion manner, although the wire is fully permittive. Thus, if $e=f(t-z/v)$ $=\epsilon^{-pz/v}f(t)$, then $f(t)$ is the value of e at $z=0$. If this is e_0, constant for positive t, then the Normal Expansion theorem gives immediately

$$E_1=\epsilon^{-pz/v}\frac{I_{0r}}{I_{0a}}e_0=e_0\epsilon^{-pz/v}\left(1+\Sigma\frac{I_{0r}\epsilon^{pt}}{\frac{1}{2}qaI_{1a}}\right)$$

$$=e_0\left[1-\Sigma\frac{J_0(sr)}{\frac{1}{2}saJ_1(sa)}\epsilon^{-s^2(t-z/v)/\mu k}\right], \tag{36}$$

where q^2 becomes $-s^2$, and the summation ranges over the positive roots of $J_0(sa)=0$. In a similar manner

$$H_{1k}=\tfrac{1}{2}kre_0-ke_0\Sigma\frac{J_{1r}}{\frac{1}{2}s^2aJ_{1a}}\epsilon^{p(t-z/v)} \tag{37}$$

This completes the solution in convergent terms. With an exception, the series are quite practical for calculation, because the exponential factors decay rapidly as $t-z/v$ increases, leaving only a few, and, finally only one significant term in the series. The exception is when $t-z/v$ is small, that is, in the region near the wave front in the wire. Then the expansions are entirely unpractical, and require the substitution of equivalent divergent series. How to find these will come a little later. By differentiation to t we obtain the results for a travelling pulse of e_0. These do not want separate writing out.

The travelling e must have been travelling already for some time before the above quasi-steady state is settled down to. In time it tends to be practically true right up to the wave front in the wire. Think of a travelling disc or pulse of e_0, and the result on the axis at the wave front. The diffusion formulæ indicate that E_1 is zero there, but that it is finite at a point close behind, even though it is too far away from the true source to have had time to receive any disturbance direct from it. And it would be true that there would be no disturbance there if the circle of curl e were only just started in motion. But when it has been moving for some time, as is sup posed in the formula, then the point in question is receiving disturb ance from the adjacent points around and behind it, though not from the front, which points received their disturbances from others around, and so back to the real source on the surface at previous moments. So the use of diffusion formulæ is not in real conflict with the finite speed of propagation. There is no disturbance at all in advance of the wave front.

(21). In the above example e was given, and the electromagnetic wave, which is spread out, is determined to correspond. But it may be more convenient to make a given outer wave be the datum, and then determine the state of the wire, and e, to suit it. Then we can make the outer wave be a mere pulse, for example Since the value of H_a fixes the nature of the outer wave, the results (33), (34) may be expressed in terms of H_a. Thus,

$$E_1 = \frac{q}{k+cp}\frac{I_{0r}}{I_{1a}}H_a, \quad H_1 = \frac{I_{1r}}{I_{1a}}H_a, \quad F_1 = -\frac{I_{1r}}{I_{1a}}\frac{dH_a/dz}{k+cp}, \tag{38}$$

give the internal state, and e is the boundary value of E_1. Now let H_a be zero when $t-z/v$ is negative, and constant when it is positive. The Normal Expansion theorem makes

$$H_1 = H_a\frac{r}{a} + H_a\Sigma\frac{J_{1r}}{\frac{1}{2}saJ_{0a}}\epsilon^{p(t-z/v)} \tag{39}$$

subject to $J_1(sa)=0$, not counting the zero or negative roots, and $p=-s^2/\mu k$ in the summation. And by the same method,

$$E_1 = H_a\left[\frac{2}{ak} - \mu v\frac{J_0(g)}{J_1(g)}\epsilon^{-k(t-z/v)/c} + \frac{2}{ak}\Sigma\frac{J_{0r}\,\epsilon^{p(t-z/v)}}{[1-(sa/g)^2]J_{0a}}\right], \tag{40}$$

where $g=\mu vka$, and sa is as before. The variable outside term arises from the root of $k+cp=0$. The quantity g varies from 0 to ∞, as we pass from a non-conducting to a perfectly conducting wire.

What is highly exceptional here is that the variable outside term may be infinite for any number of special values of g. But there is a compensation, for the same values of g each make one term of the following summation in (40) be infinite. This will be found to depend upon the expansion

$$\frac{J_0(g)}{J_1(g)}=\frac{2}{g}+\Sigma\frac{2g}{g^2-g_n^2}, \tag{41}$$

where g_n is the nth root of $J_1(g)=0$, or the sa in the previous.

The constancy of H_a behind the wave front is the easy explanation of why E_1 and e tend to constancy as we recede from the wave front. But if H_a is a pulse, say $=e_0p1$, then e tails off to nothing behind the pulse. It is given by

$$\frac{e}{e_0}=\frac{g}{ca}\frac{J_0(g)}{J_1(g)}+\frac{2}{a}\Sigma\frac{p\,\epsilon^{p(t-z/v)}}{k+cp}, \tag{42}$$

where $p=-s^2/\mu k$. This e is required to prevent the pulse wave outside the wire tailing off itself by the resistance of the wire.

(22). Closely connected with the above is the problem of the travelling of a distribution of longitudinal h, impressed magnetic force, along the wire at speed v. If it is a disc of h, then the problem becomes the interesting one of the effects produced by sliding a single circular current along the wire closely embracing it. The work is rather easier than for the travelling e, because in making the changes from E to H, etc., we do not have to alter electric conductivity to magnetic, but keep it the same.

Let H be now the longitudinal and G the radial component of magnetic force, whilst E specifies the circular electric force. The circuital equations are

$$\frac{dH}{dr}-\frac{dG}{dz}=-(k+cp)E, \qquad \frac{1}{r}\frac{d}{dr}rE=-\mu pH, \qquad \frac{dE}{dz}=\mu pG, \tag{43}$$

and the solutions due to h (subject to $p^2/v^2=d^2/dz^2$, as before) are

$$H_1=\frac{I_{0r}}{I_{0a}}h, \qquad E_1=-\mu vG_1=-\frac{\mu p}{q}\frac{I_{1r}}{I_{0a}}h, \qquad \text{(in)} \tag{44}$$

$$H_2=0, \qquad E_2=-\mu vG_2=\frac{a}{r}E_a. \qquad \text{(out)} \tag{45}$$

Here H_1 is distributed in the wire in the same way that E_1 was with travelling e. In other respects it is simpler. As before, $q^2=\mu kp$ in the wire.

If $h=\epsilon^{-pz/v}ph_0$, then it is a disc of h, or it may be a circular current, whose time integral in passing any point is h_0. Now consider. The induction must all pass through the current, and there is none in advance of the current. Also $H_2=0$. So the induction enters the wire perpendicularly behind the current at the wave front in a spread-out fashion, accumulates in the wire, and turns round towards the current. But on nearing the current it must concentrate itself towards the boundary of the wire so as to slip out round the square corner into the plane of the wave front outside the wire. The result is a pulse of outward G_2 at the wave front, followed by a distributed tail of inward G_2, connected together through the wire in a way to be found by expanding (44).

With h constant, the results are

$$E_1=-\frac{2h}{ka}\Sigma\frac{J_{1r}}{J_{1a}}\epsilon^{p(t-z/v)}, \qquad H_1=h-h\Sigma\frac{J_{0r}}{\frac{1}{2}saJ_{1a}}\epsilon^{p(t-z/v)}. \tag{46}$$

subject to $J_0(sa)=0$, $p=-s^2/\mu k$. For the pulse, turn h to h_0, multiply by p, and destroy the outside term.

(23). In the inverse problem, when we want to specify the outside wave to be of a given type, we have

$$E_1=\frac{I_{1r}}{I_{1a}}E_a, \qquad H_1=-\frac{k}{q}\frac{I_{0r}}{I_{1a}}E_a, \qquad h=H_a \tag{47}$$

where E_a defines the outside wave. If it is a pulse, and $E_a=\epsilon^{-pz/v}pf_0$, where f_0 is constant, then the normal expansions are

$$E_1=f_0\Sigma\frac{J_{1r}p\epsilon^{p(t-z/v)}}{\frac{1}{2}saJ_{0a}}, \qquad H_1=-\frac{2f_0}{\mu a}\left[1+\Sigma\frac{J_{0r}}{J_{0a}}\epsilon^{p(t-z/v)}\right], \tag{48}$$

subject to $J_1(sa)=0$ in the summations. No zero root.

But if E_a is constant behind the wave front, the zero root comes into force if the normal expansion theorem is used to find H_1. Or we may perform a time integration on (48) from 0 to t. Either way, the result is

$$H_1=-\frac{2E_a}{\mu a}\left(t-z/v+\frac{\mu ka^2}{8}\right)-\frac{2E_a}{\mu a}\Sigma\frac{J_{1r}\epsilon^{p(t-z/v)}}{\frac{1}{2}saJ_{0a}}. \tag{49}$$

But in finding E_1 there is no zero root, and the result is

$$E_1=\frac{r}{a}E_a+E_a\Sigma\frac{J_{1r}\epsilon^{p(t-z/v)}}{\frac{1}{2}saJ_{0a}}. \tag{50}$$

The reason of the unlimited increase of H_1 behind the wave front is to be seen in the E_1 formula. E_1 tends to the value $(r/a)E_a$ in the wire. Remember that the electric force is circular, and is continuous at the boundary. The existence of this circular electric force implies continuous increase in the internal H_1 at a uniform rate.

We may also see the meaning of the value $H_1=-2f_0/\mu a$ tended to in (48). It makes the induction along the wire amount to $-2\pi af_0$, that is, the time integral of the circuitation of $-E_a$, or the time integral of $2\pi a\mu vH_a$, or the length integral (along the wire) of $\mu H_a\times 2\pi a$. But this is the amount of induction leaving the wire radially in the pulse. There is no induction outside the wire anywhere else. So the induction in the wire, which is uniform in distribution a long way behind the pulse, gradually condenses towards the surface, and is wholly condensed there when it reaches the pulse, where it turns round the square corner into the pulse wave outside.

(24). As before remarked, these normal expansions are, on account of the rapid convergency produced by the time factors, thoroughly practical for calculation save near the wave front, where they are, on the contrary, quite the reverse. But in every such case there is a Providence in the form of an equivalent divergent formula, which is exactly suitable just when the convergent formula is of impossible utility practically. And the Providence is so good as to arrange matters secretly so that there is an overlapping region in which either formula may be employed, so that we may test that the one curve joins on properly to the other. It is wonderful that things should work out in this way. Logic has nothing to do with it, either with the fact, its discovery, or its use. At the same time it must be said that a sufficiently profound study of the subject would ultimately lead to the logic of its laws, as a final result. What I do strongly object to is the idea that the logic should come first, or else you prove nothing. Yet perhaps the majority of academical mathematical works are written under this idea. In reality the logic is the very last thing, and that is not final.

Physically, it is quite easy to see that in general all expansions in normal series of functions must have a second form of expression. Consider, for example, the diffusion of heat in a conductor of any shape of boundary due to surface sources. According to the shape of the boundary, there is some particular sort of normal functions concerned, and the solutions can be expressed in terms of them. But we may also express our results in terms of the waves of diffusion emanating from the individual sources themselves, which are of an entirely different type from the normal distributions. So there you are, with an equivalence between one type of formula and another. That one of these should be sometimes divergent is an observational fact; that it should be numerically equivalent to the other when calculated in a certain way is an experimental fact; so is the different range of the practicability of the two solutions in general, and the overlapping region. And no doubt the logic of it all will have to be found out experimentally. And then, finally, I suppose "rigorous" mathematicians will put the logic at the beginning, and pretend they knew all about it before they began.

(25). Let us try for a few divergent solutions of the preceding. First, to find the h needed to keep a pulse wave outside the wire from tailing behind. In (47), substitute the divergent operators for the convergent. Then

$$h = -f_0 \epsilon^{-pz/v} \frac{q}{\mu} \frac{H_{0a}}{H_{1a}}. \tag{51}$$

Expand by division. Then, writing the coefficients only,

$$\frac{H_{0a}}{H_{1a}} = 1 + \frac{1}{2} + \frac{3}{8} + \frac{3}{8} + \frac{63}{128} + \frac{54}{64} + \frac{949\frac{1}{2}}{512} + \ldots, \tag{52}$$

where the nth number is the factor of $(1/qa)^n$. Use this in (51), then multegration makes, if P is the differentiator with respect to $(t-z/v)/\mu ka^2 = T$,

$$h = -\frac{f_0}{\mu a}\left[P^{\frac{1}{2}} + \frac{1}{2} + \frac{3}{8}P^{-\frac{1}{2}} + \frac{3}{8}P^{-1} + \frac{63}{128}P^{-1\frac{1}{2}} + \frac{54}{64}P^{-2} + \frac{949\frac{1}{2}}{512}P^{-2\frac{1}{2}} + \right]$$

$$= -\frac{f_0}{\mu a}\left[\frac{1}{(\pi T)^{\frac{1}{2}}} + \frac{1}{2} + \frac{3}{8}\frac{T^{\frac{1}{2}}}{\lfloor\frac{1}{2}} + \frac{3}{8}T + \frac{63}{128}\frac{T^{1\frac{1}{2}}}{\lfloor 1\frac{1}{2}} + \frac{54}{64}\frac{T^2}{2} + \frac{949\frac{1}{2}}{512}\frac{T^{2\frac{1}{2}}}{\lfloor 2\frac{1}{2}} + \right]. \tag{53}$$

So, when T is small the practical result is given by the first term only, and is very large. After that, as the first term becomes smaller, additional terms must be added, taking care, however, to stop at the point of initial convergence, for it is a divergent series. But as with increase of T the point of convergence moves in to the beginning, we should have recourse to the other or convergent formula. This is, by (48),

$$h = -\frac{2f_0}{\mu a}\left[1 + \epsilon^{-g_1^2 T} + \epsilon^{-g_2^2 T} + \ldots\right], \tag{54}$$

when g_1, g_2, etc., are the roots of $J_1(g)=0$, of which the first three are roughly, by inspection of a graph, 3·85, 7·0, 10·15. It is when T is less than 1 that the divergent series is usable, and the smaller T the more useful is this series, because the convergent one becomes impracticable.

When $T=7/9$, the fifth term of the divergent series equals the sixth. So stop at end of fifth term. The total is 1·946. The real result is a little over 2. This is, however, nearly at the end of the range of use of the divergent series, and the convergent has not become troublesome.

(26). For a second example, find the divergent formula for the wave due to a disc of h. We have, by (44),

$$E_a = -\frac{\mu p}{q}\frac{I_{1a}}{I_{0a}}h = -\frac{\mu p}{q}\frac{H_{1a}}{H_{0a}}h, \tag{55}$$

by substituting the divergent operators. Expand by division. Then

$$\frac{H_{1a}}{H_{0a}} = 1 - \frac{1}{2} - \frac{1}{8} - \frac{1}{8} - \frac{25}{128} - \frac{13}{32} - \frac{69}{64} - \ldots, \tag{56}$$

where the nth number is the factor of $(qa)^{-n}$. So, by multegration, if h=constant,

$$\mathrm{E}_a = -\frac{h}{ka}\left[\mathrm{P}^{\frac{1}{2}} - \frac{1}{2} - \frac{1}{8}\mathrm{P}^{-\frac{1}{2}} - \frac{1}{8}\mathrm{P}^{-1} - \frac{25}{128}\mathrm{P}^{-1\frac{1}{2}} - \frac{13}{32}\mathrm{P}^{-2} - \frac{69}{64}\mathrm{P}^{-2\frac{1}{2}} - \right] \tag{57}$$

$$= -\frac{h}{ka}\left[\frac{1}{(\pi \mathrm{T})^{\frac{1}{2}}} - \frac{1}{2} - \frac{1}{8}\frac{\mathrm{T}^{\frac{1}{2}}}{\underline{|\tfrac{1}{2}}} - \frac{1}{8}\mathrm{T} - \frac{25}{128}\frac{\mathrm{T}^{1\frac{1}{2}}}{\underline{|1\tfrac{1}{2}}} - \frac{13}{32}\frac{\mathrm{T}^2}{2} - \frac{69}{64}\frac{\mathrm{T}^{2\frac{1}{2}}}{\underline{|2\tfrac{1}{2}}} - \right]. \tag{58}$$

This is suitable for use near the wave front, and further, until the point of convergence comes too near the beginning. The equivalent convergent formula is

$$\mathrm{E}_a = -\frac{2h}{ka}\left[\epsilon^{-g_1^2\mathrm{T}} + \epsilon^{-g_2^2\mathrm{T}} + \dots\right], \tag{59}$$

where the g's are the roots of $\mathrm{J}_0(g)=0$, that is, 2·40, 5·52, 8·65, etc. Differentiate (58) and (59) to t and turn h to h_0 to get the results for a pulse h_0.

As a numerical test of the error when T is large enough to reduce the convergent series to its first term nearly, say T = ·5022, then the convergent first term comes to $2\epsilon^{-2\cdot892} = 111$. The corresponding divergent series makes

$$\cdot781 - \cdot5 - \cdot071 - \cdot0627 - \cdot0512 = \cdot096.$$

The error is less than one-third of the l.c.t. This will pass, but should be less still. It is to be noted that if we carry on the divergent series further, and use it for smaller values of T, it becomes practically a convergent series.

Now as regards the pulse wave. The formulæ are

$$\mathrm{E}_a = \frac{2h_0}{ka}\left[g_1^2\epsilon^{-g_1^2\mathrm{T}} + g_2^2\epsilon^{-g_2^2\mathrm{T}} + g_3^2\epsilon^{-g_3^2\mathrm{T}} + \dots\right], \tag{60}$$

$$\mathrm{E}_a = \frac{h_0}{ka}\frac{1}{2\mathrm{T}}\left[\frac{1}{(\pi\mathrm{T})^{\frac{1}{2}}} + \frac{1}{8}\frac{\mathrm{T}^{\frac{1}{2}}}{\underline{|\tfrac{1}{2}}} + \frac{\mathrm{T}}{4} + \frac{75}{128}\frac{\mathrm{T}^{1\frac{1}{2}}}{\underline{|1\tfrac{1}{2}}} + \frac{13}{8}\frac{\mathrm{T}^2}{2} + \dots\right]. \tag{61}$$

If we use the same value of T as before to test the fit, it goes all wrong, because the point of convergence is brought right up to the beginning. A smaller value of T must be used, say T=0·2941, which makes the fourth and fifth terms of the divergent series equal. Then the true value, by the first term of the convergent series, is 2·12, leaving out the factor h_0/ka, whilst the divergent series makes

$$1\cdot04 \times (1\cdot699 + \cdot125 + \cdot122 + \cdot114) = 2\cdot15.$$

The error is now one-fourth of the l.c.t.

The above divergent solutions were obtained easily. Others, involving variable distance from the axis, are too complicated for present treatment.

(27). Returning to (30), (31), relating to an impressed force travelling along a wire at speed v, if we make $k=0$, we do away with the back effects due to a pulse of e, or with the back e required to maintain an external electromagnetic pulse. The reduced equations are simply

$$E_1=e, \qquad E_2=0,$$
$$F_1=-\frac{r}{a}\frac{de}{dz}, \quad H_1=\frac{cpr}{a}e; \qquad F_2=-\frac{a}{r}\frac{de}{dz}, \quad H_2=\frac{cpa}{r}e; \qquad (62)$$

subject to $e=f(t-z/v)$, or any other arrangement making $p^2/v^2=d^2/dz^2$. The meaning can be seen at once by letting e be in a mere disc, and constant therein. There are just two external plane pulses, one at the front, the other at the back of the disc, together with supplementary internal pulses also on the back and front. The displacement converges to the back of the disc radially from outside, then goes straight through the disc uniformly to the front, from which it diverges radially in the front plane. The whole is in motion along z at speed v. The magnetic force is as usual in plane waves.

For the travelling e we may substitute equivalent travelling electrification, namely, the convergence of ce. We do away with E_1 also in this way. Else it is the same. In the example of a disc of e, this is replaced by positive electrification on its front, and negative on its back.

(28). If, still keeping $k=0$, we alter the speed of e from v to u, we come to the case of impressed e travelling through a uniform non-conducting ether at a speed which may be either greater or less than the natural speed of propagation. The equations are now subject to $d^2/dz^2=p^2/u^2$, so that e may be $f(t-z/u)$, and $q^2=p^2/v^2-p^2/u^2$, or $q=\lambda p/v$ where $\lambda=\sqrt{1-v^2/u^2}$, which is real positive if $u>v$, and unreal if $u<v$. In the case $u<v$, the wave accompanying e extends over all space. If $u>v$, it will only do so if the region of e does so. If e has a front, then the disturbance is non-existent in advance of it, and also outside the cone of semi-vertical angle given by $\theta=\sin^{-1}(v/u)$, with axis coincident with the axis of z, and touching the circular boundary of the front of e. The equations are

$$E_1=\tfrac{1}{2}\pi qaI_{0r}K_{1a}e, \qquad H_1=\tfrac{1}{2}\pi acpI_{1r}K_{1a}e=cuF_1, \qquad (63)$$
$$E_2=-\tfrac{1}{2}\pi quK_{0r}I_{1a}e, \qquad H_2=\tfrac{1}{2}\pi acpK_{1r}I_{1a}e=cuF_2. \qquad (64)$$

Particular cases were considered before, p. 81, relating, however, to a travelling line of e or of electrification. Here the extension is made to a cylinder.

Say that $e=e_0\epsilon^{-pz/u}f(t)$, then the simply periodic case is easily treatable by the transformation $p=ni$, remembering that q is a simple multiple of p, and is real when $u>v$, and unreal when $u<v$. For example, if $f(t)=\cos nt$,

$$E_1=\tfrac{1}{2}\pi a\frac{\lambda n}{v}J_0\left(\frac{\lambda nr}{v}\right)\left[G_1\left(\frac{\lambda na}{v}\right)-iJ_1\left(\frac{\lambda na}{v}\right)\right]e_0\cos n(t-z/u), \quad (65)$$

where i means $d/d(nt)$, is the result when $u>v$, making λ positive real. In the other case, put $\lambda = si$, then s is real positive and $q = sn/v$, with the result

$$E_1 = +\tfrac{1}{2}\pi a \frac{sn}{v} I_0\left(\frac{snr}{v}\right) K_1\left(\frac{sna}{v}\right) e_0 \cos n(t - z/u). \qquad (66)$$

Similarly for the others. Observe that the external travelling field of **E** and **H** may disappear only when $u > v$, at particular frequencies, but not when $u < v$.

(29). If we take $f(t) = pt$, it means a disc or pulse of e. For this we may substitute a disc positively electrified on one side, negatively on the other, bearing in mind the equivalence $\rho u = cpe$ of electrification density ρ in motion at speed u, and impressed force e in producing the same electromagnetic effects. Or, if we take $f(t) = t^0$, or positive 1 when t is positive, and zero when t is negative, then e is uniform from $z = -\infty$ up to $z = ut$, and zero beyond. So here we may substitute a single disc of electrification, travelling at speed u perpendicular to its plane. Then

$$E_1 = -\tfrac{1}{2}\pi a e_0 \epsilon^{-pz/u} \frac{d}{da} I_{0r} K_{0a}, \quad \lambda\mu v H_1 = -\tfrac{1}{2}\pi e_0 \frac{a}{q} \epsilon^{-pz/u} \frac{d^2}{da\,dr} I_{0r} K_{0a}, \qquad (67)$$

$$E_2 = \ldots\ldots\ldots\ldots\ldots\ldots\ldots I_{0a} K_{0r} \quad \lambda\mu v H_2 = \ldots\ldots\ldots\ldots\ldots\ldots\ldots\ldots I_{0a} K_{0r}. \qquad (68)$$

This being for e constant as described, we see that the internal and external E are determined when e is impulsive by a single differentiation upon the known function $I_{0r} K_{0a} q$, the double branched first elliptic curve, with the further change of t to $(t - z/u)/\lambda$, or, which is the same, of vt to $(ut - z) \tan\theta$. So this result is known in all respects, and is the electric force for a double sheet of electrification. But a time integration is required as well to functionize H, together with a double differentiation. As the differentiations complicate the formulæ, it will be as well to modify the problem in such a way that we can use the first elliptic function unchanged, and then we can obtain easily an expression for the force required to drive a cylindrical surface distribution of electrification through the ether.

(30). Let there be surface electrification, density σ, on the cylinder $r = a$, moving at speed u along its axis. The equations are

$$E_1 = -\tfrac{1}{2}\pi q a \lambda \mu v I_{0r} K_{0a} \sigma u, \qquad H_1 = -\tfrac{1}{2}\pi q a I_{1r} K_{0a} \sigma u, \qquad (69)$$

$$E_2 = -\tfrac{1}{2}\pi q a \lambda \mu v I_{0a} K_{0r} \sigma u, \qquad H_2 = +\tfrac{1}{2}\pi q a I_{1a} K_{0r} \sigma u, \qquad (70)$$

together with $H_1 = cuF_1$, $H_2 = cuF_2$, to find the radial component of the electric force. To prove, test that $E_1 = E_2$, and $H_2 - H_1 = \sigma u$ at $r = a$, and that the circuital equations are satisfied in the manner of (15), subject to $q = \lambda p/v$. The surface density σ may be any function of $z - ut$. We now see at once that the E results are of the first elliptic type without any differentiations or integrations, when we make σ be constant. That is, in general, $\sigma = \epsilon^{-pz/u} f(t)$, and we now make $f(t)$ be zero before and constant after $t=0$. Thus,

$$E=\frac{-a\lambda\sigma u\mu v}{\sqrt{4ar}}\left[1+\frac{1^2}{2^2}\frac{(ut-z)^2\tan^2\theta-(a-r)^2}{4ar}+\dots\right], \qquad (71)$$

when the fraction after $1^2/2^2$ is <1. Or else, if it is >1, then

$$E=\frac{-a\lambda\sigma u\mu v}{\sqrt{(ut-z)^2\tan^2\theta-(a-r)^2}}\left[1+\frac{1^2}{2^2}\frac{4ar}{(ut-z)^2\tan^2\theta-(a-r)^2}+\dots\right] \qquad (72)$$

The same formula is valid for E_1 and E_2.

To interpret, a diagram will assist. We have $\sin\theta=v/u$, and $\cos\theta=\lambda$. This θ is the semi-vertical angle of the cones in the figure. The electrified cylinder itself is shown, in section, by the horizontal lines AB, AB. The motion is from left to right. The wave front is shown by EACAE. It consists of a big cone from which a little cone has been cut straight off its tip, and then another

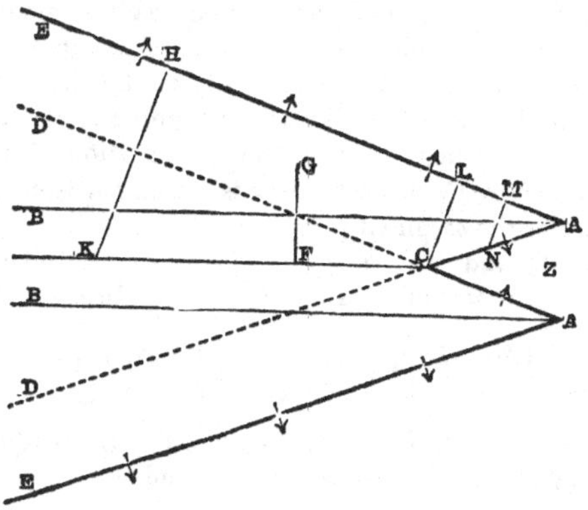

little cone ACA cut out of the remainder. There is no disturbance inside the little cone ACA. The large internal cone DCD bounded by the dotted lines is the region in which the formula (72) is valid. Outside it, beyond the dotted lines, and up to the wave fronts, the other formula (71) is valid.

Now as regards the magnitude of E. It is finite on the wave fronts and everywhere else, save on the conical surface dotted, which is the boundary between the outer and inner regions occupied by the conico-elliptic waves. On this surface E is everywhere infinite. This infiniteness does not interfere with the propagation according to the circuital laws, and has no energetic significance. If we substitute a thin layer of volume electrification for the surface electrification, the infinity will be replaced by a finite rounded peak. Now see the diagram on p 260. If we pass from M to N in the present figure, E varies in size like the ordinate of the curve $\frac{1}{2}, \frac{1}{2}$ in the former figure. It rises from a finite value at M to a greater finite value at N. This is true all the way between A and the line LC. The rise is then from finiteness at L to infinity at C. After that, going further to the left, the second or inner branch of the elliptic curve on p. 260 comes into existence. On HK for example, E rises from finiteness at H to infinity at the dotted line, and then falls to finiteness on the other side, the least value being on the axis.

(31). The quantity $-\mathrm{E}_a\sigma$, if E_a is the value of E at $r=a$, is the impressed force on σ in the direction of its motion needed to maintain the motion. By integrating this from A to B we obtain the force on a strip of length AB, and of unit breadth. Multiply by $2\pi a$ to get the total force on the cylinder AB. It is the same when the continuation of the cylinder to the left is removed. That will make no difference to the force on AB, though it will alter the distribution of **E** and **H** behind. So we can readily calculate the impressed force upon any piece, long or short An important quantity in this connection is $b=(ut-z)\tan\theta$. It is the distance from the axis to midway between the dotted line and the wave front, say from F to G; or from the axis to the continuation of this median line in the region ACA, because $ut-z$ is measured from the point Z, in the centre of the terminal plane.

Put $2\pi a\sigma=\mathrm{Q}$, and let l be the length of the cylinder. Then, if $b<2a$, we use the formula (71), and the total force on the cylinder comes to

$$\mathrm{X}=\frac{\mathrm{Q}^2\mu vu}{2\pi}\frac{l\lambda}{2a}\left[1+\frac{1}{3}\frac{1^2}{2^2}\left(\frac{b}{2a}\right)^2+\frac{1}{5}\frac{1^23^2}{2^24^2}\left(\frac{b}{2a}\right)^4+\dots\right]. \qquad (73)$$

In this, $b=l\tan\theta$. As b increases from 0 to $2a$, the quantity $[\dots]$ does not vary much. Beginning at 1, it ends at about 1·15 or 1·2, by rough calculation.

But after this, if $b>2a$, we must add on the force on the excess length calculated by the other formula (72). The total force becomes

$$\mathrm{Y}=\frac{\mathrm{Q}^2\mu vu}{2\pi}\lambda\cot\theta\left[1{\cdot}2+\log\frac{b}{2a}-\frac{1}{2}\frac{1^2}{2^2}\left(\frac{2a}{b}\right)^2-\frac{1}{4}\frac{1^23^2}{2^24^2}\left(\frac{2a}{b}\right)^2-\dots\right.$$
$$\left.+\frac{1}{2}\frac{1}{2^2}+\frac{1}{4}\frac{1^23^2}{2^24^2}+\frac{1}{6}\frac{1^23^25^2}{2^24^26^2}+\dots\right]. \qquad (74)$$

Here the series in the second line comes to about ·25. So, if l is very great, the impressed force is

$$\mathrm{Y}=\frac{\mathrm{Q}^2\mu vu}{2\pi}\left(\frac{u}{v}-\frac{v}{u}\right)\left[1{\cdot}45+\log\frac{b}{2a}\right], \qquad (75)$$

provided $2a/b$ is small. The outer factor varies as the square of the speed when u/v is large. But this must not be pushed too far, because by sufficient increase of u/v, $2a/b$ will cease to be a small quantity.

It is always possible. with any length of cylinder, to make u/v be so little over 1 that $b/2a$ is a large number. Then Y is the proper formula, and it makes $Y=0$ when $u=v$ That is to say, when the speed of motion is made to exceed that of light, the resisting force comes on gradually. Under any circumstances the resistance is finite, provided the electrification remains a surface distribution. But if it is condensed to a linear distribution, by shortening l to 0, keeping Ql finite, then the X formula shows that the force becomes infinite.

What is striking, and interesting, is the wide range within which the resisting force can be proportional to the speed $x \cos\theta$. Here we use the X formula. Make $b/2a$ small. Then

$$X=\frac{Q^2\mu vul\lambda}{4\pi a}, \qquad Xu=\frac{\mu vl\lambda}{4\pi a}(Qu)^2=\tfrac{1}{2}\mu v\lambda \, . \, 2\pi al\left(\frac{Qu}{2\pi a}\right)^2. \tag{76}$$

Here the cylinder is a ring of depth l and length $2\pi a$, and surface $2\pi al$, whilst $Qu/2\pi a$ is the convection current per unit surface Comparing with Joule's law, and with Ohm's law, we see that the "resistance" per unit surface is $\frac{1}{2}\mu v\lambda$, which is 15λ ohms. We may also write

$$Xu=\tfrac{1}{2}\mu v\lambda \, . \, \frac{(Qul)^2}{2\pi al}=RC^2. \tag{77}$$

Here C is the total convection current, and $R=\frac{1}{2}\mu v\lambda/2\pi al$ is the total resistance. It varies inversely as the area. Though not exactly true, the formula is nearly true within wide limits for the speed and the size of the cylinder.

The meaning of $\frac{1}{2}\mu v\lambda$ may be seen in the calculation on p. 17, relating to a plane strip of electrification moving in its own plane. The resistance per unit area comes out the same, as an exact result in that case. There is also another calculation, on p. 64, of the force needed to drive an electrified line in its own line. The result there, equation (17), comes from the present formula (72) by turning the present σ to $\sigma/2\pi a$ (because the present σ is surface density), and then putting $a=0$, $r=0$.

(32). It is much simpler when a plane sheet of electrification moves perpendicularly to its plane. Let it have any shape of boundary, and let σ be the uniform surface density, and Q the total charge. Then the driving force per unit area is simply=the electric tension behind the sheet, or $\frac{1}{2}cE^2$, or $\frac{1}{2}E\sigma$, or $\frac{1}{2}\sigma^2/c$. This makes

$$Xu=\tfrac{1}{2}(\sigma u)^2\mu v(v/u)=RC^2 \tag{78}$$

Therefore, $R=\frac{1}{2}\mu v \sin\theta$. Multiply by the area to obtain the total driving force. Or divide by the area to obtain the equivalent resistance to suit the total current. But it will be seen that the results are simpler in terms of the driving force itself.

The proof of this general result will be found by inspecting fig. 19 on p. 22, relating to the motion of a plane strip, and thinking about it. In the overlap there is no magnetic force, and the electric force is perpendicular to the strip. This makes $X=\frac{1}{2}\sigma^2/c$ per unit area, obviously. Now transfer attention to a plane sheet of any shape. From any point of the boundary draw a cone backward, of the semi-vertical angle given by $\sin\theta=v/u$. Move this cone all round the boundary, keeping its axis parallel to itself. It will in its motion touch the boundary of a region behind the plane sheet which may be of various shapes, but always such that one side thereof is the plane sheet itself. Any point P within this region is so situated that if it be the apex of a similar secondary cone pointing the other way, the base of this cone will be fully upon the plane sheet. The disturbance at P is, therefore, due to the σ on the base of this secondary cone only. It is, therefore, the same for all points within the region of overlap already described. It has been calculated to be a state of uniform electric force in a particular case, and by the present argument it is the same in all cases, or is independent of the shape of the boundary. Even when u/v is very slightly over unity the overlap exists, close to the plane sheet, and the property is true.

But it is also true when the speed u varies, provided it is not allowed to become smaller than v. For I have shown (p. 120) that the sudden starting of a point charge instantly establishes the steady conical state close to the charge. This being true for every element of charge on the plane sheet, it follows that immediately behind the sheet the electric force is always perpendicular to it. The state of things further away behind will vary in all sorts of ways, when u varies, but the sheet itself is the only place where the driving force is required, and there is no alteration there. When u is allowed to fall exactly to v, the driving force ceases. The slightest increase in u will bring it on again.

In the above the sheet always moves perpendicularly to itself, and the force is constant, however u varies, provided it is not $<v$. It is clear also that the straight path of the centre of the sheet might be curved considerably without producing a very great difference. But if the sheet also rotates, then obviously there will be important changes in the phenomena.

If the sheet has holes in it, the overlap is reduced in size. This makes no difference in the driving force on the rest. Now, a hole means that $\sigma=0$ over a part of the sheet. It follows that σ may vary anyhow over the sheet without altering the driving force on any particular σ, and the displacement will always be $D=\sigma$, at the place of σ, and perpendicular to the sheet. But it will be only at the sheet itself, in general, that this is true. The mixing up will begin close behind it, when σ varies.

(33). In the above the expressions for the driving force on a plane sheet of electrification moving either in its own plane, or else perpendicular to the same, have turned up. But we cannot derive from them the driving force when the plane sheet slants. This requires separate doing, and is worth it, on account of a striking peculiarity. Let ϕ be the angle of slant, meaning the angle the plane electrified strip makes with the direction of motion. The theory changes somewhat when the slant passes through the value θ. When there is no slant, the diverging plane slab waves are in every way similar, and one is on each side of the strip. Now increase ϕ to the value θ. One of the slab waves gets thinner, and becomes ultimately of no depth at all. It is then a condensed pulse wave. Further increase of slant makes this wave pass over to the other side of the strip, and increase in depth. Both slab waves being now upon one side of the strip, one thin and the other thick, further increase of slant brings them to equality again, when the strip moves perpendicularly to its plane. The figure here will illus-

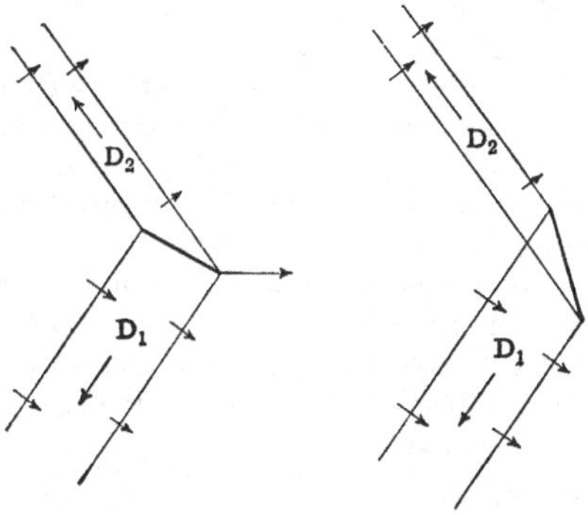

trate the two cases. At the critical angle $\phi=\theta$, the driving force becomes infinite. Besides the driving force in the direction of motion, there is also usually transverse force needed to prevent deflection of the strip up or down. It is not necessary to consider this specially.

To calculate the straightforward driving force it is only necessary to find the rate of increase of energy in the two slab waves. Let D_1 and D_2 be the displacements in the waves, U_1 and U_2 the densities of electric energy, b_1 and b_2 the depths of the waves, X the driving force on the strip per unit length perpendicular to the paper, and u the speed. Then

$$Xu=(2U_1b_1+2U_2b_2)u\cos\theta=(D_1^2b_1/c+D_2^2b_2/c)u\cos\theta \qquad (79)$$

expresses the activity of the driving force. Let $\phi<\theta$, then

$$\begin{aligned}b_1&=a\sin(\theta+\phi),\\ b_2&=a\sin(\theta-\phi),\end{aligned}\qquad D_1=\frac{\frac{1}{2}\sigma}{\sin(\theta+\phi)},\qquad D_2=\frac{\frac{1}{2}\sigma}{\sin(\theta-\phi)}, \qquad (80)$$

where a is the breadth of the strip, and σ the surface density of electrification, counting both sides. Using these in (79) we come to

$$X=\frac{\sigma^2 a}{4c}\cos\theta\left(\frac{1}{\sin(\theta+\phi)}+\frac{1}{\sin(\theta-\phi)}\right), \tag{81}$$

when $\phi<\theta$. But when $\phi>\theta$, $b_2=a\sin(\phi-\theta)$, and the result is to interchange θ and ϕ in the second fraction in (81), so that it continues positive. It is easily seen that X increases infinitely at the critical angle. It is also verifiable that the values assumed by X in the extreme cases $\phi=0$, and $\phi=\frac{1}{2}\pi$, agree with those already found.

This infinite driving force at a certain speed may be reduced to finiteness by substituting a thin layer of volume electrification for the surface distribution. The same remark applies to the motion of a surface charge of any shape, if at a particular speed, or more than one, or at any speed $>v$, there should be a similar local or general intensification of force. Substitute finite volume density where it may be required.

(34). Now consider briefly the waste of energy from a wire subjected to the impressed force $e=e_0\sin(nt-mz)$, or similar, making $p^2=-n^2$, $d^2/dz^2=-m^2$. This is sufficiently done through the resistance operators, equation (20). The waste is partly internal, in the wire itself, and partly external, by outward flux of energy without return, which occurs sometimes, though the energy may be kept in connection with the wire without waste in this way. It will be convenient to refer the external and internal resistance operators to unit length of wire instead of unit surface, as in (20). Thus changed,

$$Z_2=\frac{q}{2\pi acp}\frac{K_0(qa)}{K_1(qa)}, \qquad q^2=\frac{p^2}{v^2}-\frac{d^2}{dz^2}=m^2-\frac{n^2}{v^2}=-s^2, \tag{82}$$

is the external resistance operator. Here we see immediately that if $m>n/v$, q is real, so there is no external waste at all. The wave speed of e along the wire is $n/m=u$. So the condition of no waste is that $u<v$. This may be compared with the steady straight motion of a charge. If $u<v$, there is no waste, but if $u>v$, there is. The form of Z_2 shows that it represents negative effective inductance, or equivalent positive permittance. That is,

if $Z_2=R_2+L_2p$, then $R_2=0$, and

$$L_2n=\frac{-q}{2\pi acn}\frac{K_0(qa)}{K_1(qa)}, \qquad q=\sqrt{m^2-n^2/v^2}. \tag{83}$$

When qa is very large, this reduces to $L_2n=-q/2\pi acn$. When qa is small, it makes

$$L_2n=\frac{q^2}{2\pi cn}\log\tfrac{1}{2}qa. \tag{84}$$

The case $q=0$ means $R_2=0$, $L_2=0$.

(35). Passing next to the more interesting case $u>v$, we should use s instead of q. This makes

$$Z_2=\frac{s}{2\pi acn}\frac{G_0(sa)-iJ_0(sa)}{i^{-1}[G_1(sa)-iJ_1(sa)]}=\frac{s}{2\pi acn}\frac{2/\pi sa+i(J_0J_1+G_0G_1)}{J_1^2+G_1^2}, \quad (85)$$

of which the real part is the value of R_2, and the rest the value of L_2ni.

When sa is large

$$R_2=\frac{(n^2/v^2-m^2)^{\frac{1}{2}}}{2\pi acn}, \quad (86)$$

which approximates to $\mu v/2\pi a$ when mv is small compared with n. That is, the tendency is towards external resistance μv per unit area when the frequency is very great, unless counteracted by large m.

But when sa is small,

$$Z_2=\frac{s}{2\pi acn}\frac{s\pi a}{2}[1-i\,.\,2/\pi\,.\,\log\tfrac{1}{2}sa]. \quad (87)$$

This makes

$$R_2=\frac{1}{4cn}\left(\frac{n^2}{v^2}-m^2\right), \quad (88)$$

independent of a, and an important special case is $R_2=\frac{1}{4}\mu n$, when mv/n is small. Compare with p. 90, where the same result was found for an infinitely fine line of electrification. The corresponding inductance formula is

$$L_2n=-\tfrac{1}{4}\mu n\,.\,\frac{2}{\pi}\log\frac{na}{2v}, \quad (89)$$

which is, of course, very far from being independent of the radius. If $m=0$, the external resistance R_2 passes from the value $\frac{1}{4}\mu n$ at moderate frequencies to the large value $\mu v/2\pi a$ at excessive frequencies.

(36). The corresponding internal resistance operator is

$$Z_1=\frac{q}{2\pi a(k+cp)}\frac{I_0(qa)}{I_1(qa)}, \qquad q^2=m^2-\frac{n^2}{v^2}+\mu kni. \quad (90)$$

Here, in q, allowance is made for the permittivity of the wire, using the proper value of v, in order to compare with the external results in certain cases. For instance, n may be so large as to swamp the effects of k and m. Then Z_1 will reduce to $Z_1=\mu v/2\pi a$, using the proper value of v. This is similar to the external behaviour. But there is likely to be a want of accordance with fact if we use the value of v appropriate to low frequency, for pulses and very short waves go through a material dielectric at the etherial speed. That, however, belongs to the subject of dispersion.

When qa is large enough,

$$Z_1=\frac{(m^2-n^2/v^2+\mu kni)^{\frac{1}{2}}}{2\pi a(k+cp)}. \quad (91)$$

Here we can see that m may be so great as to be paramount in the numerator. Then Z_1 tends to become $Z_1=m/2\pi a(k+cp)$, or to the resistance $m/2\pi ak$, if cn/k is small.

If $m=n/v$ in (91), then n must be large in general to make qa be large. We now have

$$Z_1=\frac{1}{2\pi a}\left(\frac{\mu kn}{2}\right)^{\frac{1}{2}}\frac{1+i}{k+cni}=\frac{1}{2\pi a}\left(\frac{\mu kn}{2}\right)^{\frac{1}{2}}\frac{(k+cn)+i(k-cn)}{k^2+c^2n^2}. \qquad (92)$$

On the other hand, if qa is sufficiently small

$$Z_1=\frac{1}{\pi a^2(k+cp)}=\frac{k-cp}{\pi a^2(k^2+c^2n^2)}, \qquad (93)$$

to the first approximation.

In the usual estimation of the resistance and inductance of a round straight wire subjected to simply periodic impressed voltage, we make $m=0$, and $c=0$, or $v=\infty$, so that the conductivity is the controlling factor, as it certainly is, in metals. Then $q^2=\mu kp$, and $q=(\frac{1}{2}\mu kn)^{\frac{1}{2}}(1+i)$. The use of this q in (90), and reduction to the form R_1+L_1p will lead to the known formulæ in different forms (which are, however, not always attributed to their first authors), showing increased resistance and reduced inductance as the frequency increases.

It will be of interest to see how the external resistance $R_2=\frac{1}{4}\mu n$ behaves in relation to the internal resistance. Say $n=2,000$, then $R_2=500$ per cm., or 5×10^7 per kilom. Now this is only a small fraction of an ohm, whereas the resistance of a practical telephonic wire would be several ohms per kilom. So the external resistance is not of any importance in telephony, even with much larger values of n.

If, however, we pass to such frequencies as are concerned in long Hertzian waves the external resistance mounts up rapidly, and far faster than the internal resistance. The external may become a considerable multiple of the internal resistance. But it has to be remembered here that the calculations apply to a solitary wire in the ether. If there is a pair of wires the case is different. The possible great external waste of energy is abolished. One wire tends to neutralise the other in external radiative power, and the energy is kept in their neighbourhood. The same property is true in a lesser degree when the second conductor is the earth, whether the wire is horizontal or vertical. There is practically the waste in the wire, and the waste in the earth, but no important radiative waste in addition. By "earth," we should understand, in the present application, not only the real earth, but any other conductors that may be present. The radiative resistance R_2, on the other hand, is purely an etherial matter.

Plane Waves in a Dielectric Loaded in a Certain Way.

§ 530. (1). In my discussion of some features of Helmholtz's theory of dispersion, (vol. 2, p 507), these forms of the circuital equations were arrived at,

$$-\frac{d\mathrm{H}}{dx}=\left\{cp+\frac{1}{1/sp+r+mp}\right\}\mathrm{E}, \qquad -\frac{d\mathrm{E}}{dx}=\mu p\mathrm{H}. \tag{1}$$

Here E and H are the electric and magnetic forces, c and μ the permittivity and inductivity of the ether. The second law is as usual, but in the expression of the first law occurs an extra term containing an inverse operator. It may be interpreted as the extra current density due to electricity moved under the action of E, when its motion is resisted elastically, frictionally, and inertially. Of the three constants s, r, and m, if s alone is finite, its action is to increase the permittivity to $c+s$. If r alone is finite, its reciprocal is electric conductivity. In the case of s alone the speed of propagation is reduced from the v in $\mu cv^2=1$ to the v_1 in $\mu(c+s)v_1{}^2=1$. In the case of r alone, the speed of propagation is still the v of ether, but there is distortion of a known kind due to the conductivity.

The inverse operator may also be interpreted in terms of the electrons in atoms, their free periods and effective masses, as in Drude's theory. Put $r=0$, and let there be any number of inverse operators of the same type with different constants. But one will be enough here, for the purpose of considering some special cases. Let $r=0$. Then the characteristic of propagation is

$$\frac{d^2\mathrm{H}}{dx^2}=q^2\mathrm{H}, \qquad q^2=\frac{p^2}{v^2}+\frac{\mu sp^2}{1+msp^2}; \tag{2}$$

and we have also

$$\frac{\mathrm{E}}{\mathrm{H}}=\frac{\mu p}{q}, \qquad \mathrm{E}=\epsilon^{-qx}\mathrm{E}_0, \qquad \mathrm{H}=\epsilon^{-qx}\mathrm{H}_0, \tag{3}$$

if E_0, H_0 are the values at $x=0$. Carrying out the operations will show how E and H are propagated, in a pulse, for example.

If we put $p=\infty$, we obtain $\mathrm{E}=\mu v\mathrm{H}$. This shows that a discontinuity, say the wave front due to a sudden change in H_0, is propagated at the etherial speed v, and unattenuated. But it does not follow that there is no tailing after a pulse, even though there is no conductivity. It is the presence of the inertia embodied in m that allows full etherial speed of a wave front.

(2). The speed of a train of simply periodic waves is something very different. It is the resultant of actions proceeding both ways, after the wave train has been established. Say $\mathrm{H}_0 \sin nt$ is impressed at $x=0$ Then put $p^2=-n^2$ in q, and we obtain $q=bp/v$, and

$$\mathrm{H}=\epsilon^{-pxb/v}\mathrm{H}_0 \sin nt, \qquad b=\left[1+\frac{\mu sv^2}{1-msn^2}\right]^{\frac{1}{2}}, \tag{4}$$

where b is real positive when n lies between 0 and n_1, or between n_2 and ∞, given by

$$1=\mu s n_1^2, \qquad 1=\frac{\mu s v^2}{m s n_2^2-1}. \tag{5}$$

Under these circumstances,

$$H=H_0 \sin n\,[t-xb/v]=bE/\mu v. \tag{6}$$

This b is also the index of refraction in optical applications, and the wave speed is $w=v/b$.

But when n lies between n_1 and n_2, then

$$q=\frac{na}{v}, \qquad a=\left(-1+\frac{\mu s v^2}{m s n^2-1}\right)^{\frac{1}{2}}, \tag{7}$$

where a is real positive; so

$$H=H_0\epsilon^{-nxa/v}\sin nt, \qquad E=(\mu v/a)H_0\epsilon^{-nxa/v}\cos nt, \tag{8}$$

which represent standing vibrations, attenuated along x.

Now consider how w, b, and a, vary with n. At zero frequency, w is the same as v_1, the speed of long waves, the same as if $m=0$. As n increases to n_1, w falls to zero, whilst b rises to ∞. So we see by (6), that E gets smaller relatively to H, and ultimately vanishes. At the same time the wave-length in the material gets shorter and ultimately vanishes. So the remanent vibrating H counts for nothing over any small finite depth of the material.

Next jump to the frequency n_2, and then increase n to ∞. The wave speed falls from ∞ at n_2 to v at $n=\infty$ Or b rises from 0 to 1. At the lower limit n_2, the infinite w means infinite wave-length. Also it is now H that vanishes relatively to E. This shows that the impressed datum H_0 requires emendation. There is no discontinuity in passing through n_2 to the lower frequency, however. What happens is that a change of phase begins as the vanished H comes on again. Also the attenuation comes on gradually. But when n falls to n_1, the attenuation increases to ∞; so, in a sense, there is no discontinuity at n_1 either.

(3). But it is neccessary to revise the terminal datum for a better understanding. Let $x=0$ be the plane face of the dispersive material, ether being on the other side. Let radiation specified by $h_0 \sin nt$ be incident upon the boundary. Find the transmitted wave. By continuity of electric and magnetic force at the boundary, it is easily found that

$$\frac{H}{h}=\frac{2\mu v}{Z+\mu v}, \qquad Z=\frac{E}{H}=\frac{\mu v}{b}, \tag{9}$$

E and H being the values in the material, and h the incident magnetic force, all being boundary values. So we now have

$$H=\frac{2h_0 b}{1+b}\sin n(t-bx/v)=\frac{bE}{\mu v}, \tag{10}$$

when b is real, and this shows that H vanishes at $n=n_2$.

We may conclude that there is total reflection of the incident radiation at both limits, though of different sorts. But the reflection is total all the way between as well. Thus, if h' is the reflected magnetic force corresponding to the incident h, we have

$$\frac{h'}{h}=\frac{\mu v-Z}{\mu v+Z}=\frac{b-1}{b+1}=\frac{ai-1}{ai+1}. \tag{11}$$

At the n_1 end of the band, $b=\infty$, and $h'/h=1=-e'/e$, if e and e' are the incident and reflected electric forces. This is like reflection from the surface of a perfect electric conductor. And at the other n_2 end, $b=0$, making $h'/h=-e'/e=-1$, which is like reflection from a perfect magnetic conductor. Between the limits, the size (or modulus) of h'/h is always 1, so the reflection is always total, with change of phase, which varies continuously between n_1 and n_2. After that, from n_2 to ∞, there is again no change of phase in reflection.

As for the absorption in the band, there is not any. The incident energy is returned on reaching the boundary. The existence of internal vibration does not contradict this. It implies a preliminary entry of energy to set up the standing vibration; but that ceases (in the average), and the permanent state at any frequency in the band is then associated with total reflection in a time average. It is not possible to change suddenly from the permanent state of one frequency to that of another, whether in the band or outside it.

The above rudimentary system furnishes a rough imitation of ordinary dispersion in glass, &c., as well as of some cases of anomalous dispersion. For let there be a band of total reflection with n_1 beyond the detectable ultra-violet. Then from red to violet the wave speed will fall continuously, and will fall faster the higher the frequency. On the other hand, let the band n_1 to n_2 be in the green. Then the same thing will happen from red to near the green with a great spreading out and fading away of the spectrum. The green will be absent. As for the blue and violet, they will be deviated the other way, and be in the infra-red region. This roughly imitates the behaviour of fuchsine in alcohol, except that the index of refraction is then not less than unity for the blue and violet.

Now it may perhaps not be difficult to change the form of the inverse operator so as to make b be greater than 1 beyond n_2, or in a suitable portion of the region. But it does not seem possible to make the operator of the above type, or a succession of the same, represent the state of things when there are bands of total reflection in the visible spectrum. For since w goes down to 0 at the beginning of the first band, the deviation becomes unlimited in nearing it, whereas the state of things referred to only indicates an abnormally increased deviation on one side of the band, followed on the other side by abnormally reduced deviation.

(4). To find the way of propagation of a pulse through the medium requires that the functionization of ϵ^{-qx} 1 shall be done. It can be done in a power series, but as that is of a very complicated form, its interpretation is not easy. Conversion to a series of functions of x and t is desirable. There is one case in which the meaning of ϵ^{-qx} can be readily determined by a slight modification of the analysis in the telegraph theory. This case is $s=\infty$, m finite. The elastic resistance to the motion of the auxiliary charges is done away with, leaving them free to move except as regards their inertia. The equations are

$$-\frac{dH}{dx}=\left(cp+\frac{1}{mp}\right)E, \qquad -\frac{dE}{dx}=\mu pH, \qquad q^2=\frac{p^2}{v^2}+\frac{\mu}{m}. \tag{12}$$

Now in the development of my telegraph theory, the algebrisation of ϵ^{-qx} was done for

$$q^2=[(p+\rho)^2-\sigma^2]/v^2, \tag{13}$$

when ρ was a positive constant, and σ either positive or negative. To meet the present case we want $\rho=0$, $\sigma^2=-\mu v^2/m=-1/cm$. The negativity of σ^2 shows that the present case does not belong to the telegraph theory at all. It is an impossible application. But since q^2 in (13) is a function of σ^2, and the corresponding development of ϵ^{-qx} is also, as in equation (57), vol. 2, p. 337, we may use the same formulæ precisely, with the sign of σ^2 changed and $\rho=0$. This done, we have

$$\epsilon^{-qx}=\left(P_0-2\frac{w^2}{\lfloor 2}P_2+2\frac{w^4}{\lfloor 4}P_4-2\frac{w^6}{\lfloor 6}P_6+\dots\right)(z), \tag{14}$$

where $P_n(z)$ is $I_n(z)$ divided by its first term, and

$$w^2=\tfrac{1}{4}\sigma^2(t-x/v)^2, \qquad z=\sigma(t^2-x^2/v^2)^{\frac{1}{2}}, \qquad \sigma^2=-1/mc. \tag{15}$$

Consequently, if H_0 constant is impressed at $x=0$ from the left side, beginning $t=0$, the resulting H at x, t is

$$\frac{H}{H_0}=J_0(z')+2\frac{vt-x}{vt+x}J_2(z')+2\left(\frac{vt-x}{vt+x}\right)^2 J_4(z')+\dots, \tag{16}$$

from $x=0$ up to $x=vt$, and zero beyond. Here $z'=(\mu/m)^{\frac{1}{2}}(v^2t^2-x^2)^{\frac{1}{2}}$. The value of H is H_0 at the wave front as well as at the origin. The same formula also expresses the value of E/E_0, when E_0 is impressed at the origin.

To find the formula for E to match (16), we have

$$\frac{E_0}{H_0}=\frac{\mu p}{q}=\frac{\mu vp}{(p^2-\sigma^2)^{\frac{1}{2}}}, \qquad \frac{E}{H_0}=\epsilon^{-qx}\frac{\mu vp}{(p^2-\sigma^2)^{\frac{1}{2}}}. \tag{17}$$

To expand, use the a differentiator, as in vol. 2, p. 347; thus,

$$\frac{E_0}{H_0}=\mu v\frac{1+a^2}{1-a^2}\frac{1-a^2}{1+a^2}=\mu va^0. \tag{18}$$

The complete solution for E is therefore

$$E/H_0=\mu vP_0(z)=\mu vJ_0(z'), \tag{19}$$

consisting of a single term. At the wave front, $E=\mu v H_0=\mu v H$. Behind the wave front E is wavy and oscillatory. At the origin, $E_0=\mu v H_0 J_0[t/(cm)^{\frac{1}{2}}]$. It is satisfactory to obtain so simple a result as (19). But when it is E_0 that is impressed, the result is an infinite series. Thus,

$$\frac{\mu v H_0}{E_0}=\frac{1-a^2}{1+a^2}\frac{1-a^2}{1+a^2}=1-4a^2+8a^4-12a^6+\dots,$$

therefore $$\frac{\mu v H}{E_0}=J_0(z')+4\left(\frac{vt-x}{vt+x}\right)J_2(z')+8\left(\frac{vt-x}{vt+x}\right)^2 J_4(z')+\dots. \quad (20)$$

The formulæ for impulsive impressed H_0 or E_0 follow by a time differentiation. Considering the simplest case, which is (19), it is readily seen that the E due to impressed ph_0 consists of a pulse at distance vt followed by a wavy tail. Thus, h_0 being constant,

$$\frac{E}{\mu v h_0}=-\frac{(\mu/m)^{\frac{1}{2}}v^2 t}{(v^2t^2-x^2)^{\frac{1}{2}}}J_1(z'),\text{ in the tail.} \quad (21)$$

The pulse, of the type $E=\mu v H$, does not attenuate, but the amplitude of the tail does to any extent if we allow t to increase whilst keeping x constant. But if we keep at a constant distance y from the head, so that $vt=x+y$, then

$$\frac{E}{\mu v H}=-\frac{(1/cm)^{\frac{1}{2}}(x+y)}{[y(2x+y)]^{\frac{1}{2}}}J_1[(\mu/m)y(2x+y)]^{\frac{1}{2}}, \quad (22)$$

which, when x is very large, tends to constancy of amplitude.

(5). The speed of propagation is v in the above, and remains v if we increase m to ∞, when of course the theory reduces to that of plane waves in the ether. But the speed is also v when s is finite, provided m is finite, too. Yet if we now make $m=0$, the speed is reduced to the smaller value v_1. The only way I can think of for this to occur is by supposing that when m is decreased, the portion of the wave (due to impressed H_0, say) between $x=v_1t$ and $x=vt$ gets smaller, and in the limit, when $m=0$, vanishes altogether, leaving only the part up to $x=v_1t$. It would be interesting to have the full theory worked out to show the two wave fronts when m is small, and the ultimate disappearance of the forward portion of the complete wave. The beginning of the theory is the determination of E or H at $x=0$ due to H_0 or E_0 impressed there Thus,

$$\frac{E_0}{\mu v H_0}=\left(\frac{p^2+1/ms}{p^2+(1+\mu s v^2)/ms}\right)^{\frac{1}{2}}=\left(\frac{p^2-\sigma^2}{p^2-\rho^2}\right)^{\frac{1}{2}}\text{ say.} \quad (23)$$

Here $\rho^2>\sigma^2$. Or

$$\frac{E_0}{\mu v H_0}=\left(1-\frac{\sigma^2}{p^2}\right)^{\frac{1}{2}}I_0(\rho t). \quad (24)$$

It is also quite easy, by the use of the a differentiator, to convert (23) or (24) to the sum of the form $\Sigma A\, I_n(\rho t)$, or the same with σ instead of ρ, n having the even values. But we do not obtain the proper generalisation by turning t^2 to $t^2 - x^2/v^2$ in the formula for $x=0$. It remains to discover the proper functions to employ at a distance from the origin. The direct application of the ϵ^{-qx} operator gives a series in powers of x, with complicated functions of t for coefficients.

The Generation of Spherical Pulses in an Elastic Solid.

§ 531. (1). The investigation of the manner of generation and progression of spherical electromagnetic pulses in the ether will be found in my *Electrical Papers*, vol. 2, p. 409 and after. Only one speed of propagation is concerned. But in the elastic solid there are two speeds of propagation. No essential change in the analysis there used, viz., the multegrational process, is required to adapt it to two speeds, but only an extension. It will be found to greatly simplify the subject, by enabling us to represent results by a single equation, containing in itself all the various discontinuous details.

The well-known equation of motion may be written in the form

$$\mathbf{F}/\rho + (v_2^2 - v_1^2)\,\nabla\,\text{div}\,\mathbf{G} + v_1^2\nabla^2\mathbf{G} = \ddot{\mathbf{G}}, \tag{1}$$

where $\mathbf{G}$ is the displacement, ρ the density, and v_1, v_2 the two speeds, of which, in the elastic solid, v_2 is the greater, being the speed of condensational waves. Also, $\mathbf{F}$ is impressed force per unit volume, and is the source of the disturbances to be found. If we write $\rho c_1 v_1^2 = 1$, $\rho c_2 v_2^2 = 1$ then c_1 and c_2 are the elastic compliancies concerned in the two sorts of waves.

Dividing (1) into circuital and divergent parts, we obtain, if $q_1 = d/dv_1 t$, $q_2 = d/dv_2 t$,

$$c_1\mathbf{F}_1 = (q_1^2 - \nabla^2)\mathbf{G}_1, \qquad c_2\mathbf{F}_2 = (q_2^2 - \nabla^2)\mathbf{G}_2, \tag{2}$$

where $\mathbf{F}_1$, $\mathbf{G}_1$ are circuital, and $\mathbf{F}_2$, $\mathbf{G}_2$ divergent. The separate solutions are, therefore, by p. 165 above, or p. 256,

$$\mathbf{G}_1 = \text{pan}_1\, c_1\mathbf{F}_1 = \Sigma\, p_1 c_1 \mathbf{F}_1, \qquad \mathbf{G}_2 = \text{pan}_2\, c_2\mathbf{F}_2 = \Sigma\, p_2 c_2\mathbf{F}_2, \tag{3}$$

where
$$p_1 = \frac{\epsilon^{-q_1 r}}{4\pi r}, \qquad p_2 = \frac{\epsilon^{-q_2 r}}{4\pi r}. \tag{4}$$

Also
$$\text{curl}\,\mathbf{G} = \text{pan}_1\,\text{curl}\,c_1\mathbf{F}, \qquad \text{div}\,\mathbf{G} = \text{pan}_2\,\text{div}\,c_2\mathbf{F}, \tag{5}$$

where pan and curl or pan and div may be interchanged if desired. The distinction between pan_1 and pan_2 is only in the use of q_1 or q_2. Thus, the curl and div of $\mathbf{G}$ are given directly in terms of the curl and div of $\mathbf{F}$.

(2). But for $\mathbf{G}$ itself we need to make the separation of $\mathbf{F}$ into $\mathbf{F}_1 + \mathbf{F}_2$, or do something equivalent. Thus,

$$\mathbf{F}_2 = -\nabla\,\text{pot}\,\text{div}\,\mathbf{F} = -\nabla\,\text{div}\,\text{pot}\,\mathbf{F} = \frac{\nabla\cdot\nabla}{\nabla^2}\mathbf{F}, \tag{6}$$

where pot is the instantaneous potential, whereas pan is the progressive potential. So we come to

$$\mathbf{G} = c_1 \operatorname{pan}_1 \mathbf{F} + (c_2 \operatorname{pan}_2 - c_1 \operatorname{pan}_1)\mathbf{F}_2$$
$$= c_1 \operatorname{pan}_1 \mathbf{F} + (c_2 \operatorname{pan}_2 - c_1 \operatorname{pan}_1)\frac{\nabla.\nabla}{\nabla^2}\mathbf{F}. \qquad (7)$$

Thus if we find pan **F**, the rest follows by introducing c_1 and c_2, and doing $\nabla.\nabla$ pot.

But by using (2) above, it can be shown that (7) is also represented by

$$\mathbf{G} = c_1 \operatorname{pan}_1 \mathbf{F} + (\operatorname{pan}_2 - \operatorname{pan}_1)\frac{\nabla.\nabla\mathbf{F}}{p^2\rho}, \qquad (8)$$

where $p = d/dt$. Thus,

$$(q_1^2 - \nabla^2) \operatorname{pan}_1 \mathbf{f} = \mathbf{f}, \qquad (q_2^2 - \nabla^2) \operatorname{pan}_2 \mathbf{f} = \mathbf{f},$$

are identical relations. Pot both of them. Then

$$(q_1^2 \operatorname{pot} + 1) \operatorname{pan}_1 \mathbf{f} = (q_2^2 \operatorname{pot} + 1) \operatorname{pan}_2 \mathbf{f};$$

therefore, $(\operatorname{pan}_2 - \operatorname{pan}_1)\,\mathbf{f} = (q_1^2 \operatorname{pan}_1 - q_2^2 \operatorname{pan}_2) \operatorname{pot} \mathbf{f}$,

or, $$\frac{\operatorname{pan}_2 - \operatorname{pan}_1}{\rho p^2}\mathbf{f} = (c_1 \operatorname{pan}_1 - c_2 \operatorname{pan}_2) \operatorname{pot} \mathbf{f}. \qquad (9)$$

Now use this in (7) and we obtain (8), without the pot. The special **f** is to be $\nabla.\nabla\mathbf{F}$.

There is more in this transformation than may appear at first sight. For we may obtain (8) from (7) by simply putting $\nabla^{-2} = q_1^{-2}$ in the pan_1 term, and $\nabla^{-2} = q_2^{-2}$ in the pan_2 term. But these substitutions are not true, although, by (2), $\nabla^2 = q_1^2$ or q_2^2 outside the region of $\mathbf{F}_1$ or of $\mathbf{F}_2$. Equation (8) is nevertheless true everywhere, whether inside or outside the region of **F**.

(3). In the above **F** is any distribution. For the examination of pulses of course specially simple distributions are taken. The most instructive is when **F** is unidirectional, and is confined to a spherical space. Then there are two distinct solutions, one inside, the other outside the sphere of **F**, because the waves come from its boundary, and progress both inwardly and outwardly. But if we condense the sphere to a point, the internal solution goes out of existence. Let then f_0 be the volume integral of F, the direction of **F** being parallel to z. Then (8) becomes

$$\mathbf{G} = c_1 p_1 \mathbf{f}_0 - \nabla\frac{d}{dz}(p_1 - p_2)\frac{f_0}{p^2\rho}, \qquad (10)$$

outside the point $r=0$, where p_1 and p_2 are as above defined. Or, if f_0 is steady, beginning $t=0$,

$$\mathbf{G} = \frac{c_1\mathbf{f}_0}{4\pi r} - (c_1 - c_2)\,\nabla\,\frac{f_0\cos\theta}{8\pi}, \quad \text{from } r=0 \text{ to } v_1 t, \qquad (11)$$

$$\mathbf{G} = -c_1\nabla\frac{f_0 v_1^2 t^2}{8\pi r^2}\cos\theta + c_2\,\nabla\frac{f_0\cos\theta}{8\pi}, \text{ from } r = v_1 t \text{ to } v_2 t. \qquad (12)$$

Here θ is the angle between $\mathbf{z}$ and $\mathbf{r}$. Outside $r=v_2t$ is no disturbance. Inside v_1t, the displacement is steady. At these two places are condensed pulses, the outer being condensational, the inner rotational. These will be noticed later.

(4). Now take the case of a solid sphere of impressed force. The forms of pan F for a spherical surface ($r=a$) distribution of unidirectional force must be of the types $A\epsilon^{-qr}/4\pi r$ outside $r=a$, and $B\epsilon^{qr}/4\pi r+C\epsilon^{-qr}/4\pi r$ inside. At $r=a$ these must be equal, and at $r=0$ finite. So $B=-C$. It follows that the forms of pan F are

$$\text{(out)}\ f_0\frac{\text{shin}\,qa}{qa}\frac{\epsilon^{-qr}}{4\pi r},\qquad \text{(in)}\ f_0\frac{\text{shin}\,qr}{qr}\frac{\epsilon^{-qa}}{4\pi a},\tag{13}$$

where $f_0=4\pi a^2 f$, and f is the force per unit area. To check, $q=0$ makes the inside value be $f_0/4\pi a$.

By insertion of these values of pan F in (8), we obtain the internal and external solutions in all their details, and the algebrisation is quite easy. For greater comprehensiveness pass to the more developed case of F uniform throughout the sphere $r=a$. We must then integrate (13) throughout the sphere. This is the usual drudgery, and details need not be given. The results are, if f is now the force per unit volume, and f_0 its volume integral,

$$\text{pan}\,f=f_0\frac{\epsilon^{-qr}}{4\pi r}\frac{3}{q^3a}\frac{d}{da}\frac{\text{shin}\,qa}{a},\qquad\text{(out)}\tag{14}$$

and

$$\text{pan}\,f=\frac{f}{q^2}\left[1+\frac{a^2}{qr}\frac{d}{da}\frac{\epsilon^{-qa}}{a}\,\text{shin}\,qr\right].\qquad\text{(in)}\tag{15}$$

Using these in (8), we find first that the external solution is

$$\begin{aligned}\mathbf{G}=&\frac{c_1\mathbf{f}_0\epsilon^{-q_1r}}{4\pi r}\frac{3}{aq_1^3}\frac{d}{da}\frac{\text{shin}\,q_1a}{a}\\ &-\nabla\frac{d}{dz}\frac{1}{p^2\rho}\left[\frac{f_0\epsilon^{-q_1r}}{4\pi r}\frac{3}{aq_1^3}\frac{d}{da}\frac{\text{shin}\,q_1a}{a}-f_0\frac{\epsilon^{-q_2r}}{4\pi r}\frac{3}{aq_2^3}\frac{d}{da}\frac{\text{shin}\,q_2a}{a}\right]\end{aligned}\tag{16}$$

$$\begin{aligned}=&\frac{c_1\mathbf{f}_0}{4\pi r}\frac{3}{a}\frac{d}{da}\frac{\epsilon^{q_1(a-r)}-\epsilon^{-q_1(a+r)}}{2a}\frac{1}{q_1^3}\\ &-\nabla\frac{d}{dz}\frac{3}{a\rho}\frac{f_0}{4\pi r}\frac{d}{da}\left[\frac{\epsilon^{q_1(a-r)}-\epsilon^{-q_1(a+r)}}{2ap^2q_1^3}-\frac{\epsilon^{q_2(a-r)}-\epsilon^{-q_2(a+r)}}{2ap^2q_2^3}\right].\end{aligned}\tag{17}$$

By immediate multegration, f_0 being steady after $t=0$, this is turned to

$$\mathbf{G}=\frac{c_1\mathbf{f}_0}{4\pi r}\frac{3}{a}\frac{d}{da}\frac{y_1^3-y_2^3}{2a\,\underline{|3}}-\nabla\frac{d}{dz}\frac{3f_0}{4\pi ra}\frac{d}{da}\left[c_1\frac{y_1^5-y_2^5}{2a\,\underline{|5}}-c_2\frac{z_1^5-z_2^5}{2a\,\underline{|5}}\right],\tag{18}$$

where $y_1=v_1t+a-r$, $y_2=v_1t-a-r$, $z_1=v_2t+a-r$, $z_2=v_2t-a-r$. (19)

This equation (18) gives $\mathbf{G}$ explicitly from $r=a$ to ∞, and $t=0$ to ∞. The quantities y_1, y_2, z_1, z_2 are, by construction, positive only. For instance, y_1 is zero when $v_1t<(r-a)$, and similarly for the others. So we obtain several different solutions in different parts of space at different times.

(5). We can now give a general description of the different stages. At the moment $t=0$, two waves of displacement start from the surface of the sphere of f, progressing outward at speeds v_1 and v_2. At the same time two similar waves progress inward to the centre of the sphere. The time taken by the faster wave to traverse the sphere is $2a/v_2$. At any moment t not exceeding $2a/v_2$, the outer displacement is given by term No. 5 in (18) for points between the front of the condensational wave and the front of the rotational wave, and by the sum of Nos. 5, 3, and 1 for points between the front of the rotational wave and the surface of the sphere. There are 6 terms in (18), which are referred to as No. 1, etc., in the order they occur. The first region is entirely occupied by the condensational wave, the second by both the condensational and rotational, superposed.

Now, to be distinct, suppose $v_2=2v_1$. This will allow us to say definitely when the rotational shell ceases to overlap the condensational shell. It will be at the moment $v_2t=4a$. Between $v_2t=2a$ and $4a$, there are three regions to be considered. From $r=a+v_2t$ down to $r=a+v_1t$, the formula for the displacement is No. 5. From $r=a+v_1t$ down to $r=v_2t-a$ it is the sum of Nos. 5, 3, and 1. From $r=v_2t-a$ down to $r=a$ it is the sum of Nos. 5, 3, 1, and 6.

This is the state up to the moment $v_2t=4a$, and $v_1t=2a$. The back of the condensational shell then coincides with the front of the rotational shell, and the two shells are beginning to get clear of one another. For larger values of t, the formula is No. 5 for the complete condensational shell, that is, from $r=v_2t+a$ down to $r=v_2t-a$. From the latter place (the back of the condensational shell) down to $r=v_1t+a$ (the front of the rotational shell), the formula is the sum of Nos. 5 and 6. In the rotational shell itself, from $r=v_1t+a$ down to v_1t-a, the formula is the sum of Nos. 5, 6, 3, and 1. And between the back of the rotational shell (which is, like the other, of depth $2a$), and the surface of the sphere, the formula for the displacement is complete, the sum of all terms in (18). The displacement is here steady, as also in the interior of the sphere. The two shells then go out to infinity, leaving behind the steady state.

In this last stage, when the two shells do not overlap, and the rotational shell is clear of the sphere of impressed force, there are five formulæ for the displacement. Thus, from $r=0$ to a,

$$\mathbf{G}=\frac{c_1\mathbf{f}}{6}(3a^2-r^2)-\frac{c_1-c_2}{6}f\nabla\left(a^2r-\frac{1}{5}r^3\right)\cos\theta. \tag{20}$$

This is not contained in (18), of course, but being steady, is easily got from the internal solution. Then from $r=a$ to v_1t-a,

$$\mathbf{G}=\frac{c_1\mathbf{f}_0}{4\pi r}-(c_1-c_2)\nabla\frac{f_0\cos\theta}{8\pi}\left(1-\frac{a^2}{5r^2}\right). \tag{21}$$

Next, from $r=v_1t-a$ to v_1t+a, that is, in the rotational shell,

$$\mathbf{G}=\frac{c_1\mathbf{f}_0}{4\pi r}\frac{y_1^2(3a-y_1)}{4a^3}-c_2\nabla\frac{f_0v_2^2t^2\cos\theta}{8\pi r^2}$$

$$+c_1\nabla\frac{f_0\cos\theta}{16\pi a^3r^2}y_1^3\Big[ar+y_1\Big(\frac{y_1}{5}-\frac{v_1t}{4}\Big)\Big]+c_2\nabla\frac{f_0\cos\theta}{8\pi}\Big(1-\frac{a^2}{5r^2}\Big). \quad (22)$$

Then from $r=v_1'+a$ to v_2t-a, that is, between the shells,

$$\mathbf{G}=-c_2\nabla\frac{f_0v_2^2t^2\cos\theta}{8\pi r^2}+c_2\nabla\frac{f_0\cos\theta}{8\pi}\Big(1-\frac{a^2}{5r^2}\Big). \quad (23)$$

Finally, in the condensational shell, or from $r=v_2t-a$ to v_2t+a,

$$\mathbf{G}=-c_2\nabla\frac{f_0\cos\theta}{16\pi a^3r^2}z_1^3\Big\{ar+z_1\Big(\frac{z_1}{5}-\frac{v_2t}{4}\Big)\Big\}. \quad (24)$$

The last four formulæ come from (18) by carrying out the d/da and d/dz differentiations and reducing.

(6). The internal operational solution corresponding to (18) is got by using (15) in (8), and then, by multegrating it, we get the algebraical solution. It is

$$\mathbf{G}=c_1\mathbf{f}\Big[\frac{v_1^2t^2}{2}-\frac{1}{12r}y_1^2(3a+y_1)+\frac{1}{12r}y_2^2(3a+y_2)\Big]$$

$$-c_1\nabla\frac{\cos\theta}{12r^2}ay_1^3\Big[\Big(\frac{y_1}{4}+\frac{y_1^2}{20a}\Big)-a\Big(1+\frac{y_1}{4a}\Big)\Big]$$

$$+c_1\nabla\frac{\cos\theta}{12r^2}ay_2^3\Big[\Big(\frac{y_2}{4}+\frac{y_2^2}{20a}\Big)+a\Big(1+\frac{y_2}{4a}\Big)\Big]$$

$$+c_2\nabla\frac{\cos\theta}{12r^2}az_1^3\Big[\Big(\frac{z_1}{4}+\frac{z_1^2}{20a}\Big)-a\Big(1+\frac{z_1}{4a}\Big)\Big]$$

$$-c_2\nabla\frac{\cos\theta}{12r^2}az_2^3\Big[\Big(\frac{z_2}{4}+\frac{z_2^2}{20a}\Big)+a\Big(1+\frac{z_2}{4a}\Big)\Big], \quad (25)$$

where $y_1=v_1t-a+r$ = distance behind first front of rotational wave when going in; $y_2=v_1t-a-r$ = distance behind second front of rotational wave when going out, after centre has been reached; $z_1=v_2t-a+r$ = distance behind first front of condensational wave when going in; $z_2=v_2t-a-r$ = distance behind second front of condensational wave when going out after centre has been reached. The two y's and the two z's being essentially positive, otherwise zero, make (25) complete in all the stages, if it has been properly worked out. But it is not necessary to elaborate the various separate formulas, because the variable internal state is of little importance compared with the corresponding external state which has been done. It is rather more complicated in detail than the external state, because there are additional stages due to the fact that when the front of the condensational wave has reached and passed the centre, there is a crossing of the second or return front going out and the first front of the rotational wave going in. What is of importance is the steady state of displacement which begins at $r=0$ and spreads outward, beginning at the moment the front of the rotational wave reaches the centre. It is given by (20) above, and its range is from $r=0$ up to the second front of the rotational wave, which is the same as the back of the rotational shell which goes out to infinity.

It is also to be noticed that displacement begins at the centre, and throughout the sphere of impressed force not yet occupied by the fastest wave, instantly, being given by $\mathbf{G}=\frac{1}{2}c_1\mathbf{f}v_1^2t^2$, the first term in (25). It is the same as $\rho\mathbf{G}=\frac{1}{2}\mathbf{f}t^2$, or $\rho\dot{\mathbf{G}}=\mathbf{f}t$, and $\rho\ddot{\mathbf{G}}=\mathbf{f}$ simply for the acceleration of momentum. That is, the elastic forces of the solid are quite inoperative until the wave disturbance arrives.

(7). The activity of $\mathbf{f}$ is $\mathbf{f}\dot{\mathbf{G}}$, and ceases when $\mathbf{G}$ becomes steady at the moment $t=2a/v_1$. The total work done by the impressed force is therefore $\Sigma\ \mathbf{f}\mathbf{G}$, or $f\ \Sigma\ G_z$, throughout the sphere. The value of G_z may be derived from (20). The result of the integration is

$$\text{Work}=f_0^2(2c_1+c_2)/10\pi a. \tag{26}$$

One half of this remains as the stored energy of the steady displacement, ignoring the shells altogether. The other half is wasted, or radiated away in the two shells. To see how it divides between the shells (which may be guessed from (26) itself), find the simplified forms assumed by the displacement, velocity, &c., when the shells are at a great distance. In the outer condensational shell, it will be found that when v_2t/a is very large, the tangential velocity tends to become infinitesimal compared with the normal velocity, say u_2, which is given by

$$u_2=\frac{3c_2v_2f_0\cos\theta}{16\pi a^3r}z_1(2a-z_1), \tag{27}$$

where $z_1=v_2t+a-r$. It will also be found that the expression for the condensation tends to be given by

$$u_2=v_2\times\text{condensation}\,; \tag{28}$$

the same as in a simple plane wave of sound, in fact, by reason of the very small curvature. The distribution of intensity through the depth of the shell is parabolic, zero at both boundaries, maximal in the middle. In reality it is not quite zero on the inner side, though zero on the outer side.

The kinetic energy is, by volume integration throughout the shell,

$$T_2=\Sigma\ \tfrac{1}{2}\rho u_2^2=c_2f_0^2/40\pi a=U_2, \tag{29}$$

and the energy of condensation U_2 has the same value. The energy wasted in the condensational shell is therefore $c_2f_0^2/20\pi a$.

Similarly, the velocity in the rotational shell tends to become tangential, the normal component becoming relatively insensible. Let u_1 be the tangential velocity. It is given by

$$u_1=\frac{3c_1v_1f_0}{16\pi a^3r}\ y_1(2a-y_1)\sin\theta, \tag{30}$$

where $y_1 = v_1 t + a - r$, when $v_1 t/a$ is very large. It will also be found that the rotation is given by

$$u_1 = v_1 |\text{curl } \mathbf{G}|, \tag{31}$$

which is analogous to (28). Now in integrating the square of (30) throughout the volume of the shell, it will be observed that (27) has the factor $\cos\theta$, whilst (30) has $\sin\theta$. But the integral of $\sin^2\theta$ is twice that of $\cos^2\theta$. So the result is that

$$T_1 = \Sigma\ \tfrac{1}{2}\rho u_1^2 = 2c_1 f_0^2/40\pi a = U_1 \tag{32}$$

is the kinetic energy in the rotational shell. The potential energy U_1 has the same value, so the total energy in this shell is $2c_1 f_0^2/20\pi a$. The energy in the two shells is thus verified to be one half of the work done by the impressed force, as in (26).

(8). We may next calculate how the momentum is disposed of. First, observe that f is generating momentum (or transferring it to the body) at uniform rate, so that the amount taken in is $f_0 t$ at the time t. Now this goes on when the body comes to rest. There is then no place for the steadily increasing momentum save the two shells, or else between them. But it is easily found that the momentum between them (like the energy) becomes insensible. How then is $f_0 t$ of momentum divided between the shells?

If we use (27) for the normal speed, then the momentum in the condensational shell is

$$M_2 = \Sigma\ \rho u_2 \cos\theta = \tfrac{1}{3} f_0 t, \tag{33}$$

by integration throughout the shell. Similarly, in the rotational shell, it is, using (30),

$$M_1 = \Sigma\ \rho u_1 \sin\theta = \tfrac{2}{3} f_0 t. \tag{34}$$

Space is too valuable for any details concerning these and various other integrations in the present article. So the rotational shell carries twice as much momentum as the condensational, and both amounts increase at uniform rate. A comparison with the energies is interesting. The energy in the rotational shell divided by c_1 is twice the energy in the condensational shell divided by c_2.

If the force ceases at the moment t_1, two new shells are generated which follow the first pair. The first pair has steadily increasing momentum just as before, but the new condensational shell has momentum $-\frac{1}{3} f_0(t - t_1)$ and the new rotational shell has momentum $-\frac{2}{3} f_0(t - t_1)$. The total momentum in the four shells is therefore finite and constant, namely $f_0 t_1$. It is assumed here that t_1 is long enough (at least $2a/v_1$) to allow of the full formation of the shells. As for the energies in the two new shells, they are, when a long way out, the same as the energies in the two old shells. The new shells pick up the static energy as they expand, leaving a state of no displacement behind them.

(9). There are two principal interpretations of a physical nature of the above theory. The medium in question may be the ordinary elastic solid possessing rigidity. Or this rigidity may be done away with and an equivalent amount of rotational elasticity put in its place. Then the medium is what Lord Kelvin at one time called Ether, save that it is compressible as well. Lord Kelvin's Ether was incompressible. But the results are the same whether the elasticity is rigidity or is rotational. There is, however, a notable difference in the distribution of stored potential energy, other than that due to the condensation, in which there is no change. The rest of the stored potential energy is (per unit volume) proportional to the square of the rotation in the rotational ether, whereas it has a different distribution, of a more complicated expression, in the elastic solid. Correspondingly, the stress is different. So, if we enter into further detail regarding the distribution of energy, the rotational ether is much more manageable than the elastic solid. Nevertheless, no difference is made in the above energetic results relating to the total energy, because the total stored energy other than condensational in the elastic solid, either steady, or in the rotational shell when far out, may be precisely calculated by the same formula as in the case of the rotational ether.

Although the rotational ether is an admirable mechanical contrivance, yet, as regards its suitability for representing Maxwell's electromagnetic ether, I think it really never had a proper chance. For at the very beginning, as pointed out in vol. 1, p. 131, it fails to account for the mechanical forces concerned in electromagnetics. There are other objections, but that one is fatal. Owing to the very abstract nature of the electric and magnetic quantities, not one of which has been identified mechanically as Lord Kelvin understood a mechanical medium, there is a natural tendency to overlook the fact that the foundations of electromagnetics are purely dynamical, in terms of really observed mechanical forces of the Newtonian kind.

(10). The incompressible case deserves notice. Here $v_2=\infty$, because $c_2=0$. The outer shell goes out to infinity instantly when the impressed force is suddenly put on. It is then of zero energy. But its momentum is not zero, but is $\frac{1}{3}f_0t$ as before. That is, there is an instantaneous (kinematic) transference of momentum to an infinite distance (in an unbounded medium), always going on when the medium is incompressible, and is subjected to impressed force anywhere. The reduced formulas are, from $r=0$ to a,

$$\mathbf{G}=\tfrac{1}{6}c_1\mathbf{f}\,[2a^2-r^2+\tfrac{1}{5}z^2]+\tfrac{1}{15}c_1fz\mathbf{r}. \qquad (35)$$

From $r=a$ to v_1t-a,

$$\mathbf{G}=\frac{c_1\mathbf{f}_0}{4\pi r}-c_1\nabla\ \frac{f_0\cos\theta}{8\pi}\left(1-\frac{a^2}{5r^2}\right). \qquad (36)$$

From $r=v_1t-a$ to v_1t+a,

$$\mathbf{G}=\frac{c_1\mathbf{f}_0}{4\pi r}\frac{y_1^2(3a-y_1)}{4a^3}-c_1\nabla f_0\frac{v_1^2t^2\cos\theta}{8\pi r^2}$$
$$+c_1\nabla f_0\frac{\cos\theta}{16\pi a^3r^2}y_1^3\left\{ar+y_1\left(\frac{y_1}{5}-\frac{v_1t}{4}\right)\right\}. \qquad (37)$$

From v_1t+a to ∞,

$$\mathbf{G}=-\nabla\frac{f_0t^2\cos\theta}{8\pi r^2\rho}=-\frac{t^2}{2\rho}\nabla\mathrm{P}, \qquad (38)$$

where $\mathrm{P}=f_0\cos\theta/4\pi r^2$ is the pressure, and $-\nabla\mathrm{P}=\rho\ddot{\mathbf{G}}$.

The displacement outside the rotational shell is now entirely irrotational, and is like the induction outside a spherical uniform magnet. It begins instantly, at $r=a$, and at any fixed point varies as t^2 until the rotational shell arrives.

The kinetic energy outside the sphere r (greater than the outer radius of the shell) comes to

$$\mathrm{T}_r=f_0^2t^2/12\pi r^3\rho. \qquad (39)$$

It is not accompanied by any potential energy. This energy comes through the shell. It may be easily verified that the value of $\dot{\mathrm{T}}_r$ is also the value of Σ P$\dot{\mathbf{G}}$**N**, where P$\dot{\mathbf{G}}$ is the flux of energy, and **N** is unit normal outward, the summation extending over the surface r. But the moment the shell arrives at r, then $r=v_1t+a$, and T_r stops increasing. In fact, the values of T and $\dot{\mathrm{T}}$ become

$$\mathrm{T}=\frac{f_0^2t^2}{12\pi\rho(v_1t+a)^3}, \qquad \dot{\mathrm{T}}=\frac{f_0^2t}{12\pi\rho(v_1t+a)^3}\left[\frac{2a-v_1t}{v_1t+a}\right]. \qquad (40)$$

So the total T outside the shell decreases all the time, after $v_1t=2a$. This is the moment the shell is clear of the sphere of force, and extends from $r=a$ to $3a$, and is also the moment the force ceases to work. The kinetic energy of the irrotational motion is then at its greatest, and equals $c_1f_0^2/81\pi a$.

(11). At the other extreme we have the case $v_2=0$, $c_2=\infty$, or no elastic resistance to compression. This is best understood by making v_2 very small first, though finite. Then the rotational shell is on the outside, and beyond it is no disturbance, whilst the condensational shell is, if formed, close to the sphere of force, and is expanding very slowly. The formulas (16) and (17) are still valid, though the order of commencement of some of the terms is changed. But it will be found that the displacement, velocity, condensation, kinetic energy and potential energy in the condensational shell all increase without limit as v_2 gets smaller. In the limit, with $v_2=0$, they are infinite. Of course a medium of this sort, when subjected to the simple impressed force employed above to move it, is out of all question for practical consideration. The remedy is to do away with the divergent part of the impressed force, and employ only circuital distributions. Then the infinite condensational results do not occur, because the displacement is circuital. But then the same result is attained with finite or zero compressibility. Let the force be circuital, then the displacement is also circuital. So it looks as though the case $v_2=0$ were of no special use.

I should remark that it simplifies matters somewhat to do away with the rigidity altogether (as in vol. 1, p. 244), so that the condensational shell may be inside the rotational naturally, before making $v_2=0$.

(12). Returning now to the general case of finite v_2 and v_1, what happens when momentum is given to the medium when it is unbounded is, after preliminary irregularity, a state of rest everywhere save in the two shells. But let the medium be of finite size. Then its boundary is the place of origin of reflected waves. A part of the motion which arrives there may escape outside, into the air, for example. Ignoring that, the shells will be reflected to and fro any number of times, crossing and recrossing it may be in an increasingly distorted and mixed up manner, depending upon the shape of the boundary and the situation of the seat of impressed force. Moreover, every time a shell crosses the impressed force, more work is done; so the tendency is for the impressed force to get into a state of continuous, though fluctuating activity, whilst motion parallel to the force tends to cover the whole medium, though again in a fluctuating manner. What will be the ultimate result? Will the final result be in accordance with $\mathbf{f}_0=\mathrm{M}\dot{\mathbf{u}}$, where M is the total mass of the body, and **u** its velocity, the elementary law for a perfectly rigid body, with appropriate rotation as well should that be necessary? It does not seem likely that this result will be reached precisely. There is nothing to show that the vibrations will wholly disappear, unless of course we introduce internal frictional forces, which are not considered in the above. Another way of doing away with, or greatly reducing the importance of the vibrations is to distribute the impressed force uniformly over the whole body, as in the action of gravity.

(13). It may be useful to add the complete formulas for the condensation and rotation. The axes of rotation make circles round the z axis, that is, the circles of latitude. When the rotational shell has left the sphere of force, then there are three formulas for curl **G**. The tensors are

$$\tfrac{1}{3}c_1 fr\sin\theta,\qquad \frac{c_1 f_0}{4\pi r^2}\sin\theta,\qquad -\sin\theta\frac{d}{dr}\frac{y_1^2(3a-y_1)}{4a^3}\frac{c_1 f_0}{4\pi r}.\tag{41}$$

$$r=0 \text{ to } a,\qquad r=a \text{ to } v_1t-a,\qquad r=v_1t-a \text{ to } v_1t+a.$$

The total of $|\text{curl } \mathbf{G}|$ in the shell, got by integration to r throughout its depth, $y=2a$ to 0, makes

$$\frac{c_1 f_0 \sin\theta}{4\pi(v_1 t - a)}. \tag{42}$$

Put $a=0$ to make the value when the source is condensed to a point. This may also be reckoned by the discontinuity in the tangential **G** which occurs then, the shell becoming condensed to a surface. The value in (30) is zero at the outside of the shell, and there is no discontinuity at the inside of the shell when the full formula is used.

The corresponding condensation formulas are, if $z_1 = v_2 t + a - r$,

$$\tfrac{1}{3}c_2 fr\cos\theta, \qquad \frac{c_2 f_0}{4\pi r^2}\cos\theta, \qquad -\cos\theta\frac{d}{dr}\frac{z_1^2(3a-z_1)}{4a^3}\frac{c_2 f_0}{4\pi r}. \tag{43}$$

$$r=0 \text{ to } a, \qquad r=a \text{ to } v_2 t - a, \qquad r = v_2 t - a \text{ to } v_2 t + a.$$

The total condensation in the depth of the shell comes to

$$\frac{c_2 f_0 \cos\theta}{4\pi(v_2 t - a)}. \tag{44}$$

(14). When the impressed force is simply periodic, put $q=si$ in the operational solutions, and $f = \mathrm{F}\sin svt$, to obtain the wave trains. At a set of particular frequencies either of the two trains may vanish. Thus, the relation $\tan sa = sa$ makes pan $f=0$. This applies to either train, using v_1 or v_2 respectively. The corresponding vibrations are then confined to the sphere of force. But in the incompressible case, with $c_2=0$, then the whole external disturbance will vanish.

(15). Lastly, a few words on the solutions separately for $\mathbf{G}_1$ and $\mathbf{G}_2$. Inasmuch as the impressed forces $\mathbf{f}_1$ and $\mathbf{f}_2$ extend over all space, the separate solutions cannot be obtained so directly and with so little work as when they act together. Some kind of further integration is required. Here is one way. I give the steps and the results only. First find pot pan f. Keeping to the case of complete formation of the shells, there are four results,

$$(a).\ \text{pot pan } f = \frac{f}{6}\left(\frac{5}{4}a^4 - \frac{1}{2}a^2r^2 + \frac{1}{20}r^4\right) + \frac{f_0}{4\pi}(vt-a), \tag{45}$$

$$(b).\ \ ,, \quad ,, \quad = \frac{2}{15}\frac{fa^5}{r} - \frac{f_0 r^2}{8\pi r} - \frac{f_0 r}{8\pi} + \frac{f_0 vt}{4\pi}, \tag{46}$$

$$(c).\ \ ,, \quad ,, \quad = \frac{f_0 v^2 t^2}{8\pi r} - \frac{f_0 y^4}{64\pi a^3 r}\left(a - \frac{y}{5}\right), \tag{47}$$

$$(d).\ \ ,, \quad ,, \quad = \frac{f_0 v^2 t^2}{8\pi r}, \qquad y = vt + a - r. \tag{48}$$

Here y is the distance inward of r from outer boundary of either shell, whilst (a) applies from $r=0$ to a, (b) from $r=a$ to $vt-a$, (c) from $r=vt-a$ to $vt+a$, and (d) from $r=vt+a$ to ∞.

Now d/dz these results. Thus,

$$(a)_1.\quad \frac{d}{dz}\,\text{pot pan}\,f=\frac{f}{6}\left(\frac{r^3}{5}-a^2r\right)\cos\theta, \tag{49}$$

$$(b)_1.\quad ,,\quad ,,\quad =-\frac{f_0\cos\theta}{8\pi}\left(1-\frac{a^2}{5r^2}\right), \tag{50}$$

$$(c)_1.\quad ,,\quad ,,\quad =\frac{f_0y^3\cos\theta}{16\pi a^3r^2}\left(ar+\frac{y}{5}(a-r-\tfrac{1}{4}vt)\right)-\frac{f_0v^2t^2\cos\theta}{8\pi r^2}, \tag{51}$$

$$(d)_1.\quad ,,\quad ,,\quad =-\frac{f_0v^2t^2}{8\pi r^2}\cos\theta. \tag{52}$$

These apply to either f_1 or f_2, with the proper value of v.

Now $\mathbf{G}_2=-c_2\nabla$ the last four expressions, using v_2. So

$$(a)_2.\quad \mathbf{G}_2=-c_2\nabla\frac{f}{6}\left(\frac{r^3}{5}-a^2r\right)\cos\theta, \tag{53}$$

$$(b)_2.\quad ,,\quad =c_2\nabla\frac{f_0\cos\theta}{8\pi}\left(1-\frac{a^2}{5r^2}\right), \tag{54}$$

$$(c)_2.\quad ,,\quad =-c_2\nabla\frac{f\cos\theta}{16\pi a^3r^2}z_1^3\left(ar+\frac{z_1}{5}(a-r-\tfrac{1}{4}v_2t)\right)+\nabla\frac{f_0t^2\cos\theta}{8\pi r^2\rho}, \tag{55}$$

$$(d)_2.\quad ,,\quad =\nabla\frac{f_0t^2\cos\theta}{8\pi r^2\rho}. \qquad z_1=v_2t+a-r. \tag{56}$$

These are the displacements over all space at the moment t due to $\mathbf{f}_2$. Similarly, we have

$$\mathbf{G}_1=c_1\,\text{pan}\,\mathbf{f}+\nabla\frac{d}{dz}c_1\,\text{pot pan}\,f, \tag{57}$$

so we use the last set again, but with the factor $-c_2\nabla$ turned to $+c_1\nabla$, and v_2 turned to v_1, and then with the auxiliary values of c_1 pan $\mathbf{f}$ added. So the $\mathbf{G}_1$ set is represented by

$$(a)_3.\quad \mathbf{G}_1=\frac{c_1\mathbf{f}}{6}(3a^2-r^2)+\frac{c_1f}{6}\nabla\cos\theta\,(\tfrac{1}{5}r^3-a^2r), \tag{58}$$

$$(b)_3.\quad ,,\quad =\frac{c_1\mathbf{f}_0}{4\pi r}-c_1\nabla\frac{f_0\cos\theta}{8\pi}\left(1-\frac{a^2}{5r^2}\right), \tag{59}$$

$$(c)_3.\quad ,,\quad =\frac{c_1\mathbf{f}_0}{4\pi r}\frac{y_1^2(3a-y_1)}{4a^3}-\nabla\frac{f_0t^2\cos\theta}{8\pi r^2\rho}$$
$$+c_1\nabla\frac{f_0\cos\theta}{16\pi a^3r^2}y_1^3\left\{ar+y_1\left(\frac{y_1}{5}-\frac{v_1t}{4}\right)\right\}, \tag{60}$$

$$(d)_3.\quad ,,\quad =-\nabla\frac{f_0t^2\cos\theta}{8\pi r^2\rho}, \qquad y_1=v_1+ta-r. \tag{61}$$

These are the displacements due to $\mathbf{f}_1$, with the ranges, $(a)_3$ from $r=0$ to a, $(b)_3$ from $r=a$ to v_1t-a, $(c)_3$ from $r=v_1t-a$ to v_1t+a, $(d)_3$ from $r=v_1t+a$ to ∞.

As a check upon the accuracy of these results, we may combine the condensational with the rotational formulas to make up the combinational, equations (20) to (24). But it must be remembered that the ranges of $(b)_2$, $(c)_2$, $(d)_2$ are not the same as those of $(b)_3$, $(c)_3$, $(d)_3$. The proper way of combining them is this:—

$$\left.\begin{array}{ll}\text{From } r=0 \text{ to } a, & \mathrm{G}=(a)_2+(a)_3, \\ \text{From } r=a \text{ to } v_1t-a, & \mathrm{G}=(b)_2+(b)_3, \\ \text{From } r=v_1t-a \text{ to } v_1t+a, & \mathrm{G}=(b)_2+(c)_3, \\ \text{From } r=v_1t+a \text{ to } v_2t-a, & \mathrm{G}=(b)_2+(d)_3, \\ \text{From } r=v_2t-a \text{ to } v_2t+a, & \mathrm{G}=(c)_3+(d)_3, \\ \text{From } r=v_2t+a \text{ to } \infty, & \mathrm{G}=(d)_3+(d)_4.\end{array}\right\} \quad (62)$$

If the reader will take the trouble to make these combinations, then, after cancellations and reductions, the previous results (20) to (24) will be arrived at.

(16). Another case which I can recommend for working out by a serious minded worker is that of a radial distribution of force in the sphere $r=a$, or over its surface only to begin with. It is not difficult. The seat of force may be condensed to a point. It is then an outward acting pressure. When this has been done, the results may be generalized to suit the case of a source of the same nature moving about anyhow, by employing the "dopplerized potential." Then there will be two kinds of resultant formulas, corresponding to those I have found for the motion of electric charges through Maxwell's ether, according as the speed of motion of the source is less or greater than the speed of propagation. When that has been done, more advanced cases can be elaborated.

(**16A**). When describing to Mr. Searle the nature of the pulses in the above investigation, and the waste of momentum therein, he called my attention to the elementary theoretical problem of a solid sphere moving in an unbounded incompressible liquid. The momentum of the liquid is zero, by calculation. My interpretation is that there is an instantaneous flux of momentum to infinity whenever the velocity of the sphere changes. The momentum is wasted.

Thus, let M be the mass of the sphere of radius a, and density ρ, moving at speed u; and let σ be the density of the liquid. The velocity potential is $\frac{1}{2}(a^3u/r^2)\cos\theta$, and it is well known that the kinetic energy of the system may be calculated to be

$$\mathrm{T}=\tfrac{1}{2}(\mathrm{M}+m)u^2, \quad \text{where} \quad m=\mathrm{M}\sigma/2\rho.$$

But the visible momentum (as calculated) is only Mu, that of the sphere, whereas the true momentum is $(\mathrm{M}+m)u$. The amount mu is invisible. We might say that it did not exist at all, but the theory of pulses explains the discrepancy. Let the fluid be compressible; then any change of velocity of the sphere produced by impressed force acting upon it is propagated outward at finite speed. The energy and the momentum are then fully visible at a finite distance, and may be calculated. Now gradually reduce the compressibility to zero. This will increase the speed of propagation infinitely. The energy in the condensational shell (due, say, to impulsive communication of velocity to the sphere) tends to zero, but its momentum does not. So the kinetic energy of the liquid is visible, no part being wasted, whilst the momentum of the liquid is wasted, by going out of range instantaneously.

Here we have a case of instantaneous flux of momentum through a moving medium. But in the above article, it takes place through a stationary medium. The velocity of the medium inside the inner shell becomes zero by the establishment of the appropriate stress to suit the impressed force, and thenafter the solid behaves as though it were infinitely rigid.

A curious case in the hydrodynamic theory is got when $\rho=0$, meaning a rigid spherical shell of no mass. Then $T=\frac{1}{2}mu^2$, where m is half the mass of liquid which would fill the sphere. The visible momentum is zero always. All the momentum produced by impressed force is wasted, but none of the energy. Stopping the sphere restores the kinetic energy to the sphere in effect, but at the same time momentum of amount $-mu$ is sent out to infinity. The dynamical system appears to be at once conservative (for energy) and dissipative for momentum. It is remarkable that such elementary cases of Newton's dynamics should require abstruse considerations for their explanation. But it is far worse in the more modern dynamics, with ignoration of co-ordinates, and modified Lagrangean functions. Dynamics as visible to the naked eye seems to disappear altogether sometimes, leaving nothing but complicated algebra.

The impressed source need not be of momentum, that is, a moving force, as in the above. A source of matter, that is, of condensation, has similar properties, varied in detail. For example, see vol. 2, p. 500. There is also an electromagnetic example in this volume. See p. 36, equations (9), (10), and remarks on p. 38. The source is a linear source of magnetic induction. This induction is introduced at constant rate, and remains in the field. Yet the steady state tended to is one of zero induction and steady displacement. The increasing induction may then be looked for and found near the wave front. This is also the case when the source is stopped. The constancy of the total induction is then secured at a new wave front.

Plane Waves in moving Mediums. The Energy and Forces.

§ 532. (1). Some questions relating to the mechanical interaction of matter and ether are interestingly shown in the behaviour of plane waves. We also secure simplicity of treatment from the physical point of view. On p. 190 of this volume some results were described which demand a fuller treatment Beginning, for comparison, with a single medium, say the ether, the circuital equations are

$$\text{curl}\,(\mathbf{H}-\text{V}\mathbf{D}\mathbf{q})=\dot{\mathbf{D}}, \qquad -\text{curl}\,(\mathbf{E}-\text{V}\mathbf{q}\mathbf{B})=\dot{\mathbf{B}}; \tag{1}$$

where $\mathbf{D}=c\mathbf{E}$, $\mathbf{B}=\mu\mathbf{H}$, and $\mathbf{q}$ is the velocity of the medium. For plane waves take $\mathbf{E}\parallel\mathbf{y}$, $\mathbf{H}\parallel\mathbf{z}$, and $\mathbf{q}\parallel\mathbf{x}$. Then we get

$$-\Delta\,(\text{H}+\text{D}q)=p\text{D}, \qquad -\Delta\,(\text{E}+\text{B}q)=p\text{B}; \tag{2}$$

or, if q is constant everywhere,

$$-\nabla\text{H}=(p+q\Delta)c\text{E}, \qquad -\Delta\text{E}=(p+q\Delta)\mu\text{H}; \tag{3}$$

where $p=d/dt$, and $\Delta=d/dx$.

Now let $p=-w\Delta$ when applied to E or H. This means a solitary wave moving at speed w referred to the (assumed) fixed space in which the ether is moving. Then (3) become

$$\text{H}=(w-q)c\text{E}, \qquad \text{E}=(w-q)\mu\text{H}, \tag{4}$$

whence

$$\frac{\text{E}}{\text{H}}=\frac{1}{c(q-w)}=\mu(q-w), \tag{5}$$

which give $(w-q^{\,2}=v^2$, and $\text{E}^2=(\mu v\text{H})^2$, if $\mu cv^2=1$. This v is the wave speed when $q=0$. Two important quantities are determined, the ratio E/H and the speed w. In a positive wave $\text{E}_1=\mu v\text{H}_1$, and $w_1=v+q$. In a negative wave $\text{E}_2=-\mu v\text{H}_2$, and $w_2=v-q$. It follows that

$$\text{E}=\tfrac{1}{2}\epsilon^{-w_1t\Delta}(\text{E}_0+\mu v\text{H}_0)+\tfrac{1}{2}\epsilon^{w_2t\Delta}(\text{E}_0-\mu v\text{H}_0 \tag{6}$$

shows how an initial state of E_0, H_0 splits into two waves travelling in opposite directions. It is easy to see that the electric energy density=magnetic, in either wave, say U=T, and that the value of U or T is the same in the two waves. Also that when they are superposed, the sum of the energies in the two separate waves =sum of the energies of E_0 and H_0. There is no gain or loss of energy when the waves cross one another. The impedance μv is independent of q.

We see that everything goes on in the same way as if the uniformly moving ether were at rest. This is merely an example of *the* principle of relativity of motion. The motion of the ether merely carries forward the phenomena going on in it.

(2). Now let there be two interpenetrating media, matter and ether; the ether to be at rest, the matter to move through it at speed w. Let $\text{D}=\text{D}_0+\text{D}_1$, where $\text{D}_0=c_0\text{E}$ for the ether, and $\text{D}_1=c_1\text{E}_1$ for the matter, which is to be unmagnetisable in the first place. D_1 is the electric polarisation. E_1 need not be the same as E. The circuital equations are now (for a plane wave)

$$-\Delta\text{H}=p\text{D}_0+p'\text{D}_1, \qquad -\Delta\text{E}=p\text{B}, \tag{7}$$

where the new p is the moving time differentiator, or $p' = p + u\Delta$, it being assumed that u is uniform.

If we assume that E_1 is the same as E, then the polarisation varies as the etherial electric force, and so

$$-\Delta H = (cp + uc_1\Delta)E, \qquad -\Delta E = \mu p H; \tag{8}$$

which lead to

$$\Delta^2 = \mu p(cp + c\,u\Delta), \tag{9}$$

$$\frac{p}{\Delta} = -\frac{c_1 u}{2c} \pm \sqrt{v^2 + \left(\frac{c_1 u}{2c}\right)^2}. \tag{10}$$

Putting $p = -w\Delta$, there are two values of w, possessing the property $w_1 w_2 = v^2$. Note that $c = c_0 + c_1$, and $\mu c v^2 = 1$ defines v. The etherial speed by itself is the v_0 in $\mu c_0 v_0^2 = 1$. As before shown, this system does not agree with optical results. The fault is in making E_1 be the same as E.

(3). That the polarisation of the matter consists effectively in the separation of positive and negative charges is an old idea, independent of electronic hypotheses. But a change in this polarisation constitutes electric current, of the convectional kind. Then we must remember that the moving force on a charge ρ moving at speed u is not $\mathbf{E}\rho$ in the field $\mathbf{E}$, but is $(\mathbf{E} + \mathrm{V}\mathbf{uB})\rho$. In the present case, this defines $\mathbf{E}_1$ to be $\mathbf{E} + \mathrm{V}\mathbf{uB}$. That is,

$$\begin{aligned} \text{curl}\,(\mathbf{H} - \mathrm{V}\mathbf{D}_1\mathbf{u}) &= \dot{\mathbf{D}}, \\ -\text{curl}\,\mathbf{E} &= \dot{\mathbf{B}}, \end{aligned} \qquad \mathbf{D}_1 = c_1\mathbf{E}_1 = c_1(\mathbf{E} + \mathrm{V}\mathbf{uB}). \tag{11}$$

Or, in a plane wave,

$$-\Delta H = c_0 pE + (p + u\Delta)c_1(E - u\mu H), \qquad -\Delta E = \mu p H. \tag{12}$$

The assumption $p = -w\Delta$ now leads to the quadratic

$$1 + u\mu c_1(w - u) = \mu w(wc - uc_1), \tag{13}$$

making

$$w = \frac{c_1 u}{c} \pm \sqrt{v^2 - \frac{c_0 c_1}{c^2}u^2} = \frac{uv^2}{v_1^2} \pm \sqrt{v^2 - u^2 v^4 / v_0^2 v_1^2}. \tag{14}$$

We also find that $E_1 = \mu w_1 H_1$ in a positive wave, and $E_2 = -\mu w_2 H_2$ in a negative wave. This new w does agree with a result of Fresnel, as shown by Lorentz.

Since w_1 and w_2 are the speeds through the ether, the wave speeds through the matter are

$$w_1 - u = \sqrt{v^2 - c_0 c_1 u^2 / c^2} - c_0 u / c, \tag{15}$$

$$w_2 + u = \quad \ldots\ldots\ldots + \ldots\ldots \tag{16}$$

The wave speed through the matter is greater against the motion than with it.

(4). If an initial state is E_0 H_0, then at time t

$$E = \epsilon^{-w_1 t\Delta} \frac{w_1}{w_1 + w_2}(E_0 + \mu w_2 H_0) + \epsilon^{w_2 t\Delta} \frac{w_2}{w_1 + w_2}(E_0 - \mu w_1 H_0), \tag{17}$$

$$H = \frac{E_1}{\mu w_1} - \frac{E_2}{\mu w_2} \tag{18}$$

show E and H as the moment t, the exponential factors being the space shifters.

It is easy to see that the above form of solution is not confined so the special values of w_1, w_2 there concerned. For if

$$H=\epsilon^{-w_1t\Delta}A+\epsilon^{w_2t\Delta}B=H_1+H_2 \tag{19}$$

is given, and also that $-\Delta E=\mu pH$, then it follows that

$$E=\mu w_1H_1-\mu w_2H_2. \tag{20}$$

What determines the validity of (17), (18) is not the precise form of the first circuital law as modified, but that there is no change made in the second circuital law.

(5). A striking case is got by raising u to v_1, the speed given by $\mu c_1v_1^2=1$. Then

$$w_1=2v^2/v_1, \qquad w_2=0, \qquad w_1-u=2v^2/v_1-v_1, \qquad w_2+u=v_1. \tag{21}$$

Here the negative wave stands still in the ether, or moves backward in the matter at speed v_1. But inspection of (17) shows that $E_2=0$, or $E_1=E_0$ with a positive shift. Also that $H_1=E_1/\mu w_1$ and $H_2=-(E_0-\mu w_1H_0)/\mu w_1=H_0-E_1/\mu w_1$ So we see that the full E_0 goes off positively, taking with it as much of H_0 as will make a positive wave; whilst the rest of H_0, without any of E_0, stands still in the ether. We may verify this simply by making the initial state be H_0 only.

(6). Another striking case is $u=v_0v_1/v$. This destroys the radical, so that

$$w_1=vv_0/v_1=-w_2, \qquad w_1-u=v_0(v/v_1-v_1/v)=-(w_2+u). \tag{22}$$

Here the negativity of the negative wave speed means that the wave travels positively. So (22) alone shows that the positive and negative waves, if existent, travel positively together without any splitting. And, in fact, (17), (18) show that $E=E_0$ with positive shift, and $H=H_0$ with the same shift, whatever E_0 and H_0 may be. So it is really a case of an undistorted electric wave without any H, or of an undistorted magnetic wave, without any E, or of any combination of the two, travelling at speed vv_0/v_1 through the ether, in the same direction as the matter, whose speed is v_0v_1/v.

Say $c_0=9$, $c_1=16$, then $c=25$. Also, $v_0=\frac{1}{3}$, $v_1=\frac{1}{4}$, $v=\frac{1}{5}$. These make $u=\frac{5}{12}$, $w_1=\frac{4}{15}$, $w_1-u=-\frac{3}{20}$, which are the speed of the matter, and the wave speed through the ether and through the matter.

(7). If the matter is magnetizable as well as electrizable, and we apply the same principles of modification to the second as was done to the first circuital law (to be done in three dimensions later), then the equations for plane waves are

$$-\Delta H=c_0pE+c_1(p+u\Delta)(E-u\mu_0H), \tag{23}$$

$$-\Delta E=\mu_0pH+\mu_1(p+u\Delta)(H-uc_0E), \tag{24}$$

where $\mu=\mu_0+\mu_1$, $B_0=\mu_0 H$ for the ether, $B_1=\mu_1(H-uc_0E)$ for the matter; $B=B_0+B_1$ is the induction, B_1 the polarisation.

Putting $p=-w\Delta$ as before, we get

$$-H=-c_0wE+c_1(-w+u)(E-u\mu_0H), \tag{25}$$

$$-E=-\mu_0wH+\mu_1(-w+u)(H-uc_0E); \tag{26}$$

and these lead to

$$\frac{E}{H}=\frac{1+c_1u\mu_0(w-u)}{c_0w+c_1(w-u)}=\frac{\mu_0w+\mu_1(w-u)}{1+\mu_1c_0u(w-u)}, \tag{27}$$

giving the two values of w, and the corresponding values of E/H. The quadratic for w is

$$1+(w-u)^2(\mu_0c_0\mu_1c_1u^2-\mu c+\mu_0c_0)=\mu_0c_0w^2, \tag{28}$$

making

$$w=u\left(1-\frac{\mu c_0}{\mu c-\mu_0c_0\mu_1c_1u^2}\right)\pm\frac{\sqrt{X}}{\mu c-\mu_0c_0\mu_1c_1u^2}, \tag{29}$$

$$X=(1/u^2-\mu_0c_0)(\mu c-\mu_0c_0\mu_1c_1u^2)+\mu_0^2c_0^2. \tag{30}$$

The first approximation is $w_1=v+u(1-v^2/v_0^2)$.

If either or both $c_0=0$, $\mu_0=0$, then $(w-u)^2=v^2$. Here $v_0=\infty$, and $v=(\mu c)^{-\frac{1}{2}}=(\mu_1c)^{-\frac{1}{2}}$ or else $(c_1\mu)^{-\frac{1}{2}}$

More practically, we have U=T and $E=wB$ in a solitary wave (either way, with the proper value of w), when either $\mu_1=0$, or else $c_1=0$. Thus, the values of U and T are

$$U=\tfrac{1}{2}c_0E^2+\tfrac{1}{2}c_1(E-uB_0)^2, \qquad T=\tfrac{1}{2}\mu_0H^2+\tfrac{1}{2}\mu_1(H-uD_0)^2. \tag{31}$$

Taking $\mu_1=0$ in these, the result is

$$U/T=\mu_0[c_0w^2+c_1(w-u)^2], \tag{32}$$

and this reduces, by (27), to U/T=1 simply. Similarly, when $c_1=0$. But this equality of the energies in a solitary wave does not mean in general that the energy densities of two oppositely travelling waves are equal, even though $E_1=-E_2$, or $H_1=-H_2$.

The arrangement is like two electric condensers put in parallel. Both are exposed to the force E, but one of them in addition contains the force $-uB_0$. Similarly as regards H and uD_0 applied to two magnetic condensers in parallel.

(8). Now, keeping to an unmagnetizable moving matter, consider what happens when two waves travel in opposite directions and cross one another. First, let an initial state be $E=E_0$ simply, and no H. The most convenient arrangement is a slab of unit depth. The two resulting waves are, by (17), (18),

$$E_1=\mu w_1H_1=\frac{w_1}{w_1+w_2}E_0, \qquad E_2=-\mu w_2H_2=\frac{w_2}{w_1+w_2}E_0. \tag{33}$$

If these waves are going towards one another, they will at a certain moment coincide. Then $E=E_0$, $H=0$. This state, a slab containing a uniform E_0, but without any H_0, may be regarded as an initial state. It immediately splits into the two slabs separating from one another. The equations serve equally well when they are advancing towards one another. It is not necessary to write the space shifters.

Let U_1 be the density of the electric energy in the positive wave. Its amount is

$$U_1=\tfrac{1}{2}c_0E_1^2+\tfrac{1}{2}c_1(E_1-uB_1)^2$$
$$=\tfrac{1}{2}c_0\frac{w_1^2}{(w_1+w_2)^2}E_0^2+\tfrac{1}{2}c_1\frac{(w_1-u)^2}{(w_1+w_2)^2}E_0^2. \tag{34}$$

Putting in the values of w_1 and w_2, this reduces to

$$U_1=\frac{\frac{1}{2}E_0^2}{\mu(w_1+w_2)^2}=\frac{cE_0^2}{8}\,\frac{v^2}{v^2-c_0c_1u^2/c}. \tag{35}$$

As already shown, this is also the value of T_1, and the form of (35) shows that it is the value of U_2, and of T_2. The total energy density of the waves when separate is $4U_1$. But when they coincide, the mutual energy must also be counted. This is

$$U_{12}=c_0E_1E_2+c_1(E_1-uB_1)(E_2-uB_2)=\frac{cE_0^2}{4}\,\frac{v^2-2c_0c_1u^2/c^2}{v^2-c_0c_1u^2/c^2}, \tag{36}$$

making altogether

$$U_1+U_2+U_{12}=\tfrac{1}{2}c_0E_0^2=U_0. \tag{37}$$

Comparing with (35) and (36), we see that the excess of the wave energy over the energy when the waves coincide is

$$\text{Loss}=U_0-2U_{12}=\frac{U_0c_0c_1u^2/c^2}{v^2-u^2c_0c_1/c^2}. \tag{A}$$

In old days, say 40 or 50 years ago, it might have been said that the theory was disproved by this result, because it was shown to violate the grand Principle of the Conservation of Energy and its Scholium. That way of talking has gone out of fashion. It is not done now in any respectable book. A better understanding of the grand Principle, as contained in Newton's Third Law, has come about. The Principle cannot be violated. It conserves itself automatically in any self-consistent dynamical connections. The question nowadays is, How are we to account for the energy lost or gained? What force is concerned? It must be implicitly contained in the connections already given, though it is not yet displayed. Reserving details, it may be said at once that the force concerned is that between the electromagnetic field and the matter. The ether rests, so there can be no loss of energy in it. The matter moves, so any mechanical force brought into being may cause a loss or gain of energy.

(9). Do the magnetic case similarly. Let there be an initial state H_0. Then

$$H_1=\frac{w_2H_0}{w_1+w_2},\quad H_2=\frac{w_1H_0}{w_1+w_2},\quad E_1=\mu w_1H_1,\quad E_2=-\mu w_2H_2. \tag{38}$$

Here note that H_1 and H_2 are unequal, although $E_1=-E_2$. Also,

$$T_1=T_0\frac{w_2^2}{(w_1+w_2)^2},\qquad T_2=T_0\frac{w_1^2}{(w_1+w_2)^2},\qquad T_0=\tfrac{1}{2}\mu H_0^2, \tag{39}$$

where note that T_1 and T_2 are unequal.

As regards the electric energy, since

$$E_1-uB_1=(w_1-u)B_1,\qquad E_2-uB_2=-(w_2+u)B_2, \tag{40}$$

the value of U_1 is

$$U_1=\tfrac{1}{2}c_0E_1^2+\tfrac{1}{2}c_1(w_1-u)^2B_1^2=\tfrac{1}{2}B_1^2/\mu=T_1; \tag{41}$$

and similarly, $U_2=T_2$. The total wave energy is therefore

$$2(T_1+T_2)=\frac{w_1^2+w_2^2}{(w_1+w_2)^2}2T_0, \tag{42}$$

and the excess of the wave energy over T_0 is

$$T_0\left(\frac{w_1-w_2}{w_1+w_2}\right)^2=T_0\frac{c_1^2u^2/c^2}{v^2-c_0c_1u^2/c^2}, \tag{43}$$

which is positive, and as before indicates a loss of energy in uniting.

But there is another consideration. Although $E_1=-E_2$, which cancel when coincident, there is still electric energy during coincidence, namely $U_0=\frac{1}{2}c_1(uB_0)^2$. This could easily be overlooked. The energy when coincident is U_0+T_0. This is also the energy of the initial state producing the two waves. We must therefore deduct U_0 from (43) to find the loss of energy in transit. We have $U_0=\mu c_1u^2T_0$; and the real loss is

$$\text{Loss}=2(T_1+T_2)-T_0-U_0=-T_0\frac{\mu u^2(c_0c_1/c)\,(v^2-u^2c_1/c)}{v^2-u^2c_0c_1/c^2}. \tag{B}$$

Taking the slabs to be of unit depth does away with a lot of needless symbolism.

Comparing (B) with (A), we see another difference. (A) is positive, whilst (B) is negative. The same mechanical force must account for both results. Also remember that the energy lost on union equals the energy gained in separation, so that there is no permanent alteration of energy according to the data assumed.

(10). Formulas (A) and (B) are both contained in

$$\text{Loss}=-2(c_0c_1u^2/c)B_1B_2. \tag{C}$$

For, $B_1=-B_2$ reduces (C) to (A), and $E_1=-E_2$ reduces (C) to (B). Verify this. Space is too valuable for all the details.

Now the time taken by the waves in crossing is $2(w_1+w_2)^{-1}$, and the speed of motion is u. Therefore, since (C)=Force×time×speed, the resultant force on the matter between first touch and coincidence is

$$(C)\times\frac{w_1+w_2}{u}=-\frac{c_0c_1}{c}\,\frac{2u}{w_1+w_2}B_1B_2. \tag{44}$$

But this by no means represents the real force either in time or space.

(11). Only a simplified form of the general activity equation is needed, when plane disturbances are in question, to find the mechanical force. If W is the flux of energy,

$$-W' = \dot{U} + \dot{T} + Fu \tag{45}$$

is the activity equation. The convergence of **W** is the rate of loss of energy per unit volume, here employed in increasing the stored energy, and in working upon the matter. Let the polarisation be now called I, to distinguish it from the D_2 in a negative wave. Thus, $I = c_1(E - uB)$. Then we have

$$-H' = \dot{D} + (Iu)', \qquad -E' = \dot{B}. \tag{46}$$

Multiply the first by E, and the second by H, and add. Then

$$-(EH)' = \dot{U} + \dot{T} + \dot{U}_{Ic_1} - e\dot{I} + E(Iu)', \tag{47}$$

where $\dot{U}_{Ic_1}$ is the rate of increase as dependent upon variation of c_1, since c_0 is here constant, and $e = -uB$. Variability of u will also be allowed for in the first place. But the complete flux of energy consists of the electromagnetic EH, the convective flux $U_I u$, and the stress flux, which is also $U_I u$. So add $-(2U_I u)'$ to both sides of (47). Then

$$-W' = \dot{U} + \dot{T} + \dot{U}_{Ic_1} - e\dot{I} + E(Iu)' - (c_1^{-1} I^2 u)', \tag{48}$$

which easily reduces to

$$-W' = \dot{U} + \dot{T} - eJ - uU'_I + \delta U_c/\delta t, \tag{49}$$

where $J = \dot{I} + (Iu)'$ = current density in the matter, with allowance made for its motion. Also $\delta/\delta t = p + u\Delta$, the moving time differentiation. Now the last term in (49) is zero, obviously. Therefore

$$F = -eJ/u - U'_I \tag{50}$$

is the force required, the force acting on the matter. A more convenient form is got by putting $E + e$ for I/c_1, and $\dot{B}$ for $-E'$, in

$$F = B[\dot{I} + (Iu)'] - u[(I/c_1)'I + U'_{Ic_1}], \tag{51}$$

which is merely an expansion of (50). Then (51) reduces to

$$F = (\delta/\delta t + 2u')BI - U'_{Ic_1}. \tag{52}$$

So far u and c_1 may vary with x. Let them be constant, then we get

$$F = \frac{\delta}{\delta t} BI = B\frac{\delta I}{\delta t} - U'_I. \tag{53}$$

Of the two forms in (53) the first is the simpler, but the second may be more useful. It is equivalent to (50). Mark the difference between U_{Ic_1} and U_I.

(12). Now apply (53) to a solitary wave, a slab. We have

$$\delta/\delta t = -(w_1-u)d/dx \text{ in a positive wave,}$$

or

$$= (w_2+u)d/dx \quad \text{in a negative wave.}$$

Also

$$\mathrm{I}_1 = c_1(w_1-u)\mathrm{B}_1, \qquad \mathrm{I}_2 = -c_1(w_2+u)\mathrm{B}_2. \tag{54}$$

Further, Δ is negative at the front and positive at the back of a positive wave, whilst it is negative at the back and positive at the front of a negative wave. It follows that there is no bodily force in either wave when solitary, but only a surface force at its front and back, which is always outward from the wave to the undisturbed region, whether at front or back. This terminal force is of size I_1^2/c_1 in the positive wave, and I_2^2/c_1 in the negative wave. There is no change in the energy, because of the opposite directions of the two pulls on a wave. In reality this would somewhat distort the matter, but that is prevented here by the constancy of u.

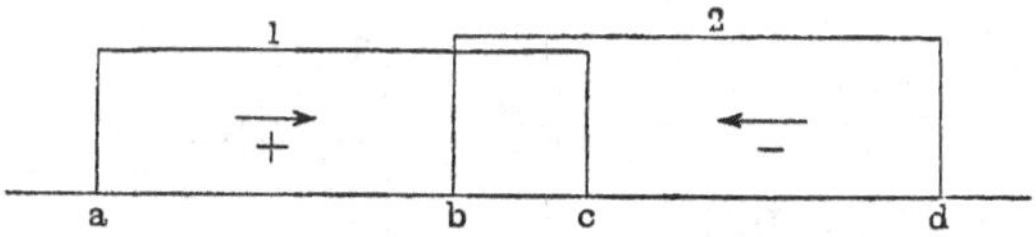

But when the waves cross, the forces at b and c do not cancel. Thus, at b, the front of the negative wave, IB jumps from $\mathrm{I}_1\mathrm{B}_1$ to IB, increasing by the amount $\mathrm{I}_2\mathrm{B}_2+\mathrm{I}_2\mathrm{B}_1+\mathrm{I}_1\mathrm{B}_2$. By (53), (54) the

$\mathrm{I}_2\mathrm{B}_2$ increment makes $(w_2+u)\,\mathrm{I}_2\mathrm{B}_2 = -\mathrm{I}_2^2/c_1$,
$\mathrm{I}_2\mathrm{B}_1$ " " $(w_2+u)\,\mathrm{I}_2\mathrm{B}_1 = \mathrm{I}_1\mathrm{I}_2\,(w_2+u)/c_1(w_1-u)$,
$\mathrm{I}_1\mathrm{B}_2$ " " $(w_2+u)\,\mathrm{I}_1\mathrm{B}_2 = -\mathrm{I}_1\mathrm{I}_2/c$,

their sum is the force (left to right) at b per unit area.

Similarly, at c, there is a jump from $\mathrm{I}_2\mathrm{B}_2$ to IB, and the

$\mathrm{I}_1\mathrm{B}_1$ increment makes I_1^2/c_1.
$\mathrm{I}_2\mathrm{B}_1$ " " $\mathrm{I}_1\mathrm{I}_2/c_1$,
$\mathrm{I}_1\mathrm{B}_2$ " " $-(\mathrm{I}_1\mathrm{I}_2/c_1)(w_1-u)/(w_2+u)$,

and their sum is the force (left to right) at c.

But the $\mathrm{I}_2\mathrm{B}_2$ term at b is cancelled by its negative at d, and the $\mathrm{I}_1\mathrm{B}_1$ term at c by its negative at a. The total force is therefore

$$\mathrm{F} = \frac{\mathrm{I}_1\mathrm{I}_2}{c_1}\left(\frac{w_2+u}{w_1-u} - \frac{w_1-u}{w_2+u}\right), \text{ where } \begin{matrix} w_1 \\ w_2 \end{matrix} = \sqrt{v^2 - c_0c_1u^2/c^2} \pm c_1u/c, \tag{55}$$

which makes

$$\mathrm{F} = -c_1\mathrm{B}_1\mathrm{B}_2 \times 4(c_0u/c)\sqrt{\ldots} = -c_1\mathrm{B}_1\mathrm{B}_2 \times (4uc_0/c)\sqrt{\ldots}, \tag{56}$$

which is the complete force acting from the moment of first meeting up to the moment of coincidence of the waves. The distance through which it works $=$ velocity $\times$ time $= u/(w_1+w_2) = u/2\sqrt{\ldots}$.

Multiply (56) by this, and we get

$$\text{Loss} = -2(c_0 c_1/c)u^2 B_1 B_2. \qquad \text{(CC)}$$

This is the proof of (C), and therefore the results (A) and (B) are reconciled. But directly the waves begin to emerge, the total force is reversed in direction, so the energy lost by the waves is restored to them in the same manner, and after emergence they are in the same state as before they met. It may be remarked that the initial calculation of the loss of energy from the waves was much simpler than the later work. On the other hand, it did not tell us the distribution of the forces acting, but only the mean value of the total.

(13). The crossing of waves in the above took place in the interior of a moving dielectric. It did not, therefore, involve any reflection. Reflection occurs when a wave passes from one medium to another. The transmitted wave differs from the incident, and there is in addition a reflected wave. In the figure XX is the inter

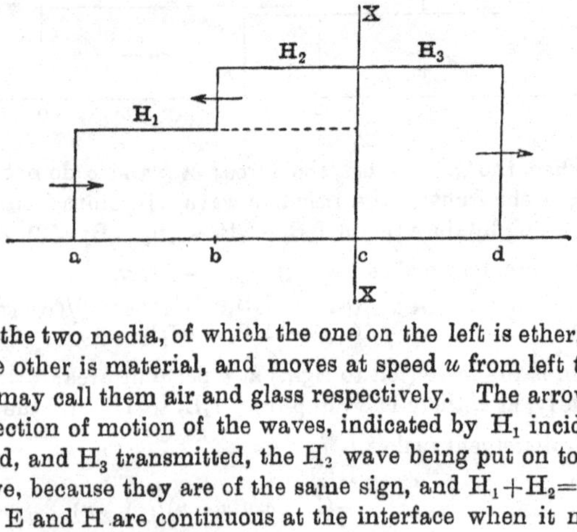

face of the two media, of which the one on the left is ether, at rest, and the other is material, and moves at speed u from left to right. Or, we may call them air and glass respectively. The arrows show the direction of motion of the waves, indicated by H_1 incident, H_2 reflected, and H_3 transmitted, the H_2 wave being put on top of the H_1 wave, because they are of the same sign, and $H_1+H_2=H_3$.

Now E and H are continuous at the interface when it moves as well as when it rests, therefore

$$E = E_1 + E_2 = E_3, \qquad H = H_1 + H_2 = H_3, \qquad (57)$$

$$E_1 = \mu v_0 H_1, \qquad -E_2 = \mu v_0 H_2, \qquad E_3 = \mu w H_3, \qquad (58)$$

where v_0 belongs to the ether, and w is the wave speed in the glass. Let

$$P_1 = 2T_1, \qquad P_2 = 2T_2, \qquad P_3 = 2T_3, \qquad (59)$$

be the longitudinal pressures in the incident, reflected, and transmitted waves. We know that $U_1=T_1$, $U_2=T_2$ in the ether, and also $T_3=U_3$ in Lorentz's modified form of theory of a moving dielectric, wherein the polarisation I_3 is given by $I_3=c_1(E_3-uB_3)$, where $c_1=c-c_0$, the excess of the permittivity of the glass over that of the ether.

It follows from (57) (58) that

$$\frac{H_2}{H_1}=\frac{v_0-w}{v_0+w}, \qquad \frac{H_3}{H_1}=\frac{2v_0}{v_0+w}. \tag{60}$$

The rate of loss of energy from the waves, by consideration of their speeds, and lengths, is

$$\begin{aligned}\text{Waste}&=P_1(v_0-u)-P_2(v_0+u)-P_3(w-u)\\&=(P_1-P_2)v_0-P_3w+(P_3-P_1-P_2)u.\end{aligned} \tag{61}$$

But here it may be shown that $(P_1-P_2)v_0=P_3w$, meaning the same as $EH=E_3H_3$, so we have

$$\text{Waste}=(P_3-P_1-P_2)u. \tag{62}$$

A remarkable thing about this formula is that it is in form exactly the negative of what we would naturally expect from the pressures at the interface, that is, the resultant force $P_3-P_1-P_2$ in the positive direction indicates that it acts in the direction of the rise of pressure. But, actually, (62) does not show the distribution of the force. There is a force at the wave front of the H_3 wave, and another at the interface, and (62) shows the resultant.

Putting P_3 and P_2 in terms of P_1 we can get various forms of expression. Thus,

$$\begin{aligned}\text{Waste}&=2P_1u\frac{v_0-w}{v_0+w}=2P_1u\frac{H_2}{H_1}=2\mu uH_1H_2\\&=\tfrac{1}{2}P_3u\left(1-\frac{w^2}{v_0^2}\right)=(\tfrac{1}{2}I_3^2/c_1)u.\end{aligned} \tag{63}$$

The last form is noteworthy. The only convection of energy is that of the polarisation (electric), and the waste of energy from the wave=convection current of polarisation energy. The first form gives the waste in terms of the incident energy. Then the only trouble is with w. Its formula is

$$w=\sqrt{v^2-u^2c_0c_1/c^2}+uc_1/c. \tag{64}$$

The distribution of force will be considered in the next section.

(14). Consider the passage of a pulse from one medium to another, when both are polarisable, though to different extents, and both move together through the ether at the same speed. The same figure will do for reference, but we must modify the notation somewhat. Thus,

$$H_1+H_2=H_3, \qquad w_1H_1-w_2H_2=w_3H_3, \tag{65}$$

express the continuity of H and E at the wave-front, combined with $E_1=\mu w_1H_1$, $E_2=-\mu w_2H_2$ in the left medium, and $E_3=\mu w_3H_3$ in the right medium.

The rate of loss of energy from the waves may be estimated in the same way as before, with one change ; namely, that in the region from b to c which is occupied by the incident and reflected, the sum of the mutual electric and mutual magnetic energy of the two waves is no longer zero, as it was when the left medium was unpolarisable. The waste is now

$$\text{Waste}=(w_1+w_2)P_1-(w_2+u)P_2-(w_3-u)P_3, \tag{66}$$

where $P_1=U_1+T_1=2T_1, \qquad P_3=U_3+T_3=2T_3,$

$$P_2=(U_1+U_2+U_{12})_e+(U_1+U_2+U_{12})_m+(T_1+T_2+T_{12}). \tag{67}$$

The quantity with the suffix e is the electric energy of the ether, and with the suffix m that of the matter. Equation (66) is now reduced to

$$\text{Waste}=2T_1(w_1-u)-2T_2(w_2+u)-2T_3(w_3-u)$$
$$-(w_2+u)\,(U_{12}+T_{12}), \tag{68}$$

and here the value of the mutual terms is

$$T_{12}+U_{12}=\mu H_1H_2+c_0E_1E_2+c_1(E_1-uB_1)\,(E_2-uB_2), \tag{69}$$

which reduces to

$$T_{12}+U_{12}=\mu H_1H_2\times 2\mu u^2c_0c_1/c. \tag{70}$$

The next question is the value of $2T_1w_1-2T_2w_2-2T_3w_3$, which was zero when the left medium was unpolarisable. Examination shows that the value is now

$$\frac{P_1(w_1-w_2)\,(w_3-w_1)}{w_2+w_3}=-\mu H_1H_2(w_1-w_2). \tag{71}$$

Lastly, the remaining terms in (68) are

$$-(2T_1+2T_2-2T_3)u=2\mu H_1H_2u=2P_1u\frac{w_1-w_3}{w_2+w_3}. \tag{72}$$

Adding together (72), (71) and (70) multiplied by $-(w_2+u)$, the result is

$$\text{Waste}=2T_{12}(uc_0/c)\,[1-\mu c_1u(w_2+u)] \tag{73}$$
$$=2T_{12}(uc_0/c)\,[1-\mu c_0c_1u^2/c-\tfrac{1}{2}\mu c_1u(w_1+w_2)], \tag{74}$$

where $T_{12}=\mu H_1H_2$, and the three c's and two w's refer to the left medium. The right medium comes in through T_{12} involving the ratio H_1/H_2 given above, which contains w_3.

(15). If we make $c_1=0$, we come to Waste$=2T_{12}u$, as before found, equation (63). We can now confirm (74) by examination of the distribution of force on the matter, and this will of course cover the simpler case. Use the force formula (52), or

$$F=(\delta/\delta t)IB-\tfrac{1}{2}I^2(dc_1/dx), \tag{75}$$

which is valid because u does not vary with x. Here I is the polarisation, and c_1 the excess permittivity. The only place where c_1 varies is the interface.

The force at (a) in the figure is

$$-(w_1-u)I_1B_1=-I_1^2/c_1, \quad \text{because} \quad I_1=(w_1-u)B_1c_1. \tag{76}$$

Similarly, the force at (d) is $+I_3^2/c_{13}$, where c_{13} is the right hand c_1. At (c) the corresponding terms are zero. But the variation of c_1 gives the force

$$\tfrac{1}{2}c_1(E-uB)^2-\tfrac{1}{2}c_{13}(E_3-uB_3)^2=(c_1/c_{13}-1)\tfrac{1}{2}I_3^2/c_{13}. \tag{77}$$

Lastly, there is (b), where there is a jump from I_1B_1 to IB. Here $\delta/\delta t=(w_2+u)d/dx$. So the I_2B_2 part of the jump gives

$$(w_2+u)I_2B_2=-I_2^2/c_1. \qquad (78)$$

The I_2B_1 part gives

$$(w_2+u)I_2B_1=(w_2+u)I_1I_2/[c_1(w_1-u)]. \qquad (79)$$

The I_1B_2 part gives

$$-(w_2+u)I_1I_2/[c_1(w_2+u)]=-I_1I_2/c_1. \qquad (80)$$

Collecting all these partial forces, we get

$$-2U_1+2U_3+U_1+U_2+U_{12}-U_3-2U_2$$
$$+U_{12}(w_2+u)/(w_1-u)-U_{12}, \qquad (81)$$

on the temporary understanding that these U's are all polarisation energies,

$$=-U_1-U_2+U_3+U_{12}(w_2+u)(w_1-u)^{-1}. \qquad (82)$$

This may be reduced to

$$T_{12}[1+\mu c_0w_1w_2-\mu c_1(w_2+u)^2] \qquad (83)$$
$$=T_{12}(2c_0/c)[1-\mu u^2c_0c_1/c-\tfrac{1}{2}\mu uc_1(w_1+w_2)]. \qquad (84)$$

This being the resultant force, agrees with the formula (74) for the waste. The reductions were rather troublesome, but the tribulation is at an end.

We can check by the alternative formula (53), or, with u constant,

$$F=B\delta I/\delta t-(d/dx)U_I. \qquad (85)$$

For the second term goes out by summation, so that the total force is $\Sigma\, B(\delta I/\delta t)$. The term at ($a$) makes $-I_1^2/2c_1$; and that at (d) makes $+I_3^2/2c_{13}$. But there is nothing at the interface, whilst at (b), B jumps from B_1 to B_1+B_2, and $\delta I/\delta t=(w_2+u)dI/dx$, where I jumps from I_1 to I_1+I_2. Now

$$(w_2+u)\int\Big(B_1+B_2\frac{x}{l}\Big)I_2dx=(w_2+u)(B_1I_2+\tfrac{1}{2}B_2I_2)$$
$$=\frac{I_1I_2}{c_1}\frac{w_2+u}{w_1-u}-\frac{I_2^2}{2c_1}; \qquad (86)$$

so the total comes to

$$-I_1^2/2c_1+I_3^2/2c_{13}-I_2^2/2c_1+(I_1I_2/c_1)(w_2+u)(w_1-u)^{-1}. \qquad (87)$$

Compare with (81). They agree. Remember, however, that the distribution of force is not correctly given by the quantity summed up here. Only the total is correct.

If we put $c_1=0$ and destroy the polarisation on the left side, the matter on the left side becomes electrically non-existent; so, whether it moves with the polarisable matter on the right side or not is immaterial. This is, for the man in the street, the explanation why $u=0$ on the left side along with $c_1=0$, gives the same results as $u=$constant on both sides.

(16). A rather striking property appears when a plane pulse is sent through a moving dielectric, according to the above specification, which is suitable for waves which are not too short. Let the "glass" be of finite depth, and the surrounding dielectric be ether on both sides. The to and fro reflections within the glass, combined with partial transmission at each boundary, make the incident pulse be the source of two trains of pulses outside the glass, one train transmitted in the further medium, the other reflected back into the first medium. If the depth of the pulse is less than the depth of the glass, the pulses in the two trains are separate from one another. The intensity of the pulses in either train decreases in geometrical progression. Now the striking thing is that if all the transmitted pulses were superposed, the result would be the original incident pulse. On the other hand, if the pulses in the train on the other side were superposed, the result would be zero. I have some vague recollection that this property is known when the glass is at rest. It is also true when the glass moves, as may be easily shown.

Let r be the coefficient of transmission of H from air to glass on the left side; s the coefficient of reflection from glass to glass on the right side; σ the coefficient of reflection from glass to glass on the left side, τ the coefficient of transmission from glass to air on the left side; ρ the coefficient of transmission from glass to air on the right side; and q the coefficient of reflection from air to air on the left side. Here "air" represents ether, and the incident pulse comes from the left side. The wave speeds in the glass are w_1 to the right, and w_2 to the left.

The fraction of the incident H_1 which is transmitted is

$$r\rho+rs\sigma\rho+rs\sigma s\sigma\rho+\ldots=r\rho(1-s\sigma)^{-1}; \tag{88}$$

and the fraction of the incident H_1 which is turned back is

$$q+rs\tau+rs\sigma s\tau+rs\sigma s\sigma s\tau+\ldots=q+rs\tau(1-s\sigma)^{-1}. \tag{89}$$

The six coefficients are to be found by the continuity of H and E at the two interfaces, together with $E=\mu w_1 H$ or $-\mu w_2 H$ inside the glass, and $E=\pm\mu v_0 H$ in the air. Thus,

$$r=\frac{2v_0}{v_0+w_1}, \qquad q=\frac{v_0-w_1}{v_0+w_1}, \qquad s=-\frac{v_0-w_1}{v_0+w_2}, \tag{90}$$

$$\sigma=-\frac{v_0-w_2}{v_0+w_1}, \qquad \tau=\frac{w_1+w_2}{v_0+w_1}, \qquad \rho=\frac{w_1+w_2}{v_0+w_2}. \tag{91}$$

From these values, the result in (88) comes to 1, and the result in (89) comes to 0, like life, as was remarked before. This proves the property described above.

The pulses in the glass of course attenuate to 0, but the two trains of pulses in the air persist until some change of conditions occurs. We may also use the above coefficients to find in what ratio the energy of the incident pulse divides between the two trains. We must sum up the squares of the terms in (88) and (89).

(17). The moving force exerted by radiation on an interposed plate of glass requires the consideration of the to and fro reflections in the plate itself, and this means a study of the complete solutions on the left and right sides. A synthetical construction is the most thorough and illuminating, but the same effect is reached, in explaining to others, by the shorter process of writing down the results obtained, and then examining their meaning, and testing that they satisfy all the conditions imposed. So let H_1 be the magnetic force in the radiation incident upon the left side of the glass, H_2 that in the reflected radiation, and H_3 in the transmitted radiation on the right side of the plate. Then first,

$$H_3=(1-m)\frac{\epsilon^{-pt_1}}{1-m\epsilon^{-pt_3}}H_1, \qquad t_1=\frac{x-l}{v}+\left(1-\frac{u}{v}\right)\frac{l}{w_1-u}, \tag{92}$$

$$t_2=\frac{l}{w_1-u}+\frac{l}{w_2+u}, \qquad t_3=\left(1-\frac{u}{v}\right)t_2, \qquad m=\frac{(v-w_1)(v-w_2)}{(v+w_1)(v+w_2)}, \tag{93}$$

give H_3 transmitted in terms of H_1 incident. Here l is the depth of the glass moving at speed u along x with the incident wave H_1, which is of the type $f(t-x/v)$. Also v is the wave speed in the air, and w_1, w_2, the two speeds in the glass, with and against the direction of motion of the incident wave; and p is the time differentiator.

To interpret, expand H_3.

$$H_3=(1-m)\,\epsilon^{-pt_1}[1+m\epsilon^{-pt_3}+m^2\epsilon^{-2pt_3}+m_3\epsilon^{-3pt_3}+\ldots]H_1. \tag{94}$$

Take H_1 to be a pulse which passes the origin of x at the moment $t=0$, when the left side of the glass is at distance y from the origin. Then (94) asserts that H_3 consists of an infinite series of pulses passing x on the right side of the glass at the moments t_1, t_1+t_3, t_1+2t_3, etc.; and of the sizes $1-m$, $m(1-m)$, $m^2(1-m)$, etc. This is correct, of course, provided that t_1 is the time of transit of a pulse from 0 to x direct, and that t_3 is the additional retardation due to one pair of internal reflections, $2t_3$ that due to two pairs, and so on; and provided the sizes are correct.

The sizes are verified by equations (91). It will be found that $1-m$ is the product $r\rho$, showing the attenuation due to the passage from air to glass on the left side, followed by glass to air on the right side, and that m itself is the product $s\sigma$, showing the attenuation due to reflection from glass to glass on the right side followed by glass to glass on the left side.

As regards t_1 and t_3, more consideration is required. Let A be the origin,

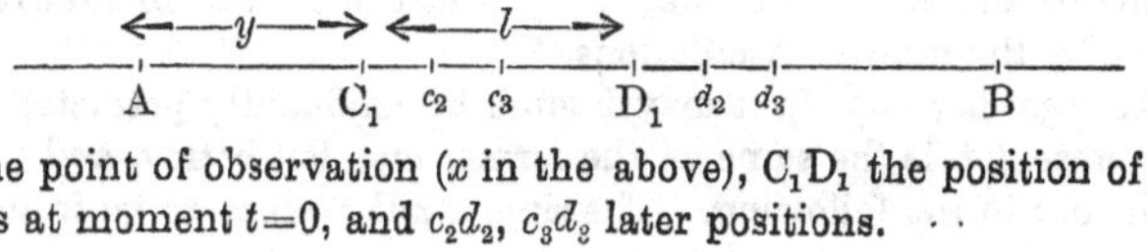

B the point of observation (x in the above), C_1D_1 the position of the glass at moment $t=0$, and c_2d_2, c_3d_3 later positions.

Now a pulse would take the time y/v to go from A to the glass if it were at rest, but its motion makes the pulse take the time $y/(v-u)$, and in this time the glass advances to c_2d_2, by moving forward the distance $yu/(v-u)$, when its distance from A becomes $yv/(v-u)$. The pulse then takes the time $l/(w_1-u)$ to traverse the glass, which has then the position c_3d_3, having moved through the distance $lu/(w_1-u)$. The time taken from A to the right side of the glass is therefore $y/(v-u)+l/(w_1-u)$, and the right side is then at distance $yv/(v-u)+lu/(w_1-u)$ from A. Finally, if x is the distance AB, we have

$$t_1=\frac{y}{v-u}+\frac{l}{w_1-u}+\frac{1}{v}\left[x-l-y-\frac{yu}{v-u}-\frac{lu}{w_1-u}\right]$$
$$=\frac{x-l}{v}+\frac{l}{w_1-u}\left(1-\frac{u}{v}\right). \qquad (95)$$

This settles t_1. But if there is a pair of internal reflections, the time becomes

$$\frac{y}{v-u}+\frac{l}{w_1-u}+\frac{l}{w_2+u}+\frac{l}{w_1-u}+\frac{1}{v}\left[x-l-y-u\left(\frac{y}{v-u}+\frac{2l}{w_1-u}+\frac{l}{w_2+u}\right)\right]$$
$$=\frac{x-l}{v}+\left(1-\frac{u}{v}\right)\left(\frac{l}{w_1-u}+t_2\right)=t_1+t_3, \qquad (96)$$

where t_3 is as defined in (93), and t_1 in (92). Similarly, if there are n_0 oscillations of the pulse in the glass, the time from A to B is $t_1+n_0t_3$. Thus (94) is fully proved, so long as B is on the right side of the glass. It is most convenient to have the origin always on the left side, so that y is positive, and the disturbance at the origin, *i.e.*, $H_1=f(t)$, commences after the glass passes the origin.

(18). The corresponding reflected wave is of the form $H_2=g(t+x/v)$, so we need only find H_{20}, the value of the reflected disturbance at the origin itself, in terms of H_{10}, the incident disturbance at the origin. Thus,

$$H_{20}=\frac{v-w_1}{v+w_1}\epsilon^{-pt_4}\frac{1-\epsilon^{-pt_3}}{1-m\epsilon^{-pt_3}}H_{10}, \qquad t_4=\frac{2z}{v+u}. \qquad (97)$$

Here t_3 is as before, and z in t_4 is the distance of the left side of the glass from the origin at the moment t. To prove, expand. Thus

$$H_{20}=\frac{v-w_1}{v+w_1}\epsilon^{-pt_4}\left[1-(1-m)\epsilon^{-pt_3}-m(1-m)\epsilon^{-2pt_3}-\dots\right]H_{10}, \qquad (98)$$

which means an infinite series of pulses, when H_1 is a pulse. The coefficients may be checked by equations (91). The first one is that of air to air on the left side, and the rest show the modifications made by the internal oscillations.

As regards t_3 and t_4, it may I think be confidently predicted that the present t_3 is the same as the former one, but both t_3 and t_4 will come out in the following. The time a pulse takes to go from the origin to the glass is $y(v-u)^{-1}$, and when it is at once reflected it takes the same time to go back to the origin; so $t_4=2y(v-u)^{-1}$. But in this time the glass has moved on the distance $2yu(v-u)^{-1}$, therefore $z=y+2yu(v-u)^{-1}=y(v+u)(v-u)^{-1}$; therefore the time from A to the glass and back to A is $t_4=2z(v+u)^{-1}$. But suppose that n_0 internal oscillations of the pulse intervene between its first arrival at the glass and its final departure. In every oscillation the glass advances the distance ut_2, so that the time from A to A is

$$\frac{2y}{v-u}+n_0t_2\left(1+\frac{u}{v}\right), \tag{99}$$

where the additional term n_0t_2u/v is the extra retardation due to the increased distance the reflected pulse has to go. We must now put y in terms of z, which is the same for all the pulses, whereas y varies from pulse to pulse. We have

$$z=\left(\frac{yv}{v-u}+n_0ut_2\right)+\left(\frac{yu}{v-u}+\frac{n_0u^2t_2}{v}\right)=y\frac{v+u}{v-u}+n_0ut_2\left(1+\frac{u}{v}\right), \tag{100}$$

therefore
$$\frac{y}{v-u}=\frac{z}{v+u}-\frac{n_0ut_2}{v}. \tag{101}$$

Put this value of y in (99) and we get

$$\frac{2z}{v+u}+n_0t_2\left(1-\frac{u}{v}\right)=t_4+n_0t_3. \tag{102}$$

This completes the proof of (97).

(19). The final simply periodic solutions are interesting. Take $H_{10}=h\sin nt$, then $p=ni$ in (92) and (97), using the condensed forms for H_3 and H_{20}. We get

$$H_{20}=\frac{v-w_1}{v+w_1}h\frac{(1+m)(1-\cos nt_3)\sin+(1-m)\sin nt_3\cos}{1+m^2-2m\cos nt_3}[n(t-t_4)], \tag{103}$$

$$H_3=(1-m)h\frac{(1-m)\cos nt_3\sin-m\sin nt_3\cos}{1+m^2-2m\cos nt_3}[n(t-t_1)], \tag{104}$$

of which the amplitudes are

$$\text{amp } H_2=\frac{v-w_1}{v+w_1}h\left[\frac{2(1-\cos nt_3)}{1+m^2-2m\cos nt_3}\right]^{\frac{1}{2}}, \tag{105}$$

$$\text{amp } H_3=\frac{(1-m)h}{(1+m^2-2m\cos nt_3)^{\frac{1}{2}}}. \tag{106}$$

By (105) we see that the reflected wave vanishes when $nt_3=2\pi\times$ integer, or 0. When this happens (106) shows, as it must, that amp $H_3=h$. At these frequencies, the transmitted wave copies the incident wave, unattenuated. The same is true for any periodic incident wave, when the period is t_3, which is the time of a single oscillation in the glass, modified through its motion by the factor $(1-u/v)$.

This property is best understood by making the incident wave be a series of equal pulses of period t_3. The reflected wave due to the first, or leading pulse, is a series of pulses of which the leader is positive, and all the rest are negative, the sum of their intensities equalling that of the leader. The same is true for all the following incident pulses. Superpose them. Then we have

$$\begin{array}{llllll} 1, & -(1-m), & -m(1-m), & -m^2(1-m), & -m^3(1-m), & \ldots\ldots \\ & 1, & -(1-m), & -m(1-m), & -m^2(1-m), & \ldots\ldots \\ & & 1, & -(1-m), & -m(1-m), & \ldots\ldots \\ & & & 1, & -(1-m), & \ldots\ldots \\ & & & & 1, & \ldots\ldots \end{array}$$

Add up the columns. The result is

$$1, \quad m, \quad m^2, \quad m^3, \quad m^4, \quad \ldots\ldots$$

which, multiplied by $(v-w_1)\,(v+w_1)^{-1}$, show the intensities of the reflected pulses due to the incident pulses of unit intensity. Since m is less than 1, the final result is 0. That is, any periodic incident wave of period t_3 produces a reflected wave which attenuates to zero, whilst the transmitted wave rises to full strength.

It is not so simple when the period is not such as to make $H_2=0$ finally, but in any case the solutions (103), (104) are only what is tended to finally. The full history of the establishment of the periodic state is given by the expanded operational or differential solutions previously given, and may be followed term by term.

The frequency of H_1 and H_3 is the same, viz., $n/2\pi$, but this is not the frequency of H_2, and that in the glass itself is different again. Thus, for H_2, let $z=a+ut$, then

$$n(t-t_4)=n\left[t-\frac{2(a+ut)}{u+v}\right]=n\left[\frac{v-u}{v+u}t-\ldots\right], \tag{107}$$

showing that the frequency is lowered to $(n/2\pi)\,(v-u)\,(v+u)^{-1}$. But in the glass itself, first turn t to $t+x/v$, to make the reflected wave, and then give x the value z; this makes

$$n\left[\frac{v-u}{v+u}\left(t+\frac{a+ut}{v}\right)-\ldots\right]=n\left(1-\frac{u}{v}\right)t, \tag{108}$$

showing $(n/2\pi)\,(1-u/v)$ to be the frequency in the glass. This is to be confirmed by making $x=z+l$ in the formula for H_3. Thus,

$$n(t-t_1)=n\left(t-\frac{z}{v}+\ldots\right)=n\left[t\left(1-\frac{u}{v}\right)+\ldots\right]. \tag{109}$$

It is curious to see how the change from n to $n(1-u/v)$, and then again to $n(1-u/v)\,(1+u/v)^{-1}$ takes place.

(20). Now as regards the moving force on the glass. This is to be estimated by the rate of loss of energy from the waves. The value of

$$(E_1+E_2)\,(H_1+H_2)-E_3H_3=\mu v(H_1^2-H_2^2-H_3^2), \tag{110}$$

at the glass surfaces is the rate of supply of energy to the glass, to be accounted for electromagnetically and mechanically. But in the final periodic state, the electromagnetic state of the glass becomes regular and periodic, so that the mean value of (110) then represents the mean activity of the force on the glass. With frequency $n/2\pi$, the given formulas (105), (106) for the amplitudes, when squared, and divided by 2, make

$$\tfrac{1}{2}\mu v h^2\left[1-\left(\frac{v-w_1}{v+w_1}\right)^2\frac{2(1-\cos nt_3)}{1+m^2-2m\cos nt_3}-\frac{(1-m)^2}{1+m^2-2m\cos nt_3}\right]$$

$$=\mu v h^2\left(\frac{v-w_1}{v+w_1}\right)\left(\frac{v-w_2}{v+w_2}-\frac{v-w_1}{v+w_1}\right)\frac{1-\cos nt_3}{1+m^2-2m\cos nt_3}$$

$$=\mu h^2\frac{v-w_1}{(v+u_1)^2}\frac{4u(v^2-w^2)}{v+w_2}\frac{1-\cos nt_3}{1+m^2-2m\cos nt_3}, \tag{111}$$

where w is the speed in the glass at rest, or $\mu c w^2=1$. Divide (111) by u to obtain the mean force. Owing to the syntonic factor, it varies widely with the frequency. It is zero when there is no reflected wave.

When u/w is small enough, we may put $w_1=w_2=w$. Then

$$\text{Mean force}=4\mu h^2\left(\frac{v-w}{v+w}\right)^2\frac{1-\cos nt_3}{1+m^2-2m\cos nt_3}. \tag{112}$$

Take $\cos nt_3=0$, and then again $=-1$. The results are

$$2\mu h^2\frac{(v^2-w^2)^2}{v^4+6v^2w^2+w^4} \quad\text{and}\quad 2\mu h^2\frac{(v^2-w^2)^2}{(v^2+w^2)^2}, \tag{113}$$

where $\frac{1}{2}\mu h^2=$ mean density of the incident energy, say P. The value of the force rises very fast as w falls off from v, when it is zero. When $w=\frac{1}{2}v$, the value of the second of (113) is 1·44P. It is less than P when $w=\frac{3}{4}v$. It is P when $3w^2=v^2$.

If, on the other hand, the glass is so deep that only the left side need be considered in estimating the energy, then the mean value of $uv\,(H_1^2-H_2^2)$ comes to

$$Pv\left\{1-\left(\frac{v-w_1}{v+w_1}\right)^2\right\}=P\frac{4v^2w_1}{(v+w_1)^2}, \tag{114}$$

in which the value of w_1 may be inserted, and the result divided by u to give the mean force. When w_1 has its least value u, then the force is 4P. This may be contrasted with the result (11), p. 188.

(21). It will also be of interest to find the total force on a glass plate of finite depth due to a single incident pulse of depth small enough not to cause any overlapping in the train of reflected pulses. If H_1 is the magnetic force in the incident pulse of unit depth, its energy is μH_1^2, and the energy in the transmitted train of pulses is the space integral

$$\Sigma\mu H_3^2=\mu H_1^2(1-m)^2(1+m^2+m^4+\ldots)=\mu H_1^2(1-m)(1+m)^{-1} \tag{115}$$

But in the reflected train the pulses are of depth $(v-u)(v+u)^{-1}$. So the sum of the energy densities in the pulses must be multiplied by this factor. This makes

$$\begin{aligned}\Sigma\mu H_2^2 &= \mu H_1^2 \frac{v+u}{v-u}\left(\frac{v-w_1}{v+w_1}\right)^2\left\{1+(1-m)^2(1+m^2+m^4+\ldots)\right\}\\ &= \mu H_1^2 \frac{v+u}{v-u}\frac{2}{1+m}\left(\frac{v-w_1}{v+w_1}\right)^2.\end{aligned} \tag{116}$$

The loss of energy is the excess of μH_1^2 over the sum of (115) and (116). The result may be reduced to

$$\mu H_1^2 \frac{2}{1+m}\frac{v-w_1}{v+w_1}\left(\frac{v-w_2}{v+w_2}-\frac{v+u}{v-u}\frac{v-w_1}{v+w_1}\right). \tag{117}$$

It will be seen that the force exerted by radiation upon an interposed body cannot be definitely specified without a knowledge of the electrical properties of the body, and also the nature of the radiation. Moreover, it is not necessary for there to be any "absorption" of energy in the body, as in black bodies. On the other hand, in Maxwell's electromagnetics, the "pressure of radiation" is a quite distinct idea, being a particular form of the stress, namely, the eolotropic pressure which corresponds to the mechanical forces exerted upon electrification and electric current and magnetic current.

(22). In connection with the above, the action of radiation upon electrons free to move in the ether is deserving of consideration briefly. When a plane pulse passes an electron which cannot move, nothing happens. The pulse goes on its way unchanged, and the electric field of the electron remains the same. But if the electron is free to move, move it will, and its field will be joined on to that of the plane pulse by a spherical pulse for ever after, and the electron itself will be in permanent motion.

The force on the electron at first is $\mathbf{E}\rho$, if $\mathbf{E}$ is the electric force of the pulse, and ρ the charge of the electron. This force moves it in the plane of the pulse. But once in motion, it also represents an electric current $\rho\mathbf{u}$, if $\mathbf{u}$ is its velocity. So there is an additional force $\rho \mathrm{V}\mathbf{u}\mathbf{B}$ acting on the electron, if $\mathbf{B}$ is the induction of the pulse. This new force is at first parallel to the direction of motion of the pulse. So the path becomes curved. The equation of motion being

$$\rho(\mathbf{E}+\mathrm{V}\mathbf{u}\mathbf{B}) = \dot{\mathbf{M}}, \tag{118}$$

where $\mathbf{M}$ is the "momentum" of the electron, if we could write $\mathbf{M}=m\mathbf{u}$, with m a constant, we should have

$$\rho\mathrm{V}\mathbf{B}(\mathbf{v}-\mathbf{u}) = m\dot{\mathbf{u}} = m(\dot{\mathbf{u}}-\dot{\mathbf{v}}), \tag{119}$$

and by multiplying this by $\mathbf{v}-\mathbf{u}$, it would follow that $\frac{1}{2}m(\mathbf{v}-\mathbf{u})^2$ was constant. So the speed of the electron referred to origin moving with the wave front of a very deep pulse would be constant, and the path would be a circle with centre at the moving origin. The speed would vary from 0 to $2v$. But this would be electromagnetic nonsense, because $\mathbf{M}=m\mathbf{u}$ with m constant, is only approximately true when the velocity and the acceleration are both small. So only the beginning of the path is obtainable in this way, the first small arc of the circle, nearly a straight line. The solution from the moment $t=0$ when the pulse first reaches the electron is, if $p=d/dt$,

$$u=\frac{\mathrm{V}\rho\mathbf{B}}{mp+\mathrm{V}\rho\mathbf{B}}\mathbf{v}=\frac{\mathrm{V}\rho\mathbf{B}/mp}{1+\mathrm{V}\rho\mathbf{B}/mp\mathbf{v}} \qquad (120)$$

which, expanded by division, and noting that $\mathbf{VBVB}=-\mathrm{B}^2$ when the operand is $\mathbf{v}$, makes

$$\mathbf{u}=\mathbf{v}+v(\mathrm{i}_1\sin-\mathrm{i}_3\cos)(\rho\mathrm{B}t/m), \qquad (121)$$

if i_1 and i_3 are unit vectors parallel to $\mathbf{E}$ and $\mathbf{v}$. The circular speed is $\mathrm{E}\rho/m$, but as before said, this only applies to the beginning of the first circle.

But if the pulse is not deep, and has passed the electron before it has sensibly moved, the force-impulse on it is $\mathrm{E}\rho\tau$, if τ is the depth of the pulse. So

$$\mathrm{E}\rho\tau=\mathbf{M}. \qquad (122)$$

Given $\mathbf{u}$, $\mathbf{M}$ can be calculated for the particular distribution of electrification assumed for the electron. Say $2\mathbf{M}=m\mathbf{u}$ results, then u/v may be about 0·9. If $\rho/m=1{\cdot}75\times10^7$, the voltage in the pulse is 923,000 volts. But calculations done this way are not complete, for the validity of the $\mathbf{M}$ formula depends upon smallness of acceleration, even when the restriction to small velocity has been removed, unless $\mathbf{M}$ is specially calculated for impulses or large acceleration.

The force on an electron ρ in the progressive field $\mathrm{E}=v\mathrm{B}$ being

$$\mathbf{VB}(\mathbf{v}-\mathbf{u})\rho=\mathrm{i}_1\mathrm{B}(v-u_3)+\mathrm{i}_3\mathrm{B}u_1, \qquad (123)$$

where $_1$ refers to the $\mathbf{E}$ axis, and $_3$ to the $\mathbf{v}$ axis, it is visible at once that if $\mathbf{v}=\mathbf{u}$, there is no force on the electron. This is a sort of dynamic equilibrium. It applies to any sort of one-way radiation. Swarms of electrons may be travelling with the radiation, and might indeed be regarded as a part thereof, only then we should have to distinguish between radiation without and with electrons. It may be inferred that electrons placed in radiation tend to assume the state $\mathbf{v}=\mathbf{u}$, by the action upon them of the radiation itself. But if $\mathbf{u}$ is made to exceed $\mathbf{v}$, then, as shown before, a resisting force of a dissipative nature comes into play, and sends $\mathbf{u}$ back to $\mathbf{v}$ again. Thus the equilibrium is of a peculiar nature. Note also that $v=u_3$ makes the force $\mathrm{B}u_1$ in the direction of $\mathbf{v}$. But this implies that $u>v$, if the transverse u_1 has any existence, and so there is a resisting force again bringing u down to v. Also note that $v=u_3$ even momentarily, has the effect, through the i_3 term, of making $v-u_3$ be negative, so that the supposed i_1 velocity falls.

The Electromagnetic Circuital Equations and Connected Matter.

§533. (1). By far the best way of treating the Faraday or second circuital law is to take it as an experimental fact, proved roughly by Faraday, and capable of much closer proof by telephones, induction balances, and general consistency of results. For a fixed medium the mathematical expression is $-\text{curl}\ \mathbf{E}=\dot{\mathbf{B}}$ for an infinitesimal circuit, and the meaning is quite simple, as explained in vol. 1. But it is by no means so clear when a circuit through which the induction varies is in motion. For in a comprehensive theory the ether must be taken into consideration as well as the moving matter, and the question at once presents itself what their connection is. The ether may rest or move, and so many different forms of resultant circuital equation may be imagined. By the construction, not of the circuital equation itself, but of the term VqB in Maxwell's equation of E.M.F., it may be inferred that he supposed the ether to move with the matter. At any rate, by curling his equation we remove the indistinctness and obtain the equation $-\text{curl}\ (\mathbf{E}-\mathbf{e})=\dot{\mathbf{B}}$, in which $\mathbf{e}=\mathrm{V}\mathbf{qB}$, and this form does really express Faraday's law for a single medium, meaning thereby that the ether and the matter go together. Along with this, there comes naturally a similar alteration in the first circuital law, curl $\mathbf{H}=\dot{\mathbf{D}}$ for rest becoming curl $(\mathbf{H}-\mathbf{h})=\dot{\mathbf{D}}$, where $\mathbf{h}=\mathrm{V}\mathbf{Dq}$, when there is motion with velocity $\mathbf{q}$. This term was overlooked altogether in his treatise by Maxwell. More for the sake of definiteness than anything else, this system was followed in vol. 1, having previously used the same definitions of $\mathbf{e}$ and $\mathbf{h}$ in 1885.

(2). At the other extreme is the alteration in the motional magnetic force $\mathbf{h}$ made by H. A. Lorentz. Assuming that the ether is entirely at rest, he arrived at $\mathbf{h}_1=\mathrm{V}\mathbf{D}_1\mathbf{u}$, if $\mathbf{u}$ is the velocity of the matter, and $\mathbf{D}_1$ the polarisation, or that part of the displacement which depends on the matter only. This was done by reasoning relating to ions and their spacial displacement under the influence of electric force. At the present time electrons would be spoken of rather than ions. It is some sort of electric separation that is involved, at any rate. But if the electronic idea were really necessary in the argument, it would follow, since no magnetons to match have been found, that there would be no similar change in the first circuital law, and, in fact, the author of this revised $\mathbf{h}_1$ (instead of $\mathbf{h}$) made no such change.

But if we can clearly come to $\mathbf{h}_1=\mathrm{V}\mathbf{D}_1\mathbf{u}$ without the electronic idea, but as a necessary term in the first circuital equation to express the Maxwell law if the ether rests, then the absence of magnetons becomes of no importance in the argument, and the same reasoning applied to magnetic instead of to electric polarisation will show that e in the second circuital law must be altered from $\mathrm{V}\mathbf{qB}$ to $\mathbf{e}_1=\mathrm{V}\mathbf{uB}_1$, where $\mathbf{B}_1$ is that part of the induction which is associated with the moving matter.

Moreover, the more general case of both ether and matter moving can then be treated on the same lines. This is convenient even if the ether does not move, on account of the convenience in theoretical arguments sometimes due to the relativity of motion. But this more general case demands notice also because $\mathbf{e}_1=\mathrm{V}\mathbf{uB}_1$ would make $\mathbf{e}_1=0$ in an unmagnetisable medium. So there is room for some speculation here, and a closer examination of results is demanded.

(3). The most convenient way of making the transformations referred to above is the vectorial. Say that $m\mathbf{sn}=1$, where $\mathbf{n}$ is an infinitesimal area defined by its normal, and $\mathbf{s}$ an infinitesimal length, so that $\mathbf{sn}$ is the volume of the element of space. As for m, it may be regarded as the density of matter in the space. Differentiate with respect to t, using the moving time-differentiator $\partial/\partial t$. Then

$$m\mathbf{s}\frac{\partial\mathbf{n}}{\partial t}+\mathbf{ns}\frac{\partial m}{\partial t}+m\mathbf{n}\frac{\partial\mathbf{s}}{\partial t}=0, \tag{1}$$

or

$$\mathbf{s}\frac{\partial\mathbf{n}}{\partial t}+\mathbf{n}\frac{\partial\mathbf{s}}{\partial t}=-\frac{\mathbf{ns}}{m}\frac{\partial m}{\partial t}=\frac{\nabla\mathbf{q}}{m}=\frac{\partial}{\partial t}\mathbf{ns}, \tag{2}$$

if $\mathbf{q}$ is the velocity of the matter.

Now here

$$\frac{\partial\mathbf{s}}{\partial t}=\mathbf{s}\nabla.\mathbf{q}, \tag{3}$$

a proposition to be proved by inspection of a short line in motion. It follows by (2) that

$$\frac{\partial\mathbf{n}}{\partial t}=\mathbf{n}.\nabla\mathbf{q}-\nabla(\mathbf{n}\dot{\mathbf{q}})=\mathrm{V}.\mathrm{V}\nabla\mathbf{n}.\dot{\mathbf{q}} \tag{4}$$

where the under dots indicate the vectors on which ∇ operates. The two equations (3) and (4), expressing the rates of time variation of a moving linear and surface element, are very useful formulas.

(4). Now let $\mathbf{B}_0$ be the induction (density) in a medium, say the ether, supposed to be moving at velocity $\mathbf{q}$. Then $\mathbf{B}_0\mathbf{n}$ is the induction through the area $\mathbf{n}$, and its variation is given by

$$\frac{\partial\mathbf{B}_0\mathbf{n}}{\partial t}=\mathbf{n}\frac{\partial\mathbf{B}_0}{\partial t}+\mathbf{B}_0\mathbf{n}.\nabla\mathbf{q}-\mathbf{n}.\mathbf{B}_0\nabla.\mathbf{q}-\mathbf{nq}.\nabla\mathbf{B}_0+\mathbf{nq}.\nabla\mathbf{B}_0$$

$$=\mathbf{n}[\dot{\mathbf{B}}_0+\mathbf{q}\nabla.\mathbf{B}_0-\mathrm{V}\nabla\mathrm{V}\mathbf{q}\dot{\mathbf{B}}_0]. \tag{5}$$

So, starting from the fundamental $-\mathrm{V}\nabla\mathbf{E}=\dot{\mathbf{B}}_0$, if this is to hold good for a moving area, we come to

$$-\mathbf{n}\mathrm{V}\nabla\mathbf{E}=\frac{\partial}{\partial t}(\mathbf{n}\mathbf{B}_0)=\mathbf{n}[\dot{\mathbf{B}}_0+\mathbf{q}.\nabla\mathbf{B}_0-\mathrm{V}\nabla\mathrm{V}\mathbf{q}\mathbf{B}_0], \qquad (6)$$

or

$$-\mathrm{V}\nabla(\mathbf{E}-\mathrm{V}\mathbf{q}\mathbf{B}_0)=\dot{\mathbf{B}}_0+\mathbf{q}.\nabla\mathbf{B}_0 \qquad (7)$$

which is the transformed circuital. If a single medium is concerned, then we may put $\nabla\mathbf{B}_0=0$.

But suppose there are (effectively) two mediums, one carrying the induction $\mathbf{B}_0$, the other $\mathbf{B}_1$, and we apply the Faraday principle to both together. Let $\mathbf{n}$, $\mathbf{n}_0$, $\mathbf{n}_1$ be identical at the moment t, but that $\mathbf{n}$ becomes $\mathbf{n}_0$ for one moving medium, and $\mathbf{n}_1$ for the other. Then

$$-\mathbf{n}\mathrm{V}\nabla\mathbf{E}=\frac{\partial}{\partial t}(\mathbf{n}_0\mathbf{B}_0)+\frac{\partial}{\partial t}(\mathbf{n}_1\mathbf{B}_1) \qquad (8)$$

expresses the Faraday law, and is, by the above, the same as

$$-\mathrm{V}\nabla(\mathbf{E}-\mathrm{V}\mathbf{q}\mathbf{B}_0-\mathrm{V}\mathbf{u}\mathbf{B}_1)=\dot{\mathbf{B}}+\mathbf{q}.\nabla\mathbf{B}_0+\mathbf{u}.\nabla\mathbf{B}_1, \qquad (9)$$

which is the resultant circuital equation. To avoid magnetons, we also require $\nabla\mathbf{B}_0=0$, $\nabla\mathbf{B}_1=0$ separately, in general, to secure $\nabla\mathbf{B}=0$, if $\mathbf{B}$ is the actual induction.

If there is a real convection current $\mathbf{w}.\nabla\mathbf{B}$, we must add $(\mathbf{w}-\mathbf{q})\nabla\mathbf{B}_0+(\mathbf{w}-\mathbf{u})\nabla\mathbf{B}_1$ to the right side of (9), making it become $\mathbf{w}.\nabla(\mathbf{B}_0+\mathbf{B}_1)=\mathbf{w}.\nabla\mathbf{B}$.

It will be seen that (8), (9) may be extended to any number of interpenetrating mediums. The other circuital equation, derived from curl $\mathbf{H}=\dot{\mathbf{D}}$ is obviously

$$\mathrm{V}\nabla(\mathbf{H}-\mathrm{V}\mathbf{D}_0\mathbf{q}-\mathrm{V}\mathbf{D}_1\mathbf{u})=\dot{\mathbf{D}}+\mathbf{q}.\nabla\mathbf{D}_0+\mathbf{u}.\nabla\mathbf{D}_1, \qquad (10)$$

to match equation (9).

(5). The motional electric and magnetic forces play the part of impressed forces in the primitive circuital equations. So let

$$\mathbf{e}_0=\mathrm{V}\mathbf{q}\mathbf{B}_0, \quad \mathbf{e}_1=\mathrm{V}\mathbf{u}\mathbf{B}_1, \quad \mathbf{h}_0=\mathrm{V}\mathbf{D}_0\mathbf{q}, \quad \mathbf{h}_1=\mathrm{V}\mathbf{D}_1\mathbf{u}, \qquad (11)$$

$$\mathbf{D}_0=c_0\mathbf{E}, \quad \mathbf{D}_1=c_1\mathbf{E}_1, \quad \mathbf{B}_0=\mu_0\mathbf{H}, \quad \mathbf{B}_1=\mu_1\mathbf{H}_1, \qquad (12)$$

$$\mathbf{E}_1=\mathbf{E}+\mathbf{e}, \quad \mathbf{H}_1=\mathbf{H}+\mathbf{h}. \qquad (13)$$

Here c_0 and μ_0 are the permittivity and inductivity of the ether, c_1 and μ_1 the additions made when matter is present; and $\mathbf{E}$ and $\mathbf{H}$ are the electric and magnetic forces in the ether, according to

$$\mathbf{B}=\mathbf{B}_0+\mathbf{B}_1=\mu_0\mathbf{H}+\mu_1\mathbf{H}_1, \qquad \mathbf{D}=\mathbf{D}_0+\mathbf{D}_1=c_0\mathbf{E}+c_1\mathbf{E}_1, \qquad (14)$$

where $\mathbf{B}$ and $\mathbf{D}$ are the resultant induction and displacement. It is supposed that c_0 and μ_0 move with the ether at velocity $\mathbf{q}$, whilst c_1 and μ_1 move with the matter at velocity $\mathbf{u}$. A fixed reference space is implied, in which the ether and matter move.

The circuital equations are then

$$\mathrm{V}\nabla\mathbf{H}=\dot{\mathbf{D}}+\mathrm{V}\nabla\mathbf{h}_0+\mathrm{V}\nabla\mathbf{h}_1, \tag{15}$$

$$-\mathrm{V}\nabla\mathbf{E}=\dot{\mathbf{B}}-\mathrm{V}\nabla\mathbf{e}_0-\mathrm{V}\nabla\mathbf{e}_1, \tag{16}$$

where no convection term is added to represent electrification moving, which may be done separately. Multiply the first by **E**, the second by **H** and add. Then

$$\begin{aligned}-\nabla\mathrm{V}\mathbf{EH}&=\mathbf{E}\dot{\mathbf{D}}+\mathbf{H}\dot{\mathbf{B}}+\dots\\&=\dot{\mathrm{U}}+\dot{\mathrm{T}}+\dot{\mathrm{U}}_{c1}+\dot{\mathrm{T}}_{\mu1}-\mathbf{e}\dot{\mathbf{D}}_1-\mathbf{h}\dot{\mathbf{B}}_1\\&\quad-\nabla\mathrm{V}\mathbf{E}\mathbf{h}_0+\mathbf{h}_0\mathrm{V}\nabla\mathbf{E}-\nabla\mathrm{V}\mathbf{E}_1\mathbf{h}_1+\nabla\mathrm{V}\mathbf{e}\mathbf{h}_1+\mathbf{h}_1\mathrm{V}\nabla\mathbf{E}\\&\quad+\nabla\mathrm{V}\mathbf{H}\mathbf{e}_0-\mathbf{e}_0\mathrm{V}\nabla\mathbf{H}+\nabla\mathrm{V}\mathbf{H}_1\mathbf{e}_1-\nabla\mathrm{V}\mathbf{h}\mathbf{e}_1-\mathbf{e}_1\mathrm{V}\nabla\mathbf{H}.\end{aligned} \tag{17}$$

By rearrangement this makes

$$\begin{aligned}&-\nabla[\mathrm{V}\mathbf{EH}+\mathrm{V}\mathbf{h}_0\mathbf{E}+\mathrm{V}\mathbf{h}_1\mathbf{E}_1+\mathrm{V}\mathbf{H}\mathbf{e}_0+\mathrm{V}\mathbf{H}_1\mathbf{e}_1]\\&=\dot{\mathrm{U}}+\dot{\mathrm{T}}+\dot{\mathrm{U}}_{c1}+\dot{\mathrm{T}}_{\mu1}-\mathbf{e}\dot{\mathbf{D}}_1-\mathbf{h}\dot{\mathbf{B}}_1\\&+(\mathbf{h}_0+\mathbf{h}_1)\mathrm{V}\nabla\mathbf{E}-(\mathbf{e}_0+\mathbf{e}_1)\mathrm{V}\nabla\mathbf{H}+\nabla(\mathrm{V}\mathbf{e}\mathbf{h}_1+\mathrm{V}\mathbf{e}_1\mathbf{h}).\end{aligned} \tag{18}$$

Here U and T are the electric and magnetic energy densities, and $\dot{\mathrm{U}}_{c1}$, $\dot{\mathrm{T}}_{\mu1}$ are their time rates of increase so far as c_1 and μ_1 are concerned, that is, for example,

$$\begin{aligned}\mathbf{E}_1\dot{\mathbf{D}}_1&=\mathrm{E}_1c_1\dot{\mathrm{E}}_1+\mathrm{E}_1\dot{c}_1\mathrm{E}_1=\dot{\mathrm{U}}_1+\dot{\mathrm{U}}_{c1};\\\dot{\mathrm{U}}_{c1}&=\tfrac{1}{2}\mathrm{E}_1^2\dot{c}_1=\mathrm{U}_1\dot{c}_1/c_1.\end{aligned} \tag{19}$$

Now we have

$$\begin{aligned}\mathrm{V}\mathbf{h}_0\mathbf{E}&=\mathrm{V}\mathbf{E}\mathrm{V}\mathbf{q}\mathbf{D}_0=\mathbf{q}(\mathbf{E}\mathbf{D}_0-\mathbf{E}.\mathbf{D}_0)=\mathbf{q}(\mathrm{U}_0+\mathrm{P}_0),\\\mathrm{V}\mathbf{h}_1\mathbf{E}_1&=\mathrm{V}\mathbf{E}_1\mathrm{V}\mathbf{u}\mathbf{D}_1=\mathbf{u}(\mathbf{E}_1\mathbf{D}_1-\mathbf{E}_1.\mathbf{D}_1)=\mathbf{u}(\mathrm{U}_1+\mathrm{P}_1),\\\mathrm{V}\mathbf{H}\mathbf{e}_0&=\mathrm{V}\mathbf{H}\mathrm{V}\mathbf{q}\mathbf{B}_0=\mathbf{q}(\mathbf{H}\mathbf{B}_0-\mathbf{H}.\mathbf{B}_0)=\mathbf{q}(\mathrm{T}_0+\mathrm{Q}_0),\\\mathrm{V}\mathbf{H}_1\mathbf{e}_1&=\mathrm{V}\mathbf{H}_1\mathrm{V}\mathbf{u}\mathbf{B}_1=\mathbf{u}(\mathbf{H}_1\mathbf{B}_1-\mathbf{H}_1.\mathbf{B}_1)=\mathbf{u}(\mathrm{T}_1+\mathrm{Q}_1).\end{aligned} \tag{20}$$

These represent convection of energy and activity of stresses, if

$$\mathrm{U}_0=\tfrac{1}{2}\mathbf{E}\mathbf{D}_0,\quad \mathrm{U}_1=\tfrac{1}{2}\mathbf{E}_1\mathbf{D}_1,\quad \mathrm{T}_0=\tfrac{1}{2}\mathbf{H}\mathbf{B}_0,\quad \mathrm{T}_1=\tfrac{1}{2}\mathbf{H}_1\mathbf{B}_1, \tag{21}$$

are the energy densities, and

$$\mathrm{P}_0=\mathrm{U}_0-\mathbf{E}.\mathbf{D}_0,\ \mathbf{P}_1=\mathrm{U}_1-\mathbf{E}_1.\mathbf{D}_1,\ \mathrm{Q}_0=\mathrm{T}_0-\mathbf{H}.\mathbf{B}_0,\ \mathrm{Q}_1=\mathrm{T}_1-\mathbf{H}_1.\mathbf{B}_1, \tag{22}$$

are four eolotropic pressures, the negatives of eolotropic tensions. The left member of (18) is therefore the same as

$$-\nabla[\mathrm{V}\mathbf{EH}+\mathbf{q}(\mathrm{U}_0+\mathrm{T}_0+\mathrm{P}_0+\mathrm{Q}_0)+\mathbf{u}(\mathrm{U}_1+\mathrm{T}_1+\mathrm{P}_1+\mathrm{Q}_1)]. \tag{23}$$

Since this is the convergence of the flux of energy made up of the Poynting flux, the convection of stored energy, and the activity of the pressures, the right side of (18), after the $\dot{\mathrm{U}}_{c1}$ and $\dot{\mathrm{T}}_{\mu1}$ terms, should be the mechanical activity, though a part may be involved in $\dot{c}$ and $\dot{\mu}$, not at present in question. The remainder is the same as

$$\begin{aligned}&-\mathbf{e}\dot{\mathbf{D}}_1-\mathbf{h}\dot{\mathbf{B}}_1-(\mathbf{h}_0+\mathbf{h}_1)\dot{\mathbf{B}}-(\mathbf{e}_0+\mathbf{e}_1)\dot{\mathbf{D}}+(\mathbf{h}_0+\mathbf{h}_1)\mathrm{V}\nabla(\mathbf{e}_0+\mathbf{e}_1)\\&-(\mathbf{e}_0+\mathbf{e}_1)\mathrm{V}\nabla(\mathbf{h}_0+\mathbf{h}_1)+\nabla(\mathrm{V}\mathbf{e}\mathbf{h}_1+\mathrm{V}\mathbf{e}_1\mathbf{h}),\end{aligned} \tag{24}$$

which may be written

$$=-\mathbf{e}\dot{\mathbf{D}}_1-\mathbf{h}\dot{\mathbf{B}}_1-(\mathbf{e}_0+\mathbf{e}_1)\dot{\mathbf{D}}-(\mathbf{h}_0+\mathbf{h}_1)\dot{\mathbf{B}}$$
$$+\nabla[\mathrm{V}\mathbf{e}\mathbf{h}_1+\mathrm{V}\mathbf{e}_1\mathbf{h}+\mathrm{V}(\mathbf{e}_0+\mathbf{e}_1)(\mathbf{h}_0+\mathbf{h}_1)]. \tag{25}$$

In this, put the e's and h's in terms of the fluxes and velocities, by (11) to (13). The first line easily reduces to

$$\mathbf{u}[\overline{\mathrm{V}\dot{\mathbf{D}}\mathbf{B}}-\overline{\mathrm{V}\mathbf{D}_0\mathbf{B}_0}]+\mathbf{q}\overline{\mathrm{V}\dot{\mathbf{D}}_0\mathbf{B}_0}. \tag{26}$$

Similarly, using

$$\mathbf{e}=\mathrm{V}(\mathbf{u}-\mathbf{q})\mathbf{B}_0, \qquad \mathbf{h}=\mathrm{V}\mathbf{D}_0(\mathbf{u}-\mathbf{q}), \tag{27}$$

the second line in (25) reduces to

$$\nabla[\mathbf{u}.\mathbf{u}(\mathrm{V}\mathbf{D}\mathbf{B}-\mathrm{V}\mathbf{D}_0\mathbf{B}_0)]+\nabla[\mathbf{q}.\mathbf{q}\mathrm{V}\mathbf{D}_0\mathbf{B}_0]. \tag{28}$$

So, finally, the mechanical activity is

$$+\left\{\mathbf{u}\frac{d}{dt}+\nabla(\mathbf{u}.\mathbf{u})\right\}(\mathbf{M}-\mathbf{M}_0)+\left\{\mathbf{q}\frac{d}{dt}+\nabla(\mathbf{q}.\mathbf{q})\right\}\mathbf{M}_0, \tag{29}$$

where $\mathbf{M}=\mathrm{V}\mathbf{D}\mathbf{B}$, $\mathbf{M}_0=\mathrm{V}\mathbf{D}_0\mathbf{B}_0$ Here $\mathbf{M}$ is the complete *quasi*-momentum, and $\mathbf{M}_0$ the etherial part. Or, as on p. 191,

$$=\mathbf{u}\mathbf{F}_1+\mathbf{q}\mathbf{F}_0, \tag{30}$$

if

$$\mathbf{F}_1=\left(\frac{d}{dt}+\mathbf{u}\nabla+\nabla\mathbf{u}+\nabla.\dot{\mathbf{u}}\right)(\mathbf{M}-\mathbf{M}_0)$$
$$\mathbf{F}_0=\left(\frac{d}{dt}+\mathbf{q}\nabla+\nabla\mathbf{q}+\nabla\,\dot{\mathbf{q}}\right)\mathbf{M}_0. \tag{31}$$

These indicate the activity of the forces per unit volume ; on the moving matter and on the moving ether. Compare with p. 148, where a similar result is obtained for a singly moving medium. The extension to any number of mediums is obvious, though the utility thereof is not.

If $\mathbf{q}=0$, the term $\mathbf{F}_0\mathbf{q}$ goes out of existence. But this is the limiting case of an ether kept rigid. It does not mean that $\mathbf{F}_0$ is zero.

If $\mathbf{u}=\mathbf{q}$ we have the extreme case of ether carried with matter, the velocity of the ether away from the matter being left arbitrary. In any case we require a standard fixed space for reference, not that actually defined by the configuration of the ether at a given moment, because it cannot be at rest in its innermost parts, even though stagnant on the whole. But the standard fixed space is only necessary for convenience and definiteness.

(6). The above forces are not the same as the pressure-forces, meaning the forces derived from the four eolotropic pressures, two electric, two magnetic. For take the case of Q_0. We have

$$-\nabla[\mathbf{q}(\mathrm{T}_0-\mathbf{H}.\mathbf{B}_0)]=\mathbf{H}.\mathbf{B}_0\nabla.\mathbf{q}-\mathrm{T}_0\nabla\mathbf{q}$$
$$+\mathbf{q}[\mathbf{H}.\nabla\mathbf{B}_0+\mathrm{V}.\mathrm{V}\nabla\mathbf{H}\,\mathbf{B}_0-\nabla_0\mathrm{T}_{\mu 0}], \tag{32}$$

so the activity of the pressure-forces is

$$\mathbf{q}[\mathrm{V}.\mathrm{V}\nabla\mathbf{E}.\mathbf{D}_0+\mathrm{V}.\mathrm{V}\nabla\mathbf{H}.\mathbf{B}_0]+\mathbf{u}[\mathrm{V}.\mathrm{V}\nabla\mathbf{E}_1.\mathbf{D}_1+\mathrm{V}.\mathrm{V}\nabla\mathbf{H}_1.\mathbf{B}_1]$$
$$+\mathbf{q}[-\nabla_{\mu 0}\mathrm{T}_0-\nabla_{c0}\mathrm{U}_0+\mathbf{E}.\nabla\mathbf{D}_0+\mathbf{H}.\nabla\mathbf{B}_0]$$
$$+\mathbf{u}[-\nabla_{\mu 1}\mathrm{T}_1-\nabla_{c1}\mathrm{U}_1+\mathbf{E}_1.\nabla\mathbf{D}_1+\mathbf{H}_1.\nabla\mathbf{B}_1]. \tag{33}$$

Disregarding the second and third lines, which are concerned with variations in μ and c and with divergences of **D** and **B**, the first line may be transformed to

$$-(\mathbf{h}_0+\mathbf{h}_1)\dot{\mathbf{B}}-(\mathbf{e}_0+\mathbf{e}_1)\dot{\mathbf{D}}+\mathbf{h}_1\dot{V}\nabla\mathbf{e}-\mathbf{e}_1V\nabla\mathbf{h}+\nabla V(\mathbf{e}_0+\mathbf{e}_1)(\mathbf{h}_0+\mathbf{h}_1). \quad (34)$$

Taking this to be the complete pressure-force, deduct it from (25). The difference may be reduced to

$$=-\mathbf{e}(\dot{\mathbf{D}}_1+V\nabla\mathbf{h}_1)-\mathbf{h}(\dot{\mathbf{B}}_1-V\nabla\mathbf{e}_1)=-(\mathbf{eJ}_1+\mathbf{hG}_1), \quad (35)$$

where $\mathbf{J}_1$ and $\mathbf{G}_1$ are the polarisation currents, with allowance made for the distortion due to the motion.

The pressure-forces and the former forces are therefore the same if $\mathbf{e}=0$, $\mathbf{h}=0$. This would mean either $\mathbf{u}=\mathbf{q}$, or else that the polarisations were exactly proportional to the etherial electric and magnetic forces.

With **e** and **h** left unstated in form in the equations (13) above, we may construct the activity equation

$$\mathbf{eJ}_1+\mathbf{hG}_1-\Delta\mathbf{W}=\dot{U}+T+\frac{\partial}{\partial t}[U_{c_0}+U_{c_1}+T_{\mu_0}+T_{\mu_1}]$$
$$-\mathbf{q}\nabla(P_0+Q_0)-\mathbf{u}\nabla(P_1+Q_1), \quad (36)$$

where $\mathbf{W}=V\mathbf{EH}+\mathbf{q}(U_0+T_0+P_0+Q_0)+\mathbf{u}(U_1+T_1+P_1+Q_1)$. (37)

This is quite symmetrical, and if **e** and **h** are regarded as intrinsically impressed forces, is the standard form, **e** acting on $\mathbf{J}_1$ only, and **h** on $\mathbf{G}_1$ only. Note that $\partial/\partial t$ is the moving time differentiator, that is $d/dt+\mathbf{q}\nabla$ or else $d/dt+\mathbf{u}\nabla$; and that only c_0, c_1, μ_0, μ_1 are differentiated. Also that the co-factors of **q** and **u** are the corresponding pressure forces. They are exhibited in (33).

Reverting to Lorentz's form for **e**, and its magnetic analogue, we have

$$\mathbf{e}_0+\mathbf{e}_1+\mathbf{e}=V\mathbf{uB}, \qquad \text{curl}\,(\mathbf{H}_1-V\mathbf{Du})=\dot{\mathbf{D}}, \quad (38)$$

$$\mathbf{h}_0+\mathbf{h}_1+\mathbf{h}=V\mathbf{Du}, \qquad -\text{curl}\,(\mathbf{E}_1-V\mathbf{uB})=\dot{\mathbf{B}}; \quad (39)$$

which are rather neat, but need some expansion for practical use.

(7). The electromagnetic equations of activity are very interesting to contemplate in relation to the similar equations in an elastic solid or fluid, especially in the obtrusion of the $V\mathbf{EH}$ term, of unknown meaning in **E** and **H**.

In an isotropic solid or fluid without viscosity we have

$$-\nabla[(U+T+P,\mathbf{q}]=\dot{U}+\dot{T}, \quad (40)$$

if U and T are the stored potential and kinetic energy, **q** the velocity, and P is the pressure, isotropic or eolotropic for a fluid or solid respectively. But if a fluid, then we may eliminate either U or P, when P is a function of the density m only. For then

$$U=-\frac{1}{V}\int_{V_0}^{V}(P-P_0)dV, \qquad \text{if } V=m^{-1}=\text{vol. of unit mass}; \quad (41)$$

$$=m\int_{m_0}^{m}\frac{P-P_0}{m^2}dm, \qquad \text{if } P_0 \text{ and } m_0 \text{ are standard}. \quad (42)$$

By differentiation

$$\frac{d}{dm}\frac{\mathrm{U}}{m}=\frac{\mathrm{P}-\mathrm{P}_0}{m^2},\quad\text{or}\quad m\frac{d\mathrm{U}}{dm}=\mathrm{U}+\mathrm{P},\tag{43}$$

throwing away P_0 as unnecessary. So (40) may be written

$$-\nabla\left[\left(\mathrm{T}+m\frac{d\mathrm{U}}{dm}\right)\mathbf{q}\right]=\dot{\mathrm{U}}+\dot{\mathrm{T}}=-\nabla\left[\left(\frac{\mathrm{T}}{m}+\frac{d\mathrm{U}}{dm}\right)m\mathbf{q}\right].\tag{44}$$

But in general P is too useful to be eliminated.

The equation (40) splits into two. For

$$\frac{\partial\mathrm{U}}{\partial t}=\frac{\partial\mathrm{U}}{dm}\frac{\partial m}{\partial t}=-m\nabla\mathbf{q}\cdot\frac{d\mathrm{U}}{dm}=-(\mathrm{U}+\mathrm{P})\nabla\mathbf{q},$$

therefore

$$-\mathrm{P}\nabla.\mathbf{q}=\dot{\mathrm{U}}+\nabla(\mathbf{q}\mathrm{U}).\tag{45}$$

Similarly,

$$-\mathbf{q}\nabla.\mathrm{P}=\dot{\mathrm{T}}+\nabla(\mathbf{q}\mathrm{T}).\tag{46}$$

Adding these two, equation (40) results, or an equivalent, namely,

$$-\nabla(\mathrm{P}\mathbf{q})=\frac{\partial\mathrm{U}}{\partial t}+\frac{\partial\mathrm{T}}{\partial t}+(\mathrm{U}+\mathrm{T})\nabla\mathbf{q}.\tag{47}$$

In a solid, (40), (47), and the splits (45), (46) are just the same, with the notation for eolotropic linear transformers adopted in my Vector Algebra and Analysis.

(8). Now pass to the etherial electromagnetic equations, with μ, c, and m variable. They are

$$\mathrm{V}\nabla\mathbf{H}=\frac{\partial\mathbf{D}}{\partial t}-(\mathbf{D}\nabla-\mathbf{D}.\nabla)\mathbf{q},\qquad-\mathrm{V}\nabla\mathbf{E}=\frac{\partial\mathbf{B}}{\partial t}-(\mathbf{B}\nabla-\mathbf{B}.\nabla)\mathbf{q},\tag{48}$$

from which comes the activity equation

$$-\nabla\mathrm{V}\mathbf{EH}=\mathbf{E}\frac{\partial\mathbf{D}}{\partial t}+\mathbf{H}\frac{\partial\mathbf{B}}{dt}-(\mathbf{E}.\mathbf{D}\nabla.\mathbf{q}+\mathbf{H}.\mathbf{B}\nabla.\mathbf{q})+2(\mathrm{U}+\mathrm{T})\nabla\mathbf{q},\tag{49}$$

which may be transformed to

$$-\nabla\mathrm{V}\mathbf{EH}=\dot{\mathrm{U}}+\dot{\mathrm{T}}+\nabla[\mathbf{q}(\mathrm{U}+\mathrm{T})]+\mathrm{P}\nabla.\mathbf{q}+\frac{\mathrm{U}}{c}\frac{\partial c}{\partial t}+\frac{\mathrm{T}}{\mu}\frac{\partial\mu}{\partial t},\tag{50}$$

where U and T are now the electric and magnetic energies $\frac{1}{2}$ED and $\frac{1}{2}$HB, whilst P is the eolotropic pressure defined by

$$\mathrm{P}=\mathrm{U}+\mathrm{T}-\mathbf{E}.\mathbf{D}-\mathbf{H}.\mathbf{B}.\tag{51}$$

Noting the presence of $\mathrm{P}\nabla.\mathbf{q}$ in (50), we may convert it to

$$-\nabla[\mathrm{V}\mathbf{EH}+\mathbf{q}(\mathrm{U}+\mathrm{T}+\mathrm{P})]=\dot{\mathrm{U}}+\dot{\mathrm{T}}+\mathbf{Fq}+\frac{\mathrm{U}}{c}\frac{\partial c}{\partial t}+\frac{\mathrm{T}}{\mu}\frac{\partial\mu}{\partial t},\tag{52}$$

where F is the moving force $-\nabla\mathrm{P}$. (If the stress were rotational, we should have to write $-\mathrm{P}\nabla$.)

(9). The interpretation of the $\partial c/\partial t$ and $\partial\mu/\partial t$ terms may be various. For example, if c and μ do not vary intrinsically, these terms vanish. (But not in an eolotropic medium, even then.) Now c and μ must, or nearly must, vary with m, the density of the ether. A very remarkable case is when they are both proportional to m. Then

$$\frac{\mathrm{U}}{c}\frac{\partial c}{\partial t}+\frac{\mathrm{T}}{\mu}\frac{\partial\mu}{\partial t}=\frac{\mathrm{U}+\mathrm{T}}{m}\frac{\partial m}{\partial t}=-(\mathrm{U}+\mathrm{T})\nabla\mathbf{q}.\tag{53}$$

Then (52) becomes

$$-\nabla[\mathrm{V}\mathbf{EH}+\mathbf{q}(\mathrm{U}+\mathrm{T}+\mathrm{P})]=\dot{\mathrm{U}}+\dot{\mathrm{T}}+\mathbf{Fq}-(\mathrm{U}+\mathrm{T})\nabla\mathbf{q}; \qquad (54)$$

and (50) becomes

$$-\nabla\mathrm{V}\mathbf{EH}=\frac{\partial\mathrm{U}}{\partial t}+\frac{\partial\mathrm{T}}{\partial t}+\mathrm{P}\nabla.\mathbf{q}. \qquad (55)$$

Now comes a case of splitting. For if

$$\mathrm{P}\nabla.\mathbf{q}=(\mathrm{U}+\mathrm{T})\nabla\mathbf{q} \quad \text{or} \quad \mathbf{E}.\mathbf{D}\nabla.\mathbf{q}+\mathbf{H}.\mathbf{B}\nabla.\mathbf{q}=0, \qquad (56)$$

the electromagnetic equation of activity splits into two, namely,

$$-\mathrm{P}\nabla.\mathbf{q}=\frac{\mathrm{U}}{c}\frac{\partial c}{\partial t}+\frac{\mathrm{T}}{\mu}\frac{\partial\mu}{\partial t}, \qquad -\nabla\mathrm{V}\mathbf{EH}=\dot{\mathrm{U}}+\dot{\mathrm{T}}+\nabla[\mathbf{q}(\mathrm{U}+\mathrm{T})]. \qquad (57)$$

This is the case of plane waves in a compressible ether, in which the permittivity and inductivity vary as the density, as was discussed before. But I do not see how to bring about the distortionless state in general. Nor is there any apparent reason for the assumption (56) in the general case, though perhaps Professor Bjerknes may make something out of it.

(10). The parallelism of Riemann's solutions for fluid motion and the case at present in question, noticed on p. 158, is worth exhibiting here. The equations for plane waves are

$$-\mathrm{H}'=\dot{\overline{c\mathrm{E}}}+(qc\mathrm{E})', \qquad -\mathrm{E}'=\dot{\overline{\mu\mathrm{H}}}+(q\mu\mathrm{H})', \qquad (58)$$

where the ′ and · stand for d/dx and d/dt, and q is the speed of the medium parallel to x. Or,

$$\begin{aligned}-\mathrm{H}'&=c\dot{\mathrm{E}}+qc\mathrm{E}'+\mathrm{E}(\dot{c}+[cq]'),\\ -\mathrm{E}'&=\mu\dot{\mathrm{H}}+q\mu\mathrm{H}'+\mathrm{H}(\dot{\mu}+[\mu q]').\end{aligned} \qquad (59)$$

Now if m is the density,

$$\dot{m}+\nabla(\mathbf{q}m)=0=\left(\frac{\partial}{\partial t}+\nabla\mathbf{q}\right)m, \quad \text{or} \quad \dot{m}+[mq]'=0; \qquad (60)$$

the third form being for plane waves. If c and μ vary as m, similar equations hold for them, so (59) become

$$-\mathrm{H}'=c\dot{\mathrm{E}}+cq\mathrm{E}'=c\frac{\partial\mathrm{E}}{\partial t}, \qquad -\mathrm{E}'=\mu\dot{\mathrm{H}}+\mu q\mathrm{H}'=\mu\frac{\partial\mathrm{H}}{\partial t}. \qquad (61)$$

Further defining v by $\mu cv^2=1$, the three quantities mv, cv, and μv are made to be constants.

By adding and subtracting, we obtain

$$\left(\frac{\partial}{\partial t}+v\frac{d}{dx}\right)(\mathrm{E}+\mu v\mathrm{H})=0, \quad \left(\frac{\partial}{\partial t}-v\frac{d}{dx}\right)(\mathrm{E}-\mu v\mathrm{H})=0. \qquad (62)$$

The speed of the positive wave $\mathrm{E}+\mu v\mathrm{H}$ is $v+q$, that of the negative wave $v-q$, the q being contained in $\partial/\partial t$.

(11). Now in the fluid theory, if $v^2=d\mathrm{P}/dm$, P being the pressure,

$$-v^2m'=m\frac{\partial q}{\partial t}, \qquad -mq'=\frac{\partial m}{\partial t}, \qquad (63)$$

are the equations of connection of m and q. If we make mv=constant, we have

$$m'=(mv/v)'=-mv'/v, \quad \text{and} \quad \partial m/\partial t=-(m/v)\partial v/\partial t; \qquad (64)$$

so we can eliminate m, and get

$$v\dot{v}=\frac{\partial q}{\partial t}, \quad vq'=\frac{\partial v}{\partial t}; \quad \left(\frac{\partial}{\partial t}+v\frac{d}{dx}\right)(q-v)=0, \quad \left(\frac{\partial}{\partial t}-v\frac{d}{dx}\right)(q+v)=0; \qquad (65)$$

Compare with (62). Equations (65) are what Riemann's solutions are when mv= constant, which requires

$$m^2\frac{dP}{dm}=\text{const}, \quad \text{or} \quad P=\text{const}-\frac{\text{const}}{m}. \qquad (66)$$

A striking particular case is got by making $v=q$ in the first pair of (65). The members of the pair become identical. So $v=q=f(x)$, when mv= const. is one solution. Similarly $v=-q=g(x)$ is another. In the first case, the waves are $q_1=\frac{1}{2}u$ at speed $2v$, positive wave, and $q_2=q-\frac{1}{2}u$ at speed 0, negative wave. In the second case, $q_1=\frac{1}{2}u-q$ at speed 0, positive wave, and $q_2=-\frac{1}{2}u$, at speed $2v$, negative wave.

If we eliminate v' and $\dot{v}$ from (65) we obtain

$$q''=\frac{1}{v}\frac{\partial}{\partial t}\frac{1}{v}\frac{\partial}{\partial t}q, \qquad (67)$$

which looks well, but is not so workable as (65).

(12). The operator $\partial/\partial t+\nabla \mathbf{q}$, which occurs in fluid and solid theory, and in electromagnetics, when applied to a scalar or a vector, produces the strength of source of the operand. Denote the operator here by (s), and apply it to a few quantities. If m is the operand, then if it is scalar,

$$(s)m=\frac{\partial m}{\partial t}+m.\nabla\mathbf{q}=\dot{m}+\nabla(m\mathbf{q}). \qquad (68)$$

This applies to the density of moving matter, for example. If m is made a vector, both forms of expansion apply to the three scalar components, but only the first form to the complete vector. For example, the momentum $m\mathbf{q}$,

$$s(m\mathbf{q})=\frac{\partial m\mathbf{q}}{\partial t}+m\mathbf{q}.\nabla\mathbf{q}=m(s)\mathbf{q}+\mathbf{q}(s)m-m\mathbf{q}.\nabla\mathbf{q}. \qquad (69)$$

If $s(m)=0$, as usual in hydrodynamics, this reduces to $m\partial\mathbf{q}/\partial t$, as usual also.

Similarly, if U is the potential energy,

$$(s)\mathrm{U}=\dot{\mathrm{U}}+\nabla(\mathbf{q}\mathrm{U})=m\frac{\partial}{\partial t}\frac{\mathrm{U}}{m}=\left(\mathrm{U}-m\frac{d\mathrm{U}}{dm}\right)\nabla\mathbf{q}, \qquad (70)$$

if U is a function of m only. It is also $-\mathrm{P}.\nabla\mathbf{q}$, quite independently, if desired, from the idea of pressure. So (70) gives P in terms of U by a differential process, instead of the former integrational and differential.

If T is the kinetic energy,

$$(s)\mathrm{T}=\dot{\mathrm{T}}+\nabla(\mathbf{q}\mathrm{T})=m\frac{\partial}{\partial t}\frac{\mathrm{T}}{m}=m\mathbf{q}\frac{\partial\mathbf{q}}{\partial t}=-\mathbf{q}\nabla\mathrm{P}. \qquad (71)$$

If ab is a scalar product,

$$(s)\mathbf{ab}=\mathbf{b}(s)\mathbf{a}+\mathbf{a}(s)\mathbf{b}-\mathbf{ab}.\nabla\mathbf{q}, \qquad (72)$$

and (71) may be done by (72) also, by $\mathrm{T}=\frac{1}{2}m\mathbf{qq}$.

Again, $(s)\mathrm{P}$, by introducing U, leads to

$$(s)\mathrm{P}=\left[-\mathrm{U}+m\frac{d\mathrm{U}}{dm}-m^2\frac{d^2\mathrm{U}}{dm^2}\right]\nabla\mathbf{q}, \qquad (s)(\mathrm{P}+\mathrm{U})=-m^2\frac{d^2\mathrm{U}}{dm^2}\nabla\mathbf{q}. \qquad (73)$$

When there is impressed force f, it generates momentum and pressure. Thus,

$$\mathbf{f}=\nabla\left(m\frac{d}{dm}-1\right)\mathrm{U}+(s)m\mathbf{q}=m\left(\nabla\frac{d\mathrm{U}}{dm}+\frac{\partial\mathbf{q}}{\partial t}\right)=-\nabla\frac{(s)\mathrm{U}}{\nabla\mathbf{q}}+\frac{(s)\mathrm{T}}{\mathbf{q}}. \qquad (74)$$

(13). In vortex motion $(s)\mathrm{V}\nabla\mathbf{q}$ is important, and also $(s)\mathrm{V}\nabla\mathbf{q}/m$. Thus,

$$\frac{\mathbf{f}-\nabla\mathrm{P}}{m}=\frac{\partial\mathbf{q}}{\partial t}, \qquad (75)$$

by curling and writing $\mathbf{q}_1$ for $\mathrm{V}\nabla\mathbf{q}$ leads to

$$\mathrm{V}\nabla\frac{\mathbf{f}}{m}=\mathrm{V}\nabla(\dot{\mathbf{q}}+\mathrm{V}\mathbf{q}_1\mathbf{q})=\dot{\mathbf{q}}_1+\mathbf{q}_1.\nabla\mathbf{q}+\mathbf{q}\nabla.\mathbf{q}_1-\mathbf{q}_1\nabla.\mathbf{q}, \qquad (76)$$

$$=(s)\mathbf{q}_1-\mathbf{q}_1\nabla.\mathbf{q}. \qquad (77)$$

So

$$(s)\mathbf{q}_1=\mathrm{V}\nabla\frac{\mathbf{f}}{m}+\mathbf{q}_1\nabla.\mathbf{q}=m\frac{\partial}{\partial t}\frac{\mathbf{q}_1}{m}, \qquad (78)$$

equivalent to Helmholtz and Nanson, with impressed force added.

The interpretation is done by equation (3) above. Thus (78) and (3) are

$$\frac{1}{m}\mathrm{V}\nabla\frac{\mathbf{f}}{m}=\frac{\partial}{\partial t}\frac{\mathbf{q}_1}{m}-\frac{\mathbf{q}_1}{m}\nabla.\mathbf{q}, \qquad 0=\frac{\partial}{\partial t}\mathbf{s}-\mathbf{s}\nabla.\mathbf{q}. \qquad (79)$$

If the curl of the impressed force per unit mass is zero, $\mathbf{q}_1/m$ in the first equation becomes strictly comparable with s in the second, and this s may be taken to be an infinitesimal element of length upon a moving vortex line.

But I consider Kelvin's way superior for simplicity and clearness physically. Then we take s to be element of length on any closed circuit of particles, and consider time variation of the circulation. Thus,

$$\frac{\partial}{\partial t}\Sigma(\mathbf{qs})=\Sigma\left(\mathbf{s}\frac{\partial\mathbf{q}}{\partial t}+\mathbf{q}\frac{\partial\mathbf{s}}{\partial t}\right)=\Sigma\left(\mathbf{s}\frac{\partial\mathbf{q}}{\partial t}+\mathbf{q}.\mathbf{s}\nabla.\mathbf{q}\right)=\Sigma\mathbf{s}\frac{\partial\mathbf{q}}{\partial t}, \qquad (80)$$

by using (79), and then noting that $\nabla\frac{1}{2}q^2$ is polar,

$$=\Sigma\mathbf{s}\frac{\mathbf{f}-\nabla\mathrm{P}}{m}=\Sigma\frac{\mathbf{f}}{m}\mathbf{s}; \qquad (81)$$

and the general theorem of curl makes

$$\Sigma(\mathbf{qs})=\Sigma\mathbf{q}_1\mathbf{n}, \qquad \text{so} \qquad \Sigma\frac{\mathbf{f}}{m}\mathbf{s}=\frac{\partial}{\partial t}\Sigma(\mathbf{qs})=\frac{\partial}{\partial t}\Sigma(\mathbf{q}_1\mathbf{n}). \qquad (82)$$

The above equations contain the leading properties.

If we ask, apart from the mathematics, why rotation of the fluid elements is simply carried with the fluid, the answer of course is, because there is no tangential stress concerned. A wider view is obtained by considering waves in an elastic solid. There are two sorts, controlled by the normal and the tangential stresses, or by the compressibility and the rigidity, with different speeds. But besides these wave propagations of motion, there is convection. It is usual to disregard the convection in examining the wave propagation. This convection is analogous to the motion of vortex tubes in a fluid. In fact, if we do away with the rigidity, the corresponding speed of propagation is brought to zero, and there is only the convection of rotation left, that is, the motion of vortex tubes in a fluid.

Carrying this further, it is natural to abolish the other wave, the compressional. To do this, $d\mathrm{P}/dm=0$, or P is independent of m. Say, $\mathrm{P}=0$. Now there is simply convection of the compression left, without wave propagation. In the absence of impressed force, we have $\partial\mathbf{q}/\partial t=0$. Also U varies as m. And T or $\frac{1}{2}mq^2$ is constant for the same group of particles, like U, and $\mathbf{q}$. Also div $\mathbf{q}$. Moreover, $(s)(\mathrm{U}+\mathrm{T})=0$. The vortex properties are unaffected, though of course the character of the motion is different.

Theory of an Electric Charge in Variable Motion.

§ 534. (1). On pp. 158 to 167 of this volume, I have given the leading results concerning the motion of a point-charge, or very small spherical surface charge, when in any state of motion. All details of the mere working were omitted, for want of space. That was of little consequence to the experts who were the only persons likely to read the stuff, and who were naturally fully competent to work out and verify the results. But it came to my knowledge that some readers had difficulty in working out the problems. Even making due allowance for the difference between working an original investigation in ignorance of the results, and a verification of the results given, there was, on consideration, some difficulty in the working. I shall now, therefore, for the benefit of a larger number of readers, work out the problem practically in full, without attempting short cuts or making large omissions.

The problem is this. A point-charge, which will be called an electron for convenience, moves about in the ether in any path at any speed. Find the accompanying electromagnetic phenomena. There are different ways, and the more the better, if they help one another.

(2). The charge, say Q, in motion at velocity $\mathbf{u}$, constitutes an impressed current element of moment $\mathrm{Q}\mathbf{u}$. Now the magnetic force is the curl of the potential of the current. Here, at the very beginning, however, we meet a distinction between the potential of a current element at rest, and of the same in motion. It is not merely Q that moves through the ether, but the current Qu as well. The simplest case is that of steady straight motion. This I did in 1888 by one way, and by several other ways since. The potential P of Q is changed from $Q/4\pi R'c$ to

$$P=\frac{Q/4\pi R'c}{[1-(u^2/v^2)\sin^2\beta]^{\frac{1}{2}}}, \tag{1}$$

where β is the angle between **R**′ and u, **R**′ being the vector from Q to the point of observation, say N. In the figure, Q moves horizontally; its position momentarily determines the centre of a moving field of magnetic force, and also a moving field of electric force. They are quite symmetrical about the point Q; in fact they move with Q as though rigidly attached to it. The magnetic force is a new creation. The electric force is not. But it is not the same as if Q were at rest. The distribution of displacement is distorted. It is axially attenuated, and equatorially condensed. Full particulars have been previously given.

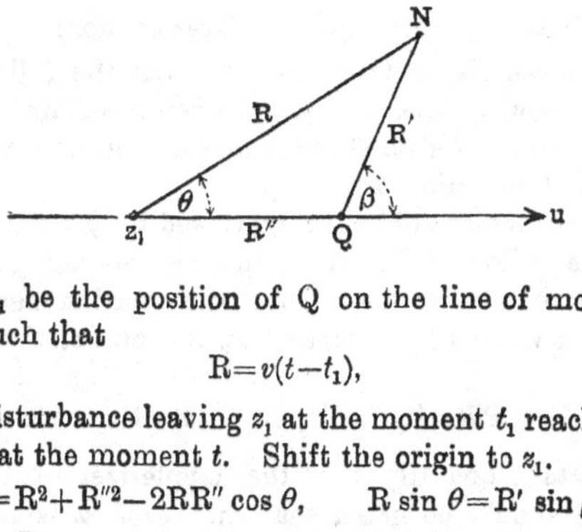

Now let z_1 be the position of Q on the line of motion at the moment t_1 such that

$$R=v(t-t_1), \tag{2}$$

so that the disturbance leaving z_1 at the moment t_1 reaches N when Q reaches z, at the moment t. Shift the origin to z_1. We have

$$R'^2=R^2+R''^2-2RR''\cos\theta, \qquad R\sin\theta=R'\sin\beta, \tag{3}$$

$$\frac{R}{v}=\frac{R''}{u}, \qquad R'^2\left(1-\frac{u^2}{v^2}\sin^2\beta\right)=R^2\left(1-\frac{u}{v}\cos\theta\right)^2, \tag{4}$$

by elementary geometry. So (1) becomes

$$P=\frac{Q/4\pi Rc}{1-(u/v)\cos\theta}. \tag{4A}$$

This looks simpler than (1), but would be less easy to work with than (1) because it is not at the centre of symmetry, if steady straight motion is in question.

Similarly, the H formula is given by the two forms

$$H=\frac{Qu[1-u^2/v^2]\sin\beta}{4\pi R'^2[1-(u^2/v^2)\sin^2\beta]^{\frac{3}{2}}}=\frac{Qu[1-u^2/v^2]\sin\theta}{4\pi R^2[1-(u/v)\cos\theta]^3}. \tag{4B}$$

(3). If these formulas are to be valid over all space, it is evident that Q must have been in straight steady motion for all past time. If the charge began to move only from z_1 at the moment t_1, then the region of disturbance is confined to the sphere of radius R, and outside this sphere is the original stationary electric force. This becomes clear when the finiteness of v is remembered. But inside this sphere, the new distorted and progressive state of electric force, accompanied by magnetic force, is fully established. If then, we make Q move along a path consisting of a series of straight lines of any lengths and directions, the speed being constant in each piece, though variable from piece to piece, the state of electric and magnetic force set up becomes fully known by the above in all the regions between the spherical surfaces whose centres are at the junctions of the separate pieces. Moreover, since the state of electric displacement changes in crossing any one of these spherical surfaces, whilst the displacement itself is perfectly continuous, in the general sense of no divergence, we can, by taking differences, find the electromagnetic state in the surfaces themselves. Every one of them constitutes an electromagnetic pulse, which is a phenomenon of radiation. This will come later in more detail.

(4). Another way is to derive results from the fully worked out problem of a growing line of electric current, as done before, pp. 102 to 117. The results for an element of moving current then follow by a terminal differentiation.

Let dz_1 be an element of length traversed by Q from the moment t_1 to t_1+dt_1, and let t and $t+dt$ be the corresponding moments at N, the point of observation. Then the relation between P the potential at N when Q moves, and P_0 the potential when it is at rest, is

$$\mathrm{P}dt=\mathrm{P}_0 dt_1, \qquad \text{or} \qquad \mathrm{P}=\mathrm{P}_0\dot{t}_1. \tag{5}$$

This important quantity $\dot{t}_1$ is the dopplerizer of the potential function. It should be noted that the speed u is not z_1, but is dz_1/dt_1. To find $\dot{t}_1$, we have, if r is the distance of N from the line of motion,

$$\mathrm{R}^2=v^2(t-t_1)^2=r^2+(z-z_1)^2, \tag{6}$$

$$v^2(t-t_1)(1-\dot{t}_1)=-(z-z_1)\dot{z}_1=-(z-z_1)u\dot{t}_1,$$

$$\mathrm{R}(1-\dot{t}_1)=-(z-z_1)u\dot{t}_1/v,$$

therefore
$$\dot{t}_1=\frac{1}{u}\dot{z}_1=\frac{1}{1-(u/v)\cos\theta}=-\frac{\dot{\mathrm{R}}}{u}. \tag{7}$$

So the scalar and vector potentials are

$$\mathrm{P}=\frac{\mathrm{Q}\dot{t}_1}{4\pi\mathrm{R}c}, \qquad \mathbf{A}=\frac{\mathrm{Q}u\dot{t}_1}{4\pi\mathrm{R}}=-\frac{\mathrm{Q}}{4\pi}\frac{\dot{\mathrm{R}}}{\mathrm{R}}, \tag{8}$$

from which the electric and magnetic forces follow by

$$\mathbf{E}=-\nabla\mathrm{P}-\mu\dot{\mathbf{A}},\qquad \mathbf{H}=\mathrm{V}\nabla\mathbf{A}, \tag{9}$$

formally in Maxwell's manner, but with specially defined potentials. The auxiliary conditions are $\nabla\mathbf{D}=0$ except at Q itself, where the total divergence must be Q, and $\nabla\mathbf{H}=0$ everywhere. These may be tested separately.

(6). The disturbance leaving Q at z_1 at moment t_1 reaches N at moment t. Describe a sphere through N, with centre at Q. Then, a little later, when Q has reached z_1+dz_1, describe about Q another sphere of radius sufficient to reach N. Now at the moment this sphere reaches N, the first sphere has expanded, and is wholly outside the second. The space between is bounded by two spherical surfaces with different centres. So the shell varies in its depth, which is

$$v\,dt_1[1-(u/v)\cos\theta]=v\,dt_1/\dot{t}_1=v\,dt. \tag{10}$$

Useful subsidiary formulas are

$$\dot{\mathrm{R}}=v(1-\dot{t}_1),\qquad \ddot{\mathrm{R}}=-v\ddot{t}_1,\qquad \dot{\mathbf{R}}=-\mathbf{u}\dot{t}_1,\qquad \ddot{\mathbf{R}}=-\dot{\mathbf{u}}\dot{t}_1-\mathbf{u}\ddot{t}_1. \tag{11}$$

$$\ddot{t}_1=\frac{\dot{t}_1}{\mathrm{R}v}[v^2\,1-t_1)^2-\dot{t}_1(u^2\dot{t}_1-\mathbf{R}\dot{\mathbf{u}})]. \tag{12}$$

$$\nabla\mathbf{R}=\frac{\mathbf{Ru}}{v}\dot{t}_1+3,\qquad \nabla\mathbf{R}_1=\frac{2}{\mathrm{R}},\qquad \nabla\dot{\mathbf{R}}_1=-\frac{2v(1-\dot{t}_1)}{\mathrm{R}^2}. \tag{13}$$

$$\nabla\frac{\dot{\mathbf{R}}}{\mathrm{R}}=-\frac{\ddot{\mathrm{R}}}{\mathrm{R}v}-\frac{\dot{\mathrm{R}}\dot{t}_1}{\mathrm{R}^2}=\frac{\ddot{t}_1}{\mathrm{R}}+\mathbf{R}_1\mathbf{u}\frac{\dot{t}_1^{\,2}}{\mathrm{R}^2}. \tag{14}$$

$$\nabla\mathbf{u}=\frac{d\mathbf{u}}{dt_1}\nabla t_1=-\frac{\dot{t}_1}{v}\mathbf{R}_1\frac{d\mathbf{u}}{dt_1}=-\frac{\mathbf{R}_1\dot{\mathbf{u}}}{v}. \tag{15}$$

Except in the last set, I have not given the working, which is plain enough by vector analysis. Note that the acceleration is not $\dot{\mathbf{u}}$, but $d\mathbf{u}/dt_1$. The symbol $\mathbf{R}_1$ means unit vector $\mathbf{R}$.

(7). Here is one derivation of the $\mathbf{H}$ formula, using some of the subsidiary results:—

$$\mathrm{V}\nabla\frac{\dot{\mathbf{R}}}{\mathrm{R}}=\frac{1}{\mathrm{R}}\mathrm{V}\nabla\dot{\mathbf{R}}+\mathrm{V}\nabla\left(\frac{1}{\mathrm{R}}\right).\dot{\mathbf{R}}=\frac{1}{\mathrm{R}}\frac{d}{dt}\mathrm{V}\nabla\mathbf{R}-\frac{\dot{t}_1}{\mathrm{R}^2}\mathrm{V}\mathbf{R}_1\dot{\mathbf{R}}$$

$$=-\frac{1}{v\mathrm{R}}\frac{d}{dt}\mathrm{V}\mathbf{R}_1\dot{\mathbf{R}}-\frac{\dot{t}_1}{\mathrm{R}^2}\mathrm{V}\mathbf{R}_1\dot{\mathbf{R}}=-\frac{1}{v\mathrm{R}}\frac{d}{dt}\frac{\mathrm{V}\mathbf{R}\dot{\mathbf{R}}}{\mathrm{R}}-\frac{\dot{t}_1}{\mathrm{R}^2}\mathrm{V}\mathbf{R}_1\dot{\mathbf{R}}$$

$$=-\frac{1}{v\mathrm{R}}\left(-\frac{\dot{\mathrm{R}}}{\mathrm{R}^2}\mathrm{V}\mathbf{R}\dot{\mathbf{R}}+\frac{1}{\mathrm{R}}\mathrm{V}\mathbf{R}\ddot{\mathbf{R}}\right)-\frac{\dot{t}_1}{\mathrm{R}^2}\mathrm{V}\mathbf{R}_1\dot{\mathbf{R}}$$

$$=-\frac{1}{\mathrm{R}v}\left\{\frac{v(1-\dot{t}_1)}{\mathrm{R}^2}\mathrm{V}\mathbf{Ru}\dot{t}_1+\frac{1}{\mathrm{R}}\mathrm{V}\mathbf{R}(-\mathbf{u}\ddot{t}_1-\dot{\mathbf{u}}\dot{t}_1)\right\}+\frac{\dot{t}_1}{\mathrm{R}^2}\mathrm{V}\mathbf{R}_1\mathbf{u}\dot{t}_1$$

$$=\frac{\dot{t}_1}{\mathrm{R}^2v}\mathrm{V}\mathbf{R}\dot{\mathbf{u}}+\frac{1}{\mathrm{R}^2v}\mathrm{V}\mathbf{Ru}\frac{\dot{t}_1}{\mathrm{R}v}\left[(v^2-u^2)\dot{t}_1^{\,2}+v^2-2v^2\dot{t}_1+\dot{t}_1\mathbf{R}\dot{\mathbf{u}}\right]+\frac{\mathrm{V}\mathbf{Ru}\dot{t}_1}{\mathrm{R}^3}(2\dot{t}_1-1)$$

$$=\frac{\dot{t}_1}{\mathrm{R}^2v}\mathrm{V}\mathbf{R}\dot{\mathbf{u}}+\frac{\dot{t}_1^{\,2}\mathbf{R}\dot{\mathbf{u}}}{\mathrm{R}^3v^2}\mathrm{V}\mathbf{Ru}+\frac{\kappa^2\dot{t}_1^{\,3}}{\mathrm{R}^3}\mathrm{V}\mathbf{Ru} \tag{16}$$

So $\mathbf{H}=\mathrm{V}\nabla\mathbf{A}$ makes, if $\kappa^2=1-u^2/v^2$,

$$\mathbf{H}=-\frac{\mathrm{Q}}{4\pi}\frac{\dot{t}_1}{\mathrm{R}^2}\left[\frac{\dot{t}_1{}^2\kappa^2}{\mathrm{R}}\mathrm{V}\mathbf{R}\mathbf{u}+\frac{1}{v}\left\{\mathrm{V}\mathbf{R}\dot{\mathbf{u}}+\frac{\mathbf{R}\dot{\mathbf{u}}\dot{t}_1}{\mathrm{R}v}\mathrm{V}\mathbf{R}\mathbf{u}\right\}\right], \tag{17}$$

showing one term depending on $\mathbf{u}$, and two more upon $\mathbf{u}$ and $\dot{\mathbf{u}}$. The interpretation will come presently. As a check upon the accuracy of the result, do it by a different way, and come to the same result. Only by that means does a feeling of confidence in fallible working arise, in complicated work.

$$\frac{d}{dx}\frac{\mathbf{u}}{\mathrm{R}-\mathbf{R}\mathbf{u}/v}=\frac{d\mathbf{u}}{dx}\frac{\dot{t}_1}{\mathrm{R}}-\frac{\mathbf{u}\dot{t}_1{}^2}{\mathrm{R}^2}\left(\frac{d\mathrm{R}}{dx}-\frac{\mathbf{u}}{v}\frac{d\mathbf{R}}{dx}-\frac{\mathbf{R}}{v}\frac{d\mathbf{u}}{dx}\right)$$

$$=-\frac{\dot{t}_1\dot{\mathbf{u}}}{\mathrm{R}v}\mathbf{R}_1\mathbf{i}_1-\frac{\mathbf{u}\dot{t}_1{}^2}{\mathrm{R}^2}\left\{\dot{t}_1\mathbf{R}_1\mathbf{i}_1-\frac{\mathbf{u}}{v}\left(\mathbf{i}_1+\frac{\mathbf{u}\dot{t}_1}{v}\mathbf{R}_1\mathbf{i}_1\right)+\frac{1}{v^2}\mathbf{R}\dot{\mathbf{u}}.\mathbf{R}_1\mathbf{i}_1\right\}. \tag{18}$$

So, operating by $\mathrm{V}\mathbf{i}_1$,

$$\mathrm{V}\mathbf{i}_1\frac{d}{dx}(\ldots)=-\frac{\dot{t}_1\mathbf{R}_1\mathbf{i}_1}{\mathrm{R}v}\mathrm{V}\mathbf{i}\dot{\mathbf{u}}$$

$$-\frac{\dot{t}_1{}^2}{\mathrm{R}^2}\mathrm{V}\mathbf{i}_1\mathbf{u}\left\{\dot{t}_1\mathbf{R}_1\mathbf{i}_1-\frac{1}{v}\left(u_1+\frac{\dot{t}_1u^2}{v}\mathbf{R}_1\mathbf{i}_1\right)+\frac{\mathbf{R}\dot{\mathbf{u}}}{v^2}\mathbf{R}_1\mathbf{i}_1\right\}. \tag{19}$$

This being the vector $\mathbf{i}_1$ component of the curl, add on the $\mathbf{i}_2$ and $\mathbf{i}_3$ components symmetrically, making

$$\mathrm{V}\nabla\frac{\mathbf{u}}{(\mathrm{R}-\mathbf{R}\mathbf{u}/v)}=-\frac{\dot{t}_1}{\mathrm{R}v}\mathrm{V}\mathbf{R}_1\dot{\mathbf{u}}-\frac{\dot{t}_1{}^2}{\mathrm{R}^2}\left(\dot{t}_1\kappa^2+\frac{\mathbf{R}\dot{\mathbf{u}}}{v^2}\right)\mathrm{V}\mathbf{R}_1\mathbf{u}. \tag{20}$$

Therefore, finally,

$$\mathbf{H}=\mathrm{V}\nabla\frac{\mathrm{Q}\mathbf{u}}{4\pi(\mathrm{R}-\mathbf{R}\mathbf{u}/v)}=-\frac{\mathrm{Q}\dot{t}_1}{4\pi\mathrm{R}}\left[\frac{\dot{t}_1{}^2\kappa^2}{\mathrm{R}}\mathrm{V}\mathbf{R}_1\mathbf{u}+\frac{1}{v}\mathrm{V}\mathbf{R}_1\dot{\mathbf{u}}+\frac{\mathbf{R}_1\dot{\mathbf{u}}\dot{t}_1}{v^2}\mathrm{V}\mathbf{R}_1\mathbf{u}\right]. \tag{21}$$

This will be seen to be the same as (17), since $\mathbf{R}=\mathrm{R}\mathbf{R}_1$.

(8). Now to find the electric force due to the moving Q. First, the $-\dot{\mathbf{A}}$ part.

$$-\dot{\mathbf{A}}=-\frac{\mathrm{Q}}{4\pi}\frac{d}{dt}\frac{\mathbf{u}}{\mathrm{R}-\mathbf{R}\mathbf{u}/v}=-\frac{\mathrm{Q}\dot{\mathbf{u}}\dot{t}_1}{4\pi\mathrm{R}}+\frac{\mathrm{Q}\mathbf{u}\dot{t}_1{}^2}{4\pi\mathrm{R}^2}\left(\dot{\mathrm{R}}-\frac{\dot{\mathbf{R}}\mathbf{u}}{v}-\frac{\mathbf{R}\dot{\mathbf{u}}}{v}\right)$$

$$=-\frac{\mathrm{Q}\dot{\mathbf{u}}\dot{t}_1}{4\pi\mathrm{R}}+\frac{\mathrm{Q}\mathbf{u}\dot{t}_1{}^2}{4\pi\mathrm{R}^2}\left\{v(1-\dot{t}_1)+\frac{\mathbf{u}}{v}\mathbf{u}\dot{t}_1-\frac{\mathbf{R}\dot{\mathbf{u}}}{v}\right\}$$

$$=-\frac{\mathrm{Q}\dot{\mathbf{u}}\dot{t}_1}{4\pi\mathrm{R}}+\frac{\mathrm{Q}\mathbf{u}\dot{t}_1{}^2}{4\pi\mathrm{R}^2}\left\{-v\kappa^2\dot{t}_1+v-\frac{\mathbf{R}\dot{\mathbf{u}}}{v}\right\}. \tag{22}$$

Next the $-\nabla\mathrm{P}$ part. We have, by differentiation,

$$\nabla\frac{\dot{t}_1}{\mathrm{R}}=-\frac{\dot{t}_1{}^2}{\mathrm{R}^2}\left\{\mathbf{R}_1\dot{t}_1\kappa^2-\frac{\mathbf{u}}{v}+\frac{\mathbf{R}_1}{v^2}.\mathbf{R}\dot{\mathbf{u}}\right\}. \tag{23}$$

So
$$-\nabla\mathrm{P}=\frac{\mathrm{Q}\dot{t}_1{}^2}{4\pi\mathrm{R}^2c}\left\{\mathbf{R}_1\dot{t}_1\kappa^2-\frac{\mathbf{u}}{v}+\frac{\mathbf{R}_1}{v^2}.\mathbf{R}\dot{\mathbf{u}}\right\}. \tag{24}$$

The electric force being $-\mu\dot{\mathbf{A}}-\nabla P$, we now add $(22)\times\mu$ to (24). Observe that the middle term of (24) cancels the v in { } term of (22). What is left makes

$$\mathbf{E}=+\frac{Q\dot{t}_1^{\,3}\kappa^2}{4\pi R^2 c}\left(\mathbf{R}_1-\frac{\mathbf{u}}{v}\right)-\frac{\mu Q\dot{t}_1}{4\pi R}\left\{\dot{\mathbf{u}}+\frac{\mathbf{R}\dot{\mathbf{u}}\dot{t}_1}{R}\left(\frac{\mathbf{u}}{v}-\mathbf{R}_1\right)\right\}. \tag{25}$$

Here it will be recognised that the first part, which does not contain $\mathbf{u}$, represents the steady progressive field due to Q moving straight at steady velocity $\mathbf{u}$; or rather, a part of the same contained in a shell between two spherical surfaces. These are infinitely close together when $\mathbf{u}$ varies. But when $\mathbf{u}$ is steady for a time, then so is $\mathbf{E}$, except for the progressive motion and the expansion of the two shell skins. When $\mathbf{u}$ varies, again, the centre of symmetry of the quasi-steady field in the infinitely thin shell is at the point which Q would occupy at the moment t if the velocity $\mathbf{u}$ were maintained steady from the moment t_1 If $\mathbf{u}$ is not steady, this centre of symmetry moves, usually in a curve, also unsteadily. A diagram will make plain better than the mathematical formulas, though I do not see how the true results could be got without mathematics, not even by Faraday.

(9). If we put $\mathbf{E}=\mathbf{E}_1+\mathbf{E}_2$, and $\mathbf{H}=\mathbf{H}_1+\mathbf{H}_2$, where $\mathbf{E}_1$, $\mathbf{H}_1$ do not contain the acceleration, and $\mathbf{E}_2$, $\mathbf{H}_2$ do, then we have these simple relations,

$$\mathbf{H}_1=\mathrm{V}\mathbf{u}\mathbf{D}_1, \qquad \mathbf{H}_2=\mathrm{V}\mathbf{v}\mathbf{D}_2, \tag{26}$$

where $\mathbf{v}$ is the velocity of the expanding shell, that is, radial from the point z_1, and of size v. So only $\mathbf{E}_1$ and $\mathbf{E}_2$ need be specified, and may be written

$$\mathbf{E}_1=\frac{Q}{4\pi R^2 c}\mathbf{S}\kappa^2\dot{t}_1^{\,3}, \qquad \mathbf{E}_2=-\frac{\mu Q\dot{t}_1^{\,2}}{4\pi R}\left\{\mathbf{w}-\frac{\mathbf{S}}{R}\dot{t}_1.\mathbf{R}_1\mathbf{w}\right\}, \tag{27}$$

where $\mathbf{S}=\mathbf{R}-\mathbf{u}R/v$, the vector from Q in the figure above to N, the point of observation, when $\mathbf{u}$ is steady. As this vector is not in general drawn from the real position of the charge at time t, the new symbol $\mathbf{S}$ is introduced. These are the formulas given on p. 162, where however the acceleration was denoted by $\mathbf{A}$. To avoid the double use of $\mathbf{A}$, which is now the vector-potential, the acceleration is now called $\mathbf{w}$. It is given by $\dot{\mathbf{u}}=\mathbf{w}\dot{t}_1$.

If we multiply $\mathbf{E}_2$ by $\mathbf{R}_1$ we get 0, because $\mathbf{R}_1\mathbf{S}\dot{t}_1=R$. So $\mathbf{E}_2$ is perpendicular to $\mathbf{R}$, and therefore lies in the shell, or is tangential. Its direction in the shell is perpendicular to $\mathrm{V}\mathbf{w}\mathbf{S}$, because if we multiply $\mathbf{E}_2$ by this vector, we get 0 again. In fact, we may write

$$\mathbf{E}_2=-\frac{\mu Q\dot{t}_1^{\,3}}{4\pi R^2}\mathrm{V}\mathbf{R}_1\mathrm{V}\mathbf{w}\mathbf{S}, \tag{28}$$

showing the two perpendicularities at a glance.

Also, $\mathbf{H}_2 = cv\mathrm{V}\mathbf{R}_1\mathbf{E}_2$; so $\mathbf{H}_2$ is also entirely in the shell, and is perpendicular to $\mathbf{E}_2$, and $\mathbf{E}_2$, $\mathbf{H}_2$ together constitute a simple electromagnetic spherical pulse, in which $\mathrm{E} = \mu v \mathrm{H}$. But it requires to be supplemented by the quasi-static $\mathbf{E}_1$, $\mathbf{H}_1$ in the shell, to produce complete satisfaction as regards divergence of $\mathbf{D}$.

Since $\mathrm{V}\mathbf{u}\mathbf{D}_1 = \mathrm{V}\mathbf{v}\mathbf{D}_1$, we may also write $\mathbf{H} = \mathrm{V}\mathbf{v}\mathbf{D}$ complete, by (26).

(10). Another way of exhibiting the $\mathbf{E}$ result is got thus.

$$\mathbf{E} = \frac{\mu Q}{4\pi}\left\{\frac{d}{dt}\frac{\dot{\mathbf{R}}}{\mathrm{R}} + v\nabla\frac{\dot{\mathrm{R}} - v}{\mathrm{R}}\right\} = \dots\left\{\frac{d}{dt}\left(\frac{\dot{\mathbf{R}}}{\mathrm{R}} + \frac{v\nabla\mathrm{R}}{\mathrm{R}}\right) - v^2\nabla\frac{1}{\mathrm{R}}\right\}. \tag{29}$$

Here $\dot{\mathbf{R}} = -\mathbf{u}\dot{t}_1$, and $\nabla\mathrm{R} = \mathbf{R}_1\dot{t}_1$, so

$$\mathbf{E} = \dots\left\{\frac{d}{dt}\frac{\dot{t}_1}{\mathrm{R}}(\mathbf{v} - \mathbf{u}) - v^2\nabla\frac{1}{\mathrm{R}}\right\} = \dots\left\{\frac{d}{dt}\frac{v\dot{t}_1\mathbf{S}}{\mathrm{R}^2} - v^2\nabla\frac{1}{\mathrm{R}}\right\}, \tag{30}$$

where $\mathbf{S} = \mathbf{R} - \mathrm{R}\mathbf{u}/v$ and $\mathrm{R}/\dot{t}_1 = \mathbf{R}_1\mathbf{S}$. Also,

$$\frac{\dot{t}_1}{\mathrm{R}}(\mathbf{v} - \mathbf{u}) = \frac{1}{\mathrm{R}}(\mathbf{R}_1 v\dot{t}_1 + \dot{\mathbf{R}}) = \frac{1}{\mathrm{R}}(\dot{\mathbf{R}} + (v - \dot{\mathrm{R}})\mathbf{R}_1)$$

$$= \frac{1}{\mathrm{R}}(\mathrm{R}\dot{\mathbf{R}}_1 + v\mathbf{R}_1) = \dot{\mathbf{R}}_1 + \frac{v\mathbf{R}_1}{\mathrm{R}}. \tag{31}$$

So
$$\mathbf{E} = \frac{\mu Q}{4\pi}\left\{\ddot{\mathbf{R}}_1 + \frac{v}{\mathrm{R}^2}(\dot{\mathbf{R}} - 3\mathbf{R}_1\dot{\mathrm{R}} + v\mathbf{R}_1)\right\}$$

$$= \frac{\mu Q}{4\pi}\left\{\ddot{\mathbf{R}}_1 + \frac{v}{\mathrm{R}^2}(\mathrm{R}\dot{\mathbf{R}}_1 - 2\dot{\mathrm{R}}\mathbf{R}_1 + v\mathbf{R}_1)\right\} \tag{32}$$

as on p. 174. From this,

$$\mathbf{H} = \mathrm{V}\mathbf{v}\mathbf{D} = \frac{1}{\mu v}\mathrm{V}\mathbf{R}_1\mathbf{E} = \frac{Q}{4\pi v}\,\mathrm{V}\mathbf{R}_1\left(\ddot{\mathbf{R}}_1 + \frac{v\dot{\mathbf{R}}_1}{\mathrm{R}}\right). \tag{33}$$

In these, remember that $\mathbf{R}_1$ is unit vector.

(11). At a great distance we may write approximately

$$\mathbf{E} = \frac{\mu Q}{4\pi}\ddot{\mathbf{R}}_1, \tag{34}$$

which is str:kingly simple, but needs some development for use. Thus,

$$\ddot{\mathbf{R}}_1 = \frac{d^2}{dt^2}\frac{\mathbf{R}}{\mathrm{R}} = \frac{d}{dt}\left(\frac{\dot{\mathbf{R}}}{\mathrm{R}} - \frac{\mathbf{R}\dot{\mathrm{R}}}{\mathrm{R}^2}\right) = \frac{\ddot{\mathbf{R}}}{\mathrm{R}} - 2\frac{\dot{\mathbf{R}}\dot{\mathrm{R}}}{\mathrm{R}^2} + 2\frac{\mathbf{R}\dot{\mathrm{R}}^2}{\mathrm{R}^3} - \frac{\mathbf{R}\ddot{\mathrm{R}}}{\mathrm{R}^2}. \tag{34A}$$

From this the R^{-2} terms should be rejected again, making

$$\mathbf{E} = \frac{\mu Q}{4\pi\mathrm{R}}(\ddot{\mathbf{R}} - \ddot{\mathrm{R}}\mathbf{R}_1). \tag{35}$$

Now suppose an electron is revolving in an orbit whose vector radius from a fixed origin is $\mathbf{s}$, whilst $\mathbf{r}$ is the vector from the origin to the observing point. Then by differentiations,

$$\mathbf{R} = \mathbf{r} - \mathbf{s}, \quad \ddot{\mathbf{R}} = -\ddot{\mathbf{s}}; \quad \mathrm{R}^2 = r^2 + s^2 - 2\mathbf{r}\mathbf{s}, \quad \mathrm{R}\dot{\mathrm{R}} = \mathbf{s}\dot{\mathbf{s}} - \dot{\mathbf{s}}\mathbf{r},$$

$$\dot{\mathrm{R}}^2 + \mathrm{R}\ddot{\mathrm{R}} = \dot{\mathbf{s}}^2 + \mathbf{s}\ddot{\mathbf{s}} - \mathbf{r}\ddot{\mathbf{s}}, \qquad \ddot{\mathrm{R}} = \frac{\dot{\mathbf{s}}^2 + \mathbf{s}\ddot{\mathbf{s}} - \mathbf{r}\ddot{\mathbf{s}}}{\mathrm{R}} - \frac{(\mathbf{s}\dot{\mathbf{s}} - \dot{\mathbf{s}}\mathbf{r})^2}{\mathrm{R}^3}. \tag{36}$$

So, at a distance, when **r** and **R** are practically the same,

$$\ddot{\mathbf{R}}=-\ddot{\mathbf{s}}\mathbf{r}_1, \qquad \mathbf{E}=-\frac{\mu Q}{4\pi R}(\ddot{\mathbf{s}}-\ddot{\mathbf{s}}\mathbf{r}_1.\mathbf{r}_1). \tag{37}$$

That is, if we describe a sphere of radius **r**, the tangential component on its surface of the vector $-(\mu Q/4\pi R)\ddot{\mathbf{s}}$ is the electric force due to Q moving in any small orbit **s**. But the formula will fail when s/R is not very small. Electronic orbits may be imagined to be excessively small, but R may be so too.

(12). Let the path be elliptic, with velocity of the spring kind. Say,

$$\mathbf{s}=\frac{1}{n}\Big\{u_2\mathbf{i}_1\cos+u_1\mathbf{i}_2\sin\Big\}nt_1, \quad \mathbf{u}=\frac{d\mathbf{s}}{dt_1}=\Big\{-u_2\mathbf{i}_1\sin+u_1\mathbf{i}_2\cos\Big\}nt_1. \tag{38}$$

Then

$$n^{-1}(1-\mathbf{r}_1.\mathbf{r}_1)\mathbf{s}=\Big[\mathbf{i}_1u_2\cos+\mathbf{i}_2u_1\sin-\mathbf{r}_1\Big(\frac{x}{r}u_2\cos+\frac{y}{r}u_1\sin\Big)\Big]nt_1 \tag{39}$$

$$=\mathbf{i}_1\Big\{u_2\cos-\Big(\frac{x^2}{r^2}u_2\cos+\frac{xy}{r^2}u_1\sin\Big)\Big\}nt_1$$

$$+\mathbf{i}_2\Big\{u_1\sin-\Big(\frac{xy}{r^2}u_2\cos+\frac{y^2}{r^2}u_1\sin\Big)\Big\}nt_1$$

$$-\mathbf{i}_3\Big\{\frac{xz}{r^2}u_2\cos+\frac{yz}{r^2}u_1\sin\Big\}nt_1, \tag{40}$$

where $\qquad x=r\sin\theta\cos\phi, \quad y=r\sin\theta\sin\phi, \quad z=r\cos\theta.$ (41)

Also, $\qquad s_\theta=-\dfrac{s_3}{\sin\theta}, \qquad s_\phi=s_2\cos\phi-s_1\sin\phi.$ (42)

The components of the projection of $\mathbf{s}/n$ on the spherical surface are therefore

$$n^{-1}s_\theta=\cos\theta(u_2\cos nt_1\cos\phi+u_1\sin nt_1\sin\phi), \tag{43}$$

$$n^{-1}s_\phi=(u_1\sin nt_1\cos\phi-u_2\cos nt_1\sin\phi)\,; \tag{44}$$

and the components of the electric force at a distance due to the orbitally revolving Q are

$$E_\theta=\mu vH_\phi=-\frac{\mu Q}{4\pi Rn}\frac{d^2}{dt^2}\cos\theta\,(u_2\cos nt_1\cos\phi+u_1\sin nt_1\sin\phi), \tag{45}$$

$$E_\phi=-\mu vH_\theta=\frac{\mu Q}{4\pi Rn}\frac{d^2}{ut^2}(u_2\cos nt_1\sin\phi-u_1\sin nt_1\cos\phi). \tag{46}$$

(13). In case of circular motion $u_1=u_2=u$. Then

$$E_\theta=-\frac{\mu Qu}{4\pi Rn}\cos\theta\frac{d^2}{dt^2}\cos(nt_1-\phi), \tag{47}$$

$$E_\phi=-\frac{\mu Qu}{4\pi Rn}\frac{d^2}{dt^2}\sin(nt_1-\phi). \tag{48}$$

But a linear oscillator had better be done with different coordinates. Let Ω be the angle between **r** and the y axis, supposed to be the axis of vibration. The projection of **s** on the sphere is simply $nu \sin \Omega \sin nt_1$. Lines of longitude. So

$$\mathrm{E}=\mu v\mathrm{H}=-\frac{\mu \mathrm{Q}u}{4\pi \mathrm{R}n}\frac{d^2}{dt^2}\sin \Omega . \sin nt_1. \tag{49}$$

These formulas are deceptively simple. The developments arising from d^2/dt^2 will follow. They are closely related to the old theory of the mean and eccentric anomalies in planetary motion. Also, the more general developments, including the state of **E** and **H** right up to the electron itself, will come in due course.

(14). To obtain the exact formulas for a circular orbit, at any distance therefrom, use the ordinary spherical coordinates. The circle is of radius a, in the plane x, y, whilst z is the axis of revolution; θ is measured from the z axis, and ϕ from the x axis. The charge Q revolves positively in the circle, at speed u, and angular speed $n=u/a$, and $\phi=0$ when $t_1=0$ fixes the epoch. So

$$\mathbf{a}=a(\mathbf{i}_1 \cos+\mathbf{i}_2 \sin)nt_1, \qquad \mathbf{u}=u(-\mathbf{i}_1 \sin+\mathbf{i}_2 \cos)nt_1, \tag{50}$$

$$\mathbf{R}=\mathbf{r}-\mathbf{a}, \qquad \mathbf{Ru}=\mathbf{ru}, \qquad d\mathbf{u}/dt_1=-n^2\mathbf{a}, \qquad \phi_1=\phi-nt_1, \tag{51}$$

$$\mathbf{Ru}=ru \sin \theta \sin(\phi-nt_1)=ru \sin \theta \sin \phi_1, \tag{52}$$

$$\mathrm{R}^2=a^2+r^2-2ar \sin \theta \cos \phi_1, \qquad \mathbf{ra}=ra \sin \theta \cos \phi_1. \tag{53}$$

From these, the formula for the vector potential is

$$\mathbf{A}=\frac{\mathrm{Q}u(\mathbf{i}_2 \cos -\mathbf{i}_1 \sin)nt_1}{4\pi[\{r^2+a^2-2ar \sin \theta \cos \phi_1\}^{\frac{1}{2}}-(ur/v) \sin \theta \sin \phi_1]}. \tag{54}$$

Now find the radiational part of **H**, or $\mathbf{H}_2$. It is, by (17),

$$\mathbf{H}_2=-\frac{\mathrm{Q}\dot{t}_1^2}{4\pi \mathrm{R}^2 v}\mathrm{V}\mathbf{R}\left[\frac{d\mathbf{u}}{dt_1}+\frac{\mathbf{u}\dot{t}_1}{\mathrm{R}v}\,\mathbf{R}\frac{d\mathbf{u}}{dt_1}\right]. \tag{55}$$

By the preceding formulas (50) to (53), we find

$$\mathrm{V}\mathbf{Ru}=\mathbf{i}_1(-zu \cos nt_1)+\mathbf{i}_2(-zu \sin nt_1)+\mathbf{i}_3[(xu \cos+yu \sin)nt_1-au] \tag{56}$$

$$\mathrm{V}\mathbf{R}\frac{d\mathbf{u}}{dt_1}=nu[\mathbf{i}_1 z \sin nt_1-\mathbf{i}_2 z \cos nt_1+\mathbf{i}_3(y \cos-x \sin)nt_1]. \tag{57}$$

Insert these in (55), and we get, after rearrangements,

$$\mathbf{H}_2=-\frac{\mathrm{Q}nu}{4\pi \mathrm{R}v}\dot{t}_1^2\Big[\mathbf{i}_1\Big\{\cos \theta \sin nt_1+\frac{ur\dot{t}_1}{v\mathrm{R}}\cos \theta \cos nt_1(\sin \theta \cos \phi_1-a/r)\Big\}$$
$$+\mathbf{i}_2\{-\cos \theta \cos nt_1+\ldots\ldots\ldots \sin nt_1(\ldots\ldots\ldots\ldots\ldots\ldots\ldots)\}$$
$$+\mathbf{i}_3\Big\{\sin \theta \sin \phi_1-\frac{ur\dot{t}_1}{v\mathrm{R}}\Big(\sin \theta \cos \phi_1-\frac{a}{r}\Big)^2\Big\}\Big] \tag{58}$$

This is exact, but requires the value of $\dot{t}_1$ and of $\dot{t}_1^2$ to be put in. We have

$$\frac{\dot{t}_1}{\mathrm{R}}=\frac{1}{r\Big[\Big(1+\frac{a^2}{r^2}-\frac{2a}{r}\sin \theta \cos \phi_1\Big)^{\frac{1}{2}}-\frac{u}{v}\sin \theta \sin \phi_1\Big]} \tag{59}$$

The first approximation is

$$\frac{\dot{t}_1}{\mathrm{R}}=\frac{1}{r}\left[1+\frac{a}{r}\sin\theta\cos\phi_1+\frac{u}{v}\sin\theta\sin\phi_1\right]. \tag{60}$$

Now a/r at a distance from the orbit may be a very small fraction of u/v. If so, we have, by (59),

$$\frac{rt_1}{\mathrm{R}}=\left(1-\frac{u}{v}\sin\theta\sin\phi_1\right)^{-1}. \tag{61}$$

This dismissal of a/r from (59) produces great simplification. Put $a/r=0$ and $r/\mathrm{R}=1$ in (58). It reduces to

$$\begin{aligned}\mathbf{H}_2=-\frac{Qnu}{4\pi rv}\dot{t}_1^{\,2}\Big[&\mathbf{i}_1\left\{+\cos\theta\sin nt_1+\frac{u}{v}\dot{t}_1\cos\theta\cos nt_1\sin\theta\cos\phi_1\right\}\\ &+\mathbf{i}_2\{-\cos\theta\cos nt_1+\;\ldots\ldots\ldots\;\sin nt_1\;\ldots\ldots\ldots\ldots\}\\ &+\mathbf{i}_3\left\{\sin\theta\sin\phi_1-\frac{u}{v}\dot{t}_1\sin^2\theta\cos^2\phi_1\right\}\Big],\end{aligned} \tag{62}$$

in which

$$\dot{t}_1=\left(1-\frac{u}{v}\sin\theta\sin\phi_1\right)^{-1}. \tag{63}$$

Divide into θ and ϕ components.

$$\mathrm{H}_{2\theta}=\frac{Qnu\dot{t}_1^{\,2}}{4\pi rv}\left\{\sin\phi_1-\frac{u\dot{t}_1}{v}\sin\theta\cos^2\phi_1\right\}, \tag{64}$$

$$\mathrm{H}_{2\phi}=\;\ldots\ldots\;\left\{1+\frac{u\dot{t}_1}{v}\sin\theta\sin\phi_1\right\}\cos\theta\cos\phi_1. \tag{65}$$

But here

$$\sin\phi_1-\frac{(u/v)\sin\theta\cos^2\phi_1}{1-(u/v)\sin\theta\sin\phi_1}=\left(\sin\phi_1-\frac{u}{v}\sin\theta\right)\dot{t}_1, \tag{66}$$

and

$$1+\frac{u\dot{t}_1}{v}\sin\theta\sin\phi_1=\dot{t}_1. \tag{67}$$

So (64), (65) are transformed to

$$\mathrm{H}_{2\theta}=\frac{Qnu}{4\pi rv}\dot{t}_1^{\,3}\left(\sin\phi_1-\frac{u}{v}\sin\theta\right), \tag{68}$$

$$\mathrm{H}_{2\phi}=\;\ldots\ldots\ldots\;\cos\theta\cos\phi_1. \tag{69}$$

Now here comes a very curious phenomenon. The value of ϕ_1 is

$$\phi_1=\phi-nt_1=\phi-n(t-\mathrm{R}/v). \tag{70}$$

In my first working out of this problem, I naturally substituted r for R in (70) to suit the formulas (62), and later forms. That it could be otherwise never occurred to me. Yet it led to results which were physically wrong. The explanation found was this. Although a/r is negligible compared with u/v, we must not turn R to r in ϕ_1. Thus,

$$\mathrm{R}=r\left\{1-\frac{2a}{r}\sin\theta\cos\phi_1+\frac{a^2}{r^2}\right\}^{\frac{1}{2}}=r\left(1-\frac{a}{r}\sin\theta\cos\phi_1\right)$$

to the first power of a/r. It makes

$$\phi_1=\phi-n\left\{t-\frac{r}{v}+\frac{a}{v}\sin\theta\cos\phi_1\right\}=\phi-n\left(t-\frac{r}{v}\right)-\frac{u}{v}\sin\theta\cos\phi_1. \tag{71}$$

So we may write

$$\phi_0=\phi_1-\frac{u}{v}\sin\theta\cos\phi_1=\phi-n\left(t-\frac{r}{v}\right) \tag{72}$$

as the equation of condition to which (68), (69) are subject, along with (63). The may-be excessively small difference between R and r turns out to be quite important enough to be counted in.

Observe that

$$\frac{a}{r}+\frac{u}{v}=\frac{a}{r}\left(1+\frac{ru}{av}\right)=\frac{a}{r}\left(1+\frac{2\pi r}{\lambda}\right)=\frac{u}{v}\left(1+\frac{\lambda}{2\pi r}\right), \tag{73}$$

if λ= wave length. So if $\lambda/2\pi r$ is large, it is u/v that is relatively negligible. But if $\lambda/2\pi r$ is small, it is a/r that is negligible (as above) in the value of $\dot{t}_1$.

(15). To analyse (68), (69) into circular functions. Let

$$\phi_1=\phi_0-\beta\cos\phi_1. \tag{74}$$

Then $$\frac{d\phi_1}{d\phi_0}=1+\frac{u}{v}\sin\theta\sin\phi_1\frac{d\phi_1}{d\phi_0}=\frac{1}{1-\beta\sin\phi_1}=\dot{t}_1, \tag{75}$$

$$\frac{d}{d\phi_0}\cos\phi_1=-\sin\phi_1.\dot{t}_1, \qquad -\frac{d^2}{d\phi_0^2}\cos\phi_1=\cos\phi_1.\dot{t}_1^3, \tag{76}$$

$$\frac{d}{d\phi_0}\sin\phi_1=\cos\phi_1.\dot{t}_1, \qquad -\frac{d^2}{d\phi_0^2}\sin\phi_1=(\sin\phi_1-\beta)\dot{t}_1^3. \tag{77}$$

Since $d/d\phi_0=-d/d(nt)$ at a fixed point, the use of differentiations to ϕ_0 is obviously appropriate. Equations (68), (69) are the same as

$$+\mathrm{H}_{2\phi}=-\frac{Qun}{4\pi rv}\cos\theta\frac{d^2}{d\phi_0^2}\cos\phi_1=\frac{Qun}{4\pi rv}\cos\theta\cos\phi_1.\dot{t}_1^3, \tag{78}$$

$$+\mathrm{H}_{2\theta}=-\frac{Qun}{4\pi rv}\frac{d^2}{d\phi_0^2}\sin\phi_1=\frac{Qun}{4\pi rv}(\sin\phi_1-\beta)\dot{t}_1^3. \tag{79}$$

Now expand by trigonometry (theorem due to Lagrange) $\sin\phi_1$ and $\cos\phi_1$ in series of sines and cosines of multiples of ϕ_0. Thus, first,

$$\phi_1=\phi_0-\beta\cos\phi_0+\frac{\beta^2}{\underline{|2}}(\cos^2\phi_0)'-\frac{\beta^3}{\underline{|3}}(\cos^3\phi_0)''+\dots, \tag{80}$$

$$\sin\phi_1=\sin\phi_0-\beta\cos^2\phi_0+\frac{\beta^2}{\underline{|2}}(\cos^3\phi_0)'-\frac{\beta^3}{\underline{|3}}(\cos^4\phi_0)''\dots, \tag{81}$$

$$\cos\phi_1=\cos\phi_0+\beta\cos\phi_0\sin\phi_0-\frac{\beta^2}{\underline{|2}}(\cos^2\phi_0\sin\phi_0)' + \frac{\beta^3}{\underline{|3}}(\cos^3\phi_0\sin\phi_0)''-\dots, \tag{82}$$

$$\dot{t}_1^3(\sin\phi_1-\beta)=-(\cos\phi_0)'+\beta(\cos^2\phi_0)''-\frac{\beta^2}{\underline{|2}}(\cos\phi_0)''' + \frac{\beta^3}{\underline{|3}}(\cos^4\phi)''''-\dots, \tag{83}$$

where the accent means $d/d\phi_0$. Now put the powers of $\cos\phi_0$ in

terms of multiples. For convenience in printing, let $c7$ mean $\cos 7\phi_0$, and $s5$ mean $\sin 5\phi_0$.
Then

$$\sin\phi_1 = s1 - \frac{\beta}{2}(1+c2) + \frac{\beta^2}{\underline{|2}}\,\frac{1}{4}(c3+3c1)' - \frac{\beta^3}{\underline{|3}}\,\frac{1}{8}(c4+4c2+3)''$$
$$+\frac{\beta^4}{\underline{|4}}\,\frac{1}{2^4}(c5+5c3+10c1)''' - \frac{\beta^5}{2^5\underline{|5}}(c6+6c4+15c2+10)^{\text{vi}}$$
$$+\frac{\beta^6}{2^6\underline{|6}}(c7+7c5+21c3+35c1)^{\text{v}} - \frac{\beta^7}{2^7\underline{|7}}(c8+8c6+28c4+56c2+35)^{\text{vi}}$$
$$+\frac{\beta^8}{2^8\underline{|8}}(c9+9c7+36c5+84c3+126c1)^{\text{vii}} - \ldots. \qquad (84)$$

Carrying out the differentiations, we come to

$$\sin\phi_1 = s1 - \frac{\beta}{2}(c2+1) - \frac{\beta^2}{2^2\underline{|2}}(-3s3-3s1) - \frac{\beta^3}{2^3\underline{|3}}(-4^2c4-16c2)$$
$$+\frac{\beta^4}{2^4\underline{|4}}(5^3s5+5.3^3s3+10s1) - \frac{\beta^5}{2^5\underline{|5}}(6^4c6+6.4^2c4+15.2^2c2)$$
$$+\frac{\beta^6}{2^6\underline{|6}}(-7^5s7-7.5^5s5-21.3^5s3-35s1). \qquad (85)$$

Here notice that the expansion of $\cos\phi_1$ may be got from that of $\sin\phi_1$ by differentiating to ϕ_0 and dividing by 1, 2, 3, ... all along And the same applies to $(\cos\phi_1)''$ by doing the same to $(\sin\phi_1)''$. So we need only develop $(\sin\phi_1)''$, and this is facilitated by doing the double differentiation to ϕ_0 required in (76), (77) along with the others. Thus, observing (77), we have

$$\dot{t}_1^3(\sin\phi_1-\beta) = -(\cos\phi_0)' + \beta(\cos^2\phi_0)'' - \frac{\beta^2}{\underline{|2}}(\cos^3\phi_0)''' + \frac{\beta^3}{\underline{|3}}(\cos^4\phi_0)^{\text{iv}} - \ldots$$
$$= -(\cos\phi_0)' + \frac{\beta}{2}(\cos 2\phi_0+1)'' - \frac{\beta^2}{2^2\underline{|2}}(\cos 3\phi_0 + 3\cos\phi_0)'''$$
$$+\frac{\beta^3}{2^3\underline{|3}}(\cos 4\phi_0 + 4\cos 2\phi_0 + 3)^{\text{iv}} - \ldots. \qquad (86)$$

Here $\quad -c' + \frac{\beta^2}{2^2\underline{|2}}3c' - \frac{\beta^4}{2^4\underline{|4}}10c' + \frac{\beta^6}{2^6\underline{|6}}\,\frac{7.6.5}{1.2.3}c' - \ldots$

$$= s\left(1 - \frac{3}{2^2\underline{|2}}\beta^2 + \frac{10}{2^4\underline{|4}}\beta^4 - \frac{7.6.5}{1.2.3}\,\frac{\beta^6}{2^6\underline{|6}} + \right) = 2s\frac{d}{d\beta}J_1(\beta) \qquad (87)$$

if, for the moment $c' = (\cos\phi_0)'$, $\quad s = \sin\phi_0$.

The next gives $-2(d/d\beta)J_2(2\beta)$; the next $-2(d/d\beta)J_3(3\beta)$, and so on. So the final result is

$$\dot{t}_1^3(\sin\phi_1-\beta) = 2\frac{d}{d\beta}[\sin\phi_0.J_1(\beta) - \cos 2\phi_0 J_2(2\beta) - \sin 3\phi_0.J_3(3\beta)$$
$$+\cos 4\phi_0.J_4(4\beta) + \sin 5\phi_0.J_5(5\beta) - \ldots - \ldots + \ldots], \qquad (88)$$

and the other one, by the process described just after (85), comes to

$$\dot{t}_1^3\cos\phi_1 = \frac{2}{\beta}[\cos\phi_0.J_1(\beta) - 3\cos 3\phi_0.J_3(3\beta) + 5\cos 5\phi_0.J_5(5\beta) - \ldots$$
$$+2\sin 2\phi_0.J_2(2\beta) - 4\sin 4\phi_0.J_4(4\beta) + \ldots]. \qquad (89)$$

These results, (88), (89), used in (78), (79), complete the explicit solutions.

(16). This brings us to Bessel's expansions of the formulas for the mean and eccentric anomaly in a planetary orbit. Very few electronic investigators know much about this old subject, so for comparison with the electromagnetic problem, I add a few notes thereon. The figure shows part of the elliptic orbit, and part of a

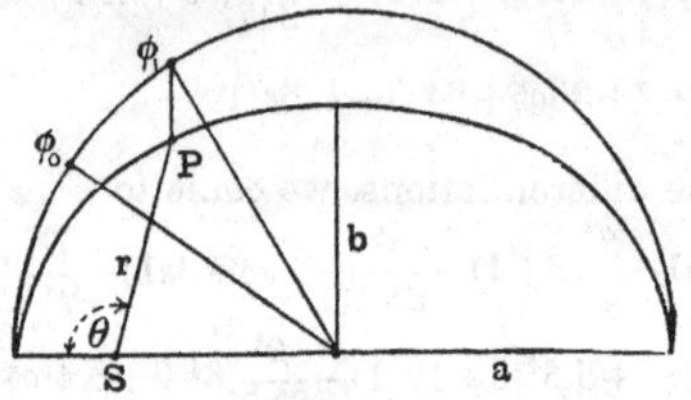

concentric circle. The polar co-ordinates of a point on the ellipse are r, θ, with the focus as origin; the circle has radius a, the major semi-axis. The minor one is b. Then

$$r=a(1-e\cos\phi_1)=\frac{a(1-e^2)}{1+e\cos\theta}, \qquad e^2=1-\frac{b^2}{a^2}, \tag{90}$$

is the polar equation referred to focus. Draw a perpendicular from the end of r to the circle at ϕ_1, and let $\phi_0=nt$. Then the angle ϕ_0 is called the mean anomally, ϕ_1 the eccentric anomaly, and θ the true anomaly, when a planet is at P, and the sun at S. Let ϕ_0 revolve uniformly in the circle. We have, by dynamics of particles,

$$\phi_1=\phi+e\sin\phi_1, \qquad \tan\tfrac{1}{2}\theta=\left(\frac{1+e}{1-e}\right)^{\frac{1}{2}}\tan\tfrac{1}{2}\phi_1. \tag{91}$$

The uniform revolution of ϕ_0 is accompanied by a continuously variable revolution of ϕ_1. Bessel showed that

$$\phi_1=\phi_0+2\{J_1(e)\sin\phi_0+\tfrac{1}{2}J_2(2e)\sin 2\phi_0+\tfrac{1}{3}J_3(3e)\sin 3\phi_0+\ldots\}$$

$$\cos\phi_1=-\tfrac{1}{2}e+(J_0-J_2)(e)\cos\phi_0+\tfrac{1}{2}(J_1-J_3)(2e)\cos 2\phi_0$$
$$+\tfrac{1}{3}(J_2-J_4)(3e)\cos 3\phi_0+\ldots$$

$$\sin\phi_1=(J_0+J_2)\,e)\sin\phi_0+\tfrac{1}{2}(J_1+J_3)(2e)\cos 2\phi_0+\tfrac{1}{3}(J_2+J_4)(3e)\sin 3\phi_0+\ldots$$

If in these, we put $e=$ my β, and reduce ϕ_0 and ϕ_1 by $\frac{1}{2}\pi$ each, we make $\phi_0=\phi_1+\beta\cos\phi_1$, and the results will correspond to some in my electromagnetic problem above, although there is no dynamical resemblance between the two problems.

(17). But it is not from these expansions, however useful for minute calculation, that we can see what the resultant effect is. For that we should examine the unexpanded **H** formulas themselves, giving special values to ϕ_1 and ϕ_0 in a complete revolution.

There is a great difference between the results on the axis and in the equatorial plane, where the Doppler effect is fully developed. In speaking of dopplerisation, it should be understood in a wider sense than in the old Doppler effect. The distortional effects due to the motion of a source of disturbance should be understood. In the uniform straight motion of a charge, for instance, the effect is equatorial compression of the tubes of displacement, and this is the same fore and aft. This is different from the old Doppler effect, which is a compression in front of the moving source of light, causing an increase of frequency in light of any one sort. Behind effects are ignored; but in plane waves the effect is the opposite, a lengthening of the waves and a lowering of frequency.

Now when a charge revolves in a fixed circle, and the point of observation is in its plane, the charge is sometimes moving towards, and sometimes moving away from the observing point. There is compression and increase of intensity in the first case, and expansion and decrease of intensity in the second case. So the sine wave is distorted, with a shifting closer of the nodes, an increase of intensity

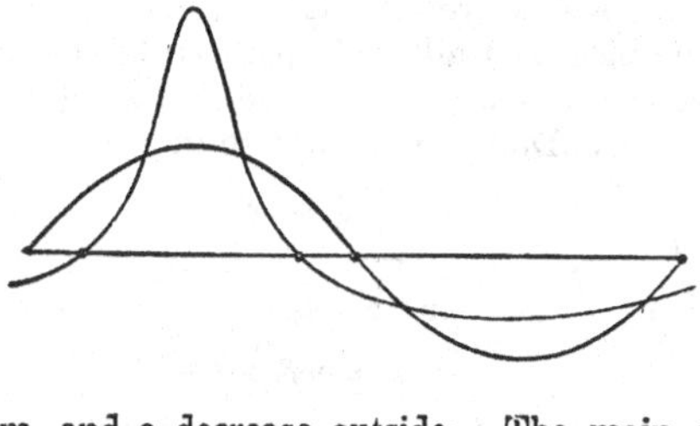

between them, and a decrease outside. The main effect at the observing point of increasing the speed gradually from 0 up to v is to gradually squeeze the nodes together once per revolution, with a great concentration of the disturbance. During the rest of the revolution, the disturbance, of the opposite sign, is widespread and weak. The final result at $u/v=1$, is mere pulses, one per revolution, together with the spread weak disturbance of the opposite sign. The figure will show the beginning of this process.

In the plane of the orbit, $\theta=\frac{1}{2}\pi$, $H_{2\phi}=0$, and

$$H_{2\theta}=\frac{Qnu}{4\pi rv}\frac{\sin\phi_1-u/v}{\{1-(u/v)\sin\phi_1\}^3},\qquad \begin{aligned}\phi_1&=\phi-nt_1\\&=\phi_0-\frac{u}{v}\cos\phi_1.\end{aligned}\tag{92}$$

The position of the nodes is given by

$$\sin\phi_1=\frac{u}{v},\qquad \phi_1=\phi_0-\frac{u}{v}\Big(1-\frac{u^2}{v^2}\Big)^{\frac{1}{2}}.\tag{93}$$

Say $u/v=\frac{1}{2}$, then $\phi_1=\pi/6$, and $\phi_0=\pi/6\pm\frac{1}{2}(\frac{3}{4})^{\frac{1}{2}}$.

Next carry ϕ_1 through a cycle. Let h be the second fractional factor in (92).

$$\phi_1=0, \qquad \phi_1=\tfrac{1}{2}\pi, \qquad \phi_1=\pi, \qquad \phi_1=\tfrac{3}{2}\pi,$$

$$h=-\frac{u}{v}, \qquad h=\frac{1}{(1-u/v)^2}, \qquad h=-\frac{u}{v}, \qquad h=-\frac{1}{(1+u/v)^2}. \tag{94}$$

Note the change from $1-u/v$ to $1+u/v$ in the second and fourth results for H.

If $u/v=\frac{1}{2}$, the variation of intensity is from $-\frac{4}{9}$ to $+4$.

If $u/v=\frac{9}{10}$, the variation is from $-\frac{100}{361}$ to $+100$.

So $u/v=9/10$ makes strongly marked pulses in the equatorial plane.

We always have $\phi_0=\phi_1$, when the charge is nearest to or furthest from the observing point, and is therefore moving perpendicularly to R. In the first quadrant, ϕ_1 leads, and in the second it lags. ϕ_0 may be regarded as the real revolving charge, and ϕ_1 as a fictitious effective charge, nearly the same as the real when u/v is very small, but in spasmodic motion when it is large.

(18). It is now desirable to give the fully expanded formulas for the complete **H**, and so remove the previous restriction to the electromagnetic part, and allow of approach to the centre of the orbit. The same coordinates x, y, z, and r, θ, ϕ will be used. The quasi-static $\mathbf{H}_1$ or $\mathbf{VuD}_1$, is represented thus.

$$\mathrm{H}_{1r}=\frac{\mathrm{Q}\kappa^2 u}{4\pi r^2}\frac{\frac{a}{r}\cos\theta}{\left\{\sqrt{1-\frac{2a}{r}\sin\theta\cos\phi_1+\frac{a^2}{r^2}}-\frac{u}{v}\sin\theta\sin\phi_1\right\}^3},$$

$$\mathrm{H}_{1\phi}=\ldots\ldots\frac{-\sin\phi_1\cos\theta}{\ldots\ldots\ldots\ldots}, \tag{95}$$

$$\mathrm{H}_{1\theta}=\ldots\ldots\frac{\cos\phi_1-\frac{a}{r}\sin\theta}{\ldots\ldots\ldots\ldots}.$$

Here H_{1r} is the radial component out from centre of circular orbit. If a/r is very small, H_{1r} varies as r^{-3}, and then also

$$\mathrm{H}_{1}\phi=\frac{\mathrm{Q}\kappa^2 u}{4\pi r^2}\frac{-\sin\phi_1\cos\theta}{\left(1-\frac{u}{v}\sin\theta\sin\phi_1\right)^3}, \qquad \mathrm{H}_{1\theta}=\ldots\frac{\cos\phi_1}{\ldots\ldots\ldots} \tag{96}$$

Comparing with the corresponding H_2 formulas, we see that $\kappa^2 v/nr$ is the multiplying factor, not counting phase change. Or $(1-u^2/v^2)/(2\pi r/\lambda)$. So if $2\pi r/\lambda$ is large, the quasi-static $\mathbf{H}_1$ is correspondingly small.

It may also be asked whether the quasi-static part can be sensible at a distance when u is nearly equal to v. The answer is that the R component of $\mathbf{H}_1$ tends to ∞/R^2 when the corresponding part of H_2 tends to ∞/R.

In vector-cartesian form (95) is the same as

$$\mathbf{H}_1 = \frac{Q\kappa^2 u}{4\pi r^2}\frac{\mathbf{i}_1\cos\theta\cos nt_1 + \mathbf{i}_2\cos\theta\sin nt_1 - \mathbf{i}_3(\sin\theta\cos\phi_1 - a/r)}{\{\mathrm{R}/r - (u/v)\sin\theta\sin\phi_1\}^3}, \quad (97)$$

the denominator being the same as in (95). At the centre of the orbit $r=0$, and $\mathbf{H}_1$ reduces to the axial component only, say

$$\mathbf{H}_1 = \mathbf{i}_3\frac{Q\kappa^2 u}{4\pi a^2}. \quad (98)$$

It varies as the speed at first, but then comes κ^2 into play and shuts it off eventually. But this is not the complete magnetic force at the centre. We must add on the electromagnetic part. Put $r=0$ in the formula (58) for $\mathbf{H}_2$. It reduces to

$$\mathbf{H}_2 = \mathbf{i}_3\frac{Qu}{4\pi a^2}\frac{u^2}{v^2}. \quad (99)$$

So the complete $\mathbf{H}$ at the centre is $\mathbf{H}=\mathbf{i}_3(Qu/4\pi a^2)$ at any speed.

The complete electromagnetic $\mathbf{H}_2$ may also be written thus,

$$\mathbf{H}_2 = -\frac{Qnu}{4\pi r v[\mathrm{R}/r - (u/v)\sin\theta\cos\phi_1]^3}$$
$$\times\Big[\mathbf{i}_1\Big\{\frac{\mathrm{R}}{r}\sin nt_1 + \frac{u}{v}\sin\theta\cos\phi - \frac{ua}{vr}\cos nt_1\Big\}\cos\theta$$
$$+\mathbf{i}_2\Big\{-\frac{\mathrm{R}}{r}\cos nt_1 + \frac{u}{v}\sin\theta\sin\phi - \frac{ua}{vr}\sin nt_1\Big\}\cos\theta$$
$$+\mathbf{i}_3\Big\{\frac{\mathrm{R}}{r}\sin\theta\sin\phi_1 - \frac{u}{v}\sin^2\theta - \frac{ua^2}{vr^2} + \frac{2ua}{vr}\sin\theta\cos\phi_1\Big\}\Big]. \quad (100)$$

If to this we add the quasi-static $\mathbf{H}_1$, equation (97), we obtain the complete $\mathbf{H}$.

$$\mathbf{H} = \frac{Qu}{4\pi r v[\mathrm{R}/r - (u/v)\sin\theta\cos\phi_1]^3}$$
$$\times\Big[\mathbf{i}_1\cos\theta\Big\{\frac{1}{r}\cos nt_1 - \frac{u}{av}\Big(\frac{\mathrm{R}}{r}\sin nt_1 + \frac{u}{v}\sin\theta\cos\phi\Big)\Big\}$$
$$+\mathbf{i}_2\cos\theta\Big\{\frac{1}{r}\sin nt_1 - \frac{u}{av}\Big(-\frac{\mathrm{R}}{r}\cos nt_1 + \frac{u}{v}\sin\theta\sin\phi\Big)\Big\}$$
$$+\mathbf{i}_3\Big\{\frac{a}{r^2} - \Big(1+\frac{u^2}{v^2}\Big)\frac{1}{r}\sin\theta\cos\phi_1 - \frac{\mathrm{R}u}{rav}\sin\theta\sin\phi_1 + \frac{u^2}{v^2 a}\sin^2\theta\Big\}\Big]; \quad (101)$$

from which come the polar formulas

$$\mathbf{H}_r = \frac{Qu}{4\pi r v[\mathrm{R}/r - (u/v)\sin\theta\cos\phi_1]^3}\Big[\frac{1}{r}\cos\theta\Big(\frac{a}{r} - \frac{u^2}{v^2}\sin\theta\cos\phi_1\Big)\Big], \quad (102)$$

$$\mathbf{H}_\phi = \ldots\ldots\ldots\ldots \Big[-\frac{1}{r}\sin\phi_1 + \frac{n\mathrm{R}}{vr}\cos\phi_1\Big)\Big], \quad (103)$$

$$\mathbf{H}_\theta = \ldots\ldots\ldots\ldots \Big[\frac{1}{r}\cos\phi_1\Big(1+\frac{u^2}{v^2}\sin^2\theta\Big)$$
$$-\frac{a}{r^2}\sin\theta + \frac{n}{v}\Big(\frac{\mathrm{R}}{r}\sin\phi_1 - \frac{u}{v}\sin\theta\Big)\Big]. \quad (104)$$

At a great distance the terms which tend to be negligible are the whole of H_r, the first term in H_ϕ, and the first and second in H_θ.

Remember in all these that $\phi_1=\phi-nt_1$, and $R=v(t-t_1)$.

In the plane of the orbit the formula is

$$\mathbf{H}=\frac{\mathrm{i}_3 Qua}{4\pi r^3[R/r-\ldots]^3}\left[1+\frac{u^2r^2}{v^2a^2}-\frac{r}{a}\cos\phi_1\left(1+\frac{u^2}{v^2}\right)-\frac{uRr}{va^2}\sin\phi_1\right]. \quad (105)$$

The transformation formulas from (101) to (102), (103), (104) cannot be remembered, so are here inserted for reference. Say for the moment that H_1, H_2, H_3 are the three scalar rectangular components of $\mathbf{H}$, then

$$H_\phi=-H_1\sin\phi+H_2\cos\phi,\qquad H_r\sin\theta+H_\theta\cos\theta=H_1\cos\phi+H_2\sin\phi,$$
$$H_\theta=(H_1\cos\phi+H_2\sin\phi)\cos\theta-H_3\sin\theta,\qquad H_r\cos\theta-H_\theta\sin\theta=H_3.$$
$$H_r=(H_1\cos\phi+H_2\sin\phi)\sin\theta+H_3\cos\theta, \quad (106)$$

(19). Now similarly for the spring elliptic orbit. Semi-axes a and b, along i_2 and i_1. Let

$$\mathbf{s}=(\mathrm{i}_1 b\cos+\mathrm{i}_2 a\sin)nt_1,\qquad \mathbf{u}=n(-\mathrm{i}_1 b\sin+\mathrm{i}_2 a\cos)nt_1, \quad (107)$$

where $\mathbf{s}$ is the path, $\mathbf{u}$ the velocity, and $\mathbf{w}=d\mathbf{u}/dt_1=-n^2\mathbf{s}$ the acceleration. Then in

$$\mathbf{H}_2=-\frac{Q\dot{t}_1^2}{4\pi R^2 v}\left\{\mathrm{V}\mathbf{R}\mathbf{w}+\frac{\dot{t}_1}{Rv}\mathbf{R}\mathbf{w}.\mathrm{V}\mathbf{R}\mathbf{u}\right\}, \quad (108)$$

we have

$$\mathbf{R}=\mathbf{r}-\mathbf{s}=\mathrm{i}_1(x-b\cos nt_1)+\mathrm{i}_2(y-a\sin nt_1)+\mathrm{i}_3 z, \quad (109)$$
$$\mathbf{R}\mathbf{u}=n(-bx\sin+ay\cos)nt_1+n(b^2-a^2)\sin nt_1\cos nt_1, \quad (110)$$
$$\mathbf{R}\mathbf{w}=-n^2(bx\cos+ay\sin-b^2\cos^2-a^2\sin^2)nt_1, \quad (111)$$
$$\mathrm{V}\mathbf{R}\mathbf{u}=n[(-\mathrm{i}_1 za\cos-\mathrm{i}_2 zb\sin)nt_1+\mathrm{i}_3\{(xa\cos+yb\sin)nt_1-ab\}], \quad (112)$$
$$\mathrm{V}\mathbf{R}\mathbf{w}=n^2\{\mathrm{i}_1 za\sin-\mathrm{i}_2 zb\cos+\mathrm{i}_3(-xa\sin+yb\cos)\}nt_1. \quad (113)$$

Using these in (108) we come to

$$\begin{aligned}\mathbf{H}_2=-\frac{Qn^2\dot{t}_1^2}{4\pi R^2 v}\Big[&\{\mathrm{i}_1 za\sin-\mathrm{i}_2 zb\cos+\mathrm{i}_3(yb\cos-xa\sin)\}\,nt_1\\ &+\frac{n\dot{t}_1}{Rv}(bx\cos+ay\sin-b^2\cos^2-a^2\sin^2)nt_1\\ &\times\{(\mathrm{i}_1 za\cos+\mathrm{i}_2 zb\sin)nt_1-\mathrm{i}_3[(xa\cos+yb\sin)nt_1-ab]\}\Big],\end{aligned} \quad (114)$$

which may be more conveniently transformed to

$$\begin{aligned}\mathbf{H}_2=-\frac{Qn^2\dot{t}_1^2}{4\pi R^2 v}\Big[&\mathrm{i}_1 ar\cos\theta\left\{\sin nt_1+\frac{n\dot{t}_1}{Rv}r\sin\theta.\mathrm{X}\cos nt_1\right\}\\ &+\mathrm{i}_2 br\cos\theta\left\{-\cos nt_1+\frac{n\dot{t}_1}{Rv}r\sin\theta.\mathrm{X}\sin nt_1\right\}\\ &+\mathrm{i}_3 r^2\sin\theta\left\{\mathrm{Y}-\frac{n\dot{t}_1}{Rv}r\sin\theta.\mathrm{XZ}\right\}\Big],\end{aligned} \quad (115)$$

where, for convenience in composing, X, Y Z are defined by

$$X=b\cos\phi\cos nt_1+a\sin\phi\sin nt_1-\frac{(b^2\cos^2+a^2\sin^2)nt_1}{r\sin\theta}, \tag{116}$$

$$Y=\frac{b}{r}\sin\phi\cos nt_1-\frac{a}{r}\cos\phi\sin nt_1, \tag{117}$$

$$Z=\frac{a}{r}\cos\phi\cos nt_1+\frac{b}{r}\sin\phi\sin nt_1-\frac{ab}{r^2\sin\theta}. \tag{118}$$

No assumptions about smallness have been made, so (115) is complete, up to $u=v$, after which a wider theory is necessary. But we still require an expression for $\dot{t}_1/R$. This is given in the general case by

$$\left(\frac{r\dot{t}_1}{R}\right)^{-1}=\Big[1+\left(\frac{a^2}{r^2}\sin^2+\frac{b^2}{r^2}\cos^2\right)nt_1$$
$$-2\sin\theta\left(\frac{b}{r}\cos\phi\cos+\frac{a}{r}\sin\phi\sin\right)nt_1\Big]^{\frac{1}{2}}$$
$$-\frac{nr}{v}\sin\theta\left(-\frac{b}{r}\cos\phi\sin nt_1+\frac{a}{r}\sin\phi\cos nt_1+\frac{b^2-a^2}{r^2\sin\theta}\sin nt_1\cos nt_1\right). \tag{119}$$

But here, for distant application, we may of course first neglect a^2/r^2 and b^2/r^2. This reduces (119) to

$$\frac{r\dot{t}_1}{R}=\Big[1-\sin\theta\Big\{\left(\frac{b}{r}\cos\phi+\frac{na}{v}\sin\phi\right)\cos nt_1$$
$$+\left(\frac{a}{r}\sin\phi-\frac{nb}{v}\cos\phi\right)\sin nt_1\Big\}\Big]^{-1}. \tag{120}$$

If we further neglect a/r and b/r against na/v and nb/v, which are u_1/v and u_2/v, we come to

$$\frac{r\dot{t}_1}{R}=\left[1-\sin\theta\left(\frac{u_1}{v}\sin\phi\cos nt_1-\frac{u_2}{v}\cos\phi\sin nt_1\right)\right]^{-1}. \tag{121}$$

The *quasi*-static part $\mathbf{H}_1$ must be counted along with (115) near the orbit, and up to the centre, if desired. Its expression is

$$\mathbf{H}_1=\frac{Q\kappa^2}{4\pi}\left(\frac{\dot{t}_1}{R}\right)^3 nr\Big[\cos\theta\,(\mathbf{i}_1 a\cos nt_1+\mathbf{i}_2 b\sin nt_1)$$
$$-\mathbf{i}_3\sin\theta\left\{a\cos\phi\cos nt_1+b\sin\phi\sin nt_1-\frac{ab}{r\sin\theta}\right\}\Big]; \tag{122}$$

and the components for spherical coordinates are

$$H_{1\phi}=\frac{Q\kappa^2}{4\pi}\left(\frac{\dot{t}_1}{R}\right)^3 nr(b\cos\phi\sin nt_1-a\sin\phi\cos nt_1)\cos\theta,$$

$$H_{1\theta}=\ldots\ldots\ldots\ldots\ldots\left\{a\cos\phi\cos nt_1+b\sin\phi\sin nt_1-\frac{ab}{r}\sin\theta\right\}, \tag{123}$$

$$H_{1r}=\ldots\ldots\ldots\ldots\ldots\frac{ab}{r}\cos\theta.$$

The dotted lines mean what is over them, from =, to and with nr. It will, of course, be remembered that there are circumstances under which the radiational part is insignificant. Then only the *quasi*-static magnetic force counts. There is then no sensible waste of energy by radiation. The estimates of the waste of energy and momentum will follow.

(20). In order to verify that the general solutions (115), (122) above, correctly reduce at a distance to the much simpler ones (45), (46) previously obtained by a vectorial process, we may employ the transformation formulas (106). By neglecting a^2/r^2 and b^2/r^2, and also making $R/r=1$, and disregarding the quasi-static part, we come after some complicated reductions to

$$H_\phi=\frac{Qn\dot{t}_1^3}{4\pi rv}\cos\theta[u_2\cos\phi\cos nt_1+u_1\sin\phi\sin nt_1],$$

$$H_\theta=\frac{Qn\dot{t}_1^3}{4\pi rv}\left[u_2\sin\phi\cos nt_1-u_1\cos\phi\sin nt_1-\frac{u_1u_2}{v}\sin\theta\right], \qquad (124)$$

in which (121) must be used, with $r=R$ on the left side. It may also be derived from

$$\begin{aligned}nt_1&=n(t-R/v)=n(t-r/v)-n(R-r)/v\\&=n(t-r/v)+v^{-1}\sin\theta(u_2\cos\phi\cos nt_1+u_1\sin\phi\sin nt_1).\end{aligned} \qquad (125)$$

We should now, by analogy with the circular orbit, bring in d^2/dt^2. This means more complicated work, though it is only to test the accuracy of the results, We come to

$$H_\phi=-\frac{Qn}{4\pi rv}\frac{d^2}{d(nt)^2}(u_2\cos\phi\cos nt_1+u_1\sin\phi\sin nt_1)\cos\theta, \qquad (126)$$

$$H_\theta=-\ldots\ldots\ldots\ldots\ldots(u_2\sin\phi\cos nt_1-u_1\cos\phi\sin nt_1), \qquad (127)$$

subject to (121) with $r=R$ on the left side.

This peculiarity of the elliptic formulas should be borne in mind, which does not present itself in the circular formulas. It may be that b/R is negligible, but that a/R is not. Then we should go back to the general formulas, and use the appropriate reduced forms. Then again one of these ratios may be actually zero, namely, in the case of a linear vib rator, whilst the other may be either negligible or not, according to the position of the point of observation. If this is at such a distance as to make r and R practically identical, then, if $u_2=0$, we have

$$H_\phi=\frac{Qn\dot{t}_1^3u_1}{4\pi rv}\cos\theta\sin\phi\sin nt_1=-\frac{Qn}{4\pi rv}\frac{d^2}{d(nt)^2}u_1\sin\phi\sin nt_1\cos\theta, \qquad (128)$$

$H_\theta=\ldots\ldots\ldots(-\cos\phi\ \sin\ nt_1)=-\ldots\ldots\ldots\ldots\ldots(-u_1\cos\phi\sin nt_1)$,

subject to $\quad nt_1=n(t-r/v)+(u_1/v)\sin\theta\sin\phi\sin nt_1. \qquad (129)$

The velocity is $\mathbf{u}=\mathbf{i}_2u_1\cos nt_1$, and the path is $\mathbf{s}=\mathbf{i}_2a\sin nt_1$.

But in the general case, to allow of approach of the point of observation right up to the vibrating charge, we should put $b=0$ in (114) and (119), and not forget to include the quasi-static part (122) of the magnetic force.

At the centre itself where $r=0$, in the case of an elliptic orbit,

$$\frac{\dot{t}_1}{\mathrm{R}}=\frac{1}{\mathrm{R}-(n/v)(b^2-a^2)\sin nt_1 \cos nt_1}, \quad \mathrm{R}^2=b^2\cos^2 nt_1+a^2\sin^2 nt_1, \tag{130}$$

$$\mathrm{H}=\frac{\mathrm{Q}n^3\dot{t}_1^3ab}{4\pi\mathrm{R}v^2}+\frac{\mathrm{Q}\kappa^2\dot{t}_1^3abn}{4\pi\mathrm{R}^3}. \tag{131}$$

The first part of H is electromagnetic, the second part quasi-static. The direction is axial. When $a=b$, we reduce to $\mathrm{Q}u/4\pi a^2$, as was before verified.

(21). No formal proof is needed that the induction in the above solutions has no divergence, because that is secured by the magnetic force being the curl of a vector. But a proof is, perhaps needed that the displacement has no divergence outside the moving charge, that is, that

$$\mathrm{div}\,\mathbf{E}=-\mathrm{div}(\mu\dot{\mathbf{A}}+\nabla\mathrm{P})=0=-\frac{\mu\mathrm{Q}}{4\pi}\,\mathrm{div}\left(\dot{\mathrm{u}}+\mathrm{u}\frac{d}{dt}-v^2\nabla\right)\frac{\dot{t}_1}{\mathrm{R}}. \tag{132}$$

This leads to a few remarks on the potentials in electromagnetics. Maxwell's potentials were of the instantaneous kind appropriate to steady states. They can, and have been, used in electromagnetic investigations. Yet the progressive scalar potential was well-known in investigations of sound vibrations. Why then was it so slow in getting into electromagnetics? It arose from the presence of two potentials. First **A**, and then P as an auxiliary to make the divergence of **E** come right. This gave trouble, and sometimes confusion and error arose even when these potentials were of the instantaneous sort. In my investigations of plane, spherical, and cylindrical waves published in 1888 I used no potentials, but the electric and magnetic forces themselves, connected together in differential equations by differential operators analogous to the resistance operator Z in V=ZC, of which a good deal was said in vol. 2. Now this work was entirely of the progressive kind, and if translated into potentials would involve the use of progressive potentials. There would be no advantage in the translation in general, in fact, the reverse; though in some special cases relating to point sources the potentials might be useful. The Z operators are more advanced. In the article in this volume on the two sorts of pulses in an elastic solid, there is a transition by integration from the operators for points to those for surfaces and solids. This rounds off the treatment, and harmonises diverse ways of working.

I think it was G. F. FitzGerald who first brought the progressive **A** and P into electromagnetics. There is an account of his investigation in Gray's "Electricity and Magnetism," vol. 1, near the end. But his potentials were not dopplerized, and were only suitable for continuous distributions of sources. Now it is possible, by elaborate integrational expedients, to use the undistorted potentials when the sources are discontinuous. But it is much better to use the distorted potentials at once. We should then show that the distorted **A** and P used above fit together properly. I should be satisfied myself by the general fitness of results, and only forced myself to do the following:—

$$\operatorname{div}\left\{\frac{\dot{t}_1^{\,3}}{\mathrm{R}^3}\left(\mathbf{R}-\frac{\mathrm{R}\mathbf{u}}{v}\right)(v^2-u^2+\mathrm{R}\mathbf{w})\frac{\dot{\mathbf{u}}\dot{t}_1}{\mathrm{R}}\right\}$$

$$=\frac{\dot{t}_1^{\,3}}{\mathrm{R}}\left(v^2-u^2+\mathrm{R}\mathbf{w}\right)\left(3+\frac{\mathrm{R}\dot{\mathbf{u}}}{v^2}\right)$$

$$+\frac{\dot{t}_1^{\,3}}{\mathrm{R}^3}\left(\mathbf{R}-\frac{\mathrm{R}\mathbf{u}}{v}\right)\left(3\mathbf{u}\mathbf{w}\dot{t}_1\frac{\mathbf{R}_1}{v}+\mathbf{w}-\mathrm{R}\dot{\mathbf{w}}.\frac{\mathbf{R}_1}{v}\right)$$

$$-\left(\mathbf{R}-\frac{\mathrm{R}\mathbf{u}}{v}\right)\left(v^2-u^2+\mathrm{R}\mathbf{w}\right)3\frac{\dot{t}_1^{\,4}}{\mathrm{R}^4}\left(\dot{t}_1\mathbf{R}_1\kappa^2-\frac{\mathbf{u}}{v}+\frac{\mathbf{R}_1}{v^2}.\mathrm{R}\dot{\mathbf{u}}\right)$$

$$+\dot{\mathbf{u}}\frac{\dot{t}_1^{\,2}}{\mathrm{R}^2}\left(\dot{t}_1\mathbf{R}_1\kappa^2-\frac{\mathbf{u}}{v}+\frac{\mathbf{R}_1}{v^2}.\mathrm{R}\dot{\mathbf{u}}\right)+\frac{\dot{t}_1}{\mathrm{R}}\left(-\frac{\mathbf{u}\dot{\mathbf{u}}\dot{t}_1}{\mathrm{R}v}+\frac{\mathrm{R}\ddot{\mathbf{u}}}{\mathrm{R}v}-\frac{\mathrm{R}\dot{\mathbf{u}}}{\mathrm{R}^2v}.v(1-\dot{t}_1)\right)$$

$$=\frac{\dot{t}_1^{\,3}}{\mathrm{R}^3}\left[\frac{3\mathbf{u}\dot{\mathbf{u}}}{v}\frac{\mathrm{R}}{\dot{t}_1}+\mathrm{R}\mathbf{w}-\frac{\mathrm{R}}{v}\mathbf{u}\mathbf{w}-\frac{\mathrm{R}}{v\dot{t}_1}\frac{\mathrm{R}\dot{\mathbf{u}}}{\dot{t}_1^{\,2}}\frac{v}{\mathrm{R}}\left(\kappa^2\dot{t}_1^{\,3}+\dot{t}_1-2\dot{t}_1^{\,2}+\dot{t}_1^{\,2}\frac{\mathrm{R}\dot{\mathbf{u}}}{v^2}\right)\right]$$

$$+\frac{\dot{t}_1^{\,3}}{\mathrm{R}^3}(v^2-u^2+\mathrm{R}\mathbf{w})\left(3+\frac{\mathrm{R}\dot{\mathbf{u}}}{v^2}\right)$$

$$-\frac{3\dot{t}_1^{\,4}}{\mathrm{R}^4}(v^2-u^2+\mathrm{R}\mathbf{w})\left\{\mathrm{R}\kappa^2+\frac{\mathrm{R}.\mathrm{R}\dot{\mathbf{u}}}{v^2\dot{t}_1}-\frac{\mathrm{R}\mathbf{u}}{v}+\frac{\mathrm{R}u^2}{v^2}\right\}$$

$$+\frac{\dot{t}_1^{\,2}}{\mathrm{R}^2}\left\{\frac{\mathrm{R}\dot{\mathbf{u}}\dot{t}_1\kappa^2}{\mathrm{R}}-\frac{\mathbf{u}\dot{\mathbf{u}}}{v}+\frac{(\mathrm{R}\dot{\mathbf{u}})^2}{\mathrm{R}v^2}\right\}+\frac{\dot{t}_1}{\mathrm{R}}\left\{-\frac{\mathbf{u}\dot{\mathbf{u}}\dot{t}_1}{\mathrm{R}v}+\frac{\mathrm{R}\ddot{\mathbf{u}}}{\mathrm{R}v}-\frac{\mathrm{R}\dot{\mathbf{u}}}{\mathrm{R}^2}+\frac{\mathrm{R}\dot{\mathbf{u}}\dot{t}_1}{\mathrm{R}^2}\right\}$$

$$=\frac{\dot{t}_1^{\,3}}{\mathrm{R}^3}(\underset{(2)}{v^2-u^2}+\underset{(1)}{\mathrm{R}\mathbf{w}})\left[3+\frac{\mathrm{R}\dot{\mathbf{u}}}{v^2}-\frac{3\dot{t}_1}{\mathrm{R}}\left(\frac{\mathrm{R}}{\dot{t}_1}+\frac{\mathrm{R}.\mathrm{R}\dot{\mathbf{u}}}{v^2\dot{t}_1}\right)=-\frac{2\mathrm{R}\dot{\mathbf{u}}}{v^2}\right]$$

$$+\frac{\dot{t}_1^{\,3}}{\mathrm{R}^3}\left[\underset{(3)}{\frac{2\mathbf{u}\dot{\mathbf{u}}\mathrm{R}}{v\dot{t}_1}}+\underset{(5)}{\frac{\mathrm{R}\dot{\mathbf{u}}}{\dot{t}_1}}-\underset{(4)}{\frac{\mathrm{R}.\mathrm{R}\ddot{\mathbf{u}}}{v\dot{t}_1^{\,2}}}+\frac{\mathrm{R}\dot{\mathbf{u}}}{\dot{t}_1^{\,3}}\left(\underset{(2)}{\kappa^3\dot{t}_1^{\,3}}+\underset{(6)}{\dot{t}_1}-\underset{(5)}{2\dot{t}_1^{\,2}}+\underset{(1)}{\frac{\mathrm{R}\dot{\mathbf{u}}\dot{t}_1^{\,2}}{v^2}}\right)\right.$$

$$\left.+\underset{(2)}{\mathrm{R}\dot{\mathbf{u}}\kappa^2}-\underset{(3)}{\frac{\mathrm{R}\mathbf{u}\dot{\mathbf{u}}}{v\dot{t}_1}}+\underset{(1)}{\frac{(\mathrm{R}\dot{\mathbf{u}})^2}{v^2\dot{t}_1}}\right]$$

$$+\frac{\dot{t}_1^{\,2}}{\mathrm{R}^2}\left(-\underset{(3)}{\frac{\mathbf{u}\dot{\mathbf{u}}}{v}}+\underset{(5)}{\frac{\mathrm{R}\dot{\mathbf{u}}}{\mathrm{R}}}+\underset{(4)}{\frac{\mathrm{R}\ddot{\mathbf{u}}}{v\dot{t}_1}}-\underset{(6)}{\frac{\mathrm{R}\dot{\mathbf{u}}}{\mathrm{R}\dot{t}_1}}\right)=0. \quad \text{[Q.E.D.]} \qquad (133)$$

Several of the before given subsidiary formulas have been used, and, to ease matters finally, undersigns, like (2), are used. All the (1)'s come to 0, and so do all the (2)'s, and so on to all the (6)'s. So all comes to nought. I dare say, and should think likely, that a much simpler proof can be given.

As regards the total divergence at Q itself coming to Q, I think this is sufficient. Start the charge from rest or steady straight motion at a certain moment, then let it be in variable motion for a certain interval, and finally bring it to steady straight motion again. Space will then be divided into three regions. In the inner and in the outer regions are quasi-static states of displacement, whilst in the middle region is electromagnetic disturbance as well. It has been proved for the quasi-static states that the total displacement outward is Q (in previous work). So Q of displacement enters the middle region from inside and leaves it on the outside, without (as just proved) any divergence in the middle region itself. Since this middle region may be extended inward and outward, the proof covers all space, and any sort of variable motion of Q, with this caution, that should **u** change quite suddenly, there is an electromagnetic pulse generated requiring special reckoning.

It will be noted that the terms $\mu\dot{\mathbf{A}}$ and ∇P in (132) are not free from divergence separately.

(22). The following figure will serve to show plainly many peculiarities in the analysis. The curved line 1 to 7 in the middle represents a portion of the path of a point charge in motion. The numbers show its position at seven moments of time (t_1, t_2, &c.), differing by the same interval of time from any one to the next. Since the points are not equidistant, the velocity varies. It increases, if the motion is smooth, from 1 up to $3\frac{1}{2}$ about, and then falls off greatly.

The sphere 1 is centred upon 1, 2 upon 2, and so on. Their radii are such that the whole figure corresponds to the state at one moment of time t. That is, the radius of 1 is $R_1 = v(t-t_1)$; of 2, $R_2 = v(t-t_2)$, and so on. Outside the sphere 1 may be imagined to be any number of similar spheres. They belong to positions of the charge passed through before the point 1 was reached. Similarly we may fill up the central space right up to the charge itself, by making the time interval smaller. Note, however, that the charge is not necessarily at the point 7 at the moment t, but is most likely further on. The exception is when the charge stops dead at the point 7 at the moment t_7. If so, the whole of the sphere 7 will be occupied by the simple electrostatic uniformly radial distribution of displacement. (In passing, I ask why has not a "spherical surface" a name of its own, reserving "sphere" for a solid? There is occasionally confusion.)

The shells in the figure may all be subdivided, of course. So we come ultimately to the whole of space being divided into shells of infinitesimal depth, corresponding to the infinitesimal elements of the path traversed by Q. This path element is $\mathbf{u}dt_1$. The shells are spherical, but not of uniform depth, because the two sides of a shell have different centres. The depth is vdt_1 transversely round $\mathbf{u}$, but elsewhere it is $vdt_1[1-(u/v)\cos\theta]$. Thus it varies from $vdt_1(1-u/v)$ in front to $vdt_1(1+u/v)$ behind, showing a compression in front and an expansion behind.

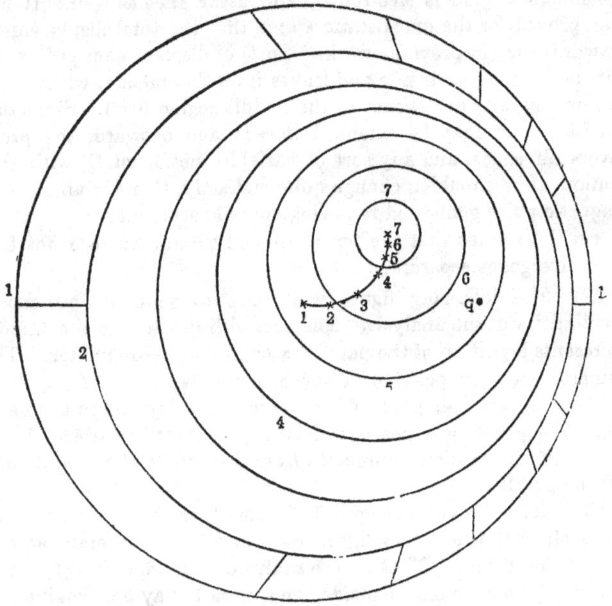

The electric and magnetic state may be described by supposing that the charge Q is continuously emitting electric and magnetic disturbance with its natural isotropic manner of spreading distorted by the motion, so that what would be in a spherical shell of uniform depth vdt_1 is really found in the shell of variable depth as above. Similarly, in the case of a shell of finite depth, as in the figure, say that bounded by the surfaces 4 and 5, the disturbance emitted by Q between t_4 and t_5 is wholly contained in that shell. Moreover, this is true not only at a particular moment, but for all time. For the two surfaces of the shell expand at speed v, each from its own centre, and the same formulas for **E** and **H** in the shells continue valid for all time, or until a change of conditions occurs. Some of the consequences are rather striking.

Suppose, for instance, that the velocity of Q is constant between the points 1 and 2. Then the shell between the 1 and 2 spheres contains only the quasi-static state called $\mathbf{E}_1$ and $\mathbf{H}_1$ in the above. The centre of symmetry is at a point q to the right of the point 2. It is exactly where the charge Q would be at the moment t if it continued moving with the velocity it had between the points 1 and 2. The actual position of the charge may, however, be anywhere else. No matter, the distribution of $\mathbf{E}_1$ in the shell 1, 2 is radial outward from the progressive virtual source, and is condensed equatorially, or up and down, and expanded laterally, or right and left. As the shell expands the electric force $\mathbf{E}_1$ alters so as to keep in touch with the new position of the virtual charge.

There need be no magnetic disturbance outside the sphere 1, or inside the sphere 2. If so, then Q was jerked into motion at the point 1, and stopped at the point 2. Then the surface 1 is an electromagnetic pulse, and 2 is another one. They are not quite pure, because of the electric displacement which leaves 2 and enters 1. What really occurs, however, is this. The electromagnetic skin 1 as it expands, lays down the electric displacement in the shell on its outer boundary, and at the same time, the skin 2 picks up the displacement on its inner boundary. At a sufficiently great distance, the displacement in the shell becomes insignificant. But the skins are left, and they are of great importance, as they contain condensed energy and momentum.

If the charge really continued moving uniformly, instead of being stopped at the point 2, it would coincide with the virtual charge, and the quasi-static field of $\mathbf{E}_1$ and $\mathbf{H}_1$ would extend right up to the charge itself. Then the whole field of force moves as if it were rigidly attached to the charge, save that the electromagnetic skin 1 is continuously enlarging the field. There is certainly no supply of energy to the field from the charge when it is not accelerated and no electromagnetic disturbance is being generated. Yet it may be seen from the above theory of the shells expanding outward from a moving charge, that it is convenient to think and speak of the shells as being generated, and to associate a particular shell with a particular element of path of the charge.

When there is continuous acceleration, the quasi-static $\mathbf{H}_1 = \mathrm{V}\mathbf{u}\mathbf{D}_1$ is accompanied by the electromagnetic $\mathbf{H}_2 = \mathrm{V}\mathbf{v}\mathbf{D}_2$, and they are superposed. No proper separation into their shells with skins is possible. But at a distance from the charge the quasi-static part of the disturbance becomes negligible. What is left is electromagnetic, not in skins, but throughout the shells. This disturbance is radiation. A shell expands indefinitely, and carries its momentum and energy with it. The shells are therefore appropriate for the calculation of the waste of energy and momentum, some details of which now follow.

(23). The spherical coordinates are best to find the energy in a shell. Thus,

$$\int_{-1}^{1}\int_{0}^{2\pi}\tfrac{1}{2}c\mathrm{E}^2 v dt_1\left(1-\frac{u}{v}\mu\right)\mathrm{R}^2 d\mu d\phi \tag{134}$$

represents the electric energy, μ standing for $\cos\theta$, as usual. The integration to r is already done, being represented by the factor $vdt_1(1-\mu u/v)$. The next thing is a convenient expression for E^2. First as regards $\mathrm{E}_2{}^2$. Let μ_1 and ϕ_1 belong to the acceleration **w**, and $\nu=\sin\theta$, and

$$\cos\phi'=\mu\mu_1+\nu\nu_1\cos(\phi-\phi_1). \tag{135}$$

Then ϕ' is the angle between **R** and **w**. Also,

$$\begin{aligned}[\mathbf{w}_1-(\mathbf{R}_1-\mathbf{u}/v)\dot{t}_1\mathbf{R}_1\mathbf{w}_1]^2 &= 1+(1+u^2/v^2-2\mu u/v)\dot{t}_1{}^2(\mathbf{R}_1\mathbf{w}_1)^2\\ &\quad -2(\mathbf{R}_1\mathbf{w}_1-\mu_1 u/v)\dot{t}_1\mathbf{R}_1\mathbf{w}_1+(u^2/v^2)\dot{t}_1{}^2(\mathbf{R}_1\mathbf{w}_1)^2\\ &= 1-\kappa^2\dot{t}_1{}^2\cos^2\phi'+2(u/v)\dot{t}_1\mu_1\cos\phi'.\end{aligned} \tag{136}$$

The ϕ integration on the last can now be done, making

$$2\pi-\kappa^2\dot{t}_1{}^2(2\pi\mu^2\mu_1{}^2+\pi\nu^2\nu_1{}^2)+4\pi(u/v)\dot{t}_1\mu\mu_1{}^2. \tag{137}$$

The electromagnetic energy in the shell is therefore, by (134),

$$\Sigma c\mathrm{E}_2{}^2=cv\left(\frac{\mu\mathrm{Q}}{4\pi}\right)^2 w^2\int\left[\dot{t}_1{}^4 2\pi+\dot{t}_1{}^5 4\pi\frac{u}{v}\mu\mu_1{}^2-\dot{t}_1{}^6\kappa^2(2\pi\mu^2\mu_1{}^2+\pi\mu^2\nu_1{}^2)\right]d\mu. \tag{138}$$

where $\dot{t}_1=(1-\mu u/v)^{-1}$. Call the integral on the right side X. Expansion of the powers of $\dot{t}_1$ by division seems the readiest way of integrating. Carried out, it makes

$$\begin{aligned}\mathrm{X}=4\pi&\left(1+\frac{1}{3}\frac{3.4}{1.2}\frac{u^2}{v^2}+\frac{1}{5}\frac{3.4.5.6}{1.2.3.4}\frac{u^4}{v^4}+\dots\right)\\ &+8\pi\mu_1{}^2\left(0+\frac{1}{3}.4.\frac{u^2}{v^2}+\frac{1}{5}\frac{4.5.6}{1.2.3}\frac{u^4}{v^4}+\dots\right)\\ &-2\pi\kappa^2\nu_1{}^2\left(1+\frac{1}{3}\frac{5.6}{1.2}\frac{u^2}{v^2}+\frac{1}{5}\frac{5\,6.7.8}{1.2\,3\,4}\frac{u^4}{v^4}+\dots\right)\\ &-2\pi\kappa^2(3\mu_1{}^2-1)\left(\frac{1}{3}+\frac{1}{5}\frac{5.6}{1.2}\frac{u^2}{v^2}+\frac{1}{7}\frac{5.6.7.8}{1.2.3.4}\frac{u^4}{v^4}+\dots\right),\end{aligned} \tag{139}$$

which, by proper distribution, gives

$$\mathrm{X}=(8\pi/3)(\kappa^{-4}+\kappa^{-6}\mu_1{}^2u^2/v^2). \tag{140}$$

This, by (138), makes

$$\mathrm{U}_2=\mathrm{T}_2=\frac{\mu\mathrm{Q}^2w^2}{12\pi v}dt_1\left(\frac{1}{\kappa^4}+\frac{\cos^2\phi_1.u^2/v^2}{\kappa^6}\right). \tag{141}$$

It will be remembered that $E_2 = \mu v H_2$. This makes $U_2 = T_2$. The distance R does not appear in the result. So the energy is wasted. It is the portion of the energy given to the field by Q in the time interval dt_1 which is not stored within range, but is irrecoverable. The rate of waste of energy at the moment t_1 is therefore

$$\text{Waste} = \frac{\mu Q^2 w^2}{6\pi v \kappa^4}\left(1 + \frac{\cos^2\phi_1 . u^2/v^2}{\kappa^2}\right), \tag{142}$$

though not till later does it unmix itself from the energy of the field in general. It is least when $\mathbf{u} \perp \mathbf{w}$. Then $\cos\phi_1 = 0$, and the waste varies as κ^{-4}. It is greatest when $\mathbf{u} \parallel \mathbf{w}$. Then it varies as κ^{-6}. Under any circumstances, it varies as the square of the acceleration. Starting and stopping are the same in this respect. Observe, too, that though this formula refers to a charge at a point, with infinite total energy, the waste is finite. (There have been similar cases previously in this volume.) We may infer then that the theory is practically the same for a charge of small finite size, when the total energy is finite. The limitation of size will be conditioned by the time taken for disturbances to cross the charge. It should be so small that there is no sensible change in the acceleration in the time named.

Another way of obtaining the waste does it for a sphere of finite size, with the same result. The theory of impulsive changes of velocity is used. There is much less elaborate calculation, like the above, but more thinking and imagination are needed, such as a physicist prefers when possible. This way will come later. At present there is the rest of the energy in a shell at any distance to be calculated. This cannot be done by mere thinking.

We have
$$\tfrac{1}{2}cE^2_1 = \tfrac{1}{2}c\left(\frac{Q\dot{t}_1{}^3\kappa^3}{4\pi R^2 c}\right)\left(1 + \frac{u^2}{v^2} - 2\frac{u\mu}{v}\right), \tag{143}$$

where the last factor is the square of the vector $\mathbf{R}_1 - \mathbf{u}/v$. Therefore

$$U_1 = \frac{Q^2\kappa^4 v\,dt_1}{16\pi R^2 c}\int\frac{1 + u^2/v^2 - 2\mu u/v}{(1-\mu u/v)^5}\,d\mu, \tag{143A}$$

where the ϕ integration has been done. Calling the integral Y, integration makes

$$\begin{aligned} Y = 2\Big\{1 &+ \frac{1}{3}\frac{5.6}{1.2}\frac{u^2}{v^2} + \frac{1}{5}\frac{5.6.7.8}{1.2.3.4}\frac{u^4}{v^4} + \frac{1}{7}\frac{6.7.8.9.10}{1.2.3.4.5}\frac{u^6}{v^6} + \dots \\ &+ 1\ .\ \frac{u^2}{v^2} + \frac{1}{3}\cdot\frac{5.6}{1.2}\cdot\frac{u^4}{v^4} + \frac{1}{5}\cdot\frac{5.6.7.8}{1.2.3.4}\frac{u^6}{v^6} + \dots \\ &- 2\,\frac{1}{3}.5\,\frac{u^2}{v^2} - \frac{2}{5}\cdot\frac{5.6.7}{1.2.3}\frac{u^4}{v^4} - \frac{2}{7}\cdot\frac{5.6.7.8.9}{1.2.3.4.5}\frac{u^6}{v^6} + \dots\Big\} \end{aligned}$$

$$= \tfrac{2}{3}\left\{3 + 2\frac{u^2}{v^2} + 2\frac{u^4}{v^4} + 2\frac{u^6}{v^6} + \dots\right\} = 2 + \frac{4}{3}\frac{u^2}{v^2\kappa^2}. \tag{144}$$

Therefore, by (143),

$$U_1=\frac{Q^2vdt_1}{4\pi R^2c}\left\{1+\frac{2}{3}\frac{u^2}{v^2\kappa^2}\right\}. \qquad (145)$$

The final part of the electric energy is $\sum cE_1E_2$. We have

$$cE_1E_2=\frac{cQ\dot{t}\,^3\kappa^2}{4\pi R^2c}\frac{\mu Qw\dot{t}_1^{\,2}}{4\pi R}\times - \left(\mathbf{R}_1-\frac{\mathbf{u}}{v}\right)\left[\mathbf{w}_1-\left(\mathbf{R}_1-\frac{\mathbf{u}}{v}\right)\dot{t}_1.\mathbf{R}_1\mathbf{w}_1\right]. \qquad (146)$$

Here it may be shown that the factor following the $\times$ is the same as

$$\frac{+(u/v)\sin^2\theta\cos\phi_1-(u/v)(\cos\theta-u/v)\sin\theta\sin\phi_1\cos\phi}{1-\mu u/v}. \qquad (147)$$

Integrating this to ϕ, the result is

$$\frac{2\pi(u/v)(1-\mu^2)\cos\phi_1}{1-\mu u/v}, \qquad (148)$$

which, used in (146), makes

$$U_{12}=\frac{\mu Q^2w}{8\pi R}\frac{u}{v}\cos\phi_1.vdt_1\int\frac{(1-\mu^2)d\mu}{(1-\mu u/v)^5}. \qquad (149)$$

Calling the integral Z, its value is

$$Z=2\left\{1+\frac{1}{3}\frac{5.6}{1.2}\frac{u^2}{v^2}+\ldots-\frac{1}{3}-\frac{1}{5}\frac{5.6}{1.2}\frac{u^2}{v^2}-\ldots\right\}=\frac{4}{3\kappa^5}. \qquad (150)$$

So, finally,

$$U_{12}=\frac{\mu Q^2w}{6\pi R}\frac{u}{v}\cos\phi_1\,vdt_1.\frac{1}{\kappa^4}. \qquad (151)$$

The magnetic energy wants less work, because part is already done in finding the electric energy. T_2 is known, $=U_2$. Then, as regards $T_{12}=\sum\mu\mathbf{H}_1\mathbf{H}_2$, we have

$$\mathbf{H}_1\mathbf{H}_2=\mathrm{V}\mathbf{v}\mathbf{D}_1\mathrm{V}\mathbf{v}\mathbf{D}_2=v^2\mathbf{D}_1\mathbf{D}_2-\mathbf{v}\mathbf{D}_1.\mathbf{v}\mathbf{D}_2=(c/\mu)\mathbf{E}_1\mathbf{E}_2, \qquad (152)$$

from which it follows that $T_{12}=U_{12}$, already calculated. And, as regards T_1, we have

$$\mu H_1^2=\mu\mathbf{H}_1\mathrm{V}\mathbf{v}\mathbf{D}_1=\mu\mathbf{v}\mathrm{V}\mathbf{D}_1\mathbf{H}_1=\mu\mathbf{v}\mathrm{V}\mathbf{D}_1\mathrm{V}\mathbf{v}\mathbf{D}_1$$
$$=\mu\mathbf{v}(\mathbf{v}D_1^2-\mathbf{D}_1.\mathbf{v}\mathbf{D}_1)=c\mathbf{E}_1^2-c(\mathbf{E}_1\mathbf{R}_1)^2. \qquad (153)$$

That is, $U_1-T_1=\frac{1}{2}c(\mathbf{E}_1\mathbf{R}_1)^2$ per unit volume. The last quantity must now be summed up through the complete shell. The integral is

$$\frac{Q^2\kappa^4vdt_1}{16\pi R^2c}\int\frac{d\mu}{(1-\mu u/v)^3}=\ldots\ldots\times 2\left(1+\frac{1}{3}\frac{3.4}{1.2}\frac{u^2}{v^2}+\frac{1}{5}\frac{3.4.5.6}{1.2.3.4}\frac{u^4}{v^4}+\right)$$
$$=\ldots\ldots\times\frac{2}{\kappa^4}=\frac{Q^2vdt_1}{8\pi R^2c}=U_1-T_1. \qquad (154)$$

Thus T_1 is known. Collecting the energy results for the shell, we have

$$U_1=\frac{Q^2vdt_1}{8\pi R^2c}\left(1+\frac{2u^2}{3v^2\kappa^2}\right),\qquad T_1=\frac{Q^2vdt_1}{12\pi R^2c}\frac{u^2}{v^2\kappa^2}, \qquad (155)$$

$$U_{12}=T_{12}=\frac{\mu Q^2w}{6\pi R}\frac{udt_1\cos\phi_1}{\kappa^4}, \qquad (156)$$

$$U_2=T_2=\frac{\mu Q^2w^2}{12\pi v}dt_1\left(\frac{1}{\kappa^4}+\cos^2\phi_1.\frac{u^2}{v^2\kappa^6}\right), \qquad (157)$$

where $\kappa^2=1-u^2/v^2$, and ϕ_1 is the angle between $\mathbf{u}$ and $\mathbf{w}$

The relation (154) attracts notice. The excess of U_1 over T_1, or of U over T, in any elementary shell equals the static energy in the same space when Q is at rest at the centre of the shell. So, if we traverse the shells, from Q outward, at any fixed moment, U and T always change at the same rate, as settled by (154).

(24). There is now the momentum in an elementary shell to be calculated. This is

$$\mathbf{M}=\Sigma \mathrm{V}\mathbf{DB}=\Sigma \mathrm{V}(\mathbf{D}_1\mathbf{B}_1+\mathbf{D}_1\mathbf{B}_2+\mathbf{D}_2\mathbf{B}_1+\mathbf{D}_2\mathbf{B}_2)=\mathbf{M}_1+\mathbf{M}_{12}+\mathbf{M}_{21}+\mathbf{M}_2$$

In these, (158)

$$\mathbf{D}_1=\frac{Q\kappa^2\dot{t}_1^{\,3}}{4\pi R^2}(\mathbf{R}_1-\mathbf{u}/v), \qquad \mathbf{H}_1=\mathrm{V}\mathbf{u}\mathbf{D}_1,$$

$$\mathbf{B}_2=\mu v\mathrm{V}\mathbf{R}_1\mathbf{D}_2=-\frac{\mu Q\dot{t}_1^{\,2}w}{4\pi Rv}[\mathrm{V}\mathbf{R}_1\mathbf{w}_1+\dot{t}_1\mathbf{R}_1\mathbf{w}_1.\mathrm{V}\mathbf{R}_1\mathbf{u}/v]. \tag{159}$$

Take $\mathbf{M}_1$ first. We have

$$\mathbf{u}\mathbf{M}_1=\mathbf{u}\mathrm{V}\mathbf{D}_1\mathbf{B}_1=\mathbf{B}_1\mathrm{V}\mathbf{u}\mathbf{D}_1=\mathrm{H}_1\mathrm{B}_1. \tag{160}$$

This is per unit volume. But also, we can see that $\mathrm{V}\mathbf{D}_1\mathbf{B}_1$ is symmetrical round $\mathbf{u}$. It follows that the total $\mathbf{M}_1$ is parallel to $\mathbf{u}$. Therefore

$$\mathrm{T}_1=\tfrac{1}{2}\mathrm{M}_1u, \qquad \mathrm{M}_1=2\mathrm{T}_1/u, \tag{161}$$

where the known value of T_1 is to be used.

Take $\mathbf{M}_{21}$ next. We have

$$\mathbf{R}_1\mathrm{V}\mathbf{D}_2\mathbf{B}_1=\mathbf{B}_1\mathrm{V}\mathbf{R}_1\mathbf{D}_2=\mathbf{B}_1\mathbf{H}_2/v=\mathrm{H}_1\mathrm{B}_2/v, \text{ or } \mathrm{V}\mathbf{D}_2\mathbf{B}_1=\mathbf{R}_1.\mathrm{H}_1\mathrm{B}_2/v, \tag{162}$$

because $\mathbf{D}_2$ and $\mathbf{B}_1$ are both tangential. We may therefore utilise the already done calculation of T_{12} or U_{12} to find $\mathbf{M}_{21}\mathbf{u}_1$, or the component of $\mathbf{M}_{21} \| \mathbf{u}$. The result is

$$\frac{\mu Q^2wu\kappa^2}{8\pi Rv}\cos\phi_1.dt_1\int\frac{(\mu-\mu^3)}{(1-\mu u/v)^5}d\mu=\ldots\ldots 2\frac{u}{v}\Big(\frac{1}{3}.5+\frac{1}{5}.\frac{5.6.7}{1.2.3}\frac{u^2}{v^2}+\ldots$$
$$-\frac{1}{5}.5-\frac{1}{7}\frac{5.6.7}{1.2.3}\frac{u^2}{v^2}-\ldots\Big)$$

$$=\ldots\ldots 2\frac{u}{v}\Big(\frac{2}{3}+2\frac{u^2}{v^2}+4\frac{u^4}{v^4}+\ldots\Big)=\ldots\ldots\frac{4}{3}\frac{u}{v}\Big(1+3\frac{u^2}{v^2}+6\frac{u^4}{v^4}+\ldots\Big). \tag{163}$$

That is, $$\mathbf{M}_{21}\mathbf{u}_1=\frac{\mu Q^2w}{6\pi R}dt_1.\cos\phi_1.\frac{u^2}{\iota^2\kappa^4}=\frac{u}{v^2}\mathrm{T}_{12}. \tag{164}$$

The $\mathbf{u}$ component of $\mathbf{M}_{12}$ is easily found. Thus,

$$\mathbf{u}\mathrm{V}\mathbf{D}_1\mathbf{B}_2=\mathbf{B}_2\mathrm{V}\mathbf{u}\mathbf{D}_1=\mathbf{H}_1\mathbf{B}_2, \quad\text{therefore}\quad \mathbf{M}_{12}\mathbf{u}_1=\mathrm{T}_{12}/u. \tag{165}$$

Lastly, the $\mathbf{u}$ component of $\mathbf{M}_2$ wants some more work, partly using the details given for U_2 or T_2. Because $\mathbf{D}_2$ and $\mathbf{B}_2$ are perpendicular, and tangential, we have

$$\mathrm{V}\mathbf{D}_2\mathbf{B}_2=\frac{\mu Q^2\dot{t}_1^{\,4}w^2}{(4\pi R)^2v^3}\mathbf{R}_1\{\mathbf{w}_1-(\mathbf{R}_1\mathbf{w}_1)\dot{t}_1(\mathbf{R}_1-\mathbf{u}/v)\}^2, \tag{166}$$

which, by the work given, (135 onward), becomes

$$......\mathbf{R}_1\left[1+\frac{u^2}{v^2}-\frac{2u}{v}\mu\right]\left[1-2\dot{t}_1\left\{\mu\mu_1^2\left(\mu-\frac{u}{v}\right)+\nu^2\nu_1^2\cos\phi\right.\right.$$

$$\left.+\mu\mu_1\nu\nu_1\cos\phi+\nu\nu_1\mu_1\left(\mu-\frac{u}{v}\right)\cos\phi\right\}$$

$$\left.+\dot{t}_1^2\left(1+\frac{u^2}{v^2}-2\frac{u}{v}\mu\right)(\mu^2\mu_1^2+\nu^2\nu_1^2\cos^2\phi+2\mu\mu_1\nu\nu_1\cos\phi)\right], \quad (167)$$

where μ and ν relate to $\mathbf{R}_1$, μ_1 and ν_1 to $\mathbf{w}_1$, whilst the direction of $\mathbf{u}$ is that of the polar axis, $\theta=0$.

To find $\mathbf{M}_2\mathbf{u}_1$, multiply (167) by $(vdt_1/\dot{t}_1)\mathrm{R}^2ud\mu d\phi$, and integrate. Thus,

$$\mathbf{M}_2\mathbf{u}_1=\frac{\mu Q^2w^2dt_1}{(4\pi v)^2}\iint d\mu d\phi\{\dot{t}_1^3-2\dot{t}_1^4(......)+\dot{t}_1^5(......)\}, \quad (168)$$

where the (......) and (......) may be seen in (167). The $d\phi$ integration may be done at sight, making

$$\mathbf{M}_2\mathbf{u}_1=\frac{\mu Q^2w^2dt_1}{(4\pi v)^2}\int\mu\left[2\pi\dot{t}_1^3+4\pi\dot{t}_1^4\mu\mu_1^2\frac{u}{v}-\kappa^2\dot{t}_1^5(2\pi\mu^2\mu_1^2+\pi\nu^2\nu_1^2)\right]d\mu. \quad (169)$$

Here the integral factor comes to

$$4\pi\left(\frac{1}{3}.3\frac{u}{v}+\frac{1}{5}.\frac{3.4.5}{1.2.3}\frac{u^3}{v^3}+...\right)+8\pi\mu_1^2\frac{u}{v}\left(\frac{1}{3}+\frac{1}{5}.\frac{4.5u^2}{1.2v^2}+...\right)$$

$$-4\pi\mu_1^2\kappa^2\left(\frac{1}{5}.5\frac{u}{v}+\frac{1}{7}.\frac{5.6.7}{1.2.3}\frac{u^3}{v^3}+...\right)$$

$$-2\pi\nu_1^2\kappa^2\left(0+\frac{1}{3}.5\frac{u}{v}+\frac{1}{5}.\frac{5.6.7}{1.2.3}\frac{u^3}{v^3}+...-\frac{1}{5}.5\frac{u}{v}-\frac{1}{7}.\frac{5.6.7}{1.2.3}\frac{u^3}{v^3}-...\right)$$

$$=4\pi\frac{u}{v}\left[\frac{2}{3}\left(1+2\frac{u^2}{v^2}+3\frac{u^4}{v^4}+...\right)+\mu_1^2\frac{u^2}{v^2}\frac{2}{3}\left(1+3\frac{u^2}{v^2}+...\right)\right]$$

$$=4\frac{\pi u}{v}\left(\frac{1}{\kappa^4}+\frac{\mu_1^2u^2}{\kappa^6v^2}\right). \quad (170)$$

Therefore $\qquad \mathbf{M}_2\mathbf{u}_1=\frac{\mu Q^2w^2udt_1}{6\pi v^3}\left(\frac{1}{\kappa^4}+\frac{\cos^2\phi_1.u^2}{v^2\kappa^6}\right)=\frac{u}{v^2}2\mathrm{T}_2. \quad (171)$

This completes the calculation of the $\mathbf{u}$ components. Altogether,

$$\mathbf{M}\mathbf{u}_1=\frac{2\mathrm{T}_1}{u}+\frac{2\mathrm{T}_{12}}{u}+\frac{u^2}{v^2}\frac{2\mathrm{T}_{12}}{u}+\frac{u^2}{v^2}\frac{2\mathrm{T}_2}{u}=(\mathbf{M}_1+\mathbf{M}_{12}+\mathbf{M}_{21}+\mathbf{M}_2)\mathbf{u}_1 \quad (172)$$

Now if $\mathbf{w}\|\mathbf{u}$, we make symmetry for $\mathbf{M}$ about the $\mathbf{u}$ axis. Then the four $\mathbf{M}$'s (partial) are parallel to $\mathbf{u}$; and the four scalars in the middle expression represent their values. But $\mathbf{M}$ cannot be parallel to $\mathbf{u}$ when $\mathbf{w}$ is not, and we must examine further to find how the component $\mathbf{M}$'s given are to be supplemented, remembering however that $\mathbf{M}_1\|\mathbf{u}$ always, as it does not depend upon $\mathbf{w}$.

Try the $\mathbf{w}$ component of $\mathbf{M}_{21}$. We must multiply by $\mathbf{w}_1$ instead of by $\mathbf{u}_1$. Now $\mathbf{R}_1\mathbf{w}_1=\cos\phi'=\mu\mu_1+\nu\nu_1\cos\phi$. This is to be used as a factor instead of μ before. So, if $\mathbf{M}_{21} \| \mathbf{u}$, the part of the integral depending on ν_1 must vanish. Test this.

$$\iint \nu \cos\phi \left[\mu \dot{t}_1^4 - \cos\phi' . \dot{t}_1^5 \left(\mu - \frac{u}{v}\right)\right] d\mu d\phi$$

$$= - \iint \nu \cos\phi . \nu\nu_1 \cos\phi' . \dot{t}_1^5 \left(\mu - \frac{u}{v}\right) d\mu$$

$$= -\frac{1}{2}\nu_1 \int (1-\mu^2)\left(\mu - \frac{u}{v}\right)\dot{t}_1^5 d\mu$$

$$= -\nu_1 \Big[\frac{1}{3}.5 + \frac{1}{5}\cdot\frac{5.6.7}{1.2.3}\frac{u^2}{v^2} + \frac{1}{7}\frac{5.6.7.8.9}{1.2.3.4.5}\frac{u^4}{v^4} + \dots$$

$$-\frac{1}{5}\dots - \frac{1}{7}\dots\dots\dots\dots - \frac{1}{9}\dots\dots\dots\dots\dots + \dots$$

$$-1 - \frac{1}{3}\cdot\frac{5.6}{1.2}\frac{v^2}{v^2} - \frac{1}{5}\cdot\frac{5.6.7.8}{1.2.3.4}\frac{u^4}{v^4} - \dots$$

$$+\frac{1}{3}+\frac{1}{5}\dots\dots\dots + \frac{1}{7}\dots\dots\dots\dots\dots + \dots\Big] = 0. \qquad (173)$$

That is, $\mathbf{M}_{21}\mathbf{w}_1 = (\mathbf{M}_{21}\mathbf{u}_1)\mu_1$. Therefore, since $\mu_1 = \mathbf{u}_1\mathbf{w}_1$, we conclude that $\mathbf{M}_{21}$ is parallel to $\mathbf{u}$, like $\mathbf{M}_1$.

Now $\mathbf{M}_{12}$. We have

$$\mathbf{VD}_1\mathbf{B}_2 = \frac{\mu Q^2 \kappa^2 u \dot{t}_1^5}{16\pi^2 R^3 v}\left[\mathbf{R}_1\left\{-\dot{t}_1 \cos\phi' . \kappa^2 + \frac{u}{v}\mu_1\right\} + \frac{\mathbf{W}_1}{\dot{t}_1} + \mathbf{u}_1\frac{u}{v}\cos\phi'\right] \qquad (174)$$

and the $\mathbf{w}$ component is

$$\mathbf{w}_1\mathbf{VD}_1\mathbf{B}_2 = \dots\dots\dots\dots\left[\cos\phi'\left\{-\kappa^2 \dot{t}_1 \cos\phi' + \frac{u}{v}\mu_1\right\} + \frac{1}{\dot{t}_1} + \mu_1\frac{u}{v}\cos\phi'\right]. \qquad (175)$$

Integration to ϕ makes

$$\dots\dots\dots\dots\left[2\frac{u}{v}\mu\mu_1^2 . 2\pi + \frac{2\pi}{\dot{t}_1} - \kappa^2 \dot{t}_1 (\mu^2\mu_1^2 2\pi + \nu^2\nu_1^2\pi)\right], \qquad (176)$$

and therefore

$$\mathbf{M}_{12}\mathbf{w}_1 = \frac{\mu Q^2 \kappa^2 w}{8\pi R} dt_1 \int \left[\dot{t}_1^3 + \frac{2u}{v}\mu\mu_1^2 \dot{t}_1^4 - \kappa^2(\mu^2\mu_1^2 + \tfrac{1}{2}\nu^2\nu_1^2)\dot{t}_1^5\right] d\mu, \qquad (177)$$

where the integral factor comes to

$$2\left\{1 + \frac{1}{3}\cdot\frac{3.4}{1.2}\frac{u^2}{v^2} + \frac{1}{5}\cdot\frac{3.4.5.6}{1.2.3.4}\frac{u^4}{v^4} + \right\} - \nu_1^2\left\{1 + \frac{1}{3}\cdot\frac{5.6}{1.2}\frac{u^2}{v^2} + \frac{1}{5}\cdot\frac{5.6.7.8}{1.2.3.4}\frac{u^4}{v^4} + \right.$$

$$+4\mu_1^2\left\{0 + \frac{1}{3}\;4\;\dots + \frac{1}{5}\cdot\frac{4\,5.6}{1.2.3}\dots + \right\} \qquad +0 - \;1\;\dots - \frac{1}{3}\cdot\frac{5.6}{1.2}\;\dots -$$

$$-2\mu_1^2\left\{\frac{1}{3} + \frac{1}{5}\cdot\frac{5.6}{1\,2}\dots + \frac{1}{7}\cdot\frac{5\,6.7.8}{1.2.3\,4}\dots + \right\} \qquad -\frac{1}{3} - \frac{1}{5}\cdot\frac{5.6}{1.2}\dots - \frac{1}{7}\frac{5.6.7.8}{1.2.3.4}\dots -$$

$$2\mu_1^2\left\{-\frac{1}{3}\;\frac{u^2}{v^2}\quad - \frac{1}{5}\quad\frac{5.6}{1.2}\frac{u^4}{v^4}\quad - \right\} \qquad \left. +0 + \frac{1}{3}\;\dots + \frac{1}{5}\cdot\;\frac{5.6}{1.2}\;\dots + \right\}$$

$$= \frac{4}{3}\left(1 + 2\frac{u^2}{v^2} + 3\frac{u^4}{v^4} + \dots\right) + \frac{4}{3}\mu_1^2\frac{u^2}{v^2}\left(1 + 3\frac{u^2}{v^2} + 6\frac{u^4}{v^4} + \dots\right). \qquad (178)$$

Therefore, by (177),

$$\mathbf{M}_{12}\mathbf{w}_1=\frac{\mu Q^2 w dt_1}{6\pi R}\left\{\frac{1}{\kappa^2}+\frac{u^2\cos^2\phi_1}{v^2\kappa^4}\right\} \tag{179}$$

is the $\mathbf{w}$ component of $\mathbf{M}_{12}$.

Lastly, test that there is no component $\perp$ $\mathbf{u}$ and to $\mathbf{w}$. The components of $\mathrm{V}\mathbf{uw}$ are 0, 0, $\sin\phi_1$. Those of $\mathbf{R}_1$ are μ, $\nu\cos\phi$, $\nu\sin\phi$. Therefore $\mathbf{R}_1\mathrm{V}\mathbf{u}_1\mathbf{w}_1=\nu\nu_1\sin\phi$. So we have to integrate

$$\nu\nu_1\sin\phi\left\{\frac{u}{v}\mu_1-\kappa^2\dot{t}_1(\mu\mu_1+\nu\nu_1\cos\phi)\right\} \tag{180}$$

with respect to ϕ. The result is zero.

Therefore $\mathbf{M}_{12}$ is in the plane of $\mathbf{u}$ and $\mathbf{w}$, with components in their directions given by (165) and (179). From them, the resultant is given by

$$\mathrm{M}_{12}=\frac{\mu Q^2 w dt_1}{6\pi R}\,\frac{\sqrt{\cos^2\phi_1+\kappa^4\sin^2\phi_1}}{\kappa^4},\qquad \tan\phi_2=\kappa^2\tan\phi_1, \tag{181}$$

where ϕ_2 is the angle between $\mathbf{M}_{12}$ and $\mathbf{u}$, and ϕ_1 that between $\mathbf{w}$ and $\mathbf{u}$. Thus M_{12} lies between $\mathbf{u}$ and $\mathbf{w}$, and moves from $\theta=0$ to $\frac{1}{2}\pi$ when $\mathbf{w}$ does so.

Finally, there is $\mathbf{M}_2$. It is $\parallel$ $\mathbf{R}$ in detail, and $\mathbf{R}_1\mathbf{w}_1=\mu\mu_1+\nu\nu_1\cos\phi$. In summing up through the shell for the second term of the last expression, we have to integrate

$$-2\dot{t}_1^{\,4}\left\{\pi\mu\mu_1\nu^2\nu_1+\pi\nu^2\nu_1\mu_1\left(\mu-\frac{u}{v}\right)\right\}+\dot{t}_1^{\,5}.2\pi\mu\mu_1\nu^2\nu_1\left(1+\frac{u^2}{v^2}-2\frac{u}{v}\mu\right) \tag{182}$$

with respect to μ, making

$$\begin{aligned}&-8\pi\mu_1\nu_1\left\{\frac{1}{3}.4+\frac{1}{5}.\frac{4.5.6}{1.2.3}\frac{u^2}{v^2}+\ldots-\frac{1}{5}.4-\frac{1}{7}.\frac{4.5.6}{1.2.3}\frac{u^2}{v^2}-\ldots\right\}\\&+4\pi\frac{u}{v}\mu_1\nu_1\left\{1+\frac{1}{3}.\frac{4.5}{1.2}\frac{u^2}{v^2}+\ldots-\frac{1}{3}-\frac{1}{5}.\frac{4.5}{1.2}\frac{u^2}{v^2}-\ldots\right\}\\&+4\pi\mu_1\nu_1\left(1+\frac{u^2}{v^2}\right)\left\{\frac{1}{3}.5+\frac{1}{5}.\frac{5.6.7}{1.2.3}\frac{u^2}{v^2}+\ldots-\frac{1}{5}.5-\frac{1}{7}.\frac{5.6.7}{1.2.3}\frac{u^2}{v^2}-\ldots\right\}\\&-8\pi\frac{u}{v}\mu_1\nu_1\left\{\frac{1}{3}+\frac{1}{5}.\frac{5.6}{1.2}\frac{u^2}{v^2}+\ldots-\frac{1}{5}-\frac{1}{7}.\frac{5.6}{1.2}\frac{u^2}{v^2}-\ldots\right\}.\end{aligned} \tag{183}$$

This comes to 0 exactly. So $\mathbf{M}_2$ has no $\mathbf{w}$ component, for $\mathbf{M}_2\mathbf{w}_1=\mathbf{M}_2\mathbf{u}_1\,.\,\mu_1$ by the above. As a check, examine the $\mathrm{V}\mathbf{u}_1\mathbf{w}_1$ component. As for $\mathbf{M}_{12}$, we must multiply by $\nu\nu_1\sin\phi$. See the full expression for $\mathbf{M}_2$. The integrals of $\sin\phi$, and $\sin\phi\cos\phi$, and $\sin\phi\,.\cos^2\phi$ all vanish. So $\mathbf{M}_2$ is parallel to $\mathbf{u}$. Only $\mathbf{M}_{12}$ is not, and then only when $\mathbf{w}$ is not.

(25). Since $\mathbf{M}_2$ is independent of R, this part of the momentum is wasted like the energy T_2 and U_2. The rate of waste of momentum is

$$\frac{\mu Q^2 w^2}{6\pi v}\left(\frac{1}{\kappa^4}+\frac{u^2\cos^2\phi_1}{v^2\kappa^6}\right)=\frac{u}{v^2}\times\text{energy waste}. \tag{208}$$

in another form, $T_2 = \frac{1}{2}M_2 v^2/u$. At the speed of light this is $T_2 = \frac{1}{2}M_2 v$, but then M_2 itself is infinite. This relates to the point-charge. There is no necessary infiniteness when the charge is distributed.

In motion at uniform speed in a circle of radius b, if we call the time rate of waste of energy dW/dt_1, and assume that the charge is a uniform spherical surface distribution, so that its energy when at rest is $U_0 = \mu v^2 Q^2 / 8\pi a$, where a is the radius of the sphere, then $\mathbf{w} \perp \mathbf{u}$, $\cos\phi_1 = 0$, and

$$\frac{dW}{dt_1} = \frac{4av}{3b^2}\frac{u^4}{v^4\kappa^4}U_0. \tag{209}$$

This U_0 presents itself naturally as a unit of energy. Using the high value of $u/v = \cdot 95$, first measured by Kaufmann (Becquerel's β "rays"), $u^2/v^2 = \cdot 9$, $\kappa^2 = \cdot 1$,

$$\frac{dW}{dt_1} = \frac{281}{10^{13}}\frac{v}{b^2}U_0 = \frac{843}{10^3 b^2}U_0, \tag{210}$$

or nearly U_0/b^2. This is a very large multiple of U_0 if b is of atomic size. But then this waste is per second. The waste per centim. is

$$\frac{dW}{dz_1} = \frac{10}{9v}\frac{dW}{dt_1} = \frac{31}{10^{12}b^2}U_0, \tag{211}$$

which is still a large multiple of U_0. The waste per revolution is $2\pi b$ times (211), or $(195/10^{12}b)U_0$. Say $b = 10^{-8}$, then this is only $(1/50)U_0$, that is 2 per cent. loss per rev., which could not go on for many revolutions, if we remember that the total energy $U+T$ is only a small multiple of U_0.

I give this numerical example to show the practical impossibility of maintaining very high revolutionary velocities, in accordance with the waste formula. It is much worse when u/v is further increased. Revolutionary velocities are probably quite small, against v, and even then there is rapid exhaustion by the radiation of energy. Maintenance by renewal is required, and here the old banging about of molecules presents itself, besides radiation from outside. The white hot poker takes some time to turn black; a very thin wire turns black at once; and carrying this on to molecular dimensions, there is nothing to wonder at in the very rapid loss of energy from a revolving charge left to itself.

But in systems of electrons in connected motion the waste is reduced. Then so much less is the radiation. The waste may, specially, be nothing at all. An elementary case is a circle of electrification revolving in its own plane round its own centre. Only when the veloc ty changes is there waste of energy. There 's a'so no waste when there is entire disappearance of the external **E** and **H** due to vibrating systems of electrification, of wh ch examples may be found in my *Elec. Pa.*, vol. 2, p. 418, and elsewhere; bearing in mind that problems relating to vibrating impressed electric force have all their corresponding representatives with vibrating electrification, in the way described before, p. 92 of this volume.

But a whole book could be filled with calculations of this sort, with various assumed data, and it is no part of my business to fill the rest of this volume with that kind of matter. There are other sources of information.

(26). When the magnetic effect of a sudden change of velocity of Q without change of position is wanted, it is desirable to exhibit the **H** formula in a different manner. We have

$$\mathbf{A}=-\frac{Q\dot{\mathbf{R}}}{4\pi R}, \qquad \frac{4\pi}{Q}\mathrm{V}\nabla\mathbf{A}=-\frac{1}{R}\frac{d}{dt}\mathrm{V}\nabla R-\mathrm{V}\left(\nabla\frac{1}{R}\right)\dot{\mathbf{R}}$$

$$=\ldots\ldots\ldots\ldots\ldots+\frac{\mathrm{V}\nabla R.\dot{\mathbf{R}}}{R^2}. \tag{212}$$

Here $\qquad \dot{\mathbf{R}}=-\mathbf{u}\dot{t}_1, \qquad \mathrm{V}\nabla R=+\dot{t}_1\mathrm{V}\mathbf{R}_1\mathbf{u}. \tag{213}$

So $\qquad \mathbf{H}=\mathrm{V}\nabla\mathbf{A}=\frac{Q}{4\pi Rv}\frac{d}{dt}\mathrm{V}\mathbf{u}\mathbf{R}_1\dot{t}_1+\frac{Q}{4\pi R^2}\mathrm{V}\mathbf{u}\mathbf{R}_1\dot{t}_1^{\,2}$

$$=\frac{Q}{4\pi R}\left(\frac{\dot{t}_1}{R}+\frac{d}{d(vt)}\right)\mathrm{V}\mathbf{u}\mathbf{R}_1\dot{t}_1. \tag{214}$$

If we carry out the time differentiation fully, the first given vectorial formula for **H**, (17), p. 436, will be come to. But from (214) itself, we can see that a sudden change of velocity produces a pulse of **H** given by

$$\mathbf{H}dt=\frac{Q}{4\pi Rv}d(\dot{t}_1\mathrm{V}\mathbf{u}\mathbf{R}_1)=+\frac{Q}{4\pi Rv}d\mathrm{V}\mathbf{R}_1\dot{\mathbf{R}}. \tag{215}$$

This is a spherical pulse of course, and is of uniform depth, when the change of velocity is quite sudden and the source is spherical. It is the time integral of **H** that is given, as the pulse passes the fixed observing point, and it depends upon the change made in the vector $\mathrm{V}\mathbf{R}_1\dot{\mathbf{R}}$, where **R** is the vector from the charge to the observing point, as in the figure on p. 433. Now $\mathbf{R}_1$ does not change, so the pulse of **H** varies as $\mathrm{V}\mathbf{R}_1 d\dot{\mathbf{R}}$, or as $-\mathrm{V}\mathbf{R}_1 d(\mathbf{u}\dot{t}_1)$.

If $2a$ is the depth of the pulse, $dt=2a/v$, so

$$2a\mathrm{H}=\frac{Q}{4\pi R}\left[\frac{u\sin\theta}{1-(u/v)\cos\theta}\right], \tag{216}$$

where only the change made in the quantity in the [ ] need be reckoned when the direction of **u** does not change. But if it does, then for $\sin\theta$ in (216) we must substitute $\mathrm{V}\mathbf{u}_1\mathbf{R}_1$. If the impulse is made from rest to velocity **u**, the [ ] may be omitted. This is the formula got before, p. 110, eq (64), p. 120, eq. (91). It should be noted that these formulæ, (212) to (216), relate to a quite sudden change of velocity. Another kind of pulse will turn up soon, required to effect the transition from the true pulse to continuous changes of velocity.

(27). Now considering Q to be uniformly distributed on a spherical surface of radius a, and to be suddenly jerked from rest to velocity **u**, the applied force impulse lasts just so long as it takes the inner of the two spherical waves which start from the surface to get entirely clear of the charge. From that moment the steady state exists right up to the inner surface of the shell, and therefore extends throughout the sphere, and partly outside. The shell as it expands leaves part of its contents (displacement and induction) behind, to complete the establishment of the steady state. Ultimately the shell tends to become purely electromagnetic, $E=\mu v H$, and then E and H are uniformly distributed in the depth of the shell. Its energy has been calculated already, p. 128. Say T_1 is the magnetic energy in the shell, then

$$T_1=\tfrac{1}{2}Q(P-P_0), \qquad P=\frac{Q}{4\pi ac}\left(1+\frac{1}{3}\frac{u^2}{v^2}+\frac{1}{5}\frac{u^4}{v^4}+\dots\right), \tag{217}$$

where P is Searle's and Morton's potential. It is related to the Newtonian force on unit of Q by (103), p. 128,

$$-\kappa^2\nabla P=\mathbf{E}+\mathrm{V}u\mathbf{B}. \tag{218}$$

The waste of energy is $2T_1$, because $T_1=U_1$ in the shell. This also equals the increase in the stored energy, a property due to the impulsiveness of the acceleration; that is,

$$2T_1=U+T-U_0, \tag{219}$$

where U and T are the electric and magnetic energies of the steady state, and U_0 the initial value of U belonging to P_0 and $u=0$. Also

$$U-T=\tfrac{1}{2}PQ\kappa^2. \tag{220}$$

From these relations follow Searle's formulas for U and T, namely

$$T=\frac{Q^2}{4\pi ac}\left(\frac{1}{3}\frac{u^2}{v^2}+\frac{2}{15}\frac{u^4}{v^4}+\frac{3}{35}\frac{u^6}{v^6}+\dots\right), \tag{221}$$

$$U-U_0=\frac{Q^2}{4\pi ac}\left(\frac{1}{15}\frac{u^4}{v^4}+\frac{2}{35}\frac{u^6}{v^6}+\frac{3}{7.9}\frac{u^8}{v^8}+\dots\right). \tag{222}$$

When u/v is small, the work done by the force-impulse is close to 2T, half stored, and half wasted.

(28). Let f be the applied force on Q. Then

$$\textstyle\int(f\,dt)=M+M_1, \qquad \mathbf{M}=\Sigma\mathrm{V}\mathbf{DB}, \tag{223}$$

where M belongs to the steady (progressive) field, and M_1 is the same for the shell. Now $HB=\mathbf{B}\mathrm{V}u\mathbf{D}=u\mathrm{V}\mathbf{DB}$; therefore

$$2T=Mu, \qquad u\textstyle\int(f\,dt)=M_1u+2T. \tag{224}$$

Also $$u\textstyle\int(f\,dt)=(U+T-U_0)+2T_1=2(U+T-U_0). \tag{225}$$

It follows that

$$2(\mathrm{U}-\mathrm{U}_0)=\mathrm{M}_1 u, \qquad \mathrm{T}-\mathrm{U}=\tfrac{1}{2}(\mathrm{M}-\mathrm{M}_1)u-\mathrm{U}_0. \tag{226}$$

The momentum stored is given by first of (224), the momentum wasted in the shell by the first of (226). The first momentum expression is expectable; the second is remarkable, being in terms of the increased electric energy of the steady field.

We may verify this by integrating the shell formula. Thus,

$$\mathrm{M}_1=\mathbf{u}_1\Sigma\mathbf{VDB}\text{ (shell)}=\frac{\mu\mathrm{Q}^2u^2}{16\pi av}\int_{-1}^{1}\frac{\mu-\mu^3}{(1-\mu u/v)^2}d\mu. \tag{227}$$

This, easily solved by power integration, makes

$$\mathrm{M}_1=\frac{\mu\mathrm{Q}^2u^2}{8\pi av}\left[\left(\frac{1}{3}-\frac{1}{5}\right)2\frac{u}{v}+\left(\frac{1}{5}-\frac{1}{7}\right)4\frac{u^3}{v^3}+\left(\frac{1}{7}-\frac{1}{9}\right)6\frac{u^5}{v^5}+\dots\right]$$

$$=\frac{2}{u}(\mathrm{U}-\mathrm{U}_0), \quad \text{according to (222), and verifying (226).}$$

(29). Next suppose that Q, in steady straight motion at velocity $\mathbf{u}_1$, has its velocity suddenly increased to $\mathbf{u}_2$. It is essentially sufficient to consider the details in the case of parallelism of $\mathbf{u}_1$ and $\mathbf{u}_2$. At the moment of the jerk a new shell belonging to $\mathbf{u}_2$ begins to be established by spherical expansion. Outside the new shell is the old steady field, which may be dismissed in considering the strength of the new shell. Inside the shell is the new steady field belonging to $\mathbf{u}_2$ along with the old one belonging to $\mathbf{u}_1$, and in the shell itself we have

$$2a\mathrm{H}=\frac{\mathrm{Q}}{4\pi\mathrm{R}}\left(\frac{u_2\sin\theta}{1-(u_2/v)\cos\theta}-\frac{u_1\sin\theta}{1-(u_1/v)\cos\theta}\right) \tag{228}$$

at ∞, that is, when the shell has expanded sufficiently.

The magnetic energy in the shell is $\mathrm{T}_1+\mathrm{T}_2-\mathrm{T}_{12}$, where T_1 is the former magnetic energy of the shell belonging to u_1, T_2 that belonging to u_2 (from rest), and $\frac{1}{2}\mathrm{T}_{12}$ depends on $u_1u_2\sin^2\theta(1-\mu u_1/v)^{-1}(1-\mu u_2/v)^{-1}$ in the same way as T_1 depends on $u_1^2\sin^2\theta(1-\mu u_1/v)^{-2}$. So we have to find

$$\mathrm{T}_{12}=\frac{\mu\mathrm{Q}^2u_1u_2}{16\pi a}\int_{1}^{-1}\frac{\sin^2\theta.d\mu}{(1-\mu u_1/v)(1-\mu u_2/v)}. \tag{229}$$

Here, by expansion,

$$(1-\mu u_1/v)^{-1}(1-\mu u_2/v)^{-1}$$

$$=(u_2-u_1)^{-1}\left[u_2-u_1+\frac{(u_2^2-u_1^2)\mu}{v^2}+\frac{(u_2^3-u_1^3)\mu^2}{v^4}+\dots\right]. \tag{230}$$

Using this in (229), and doing the integration, the result is, if $\Delta u=u_2-u_1$,

$$\mathrm{T}_{12}=\frac{\mu\mathrm{Q}^2u_1u_2}{4\pi a}\left(\frac{1}{3}+\frac{1}{3.5}\,\frac{u_2^3-u_1^3}{v^2\Delta u}+\frac{1}{5.7}\,\frac{u_2^5-u_1^5}{v^4\Delta u}+\dots\right). \tag{231}$$

So the final magnetic energy in the shell is

$$T_1+T_2-T_{12}=\frac{Q^2}{8\pi ac}\left\{\frac{u_1^2}{v^2}\left(\frac{1}{3}+\frac{1}{5}\frac{u_1^2}{v^2}+\frac{1}{7}\frac{u_1^4}{v^4}+\right)+\frac{u_2^2}{v^2}\left(\frac{1}{3}+\frac{1}{5}\frac{u_2^2}{v^2}+\frac{1}{7}\frac{u_2^4}{v^4}+\right)\right.$$
$$\left.-\frac{2u_1u_2}{v^2}\left(\frac{1}{3}+\frac{1}{15}\frac{u_2^3-u_1^3}{v^2\Delta u}+\frac{1}{35}\frac{u_2^5-u_1^5}{v^4\Delta u}\right)\right\}. \qquad (232)$$

This may be put in terms of logs, with the peculiarity that whereas the series in (232) are visibly explanatory, the log expressions require some study, especially the one for T_{12},

The following will show the meaning. First put on u_1 from rest The magnetic energy wasted is T_1. Then increase to u_2. The additional magnetic waste is $T_1+T_2-T_{12}$. The total magnetic waste is $2T_1+T_2-T_{12}$. The total waste is double as much. But if u_2 is put on from rest, the magnetic waste is T_2. So, doing it in two steps saves $T_{12}-2T_1$ when u_1 comes first, and the saving is $T_{12}'-2T_2$ when u_2 comes first. Say $u_2=2u_1$. Then

$$T_1=\frac{Q^2}{8\pi ac}\left(\frac{1}{3}\frac{u_1^2}{v^2}+\frac{1}{5}\frac{u_1^4}{v^4}+\right), \qquad T_2=\ldots\left(\frac{4}{3}\frac{u_1^2}{v^2}+\frac{16}{5}\frac{u_1^4}{v^4}+\right),$$
$$T_{12}=\ldots\left(\frac{4}{3}\frac{u_1^2}{v^2}+\frac{28}{15}\frac{u_1^4}{v^4}+\right).$$

So $T_{12}<T_2$ and much $>2T_1$. If T_2 is wasted first, the additional waste $2T_2-T_{12}$ is positive and large. But if T_1 is wasted first, the additional waste $2T_1-T_{12}$ is negative, or there is a saving. Note that in T_{12} we may interchange u_1 and u_2.

We may also write

$$\frac{2T_{12}}{Q}=-P_0+\frac{u_2P_1-u_1P_2}{u_2-u_1}+\frac{u_1u_2}{v^2}\frac{u_2P_2-u_1P_1}{u_2-u_1}, \qquad (233)$$

by using the potential of (217). So the energy in the shell (from u_1 to u_2) is ultimately given by

$$2(T_1+T_2-T_{12})=-P_0Q+\frac{u_2P_2-u_1P_1}{\Delta u}\left(1-\frac{u_1u_2}{v^2}\right). \qquad (234)$$

In this, use the P formula, and write u for u_1, and $u+\Delta u$ for u_2. It will be found first that the terms independent of u vanish. Then examining what is left, it will be found that the terms involving Δu also vanish. But the $(\Delta u)^2$ terms do not vanish. Retaining them, and discarding higher powers, the result is

$$2(T_1+T_2-T_{12})=\frac{Q^2}{4\pi ca}\frac{(\Delta u)^2}{3v^2}\left(1+2\frac{u^2}{v^2}+3\frac{u^4}{v^4}+\right)=\frac{\mu Q^2(\Delta u)^2}{12\pi a\kappa}, \qquad (235)$$

when Δu is a very small impulsive increment of speed. Divide by $2a$ to get the energy in shell per unit depth, and then multiply by $v/4\pi R^2$ to get the mean flux of energy per unit area.

(30). In the above the assumption was made that **H** tends to uniformity of value in the depth of the shells due to sudden changes of velocity. This must be justified. Now I have shown on p. 99 that

$$H=\frac{\rho ua}{2r}\sin\theta\left(vt_1-\frac{v^2t_1^2}{2a}+\frac{v^2t_1^2}{2r}-\frac{v^3t_1^3}{6ar}\right),\tag{236}$$

where $vt_1=vt+a-r$, is the H wave from a solid sphere of electrification of uniform density ρ, making no allowance for the Doppler effect. The sphere is at rest when $t=0$, and then assumes the velocity **u**. The quantity vt_1 is here the distance of the point r, θ from the front of the wave. H varies from 0 at the front to $(Qu\sin\theta)/4\pi r^2$ at the back or inner boundary of the shell, if Q is the total charge. In (236) let r be very great. Then it reduces to

$$H=\frac{\rho u}{4r}\sin\theta.\ (a^2-x^2),\tag{237}$$

at distance x from the middle of the shell. The distribution of intensity is parabolic. Now increase a to $a+\Delta a$. This makes

$$H'=\frac{\rho u\nu}{4r}[(a^2-x^2)+2a\Delta a],\qquad H'-H=\frac{Q'u\nu}{8\pi ar}.\tag{238}$$

where $Q'=4\pi a^2\rho\Delta a$. This difference is the H due to the sudden starting of the differential charge of surface density $\rho\Delta a$; and it is of the same value right through the shell. The integral H' is $2aH'=Q'u\nu/4\pi r$, in agreement with the more general $Qu\nu\dot{t}_1/4\pi R$, when u is so large that $\dot{t}_1$ cannot be ignored.

(31). The shell energy (as on p. 97) is 20 per cent. more when Q is solid than when it is superficial. That is,

$$\Sigma\mu H^2=\mu\frac{9Q^2u^2}{6\pi\pi^2a^2}\iint(1-\mu^2)d\mu h^2\left(1-\frac{h}{a}+\frac{h^2}{4a^2}\right)dh.2\pi=\frac{\mu Q^2u^2}{10\pi a},\tag{239}$$

when Q is solid. Or thus, if the *same* quantity of induction in the shell is changed from a uniform to a parabolic distribution, the value of the integral of its square is increased 20 per cent.

We may infer that P, P_0, $U-U_0$, and T are all similarly increased. It is so for $U-U_0+T$ because this quantity equals the shell energy. It is so for U_0 by an easy calculation of $\Sigma\frac{1}{2}cE^2$, with the value $E=Qr/4\pi a^3c$ inside the sphere. It comes to one-fifth of the external part of U_0, which is unchanged. Then we know that the outside density of $U-T$ is $\frac{1}{2}PQ\kappa^2$, so the total $U-T$ is increased 20 per cent., if $\Sigma P\rho$ in the sphere equals one-fifth of P_aQ on its surface, P_a being the surface value of P. Now at distance r from the surface we have

$$P=\frac{a}{r}P_a\times\frac{r^3}{a^3}+\int_r^a 4\pi r^2\frac{3Q}{4\pi a^3}\frac{1}{4\pi cr}(1+\ldots)dr=\left(\frac{3}{2}-\frac{1}{2}\frac{r^2}{a^2}\right)P_a,\tag{240}$$

and the mean value of P in the sphere by this formula is $6P_a/5$. This is true whether the charge is solid or superficial. Therefore the whole $U-T$ is increased 20 per cent.; so, by the previous, this is true for U and T separately. These calculations are elementary, but those relating to the variable motion are much complicated by using a solid instead of a surface distribution of electrification. The solid sphere is a collection of shells, so there is a mixture by overlapping.

(32). The H due to a solid sphere of electrification density ρ, given the velocity u is, if $q=d/d(vt)$, by p. 99,

$$\mathrm{H}=\frac{av}{2qr}\left(1+\frac{1}{qr}\right)\left[\epsilon^{-q(r-a)}\left(1-\frac{1}{qa}\right)-\epsilon^{-q(r+a)}\left(1+\frac{1}{qa}\right)\right]\rho u. \quad (241)$$

The shell solution (236) above is the immediate algebrisation of the first part of this operational solution, in which u may be any function of the time. But u/v must be small. When the second part is added the result gives the solution between the sphere and the inner boundary of the shell. Equation (241) is the same as

$$\mathrm{H}=av\left(\cosh qa-\frac{\text{shin } qa}{qa}\right)\left(-\frac{d}{dqr}\frac{\epsilon^{-qr}}{qr}\right)\rho u, \quad (242)$$

which is more convenient when ρu is simply periodic. The corresponding A is

$$\text{(outside)} \quad \mathrm{A}=\frac{a}{q^2}\left(\cosh qa-\frac{\text{shin } qa}{qa}\right)\frac{\epsilon^{-qr}}{r}\rho u$$

$$=\left(1+\frac{3.4}{\lfloor 5}q^2a^2+\frac{3.6}{\lfloor 7}q^4a^4+\frac{3.8}{9}q^6a^6+\dots\right)\mathrm{A}_0, \quad (243)$$

$$\text{if} \quad \mathrm{A}_0=\frac{4\pi a^3}{3}\frac{\epsilon^{-qr}}{4\pi r}\rho u \quad (244)$$

This A_0 is what is due to the total charge if collected at the centre, which is a known function. The extra terms to be derived by double differentiations after the manner of (243) show the external effect of distributing the charge. Now by a single differentiation to a we have

$$\frac{d\mathrm{A}}{da}=4\pi a^2\left(1+\frac{q^2a^2}{\lfloor 3}+\frac{q^4a^4}{\lfloor 5}+\dots\right)\frac{\epsilon^{-qr}}{4\pi r}\rho u. \quad (245)$$

This is the external A due to a shell of electrification of surface density ρ; so

$$\text{(out)} \quad \mathrm{A}=\frac{\text{shin } qa}{qa}\left(\frac{\epsilon^{-qr}}{4\pi r}\mathrm{Q}u\right)=\frac{\text{shin } qa}{qa}\mathrm{A}_0 \quad (246)$$

is the external A due to the surface charge Q, A_0 being the same due to Q at the centre.

The corresponding internal potentials are

$$\text{(in)} \quad \mathrm{A}=-\frac{a}{2}\left(1+\frac{1}{qa}\right)\epsilon^{-qr}\frac{\epsilon^{qr}-\epsilon^{-qr}}{q^2r}\rho u, \quad \text{(solid)} \quad (247)$$

$$\text{(in)} \quad \mathrm{A}=\epsilon^{-qa}\frac{\text{shin } qr}{qr}\frac{\mathrm{Q}u}{4\pi a}=\frac{\text{shin } qr}{qr}\mathrm{A}_{0a}. \quad \text{(surface)} \quad (248)$$

So by (246), (248), the results over all space become known for the surface Q in terms of A_0, the potential due to Q at the centre.

The matter does no end here. For if we use the dopplerized or distorted A_0 and A_{0a}, obtained by introducing the factor $\dot{t}_1$, the same operators (shin qa), qa and (shin qr)/qr will give the corresponding results when u/v is not a small quantity. I carried this out for the second term of the complete series (246), obtaining a complicated expression whose value was a very small fraction of that of the first term. But there are circumstances under which the first term of (246), that is, the solution for a point charge at the centre, is not sufficiently representative of the result due to a surface charge. The important matters are the size of the sphere, and of the acceleration. As before said, the rate of acceleration should not change sensibly in the time light takes to traverse the sphere, if we want to treat the sphere (externally) as a point.

(33). The quantity **M**, or Σ**VDB**, called the momentum, when calculated for a single electron, is the force-impulse or time-integral of the Newtonian, sometimes called the mechanical force, on the electron. For the force on the electron due to the stress in the ether is the negative of the total force on the ether itself, which is (d/dt)**VDB** per unit volume. And the negative of the force on the electron is the applied force, say **f**. So $\mathbf{f}=\dot{\mathbf{M}}$ is the "equation of motion." Or

$$-\mathbf{f}=\Sigma(\mathbf{E}+\mathrm{V}\mathbf{uB})\rho, \tag{249}$$

when summed up throughout the electrification constituting the electron in the electromagnetic analysis.

This is an example of Newton's celebrated Action = Reaction, an impregnable fundamental principle whose neglect sometimes leads to alarming consequences.

If **M**||**u** we may write

$$\dot{\mathbf{M}}=\frac{d}{dt}\mathrm{M}\mathbf{u}_1=\mathrm{M}\dot{\mathbf{u}}_1+\dot{\mathrm{M}}\mathbf{u}_1. \qquad \text{(1st Assumption)} \tag{250}$$

And if M is a function of u, we may write

$$\dot{\mathbf{M}}=\mathrm{M}\dot{\mathbf{u}}_1+\mathbf{u}_1\frac{d\mathrm{M}}{du}\dot{u}. \qquad \text{(2nd Assumption)} \tag{251}$$

There are two main cases. If the direction of **u** does not change,

$$\dot{\mathbf{u}}_1=0, \quad \text{and} \quad \mathbf{f}=\frac{d\mathrm{M}}{du}\dot{\mathbf{u}}=m_1\dot{\mathbf{u}}. \tag{252}$$

So, subject to the assumption made, we may say that $d\mathrm{M}/du$ is the effective mass in straight motion. But if the direction changes, while the speed is constant, then

$$\dot{u}=0, \quad \text{and} \quad \mathbf{f}=\mathrm{M}\dot{\mathbf{u}}_1=\frac{\mathrm{M}}{u}\dot{\mathbf{u}}=m_2\dot{\mathbf{u}}. \tag{253}$$

So, subject to the assumption made, we may say that M/u is the effective transverse mass.

Neither of the assumptions is true in general. In straight motion, the first is true, but the second is not, owing to the waste. In circular motion, the first is not true, though the second may be true sometimes.

But though there is no definite value to be given to the effective mass of an electron (or, in general, any connected distribution of electrification), the above formulas for m_1 and m_2 are not useless. The above is plain Newtonian dynamics, and when we can go nearly enough to the assumptions made, we approach m_1 and m_2. That M should be a function of u strictly means the establishment of the steady state belonging to u, and the ignoration of previous waste of energy and momentum. This is to apply approximately when the speed varies in straight motion, and also when the direction varies. Therefore the acceleration must be infinitely small. Then m_1 and m_2 become practically the straight and the transverse effective mass.

It is rather remarkable that these m_1 and m_2 formulas, which were first exhibited by Abraham, were derived by him from expressions for the energy by the Lagrangian method. Now this method is not applicable to the problem in general. There is one coordinate and one speed, namely, u, and the energy expression is a function of u^2, u^4, u^6, etc. Yet the method goes. Only, since it is the steady energy that is counted, the acceleration must, as above, be infinitesimal. Then we come indirectly to the interesting conclusion that the Lagrangian method may be applied, provided the acceleration is infinitesimal, to cases where, at first sight, there appears to be no application.

At the other extreme, we have infinite acceleration, or the sudden establishment of velocity. This is easy. For we have

$$T=\tfrac{1}{2}Mu, \qquad U-U_0=\tfrac{1}{2}M_1u, \qquad M_0=M+M_1=m_0u; \qquad (254)$$

$$U+T-U_0=\tfrac{1}{2}M_0u=\tfrac{1}{2}m_0u^2. \qquad (255)$$

Here the two parts of the momentum are joined to make M_0, and m_0 is the corresponding "mass," the effective mass under the circumstance, or at least what might be called the impulsive mass, as it is a quite special case.

If the electron is suddenly stopped, there is superimposed on the existing field, (shell with interior steady progressive state), a new state involving phenomena exactly similar to the starting phenomena, but with **E** and **H** reversed. So there are now two shells, both expanding to infinity from different centres. Their energies and momenta are ultimately equal. That is, there is as much energy and momentum wasted in stopping as in starting. But the work is not done at the electron, but is taken from the field.

Thus, on sudden stoppage, momentum of amount $(2/u)(U+T-U_0)$ is added to the pre-existing momentum, not instantly, but during the time taken for the new wave to be created and get clear of the electrification. As this second shell traverses the field, it restores the original state of isotropic radial displacement of energy U_0, without magnetic energy, and the energy thus lost from the field is stored in the second shell. So $U+T-U_0$ of energy is lost ultimately in the second shell.

The new field momentum $-2T/u$ cancels the previous momentum $+2T/u$, and the rest, which is $+2(U-U_0)/u$, goes out to infinity in the second shell. So the total waste of momentum is $(4/u)(U-U_0)$, reckoned in time from a state of rest to a state of rest again, with two sudden changes of velocity of opposite sign. In continuous acceleration too, the waste of momentum is always positive, as is obvious in the formula involving the square of the acceleration, though perhaps not visibly evident otherwise.

What is done in increasing u to $u+\Delta u$ (a positive increment) is to increase the momentum in the field, and also to waste positive momentum. Then, on reducing the speed to u again, we reduce the momentum in the field by the same amount, and also waste the same amount of + momentum as before. Say $f_1 dt_1 = x_1 + x_2$ in the first case; then $f_2 dt_2 = -x_1 + x_2$ in the second. So the force on the electron is smaller in the reduction of speed than in the increase. And, in conformity with this,

$$u f(f dt) = +2(T+U-U_0) \quad \text{at the start,}$$
$$\text{and} \quad = -2(T-U+U_0) \quad \text{at the stop.} \qquad (256)$$

The total force-time integral therefore comes to $(4/u)(U-U_0)$, and this is accounted for by the waste of momentum in the two shells.

(34). This is perhaps an appropriate place to insert some remarks on the so-called mass of an electron; though, strictly speaking, none has been found yet, because electricity in itself has no inertia in the theory, and experiment has so far found none. All the previous in this article is made up from my investigations condensed on pp. 158 to 167, which the reader is recommended to read, in order to obtain a general idea of the subject broadly, apart from complications of calculation. A well-known German professor said there was material in it for a volume. Yes; but there were difficulties in the way of the volume.

The name electron was coined long ago by Johnstone Stoney to be given to hypothetical units of electricity. It is a capital name, and is, I think, better than Kelvin's modification "electrion," which weds electricity to the "ion." The first use of the idea of mass, (another capital name, not to be confounded with weight), in connection with a moving charge was, I believe, by J. J. Thomson, somewhen about 1877. The idea comes in very naturally and simply. A charged spherical conductor is supposed to move without altering the istropic distribution of displacement. Then comes Maxwell's doctrine that the changing displacement is an electric current, with a magnetic field. Its energy is calculated, and is found to vary as u^2, the square of the speed. So does the ordinary kinetic energy of the mass of the conductor. There is therefore a small effective increase of mass.

Several things have to be noted here from a later point of view. No increase in the electric energy is allowed for. It is implied that changes of velocity are instantly accompanied by the corresponding changes of magnetic force. It is then clear that u/v must be very small, and that the solution of the problem is the first term of the real solution. Moreover, no jerks, or rapid accelerations, are permissible. With these limitations, the idea of a constant effective mass of a moving electrified sphere, in addition to the common mass, is a most useful one.

(35). But the current was closed through the conductor, if I remember rightly, and there were analytical errors, or errors of method in the theory, which made the estimate of the magnetic energy come out wrong. I gave the corrected formula, now in general use, in 1885, and a broader theory in 1888-9. (Elec Pa., vol. 2, pp. 490 to 518.) Steady straight motion at any speed was treated. The moving charge itself completes the circuit of the electric displacement currents. Inside the electrified surface there is no magnetic field. The idea of a constant effective mass, independent of the speed, goes out of existence. Both the electric and magnetic energy even when divided by u^2 increase, at first slowly, later very fast indeed, as the speed increases. Infinite series in powers of u^2 are required in general to represent them. As regards, however, the formulas for U and T, I calculated them for the region outside a sphere which itself surrounds the charged body, and might be the charged sphere itself if the displacement were, as assumed, perpendicular to the conductor. But Searle showed me that my solution for a point-charge, when applied to a sphere, would not work. This led me to revise this part of the theory (vol. 1 of this work, p. 273). The displacement cannot be perpendicular to the conducting surface. The equlibrium surfaces are oblate ellipsoids of revolution when the source is a charge at a point, or is itself distributed over one of the same oblate spheroids. Searle himself followed this up, and gave the formulas for a charged sphere or a charged ellipsoid of revolution of any ellipticity moving parallel to its axis of symmetry. Also the energy formulas. This ellipsoidal work has since come into use in connection with the electron, which should be gratifying to the rejected one

(36). But the idea of mass in association with moving charges was apparently sleeping most of the time, for it plays only a small part in the general theory. It woke again in a practical manner when J. J. Thomson, by his tube experiments, established that Crookes's radiant matter, or matter in a fourth state, consisted of a stream of something he called corpuscles, the same in all gases. Here the elementary theory of a moving charge was the principal theoretical guide, applied to the deflections produced by external electric and magnetic fields. It appeared to be very probable that the mass of the corpuscle, in the simple Newtonian sense, was a constant independent of the speed. And it was the same in Zeeman's experiments on the effect of magnetic force on the spectrum lines of a gas, as interpreted by Lorentz. In fact, it almost came to be taken for granted soon that "m" and "e/m" were constants, when m was reckoned electromagnetically as in the elementary theory.

(37). But to discriminate between electromagnetic effective mass and material mass in the old sense requires observations at high speeds, and a number of them. They soon came. The first I saw of Kaufmann's results was a short note saying he had found by high-speed measurements that So-much of the effective mass was material, and So-much electromagnetic. This was followed by a second short note giving five numerical results of observation. Knowing nothing of the method of experiment, I tried two formulas tentatively on these figures, one involving infinitesimal, the other infinite acceleration. When plotted the results made a rather violent zigzag. But there was no positive sign of any part of the effective mass being material, by either formula. The difference made in the value of e/m in passing from one to the other was only 14 per cent. This was remarkable, but then the steepness of the theoretical curves when $u/v > 9/10$, combined with the adjustment to the zigzag, could easily explain that. But the infinite acceleration formula really gave a better fit than the other, so I selected it in the Note on p. 131 above.

But Kaufmann at this time or shortly after published much more precise experimental results, accompanied or followed by Abraham with his formulas for the above m_1 and m_2, on the assumption of rigid sphericality of an electron applied to the ellipsoidal theory. The agreement in general was very good with the m_2 deflection formula. But with such steep curves as are given by many formulas, an outsider like myself could not feel quite convinced of more than that the material mass was only a small part of the apparent mass, as estimated electromagnetically. Moreover, it should be remembered that the use of either m_1 or m_2 requires infinitesimal acceleration Was the acceleration sufficiently small to allow the formula used to be valid without sensible error at the high speeds concerned? The acceleration was large enough to be visible and measureable.

(38). Then came Lorentz's electron. This has the oblate ellipsoidal shape of the surface of equilibrium. It is squeezed from the spherical shape in the line of motion, without lateral change. It has to change its shape whenever it changes its velocity. It is attractive theoretically, on account of the simplicity of energy and mass formulas. It is also suggested by the "explanation" given by FitzGerald and by Lorentz of the Michelson-Morley experiment, that the negative result could be accounted for by a certain lateral contraction, in the line of motion, of the bodies supporting the apparatus. Here the real difficulty is to explain the explanation. Closely related to Lorentz's electron was Bucherer's modification which followed. It had the same squeeze in the line of motion, but was expanded laterally to keep the volume constant. Bucherer undertook elaborate experiments on his own account to settle the matter, with the result that he gave up his own design of electron, finding that Lorentz's agreed best with his experimental results.

I am still inclined to think that the electron may have a material basis. It might be a very small element with a negative charge of much greater effective mass. As regards its constitution, that is very speculative. It cannot be just electricity and nothing more, or it would fly to pieces and disseminate to an infinite distance, or else go to positive electrification. This would not be in the wave manner of the condensation in a condensational wave, but convectionally. This of course assumes that we define electricity by the theory of the same in its dynamical aspect, combined with the fact that when like electricities mutually repel, they actually yield to the repulsion and move through the ether, unless prevented by some other force. So we seem to want something for the negative electricity of an electron to hold on to. Or, in another form, something for the tubes of displacement to terminate upon. Besides that, the electron must have size and shape and so on, to allow it to behave as it does. This is no doubt a very difficult question, but scientific research has done apparently impossible work before now, and will no doubt do it again. As for positive electricity, that seems to have no electron, but to be always associated with the recognised chemical elements, whether stable or unstable. Is it male or female? There is no doubt that the physical basis of life, which used to be protoplasm, must be sought for much lower down, if indeed there is any limit; and the electrons, which are called negative, seem most likely to be the stimulators. Life does not stop in the animal and vegetable kingdoms. In a rudimentary form, it may be everywhere, as electromagnetic waves are. Evolution comes from within, in its first nature, however it may be guided and influenced externally.

(39). The "electrical theory of matter," which has some evidence to support it, is full of difficulties. Some of them are very evident when the limited application of the "mass" theory is considered, namely to small velocities and, more important still, to small accelerations. Nor can it be safely assumed that the effective inertia of a collection of electrons equals the sum of their separate inertias; so that the number of electrons in an atom varies as the atomic "weight." Nothing is known about the gravitation of electrons, by the way. Consider, for example, the well-known electrical problem of a steady current in a straight wire, with a distant external return conductor, (tube, for simplicity). In calculating its inductance, say the L in $\frac{1}{2}LC^2$ the magnetic energy, it is really the quantity $\frac{1}{2}\mu H^2$, the density of the magnetic energy that is summed up, and the current in the wire is supposed to be uniformly distributed. Most of the magnetic energy is outside the wire, in the space between the wire and return. We need make no assumption as to what the current in the wire is in detail; that is eliminated by the averaging which is assumed in the construction of the uniform distribution. But now substitute for the last an enormous number of electrons, of such number and speed as to make up the total current. According to a calculation made by A. Schuster on the basis of the slow speed theory, the magnetic energy is increased by a very small amount, perhaps not recognisable. We might perhaps also come to the same conclusion by considering that the external magnetic field due to the uniform current and that due to the substituted condensed currents are almost exactly the same.

Now if we start with a uniform distribution of current in a sphere, such as would arise if a solid sphere of electrification were set in motion, and calculate its magnetic energy; and then substitute for it a great number (say many thousands) of electrons distributed fairly uniformly, we can estimate the difference made in the magnetic energy. The magnetic energy of the collection of electrons is not the sum of their separate energies, and I find that this affects the reckoning of proportionality of the number of electrons to atomic weight in the case of the large atomic weights. But this is a trifle compared with the difficulties connected with positive electrification, if common material energy is to be considered to be wholly electromagnetic.

It may be worth noticing that we may have electrons of the same effective mass of any size we like, by suitably altering the distribution of the electrification. Say it is to be of atomic size. That could be done by dividing the total charge into a number of small charges, each to be the charge of a sub-electron, the sub-electrons to be arranged throughout the sphere of atomic size.

But although electrons do not behave as small constant masses in electromagnetic theory, it may be observed that constancy of mass of a given quantity of matter is not an essential part of dynamics, even though it may be true observationally to a certain degree of fineness. What I always regard as *the* fundamental principle of dynamics is Newton's celebrated Third Law. If that is not true, the result is Chaos.

I cannot help remarking how fortunate it has been in the rapid growth of the new science of Radio-activity of Becquerel and Rutherford, as well as in many other experimental ways, that the usual velocities of electrons are low, so that the elementary theory of electrons can be used as a guide in interpretation; save for shocks, or large accelerations. The theory of pulses, too, is relatively simple, if they are condensed enough. In the application to the extraordinary X-rays of Röntgen, (to the discovery of which P. Lenard seems to have been tending), there must be pulses produced when a cathode stream produces X-rays; but the resultant effect of the pulses in the stream of X-rays is something like plane waves (or may be diverging waves) of light, with extraordinary relative intensity given to very high frequencies when analysed simple harmonically. The importance of the high frequencies may be readily seen in the analysis of a discontinuous curve into simply periodic waves.

But can an electron lose its charge? If it can, then X-rays may contain a stream of the residual element along with the electromagnetic waves. Being of very small mass its speed may be excessive. The residual element provisionally assumed here would not be the same as a neutral particle formed of helium and an electron, a sort of doublet. Bragg has proposed (Roy. Inst., Jan. 1911) that "the γ or X-ray is corpuscular, being an electron which has assumed a cloak of darkness in the form of sufficient positive electricity to neutralize its charge." It would depend on the manner of neutralisation how far this would behave as a really unelectrified element, which, in virtue of very small mass could penetrate the ether unresisted at speeds far exceeding that of light, whereas Bragg's cloaked electron might not do this. But we cannot get away from the pulses, and their resultant effect. For if they don't exist we have to ask, Why not? They ought to be there; and I have no doubt they are, however it may be as regards the corpuscles.

The dependence of apparent mass upon distribution of the charge is remarkably illustrated by the elementary case of a uniform plane distribution of electrification. If of infinite extent, it has no inertia at all, when moved perpendicularly to its plane, because there is no magnetic force. (I gave this case in 1889.) Even if the electrification is condensed in electrons, the correction need not be large. If the plane is finite, say with a circular edge, the theory is known, by that of a disc of impressed electric force (Elec. Pa., vol. 2, p. 415, and E.M.T., vol. 1, p. 114). The edge is the source of the waves when the disc is moved perpendicular to itself. So, if moved from rest suddenly, the greater the charge, as with a larger disc, the longer it takes before the central part is endowed with any inertia.

(40). In the above I have made much use of the vector **M**, the time integral of the Newtonian or mechanical force on the ether. Different views have been held about this force. I have always regarded it as a reality, since I first wrote on the electric and magnetic stresses in 1886, necessitated by Newton's Third Law. But it need not be the complete force on the ether, and **M** need not therefore be the actual momentum. It might be only potential momentum, and if the ether were incompressible, and it could not move under the action of the force $d\mathbf{M}/dt$, then **M** could not be visible momentum. The ether should be compressible to make **M** be visible momentum, and waves of **E**, **H** are then accompanied by waves of compression. (*See* pp. 144 to 158 of this volume for this theory.) But in general they are distorted. We must assume some law of compressibility. Then, in cases I examined, of simply periodic progressive plane waves, the disturbance in the humps and hollows tends to accumulate in front, that is at the nodes. It is only a question of time, and completely upsets the proper propagation of light from distant stars, as generally accepted.

I have, however, shown how to prevent this, in plane solitary waves of any type. The three quantities m the density of the ether, μ its inductivity, and c its permittivity, are all referred to unit volume, as usual. Now, in the wave of compression accompanying the electromagnetic wave, m varies, of course. If μ and c do not vary, there is distortion in transit. But let them vary in the same way as the density, then the distortion disappears.

We may therefore regard **M** as being really the momentum, if the electric and magnetic properties remain always the same for the same amount of ether when it is compressed. The difficulty about light from distant stars disappears. There is still some residual error when waves cross, or are coming all ways, but this error is insignificant.

The circuital equations have to be taken strictly in more general form than for the ether at rest; in fact, this is a necessary part of the theory. It makes a complication when the speed of the ether is not very small. But if very small, the practical course is to work in terms of the simple undistorted circuital equations. In the electronic theory, for example, it would be a needless complication to allow for the compression of the ether in the mathematical work.

(41). As regards the actual density of the ether, I am not aware of any safe way of determining it at present. Consider the relation $\mu H^2 = mvq$ (vol. 3, p. 157). Here q is the speed of the ether, and mq the momentum. It is clear that $\mu H^2/v$ is only of moderate size, and cannot be very great. For, using c.g.s. magnetic units, $\mu=1$, and H is moderate in solar radiation, whilst v = nearly 3×10^{10}, if v, which is the speed of light through the ether, is nearly the same as u (vol. 3, p. 149) the speed of light through space, which will be the case when q is small compared with v and u. So mq is very small under the circumstances. If the ether has the density of water, then $m=1$, and q is very small. If the ether is a million times as dense as water, then q will be a millionth part of the previous small value. But it is easy to see that both m and q may be very small. If, however, we choose m to be so small that q comes out large, then we must go back to $\mu H^2 = mvq$, in which $q+v=u=3\times 10^{10}$. So, along this line of reasoning, something not yet found is required to settle the density of the ether. Lodge's argument in favour of an enormously dense ether is founded upon different considerations. My own preference is in favour of small density and small velocity, if convenient to the ether.

Some do not believe in the materiality of the ether. This view is thoroughly anti-Newtonian, anti-Faradaic and anti-Maxwellian. What mean action and reaction, the storage of energy, the transit of force and energy through space, &c., &c., if there is no medium in space? For space is nothing at all, save extension. Lord Kelvin used to call me a nihilist. That was a great mistake, (though I did throw a bomb occasionally, to stimulate an official humbug to learn something about electricity and how to apply it). He was most intensely mechanical, and could not accept any ether unless he could make a model of it. Without the model he did not consider electromagnetics to be dynamical. But I regard electromagnetics as being fully dynamical, a special branch of dynamics, in fact, founded upon Newton's dynamics of the visible, and resting upon it throughout. This applies to the theory of the electron, which is a particular application of electromagnetics. It is, however, striking that the method of treating dynamics in use by mathematicians in general, namely, the method of generalized coordinates and its complications, does not usually show up well in the treatment of electromagnetics. I have observed repeatedly that young to-be physicists, when they leave college, are full of generalized coordinates, and the theory of functions, and unnatural spaces, and rigour; but when they become thoroughly immersed in real physics, even though mathematical, a lot of the learning referred to fades away. Is Cambridge to blame? Perhaps not. It may be my fault.

(42). Returning to the electronic theory, a few results are desirable relating to the waste of energy due to a succession of equal sudden changes of velocity, acting in the same sense, and the transition therefrom to the waste when the acceleration is continuous. The necessary formulas have been already given; p. 467. Say $2T_1$ is the final energy of impulse u_1, and $2T_2$ that of u_2, both from rest; then $2(T_1+T_2-T_{12})$ is the final energy of the shell u_2-u_1, when u_1 is increased to u_2.

Now let $u_1/v = \cdot 1$, $u_2/v = \cdot 2$. Then, in terms of U_0, the initial electric energy at rest,

$$T_1 = (U_0/300)\times 1\cdot 006,\quad T_2 = (U_0/300)\times 4\cdot 096,\quad T_{12} = (U_0/300)\times 4\cdot 056.$$

$$2T_2 - T_{12} = (U_0/300)\times 2\cdot 068,\qquad 2T_1 - T_{12} = -(U_0/300)\times 2\cdot 044,$$

$$2(T_1+T_2-T_{12}) = (U_0/300)\times 2\cdot 092.$$

So, in terms of $U_0/300$ as unit of energy, the work done by the first shell when u goes from 0 to $v/10$ is 4·024, of which half is stored in the form of T and increased U, and half is wasted. Then comes the increase from u_1 to u_2. The field energy is (ultimately) increased to $2T_2+U_0$. The amount of increase is 6·180. At the same time the second shell wastes (ultimately) 2·092. So the total work of the second shell is 8·272, and the total work of both shells is 12·296.

But if the speed u_2 or $\cdot 2v$ were instantly set up, the work would be 16·384. So the saving by two jerks instead of one is 4·088. Also, the waste in the two shells is 4·104, whilst the waste in the single shell due to u_2 from rest is 8·192. So, finally, the waste due to the two equal jerks in succession is roughly one-half that due to a single jerk of double the size, but is actually a little more, by the amount 4·104 − 4·096 = ·008.

This "little more" is significant. The property is general. If ten equal jerks are substituted for one jerk, the final fields are the same, but the energy wasted by the 10 weak shells is only one-tenth of that wasted by the one strong shell, and a little over.

I will give the results of one more case, five pulses taking the place of one. Each increment of u is $v/10$, so that the final speed is $v/2$. The works done separately in the five instalments are, in terms of the unit of energy $U_0/150$,

2, 4·136, 6·516, 9·486, 13·238.

Adding these up, the total work done at any stage amounts to

2, 6·136, 12·652, 22·138, 35·376.

Also, the total energy stored at any stage is

1, 4·096, 9·486, 17·620, 29·375.

Subtract these from the preceding figures to get the total energy wasted at infinity at any stage; the results are

1, 2·040, 3·163, 4·518, 6·001.

Now the waste is 29·375 for the strong pulse when u jumps from 0 to $v/2$. One-fifth of this is 5·875. So by means of five equal instalments, the waste is reduced to one-fifth, and ·126 over.

The separate energies stored by the five pulses are

1, 3·096, 5·390, 8·134, 11·755,

and the separate wastes to correspond are

1, 1·040, 1·126, 1·352, 1·483.

I have also worked out the results for eight equal increments of speed, especially to compare them with those for four increments of double the size, and with two of four times the size, as well as with one of full size. The results are quite similar. Doubling the number of increments always halves the waste, with a little over. In the limit, with infinitely numerous infinitely small increments of speed, we must come to continuous waste along with continuous acceleration. Some caution, however, is required in estimating its amount.

(43). When we take account of the fact that an electron, though small, cannot be infinitely small, two sorts of impulses of acceleration come into view. One of these has duration so small that when the acceleration ceases, the pulse generated at the surface of the spherical electron extends only a little way into the electron. Then we may regard the impulse as a theoretical impulse, of no duration. The electromagnetic disturbance is in an infinitely thin spherical shell at the first moment, which expands both ways, and leads finally to a shell of depth equal to the diameter of the electron, which expands to infinity. There is no dopplerisation concerned in the production of this shell however the electron may move after the moment of the condensed acceleration. Its centre is at the place where the centre of the sphere was at that moment.

But if the period of acceleration lasts so long that the electron has sensibly moved when it ceases, then we come to a distorted pulse, whose inner and outer boundaries have different centres. It may be practically a pulse, on account of the smallness of size of the electron, and the great speed of light. Now it is this sort of pulse that plays so important a part in the theory previously given, (*see* figure on p. 454, for example, divided up into smaller pulses, if desired), and we can see the reason more plainly now why the application of the theory of continuous acceleration of a sphere cannot be extended indefinitely up to infinitely great acceleration. Because, in passing to that limit, the duration of great acceleration would also pass to that of the pure pulse of acceleration, without any dopplerisation in question.

Now see equation (235). This shows the waste due to the impulse of acceleration causing the increment of speed Δu. It varies inversely as κ^4. On the other hand, the factor κ^{-6} occurs in the continuous waste formula (142), p. 457. The following may make the transition plainer. (Done shortly on p. 160). By (228) and (235),

$$2a\mathrm{H}=\frac{\mathrm{Q}\Delta u}{4\pi\mathrm{R}}\frac{\sin\theta}{[1-(u/v)\cos\theta]^2},\qquad \text{Waste}=\frac{\mu\mathrm{Q}^2}{12\pi a}\frac{(\Delta u)^2}{\kappa^4},\qquad (257)$$

where the first formula gives the final H in the shell of depth $2a$ due to instantaneous Δu and the second gives the waste of energy by integration of U+T or 2T throughout the shell. H is uniform in the depth of the shell.

Now imagine the impulse of acceleration Δu to be spread over the short interval of time $2a/v$, the time taken to cross the electron. What has been said about Δu applies to each of the n smaller impulses $\Delta u/n$ which make up Δu. But there is obvious overlapping of the n final shells each of depth $2a$. The result is to double the

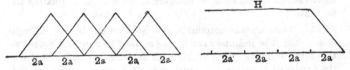

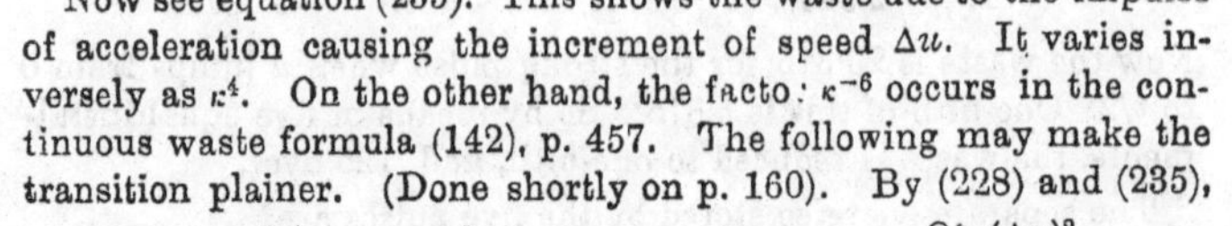

depth of the resultant single shell, and in it the final H will rise by equal steps from 0 on its front to the full value (257) in its middle, and then fall similarly by equal steps to 0 at its back. In the limit, $n=\infty$, the rise and fall will be continuous and uniform.

Next, suppose that a second Δu, of the same spread in time, is started as soon as the full H of the first shell is reached, and a third Δu when the full H of the second shell is reached, and so on. Then the rises and falls will compensate one another, and H in the resultant shell will be of the same full value given by (257), except at the beginning. This will be visible in the diagram.

Now $2a/v=dt_1$. Put this in the waste formula (257) and it makes

$$\text{Waste}=\Delta t_1\frac{\mu\mathrm{Q}^2}{6\pi v}\left(\frac{\Delta u}{\Delta t_1}\right)^2\frac{1}{\kappa^4}\qquad (258)$$

apparently be the waste in the time Δt_1, and substituting d for Δ we obtain a spurious formula for the waste during continuous acceleration. It is spurious because, being calculated for a (relatively) deep pulse, the electron has moved sensibly, and the boundaries of the resultant shells of the small elements of acceleration have different centres, and the elementary shells are of variable depth, according to the distorter $\dot{t}_1$. We must therefore allow for this in the H formula above when integrating the square of H to find the energy. This turns κ^4 to κ^6 in the continuous waste formula. Thus, as in (142), with $\phi_1=0$, we require

$$\text{Waste}=dt_1\frac{\mu Q^2}{6\pi v}\frac{w^2}{\kappa^6}. \tag{259}$$

Now the value of H × depth remains the same during the change from the undistorted to the distorted distribution. This means that the distorted H is $1/\dot{t}_1$ times the undistorted, so by (257), H in the dopplerized shell is given by

$$H=\frac{Q\Delta u}{8\pi a R}\frac{\sin\theta}{[1-(u/v)\cos\theta]^3}=\frac{Qw}{4\pi v R}\frac{\sin\theta}{[1-(u/v)\cos\theta]^3}; \tag{260}$$

that is, with the factor $\dot{t}_1^3$, if w, the rate of acceleration, is put for $\Delta u/\Delta t_1$. Squaring this, makes $\dot{t}_1^6$. Then in the integration the element of depth is $vdt_1/\dot{t}_1$ corresponding to the path element udt_1. This makes $\dot{t}_1^5$; and lastly the integration over the shell turns it to κ^{-6}.

Although (260) is only valid when R/a is large, being the final value of H in the shell, yet it has the merit of being derived from the impulse theory. Now I have given the complete formula for H at any distance due to a sphere when u receives impulsive acceleration. It would be an interesting exercise to derive therefrom in a similar way to the above the complete formulas for E and H at any distance in the case of continuous acceleration. For there is no doubt that the Maxwellian method of derivation from the scalar and vector potentials is rather cumbrous. That it is possible to do the work fully from the impulsive theory I know, because I did it that way first, and then tested the results by the potential method.

(44). I have calculated above the formulas for the energy and momentum in the elementary dopplerized shell associated with the path element $\mathbf{u}dt_1$ of a point charge. The following relates to the changes that take place when the acceleration $\Delta\mathbf{u}$ is impulsively made to previously existing $\mathbf{u}$, without change of direction. The final momentum in the shell of depth $2a$ belonging to Δu is M_2, given by

$$M_2=\sum(\mu H_2^2/v)\cos\theta=\frac{\mu Q^2(\Delta u)^2}{16\pi a v}\int_{-1}^{1}\frac{\mu-\mu^3}{[1-\mu u/v]^4}d\mu, \tag{261}$$

where T_2 is the final magnetic energy in this shell, already calculated. Here the formula for H_2 to correspond is used, namely, the first of (257). The integral comes to

$$2\left\{\frac{1}{3}\frac{4u}{v}+\frac{1}{5}\frac{4.5.6}{1.2.3}\frac{u^3}{v^3}+\frac{1}{7}\frac{4.5.6.7.8}{1.2.3.4.5}\frac{u^5}{v^5}+\dots\right.$$
$$\left.-\frac{1}{5}\dots-\frac{1}{7}\dots\dots-\frac{1}{9}\dots\dots\dots-\dots\right\}$$
$$=\frac{16u}{v}\left(\frac{1}{15}+\frac{5}{35}\frac{u^2}{v^2}+\frac{14}{63}\frac{u^4}{v^4}+\frac{30}{99}\frac{u^6}{v^6}+\dots\right); \tag{262}$$

therefore
$$M_2=\frac{\mu Q^2(\Delta u)^2}{15\pi av}\frac{u}{v}\left(1+\frac{15}{7}\frac{u^2}{v^2}+\frac{30}{9}\frac{u^4}{v^4}+\frac{50}{11}\frac{u^6}{v^6}+\dots\right),\tag{263}$$
and
$$2T_2-uM_2=\frac{\mu Q^2(\Delta u)^2}{12\pi a}\left(1+\frac{6}{5}\frac{u^2}{v^2}+\frac{9}{7}\frac{u^4}{v^4}+\frac{12}{9}\frac{u^6}{v^6}+\dots\right)$$
$$=\frac{(\Delta u)^2}{2u}\frac{d}{du}(U+T-U_0)=\frac{(\Delta u)^2}{2}\frac{dM}{du}=\tfrac{1}{2}m_1'(\Delta u)^2,\tag{264}$$
where U, T, M refer to speed u, and m_1 is the direct effective mass at that speed. See (221), (222), (224).

Now we have
$$M=\frac{2T}{u}=\frac{\mu Q^2}{2\pi a}\left\{\frac{1}{3}u+\frac{2}{15}\frac{u^3}{v^2}+\frac{3}{35}\frac{u^5}{v^4}+\dots\right\}\tag{265}$$
$$\frac{dM}{du}=\frac{\mu Q^2}{2\pi a}\left\{\frac{1}{3}+\frac{2}{5}\frac{u^2}{v^2}+\frac{3}{7}\frac{u^4}{v^4}+\dots\right\}\tag{266}$$
$$\frac{d^2M}{du^2}=\frac{2\mu Q^2}{\pi a}\frac{u}{v^2}\left\{\frac{1}{5}+\frac{3}{7}\frac{u^2}{v^2}+\frac{6}{9}\frac{u^4}{v^4}+\dots\right\}.\tag{267}$$
Therefore by the preceding relating to M_2, we come to
$$M_2=\tfrac{1}{6}(\Delta u)^2\frac{d^2M}{du^2},\qquad 2T_2=\tfrac{1}{2}(\Delta u)^2\frac{dM}{du}+\frac{1}{6}\frac{(\Delta u)^2}{u}\frac{d^2M}{du^2}.\tag{268}$$

Now from a different point of view, let F be the force on Q in turning u to $u+\Delta u=u_1$, say. It acts all the time the second shell takes to get clear of the electrification. Then the work done is
$$\int Fu_1dt=(\Delta u)\frac{d}{du}(U-U_0+T)+\tfrac{1}{2}(\Delta u)^2\frac{d^2}{du^2}(U-U_0+T)+2T_2$$
$$=u(\Delta u)\frac{dM}{du}+(\Delta u)^2\frac{dM}{du}+\tfrac{2}{3}u(\Delta u)^2\frac{d^2M}{du^2}.\tag{269}$$
Also
$$u_1\int Fdt=u_1\left\{\Delta u\frac{dM}{du}+\tfrac{1}{2}(\Delta u)^2\frac{d^2M}{du^2}\right\}+\tfrac{1}{6}u_1(\Delta u)^2\frac{d^2M}{du^2}$$
$$=u_1(\Delta M+M_2),\tag{270}$$
where ΔM is stored and M_2 wasted. It follows that the work is
$$\int Fu_1dt=u_1\int Fdt\quad\text{to second order.}\tag{271}$$
$$=\Delta u\frac{dM}{du}+\tfrac{2}{3}(\Delta u)^2\frac{d^2M}{du^2}.\tag{272}$$

Comparing with (270) and (268), we see that the waste part of the force-integral, which amounts to M_2, and the rest, which is ΔM, properly accounts for the increment of stored energy $\Delta(U-U_0+T)$ and for the waste energy M_2.

The reason for using the final M and M_2, and T and T_2 is that when the shell is not so distant that $2T_2$ may be regarded as the full shell energy, and T as the full steady magnetic energy, they are still in existence fully. For the applied force ceases working directly the back of the shell leaves the electrification. At that moment only a part of T_2 is visibly "in the field" behind the shell, the rest is in the shell itself. We make the convention then that the field is to include the shell so far as the variable part of shell energy goes. And this is right, because it is the shell itself that lays down the steady field behind it as it expands. It is continuously losing **E** and **H**, momentum and energy, in tending to its final state of constant energy and momentum.

It may be also convenient sometimes to make use of the same convention in considering continuous acceleration, after the manner of the figure on p. 454, where every infinitesimal shell is connected with its corresponding path element of a charge.

(45). When we pass from speeds less than to those greater than that of light, the coefficient 2 comes into play sometimes. I have already considered this matter (p. 27 and later). It receives a sort of physical explanation in the fact that when a charge is moving along a path at a greater speed than light, a fixed external point may be

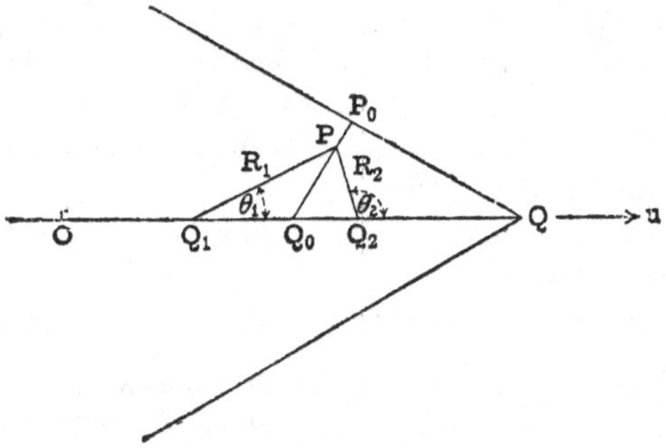

receiving disturbance from two distinct centres on the line of motion. That is, if t be the time at the observing point, and t_1 the corresponding time at the electron, meaning the moment of time when the disturbance leaves the electron which arrives at the observing point at the moment t, there may be two values of t_1 for one of t. The electron apparently becomes two electrons at different places when viewed from the observing point. It is easy to see that this must occur. For the single electron as it goes along generates and leaves behind a conical wave front. If the speed varies the angle of the cone also varies, as the cone is being continuously generated at its tip, but it is sufficient to consider the portion of the cone near the tip where we may regard the angle of the cone as constant. Now, say that $u=2v$. Then the semiangle of the cone is 30°. The charge Q being supposed to go from left to right, if the observing point is outside the cone it is not in communication with Q at all. But as soon as the wave front reaches P_0, as in the figure, it is immediately put in communication with the electron, not in its actual position at Q, but at Q_0. The disturbance going straight from Q_0 to P_0 reaches P_0 in the same interval of time as the electron takes to reach Q. That is $Q_0P_0/v = Q_0Q/u$, or $Q_0Q = 2Q_0P_0$.

Now let the observing point be moved a little way inside the cone, say to P. The corresponding point Q_0 splits into two, which separate, say Q_1 and Q_2. For Q_2P is half of Q_2Q, and Q_1P is half of Q_1Q. Moving the observing point inward to the axis makes the two sources separate further. The final positions are at O, such that Q_0 is in the middle of OQ, and another point a little to the right of Q_2, such that its distance from Q_0 is half its distance from Q. So any point on the axis is in communication at the moment t with two Q's, one in front and the other behind, whose moments are t_1 and t_2 say.

This doubleness of source applies to all points within the cone. Nor is that all. Their effects at P are equal. On p. 433 is an investigation in the case $u < v$, when there is only one correspondence between t_1 and t. Modify it thus, using the just given figure.

$$R_1^2 \sin^2\theta_1 = R_2^2 \sin^2\theta_2 = R'^2 \sin^2\beta, \tag{273}$$

where R′ is the distance QP and β is the angle between QP_0 and the direction of motion. Also

$$R'^2 = R_1^2[1+u^2/v^2-2(u/v)\cos\theta_1] = R_2^2[1+u^2/v^2-2(u/v)\cos\theta_2]; \tag{274}$$

from which, and (273),

$$R'^2[1-(u^2/v^2)\sin^2\beta] = R_1^2[1-(u/v)\cos\theta_1]^2 = R_2^2[1-(u/v)\cos\theta_2]^2. \tag{275}$$

So we prove that $\dot t_1/R_1$ and $\dot t_2/R_2$ are of the same size. As, however, squares are involved, we must attend carefully to the signs prefixed. They must be chosen so as to add effects, not to cancel. Now $\dot t_1/R_1$ is obviously negative in value if we use $\{1-(u/v)\cos\theta_1\}$. Remember that **E** and **H** and the potentials are reversed inside the cone. So for $\dot t_2/R_2$ we must use $-\{1-(u/v)\cos\theta_2\}$. Else there would be cancellation. Inspection of the last figure will show that when we add dt to t, the result is to shift Q to the right, Q_2 also to the right, but Q_1 to the left. So $\dot t_1$ is negative and $\dot t_2$ positive when t_1 and t_2 mean moments of time.

If the charge is suddenly started from rest, then we have both a sphere and a cone to consider, as in the figure on p. 102. If the observing point is in the cone, but outside the sphere, there are two correspondences as above. But if the observing point is in the sphere, there is only one.

If a single charge Q moves in any path with any variable velocity from 0 to ∞, there may be at a given moment t at the observing point, any number of corresponding sources, 0, 1, 2, 3, &c. This is quite easily to be seen by drawing a few diagrams. An easy case is circular motion. Even when the motion is steady in the circle, there may be any number of points upon it such that as the charge passes them, the waves emitted all arrive at an external observing point at the same moment. The scalar potential due to a single charge Q is therefore

$$P=\frac{Q}{4\pi c}\left\{\frac{\dot{t}_1}{R_1}+\frac{\dot{t}_2}{R_2}+\frac{\dot{t}_3}{R_3}+\dots\right\}, \tag{276}$$

when the proper signs are used.

(46). The best way to see this is to go backward in time, that is, to reverse the motion of the electron at the moment t, and at the same moment let a spherical surface begin to expand at speed v from the observing point, also going backward in time. Let this be done in the case to which the last figure applies. Then obviously the electron will go through the surface first at Q_2, and then again, when the surface has expanded more, at Q_1. There will be no more cutting after that, because the electron will be permanently outside the sphere. That is, if the electron does not change its speed. But we can make it cross the surface again merely by letting it slow down to a speed less than v. The spherical surface will then catch it up and pass it. To make a fourth passage through the surface it is only needed to raise the speed of the electron again above v. And so on to any extent, the electron oscillating from the outside to the inside of the spherical surface. In this way we determine any number of positions on the straight path which are in communication with the observing point at one moment.

In circular motion, on the other hand, the speed of the electron may be uniform to reach the same result. Exaggerate, to see this easily. Let the speed be a large multiple of that of light. As before, reverse its motion, and let a spherical surface expand from the observing point. When it cuts the circle of motion, it must obviously be passed through by the electron at least once, and therefore a first time. Then the electron will rush round the circle and pass through it a second time, before the sphere has expanded much; and then a third time, and so on. But there is a limit to this, because the sphere will ultimately leave the circle altogether. So the number of passages is always finite. Moreover, the number of passages is not the same for all positions of the observing point. The most striking case is when the observing point is at the centre of the circle. Then there is only one cutting of sphere and electron, however fast the electron may be moving. Then the magnetic force at the centre, or indeed at any point on the axis of the circle, although it varies, has a finite mean value, viz., that due to the steady circular current $Qu/2\pi r$, if r is the radius of the circle of revolution. This is only for points on the axis. Between the axis and the circle, the electron will in different places have one, two, or three, or more representative electrons on the circle, according to the speed. Generally speaking, we may divide space into a number of regions in which the H formula differs, owing to the variation in the number of equivalent electrons on the circle. This is a very curious and interesting subject, but I do not propose to go into it further here, since it is removed from practical value. There is a very good reason why electrons in practice always seem to move, so far as can be concluded from deflection observations on swarms and imperfect calculations, at speeds less than v. For though it is quite easy to think that an electron may be initially ejected and projected at a speed exceeding v, yet the dragging back to which it is subjected will very soon bring its speed down to v itself, or less.

(47). It was shown before, p. 458, that $T=U-U_0$ is true separately for all the dopplerized shells of infinitesimal depth surrounding a spherical charge in variable motion. It will be convenient for distinctness to consider that u varies continuously, although we may extend the argument to include pulses. Now here U_0 is the static energy in the shell under consideration when Q is at its centre. The shell is of variable depth, but it is provable by an easy integration that the value of U_0 in the distorted shell is the same as that in a concentric spherical whose depth is the same as that of the distorted shell at its equator, or $\theta=\frac{1}{2}\pi$. If then we transform every shell in this way, and shift their centres to a single centre, the shells will all fit together, and the total value of U_0 in them will be that due to Q at the centre chosen, in the region outside the innermost shell. It follows that the property $T=U-U_0$ is true in the sum, provided the inner boundary is the innermost dopplerized shell. Let its radius be R, and that of the charged sphere be a, then

$$\frac{R}{v}=\frac{a}{v-u}, \quad \text{or} \quad \frac{a}{R}=1-\frac{u}{v}, \tag{277}$$

if u is the speed to correspond, which may be supposed not to be varying sensibly at the moment. This innermost shell just touches the sphere a on its nose, and the region within it is excluded from the operation of $T=U-U_0$.

If we make u constant, and consider only steady straight motion, then we can go further. For then we have Searle's formula $U-T=\frac{1}{2}PQ\kappa^2$, proved on p. 129, for the region outside the sphere a, within which there is no U or T or U_0. The formula is

$$T_a-(U-U_0)_a=U_{0a}\left(\frac{2}{3}\frac{u^2}{v^2}+\frac{2}{15}\frac{u^4}{v^4}+\frac{2}{35}\frac{u^6}{v^6}+\dots\right) \tag{278}$$

for the complete field. But, by the preceding,

$$(U-T)_R=U_{0R}=(a/R)U_{0a}=(1-u/v)U_{0a}, \tag{279}$$

outside the sphere R. Therefore for the space between the sphere a and the sphere R which touches and surrounds it, we have

$$(\mathrm{U}-\mathrm{T})_{\mathrm{R}a}=\mathrm{U}_{0a}\left\{\frac{u}{v}-\frac{2}{3}\frac{u^2}{v^2}-\frac{2}{15}\frac{u^4}{v^4}-\frac{2}{35}\frac{u^6}{v^6}-\dots\right\}. \qquad (280)$$

It is obviously true when $u=0$. When $u=v$, it makes U = T. But this case must be excluded, being unrealisable, on account of the infinities involved. See p. 533, vol. 2, on this point. It is implied in (280) that the steady final state is established. But it never would be, if $u=v$. In any finite time, however great, the energy in the field, however great, would be only an infinitesimal part of the infinite value that the steady formulas assert. All effective mass formulas make nonsense when u is too close to v.

(48). There is another formula which deserves notice. On p. 2!2 I have shown that the transverse momentum of an electron revolving uniformly in a circle is given by

$$\mathrm{M}_w=\frac{\mu\mathrm{Q}^2w}{6\pi v\kappa^2}, \qquad (281)$$

if we denote by w the size of the acceleration, which is directed inward to the centre of the circle. The direction of $\mathbf{M}_w$ is outward. This is my extension of Searle's formula relating to the case $\kappa=1$, or u/v very small. Now, how is this formula (281) related to that for the momentum in an infinitesimally deep dopplerised shell? Referring to (156), p. 458, the energy formulas, we see that since $\phi_1=\frac{1}{2}\pi$, the mutual energy of the quasi-static and the electromagnetic parts are zero; that is, $\mathrm{U}_{12}=0$, $\mathrm{T}_{12}=0$. But this does not make the mutual momentum vanish. Referring to (172), (173), we see that $\mathbf{M}_{21}$ does in fact also vanish, but that by (181), p. 462, $\mathbf{M}_{12}$ does not. Putting $\phi_1=\frac{1}{2}\pi$, we get

$$\mathrm{M}_{12}=\frac{\mu\mathrm{Q}^2w}{6\pi v\kappa^2}\frac{vdt_1}{\mathrm{R}}. \qquad (282)$$

It is now clear that M_{21} and M_{12}, which were found to be unequal, ought not to be equal. Moreover, to bring (282) into harmony with (281), we require

$$\Sigma d\mathbf{R}/\mathrm{R}=1. \qquad (283)$$

For vdt_1 is the depth of the shell at its equator, or dR. If we summed up dR/R, the result would be infinite, of course. But in reality $d\mathbf{R}$, which is radial from the position of the electron, turns round just as the electron does, as we pass through the shells in succession, always keeping to their equators. So instead of an infinite result, we have only a residual finite result. I will leave it to the reader to turn (283) to a definite integral, work it out, and so obtain an independent proof of the transverse momentum formula. That the size of the electron does not appear in the formula is interesting, though not unprecedented. It does not appear in the formulas for the waste of energy and momentum either. The 1 on the right side is a unit vector transverse to the path of the electron in its actual position.

(49). In considering the electromagnetic effects of a moving point charge, considered as representative of a finite charge having a finite density throughout a very small region, or as a surface charge to the same, the rotation, if any, of the charge is left out of account for the same reason as, in the dynamics of a particle, its rotation is ignored. It is, relatively, of no consequence.

But there is another reason. If a point charge (in the above sense) revolves in a circle, it is subject to the theory of the translation of the charge, with dopplerisation effects, and a special kind of resistance of the ether to its motion should it move faster than light, different in law from the resistance equivalent to the waste of energy due to mere change of velocity. But if we distribute electrification uniformly in a circle, meaning a circular line, and set the circle spinning on its axis, there is no longer any dopplerisation. The circular current is just a circular current, with electromagnetic effects truly when it varies in strength, but without any of those troublesome peculiarities which turn up in the translation of a charge.

The same remark applies to a uniformly charged sphere, whether it is charged on its surface or throughout its volume. All the currents are circular, if the sphere is set spinning, and we may substitute for $\rho \mathbf{u}$, the convection current density, the product of the velocity $\mathbf{u}$ and the density ρ of electrification, any symbol, say $\mathbf{C}$, signifying the electric current density, and discuss the effect of $\mathbf{C}$ without reference to convection currents. The electromagnetic effects produced by varying $\mathbf{C}$, either continuously or discontinuously, were worked out a quarter of a century ago, in my work on Electromagnetic Waves, including the shell waves, and the changes of form of the functions representing the electromagnetic phenomena. So we may say roughly that there is nothing new to be elucidated in respect to the properties of convection currents by considering rotations. Moreover, we can see at once that the substitution $\mathbf{C} = \rho \mathbf{u}$, allows us to make $\mathbf{u}$ be as large as we like without any intrusion of the conical or hyperboloidal theory.

As regards electrons, whether of the round or flattened kind, these remarks apply also. I do not know whether anyone has published the theory of the shell wave sent out from a charged ellipsoid. Its general features may be readily seen, and I dare say it would not be difficult to one acquainted with ellipsoidal functions, though they are not pretty to look at. I shall not do it.

But the distortion due to dopplerisation will occur in any of the above cases if the currents, though circular, are not of the same strength all round the circles. They will then be closed outside, or may be partly inside the charged surface. This is a residual effect, though of course it may be made large in itself by very rapid or quite sudden change in the strength of current round any of the circles concerned. This would bring us back to the theory of translational effects again. At present I will work out just one problem relating to the rotation of a charged sphere, because of its relative importance and simplicity.

(50). Consider a sphere of radius a in the ether at rest. Let impressed magnetic force **h** act in it uniformly and parallel to the axis from which θ is measured. The final state when **h** is steady is simply that of the induction due to a uniformly intrinsically magnetised sphere, the intensity of magnetisation being $\mu\mathbf{h}$, where, however, μ is the same as in the outside ether. The internal induction is $\mathbf{B}=\frac{2}{3}\mu\mathbf{h}$, with the well-known complementary external induction. There is no electric force in the final state. We may substitute for **h** a surface magnetising current of density h at the equator, and $h\sin\theta$ elsewhere, the lines of current being circles of latitude round the axis. This electric current may be quite neutral as regards electrification, if attributed to moving charges. If so, there is no steady electric force outside the sphere. But if not neutral, there will be a static field. This need not be radially isotropic, because if for h we substitute σu, surface density × speed, both σ and u may be altered simultaneously to keep h constant. It is sufficient to consider u to be constant. Then σ is constant. The external electric force is radial and is isotropic, and may be left out of account.

Now let h vary. Electromagnetic disturbances are then sent out from the surface of the sphere, whether it be intrinsically magnetised or have a true magnetising current on its surface. By the second circuital law the lines of electric force produced are circles round the axis, and are quite symmetrical with respect to the equator. Let E be the electric force at distance r, and polar angle θ. The appropriate spherical functions are

$$\text{(out)}\quad \frac{\epsilon^{-qr}}{r}\left(1+\frac{1}{qr}\right)\sin\theta,\qquad\text{and}\qquad\text{(in)}\quad \frac{\epsilon^{qr}}{r}\left(1-\frac{1}{qr}\right)\sin\theta,\qquad(284)$$

for an outward and an inward wave respectively, if q stands for $d/d(vt)$. But both are concerned on account of the inward turning to an outward wave at the centre. But we need not go into the details of determination of constants, for the work has been already done. All electromagnetic problems are dual, by transformation from E to H, e to h, μ to c, etc., in a way which is now well known.

See p 95. Equation (1) there is the full differential solution of this problem of finding the circular H due to e impressed in the sphere a. So by transformation the circular E due to impressed h is

$$E_{\text{out})} = -\frac{a\sin\theta}{2cvr}\left[\epsilon^{-q(r-a)}\left(1-\frac{1}{qa}\right)+\epsilon^{-q(r+a)}\left(1+\frac{1}{qa}\right)\right]\left(1+\frac{1}{qr}\right)h. \quad (285)$$

This is outside the sphere. It consists of two waves, both outward, the one being outward from the moment of birth, the other resulting from the reflection (in effect) at the centre of the corresponding inward wave, and therefore retarded in time by the amount $2a/v$. As for h itself, it may be any function of t, limited physically to real single values, but not confined to be continuous. To obtain the $E_{\text{in})}$ to correspond to $E_{\text{(out)}}$, we have only to interchange a and r in what follows [, leaving the outside factor the same. Thus,

$$E_{\text{(in)}} = -\frac{a\sin\theta}{2cvr}\left[\epsilon^{-q(a-r)}\left(1-\frac{1}{qr}\right)+\epsilon^{q(r+a)}\left(1+\frac{1}{qr}\right)\right]\left(1+\frac{1}{qa}\right)h. \quad (286)$$

Here the first wave is inward, and the second outward.

Now H has two components, radial G and tangential H, say. The radial component is derived from E by

$$\mu vq\text{G} = \frac{1}{r}\frac{d}{d(\cos\theta)}(\text{E}\sin\theta) \quad (287)$$

So in (285) and (286) turn $\sin\theta$ to $(2/r)\cos\theta$ to produce $G_{\text{(out)}}$ and $G_{\text{(in)}}$. And the tangential H is derived from E by

$$\mu vq\text{H} = \frac{1}{r}\frac{d}{dr}(r\text{E}). \quad (288)$$

Working this out, makes

$$H_{\text{(out)}} = cv\text{E} + \frac{a\sin\theta}{2r}\left[\epsilon^{-q(r-a)}\left(1-\frac{1}{qa}\right)+\epsilon^{-q(r+a)}\left(1+\frac{1}{qa}\right)\right]\frac{h}{q^2r^2}. \quad (289)$$

$$H_{\text{(in)}} = cv\text{E} + \frac{a\sin\theta}{2r}\left[\epsilon^{-q(a-r)}\left(1-\frac{1}{qr}\right)+\epsilon^{-q(r+a)}\left(1+\frac{1}{qr}\right)\right]\frac{h}{q^2a^2}. \quad (290)$$

Now as regards the interpretation. This is given in exact detail by the above solutions themselves, in the origin of the disturbances, their spreading, attenuation, and retardation. Two practical cases present themselves of speedy algebrisation, namely, $h=+1$, or t^0, and h simple periodic. The first is physically the more interesting, in showing the origin and development and spread outwards of a pulse, and the steady state it leaves behind. The second concerns an outward permanent wave train, but also includes the initial phenomena.

(51). In (285), let h be zero before, and constant after $t=0$. The algebrisation is immediate, by inspection, and multegration. Thus,

$$E_{\text{(out)}} = -\frac{ha\sin\theta}{2cvr}\left[\left(1-\frac{vt_1}{a}+\frac{vt_1}{r}-\frac{v^2t_1^2}{ra}\right)_1+\left(1+\frac{vt_2}{a}+\frac{vt_2}{r}+\frac{v^2t_2^2}{ra}\right)_2\right], \quad (291)$$

where $\quad vt_1 = vt-(a-r), \qquad vt_2 = vt-(a+r). \quad (292)$

Similarly $E_{(in)}$ is got by interchanging a and r in the [·] in (291). Since h is zero before $t=0$, by the manner of construction t_1 and t_2 begin with the value zero, and the 1 in $(\cdot)_1$ stands for $t_1{}^0$, and in $(\cdot)_2$ for $t_2{}^0$. That is, $(\cdot)_1$ and $(\cdot)_2$ are distinct and separate, until they overlap. $(\cdot)_1$ is the initial outward wave from the surface of the sphere, and its companion in $E_{(in)}$ is the initial wave, whilst $(\cdot)_2$ and its companion have no existence at first. This primary wave is reducible to

$$E_{(out)} = -\frac{ha \sin\theta}{4cvr}\left(1 - \frac{v^2t^2 - a^2}{r^2}\right) = E_{(in)}; \qquad (293)$$

that is, this formula is valid for the whole of the shell which the two waves occupy. This lasts until $vt=a$, when the front of the inward wave reaches the centre. The shell is then of its full depth $2a$, and the reflected or outward wave begins. When $vt > a$, the wave $(\cdot)_2$ is in existence, and the resultant E in the overlapping region round the centre is obtained by adding together $(\cdot)_1$ and $(\cdot)_2$. But this does not affect the shell formula at all, because $(\cdot)_2$ is not in existence in the shell. That is, (293) is the shell formula all the time, from its initiation to its disappearance from view at ∞. It leaves the sphere at $vt=2a$, when the impressed h ceases to work. For the steady state of E, which is zero, then reaches the surface of the sphere.

The formula being simple, may be calculated and graphically illustrated with ease. The initial value of E is $-(h/2cv)\sin\theta$ on both sides of the surface. Or, at the equator, simply $-\frac{1}{2}\mu vh$. Here $-h$ is the strength of source, and cv is the reciprocal of μv, the impedance of a unit tube of ether infinitely long one way. But the disturbance (electric displacement) generated by h spreads both ways equally. This brings in the coefficient 2. See p. 8, § 454 on this point.

Spherical attenuation outwards and compression inwards makes the formula become $E = -\frac{1}{2}\mu vh(a/r)\sin\theta$ at the fronts of the two primary waves, that is on both boundaries of the shell before it reaches the centre. At the critical moment $vt=a$, there is a momentary infinite value at the centre, followed instantly by a drop to zero, which continues, and spreads out, as mentioned above. It is remarkable that at the critical moment $vt=a$, the value of E is constant all the way through the depth of the shell, being just half the initial value. That is, $E = -\frac{1}{4}\mu vh\sin\theta$. (Remember that **E** is circular.) It comes about in this way. Whilst E is running up so fast in value at and near the front of the primary inward wave, it is falling in value elsewhere in the shell, and just when it has fallen to half value, the very large value near the wave front becomes ∞ at the front alone, *i.e.*, at the centre, which is immediately destroyed by the reflected wave. In the subsequent history of the shell, E is negative in the outer part, positive in the inner part. At a great distance, these two parts tend to be each of depth a, and the formula for E at the boundaries of the shell is

$$\mathrm{E} = \mp \mu v h a / 2r \sin\theta, \tag{294}$$

using the $-$ sign outside and the $+$ sign inside. In the shell E then changes uniformly from one value to the other.

The corresponding formulas for G and H may be developed in a similar manner. By (289), (290),

$$\mathrm{H}_{(\mathrm{out})} = -cv\mathrm{E} + \frac{ha\sin\theta}{4r^3}\left[\left(v^2t_1^2 - \frac{v^3t_1^3}{3a}\right)_1 + \left(v^2t_2^2 + \frac{v^3t_2^3}{3a}\right)_2\right], \tag{295}$$

where the in and out primary waves are shown, or rather, the additions to be made to the electromagnetic $\mathrm{H}_\theta = -cv\mathrm{E}$, which is obeyed in the shell ultimately. So, by addition of $(\cdot)_1$ and $(\cdot)_2$, we obtain the steady tangential magnetic force

$$\mathrm{H}_{(\mathrm{out})} = (ha^3/3r^3)\sin\theta, \tag{296}$$

valid up to the inner boundary of the shell. And the radial component is

$$\mathrm{G}_{(\mathrm{out})} = \frac{ha\cos\theta}{r^2}\left[\left(vt_1 + \frac{v^2t_1^2}{2r} - \frac{v^2t_1^2}{2a} - \frac{v^3t_1^3}{6ra}\right)_1 + \left(\cdot\right)_2\right], \tag{297}$$

where the $(\cdot)_2$ only differs from $(\cdot)_1$ in the interchange of r and a. In the steady state,

$$\mathrm{G}_{(\mathrm{out})} = \frac{2ha^3\cos\theta}{3r^3}. \tag{298}$$

(52). In a similar manner the establishment of the internal **H** may be exhibited, leading to the steady state $\mathbf{B} = \frac{2}{3}\mu\mathbf{h}$, parallel to the axis. If **h** is considered to be uniform magnetizing force in the sphere, then the impressed activity is $\mathbf{h}\dot{\mathbf{B}}$ per unit volume throughout the sphere. But if h means the equatorial value of a surface convection current, the activity is entirely at the surface. In either case it is the same at the same moment, and the total work done is the same. It may be calculated in many ways. For example, at the moment $vt = 2a$, the activity stops, so the work done consists of the magnetic energy in the sphere, and the shell energy, electric and magnetic. Another way is to calculate the ultimate shell energy, half electric and half magnetic. The magnetic energy left behind has the same value as the shell energy. Of course now the external field energy is included. Now the total work done is, by magnetic calculation, with uniform h in the sphere, hB per unit volume, if B means the final value; or

$$\frac{4\pi a^3}{3} \times h \times \tfrac{2}{3}\mu h = \frac{2\pi a^3\mu h^2}{9} = 4\mathrm{T}_1, \quad \text{say}. \tag{299}$$

The shell energy (which is wasted) is therefore $2\mathrm{T}_1$ ultimately, T_1

being its magnetic energy, and the field energy is $2T_1$ also. Now the electric energy is in the shell and nowhere else, at any time. It follows that the value of the electric energy is constant from the moment $vt=2a$, and is of amount T_1. If this argument does not convince, the result may be verified by integration, finding the value of $\sum\frac{1}{2}cE^2$ in the shell at time t.

Convectionally considered, the equatorial activity per unit area of the surface of the sphere is $-Eh_1$ this being the applied force on the current $h=\sigma u$, and $E\sigma$ the force of the ether on σ. Here E must have the surface value. This supplies another way of finding $4T_1$. If the sphere is suddenly stopped, at the amount $vt=2a$ or later, the field energy $2T_1$ is wasted by the formation of a second shell.

If the surface is not electrically neutral, there is a steady radial isotropic outward electric displacement to be superimposed on the above considered electric and magnetic fields. But it makes no difference in the electromagnetic phenomena. The speed at the equator, denoted by u, may have any value from 0 to ∞. It will be noticed that much of the above relating to the shell is similar to that already given for the shell arising from the sudden jerking of a charged sphere into bodily motion of translation. But it is now very much easier, and I can recommend genuine students of electromagnetics to study the subject in more detail, as given in vol. 2 of my *Elec. Pa.*, p. 402 and after. I believe most of the remainder in sheets was rescued from the lumberman's cart, and sold in volumes, but I never got anything myself by the transaction, so I am not sure.

(53). The other easily treated case being the simple periodic, the sphere, if charged, must be waggled to and fro with axis fixed. Say by a crank attached to the axis. The separate waggles need not, however, be through small angles, but may comprise any number of revolutions, and the maximum speed is unlimited. Put $h=h_0 \cos nt$, and $q=ni/v$ in (285), then it becomes

$$-E_{(\text{out})}=\frac{h_0 a \sin\theta}{cvr}\left(\cos-\frac{v}{na}\sin\right)\frac{na}{v}\left(\cos-\frac{v}{nr}\sin\right)\left(\frac{nr}{v}-nt\right), \quad (300)$$

to represent the external E. From this, by interchanging a and r, except in the first factor, we derive the corresponding internal E.

In a similar manner the two components of the magnetic force may be got. The external radial component is

$$G_{(\text{out})}=-\frac{2h_0 av\cos\theta}{nr^2}\left(\cos-\frac{v}{na}\sin\right)\frac{na}{v}.\left(\sin+\frac{v}{nr}\cos\right)\left(\frac{nr}{v}-nt\right), \quad (301)$$

and the internal $G_{(\text{in})}$ comes from the last equation by the interchange of a and r, save in the first factor. The external tangential component (positive with increasing θ) is

$$\mathrm{H}_{(\mathrm{out})} = -cv\mathrm{E}_{(\mathrm{out})} - \frac{h_0 a \sin\theta}{r}\frac{v^2}{n^2r^2}\cos\left(\frac{nr}{v} - nt\right), \tag{302}$$

and the internal component is

$$\mathrm{H}_{(\mathrm{in})} = -\frac{h_0 a \sin\theta}{r}\left\{\left(1 - \frac{v^2}{n^2r^2}\right)\sin + \frac{v}{nr}\cos\right\}\frac{nr}{v}\left(\sin + \frac{v}{na}\cos\right)\left(\frac{na}{v} - nt\right). \tag{303}$$

As dependent upon the frequency and the size of the sphere, it will be observed that the external disturbances, electric and magnetic, vanish when

$$\tan(na/v) = na/v,$$

that is, at any one of an infinite series of frequencies which are nearly equally spaced, save at the beginning. Between them are the frequencies at which the external wave trains have their maximal amplitudes. The internal oscillations at the critical frequencies are entirely stationary, and involve no work being done in a period, although there is change from electric to magnetic energy and conversely. The physical reason of the external quietude is that the primary inward wave train from the source of disturbance on the surface of the sphere arrives at the surface again, at a critical frequency, in such strength and phase as to neutralize the primary external wave train where they overlap. This consideration tells us within what limits the above solutions are valid. If true over all space and time, then the oscillating h must have been going at it for all past time. But if h begins at the moment $t=0$, it is not until the moment $vt=2a$ that the E and H, G solutions are valid, and then only inside the sphere. Outside it, up to distance $r=3a$, is the uncancelled part of the primary wave outward, constituting the initial shell. So, in general, the above solutions are valid only up to the inner boundary of the shell.

But the differential solutions previously given are true completely, and include the shell. Instead of the substitution $q=ni/v$ which is valid for the established wave train, use the substitution $q^{-n}1=(vt)^n/\underline{|n}$, which is true all along. Applied to $h=h_0\cos nt$, beginning at $t=0$, it makes

$$\frac{h}{q} = h_0\frac{v}{n}\sin nt, \qquad \frac{h}{q^2} = h_0\frac{v^2}{n^2}(1-\cos nt); \tag{304}$$

and by application of these to (285) we obtain

$$\mathrm{E}_{(\mathrm{out})} = \frac{h_0 a \sin\theta}{2cvr}\left[\left\{\left(1 + \frac{v^2}{n^2ar}\right)\cos + \frac{v}{n}\left(\frac{1}{r} - \frac{1}{a}\right)\sin\right\}n\left(t + \frac{a-r}{v}\right) - \frac{v^2}{n^2ar}\right]_1$$
$$+ \frac{h_0 a \sin\theta}{2cvr}\left[\left\{\left(1 - \frac{v^2}{n^2ar}\right)\cos + \frac{v}{n}\left(\frac{1}{r} + \frac{1}{a}\right)\sin\right\}n\left(t + \frac{a+r}{v}\right) + \frac{v^2}{n^2ar}\right]_2. \tag{305}$$

This formula shows the external E from the first moment in the form of two waves, marked $_1$ and $_2$. The first line shows the primary outward wave, the second the outward retarded wave. Up to the moment $vt=2a$ the first wave alone exists outside the sphere; after that, the second must be added on in the overlap, whilst the first is still valid beyond. At a critical frequency, the internal stationary vibration and the external shell are both established at the same moment $vt=2a$. The shell then separates and expands, leaving no disturbance between it and the sphere.

The mean rate of waste of energy by the outward wave train is

$$\frac{4\pi a^2 h_0^2}{3cv}\left(\cos\frac{na}{v}-\frac{v}{na}\sin\frac{na}{v}\right)^2. \tag{306}$$

Here $na/v=2m\pi$ gives the critical frequencies nearly, if m is any integer. To see the relation to real sizes, let $n/2\pi$, which is the frequency, be of the order of visible light, say $10^{15}/2$, then $a/m=6\times10^{-5}$, and the value of m required to bring a to atomic size 10^{-8} cm. is $10^{-3}/6$, which is absurd. It is worse with assumed electronic size 10^{-13}. We want a to be of the order 6×10^{-5} to catch the lowest, twice as great to catch the next, and so on. This is of no importance as regards an electron in oscillatory rotation. But the same equation occurs in various other problems relating to vibrating or revolving electrons or systems of the same. Or else similar series of critical frequencies given by allied equations. Even the relatively very simple case of a vibrating spherical electron has its series of critical frequencies, which may be only slightly modified by the distortion due to the doppler effect.

(54). The equation $\tan na/v=na/v$ is not a determinantal equation in the ordinary sense, that is, an equation giving the frequencies of free vibrations rendered possible by the presence of boundaries causing reflection to and fro of progressive waves. For there is no boundary of that sort concerned, and there are no free vibrations. We may, however, introduce them by letting the material inside the sphere a be different from the outside ether. Then, obviously, the initial inward and outward waves from a surface source will not be precisely the same at birth, and the inward wave on its return to the surface $r=a$ will be partly reflected and partly transmitted, and the reflected part will go to the centre and return and be split again, and so on for ever. So inside the sphere is an infinite series of waves crossing one another. This allows of the existence of normal systems, in general of the subsiding kind, with a regular determinantal equation fixing an infinite series of periods and rates of subsidence, which is not the same as the series of critical frequencies referred to above. If an electron, for example, consisted of an electrified sphere of ether having a different permittivity from the external ether, then this difference would be the cause of a great complication in the effects due to a surface source, which complication may be analysed into an infinite series of to and fro waves inside the sphere, or, with mathematical equivalence, into the infinite series of normal vibrations with special frequencies. The mathematical work becomes very complicated, but has, perhaps, little physical importance in the theory of electrons.

Slanting Motion of Electrified Straight Line.

§ 535. (1). On p. 130 was briefly considered the potential of an electrified line moving steadily in its own line, in relation to the motion of a charged ellipsoid of revolution along its axis of symmetry. What follows relates to the more general case of slanting motion. The

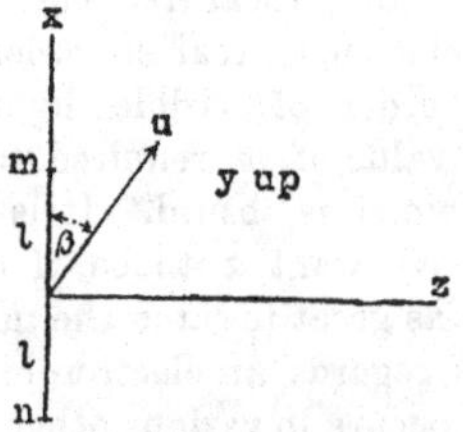

diagram will explain the coordinates. The electrified line *mn* of length $2l$ with charge Q of uniform linear density moves in the plane x, z in the direction u, and the y axis is up through the paper. Also let x_0 be the value of x of an element of the charged line. Then the eolotropic potential at the point x, y, z referred to the origin at the centre of the moving line is, by the formula for a moving point, given by the integral

$$P=\frac{Q}{8\pi lc}\int_{-l}^{+l}\frac{dx_0}{[\{(x-x_0)c+zs\}^2+\kappa^2y^2+\kappa^2\{(x-x_0)s-zc\}^2]^{\frac{1}{2}}}, \tag{1}$$

where in the integral c stands for $\cos\beta$, s for $\sin\beta$, and $\kappa^2=1-u^2/v^2$ as before. This may be transformed to

$$P=\frac{Q}{8\pi cl}\frac{1}{(c^2+\kappa^2s^2)^{\frac{1}{2}}}\int\frac{d\left\{x-x_0+\frac{zcs(1-\kappa^2)}{c^2+\kappa^2s^2}\right\}}{\left[\left\{x-x_0+\frac{zcs(1-\kappa^2)}{c^2+\kappa^2s^2}\right\}^2+\frac{\kappa^2y^2}{c^2+\kappa^2s^2}+\frac{\kappa^2z^2}{(c^2+\kappa^2s^2)^2}\right]^{\frac{1}{2}}}, \tag{2}$$

with the limits $x\pm l+\frac{zcs(1-\kappa^2)}{c^2+\kappa^2s^2}$, or $\pm x'l$,

using the + sign in the upper limit and the − in the lower. The result is

$$P=\frac{Q}{8\pi cl(c^2+\kappa^2s^2)^{\frac{1}{2}}}\log\frac{x'+l+\left[(x'+l)^2+\frac{\kappa^2y^2}{c^2+\kappa^2s^2}+\frac{\kappa^2z^2}{(c^2+\kappa^2s^2)^2}\right]^{\frac{1}{2}}}{(x'-l)+[(x'-l)^2+\ldots\ldots+\ldots\ldots]^{\frac{1}{2}}}. \tag{3}$$

When the charge moves in its own line, $s=0$, and

$$P=\frac{Q}{8\pi cl}\log\frac{x+l+[(x+l)^2+\kappa^2y^2+\kappa^2z^2]^{\frac{1}{2}}}{\text{same with }-l}. \tag{4}$$

In transverse motion $c=0$, and

$$P=\frac{Q}{8\pi cl}\frac{1}{\kappa}\log\frac{x+l+[(x+l)^2+y^2+z^2/\kappa^2]^{\frac{1}{2}}}{\text{same with }-l}. \tag{5}$$

Now let

$$B=\frac{\kappa^2y^2}{c^2+\kappa^2s^2}+\frac{\kappa^2z^2}{(c^2+\kappa^2s^2)^2}, \tag{6}$$

so that

$$P=\frac{Q}{8\pi cl}\frac{1}{(c^2+\kappa^2s^2)^{\frac{1}{2}}}\log\frac{x'+l+[(x'+l)^2+B]^{\frac{1}{2}}}{x'-l+[(x'-l)^2+B]^{\frac{1}{2}}}; \tag{7}$$

and let A be the value of the logarized quantity. It follows (by work) that

$$\frac{x'^2}{(A+1)^2}+\frac{B}{4A}=\frac{l^2}{(A-1)^2};\quad\text{or}\quad x'^2+\frac{(A+1)^2}{4A}B=\left(\frac{A+1}{A-1}\right)^2l^2, \tag{8}$$

or

$$\left[x^2+\frac{z^2c^2s^2(1-\kappa^2)^2}{(c^2+\kappa^2s^2)^2}+2\frac{xzcs(1-\kappa^2)}{c^2+\kappa^2s^2}\right]$$
$$+\frac{(A+1)^2}{4A}\left[\frac{\kappa^2y^2}{c^2+\kappa^2s^2}+\frac{\kappa^2z^2}{(c^2+\kappa^2s^2)^2}\right]=\left(\frac{A+1}{A-1}\right)^2l^2. \tag{9}$$

Let this be the same as

$$F(x\cos\beta'+z\sin\beta')^2+G(x\sin\beta'-z\cos\beta')^2+Hy^2=K; \tag{10}$$

then H and K are known, and β' is the angle the axis makes with the x-axis, measured like β. Comparing coefficients, using c' and s' with β' to match c and s with β,

$$Fc'^2+Gs'^2=1, \tag{11}$$

$$Fs'^2+Gc'^2=\frac{c^2s^2(1-\kappa^2)^2+(A+1)^2\kappa^2/4A}{(c^2+\kappa^2s^2)^2}=f,\text{ say}, \tag{12}$$

$$(F-G)c's'=\frac{cs(1-\kappa^2)}{c^2+\kappa^2s^2}; \tag{13}$$

from which find F, G, and β'. Thus,

$$F=\frac{c'^2-s'^2f}{c'^2-s'^2},\qquad G=\frac{c'^2f-s'^2}{c'^2-s'^2},\qquad F-G=\frac{1-f}{c'^2-s'^2}, \tag{14}$$

$$\frac{1-f}{c'^2-s'^2}s'c'=\frac{cs(1-\kappa^2)}{c^2+\kappa^2s^2}=\frac{1-f}{2}\tan 2\beta', \tag{15}$$

$$\tan 2\beta'=\frac{2cs(1-\kappa^2)(c^2+\kappa^2s^2)}{(c^2+\kappa^2s^2)^2-c^2s^2(1-\kappa^2)^2-(A+1)^2\kappa^2/4A}. \tag{16}$$

(2). This being the general case, let $c=s$, which is the case of equally direct and transverse motion. Then (16) reduces to

$$\tan 2\beta'=-\frac{2A(1-\kappa^4)}{(A-1)^2\kappa^2}. \tag{17}$$

Here $A=\infty$ makes $\beta=0$, and $A=1$ makes $\beta'=-\pi/4$. So $2\beta'$ is negative, and goes from 0 to $-\frac{1}{2}\pi$ as u (the speed) goes from 0 to v.

Or β' is 0 when u is 0, and the axis turns round to the left, and tends to be at right angles to **u** when u approaches v.

$$\text{Also}\qquad a'^2=\frac{(A+1)^2l^2}{(A-1)^2}\frac{\cos 2\beta'}{c'^2-s'^2f},\quad\text{displaced } x\text{-axis,}\tag{18}$$

$$b'^2=\frac{4Al^2(c^2+\kappa^2s^2)}{\kappa^2(A-1)^2},\quad\text{undisplaced } y\text{-axis,}\tag{19}$$

$$c'^2=\frac{(A+1)^2l^2}{(A-1)^2}\frac{\cos 2\beta'}{c'^2f-s'^2}.\quad\text{displaced } z\text{-axis.}\tag{20}$$

The line of Q is always the axis of the ellipsoids (prolate), when $u=0$, and is always, up to u anything $< v$, the axis of the innermost ellipsoid. It is the longest axis, and further away, where l becomes relatively small, turns round and tends to be perpendicular to the line of motion. At a sufficiently great distance, the electrified line behaves like a point charge. Then we have, at any speed, the oblate ellipsoids of a point charge, with the principal axis turned $\perp$**u**. So from the electrified line outwards there is a continuous change from very prolate to oblate ellipsoids.

(3). The above is very dreary stuff, but may be useful to those who may wish to study the distorted electric field and the magnetic of ellipsoids in motion. The case of direct motion is most important, leading to Searle's ellipsoidal formulas. They are, of course, well known now, but it may be useful to insert here some of them, for reference in relation to previous investigations in this volume.

Put $s=0$ in (2), producing (4). In this put first $r=0$, $x=a$, (where $r^2=y^2+z^2$), and secondly $x=0$, $r=b$, and equate the two P's. The result is

$$l^2=a^2-\kappa^2b^2\qquad\text{or}\qquad\frac{b^2}{a^2}=\frac{1-l^2/a^2}{\kappa^2},\tag{21}$$

which is the relation between the axes of an equipot. surface. If $a=b$, then $l/a=u/v$. Shift Q to the sphere, then P is the outside potential. The potential of the ellipsoid a, b itself is

$$P=\frac{Q}{4\pi cl}\log\frac{a+l}{a-l}=\frac{Q}{4\pi ca}\left(1+\frac{1}{3}\frac{l^2}{a^2}+\frac{1}{5}\frac{l^4}{a^4}+\dots\right).\tag{22}$$

The leading energy formula is that for the magnetic energy. For the ellipsoid a, b it is (Searle, *Phil. Mag.*, Oct., 1897).

$$T=\frac{\mu Q^2u^2}{4\pi a}\left\{\frac{1}{3}+\frac{2}{3.5}\frac{l^2}{a^2}+\frac{3}{5.7}\frac{l^4}{a^4}+\dots\right\},\tag{23}$$

subject to (21). I do not know any easy way of finding T without integration over all space outside the ellipsoid a, b; except when $a=b$. Then only a simple integration through the outside shell is needed, as done on p. 129. This finds the shell energy, and therefore also the equal stored energy. Combining with the property $U-T=\frac{1}{2}PQ\kappa^2$, of which a proof is given on p. 129, we find U and

T separately, T being (23) with u/v put for l/a. But I have given no calculation of my momentum formulas,

$$T=\tfrac{1}{2}Mu, \qquad U-U_0=\tfrac{1}{2}M_1u. \tag{24}$$

It is, however, quite easy. Here T and M belong to the steady field, M_1 to the outside shell. The density of the momentum M_1 is $2T_1/v$, directed radially outward. Therefore

$$M_1=\Sigma\frac{\mu H^2}{v}\cos\theta=\int_{-1}^{+1}\frac{\mu Q^2u^2(1-\mu^2)2\pi R^2\,2a.d\mu}{(2a)^2(4\pi R)^2(1-\mu u/v)^2}; \tag{25}$$

since H in the shell is given by

$$H=\frac{Qu\sin\theta}{8\pi aR[1-(u/v)\cos\theta]}. \tag{26}$$

So, by (25),
$$M_1=\frac{\mu Q^2u^2}{2\pi av}\left[\frac{1}{3.5}\frac{u}{v}+\frac{2}{5.7}\frac{u^3}{v^3}+\frac{3}{7.9}\frac{u^5}{v^5}+\dots\right]. \tag{27}$$

Now
$$(M+M_1)U=U-U_0+T+2T_1=4T_1, \tag{28}$$

because the shell energy equals the energy left behind, when u is instantly established. The value of $2T_1$ is known, being $\sum\mu H^2$ for the shell, calculated on p. 128 to be

$$U-U_0+T=2T_1=\frac{\mu Q^2u^2}{4\pi av}\left(\frac{1}{3}+\frac{1}{5}\frac{u^2}{v^2}+\frac{1}{7}\frac{u^4}{v^4}+\dots\right). \tag{29}$$

Also, as on p. 129,

$$U-T=\tfrac{1}{2}PQ\kappa^2, \qquad \therefore\quad 2(U-U_0)=\tfrac{1}{2}PQ\kappa^2-U_0+2T_1. \tag{30}$$

In the last use the series for P and we find $U-U_0=\frac{1}{2}M_1u$, according to (27). Insert this in (28), and we come to $T=\frac{1}{2}Mu$.

If the charge were brought from rest to speed u slowly, the waste would be less, and in the limit, zero. Here we see again the utility of the impulsive acceleration in getting useful results. So far as the particular series (27) and (29) go, the charge is rigid spherical. But (28) applies to any rigidly connected distribution of electrification suddenly set in steady motion, as is evident on considering that on sudden stoppage, the original state of things is finally restored, with $U=U_0$, $T=0$, $T_1=0$.

If the acceleration is not impulsive, we have

$$\int Fu_0dt=U-U_0+T+2T_2, \qquad u\int Fdt=(M+M_2)u, \tag{31}$$

if u_0 is the speed at time t, finishing with the value u to which U and T and M belong as before, whilst $2T_2$ is the energy wasted, and M_2 the momentum wasted. By taking the difference, we get the saving of energy due to slow working,

$$\text{Saving}=(T-U+U_0)+(M_2u-2T_2). \tag{32}$$

The first is (.) is known, the second (.) requires special calculation according to the law u_0 follows. Here $U-T$ is + and $< U_0$, so the first (.) is +. In the limit it is the whole saving when the acceleration is very slow.

(4). Now return to the electrified line, and let its motion be transverse to itself, along the z-axis. Then, by (5),

$$P=\frac{Q}{8\pi c l\kappa}\log\frac{x+l+[(x+l)^2+y^2+z^2/\kappa^2]^{\frac{1}{2}}}{\text{same with } -l}. \tag{33}$$

In this put first $x=a$, $y=0$, $z=0$; next $x=0$, $y=b$, $z=0$; and last $z=0$, $y=0$, $z=c$; and equate the three P's which result. Thus,

$$\frac{a+l}{a-l}=\frac{l+[l^2+b^2]^{\frac{1}{2}}}{-l+[l^2+b^2]^{\frac{1}{2}}}=\frac{l+[l^2+c^2/\kappa^2]^{\frac{1}{2}}}{-l+[l^2+c^2/\kappa^2]^{\frac{1}{2}}}. \tag{34}$$

This simplifies to

$$a^2-l^2=b^2=c^2/\kappa^2 \qquad \therefore \quad a>b>c. \tag{35}$$

At ∞, $a=b=c/\kappa$, the same as for a point charge.
At the line of Q itself, $a=l$, and $b=c=0$.
If $l=0$, we have $a=b=c/\kappa$ again.
Not any ellipsoid is possible, because b/c is constant throughout.

If $l^2=-\lambda^2$, then $a^2+\lambda^2=b^2=c^2/\kappa^2$, and $b>a$, and also $>c$, but a may be $>$ or $<$ c. A different interpretation is required.

The Magnetic Inertia of a Charged Conductor in a Field of Force.

[*Nature*, April 19, 1906; p. 582.]

§ 536. I think there is, in Another Place, possibly some misunderstanding concerning the inertia of a moving charged conductor due to the magnetic energy set up by its motion. It depends upon the distribution of the electrification, and may vary from a minimum up to infinity. No question of distortion due to high speed is involved, so the theory is quite simple. Say a sphere of radius a has any distribution of surface charge For simplicity, let it be symmetrical round the axis of motion, so that the surface density is representable by the sum of any number of zonal harmonic distributions. The corresponding magnetic fields follow. Their magnetic energies are all independent, so that the actual magnetic energy is the sum of the separate magnetic energies.

The really practical case, which is also very simple, is when the conductor has a charge Q and moves in a uniform electric field F. Then the surface density is

$$\sigma=\frac{Q}{4\pi a^2}+3c\text{F}\cos\theta, \tag{1}$$

where θ is the polar angle. The magnetic force is, [if u=speed],

$$\text{H}=\text{H}_1+\text{H}_2=\frac{Qu}{4\pi r^2}\sin\theta+3c\frac{\text{F}ua^3}{r^3}\sin\theta\cos\theta. \tag{2}$$

The magnetic energy is $\Sigma\frac{1}{2}\mu H^2$, and by integration comes to

$$T=\tfrac{1}{2}u^2\left[\frac{\mu Q^2}{6\pi a}+\frac{8}{5}\mu c^2F^2\pi a^3\right]. \qquad (3)$$

The magnetic inerta is therefore $m_1(1+h)$, where m_1 is the value for the uniform charge, or $m_1=\mu Q^2/6\pi a$, and

$$h=\frac{n^2}{15\pi}, \quad \text{if} \quad n=\frac{3F.4\pi a^2c}{Q}. \qquad (4)$$

This n is the ratio of the induced electric force at the pole to the undisturbed force. If $n=1$, F is just large enough to make the surface density be zero at one pole. Then $h=1/47$. This is the increased inertia due to the disturbance of the distribution of the charge. The "equation of motion" under F is

$$FQ=[m+m_1(1+h)]\dot{u}, \qquad (5)$$

where m is the mass of the body.

The whole is subject to the restriction of small u/v and small acceleration, so that the electric and magnetic fields sensibly travel with the charge. Nor need F be constant in space or in time, provided it does not vary too rapidly in relation to the size of the conductor. In slow motion the magnetic energy is the fraction u^2/v^2 of that part of the electric energy that depends upon the transverse electric force.

[**Addition.** In *Nature*, June 28, 1906, p. 198, is a continuation of the above by G. F. C. Searle. He found that if $\mathbf{u} \perp \mathbf{F}$, the magnetic energy is given by (3) above if we alter 8/5 to 16/5. This is in the steady stage with finite conductivity, and full penetration of the magnetic force into the sphere. But as I pointed out to him, there is no penetration when the conductivity is infinite, so separate calculation was required. In this case he found that the coefficient 8/5 in (3) became 6/5. In the general case of partial penetration there can be no simple formula. In electronic theory, the correction on account of redistribution of charges due to assumed conductivity is very minute, when u/v and the acceleration are small.]

Boltzmann's Interpretation of Maxwell.

[Review of Curry's "Electricity and Magnetism," *The Electrician*, September 10, 1897, p. 643.]

§ **537.** From the preface by Prof. Boltzmann, we learn that this work is based upon his "Vorlesungen über Maxwell's Theorie," though it has been much elaborated by Dr. Curry; and that Prof. Boltzmann has revised the MS. of the treatise and finds that it gives a very clear and concise exposition, not only of all his lectures, but especially of the conceptions and principles embodied in the foundation of the electromagnetic theory of light. And from Dr. Curry's preface we learn that he regards the treatise as the *philosophy* of electricity and magnetism, and that he has added two rather long chapters on Helmholtz's theory.

I am unable to agree with Dr. Boltzmann as regards the conceptions and principles, or with Dr. Curry as regards the philosophy, or that Helmholtz's theory deserves so much space, if any at all. As the subject matter is professedly Maxwell's theory we have some right to expect a following or extension of Maxwell's philosophical manner. Instead of which, the new wine is put in the old bottles. Real and neutral electricities; one fluid, two fluids; free electricity, bound electricity; two types of electricity, the current electricity and the polarised electricity; the action at a distance of the electricity due to polarisation; Maxwell's equations of action at a distance, &c., &c.! Is all that Maxwell? I never knew it before, but thought he had swept away the old fluids and given us something philosophical and dynamical.

But we should not expect too much. Germany was the breeding place and home of electrodynamic theories, so-called. They never took root in England. Indeed, Thomson and Tait severely condemned the method before Maxwell's treatise came out. Now Hertz squashed all the electrodynamic theories visibly, and continental theorists were obliged to take up Maxwell. But it is natural that they should try to fit Maxwell to their old ideas as far as possible; to have a minimum of Maxwell and a maximum of distance action. Dr. Curry says it has been his desire to embody not only the standpoint from which the theory of electricity and magnetism is studied on the Continent, but also the general methods of treatment in vogue. If Maxwell is really taught in this way generally on the Continent, I think it is a great pity. But there are notable exceptions. Hertz became quite Maxwellian after his great hit, save that, as I think, he attached rather too much importance to the mere equations, as the representation of Maxwell's theory, to the comparative exclusion of the experimental and philosophical basis. Then there is Dr. A. Föppl, whose excellent "Einfuhrung in die Maxwell'sche Theorie der Electricität" deserves to be read by all who can understand it. It is the least academical of the German works I have seen, and also the clearest and most advanced. Dr. Föppl thoroughly appreciates that the best way of exhibiting electric and magnetic relations is not by potentials, but by the electric and magnetic forces and fluxes and their reciprocal relations, and also the proper use of fluids, not electric or magnetic, but illustrative of the behaviour and distribution of the so-called electric and magnetic "forces" and "fluxes."

There is too much in this book, which is nicely got up, and is well edited, for a detailed review, especially when one does not like it. I will only notice a few points where I think Maxwell's theory is misrepresented and perverted. The starting point is not experimental, but is that particular form of the rotational ether analogy in which electric force is represented by velocity. The forms of the energies are assumed, and the equation of motion is derived through the Principle of Least Action. (There are easier ways, but they are not so learned.) The ultimate result, when, by an afterthought, the conduction current is put in, is a set of equations equivalent to the two circuital laws, with a difference. Using my own notation, these laws for a stationary conducting dielectric are

$$\text{curl}\,(\mathbf{H}-\mathbf{h})=\mathbf{C}+\dot{\mathbf{D}}=(k+cp)\mathbf{E}, \tag{1}$$

$$-\text{curl}\,(\mathbf{E}-\mathbf{e})=\quad\dot{\mathbf{B}}=\quad\mu p\mathbf{H}. \tag{2}$$

I omit the magnetic conduction current. Here $\mathbf{D}=c\mathbf{E}$, and $\mathbf{B}=\mu\mathbf{H}$, and $\mathbf{e}$ and $\mathbf{h}$ are the impressed parts of $\mathbf{E}$ and $\mathbf{H}$. Now Drs. Boltzmann and Curry (p. 34) have the second law right; but their first law is wrong in two ways. First, there is the omission of $\mathbf{h}$. As appears later this is intentional. But it is thoroughly wrong. Where is intrinsic magnetisation to come in? The other error is that in Dr. Boltzmann's exposition $\mathbf{e}$ only acts on conduction current, that is, whilst $\mathbf{E}$ produces $\mathbf{C}$; but it is only $\mathbf{E}-\mathbf{e}$ that produces (or is actively associated with) $\mathbf{D}$. This error is also intentional. Perhaps it is I that have misrepresented Maxwell. Very well, let it be I; then Dr. Boltzmann's expositions do not agree with mine in vital particulars, and this goes on pretty well all through, only getting worse later on. To prevent confusion, I may call Maxwell, my Maxwell.

Is not the energy of any condenser $\frac{1}{2}S^{-1}Q^2$, and of a unit cube condenser $\frac{1}{2}c^{-1}\mathbf{D}^2$, however $\mathbf{D}$ is caused, if elastically done and linearly? But Dr. Boltzmann says (p. 146) it is doubtful whether $\mathbf{e}$ can be generated in insulators. Excepting possible crystallic exceptions "all E.M.F.s are confined to the surface of insulators." Now is this philosophy? I think not, and regard the dielectric as the fundamental universal in my Maxwell's theory, the conductive property being occasional or frequent, but, so to speak, accidental, certainly additional. I cannot see the slightest grounds for making $\mathbf{e}$ inoperative, or else be absent altogether, unless there is electric conductivity on the spot.

In connection with this matter I see the statement (p. 224) that so-called "real electricity" (that is, Maxwell's electrification) is created or destroyed wherever impressed forces appear, on account of conductivity, but that such a variation of "real magnetism" (div. $\mathbf{B}$) is impossible, because impressed magnetic forces are "entirely unknown" in Maxwell's equations. But my Maxwell does not teach me either the fact stated about electrification, or the reason given about magnetism. My Maxwell teaches me that no **e** can produce electrification in a dielectric which is not a conductor; and that no **e** can produce electrification in a conductor which is not a dielectric; and that no **e** can produce electrification in a homogeneous conducting dielectric; but that the existence of heterogeneity is (as well as conductivity and permittivity) necessary for the production of electrification. On the other point, my Maxwell teaches me that variations in div **B** are impossible, because of the experimental absence of magnetic conductivity (not the absence of **h**), and that div **B** itself is zero as a connected experimental fact, otherwise we should have unipolar magnets. As for **h** being absent, my Maxwell teaches me it is very much in action. We cannot, indeed, *create* **h** (or μ**h**, intrinsic magnetisation density) without external agency not involved in the fundamental theory; but that is just why **h** is taken as intrinsic or impressed.

This persistent perversity about impressed forces, according to my Maxwell, takes a new form later on, in what is termed a modified form of Maxwell's equations. This remarkable assumption is made: let (my notation, &c.)

$$\mathbf{e}=\mathbf{f}+(c-c_0)k^{-1}\dot{\mathbf{f}}. \tag{3}$$

Here **e** is the old impressed force, and **f** is a new one, connected with the old one by the differential equation (3). To see its effect let **H**−**h** be $\overline{\mathbf{H}}$ and **E**−**e** be $\overline{\mathbf{E}}$ in (1) above.

Then $$\text{curl}\,\overline{\mathbf{H}}=k(\overline{\mathbf{E}}+\mathbf{e})+cp(\overline{\mathbf{E}}+\mathbf{e}) \tag{4}$$

is my Maxwell's first law, whilst Dr. Boltzmann's is

$$\text{curl}\,\overline{\mathbf{H}}=k(\overline{\mathbf{E}}+\mathbf{e})+cp\overline{\mathbf{E}}. \tag{5}$$

Now put **e** in terms of **f** according to (3). It makes (5) become

$$\text{curl}\,\overline{\mathbf{H}}=(k+cp)\mathbf{E}-c_0\dot{\mathbf{e}}, \tag{6}$$

a new form of first circuital law, which is just as wrong as the other, in another way; owing to the redundant term $-c_0$**e**.

Now our author calls (3) a "purely formal but important change," and says that since **e** and **f** are identical when steady (the only case more carefully investigated) it is impossible to decide whether **e** or **f** is impressed force! This transformation seems to me to be *hocus pocus*, and nothing more. Is there no energetic criterion of impressed force then? No one has a right to trifle with Maxwell's equations in this way. We may indeed extend, but should not violate them. There is something quite shocking about this philosophy of electricity.

But a little further on we find out why **e** goes out and **f** comes in. It is to make a transition from Maxwell's theory to Helmholtz's, and in fact, by a "slight modification" which has no justification, Helmholtz's equations, or equations resembling them, are derived. A good deal of space is devoted to Helmholtz's theory, including a so-called empirical establishment, following its originator. The theory is taken quite seriously, and is considered to harmonise with and to be an extension of Maxwell's theory, and is recommended for possible use later. I cannot agree with this view of Helmholtz's theory; but as the matter cannot be properly treated in this review, I will write a separate article about it.

[This "separate article" is in E.M.T., vol. 2, App. D., p. 493, under the title "On Compressional Electric or Magnetic Waves."]

Vectors *versus* Quaternions.

[*Nature*, April 6, 1893; p. 533.]

§ 538. Having a vivid recollection of the pleasure I derived from Prof. Gibbs's attacks upon the quaternionic system in the rather one-sided discussion that took place about two years ago in this journal, I have delayed replying to the letters of Profs. MacAulay and Tait, from an expectation that Prof. Gibbs would have something to say. In this I have not been mistaken; and, as there is a general agreement between us on the whole, I have merely to add some supplementary remarks. Prof. MacAulay refers to me as having raised the question again. I can assure him it has never been dropped. Apart from the one-sided discussion, it has been a live question with Prof. Gibbs and myself since about 1882, and is now more alive than ever. I cannot help thinking that Prof. MacAulay's letter was overhastily written, and feel sure that if he knew as much about the views and methods of those to whom he appeals as he does about Quaternions, he would have written it somewhat differently, or perhaps not have written it at all, from a conviction of the uselessness of his appeal. There is no question of suicide with us; on the contrary, quite the reverse. I am asked whether the "spoonfeeding," as he terms it, of Maxwell, FitzGerald, &c., is not good enough for me. Why, of course not. It is quaternionic, and that is the real point concerned. Again, he thinks nothing of the inscrutable negativity of the square of a vector in Quaternions; here, again, is the root of the evil. As regards a uniformity of notation amongst anti-quaternionists, I dare say that will come in time, but the proposal is premature. We have first to get people to study the matter and think about it. I have developed my system, such as it is, quite independently of Prof. Gibbs. Nevertheless, I would willingly adopt his notation (as I have adopted his dyadical notion of the linear operator) if I found it better. But I do not. I have been particularly careful in my notation to harmonise as closely as possible with ordinary mathematical ideas, processes and notation; I do not think Gibbs has succeeded so well. But that matters little now; the really important thing is to depose the quaternion from the masterful position it has so long usurped, whereby the diffusion of vector analysis has been so lamentably impeded. I have been, until lately, very tender and merciful towards quaternionic fads, thinking it possible that Prof. Tait might modify his obstructive attitude. But there is seemingly no chance of that. Whether this be so or not, I think it is practically certain that there is no chance whatever for Quaternions as a practical system of mathematics for the use of physicists. How is it possible, when it is so utterly discordant with physical notions, besides being at variance with common mathematics? A vector is not a quaternion; it never was, and never will be, and its square is not negative; the supposed proofs are perfectly rotten at the core. Vector analysis should have a purely vectorial basis, and the quaternion will then, if wanted at all, merely come in as an occasional auxiliary, as a special kind of operator. It is to Prof. Tait's devotion to his master that we should look for the reason of the little progress made in the last 20 years in spreading vector analysis.

Now I have, in my turn, an appeal to make to Prof. MacAulay. I have been much interested in his recent R. S. paper. As the heart knoweth its own wickedness, he will not be surprised when I say that I seem to see in his mathematical powers the "promise and potency" of much future valuable work of a hard-headed kind. This being so, I think it is a great pity that he should waste his talents on such an anomaly as the quaternionic system of vector analysis. I have examined a good deal of his paper, and can find nothing quaternionic about it except the language concerned in his symbols. On conversion to purely vectorial form, I find that it is greatly improved. I would suggest that he give up the quaternion. If he does not like my notation, or Prof. Gibbs's, or Prof. Macfarlane's, and will invent one for himself, it will receive proper consideration. He will greatly extend the sphere of his usefulness by the conversion. A difficulty in the way is that he has got used to quaternions. I know what it is, as I was in the quaternionic slough myself once. But I made an effort, and recovered myself, and have little doubt that Prof. MacAulay can do the same.

Passing to Prof. Tait's letter, it seems to be very significant. The quaternionic calm and peace have been disturbed. There is confusion in the quaternionic citadel; alarms and excursions, and hurling of stones and pouring of boiling water upon the invading host. What else is the meaning of his letter, and more especially of the concluding paragraph? But the worm may turn; and turn the tables.

It would appear that Prof. Tait, being unable to bring his massive intellect to understand my vectors, or Gibbs's, or Macfarlane's, has delegated to Prof. Knott the task of examining them, apparently just upon the remote chance that there might possibly be something in them that was not utterly despicable. Prof. Knott has examined them, and has made some remarkable discoveries. One of them is that those vector methods in which the quaternion is not the master lead to formulæ of the most prodigious and alarming complexity. He has counted up the number of symbols in certain equations. Admirable critic!

Now, since this discovery, and Prof. Tait's remarks, are calculated to discourage learners, I beg leave to say, distinctly and emphatically, that there is no foundation for the imputation. Prof. Knott seems to have found a mare's nest of the first magnitude; unless, indeed, he is a practical joker, and has been hoaxing his venerated friend. Speaking from a personal knowledge of the quaternionic formulæ of mathematical physics, and of the corresponding formulæ in my notation and in Prof. Gibbs's, I can say definitely that there is very little to choose between them, so far as mere length goes. Perhaps Prof. Knott has been counting the symbols in a Cartesian formula, or in a semi-cartesian one, or some kind of expanded form. I do not write for experts who delight in the most condensed symbolism. I do not even claim to be an expert myself. I have to make my readers, and therefore frequently, of set purpose, give expanded forms rather than the most condensed.

But so far as regards the brief vector formulæ, I find that the advantage is actually in my favour. I attach no importance to this, but state it merely as a fact which upsets Profs. Knott and Tait's conclusions. It is desirable that I should point out the reason, otherwise the fact may not be believed. In common algebra there is but one kind of product of a pair of quantities, say, F and v, which is denoted by Fv. In vector algebra there are two kinds of products. One of these closely resembles the usual product, whilst the other is widely different, being a vector itself. Accordingly, to harmonise with common algebra, I denote the scalar product by $\mathbf{Fv}$. It degenerates to Fv when the vectors have the same direction. Now, since the quaternionists denote this function by $-\mathrm{S}\mathbf{Fv}$, which is double as long, whilst $\pm\mathbf{Fv}$ becomes $\mp\mathrm{S}\mathbf{Fv}$, it is clear that there must be an appreciable saving of space from this cause alone, because the scalar product is usually the most frequently occurring function.

But there are other causes. The quaternionic ways of specialising formulæ are sometimes both hard to read and lengthy in execution. Look at $\mathrm{S}.\mathrm{U}\alpha\mathrm{U}\rho\mathrm{S}.\mathrm{U}\beta\mathrm{U}\rho$, which I see in Tait's book. I denote this by $(\alpha_1\rho_1)(\beta_1\rho_1)$, or else by $\alpha_1\rho_1 . \beta_1\rho_1$. Tait is twice as long. But the mere shortness is not important. It is distinctness that should be aimed at, and that is also secured by departing from quaternionic usage. Examples of shortening and clarifying by adopting my notation may be found on nearly every page of Tait's book.

Consider, for example, rotations. Quaternionists, I believe, rather pride themselves upon their power of representing a rotation by means of a quaternion. Thus, $\mathbf{b} = q\mathbf{a}q^{-1}$. The continued product of a quaternion q, a vector $\mathbf{a}$, and another quaternion q^{-1}, produces a vector $\mathbf{b}$, which is $\mathbf{a}$ turned round a certain axis through a certain angle. It is striking that it should turn out so; but is it not also a very clumsy way of representing a rotation, to have to use two quaternions, one to pull and the other to push, in order to turn round the vector lodged between them? Is it not plainer to say $\mathbf{b} = r\mathbf{a}$, where r is the rotator? Then we shall have $\mathbf{ac} = \mathbf{a}rr'\mathbf{c} = r'\mathbf{a}r'\mathbf{c} = $ &c., if r' is the reciprocal of r. Then Prof. Tait's $Vq\mathbf{a}q^{-1}q\phi(q^{-1}\mathbf{b}q)q^{-1}$ is represented by $Vr\mathbf{a}r\phi r'\mathbf{b}$. See his treatise, p. 326, 3rd edition, and note how badly the $q(\quad)q^{-1}$ system works out there and in the neighbouring pages.

What, then, is this rotator? It is simply a linear operator, like ϕ. It is, however, of a special kind, since its conjugate and its reciprocal are one, thus $rr' = 1$, or $r' = r^{-1}$. Far be it from me to follow Prof. Tait's example (see his letter) and impute to him an "imperfect assimilation" of the linear and vector operator. What I should prefer to suggest is that his admiration for the quaternionic mantle is so extreme that he will wear it in preference to a better-fitting and neater garment. If we like we can express the rotator in terms of a quaternion, in another way than above, though involving direct operations only. But I am here merely illustrating the clumsiness of the quaternionic formulæ in physical investigations, and their unnaturalness, by way of emphasising my denial and disproof of the charge made by Prof. Tait against vectorial methods. The general anti-quaternionic question I have considered elsewhere. [See the beginning and the end of the Chapter on Vectorial Algebra in vol. 1 of this work. Also p. 135 in the present volume.]

Quaternionic Innovations.

[*Nature*, January 11, 1894; p. 246.]

§ 539. That Prof. Tait should not be able to do justice to those who prefer to treat vectors as vectors, and quaternions as quaternions, instead of commingling their diverse natures, with the result, in the latter case, of confusion of physical ideas (and geometrical also, for of course geometry is itself ultimately a physical science, having an experimental foundation), is naturally to be expected. He does not know their ways, either of thinking or of working, as is abundantly evident in all that he has written adversely to Prof. Willard Gibbs and others. It is, however, a little strange, in view of Prof. Tait's often expressed conservatism regarding Quaternionics, that he should tolerate *any* innovations therein, such as Mr. MacAulay has introduced. The latter may perhaps take this as a compliment to his analytical powers, which compel the former's admiration, and toleration of his departures from quaternionic usage. For myself, I welcome any quaternionic innovations that may (ultimately) tend in the direction of the standpoint assumed by Prof. Gibbs and others, and foresaw some two years since (when a very bulky manuscript came to me for my opinion) that there would be some quaternionic upstirring.

Prof. Gibbs has already pointed out how the development of Quaternionics has involved first the elimination of the imaginary, and next the gradual elimination of the quaternion! Now there is a capital illustration of this innate tendency in Prof. Tait's review (*Nature*, December 28, 1893), where, on p. 194, he explains by an example the meaning of a startling innovation of Mr. MacAulay's. Put it, however, in vectorial form, and let us see what it comes to then. Take the case of a stress and the force to correspond (which is a little easier than Prof. Tait's example, though not essentially different). Let ϕ be a stress operator (pure, for simplicity), so that $\phi\mathbf{N}$, or $\mathbf{N}\phi$, is the stress per unit area on the $\mathbf{N}$ plane, $\mathbf{N}$ being any unit vector. Now we know, by consideration of the stresses acting upon the faces of a unit cube, that the $\mathbf{N}$ component of the force $\mathbf{F}$ per unit volume is the divergence of the stress vector for the $\mathbf{N}$ planes. That is,

$$\mathbf{FN}=\nabla\phi\mathbf{N}, \tag{1}$$

for any direction of $\mathbf{N}$. I employ my usual notation for the benefit of readers (now becoming numerous) who, though they cannot follow the obscure quaternionic processes, can understand the plainer ones of pure vector algebra. Now, may we remove the vector $\mathbf{N}$ (which is any one of an infinite number of vectors) and write

$$\mathbf{F}=\nabla\phi \quad \text{or} \quad =\phi\nabla \tag{2}$$

simply, as the complete expression for the force? Certainly we may. For, in full, we have

$$\nabla=\mathbf{i}\nabla_1+\mathbf{j}\nabla_2+\mathbf{k}\nabla_3, \tag{3}$$

$$\phi=\phi_1.\mathbf{i}+\phi_2.\mathbf{j}+\phi_3.\mathbf{k} \quad \text{or} \quad =\mathbf{i}\,\phi_1+\mathbf{j}.\phi_2+\mathbf{k}.\phi_3, \tag{4}$$

where ∇_1, &c., are the scalar components of the vector ∇ (not a quaternion, of course) and ϕ_1, &c., are the vector stresses on the planes of i, &c., so that $\phi_1 = \phi \mathbf{i}$, &c. Direct multiplication gives at once

$$\dot{\psi} = \nabla_1 \phi_1 + \nabla_2 \phi_2 + \nabla_3 \phi_3, \tag{5}$$

which is **F**. We may also write it $\phi\nabla$, because ϕ is pure.

On the other hand, when ϕ is rotational, let its conjugate be ϕ', then instead of (1) we have

$$\mathbf{FN} = \nabla\phi'\mathbf{N}, \tag{6}$$

and therefore

$$\mathbf{F} = \nabla\phi' = \phi\nabla. \tag{7}$$

Here if ϕ is given by the first expansion in (4), ϕ' is given by the second.

Now there are several things that deserve to be pointed out about the above, which should be compared with Prof. Tait on p. 194. First, that the result $\mathbf{F} = \phi\nabla$, irrespective of pureness, or $\mathbf{F} = \nabla\phi$ also when the stress is pure, when got quaternionically seems to be a great novelty to Prof. Tait, and to give him a "severe wrench," involving a "dislocation" and a "startling innovation." Perhaps, however, it is Mr. MacAulay's peculiar way of arriving at the result that Prof. Tait is alluding to. Moreover, secondly, in the vector algebra of Willard Gibbs and others the use of equation (2) or of (7) to express the force complete, by removal of the intermediate vector **N**, is neither new, nor does it involve any straining of the intellect, for it is actually a part of the system itself, done naturally and in harmony with Cartesian mathematics. See Gibbs's "Elements of Vector Analysis" (1881–4) for the direct product of ∇ and ϕ. (Also for the skew product, a more advanced idea; it, too, is a physically useful result.) Thirdly, note how very differently the same thing presents itself to Prof. Tait according as it is clothed in his favourite quaternionic garb or in vectorial vestments. In the latter case it is either unnoticed or is contemptible; in the former, it may be a novel and valuable improvement.

I do not think that Prof. Tait does justice to Mr. MacAulay in making so much of a trifle such as passes unnoticed or unappreciated in the previous work of others. There is, I know, much more in Mr. MacAulay's mathematics than Prof. Tait has yet fathomed. For my own part, I like to translate it into vectors, not merely because it is then in a form I am used to, and is plainer, but also because the true inwardness of these processes involving linear operators is properly exhibited by the dyadical way of viewing them in conjunction with vectors, without the forced and unnatural amalgamation with quaternions, and the attendant obscurities. This seems to me to be particularly true in physical applications. I should not be writing this note were it not for the misconceptions that Prof. Tait indulges in about what he does not know, viz., vector algebra apart from quaternions. At the same time, to avoid possible misunderstanding, I disclaim any hostility to Mr. MacAulay's quaternionic innovations, although I must agree with Prof. Tait as to the "singular uncouthness" of some of his expressions in their present form. I hope he may be able to see his way to do his work vectorially. It will be more amenable to innovations, I think, without mental wrenches. At any rate he is a reformer, and not afraid to innovate when he thinks fit.

The Teaching of Mathematics.

[*Nature*, October 4, 1900; p. 548.]

§ 540. Prof. John Perry has asked me to write something in criticism of the views he has lately expressed about the teaching of mathematics. I am inclined to ask, What is the use? He knows my views pretty well, and others too; and those who don't can learn them if they want to by buying my books. That is the best way, as it brings in one-and-threepences, and so does some good. I think there is a great deal to be said on both sides, and that if you are a born logic-chopper you will think differently from Faraday. The subject is too large, and I will only offer a few remarks about the teaching of geometry, based upon my own experience and observations. Euclid is the worst. It is shocking that young people should be addling their brains over mere logical subtleties, trying to understand the proof of one obvious fact in terms of something equally, or, it may be, not quite so obvious, and conceiving a profound dislike for mathematics, when they might be learning geometry, a most important fundamental subject, which can be made very interesting and instructive. I hold the view that it is essentially an experimental science, like any other, and should be taught observationally, descriptively and experimentally in the first place. The teaching should be a natural continuation of that education in geometry which every child undergoes by contact with his surroundings, only, of course, made definite and purposeful. It should be a teaching of the broad facts of geometry as they really exist, so as to impart an all-round knowledge of the subject. It should be Solid as well as Plane; the sphere and cube, &c., as well as the usual circle and square; models, sections, diagrams, compasses, rulers, &c., every aid that is useful and practical should be given. And it should be quantitative as well. The value of π should be *measured;* it may be done to a high degree of accuracy. So with the area of the circle, ellipse and all sorts of other things. The famous 47th. The boy who really measures and finds it true will have grasped the fact far better than by a logical demonstration without adequate experimental knowledge; for it happens that boys, who are generally very stupid in abstract ideas, learn a demonstration without knowing what it is all about in an intelligent manner. It may be said by logicians that you do not *prove* anything in this way. I differ. It might equally well be said that you prove nothing by *any* physical measurements. You have really proved the most important part. What a so-called rigorous proof amounts to is only this, that by limitation and substitution, arguing about abstract perfect circles, &c., replacing the practical ones, you can be as precise as you please. Now when a boy has learnt geometry, and has become competent to reason about its connections, he may pass on to the theory of the subject. Even then it should not be in Euclidean style; let the invaluable assistance of arithmetic and algebra be invoked, and the most useful idea of the vector be made prominent. I feel quite certain that I am right in this question of the teaching of geometry, having gone through it at school, where I made the closest observations on the effect of Euclid upon the rest of them. It was a sad farce, though conducted by a conscientious, hard-working teacher. Two or three followed, and were made temporarily into conceited logic-choppers, contradicting their parents; the effect upon most of the rest was disheartening and demoralising. I also feel quite certain about the experiential and experimental basis of space geometry, though that opinion has been of slow growth. If I understand them rightly, it is generally believed by mathematicians that geometry is pre-existent in the human mind, and that all we do is to look at nature and observe an approximate resemblance to the properties of the ideal space. You might assert the same pre-existence of dynamics or chemistry. I think it is a complete reversal of the natural order of ideas. It seems to me that geometry is only pre-existent in this limited sense; that since we are the children of many fathers and mothers, all of whom grew up and developed their minds (so far as they went) in contact with nature, of which they were a part, so our brains have grown to suit. So the child takes in the facts of space geometry naturally and easily. The experience of past generations makes the acquisition of present experience easier, and so it comes about that we cannot help seeing it. But it is all experience, after all; although learned philosophers, by long, long thinking over the theory of groups and other abstruse high developments, may perhaps come to what I think is a sort of self-deception, and think that their geometry is pre-existent in themselves, whilst nature's is only a bad copy. Like the old Indian pundit, whose name was something like Bhatravistra, who, after fifty years inward contemplation, discovered God:—where—it would not be polite to mention.

The Teaching of Mathematics.

[B.A. Meeting, Glasgow, 1901. Contribution to the Discussion on Prof. Perry's Address.]

§ **541.** I have read your Address. I do not doubt that in matters of detail, in practical application, many points of divergence might arise between us. For example, I may not think so much of the slide-rule as you do. And, as regards squared paper, what I object to is that the lines are not equidistant, and that is an offence to the artistic eye, besides introducing visible error. But these and similar matters are trifles in view of the most important object you are aiming at. It is fortunate that we possess an educational reformer who is so earnest, enthusiastic, and persistent, and I hope you will meet with your reward in some substantial realisation of your ideas. So far as I can see, there is a large body of educationalists in fair agreement with you at present. There have been, for long past, isolated would-be reformers. What is, I suppose, the principal difficulty, is the stupid old Toryism of the leading Universities. But even that has partly given way. If it is any satisfaction to you to be told what you know, I beg to say that I very much approve the general spirit of your Address. I agree entirely that mathematical works for the instruction of boys, sometimes including big boys, maybe up to 25 years of age, are generally written on wrong principles. Boys are not philosophers and logicians. Boys are usually exceedingly stupid in anything requiring concentrated reasoning. It is not in the nature of their soft brains that they should take kindly to Euclid and other stuff of that logic-chopping kind. But they usually possess another source of mental ability—namely, the ready acquisition of new facts and ideas—and that is what should be taken advantage of. They have also the power of learning to work processes, long before their brains have acquired the power of understanding (more or less) the scholastic logic of what they are doing. I have known boys of 14 extract cubes, fourth and fifth roots of numbers, to several places of decimals, easily. I am sure most of them would never live to understand the reason why, if they studied for a thousand years. Now, the prevalent idea of mathematical works is that you must understand the reason why first, before you proceed to practice. This is fudge and fiddlesticks. I speak with confidence in this matter, not merely from experience as a boy myself, and from knowledge of other boys, but as a grown man who has had some practice in applications of mathematics. I know mathematical processes, that I have used with success for a very long time, of which neither I nor any one else understands the scholastic logic. I have grown into them, and so understand them that way. Facts are facts, even though you do not see your way to a complete theory of them. And no *complete* theory is possible. There is always something wanting, no matter how logical people may pretend.

The fact is, there is theory and practice in mathematics just as in everything else. A man may be a good musician, by practice, and yet know nothing about the theory of music. He may arrive at that, or he may not. The same applies all round, and mathematics is no exception. The theory of mathematics is very important, but it is not the same thing as its practice, and the important matter here in relation to young persons is that the practice should come first. Then familiarity may gradually lead to some understanding of the theory, which may be studied later on in a rigorous manner, if the developing boy should be mentally fit for close reasoning. This is particularly true, I think, in geometry. It should be entirely observational and experimental at first, a natural continuation of the everyday education acquired through the senses. And even when you do come to the theory, you should put aside all old-Toryish ideas, and logical tricks and puzzles, and let the boy work more practically with the assistance of arithmetic and algebra.

Now the majority of British boys are not fit for any sort of logical theory. It is mere waste of time forcing them through Euclid, just as it is waste of time forcing Latin into them. The little they learn is soon forgotten, and gladly. What is it done for? I see nothing but old Toryism in the common arguments. Did the Romans and Greeks make their children painfully study [or, feed their young upon] scraps of ancient dead languages?

There is so much to learn nowadays, really valuable knowledge of all sorts, that it seems to me a wicked sin to go on in the old way, with Latin, Greek and Euclid. It takes such a time, involves such labour and does so little good. And we are living in the twentieth century.

The general inability of boys to study Euclid profitably is no reason why they should not learn geometry. Even stupid boys can do that, when properly directed and experimentally assisted. And as regards mathematics in general, I think it a very important assistance to have it taught in conjunction with elementary physics. That is, geometry *and* other natural facts.

As regards clever boys, I agree with you that it will do them good to go through the same process in the first place.

I seem to be running down logic. I do not mean to. But there is logic *and* logic. There is narrow-minded logic confined within narrow limits, rather conceited, and professing to be very exact, with absolutely certain premisses. And there is a broader sort of logic, more common-sensical, wider in its premisses, with less pretension to exactness, and more allowance for human error, and more room for growth.

The Pan-Potential as a Surface-Integral.

§ 542. The most practical way of treating potentials is not by using two sets of triple space-coordinates in triple or sextuple integrals, but by taking advantage of the symmetry of space, and thereby reducing the variables to a single one. When this is done, the elementary solution of $(q^2-\nabla^2)U=Q$, where U is the potential due to the source Q, comes out at once in the form

$$U=\frac{\epsilon^{-qr}}{4\pi r}Q, \quad \text{or} \quad U=PQ, \tag{1}$$

when there is a single point source of strength Q. And therefore

$$U=\Sigma\, PQ = \text{pan}\, Q, \tag{2}$$

for any distribution of sources. Here P is the potential at distance r from the point source of strength Q; or else Q is the density of source × element of volume. The source Q in the integral (2) may vary with position and time. Proper reckoning of discontinuities is assumed. But with a single fixed point source, Q in (1) only varies with the time. (*See* p. 167 and p. 256, and p. 388, above for pan-potential work.)

The form of q in electromagnetics is $q^2=(k+cp)(g+\mu p)$, where $p=d/dt$. So P may be a constant, and then U is an instantaneous potential; or when P contains p, it may be a progressive potential, without local waste of energy involved; or it may be progressive, with attenuation due to local waste of energy, but without distortion thereby; or it may be progressive with local waste of energy and with accompanying distortion, which is of two kinds, since there are two sorts of waste concerned in k and g. But although P is in general a differential operator, yet the formulas (1) and (2) may be regarded functionally as well. For the functions P1, qP1, q^{-1}P1 &c., have all been worked out algebraically, first in my *Electrical Papers*, and then in detail in vol. 2 of this work. So PQ is quite definite.

Now, as usual, it is a homogeneous medium that is considered, and (2) should include the sources over all space. We may however represent U by a surface integration partly. Thus, divide space into two regions, one bounded by a closed surface, the other the rest of space. Then we may say, if Q_1 is the internal and Q_2 the external Q,

$$U_1=U_{11}+U_{12}, \qquad U_2=U_{21}+U_{22}, \tag{3}$$

if U_{11} is the inside U due to Q_1, and U_{12} the inside U due to Q_2; and also U_{21} is the outside U due to Q_1, and U_{22} the outside U due to Q_2.

Now in (3), in the reckoning of U_1 inside, we may abolish U_{12} and substitute an integration over the boundary. Calling the last u_{12}, its value inside is precisely that due to the outside sources. Its value outside is zero. And in the reckoning of U_2 outside, we may abolish U_{21} and substitute a surface integral over the boundary, say u_{21}. Its outside value is the same as that of the inside sources. Its inside value is zero.

To prove; let

$$u = \Sigma\, \mathbf{N}(U\nabla P - P\nabla U)Q \tag{4}$$

represent the boundary integral, **N** being the unit normal. We may apply the Divergence Theorem to this in two ways, inside and outside. Doing this vectorially, first inside, then outside, we get

$$u_{(\text{in})} = U_{12} = \text{pan}\, Q_2, \qquad u_{(\text{out})} = U_{21} = \text{pan}\, Q_1. \tag{5}$$

Put ∇ for **N** in (4). Then (4) becomes the equivalent volume integral, for the inner or the outer space, according to the direction of **N** taken, and the position of the observing point. Therefore (4) is the required surface-integral. This allows us to ignore inside all the outside sources, so that U_1 inside is to be found from the inside sources Q_1, and the state of the boundary. Only the inside region need be homogeneous now. But the state of the boundary is not merely the local value of U, but of its local space variation as well.

In the special case $k=0$, $g=0$, that of ordinary optical theory for a simple medium ($\mu cv^2=1$, and v constant), I believe the formula (4) or an equivalent is due to Kirchhoff; but I only go by memory of a reference to him. An equivalent formula in which the surface is a sphere, and the observing point its centre was given by Poisson very long ago. I prefer the form (4) for analytical and vectorial convenience.

It is only necessary to consider a single point source and the equivalent integral, for ease of treatment. The last article in vol. 1, *Elec. Pa.*, may be referred to; although it refers to the instantaneous potential, the results may be extended. Eqn. (24) there represents (4) here.

In the above, the sources are limited in space. But they may also be limited in time. For if we specify Q only from the moment $t=0$, the whole previous action of Q is represented by its result at that moment. That is, the initial state of the function U may be given, instead of the function Q for all past time. This presents nothing new.

Limitations on Scientific Prediction.

§ 543. All known disturbances are conveyed either electromagnetically or gravitationally. If the first way, the speed is finite If the second way, it may also be finite, perhaps with the same value. *See* p. 463, vol. 1, on this matter. Assuming then that all disturbances are conveyed at finite speed, it follows instantaneously that the destruction of this wicked world may come at any moment without any warning. There is no possibility of foretelling this calamity (or blessing, possibly), because the cause thereof cannot give us any information till it arrives, when it will be too late to take precautions against destruction. The theological, metaphysical, legal, moral, and pecuniary consequences of this indeterminateness of knowledge, however much the state and the laws of nature may be explored, are tremendous. But practically I do not think it makes any difference. We trust by habit, guided by past experience. The thief of yesterday is a thief to-day, even though he turn pious, or fly to remote lands to escape from entomological associations.

What applies to catastrophes applies of course to smaller events, down to most minute effects, reckoned by human standards. But there is no absolute scale of size in nature, and the small may be as important, or more so, than the great. This brings us to a different kind of indeterminateness from necessary ignorance, which does not depend upon the argument employed above As the universe is boundless one way, towards the great, so it is equally boundless the other way, towards the small; and important events may arise from what is going on in the inside of atoms, and again, in the inside of electrons. There is no energetic difficulty. Large amounts of energy may be very condensed by reason of great forces at small distances. How electrons are made has not yet been discovered. From the atom to the electron is a great step, but is not finality.

Living matter is sometimes, perhaps generally, left out of consideration when asserting the well-known proposition that the course of events in the physical world is determined by its present state, and by the laws followed. But I do not see how living matter can be fairly left out. For we do not know where life begins, if it has a beginning. There may be and probably is no ultimate distinction between the living and the dead. But even if there were a clear separation, the continuous action and reaction going on between the living and the dead would make it become imperative to include the living matter. So determinateness must be applied to the universe completely, if at all.

But why is the Action and Reaction principle to be taken as fundamental? Because it is always true when proper examination can be made, and is the guide to fresh knowledge. Besides that, the untruth of the principle in practice would lead to chaos. Even the spiritual world, if it exists, must be a part of nature itself.

END OF VOL. 3.

Printed in the United States
By Bookmasters